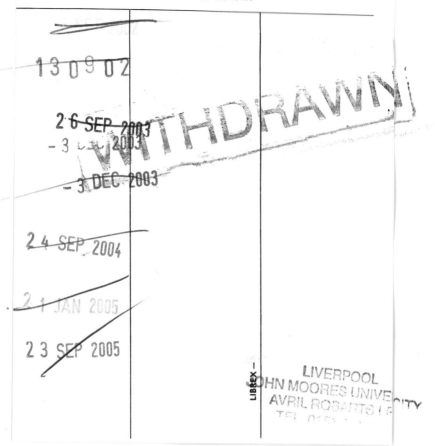

MODERN
CONTROL
ENGINEERING

CONTROL ENGINEERING

A Series of Reference Books and Textbooks

Editor

NEIL MUNRO, PH.D., D.SC.

Professor
Applied Control Engineering
University of Manchester Institute of Science and Technology
Manchester, United Kingdom

Additional Volumes in Preparation

MODERN CONTROL ENGINEERING

P. N. Paraskevopoulos

National Technical University of Athens
Athens, Greece

MARCEL DEKKER, INC.   NEW YORK · BASEL

ISBN: 0-8247-8981-4

This book is printed on acid-free paper.

Headquarters
Marcel, Dekker, Inc.
270 Madison Avenue, New York, NY 10016
tel: 212-696-9000; fax: 212-685-4540

Eastern Hemisphere Distribution
Marcel Dekker AG
Hutgasse 4, Postfach 812, CH-4001 Basel, Switzerland
tel: 41-61-261-8482; fax: 41-61-261-8996

World Wide Web
http://www.dekker.com

The publisher offers discounts on this book when ordered in bulk quantities. For more information, write to Special Sales/Professional Marketing at the headquarters address above.

Current printing (last digit):
10 9 8 7 6 5 4 3 2 1

PRINTED IN THE UNITED STATES OF AMERICA

To my wife, Mary, and our son, Nikos

Series Introduction

Many textbooks have been written on control engineering, describing new techniques for controlling systems, or new and better ways of mathematically formulating existing methods to solve the ever-increasing complex problems faced by practicing engineers. However, few of these books fully address the applications aspects of control engineering. It is the intention of this new series to redress this situation.

The series will stress applications issues, and not just the mathematics of control engineering. It will provide texts that present not only both new and well-established techniques, but also detailed examples of the application of these methods to the solution of real-world problems. The authors will be drawn from both the academic world and the relevant applications sectors.

There are already many exciting examples of the application of control techniques in the established fields of electrical, mechanical (including aerospace), and chemical engineering. We have only to look around in today's highly automated society to see the use of advanced robotics techniques in the manufacturing industries; the use of automated control and navigation systems in air and surface transport systems; the increasing use of intelligent control systems in the many artifacts available to the domestic consumer market; and the reliable supply of water, gas, and electrical power to the domestic consumer and to industry. However, there are currently many challenging problems that could benefit from wider exposure to the applicability of control methodologies, and the systematic systems-oriented basis inherent in the application of control techniques.

This series presents books that draw on expertise from both the academic world and the applications domains, and will be useful not only as academically recommended course texts but also as handbooks for practitioners in many applications domains. *Modern Control Engineering* is another outstanding entry to Dekker's Control Engineering series.

Neil Munro

v

Preface

Automatic control is one of today's most significant areas of science and technology. This can be attributed to the fact that automation is linked to the development of almost every form of technology. By its very nature, automatic control is a multi-disciplinary subject; it constitutes a core course in many engineering departments, such as electrical, electronic, mechanical, chemical, and aeronautical. Automatic control requires both a rather strong mathematical foundation, and implementation skills to work with controllers in practice.

The goal of this book is to present control engineering methods using only the essential mathematical tools and to stress the application procedures and skills by giving insight into physical system behavior and characteristics. Overall, the approach used herein is to help the student understand and assimilate the basic concepts in control system modeling, analysis, and design.

Automatic control has developed rapidly over the last 60 years. An impressive boost to this development was provided by the technologies that grew out of space exploration and World War II. In the last 20 years, automatic control has undergone a significant and rapid development due mainly to digital computers. Indeed, recent developments in digital computers—especially their increasingly low cost—facilitate their use in controlling complex systems and processes.

Automatic control is a vast technological area whose central aim is to develop control strategies that improve performance when they are applied to a system or a process. The results reported thus far on control design techniques are significant from both a theoretical and a practical perspective. From the theoretical perspective, these results are presented in great depth, covering a wide variety of modern control problems, such as optimal and stochastic control, adaptive and robust control, Kalman filtering, and system identification. From the practical point of view, these results have been successfully implemented in numerous practical systems and processes—for example, in controlling temperature, pressure, and fluid level; in electrical energy plants; in industrial plants producing paper, cement, steel, sugar, plastics, clothes, and food; in nuclear and chemical reactors; in ground, sea, and air transportation systems; and in robotics, space applications, farming, biotechnology, and medicine.

I should note that *classical* control techniques—especially those using proportional-integral-derivative (PID) controllers, which have existed since 1942—predominate in the overall practice of control engineering today. Despite the impressive progress since the 1940s, practical applications of *modern* control techniques are limited. This is indeed a serious gap between theory and practice. To reduce this gap, techniques of modern control engineering should be designed with an eye toward applicability, so as to facilitate their use in practice. To this end, modern control techniques must be presented in a simple and user-friendly fashion to engineering students in introductory control courses, so that these techniques may find immediate and widespread application. In turn, control engineering could serve human needs better and provide the same breadth of technological application found in other, related areas, such as communications and computer science. This book has been written in this spirit.

Modern Control Engineering is based on the introductory course on control systems that I teach to junior undergraduate students in the Department of Electrical and Computer Engineering at the National Technical University of Athens. It begins with a description and analysis of linear time-invariant systems. Next, classical (Bode and Nyquist diagrams, the root locus, compensating networks, and PID controllers) and modern (pole placement, state observers, and optimal control) controller design techniques are presented. Subsequent chapters cover more advanced techniques of modern control: digital control, system identification, adaptive control, robust control, and fuzzy control. This text is thus appropriate for undergraduate and first-year graduate courses in modern control engineering, and it should also prove useful for practicing engineers. The book has 16 chapters, which may be grouped into two parts: Classical Control (Chapters 1 through 9) and Modern Control (Chapters 10 through 16). (Please note that, throughout the book, bold lowercase letters indicate vectors and bold capital letters indicate matrices.)

CLASSICAL CONTROL

Chapter 1 is an introduction to automatic control systems. Chapter 2 presents the Laplace transform and matrix theory, which is a necessary mathematical background for studying continuous-time systems. Chapter 3 describes and analyzes linear time-invariant systems by using the following mathematical models: differential equations, transfer functions, impulse response, and state-space equations; the topics of block diagrams and signal-flow graphs are also covered.

Chapter 4 describes classical time-domain analysis, covering topics such as time response, model simplification, comparison of open- and closed-loop systems, model reduction, sensitivity analysis, steady-state errors, and disturbance rejection. Chapter 5 describes state-space analysis of linear systems and discusses the important concepts of controllability and observability, along with their relation to the transfer function. Chapter 6 discusses the important problem of stability. It covers the algebraic criteria of Ruth, Hurwitz, and continuous fraction, and provides an introduction to the stability of nonlinear and linear systems using the Lyapunov methods.

Chapter 7 covers the popular root locus method. Chapter 8 describes the frequency response of linear time-invariant systems, introducing the three well-known frequency domain methods: those of Nyquist, Bode, and Nichols. Chapter 9 is devoted to the classical design techniques, emphasizing controller design methods using controllers of the following types: PID, phase-lead, phase-lag, and phase lead-lag. The chapter also presents an introduction to classical optimal control.

MODERN CONTROL

Chapters 10 and 11 focus on modern controller design techniques carried out in state-space. Chapter 10 covers the design problems of pole assignment, input-output decoupling, model matching, and state observers. Closed-loop system design using state observers is also explained. Chapter 11 elucidates the problem of optimal control, as illustrated in the optimal regulator and servomechanism problems.

Chapter 12 is an introduction to digital control that provides extensive coverage of basic problems in discrete-time system description, analysis, stability, controllability, observability, and classical control techniques. Chapter 13 explains discrete-time system identification and gives the basic algorithms for off-line and on-line parametric identification.

Chapter 14 covers discrete-time system adaptive control. The following four adaptive schemes are presented: the gradient method (MIT rule), model reference, adaptive control, and self-tuning regulators. Chapter 15 is an introduction to robust control, focusing on topics such as model uncertainty, robust stability, robust performance, and Kharitonov s theorem. Chapter 16 is an introduction to fuzzy control, emphasizing the design of fuzzy controllers.

The book concludes with three appendixes that provide useful background information. Appendix A presents the Laplace transform tables, Appendix B demonstrates the Z-transform technique necessary for analyzing and designing the discrete-time (or digital) control systems presented in Chapter 12, and Appendix C gives the Z-transform tables.

ACKNOWLEDGMENTS

I would like to thank very much my undergraduate students A. Dimeas and D. Kazizis for preparing the figures and tables, my graduate student A. Vernardos for proofreading the typewritten manuscript, and Dr. Iliana Gravalou for her help in checking most solved examples and in preparing the Solutions Manual. Many thanks to Dr. Argiris Soldatos for carefully reading several chapters. Special thanks are also due Dr. K. Arvanitis for his assistance in formulating the material of Chapter 15 and to Professors R. E. King and G. Bitsoris, my colleagues, for their numerous suggestions.

P. N. Paraskevopoulos

Contents

1

Introduction to Automatic Control Systems

1.1 INTRODUCTION

An automatic control system is a combination of components that act together in such a way that the overall system behaves automatically in a prespecified desired manner.

A close examination of the various machines and apparatus that are manufactured today leads to the conclusion that they are partially or entirely automated, e.g., the refrigerator, the water heater, the clothes washing machine, the elevator, the TV remote control, the worldwide telephone communication systems, and the Internet. Industries are also partially or entirely automated, e.g., the food, paper, cement, and car industries. Examples from other areas of control applications abound: electrical power plants, reactors (nuclear and chemical), transportation systems (cars, airplanes, ships, helicopters, submarines, etc.), robots (for assembly, welding, etc.), weapon systems (fire control systems, missiles, etc.), computers (printers, disk drives, magnetic tapes, etc.), farming (greenhouses, irrigation, etc.), and many others, such as control of position or velocity, temperature, voltage, pressure, fluid level, traffic, and office automation, computer-integrated manufacturing, and energy management for buildings. All these examples lead to the conclusion that automatic control is used in all facets of human technical activities and contributes to the advancement of modern technology.

The distinct characteristic of automatic control is that it reduces, as much as possible, the human participation in all the aforementioned technical activities. This usually results in decreasing labor cost, which in turn allows the production of more goods and the construction of more works. Furthermore, automatic control reduces work hazards, while it contributes in reducing working hours, thus offering to working people a better quality of life (more free time to rest, develop hobbies, have fun, etc.).

Automatic control is a subject which is met not only in technology but also in other areas such as biology, medicine, economics, management, and social sciences. In particular, with regard to biology, one can claim that plants and animals owe their very existence to control. To understand this point, consider for example the human body, where a tremendous number of processes take place automatically: hunger, thirst, digestion, respiration, body temperature, blood circulation, reproduction of cells, healing of wounds, etc. Also, think of the fact that we don't even decide when to drink, when to eat, when to go to sleep, and when to go to the toilet. Clearly, no form of life could exist if it were not for the numerous control systems that govern all processes in every living organism.

It is important to mention that modern technology has, in certain cases, succeeded in replacing body organs or mechanisms, as for example in replacing a human hand, cut off at the wrist, with an artificial hand that can move its fingers automatically, as if it were a natural hand. Although the use of this artificial hand is usually limited to simple tasks, such as opening a door, lifting an object, and eating, all these functions are a great relief to people who were unfortunate enough to lose a hand.

1.2 A BRIEF HISTORICAL REVIEW OF AUTOMATIC CONTROL SYSTEMS

Control systems have been in existence since ancient times. A well-known ancient automatic control system is the regulator of Heron of Alexandria (Figure 1.1). This control system was designed to open the doors of a temple automatically when a fire was lit at the altar located outside the temple and to close the doors when the fire was put out. In particular, the regulator operated in the following way: the fire, acting as the input to the system, heated the air underneath the altar and the warm (expanded)

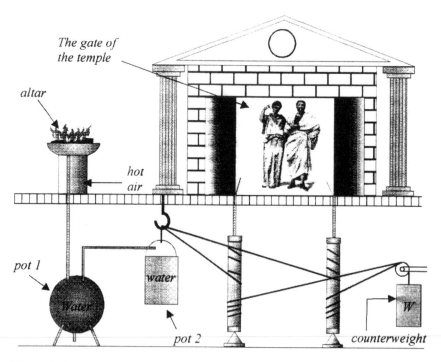

Figure 1.1   The regulator of Heron of Alexandria.

air pushed the water from the water container (pot 1) to the bucket (pot 2). The position of the water container was fixed, while the bucket was hanging from ropes wrapped around a mechanism (the door spindles) with a counterweight W. When pot 2 was empty, this mechanism, under the pull of the counterweight W, held the doors closed. When pot 2 was filled with adequate amount of water from pot 1, it moved downwards, while the counterweight W moved upwards. As a result of the downward motion of pot 2, the door spindles turned and the doors opened. When the fire was put out, water from pot 2 returned to pot 1, and the counterweight W moved downwards forcing the gates to close. Apparently, this control system was used to impress believers, since it was not visible or known to the masses (it was hidden underground).

Until about the middle of the 18th century, automatic control has no particular progress to show. The use of control started to advance in the second half of the 18th century, due to James Watt, who, in 1769, invented the first centrifugal speed regulator (Figure 1.2) which subsequently has been widely used in practice, most often for the control of locomotives. In particular, this regulator was used to control the speed of the steam engine. This is accomplished as follows: as the angular velocity of the steam engine increases, the centrifugal force pushes the masses m upwards and the steam valve closes. As the steam valve closes, the steam entering the engine from the boiler is reduced and the steam engine's angular velocity decreases, and vice versa: as the angular velocity of the steam engine decreases, the masses m go down, the steam valve opens, the amount of steam entering the engine increases, resulting in an increase of the angular velocity. This way, one can regulate the speed of the engine.

The period until about the middle of the 19th century is characterized by developments based on intuition, i.e., there was no mathematical background for control design. Maxwell in 1868 [82, 83] and Vyshnegradskii in 1877 [52] set the first

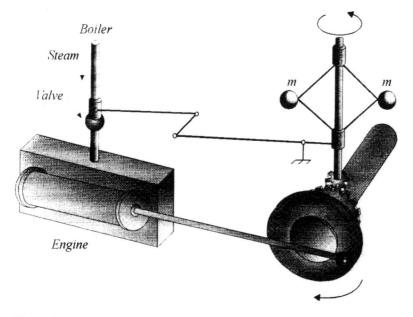

Figure 1.2 Watt's centrifugal speed regulator.

mathematical background for control design for applying their theoretical (mathematical) results on Watt's centrifugal regulator. Routh's mathematical results on stability presented in 1877 [47] were also quite important.

Automatic control theory and its applications have developed rapidly in the last 60 years or so. The period 1930–1940 was important in the history of control, since remarkable theoretical and practical results, such as those of Nyquist [84, 85] and Black [60, 61], were reported.

During the following years and until about 1960, further significant research and development was reported, due mainly to Ziegler and Nichols [92], Bode [11], Wiener [53] and Evans [18, 64]. All the results of the last century, and up to about 1960, constitute what has been termed *classical control*. Progress from 1960 to date has been especially impressive, from both the theoretical and the practical point of view. This last period has been characterized as that of *modern control*, the most significant results of which have been due to Astrom [3–5], Athans [6, 57–59], Bellman [7, 8], Brockett [12, 62], Doyle [63, 66], Francis [63, 66], Jury [24, 25], Kailath [26, 67], Kalman [27, 28, 68–79], Luenberger [33, 80, 81], MacFarlane [34], Rosenbrock [45, 46], Saridis [48], Wonham [54, 89, 90], Wolovich [55], Zames [91], and many others. For more on the historical development of control the reader can refer to [35] and [41].

A significant boost to the development of classical control methods was given by the Second World War, whereas for modern control techniques the launch of Sputnik in 1957 by the former Soviet Union and the American Apollo project, which put men on the moon in 1969, were prime movers. In recent years, an impressive development in control systems has taken place with the ready availability of digital computers. Their power and flexibility have made it possible to control complex systems efficiently, using techniques which were hitherto unknown.

The main differences between the classical and the modern control approaches are the following: classical control refers mainly to single input–single output systems. The design methods are usually graphical (e.g., root locus, Bode and Nyquist diagrams, etc.) and hence they do not require advanced mathematics. Modern control refers to complex multi-input multi-output systems. The design methods are usually analytical and require advanced mathematics. In today's technological control applications, both classical and modern design methods are used. Since classical control is relatively easier to apply than modern control, a control engineer may adopt the following general approach: simple cases, where the design specifications are not very demanding, he uses classical control techniques, while in cases where the design specifications are very demanding, he uses modern control techniques.

Today, automatic control systems is a very important area of scientific research and technological development. Worldwide, a large number of researchers aim to develop new control techniques and apply them to as many fields of human activity as possible. In Sec. 1.4, as in other parts of this book, we present many practical control examples that reflect the development of modern control engineering.

1.3 THE BASIC STRUCTURE OF A CONTROL SYSTEM

A system is a combination of components (appropriately connected to each other) that act together in order to perform a certain task. For a system to perform a certain task, it must be excited by a proper input signal. Figure 1.3 gives a simple

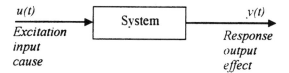

$u(t)$
Excitation
input
cause

System

$y(t)$
Response
output
effect

Figure 1.3 Schematic diagram of a system with its input and output.

view of this concept, along with the scientific terms and symbols. Note that the response $y(t)$ is also called system's behavior or performance.

Symbolically, the output $y(t)$ is related to the input u(t) by the following equation

$$y(t) = Tu(t) \qquad (1.3\text{-}1)$$

where T is an operator. There are three elements involved in Eq. (1.3-1): the input $u(t)$, the system T, and the output $y(t)$. In most engineering problems, we usually know (i.e., we are given) two of these three elements and we are asked to find the third one. As a result, the following three basic engineering problems arise:

1. *The analysis problem.* Here, we are given the input $u(t)$ and the system T and we are asked to *determine* the output $y(t)$.
2. *The synthesis problem.* Here, we are given the input $u(t)$ and the output $y(t)$ and we are asked to *design* the system T.
3. *The measurement problem.* Here, we are given the system T and the output $y(t)$ and we are asked to *measure* the input $u(t)$.

The control design problem does not belong to any of these three problems and is defined as follows.

Definition 1.3.1

Given the system T under control and its *desired response* $y(t)$, find an appropriate input signal $u(t)$, such that, when this signal is applied to system T, the output of the system to be the desired response $y(t)$. Here, this appropriate input signal $u(t)$ is called *control signal*.

From Definition 1.3.1 it appears that the control design problem is a signal synthesis problem: namely, the synthesis of the control signal $u(t)$. However, as it will be shown later in this section, in practice, the control design problem is reduced to that of designing a controller (see Definition 1.3.4).

Control systems can be divided into two categories: the *open-loop* and the *closed-loop* systems.

Definition 1.3.2

An open-loop system (Figure 1.4a) is a system whose input $u(t)$ does not depend on the output $y(t)$, i.e., $u(t)$ is not a function of $y(t)$.

Definition 1.3.3

A closed-loop system (Figure 1.4b) is a system whose input $u(t)$ depends on the output $y(t)$, i.e., $u(t)$ is a function of $y(t)$.

In control systems, the control signal $u(t)$ is not the output of a signal generator, but the output of another new additional component that we add to the

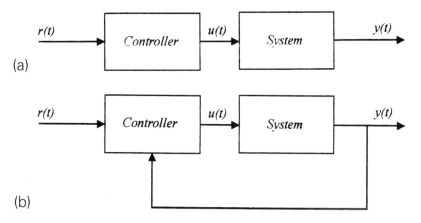

(a)

(b)

Figure 1.4   Two types of systems: (a) open-loop system; (b) closed-loop system.

system under control. This new component is called *controller* (and in special cases *regulator* or *compensator*). Furthermore, in control systems, the controller is excited by an external signal $r(t)$, which is called the *reference* or *command* signal. This reference signal $r(t)$ specifies the desired performance (i.e., the desired ouput $y(t)$) of the open- or closed-loop system. That is, in control systems, we aim to design an appropriate controller such that the output $y(t)$ follows the command signal $r(t)$ as close as possible. In particular, in open-loop systems (Figure 1.4a) the controller is excited only by the reference signal $r(t)$ and it is designed such that its output $u(t)$ is the appropriate input signal to the system under control, which in turn will produce the desired output $y(t)$. In closed-loop systems (Figure 1.4b), the controller is excited not only by reference signal $r(t)$ but also by the output $y(t)$. Therefore, in this case the control signal $u(t)$ depends on both $r(t)$ and $y(t)$. To facilitate better understanding of the operation of open-loop and closed-loop systems we present the following introductory examples.

A very simple introductory example of an open-loop system is that of the clothes washing machine (Figure 1.5). Here, the reference signal $r(t)$ designates the various operating conditions that we set on the "programmer," such as water temperature, duration of various washing cycles, duration of clothes wringing, etc. These operating conditions are carefully chosen so as to achieve satisfactory clothes washing. The controller is the "programmer," whose output $u(t)$ is the control signal. This control signal is the input to the washing machine and forces the washing machine to execute the desired operations preassigned in the reference signal $r(t)$, i.e., water heating, water changing, clothes wringing, etc. The output of the system $y(t)$ is the "quality" of washing, i.e., how well the clothes have been washed. It is well known that during the operation of the washing machine, the output (i.e., whether the

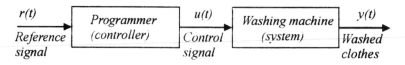

Figure 1.5   The clothes washing machine as an open-loop system.

clothes are well washed or not) it not taken into consideration. The washing machine performs only a series of operations contained in $u(t)$ without being influenced at all by $y(t)$. It is clear that here $u(t)$ is not a function of $y(t)$ and, therefore, the washing machine is a typical example of an open-loop system. Other examples of open-loop systems are the electric stove, the alarm clock, the elevator, the traffic lights, the worldwide telephone communication system, the computer, and the Internet.

A very simple introductory example of a closed-loop system is that of the water heater (Figure 1.6). Here, the system is the water heater and the output $y(t)$ is the water temperature. The reference signal $r(t)$ designates the desired range of the water temperature. Let this desired temperature lie in the range from 65 to 70°C. In this example, the water is heated by electric power, i.e., by a resistor that is supplied by an electric current. The controller of the system is a thermostat, which works as a switch as follows: when the temperature of the water reaches 70°C, the switch opens and the electric supply is interrupted. As a result, the water temperature starts falling and when it reaches 65°C, the switch closes and the electric supply is back on again. Subsequently, the water temperature rises again to 70°C, the switch opens again, and so on. This procedure is continuously repeated, keeping the temperature of the water in the desired temperature range, i.e., between 65 and 70°C.

A careful examination of the water heater example shows that the controller (the thermostat) provides the appropriate input $u(t)$ to the water heater. Clearly, this input $u(t)$ is decisively affected by the output $y(t)$, i.e., $u(t)$ is a function of not only of $r(t)$ but also of $y(t)$. Therefore, here we have a typical example of a closed-loop system.

Other examples of closed-loop systems are the refrigerator, the voltage control system, the liquid-level control system, the position regulator, the speed regulator, the nuclear reactor control system, the robot, and the guided aircraft. All these closed-loop systems operate by the same principles as the water heater presented above.

It is remarked that in cases where a system is not entirely automated, man may act as the controller or as part of the controller, as for example in driving, walking, and cooking. In driving, the car is the system and the system's output is the course and/or the speed of the car. The driver controls the behavior of the car and reacts accordingly: he steps on the accelerator if the car is going too slow or turns the steering wheel if he wants to go left or right. Therefore, one may argue that driving a car has the structure of a closed-loop system, where the driver is the controller. Similar remarks hold when we walk. When we cook, we check the food in the oven and appropriately adjust the heat intensity. In this case, the cook is the controller of the closed-loop system.

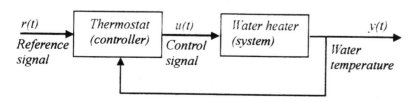

Figure 1.6 The water heater as a closed-loop system.

From the above examples it is obvious that closed-loop systems differ from open-loop systems, the difference being whether or not information concerning the system's output is fed back to the system's input. This action is called *feedback* and plays the most fundamental role in automatic control systems.

Indeed, it is of paramount importance to point out that in open-loop systems, if the performance of the system (i.e., $y(t)$) is not satisfactory, the controller (due to the lack of feedback action) does nothing to improve it. On the contrary, in closed-loop systems the controller (thanks to the feedback action) acts in such a way as to keep the performance of the system within satisfactory limits.

Closed-loop systems are mostly used when the control specifications are highly demanding (in accuracy, in speed, etc.), while open-loop systems are used in simple control problems. Closed-loop systems are, in almost all cases, more difficult to design and implement than open-loop systems. More specific comparisons between open- and closed-loop systems are made in several parts of teh book (e.g., see Sec. 4.5).

The complexity in implementing controllers for open- or closed-loop systems increases as the design requirements increase. We can have simple controllers, e.g., thermostats or programmers, but we can also have more complex controllers like an amplifier and/or an RC or an RL network to control a system or process, a computer to control an airplane, or even a number of computers (a computer centre) to control the landing of a spacecraft on Mars. Furthermore, depending mainly upon the design requirements and the nature of the system under control, a controller may be electronic, electrical, mechanical, pneumatic, or hydraulic, or a combination of two or more of these types of controllers.

On the basis of all the above material, we can now give the well-known definition of the control design problem.

Definition 1.3.4

Given the system T under control and the desired response $y(t)$, find a controller whose output $u(t)$ is such that, when applied to system T, the output of the system is the desired response $y(t)$.

It is obvious that Definitions 1.3.1 and 1.3.4 are equivalent. In practice, only Definition 1.3.4 is used, which reduces the control problem to that of designing a controller. Many controller design methods have been developed that give satisfactory practical results; however, as technology advances, new control design problems appear, which in turn require new research and development techniques.

In closing this section, we present a more complete schematic diagram of open- and closed-loop systems. Open-loop systems have the structure of Figure 1.7 and closed-loop systems have the structure of Figure 1.8. In both cases, the control problem is to have $y(t)$ follow, as close as possible, the reference signal $r(t)$. This is clearly demonstrated in the many practical control systems presented in Sec. 1.4, which follows. The term *disturbances* refer to changes in the system's environment or in the system itself, which result in a deviation of the actual system's output from its desired form. Based on the material presented thus far, it is obvious that when the output of an open-loop system deviates from its desired form due to disturbances, the controller (due to the lack of feedback action) does nothing to bring it back to its desired form. On the contrary, in a closed-loop system, if the output deviates from its desired form due to disturbances, then (thanks to the feedback action) the controller acts in such a way so as to restore the output to its desired form.

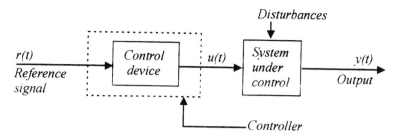

Figure 1.7 Schematic diagram of an open-loop system.

1.4 PRACTICAL CONTROL EXAMPLES

In this section we describe several well-known practical control examples (both open- and closed-loop systems) that are certainly more complex than those described in Sec. 1.3. These examples give an overall picture of the wide use of control in modern technology. Furthermore, some of these examples show how the principles of control can be used to understand and solve control problems in other fields, such as economics, medicine, politics, and sociology. Some of the examples given below are studied further in Sec. 3.13, as well as in other parts of this book.

From the examples that follow, it will become obvious that many control systems are designed in such a way as to control automatically certain variables of the system (e.g., the voltage across an element, the position or velocity of a mass, the temperature of a chamber, etc.). It is remarked that for the special category of control systems where we control a mechanical movement—e.g., the position or velocity of a mass—the term *servomechanism* is widely used.

1 Position Control System or Position Servomechanism (Figure 1.9)

The desired angular position $r(t)$ of the steering wheel is the reference input to the system and the angular position $y(t)$ of the small gear is the output of the system. Here, the system is designed such that $y(t)$ follows $r(t)$ as closely as possible. This is accomplished as follows: the angular positions $r(t)$ and $y(t)$ are transformed into

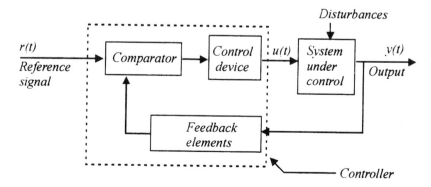

Figure 1.8 Schematic diagram of a closed-loop system.

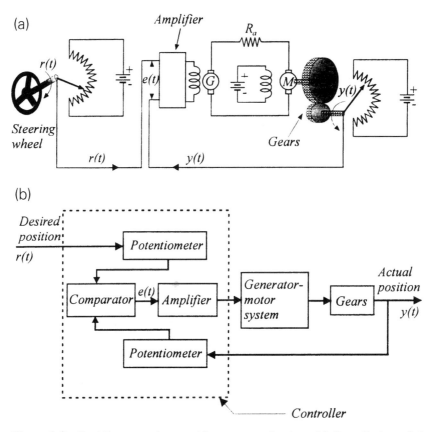

Figure 1.9   Position control or position servomechanism. (a) Overall view of the position control system; (b) schematic diagram of the position control system.

voltages by using potentiometers (see Subsec. 3.12.4). The *error* $e(t) = r(t) - y(t)$ between these two voltages is driven into the amplifier. The output of the amplifier excites the system generator motor (see Subsec. 3.12.1). As a result, the motor turns the gears in one or the other direction, depending on the sign of the error $e(t)$, thus reducing (and finally completely eliminating) the error $e(t) = r(t) - y(t)$. This way, the actual output $y(t)$ follows the reference input $r(t)$, i.e., $y(t) = r(t)$. In figure 1.9b, a schematic diagram of the system is given, where one can clearly understand the role of feedback and of the controller. A similar system is described in more detail in Subsec. 3.13.2.

2 Metal Sheet Thickness Control System (Figure 1.10)

The desired thickness $r(t)$ is the reference input to the system and the actual thickness $y(t)$ of the metal sheet is the otuput of the system. Here, the system is designed such that $y(t)$ follows $r(t)$ as closely as possible. This is accomplished as follows: the desired thickness is secured by the appropriate choice of the pressure $p(t)$ of the cylinders applied to the metal sheet. This pressure is measured indirectly via the thickness meter which measures the thickness $y(t)$. Let $b(t)$ be the indication of this meter. Then, when the error $e(t) = r(t) - b(t)$ is not zero, where $r(t)$ is the desired

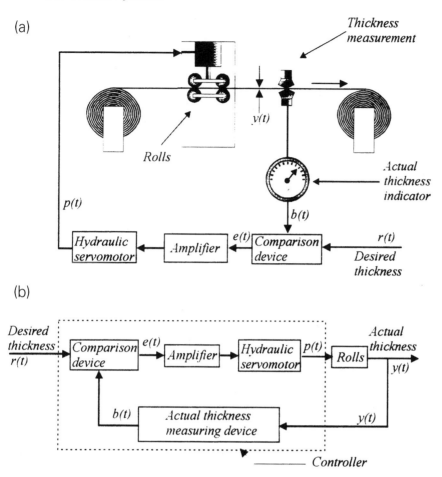

Figure 1.10 Metal sheet thickness control system. (a) Overall view of the metal sheet thickness control system; (b) schematic diagram of the metal sheet control system.

thickness, the hydraulic servomotor increases or decreases the pressure $p(t)$ of the cylinders and thus the thickness $y(t)$ becomes smaller or greater, respectively. This procedure yields the desired result, i.e., the thickness $y(t)$ of the metal sheet is, as close as possible, to the desired thickness $r(t)$. In Figure 1.10b, a schematic diagram of the closed-loop system is given.

3 Temperature Control of a Chamber (Figure 1.11)

This system is designed so that the temperature of the chamber, which is the system's output, remains constant. This is accomplished as follows: the temperature of the chamber is being controlled by a bimetallic thermostat, appropriately adjusted to deactivate the circuit of the magnetic valve whenever the chamber temperature is higher than the desired one. The valve closes and the supply of fuel gas into the burner stops. When the temperature of the chamber is lower than the desired temperature, the circuit of the magnetic valve opens and the supply of fuel gas into the

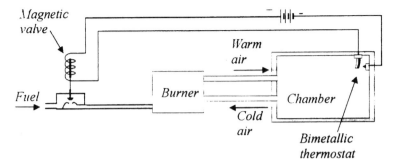

Figure 1.11   Temperature control of a chamber.

burner starts again. This way, the temperature in the chamber remains close to constant. A similar system is described in more detail in Subsection 3.13.5.

4 Liquid-Level Control (Figure 1.12)

This system is used in chemical and other industries and is designed such that the height $y(t)$ of the surface of a liquid remains constant. This is accomplished as follows: the cork floating on the surface of the liquid is attached to the horizontal surface of the flapper in such a way that when the height $y(t)$ increases or decreases, the distance $d(t)$ between the end of the nozzle and the vertical surface of the flapper decreases or increases, respectively. When the distance $d(t)$ decreases or increases, subsequently the pressure of the compressed air acting upon the surface A of the valve increases or decreases. As a result, the distance $q(t)$ between the piston and the base of the container, decreases or increases, respectively. This control system can be considered as a system whose input is the distance $d(t)$ and the output is the pressure of the compressed air acting upon the surface A of the valve. This system is actually a pneumatic amplifier, since although changing $d(t)$ does not demand a great amount

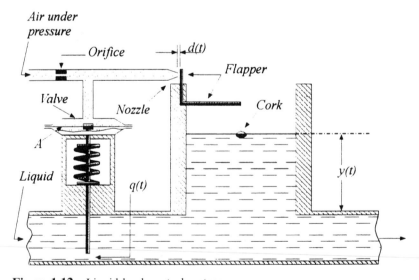

Figure 1.12   Liquid-level control system.

of pressure, the corresponding pressure on the surface A is indeed very big. Finally, knowing that a decrease or increase in the distance $q(t)$ corresponds to a decrease or increase in the height $y(t)$, it is obvious that in this way the liquid level will remain constant. A similar system is described in more detail in Subsec. 3.13.4.

5 Aircraft Wing Control System (Figure 1.13)

This system is designed such that the slope (or angle or inclination) of the wings of the aircraft is controlled manually by the pilot using a control stick. The system works as follows: when the control stick is moved to a new position, the position A of the potentiometer P changes, creating a voltage across the points A and B. This voltage activates the electromagnet and the piston of the valve of the hydraulic servomotor is moved (see Subsec. 3.12.6). The movement of the valve will allow the oil under pressure to enter the power cylinder and to push its piston right or left, moving the wings of the aircraft downwards or upwards. This way, the pilot can control the inclination of the wings.

6 Missile Direction Control System (Figure 1.14)

This system directs a missile to destroy an enemy aircraft. The system works as follows: the guided missile, as well as its target, are monitored by a radar system. The information acquired by the radar is fed into a computer, which estimates the possible course of the enemy aircraft. The missile's course changes, as new data of the aircraft's course is received. The computer constantly compares the two courses and makes the necessary corrections in the direction of the missile so that it strikes the target.

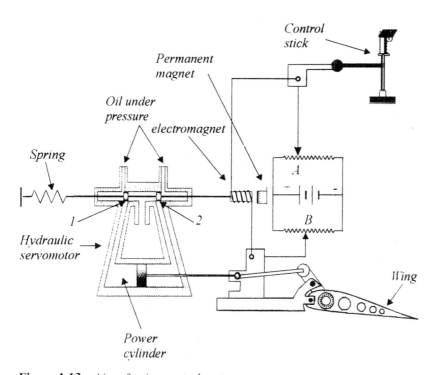

Figure 1.13 Aircraft wing control system.

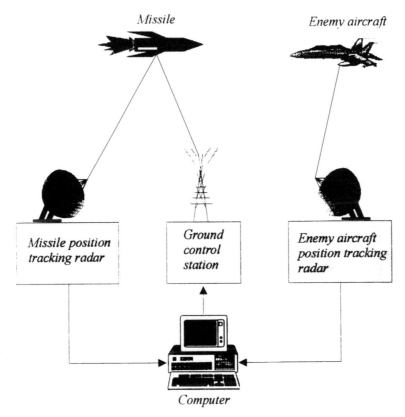

Figure 1.14   Missile direction control system.

7 Paper-Making Control System (Figure 1.15)

This system is designed such that the output $y(t)$ of the system, i.e., the consistency of the dilution of the thick pulp, remains constant. This is accomplished as follows: the pulp is stored in a large container, it is constantly rotated by a mixing mechanism to maintain pulp uniformity. Subsequently, as the pulp is driven into the drying and rolling stations of the paper-making industrial plant, water is added, which dilutes the thick pulp to a desired consistency $r(t)$. The actual consistency $y(t)$ is measured by an appropriate device and is compared with the desired consistency $r(t)$. The controller compares $r(t)$ and $y(t)$. If $y(t) \neq r(t)$, then the output of the controller $u(t)$ adjusts the water valve such that $y(t) = r(t)$.

8 Nuclear Reactor Control System (Figure 1.16)

The major control objective of a nuclear reactor is to maintain the output power within specified limits. This can be achieved as follows. The nuclear reaction releases energy in the form of heat. This energy is used for the production of steam. The steam is subsequently used to drive a turbine and, in turn, the turbine drives a generator, which finally produces electric power.

   The reference signal $r(t)$ corresponds to the desired output power, whereas $y(t)$ is the actual output power. The two signals $r(t)$ and $y(t)$ are compared and their

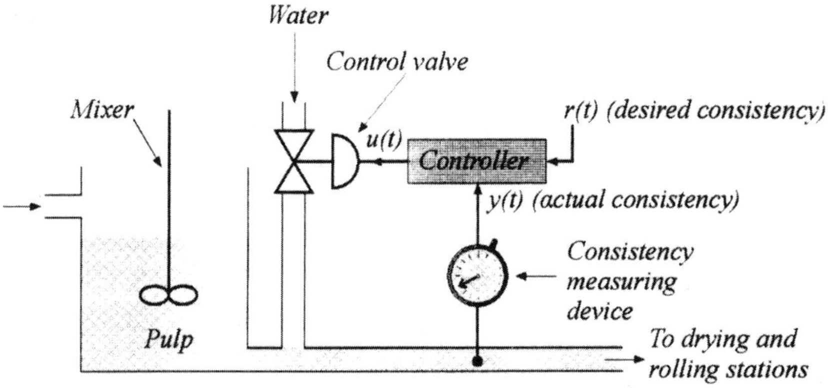

Figure 1.15 Paper-making control system.

difference $e(t) = r(t) - y(t)$ is fed into the control unit. The control unit consists of special rods which, when they move towards the point where the nuclear reaction takes place, result in an increase of the output power $y(t)$, and when they move away, result in a decrease of the output power $y(t)$. When $y(t) > r(t)$, the error $e(t)$ is negative, the rods move away from the point of the nuclear reaction and the output $y(t)$ decreases. When $y(t) < r(t)$, the error $e(t)$ is positive, the rods move towards the point of the nuclear reaction and the output $y(t)$ increases. This way, the power output $y(t)$ follows the desired value $r(t)$.

9 Boiler–Generator Control System (Figure 1.17)

The boiler–generator control system operates as follows: the steam produced by the boiler sets the shaft in rotation. As the shaft rotates, the generator produces electric power. Here, we have a system with many inputs (water, air, and liquid fuel) and one output (electric power). The electric power is automatically controlled as follows: the output power, along with intermediate variables or states of the system, such as oxygen, temperature, and pressure, are fed back to the controller, namely, to the computer. The computer regulates automatically the amount of water, air, and liquid fuel that should enter the boiler, as well as the angular velocity of the shaft, depending on the desired and real (measured) values of temperature, pressure, oxygen, and electric power, such that the electric power output is the desired one. Clearly, the controller here is the digital computer. Systems that are controlled by a computer are usually called *computer-controlled systems* or *digital control systems* and they are studied in Chapter 12.

10 Remote Robot Control (Figure 1.18)

Here, we consider a remote control system that can control, from the earth, the motion of a robot arm on the surface of the moon. As shown in Figure 1.18a, the operator at earth station watches the robot on the moon on a TV monitor. The system's output is the position of the robot's arm and the input is the position of the control stick. The operator compares the desired and the real position of the robot's arm, by looking at the position of the robot's arm on the monitor and decides on how to move the control stick so that the position of the robot arm is the desired one.

(a)

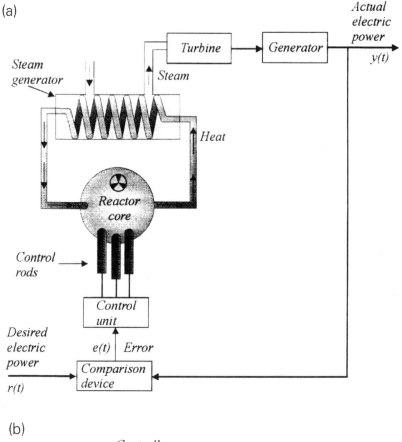

(b)

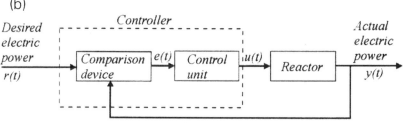

Figure 1.16   Control system of a nuclear reactor. (a) Overall picture of the nuclear reactor; (b) schematic diagram of the system.

In Figure 1.18b a schematic diagram of this system is given. In this example, the operator is part of the controller.

11 Machine Tool Control (Figure 1.19)

A simplified scheme of a machine tool for cutting (or shaping or engraving) metals is shown in Figure 1.19. The motion of the cutting tool is controlled by a computer. This type of control is termed *numerical control*. When there is a difference between the desired position $r(t)$ and the actual position $y(t)$ of the cutting tool, the amplifier amplifies this difference so that the output current of the amplifier is large enough to

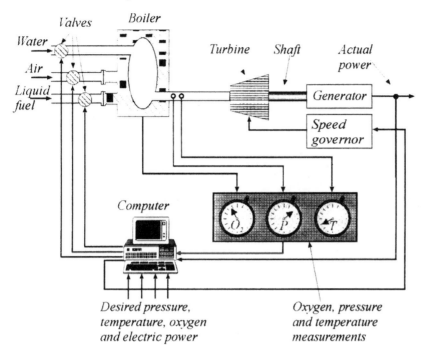

Figure 1.17 Boiler–generator control system [16].

activate the coil. The magnetic field produced around the coil creates a force on the piston of the valve of the hydraulic servomotor, moving it to the left or to the right. These small movements of the piston result in controlling the position of the cutting tool in such a way that $y(t) = r(t)$.

12 Ship Stabilization (Figure 1.20)

This example refers to the stabilization of ship oscillations due to waves and strong winds. When a ship exhibits a deviation of $\theta°$ from the vertical axis, as shown in Figure 1.20, then most ships use fins to generate an opposite torque, which restores the ship to the vertical position. In Figure 1.20b the block diagram of the system is given, where, obviously, $\theta_r(t) = 0$ is the desired position of the ship. The length of the fins projecting into the water is controlled by an actuator. The deviation from the vertical axis is measured by a measuring device. Clearly, when the error $e(t) = \theta_r(t) - \theta_y(t) \neq 0$, then the fin actuator generates the proper torque, such that the error $e(t)$ goes to zero, i.e., the ship position is restored to normal $(\theta_r(t) = 0)$.

13 Orientation Control of a Sun-Seeker System (Figure 1.21)

The sun-seeker automatic control system is composed of a telescope, two light-sensing cells, an amplifier, a motor, and gears. The two light-sensing cells are placed on the telescope in such a way that when the telescope is not aligned with the sun, one cell receives more light than the other. The two cells behave as current sources and are conected with opposite polarity, so that when one of the

(a)

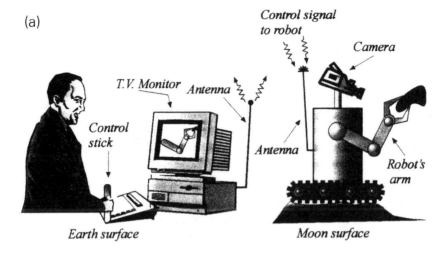

(b)

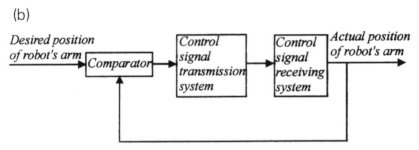

Figure 1.18 Remote robot control system. (a) Overall picture of remote robot control system; (b) schematic diagram of remote robot control.

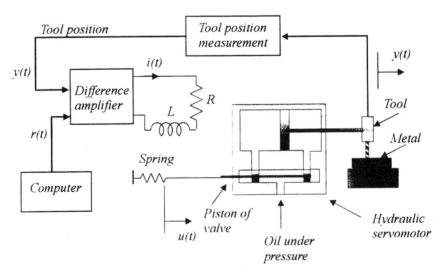

Figure 1.19 Machine tool control [16].

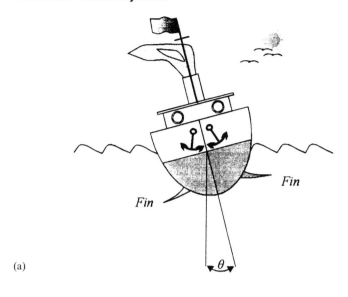

(a)

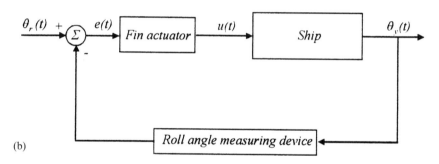

(b)

Figure 1.20 Ship stabilization control system. (a) Ship in roll position; (b) simplified block diagram of ship stabilization control system.

two cells gets more light, a current I_s is produced, which is equal to the difference of the two currents of the cells. This current is subsequently driven into the amplifier. The output of the amplifier is the input to the motor, which in turn moves the gears in such a way as to align the telescope with the sun, i.e., such that $\theta_y(t) = \theta_r(t)$, where $\theta_r(t)$ is the desired telescope angle and $\theta_y(t)$ is the actual telescope angle.

14 Laser Eye Surgery Control System (Figure 1.22)

Lasers can be used to "weld" the retina of the eye in its proper position inside the eye in cases where the retina has been detached from its original place. The control scheme shown in Figure 1.22 is of great assistance to the ophthalmologist during surgery, since the controller continuously monitors the retina (using a wide-angle video camera system) and controls the laser's position so that each lesion of the retina is placed in its proper position.

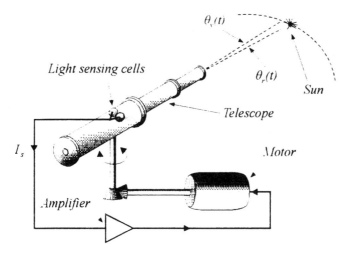

Figure 1.21 Orientation control of a sun-seeker system [31].

15 Wheelchair

The automatic wheelchair is especially designed for people disabled from their neck down. It is actually a system which the disabled person activates by moving his head. In so doing, he determines both the direction and the speed of the wheelchair. The direction is determined by sensors placed on the person's head 90° apart, so that he may choose one of the following four movements: forward, backward, left, or right. The speed is determined by another sensor whose output is proportional to the speed of the head movement. Clearly, here, the man is the controller.

16 Economic Systems (Figure 1.23)

The concept of closed-loop control systems also appears in economic and social systems. As an example, consider the inflation control system presented in Figure

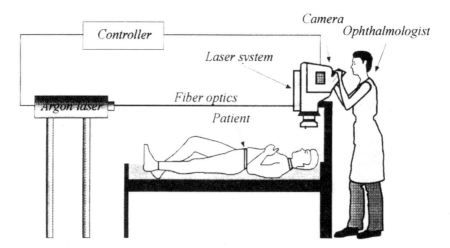

Figure 1.22 Laser eye surgery control system.

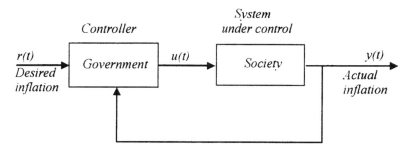

Figure 1.23   Schematic diagram of inflation control system.

1.23. Here, the input $r(t)$ to the system is the desired level of inflation. The system under control is the society, and $y(t)$ is the actual inflation. The government operates as a controller, comparing the desired inflation $r(t)$ and the actual inflation $y(t)$. If $y(t) \leq r(t)$, no action takes place. If $y(t) > r(t)$, the controller (i.e., the government) takes the necessary decisions so as to keep $y(t) \leq r(t)$. The same closed-loop scheme may be used to describe a variety of economic systems, as for example unemployment and national income. In example 4.8.1, the operation of a company is also described as a closed-loop system, wherein its mathematical model is used to study the performance of the company.

17 Human Speech (Figure 1.24)

As we all know, we use our ears not only to hear others but also to hear ourselves. Indeed, when we speak, we hear what we are saying and, if we realize that we didn't say something the way we had in mind to say it, we immediately correct it. Thus, human speech operates as a closed-loop system, where the reference input $r(t)$ is what we have in mind to say and want to put into words, the system is the vocal cords, and its output $y(t)$ is our voice. The output $y(t)$ is continuously monitored by our ears, which feed back our voice to our brain, where comparison is made between our-intended (desired) speech $r(t)$ and the actual speech $y(t)$ that our own ears hear (measure). If the desired speech $r(t)$ and the "measured" speech $y(t)$ are the same, no correction is necessary, and we keep on talking. If, however, an error is realized, e.g., in a word or in a number, then we immediately make the correction by saying

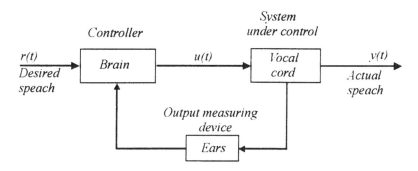

Figure 1.24   Block diagram of human speech.

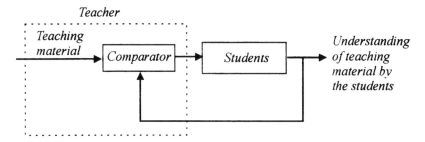

Figure 1.25   Schematic diagram of teaching.

the correct word or number. The reason that some people talk very loud is that their hearing is not very good and, in order to be able to hear themselves (so as to make the necessary speech corrections), they talk louder than normal.

18 Teaching (Figure 1.25)

The proper procedure for teaching has the structure of a closed-loop system. Let the students be the system, the teaching material presented by the teacher the input, and the "degree" of understanding of this material by the students the system's output. Then, teaching can be described with the schematic diagram of Figure 1.25. This figure shows that the system's output, i.e., the degree of understanding by students of the material taught, is fed back to the input, i.e., to the teacher. Indeed, an experienced teacher should be able to "sense" (measure) if the students understood the material taught. Subsequently, the teacher will either go on teaching new material, if the students understood the material taught, or repeat the same material, if they did not. Therefore, proper teaching has indeed the structure of a closed-loop system. Clearly, if teachers keep on teaching new material without checking whether or not the students understand what they are saying, this is not proper teaching.

BIBLIOGRAPHY

Books

1. J Ackermann. Sampled Data Control Systems. New York: Springer Verlag, 1985.
2. PJ Antsaklis, AN Michel. Linear Systems. New York: McGraw-Hill, 1997.
3. KJ Astrom. Introduction to Stochastic Control Theory. New York: Academic Press, 1970.
4. KJ Astrom, B Wittenmark. Computer Controlled Systems: Theory and Design. 3rd edn. Englwood Cliffs, New Jersey: Prentice Hall, 1997.
5. KJ Astrom, B Wittenmark. Adaptive Control. 2nd edn. New York: Addison-Wesley, 1995.
6. M Athans, PL Falb. Optimal Control. New York: McGraw-Hill, 1966.
7. R Bellman. Dynamic Programming. Princeton, New Jersey: Princeton University Press, 1957.
8. R Bellman. Adaptive Control Processes, A Guided Tour. Princeton, New Jersey: Princeton University Press, 1961.
9. S Bennet. Real-Time Computer Control, An Introduction. New York: Prentice Hall, 1988.
10. J Billingsley. Controlling with Computers. New York: McGraw-Hill, 1989.

11. HW Bode. Network Analysis and Feedback Amplifier Design. New York: Van Nostrand, 1945.
12. RW Brockett. Finite Dimensional Linear Systems. New York: John Wiley, 1970.
13. JA Cadzow. Discrete-Time Systems, An Introduction with Interdisciplinary Applications. Englewood Cliffs, New Jersey: Prentice Hall, 1973.
14. CT Chen. Linear System Theory and Design. New York: Holt, Rinehart and Winston, 1984.
15. JJ D'Azzo, CH Houpis. Linear Control System Analysis and Design, Conventional and Modern. New York: McGraw-Hill, 1975.
16. RC Dorf, RE Bishop. Modern Control Analysis. London: Addison-Wesley, 1995.
17. JC Doyle, BA Francis, A Tannenbaum. Feedback Control theory. New York: Macmillan, 1992.
18. WR Evans. Control Systems Dynamics. New York: McGraw-Hill, 1954.
19. GF Franklin, JD Powell, A Emami-Naeini. Feedback Control of Dynamic Systems. Reading, MA: Addison-Wesley, 1986.
20. GF Franklin, JD Powell, ML Workman. Digital Control of Dynamic Systems. 2nd edn. London: Addison-Wesley, 1990.
21. M Santina, A Stubbersud, G Hostetter. Digital Control System Design. Orlando, Florida: Saunders College Publishing, 1994.
22. CH Houpis, GB Lamont. Digital Control Systems. New York: McGraw-Hill, 1985.
23. R Iserman. Digital Control Systems, Vols I and II. Berlin: Springer Verlag, 1989.
24. EI Jury. Theory and Application of the Z-Transform Method. New York: John Wiley, 1964.
25. EI Jury. Sampled Data Control Systems. New York: John Wiley, 1958. Huntington, New York: Robert E Krieger, 1973 (2nd edn).
26. T Kailath. Linear Systems. Englewood Cliffs, New Jersey: Prentice Hall, 1980.
27. RE Kalman. The theory of optimal control and the calculus of variations. In: R Bellman, ed. Mathematical Optimization Techniques, Berkeley, California: University of California Press, 1963.
28. RE Kalman, PL Falb, MA Arbib. Topics in Mathematical System Theory. New York: McGraw-Hill, 1969.
29. P Katz. Digital Control Using Microprocessors. London: Prentice Hall, 1981.
30. V Kucera. Discrete Linear Control, The Polynomial Equation Approach. New York: John Wiley, 1979.
31. BC Kuo. Automatic Control Systems. London: Prentice Hall, 1995.
32. BC Kuo. Digital Control Systems. Orlando, Florida: Saunders College Publishing, 1992.
33. DG Luenberger. Optimatization by Vector Space Methods. New York: John Wiley, 1969.
34. AGC MacFarlane. Dynamic Systems Models. London: George H Harrap, 1970.
35. O Mayr. The Origins of Feedback Control. Cambridge, Massachusetts: MIT Press, 1970.
36. RH Middleton, GC Goodwin. Digital Control and Estimation: A Unifieid Approach. New York: Prentice Hall, 1990.
37. G Newton, L Gould, J Kaiser. Analytical Design of Linear Feedback Controls. New York: John Wiley & Sons, 1957.
38. NS Nise. Control Systems Engineering. New York: Benjamin and Cummings, 1995.
39. K Ogata. Modern Control Systems. London: Prentice-Hall, 1997.
40. K Ogata. Discrete-Time Control Systems. Englewood Cliffs, New Jersey: Prentice Hall, 1987.
41. M Otto. Origins of Feedback Control. Cambridge, Massachusetts: MIT Press, 1971.
42. PN Paraskevopoulos. Digital Control Systems. London: Prentice Hall, 1996.
43. CL Phillips, HT Nagle Jr. Digital Control Systems Analysis and Design. Englewood Cliffs, New Jersey, 1984.

44. JR Ragazzini, GF Franklin. Sampled Data Control Systems. New York: McGraw-Hill, 1958.
45. HH Rosenbrock. State Space and Multivariable Theory. London: Nelson, 1970.
46. HH Rosenbrock. Computer Aided Control System Design. New York: Academic Press, 1974.
47. EJ Routh. A Treatise on the Stability of a Given State of Motion. London: Macmillan, 1877.
48. GN Saridis. Self-Organizing Control of Stochastic Systems. New York: Marcel Dekker, 1977.
49. V Strejc. State Space Theory of Discrete Linear Control. New York: Academia Prague and John Wiley, 1981.
50. GJ Thaler. Automatic Control Systems. St. Paul, Minn.: West Publishing, 1989.
51. AIG Vardulakis. Linear Multivariable Control. Algebraic Analysis and Synthesis Methods. New York: Wiley, 1991.
52. IA Vyshnegradskii. On Controllers of Direct Action. Izv.SPB Teckhnolog. Inst., 1877.
53. N Wiener. The Extrapolation, Interpolation and Smoothing of Stationary Time Series. New York: John Wiley, 1949.
54. WM Wohnam. Linear Multivariable Control: A Geometric Approach. 2nd edn. New York: Springer Verlag, 1979.
55. WA Wolovich. Linear Multivariable Systems. New York: Springer Verlag, 1974.
56. LA Zadeh, CA Desoer. Linear System Theory – The State Space Approach. New York: McGraw-Hill, 1963.

Articles

57. M Athans. Status of optimal control theory and applications for deterministic systems. IEEE Trans Automatic control AC-11:580–596, 1966.
58. M Athans. The matrix minimum principle. Information and Control 11:592–606, 1968.
59. M Athans. The role and use of the stochastic linear quadratic-gaussian problem in control system design. IEEE Trans Automatic Control AC-16:529–551, 1971.
60. HS Black. Stabilized feedback amplifiers. Bell System Technical Journal, 1934.
61. HS Black. Inverting the negative feedback amplifier. IEEE Spectrum 14:54–60, 1977.
62. RW Brockett. Poles, zeros and feedback: state space interpretation. IEEE Trans Automatic control AC-10:129–135, 1965.
63. JC Doyle, K Glover, PP Khargonekar, BA Francis. State space solutions to standard H_2 and H_∞ control problems. IEEE Trans Automatic Control 34:831–847, 1989.
64. WR Evans. Control system synthesis by root locus method. AIEE Trans, Part II, 69:66–69, 1950.
65. AT Fuller. The early development of control theory. Trans ASME (J Dynamic Systems, Measurement & Control) 96G:109–118, 1976.
66. BA Francis, JC Doyle. Linear control theory with an H_∞ optimally criterion. SIAM J Control and Optimization 25:815–844, 1987.
67. T Kailath. A view of three decades of linear filtering theory. IEEE Trans Information Theory 20:146–181, 1974.
68. RE Kalman. A new approach to linear filtering and prediction problems. Trans ASME (J Basic engineering) 82D:35–45, 1960.
69. RE Kalman. On the general theory of control systems. Proceedings First International Congress, IFAC, Moscow, USSR, 1960, pp 481–492.
70. RE Kalman. Contributions to the theory of optimal control. Proceedings 1959 Mexico City Conference on Differential Equations, Mexico City, 1960, pp 102–119.
71. RE Kalman. Canonical structure of linear dynamical systems. Proceedings National Academy of Science 48:596–600, 1962.

72. RE Kalman. Mathematical description of linear dynamic systems. SIAM J Control 1:152–192, 1963.
73. RE Kalman. When is a linear control system optimal? Trans ASME (J Basic Engineering) 86D:51–60, 1964.
74. RE Kalman, JE Bertram. A unified approach to the theory of sampling systems. J Franklin Institute 405–524, 1959.
75. RE Kalman, JE Bertram. Control systems analysis and design via second method of Liapunov: II discrete-time systems. Trans ASME (J Basic Engineering) D:371–400, 1960.
76. RE Kalman, RS Bucy. New results in linear filtering and prediction theory. Trans ASME (J Basic Engineering) 83D:95–108, 1961.
77. RE Kalman, RW Koepcke. Optimal synthesis of linear sampling control systems using generalized performance indices. Trans ASME 80:1820–1826, 1958.
78. RE Kalman, YC Ho, KS Narenda. Controllability of linear dynamic systems. Contributions to Differential Equations 1:189–213, 1963.
79. RE Kalman et al. Fundamental study of adaptive control systems. Wright-Patternson Air Force Base Technical Report, ASD-TR-61-27, April 1962.
80. DG Luenberger. Observingt the state of a linear system. IEEE Trans Military Electronics MIL-8:74–80, 1964.
81. DG Luenberger. An introduction to obsevers. IEEE Trans Automatic Control AC-16:596–602, 1971.
82. JC Maxwell. On governors. Proceedings of the Royal Society of London Vol. 16, 1868. See also Selected Papers on Mathematical Trends in Control Theory. New York: Dover, 1964, pp 270–283.
83. JC Maxwell. On governors. Philosophical Magazine 35:385–398, 1868.
84. H Nyquist. Certain topics in telegraph transmission theory. Trans AIEE 47:617–644, 1928.
85. H Nyquist. Regeneration theory. Bell System Technical J 11:126–147, 1932.
86. Special issue on linear-quadratic-gaussian problem. IEEE Trans Automatic Control AC-16: December 1971.
87. Special issue on identification and system parameter estimation. Automatica 17, 1981.
88. Special issue on linear multivariable control systems. IEEE Trans on Automatic Control AC-26, 1981.
89. WM Wonham. On pole assignment in multi-input controllable systems. IEEE Trans Automatic Control AC-12:660–665, 1967.
90. WM Wonham. On the separation theorem of stochastic control. SIAM J Control 6:312–326, 1968.
91. G Zames. Feedback and optimal sensitivity: model reference transformations, multiplicative seminorms and approximate inverses. IEEE Trans Automatic Control 26:301–320, 1981.
92. JG Ziegler, NB Nichols. Optimum settings for automatic controllers. Trans ASME 64:759–768, 1942.

2

Mathematical Background

2.1 INTRODUCTION

This chapter covers certain mathematical topics necessary for the study of control systems: in particular, it aims to offer the appropriate mathematical background on subjects such as basic control signals, Laplace transform, and the theory of matrices. This background is very useful for most of the material of the book that follows.

2.2 THE BASIC CONTROL SIGNALS

In this section we present definitions of the following basic control signals: the step function, the gate function, the impulse function, the ramp function, the exponential function, and the sinusoidal function. These signals are of major importance for control applications.

1 The Unit Step Function

The unit step function is designated by $u(t - T)$ and is defined as follows:

$$u(t - T) = \begin{bmatrix} 1 & \text{for } t > T \\ 0 & \text{for } t < T \\ \text{undefined for } t = T \end{bmatrix} \tag{2.2-1}$$

The graphical representation of $u(t - T)$ is shown in Figure 2.1. The amplitude of $u(t - T)$, for $t > T$, is equal to 1. This is why the function $u(t - T)$ is called the "unit" step function.

A physical example of a unit step function is the switch of the circuit shown in Figure 2.2. It is obvious that the voltage $v_R(t)$ is given by:

$$v_R(t) = \begin{bmatrix} v(t) & \text{for } t > T \\ 0 & \text{for } t < T \\ \text{undefined} & \text{for } t = T \end{bmatrix}$$

or

$$v_R(t) = v(t)u(t - T)$$

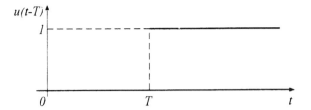

Figure 2.1   The unit step function.

Here, the role of the switch is expressed by the unit step function $u(t - T)$.

2 The Unit Gate Function

The unit gate (or window) function is denoted by $g_\pi(t) = u(t - T_1) - u(t - T_2)$, where $T_1 < T_2$, and is defined as follows:

$$g_\pi(t) = \begin{bmatrix} 1 & \text{for } t \in (T_1, T_2) \\ 0 & \text{for } t \neq (T_1, T_2) \\ \text{undefined} & \text{for } t = T_1 \text{ and } t = T_2 \end{bmatrix} \tag{2.2-2}$$

The graphical representation of $g_\pi(t)$ is given in Figure 2.3. The unit gate function is usually used to zero all values of another function, outside a certain time interval. Consider for example the function $f(t)$. Then, the function $y(t) = f(t)g_\pi(t)$ is as follows:

$$y(t) = f(t)g_\pi(t) = \begin{bmatrix} f(t) & \text{for } T_1 \leq t \leq T_2 \\ \\ 0 & \text{for } t < T_1 \text{ and for } t > T_2 \end{bmatrix}$$

3 The Unit Impulse Function

The unit impulse function, which is also called the Dirac function, is designated by $\delta(t - T)$ and is defined as follows:

$$\delta(t - T) = \begin{bmatrix} 0 & \forall t, & \text{except for } t = T \\ \infty & & \text{for } t = T \end{bmatrix} \tag{2.2-3}$$

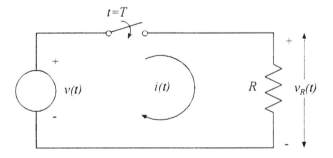

Figure 2.2   The switch as the unit step function.

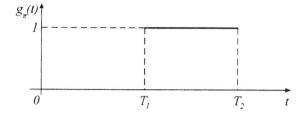

Figure 2.3 The unit gate function.

The graphical representation of $\delta(t - T)$ is given in Figure 2.4. In Figure 2.5 $\delta(t - T)$ is defined in a different way as follows: the area $c(t)$ of the parallelogram is $c(t) = (1/a)a = 1$. As a becomes larger, the base of the parallelogram $1/a$ becomes smaller. In the limit, as the height a tends to infinity, the base $1/a$ tends to zero, i.e.,

$$\delta(t - T) \text{ occurs when } \lim_{a \to \infty} c(t) \tag{2.2-4}$$

From definition (2.2-4) we readily have

$$\int_{-\infty}^{\infty} \delta(t - T) \, dt = 1 \tag{2.2-5}$$

Relation (2.2-5) shows that the area of the unit impulse function is equal to 1 (this is why it is called the "unit" impulse function).

The functions $u(t - T)$ and $\delta(t - T)$ are related as follows:

$$\delta(t - T) = \frac{du(t - T)}{dt} \quad \text{and} \quad u(t - T) = \int_{-\infty}^{t} \delta(\lambda - T) \, d\lambda \tag{2.2-6}$$

Finally, we have the following interesting property of $\delta(t - T)$: consider a function $x(t)$ with the property $|x(t)| < \infty$, then

$$\int_{-\infty}^{\infty} x(t)\delta(t - T) \, dt = x(T) \tag{2.2-7}$$

4 The Ramp Function

The ramp function is designated by $r(t - T)$ and is defined as follows:

$$r(t - T) = \begin{bmatrix} t - T & \text{for } t > T \\ 0 & \text{for } t \leq T \end{bmatrix} \tag{2.2-8}$$

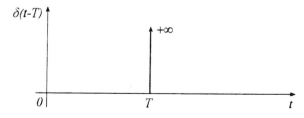

Figure 2.4 The unit impulse function.

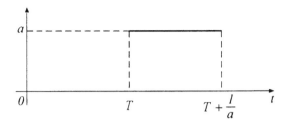

Figure 2.5   The area function $c(t)$.

The graphical representation of $r(t - T)$ is shown in Figure 2.6. It is obvious that $u(t - T)$ and $r(t - T)$ are related as follows:

$$u(t - T) = \frac{dr(t - T)}{dt} \qquad \text{and} \qquad r(t - T) = \int_{-\infty}^{t} u(\lambda - T)\,d\lambda$$

Remark 2.2.1

All the above functions are usually applied when $T = 0$. In cases where $T > 0$, then the function is delayed by T units of time, whereas when $T < 0$ the function is preceding by T units of time.

5 The Exponential Function

The exponential function is the function $f(t) = Ae^{at}$ and its graphical representation is shown in Figure 2.7.

6 The Sinusoidal Function

The sinsoidal function is the function $f(t) = A\sin(\omega t + \theta)$ and its graphical representation is shown in Figure 2.8.

Remark 2.2.2

All functions presented in this section can be expressed in terms of exponential functions or derived from the exponential function, a fact which makes the exponential function very interesting. This can easily be shown as follows: (a) the sinusoidal function is a linear combination of two exponential functions, e.g., $\sin\theta = (1/2j)(e^{j\theta} - e^{-j\theta})$; (b) the unit step function for $T = 0$ is equal to the exponential function when $A = 1$ and $a = 0$, i.e., $u(t) = f(t) = Ae^{at} = 1$, for $A = 1$ and $a = 0$; (c) the functions $\delta(t - T)$ and $r(t - T)$ can be derived from the unit step

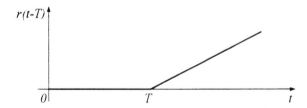

Figure 2.6   The ramp function.

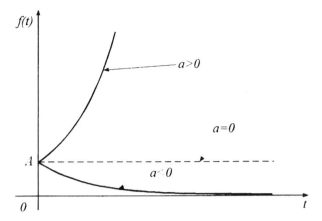

Figure 2.7 The exponential function.

function $u(t - T)$, while $u(t - T)$ may be derived from the exponential function. Furthermore, a periodic function can be expressed as a linear combination of exponential functions (Fourier series). Moreover, it is worth mentioning that the exponential function is used to describe many physical phenomena, such as the response of a system and radiation of nuclear isotopes.

2.3 THE LAPLACE TRANSFORM

To study and design control systems, one relies to a great extent on a set of mathematical tools. These mathematical tools, an example of which is the Laplace transform, facilitate the engineer's work in understanding the problems he deals with as well as solving them.

For the special case of linear time-invariant continuous time systems, which is the main subject of the book, the Laplace transform is a very important mathematical tool for the study and design of such systems. The Laplace transform is a special case of the generalized integral transform presented just below.

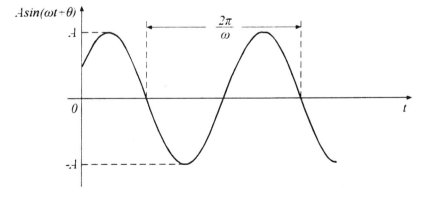

Figure 2.8 The sinusoidal function.

2.3.1 The Generalized Linear Integral Transform

The generalized linear integral transform of a function $f(t)$ is defined as follows:

$$F(s) = \int_a^b f(t)k(s, t)\,\mathrm{d}t \tag{2.3-1}$$

where $k(s, t)$ is known as the kernel of the transform. It is clear that the main feature of Eq. (2.3-1) is that it transforms a function defined in the t domain to a function defined in the s domain. A particular kernel $k(s, t)$, together with a particular time interval (a, b), define a specific transform.

2.3.2 Introduction to Laplace Transform

The Laplace transform is a linear integral transform with kernel $k(s, t) = e^{-st}$ and time interval $(0, \infty)$. Therefore, the definition of the Laplace transform of a function $f(t)$ is as follows:

$$L\{f(t)\} = \int_0^\infty f(t)e^{-st}\,\mathrm{d}t = F(s) \tag{2.3-2}$$

where L designates the Laplace transform and s is the complex variable defined as $s = \sigma + j\omega$. Usually, the time function $f(t)$ is written with a small f, while the complex variable function $F(s)$ is written with a capital F.

For the integral (2.3-2) to converge, $f(t)$ must satisfy the condition

$$\int_0^\infty |f(t)|e^{-\sigma t}\,\mathrm{d}t \leq M \tag{2.3-3}$$

where σ and M are finite positive numbers.

Let $L\{f(t)\} = F(s)$. Then, the inverse Laplace transform of $F(s)$ is also a linear integral transform, defined as follows:

$$L^{-1}\{F(s)\} = \frac{1}{2\pi j} \int_{c-j\infty}^{c+j\infty} F(s)e^{st}\,\mathrm{d}s = f(t) \tag{2.34}$$

where L^{-1} designates the inverse Laplace transform, $j = \sqrt{-1}$, and c is a complex constant.

Clearly, the Laplace transform is a mathematical tool which transforms a function from one domain to another. In particular, it transforms a time-domain function to a function in the frequency domain and vice versa. This gives the flexibility to study a function in both the time domain and the frequency domain, which results in a better understanding of the function, its properties, and its time-domain frequency-domain properties.

A popular application of the Laplace transform is in solving linear differential equations with constant coefficients. In this case, the motivation for using the Laplace transform is to simplify the solution of the differential equation. Indeed, the Laplace transform greatly simplifies the solution of a constant coefficient differential equation, since it reduces its solution to that of solving a linear algebraic equation. The steps of this impressive simplification are shown in the bottom half of Figure 2.9. These steps are analogous to the steps taken in the case of multiplying numbers using logarithms, as shown in the top half of Figure 2.9. The analogy here is

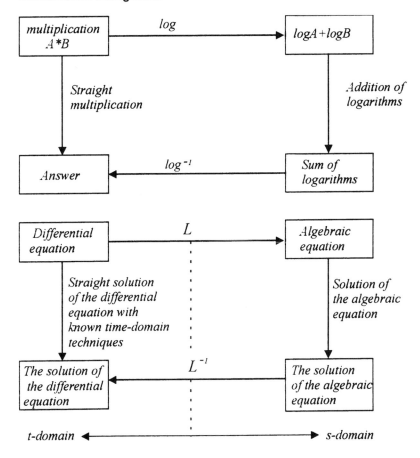

Figure 2.9   Comparison of logarithms with the Laplace transform.

that logarithms reduce the *multiplication* of two numbers to the *sum* of their logarithms, while the Laplace transform reduces the solution of a *differential* equation to an *algebraic* equation.

2.3.3   Properties and Theorems of the Laplace Transform

The most important properties and theorems of the Laplace transform are presented below.

1 Linearity

The Laplace transform is a linear transformation, i.e., the following relation holds

$$L\{c_1 f_1(t) + c_2 f_2(t)\} = L\{c_1 f_1(t)\} + L\{c_2 f_2(t)\} = c_1 F_1(s) + c_2 F_2(s) \qquad (2.3\text{-}5)$$

where c_1 and c_2 are constants, $F_1(s) = L\{f_1(t)\}$ and $F_2(s) = L\{f_2(t)\}$.

2 The Laplace Transform of the Derivative of a Function

Let $f^{(1)}(t)$ be the time derivative of $f(t)$, and $F(s)$ be the Laplace transform of $f(t)$. Then, the Laplace transform of $f^{(1)}(t)$ is given by

$$L\left\{f^{(1)}(t)\right\} = sF(s) - f(0) \tag{2.3-6}$$

Proof

From definition (2.3-2) we have that

$$L\left\{f^{(1)}(t)\right\} = \int_0^\infty f^{(1)}(t)e^{-st}\,dt = \int_0^\infty e^{-st}\,df(t) = f(t)e^{-st}\bigg]_0^\infty + s\int_0^\infty f(t)e^{-st}\,dt$$

$$= sF(s) - f(0)$$

where use was made of the integration-by-parts method. Working in the same way for the Laplace transform of the second derivative $f^{(2)}(t)$ of $f(t)$, we have that

$$L\left\{f^{(2)}(t)\right\} = s^2 F(s) - sf(0) - f^{(1)}(0) \tag{2.3-7}$$

For the general case we have

$$L\left\{f^{(n)}(t)\right\} = s^n F(s) - s^{n-1}f(0) - s^{n-2}f^{(1)}(0) - \ldots - f^{(n-1)}(0)$$

$$= s^n F(s) - \sum_{k=0}^{n-1} s^k f^{(n-k-1)}(0) \tag{2.3-8}$$

where $f^{(m)}(t)$ is the mth time derivative of $f(t)$.

3 The Laplace Transform of the Integral of a Function

Let $\int_{-\gamma}^t f(\lambda)\,d\lambda$ be the integral of a function $f(t)$, where γ is a positive number and $F(s)$ is the Laplace transform of $f(t)$. Then, the Laplace transform of the integral is given by

$$L\left\{\int_{-\gamma}^t f(\lambda)\,d\lambda\right\} = \frac{F(s)}{s} + \frac{f^{(-1)}(0)}{s} \tag{2.3-9}$$

where $f^{(-1)}(0) = \int_{-\gamma}^0 f(t)\,dt$.

Proof

From definition (2.3-2) we have that

$$L\left\{\int_{-\gamma}^t f(\lambda)\,d\lambda\right\} = \int_0^\infty \left[\int_{-\gamma}^t f(\lambda)\,d\lambda\right]e^{-st}\,dt = -\frac{1}{s}\int_0^\infty \left[\int_{-\gamma}^t f(\lambda)\,d\lambda\right]d[e^{-st}]$$

$$= -\frac{1}{s}\left\{\left[\int_{-\gamma}^t f(\lambda)\,d\lambda\right]e^{-st}\bigg]_0^\infty - \int_0^\infty e^{-st}f(t)\,dt\right\}$$

$$= \frac{1}{s}\int_0^\infty f(t)e^{-st}\,dt + \frac{1}{s}\int_{-\gamma}^0 f(t)\,dt = \frac{F(s)}{s} + \frac{f^{(-1)}(0)}{s}$$

where use was made of the integration-by-parts method.

Working in the same way, we may determine the Laplace transform of the double integral $\int_{-\gamma}^t \int_{-\gamma}^t f(\lambda)(d\lambda)^2$ to yield

$$L\left\{\int_{-\gamma}^t \int_{-\gamma}^t f(\lambda)(d\lambda)^2\right\} = \frac{F(s)}{s^2} + \frac{f^{(-1)}(0)}{s^2} + \frac{f^{(-2)}(0)}{s} \tag{2.3-10}$$

where

$$f^{(-2)}(0) = \int_{-\gamma}^{0} \int_{-\gamma}^{0} f(t)(\mathrm{d}t)^2.$$

For the general case we have

$$L\left\{\underbrace{\int_{-\gamma}^{t} \cdots \int_{-\gamma}^{t} f(\lambda)(\mathrm{d}\lambda)^n}_{n\,\text{times}}\right\} = \frac{F(s)}{s^n} + \frac{f^{(-1)}(0)}{s^n} + \frac{f^{(-2)}(0)}{s^{n-1}} + \cdots + \frac{f^{(-n)}(0)}{s} \quad (2.3\text{-}11)$$

where

$$f^{(-k)}(0) = \underbrace{\int_{-\gamma}^{0} \cdots \int_{-\gamma}^{0} f(t)(\mathrm{d}t)^k}_{k\,\text{times}}$$

Remark 2.3.1

In the special case where $f^{(k)}(0) = 0$, for $k = 0, 1, \ldots, n-1$ and $f^{(-k)}(0) = 0$, for $k = 1, 2, \ldots, n$, relations (2.3-8) and (2.3-11) reduce to

$$L\left\{f^{(n)}(t)\right\} = s^n F(s) \quad (2.3\text{-}12a)$$

$$L\left\{\underbrace{\int_{-\gamma}^{t} \cdots \int_{-\gamma}^{t} f(\lambda)(\mathrm{d}\lambda)^n}_{n\,\text{times}}\right\} = \frac{F(s)}{s^n} \quad (2.3\text{-}12b)$$

Relation (2.3-12) points out that the important feature of the Laplace transform is that it greatly simplifies the procedure of taking the derivative and/or the integral of a function $f(t)$. Indeed, the Laplace transform "transforms" the *derivative* of $f(t)$ in the time domain into *multiplying* $F(s)$ by s in the frequency domain. Furthermore, it "transforms" the integral of $f(t)$ in the time domain into *dividing* $F(s)$ by s in the frequency domain.

4 Time Scaling

Consider the functions $f(t)$ and $f(at)$, where a is a positive number. The function $f(at)$ differs from $f(t)$, in time scaling, by a units. For these two functions, it holds that

$$L\{f(at)\} = \frac{1}{a} F\left(\frac{s}{a}\right), \qquad \text{where } F(s) = L\{f(t)\} \quad (2.3\text{-}13)$$

Proof
From definition (2.3-2), we have

$$L\{f(at)\} = \int_{0}^{\infty} f(at)e^{-st}\,\mathrm{d}t = \frac{1}{a} \int_{0}^{\infty} f(at)e^{-\frac{s}{a}(at)}\,\mathrm{d}(at)$$

Setting $\lambda = at$, we arrive at the relation

$$L\{f(at)\} = L\{f(\lambda)\} = \frac{1}{a}\int_0^\infty f(\lambda)e^{-\frac{s\lambda}{a}}\,d\lambda = \frac{1}{a}F\left(\frac{s}{a}\right)$$

5 Shift in the Frequency Domain

It holds that

$$L\{e^{-at}f(t)\} = F(s+a) \tag{2.3-14}$$

Relation (2.3-14) shows that the Laplace transform of the product of the functions e^{-at} and $f(t)$ leads to shifting of $F(s) = L\{f(t)\}$ by a units.

Proof

From definition (2.3-2) we have

$$L\{e^{-at}f(t)\} = \int_0^\infty f(t)e^{-at}e^{-st}\,dt = \int_0^\infty f(t)e^{-(s+a)t}\,dt = F(s+a)$$

6 Shift in the Time Domain

Consider the function $f(t)u(t)$. Then, the function $f(t-T)u(t-T)$ is the same function shifted to the right of $f(t)u(t)$ by T units (Figure 2.10). The Laplace transform of the initial function $f(t)u(t)$ and of the shifted (delayed) function $f(t-T)u(t-T)$, are related as follows:

$$L\{f(t-T)u(t-T)\} = e^{-sT}F(s) \tag{2.3-15}$$

Proof

Setting $\lambda = t - T$, we have

$$L\{f(t-T)u(t-T)\} = L[f(\lambda)u(\lambda)] = \int_0^\infty f(\lambda)u(\lambda)e^{-s(\lambda+T)}d\lambda$$

$$= e^{-sT}\int_0^\infty f(\lambda)u(\lambda)e^{-s\lambda}\,d\lambda$$

$$= e^{-sT}F(s)$$

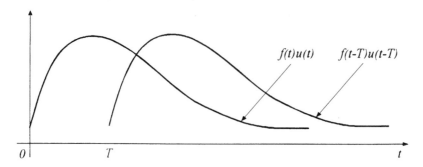

Figure 2.10 Time-delayed function.

7 The Initial Value Theorem

This theorem refers to the behavior of the function $f(t)$ as $t \to 0$ and, for this reason, is called the initial value theorem. This theorem is given by the relation

$$\lim_{t \to 0} f(t) = \lim_{s \to \infty} sF(s) \tag{2.3-16}$$

assuming that the Laplace transform of $f^{(1)}(t)$ exists.

Proof
Taking the limit of both sides of Eq. (2.3-6) as $s \to \infty$, the left-hand side of this relation becomes

$$\lim_{s \to \infty} L\left\{f^{(1)}(t)\right\} = \lim_{s \to \infty} \int_0^\infty f^{(1)}(t)e^{-st}\,dt = 0$$

while the right-hand side becomes

$$\lim_{s \to \infty} [sF(s) - f(0)] = 0 \qquad \text{and hence} \qquad \lim_{t \to 0} f(t) = \lim_{s \to \infty} sF(s)$$

8 The Final Value Theorem

This theorem refers to the behavior of the function $f(t)$ as $t \to \infty$ and, for this reason, it is called the final value theorem. This theorem is given by the relation

$$\lim_{t \to \infty} f(t) = \lim_{s \to 0} sF(s) \tag{2.3-17}$$

assuming that the Laplace transform of $f^{(1)}(t)$ exists and that the denominator of $sF(s)$ has no roots on the imaginary axis or in the right-half complex plane.

Proof
Taking the limit of both sides of Eq. (2.3-6) as $s \to 0$, the left-hand side of this relation becomes

$$\lim_{s \to 0} \int_0^\infty f^{(1)}(t)e^{-st}\,dt = \int_0^\infty f^{(1)}(t)\,dt = \lim_{t \to \infty} \int_0^t f^{(1)}(\lambda)\,d\lambda$$

$$= \lim_{t \to \infty} [f(t) - f(0)] = \lim_{t \to \infty} f(t) - f(0)$$

while the right-hand side becomes

$$\lim_{s \to 0} [sF(s) - f(0)] = \lim_{s \to 0} sF(s) - f(0)$$

Equating the resulting two sides, we readily have relation (2.3-17).

Remark 2.3.2

Clearly, given $F(s)$, one can find the behavior of $f(t)$ as $t \to 0$ and as $t \to \infty$, by first determining the inverse Laplace transform of $F(s)$, i.e., by determining $f(t) = L^{-1}F(s)$ and subsequently determining $f(0)$ and $f(\infty)$ using directly the function $f(t)$. The initial and final value theorems greatly simplify this problem, since they circumvent the rather cumbersome task of determining $L^{-1}F(s)$, and yield the values of $f(0)$ and $f(\infty)$ by directly applying the relations (2.3-16) and (2.3-17), respectively, which are simple to carry out.

9 Multiplication of a Function by t

The following relation holds

$$L\{tf(t)\} = -\frac{\mathrm{d}}{\mathrm{d}s}F(s) \tag{2.3-18}$$

Proof

Differentiating Eq. (2.3-2) with respect to s, we have

$$\frac{\mathrm{d}}{\mathrm{d}s}F(s) = -\int_0^\infty tf(t)\mathrm{e}^{-st}\,\mathrm{d}t = -L\{tf(t)\}$$

Thus relation (2.3-18) is established. In the general case, the following relation holds

$$L\{t^n f(t)\} = (-1)^n \frac{\mathrm{d}^n}{\mathrm{d}s^n}F(s) \tag{2.3-19}$$

10 Division of a Function by t

The following relation holds:

$$L\left\{\frac{f(t)}{t}\right\} = \int_s^\infty F(s)\,\mathrm{d}s \tag{2.3-20}$$

Proof

Integrating Eq. (2.3-2) from s to ∞, we have that

$$\int_s^\infty F(s)\,\mathrm{d}s = \int_s^\infty \left[\int_0^\infty f(t)\mathrm{e}^{-\sigma t}\,\mathrm{d}t\right]\mathrm{d}\sigma = \int_0^\infty \left[\int_s^\infty f(t)\mathrm{e}^{-\sigma t}\,\mathrm{d}\sigma\right]\mathrm{d}t$$

$$= \int_0^\infty \left[-\frac{f(t)}{t}\int_s^\infty \mathrm{d}\mathrm{e}^{-st}\right]\mathrm{d}t = \int_0^\infty \frac{f(t)}{t}\mathrm{e}^{-st}\,\mathrm{d}t = L\left\{\frac{f(t)}{t}\right\}$$

In the general case, the following relation holds

$$L\left\{\frac{f(t)}{t^n}\right\} = \underbrace{\int_s^\infty \cdots \int_s^\infty}_{n\text{ times}} F(\sigma)(\mathrm{d}\sigma)^n \tag{2.3-21}$$

11 Periodic Functions

Let $f(t)$ be a periodic function with period T. Then, the Laplace transform of $f(t)$ is given by

$$L\{f(t)\} = \frac{F_1(s)}{1 - \mathrm{e}^{-sT}}, \qquad F_1(s) = L\{f_1(t)\} \tag{2.3-22}$$

where $f_1(t)$ is the function $f(t)$ during the first period, i.e., for $t \in [0, T]$.

Proof

The periodic function $f(t)$ can be expressed as a sum of time-delay functions as follows:

$$f(t) = f_1(t)u(t) + f_1(t - T)u(t - T) + f_1(t - 2T)u(t - 2T) + \cdots$$

Taking the Laplace transform of $f(t)$, we have

$$Lf(t)\} = F_1(s) + F_1(s)e^{-st} + F_1(s)e^{-2sT} + \cdots = F_1(s)[1 + e^{-sT} + e^{-2sT} + \cdots]$$
$$= \frac{F_1(s)}{1 - e^{-sT}}$$

where use was made of the property (2.3-15).

2.4   THE INVERSE LAPLACE TRANSFORM

The inverse Laplace transform of a function $F(s)$ is given by the relation (2.3-4). To avoid the calculation of the integral (2.3-4), which it is often quite difficult and time consuming, we usually use special tables (see Appendix A) which give the inverse Laplace transform directly. These tables cover only certain cases and therefore they cannot be used directly for the determination of the inverse Laplace transform of any function $F(s)$. The way to deal with cases which are not included in the tables is, whenever possible, to convert $F(s)$ by using appropriate methods, in such a form that its inverse Laplace transform can be found directly in the tables. A popular such method is, when $F(s)$ is a rational function of s (and this is usually the case), to expand $F(s)$ in partial fractions, in which case the inverse Laplace transform is found directly in the tables.

Therefore, our main interest here is to develop a method for expanding a rational function to partial fractions. To this end, consider the rational function

$$F(s) = \frac{b(s)}{a(s)} = \frac{b_m s^m + b_{m-1} s^{m-1} + \cdots + b_1 s + b_0}{s^n + a_{n-1} s^{n-1} + \cdots + a_1 s + a_0}, \qquad m < n \qquad (2.4\text{-}1)$$

Let $\lambda_1, \lambda_2, \ldots, \lambda_n$ be the roots of the polynomial $a(s)$, i.e., let $a(s) = \prod_{i=1}^{n}(s - \lambda_i)$. We distinguish three cases: (a) all roots are real and distinct, (b) all roots are real but not all distinct, and (c) certain or all roots are complex conjugates.

1 Distinct Real Roots

In this case the roots $\lambda_1, \lambda_2, \ldots, \lambda_n$ of the polynomial $a(s)$ are real and distinct, i.e., $\lambda_1 \neq \lambda_2 \neq \lambda_3 \neq \ldots \neq \lambda_n$. Here, $F(s)$ can be expanded into a sum of n partial fractions, as follows:

$$F(s) = \frac{b(s)}{a(s)} = \frac{c_1}{s - \lambda_1} + \cdots + \frac{c_k}{s - \lambda_k} + \cdots + \frac{c_n}{s - \lambda_n} = \sum_{i=1}^{n} \frac{c_i}{s - \lambda_i} \qquad (2.4\text{-}2)$$

The inverse Laplace transform of each term of $F(s)$ can be found in the Laplace transform tables (Appendix A). From the tables, we have

$$f(t) = L^{-1}\{F(s)\} = c_1 e^{\lambda_1 t} + c_2 e^{\lambda_2 t} + \cdots + c_n e^{\lambda_n t}$$

The constants $c_1, c_2, \ldots, c_n$, are determined as follows: multiply both sides of (2.4-2) by the factor $s - \lambda_k$ to yield

$$(s - \lambda_k)F(s) = \frac{s - \lambda_k}{s - \lambda_1} c_1 + \frac{s - \lambda_k}{s - \lambda_2} c_2 + \cdots + c_k + \cdots + \frac{s - \lambda_k}{s - \lambda_n} c_n$$

Taking the limit as s approaches the root λ_k, we have

$$c_k = \lim_{s \to \lambda_k} (s - \lambda_k)F(s), \qquad k = 1, 2, \ldots, n \qquad (2.4\text{-}3)$$

Relation (2.4-3) is a very simple procedure for determining the constants $c_1, c_2, \ldots, c_n$.

2 Nondistinct Real Roots

In this case the roots $\lambda_1, \lambda_2, \ldots, \lambda_n$ of the polynomial $a(s)$ are real but not distinct. For simplicity, let the polynomial $a(s)$ have only one repeated root λ_1, with multiplicity r, i.e., let

$$a(s) = (s - \lambda_1)^r \prod_{i=r+1}^{n} (s - \lambda_i) \tag{2.4-4}$$

Then, $F(s)$ is expanded into partial fractions, as follows:

$$F(s) = \frac{b(s)}{a(s)} = \frac{c_1}{s - \lambda_1} + \frac{c_2}{(s - \lambda_1)^2} + \cdots + \frac{c_k}{(s - \lambda_1)^k} + \cdots + \frac{c_r}{(s - \lambda_1)^r}$$
$$+ \frac{c_{r+1}}{s - \lambda_{r+1}} + \cdots + \frac{c_n}{s - \lambda_n}$$

Using the Laplace transform tables, the inverse Laplace transform of $F(s)$ is

$$f(t) = L^{-1}\{F(s)\} = c_1 e^{\lambda_1 t} + c_2 t e^{\lambda_1 t} + \cdots + \frac{c_k}{(k-1)!} t^{k-1} e^{\lambda_1 t} + \cdots$$
$$+ \frac{c_r}{(r-1)!} t^{r-1} e^{\lambda_1 t} + c_{r+1} e^{\lambda_{r+1} t} + \cdots + c_n e^{\lambda_n t}$$

The coefficients $c_{r+1}, c_{r+2}, \ldots, c_n$ are determined according to relation (2.4-3). To determine the coefficients c_k, $k = 1, 2, \ldots, r$, we work as follows: multiply both sides of Eq. (2.4-4) by the factor $(s - \lambda_1)^r$ to yield

$$(s - \lambda_1)^r F(s) = (s - \lambda_1)^{r-1} c_1 + \cdots + (s - \lambda_1)^{r-k} c_k + \cdots$$
$$+ c_r + \frac{(s - \lambda_1)^r}{s - \lambda_{r+1}} c_{r+1} + \cdots + \frac{(s - \lambda_1)^r}{s - \lambda_n} c_n$$

Differentiating by s, $(r - k)$ times, both sides of the above relation and taking the limit as $s \to \lambda_1$, all terms of the right-hand side go to zero, except for the term $(s - \lambda_1)^{r-k} c_k$, which becomes $(r - k)! c_k$. Hence

$$c_k = \lim_{s \to \lambda_1} \frac{1}{(r-k)!} \frac{d^{r-k}}{ds^{r-k}} [(s - \lambda_1)^r F(s)], \qquad k = 1, 2, \ldots, r \tag{2.4-5}$$

Clearly, if more than one root is repeated, one may apply the above procedure for each repeated root.

3 Complex Roots

Since all coefficients of $a(s)$ are real numbers, it follows that all complex roots of $a(s)$ appear in complex conjugate pairs. For simplicity, let the polynomial $a(s)$ have only one such pair, say $\lambda_1 = \alpha + j\omega$, and $\lambda_2 = \bar{\lambda}_1 = \alpha - j\omega$, where the symbol "-" indicates the complex conjugate number. Here, $F(s)$ may be expanded in partial fractions, as in the case of distinct real roots, as follows:

$$F(s) = \frac{c_1}{s - \lambda_1} + \frac{c_2}{(s - \bar{\lambda}_1)} + \frac{c_3}{s - \lambda_3} + \cdots + \frac{c_n}{s - \lambda_n} \tag{2.4-6}$$

Using the tables, the inverse Laplace transform is

$$f(t) = L^{-1}\{F(s)\} = c_1 e^{\lambda_1 t} + c_2 e^{\bar{\lambda}_1 t} + \sum_{i=3}^{n} c_i e^{\lambda_i t}$$

Using relation (2.4-3) we conclude that $c_2 = \bar{c}_1$. If the complex roots are repeated, we use relation (2.4-5). Consequently, the results of cases (a) and (b) can be used in the present case, with the difference that the coefficients c_i of complex roots appear always in conjugate pairs, a fact which simplifies their calculation.

Consider the case where $a(s)$ has the form $a(s) = [(s+a)^2 + \omega^2]q(s)$, where $q(s)$ has only real roots. This special form of $a(s)$ is quite frequently met in practice. In this case and when the polynomial $q(s)$ is of a low degree (e.g., first or second degree), the following partial fractions expansion of $F(s)$ is recommended:

$$F(s) = \frac{b(s)}{a(s)} = \frac{b(s)}{[(s+a)^2 + \omega]q(s)} = \frac{\gamma_1 s + \gamma_0}{(s+a)^2 + \omega^2} + e(s) \tag{2.4-7}$$

where $e(s)$ consists of all fractions which have as denominators the factors of $q(s)$. Consequently, $e(s)$ can be determined using the results of this section. The coefficients γ_0 and γ_1 are determined from the following relation:

$$\frac{\gamma_1 s + \gamma_0}{(s+a)^2 + \omega^2} = F(s) - e(s)$$

The motivation for presenting the method (2.4-7) is that it has computational advantages over relation (2.4-6), since the complex conjugate roots do not appear in determining γ_0 and γ_1.

Remark 2.4.1

In $F(s)$ of relation (2.4-1), let $m \geq n$, which means that the degree of the polynomial of the numerator is equal or greater than the degree of the polynomial of the denominator. In this case, the method of partial fractions expansion cannot be applied. To circumvent this difficulty, divide the numerator by the denominator to yield

$$F(s) = \frac{b(s)}{a(s)} = \pi(s) + \frac{\gamma(s)}{a(s)} \tag{2.4-8}$$

where $\pi(s)$ is a polynomial of degree $m - n$ and $\gamma(s)$ is a polynomial of degree $n - 1$. Clearly, the inverse Laplace transform of $F(s)$ may now be found from the tables.

2.5 APPLICATIONS OF THE LAPLACE TRANSFORM

This section presents certain applications of the Laplace transform in the study of linear systems.

Example 2.5.1

Determine the voltage across the capacitor of the circuit shown in Figure 2.11. The switch S closes when $t = 0$. The initial condition for the voltage capacitor is zero, i.e. $v_c(0) = 0$.

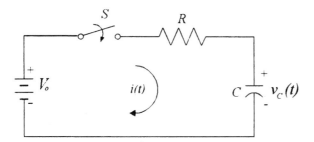

Figure 2.11   RC network.

Solution

From Kirchhoff's voltage law we have

$$Ri(t) + \frac{1}{C}\int_{-\infty}^{t} i(t)\,dt = V_0$$

Applying the Laplace transform to both sides of the integral equation, we get the following algebraic equation

$$RI(s) + \frac{1}{C}\left[\frac{I(s)}{s} + \frac{i^{(-1)}(0)}{s}\right] = \frac{V_0}{s}$$

where $I(s) = L\{i(t)\}$ and $i^{(-1)}(0) = \int_{-\infty}^{0} i(t)\,dt = Cv_c(0) = 0$. Replacing $i^{(-1)}(0) = 0$ in the above equation, we have

$$I(s)\left[\frac{1}{Cs} + R\right] = \frac{V_0}{s} \qquad \text{or} \qquad I(s) = \frac{V_0/R}{s + 1/RC}$$

The inverse Laplace transform of $I(s)$ is found in Appendix A and is as follows

$$i(t) = L^{-1}\{I(s)\} = \frac{V_0}{R}\,e^{-t/RC}$$

Hence, the voltage $v_c(t)$ across the capacitor will be

$$v_c(t) = V_0 - Ri(t) = V_0 - V_0 e^{-t/RC} = V_0[1 - e^{-t/RC}]$$

Example 2.5.2

Consider the mechanical system shown in Figure 2.12, where y, K, m, and B are the position of the mass, the spring's constant, the mass, and the friction coefficient, respectively. The initial conditions are $y(0) = 0$ and $y^{(1)}(0) = 2$. Let the applied force $f(t) = u(t)$; here $u(t)$ is the unit step function. Determine the response $y(t)$ of the mechanical system, where for simplicity, let $m = 1$, $B = 3$, and $K = 2$.

Solution

Using d'Alemberts law of forces, the following differential equation is obtained:

$$m\frac{d^2 y}{dt^2} + B\frac{dy}{dt} + Ky = f(t)$$

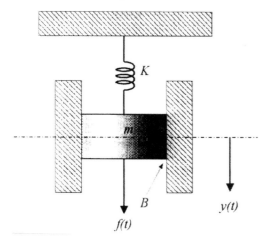

Figure 2.12 A spring and a mass.

Using the particular values for m, B, K, and $f(t)$, the differential equation becomes

$$\frac{d^2y}{dt^2} + 3\frac{dy}{dt} + 2y = u(t)$$

Applying the Laplace transform to both sides of the differential equation, we arrive at the following algebraic equation

$$s^2 Y(s) - sy(0) - y^{(1)}(0) + 3[sY(s) - y(0)] + 2Y(s) = \frac{1}{s}$$

Solving for $Y(s)$ we have

$$Y(s) = \frac{sy(0) + 3y(0) + y^{(1)}(0)}{s^2 + 3s + 2} + \frac{1}{s(s^2 + 3s + 2)}$$

$$= \frac{2}{s^2 + 3s + 2} + \frac{1}{s(s^2 + 3s + 2)}$$

where use was made of the given values of the initial conditions. Next, using the method of partial fraction expansion, we have

$$Y(s) = \frac{2}{(s^2 + 3s + 2)} + \frac{1}{s(s^2 + 3s + 2)} = \frac{2}{(s+1)(s+2)} + \frac{1}{s(s+1)(s+2)}$$

or

$$Y(s) = \left[\frac{2}{s+1} - \frac{2}{s+2}\right] + \left[\frac{1/2}{s} - \frac{1}{s+1} + \frac{1/2}{s+2}\right]$$

Taking the inverse Laplace transform, yields that the position $y(t)$ of the mass m is given by

$$y(t) = \left[2e^{-t} - 2e^{-t}\right] + \left[\tfrac{1}{2} - e^{-t} + 0.5e^{-2t}\right] = \tfrac{1}{2} - e^{-t} + \tfrac{3}{2}e^{-2t}$$

Example 2.5.3

Determine the voltage across the resistor R of the circuit shown in Figure 2.13. The capacitor C is initially charged at 10 V and the switch S closes when $t = 0$.

Solution

From Kirchhoff's voltage law we have

$$L\frac{di}{dt} + Ri + \frac{1}{C}\int_{-\infty}^{t} idt = 0$$

Applying the Laplace transform to both sides of the integrodifferential equation, we get the following algebraic equation

$$L[sI(s) - i(0)] + RI(s) + \frac{1}{C}\left[\frac{I(s)}{s} + \frac{i^{(-1)}(0)}{s}\right] = 0$$

or

$$\left[Ls + R + \frac{1}{Cs}\right]I(s) = Li(0) - \frac{i^{(-1)}(0)}{Cs}$$

Replacing the values of the parameters of the circuit and the initial conditions $v_c(0) = i^{(-1)}(0)/C = -10\,\text{V}$ and $i(0) = 0$, we have

$$\left[5s + 20 + \frac{20}{s}\right]I(s) = \frac{10}{s}$$

Solving this equation for $I(s)$, we get

$$I(s) = \frac{2}{s^2 + 4s + 4} = \frac{2}{(s + 2)^2}$$

Hence

$$i(t) = L^{-1}[I(s)] = L^{-1}\left[\frac{2}{(s + 2)^2}\right] = 2te^{-2t}$$

Therefore, the voltage $v_R(t)$ across the resistor will be $v_R(t) = 20i(t) = 40te^{-2t}$.

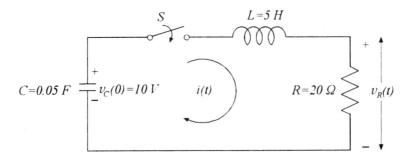

Figure 2.13   RLC network.

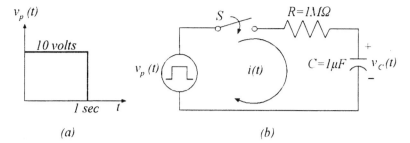

Figure 2.14   RC network with a pulse function as an input voltage. (a) Pulse of 1 sec duration; (b) RC network.

Example 2.5.4

Determine the voltage across the capacitor of the circuit shown in Figure 2.14. The
input voltage is a pulse of 1 sec duration. The switch S closes when $t = 0$.

Solution

From Kirchhoff's voltage law we have

$$10^6 i(t) + 10^6 \int_0^t i(t)\,dt = 10[u(t) - u(t-1)]$$

Applying the Laplace transform to the above equation, we get the following algebraic equation

$$10^5 I(s) + \frac{10^5}{s} I(s) = \frac{1}{s} - \frac{e^{-s}}{s}$$

Solving for $I(s)$, we get

$$I(s) = 10^{-5}\left[\frac{1 - e^{-s}}{s+1}\right] = 10^{-5}\left[\frac{1}{s+1} - \frac{e^{-s}}{s+1}\right]$$

Hence

$$i(t) = L^{-1}\{I(s)\} = 10^{-5}[e^{-t}u(t) - e^{-(t-1)}u(t-1)]$$

where relation (2.3-15) was used. The voltage $v_c(t)$ across the capacitor will be

$$v_c(t) = \frac{1}{C}\int_0^t i(t)\,dt = 10\int_0^t\left[e^{-t}u(t) - e^{-(t-1)}u(t-1)\right]dt$$

$$= 10\left[[1 - e^{-t}]u(t) - [1 - e^{-(t-1)}]u(t-1)\right]$$

The plot of the capacitor voltage is given in Figure 2.15.

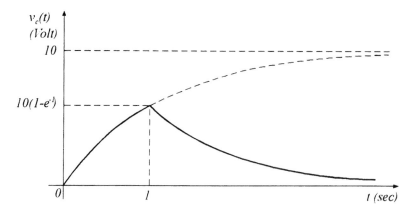

Figure 2.15 The voltage capacitor of Example 2.5.4.

2.6 MATRIX DEFINITIONS AND OPERATIONS

2.6.1 Matrix Definitions

Consider the system of linear algebraic equations

$$
\begin{aligned}
a_{11}x_1 + a_{12}x_2 + \cdots a_{1n}x_n &= b_1 \\
a_{21}x_1 + a_{22}x_2 + \cdots + a_{2n}x_n &= b_2 \\
\vdots \qquad \vdots \qquad \vdots \qquad \quad \vdots \\
a_{n1}x_1 + a_{n2}x_2 + \cdots + a_{nn}x_n &= b_n
\end{aligned}
$$

(2.6-1)

This system can be written compactly as:

$$ \mathbf{Ax} = \mathbf{b} \tag{2.6-2}$$

where

$$
\mathbf{A} = \begin{bmatrix} a_{11} & a_{12} & \cdots & a_{1n} \\ a_{21} & a_{22} & \cdots & a_{2n} \\ \vdots & \vdots & & \vdots \\ a_{n1} & a_{n2} & \cdots & a_{nn} \end{bmatrix}, \quad \mathbf{x} = \begin{bmatrix} x_1 \\ x_2 \\ \vdots \\ x_n \end{bmatrix}, \quad \text{and } \mathbf{b} = \begin{bmatrix} b_1 \\ b_2 \\ \vdots \\ b_n \end{bmatrix}
$$

The rectangular array, which has n columns and n rows, designated by the upper case letter $\mathbf{A}$, is called a *matrix*. Matrices which have only one column, designated by lower case letters (i.e., $\mathbf{x}$ and $\mathbf{b}$) are called *vectors* and have $n \times 1$ dimensions. A can also be written as $\mathbf{A} = [a_{ij}]$, $i, j = 1, 2, \ldots, n$. The parameters a_{ij} are called the *elements* of matrix $\mathbf{A}$.

One of the basic reasons which lead to the use of matrices is that they provide a concise description of multivariable systems. For example, relation (2.6-2) is a concise description of the algebraic equations (2.6-1). Furthermore, investigation of the solution of Eq. (2.6-1) is simplified when using Eq. (2.6-2). For example, we say that Eq. (2.6-2) has a unique solution if the determinant (see Section 2.6-2) of $\mathbf{A}$ is nonzero.

We present below certain useful types of matrices.

1 The Column Vector

This matrix is composed of a single column, i.e., it has the form

$$
\mathbf{a} = \begin{bmatrix} a_1 \\ a_2 \\ \vdots \\ a_n \end{bmatrix}
\tag{2.6-3}
$$

2 The Row Vector

This matrix is composed of a single row, i.e., it has the form

$$
\mathbf{a}^\mathrm{T} = [a_1 \quad a_2 \quad \cdots \quad a_n]
\tag{2.6-4}
$$

3 The Square and Nonsquare Matrices

A square matrix is one which has an equal number of rows and columns, whereas a nonsquare matrix has an unequal number of rows and columns.

4 The Diagonal Matrix

The diagonal matrix is a square matrix whose elements are all zero, except those that lie on the main diagonal and has the form

$$
\mathbf{A} = \begin{bmatrix} a_{11} & 0 & \cdots & 0 \\ 0 & a_{22} & \cdots & 0 \\ \vdots & \vdots & & \vdots \\ 0 & 0 & \cdots & a_{nn} \end{bmatrix} = \mathrm{diag}\{a_{11}, a_{22}, \ldots, a_{nn}\}
\tag{2.6-5}
$$

5 The Identity Matrix

A diagonal matrix that has ones along the main diagonal and zeros elsewhere is called an $(n \times n)$ identity matrix. This matrix is denoted by $\mathbf{I}$ and has the form

$$
\mathbf{I} = \begin{bmatrix} 1 & 0 & \cdots & 0 \\ 0 & 1 & \cdots & 0 \\ \vdots & \vdots & & \vdots \\ 0 & 0 & \cdots & 1 \end{bmatrix}
\tag{2.6-6}
$$

6 The Zero Matrix

This is a matrix whose elements are all zero.

7 The Singular and Nonsingular Matrices

If the determinant of a square matrix is zero, the matrix is called *singular*, while if the determinant is nonzero, the matrix is called *nonsingular*.

8 The Transpose of a Matrix

The matrix $\mathbf{A}^\mathrm{T}$ is the transpose of $\mathbf{A}$ if the rows of the first matrix are the columns of the second. Hence, if $\mathbf{A} = [a_{ij}]$, then $\mathbf{A}^\mathrm{T} = [a_{ji}]$. Therefore, if $\mathbf{A}$ has dimensions $n \times m$, then $\mathbf{A}^\mathrm{T}$ has dimensions $m \times n$. The superscript T denotes transposition.

9 The Symmetric Matrix

The symmetric matrix is a square matrix for which $\mathbf{A} = \mathbf{A}^T$ or $a_{ij} = a_{ji}, \forall i, j$.

10 The Triangular Matrix

The triangular matrix is a square matrix which has one of the following two forms

$$\mathbf{A} = \begin{bmatrix} a_{11} & a_{12} & \cdots & a_{1n} \\ 0 & a_{22} & \cdots & a_{2n} \\ \vdots & \vdots & & \vdots \\ 0 & 0 & \cdots & a_{nn} \end{bmatrix} \quad \text{or} \quad \mathbf{B} = \begin{bmatrix} a_{11} & 0 & \cdots & 0 \\ a_{21} & a_{22} & \cdots & 0 \\ \vdots & \vdots & & \vdots \\ a_{n1} & a_{n2} & \cdots & a_{nn} \end{bmatrix} \quad (2.6\text{-}7)$$

Matrix $\mathbf{A}$ is called *upper triangular*, whereas matrix $\mathbf{B}$ is called *lower triangular*.

11 The Conjugate Matrix

The matrix $\bar{\mathbf{A}}$ is called the conjugate matrix of $\mathbf{A}$ and its elements are the conjugate elements of $\mathbf{A}$: that is, if $\mathbf{A} = [a_{ij}]$, then $\bar{\mathbf{A}} = [\bar{a}_{ij}]$.

12 The Hermitian Matrix

If $\mathbf{A} = \bar{\mathbf{A}}^T$, then the matrix $\mathbf{A}$ is called Hermitian.

13 The Orthogonal Matrix

A matrix $\mathbf{A}$ is called orthogonal if it is a square matrix with real elements and the following relation holds:

$$\mathbf{A}^T \mathbf{A} = \mathbf{A} \mathbf{A}^T = \mathbf{I} \quad (2.6\text{-}8)$$

Other useful definitions regarding matrices are:

a. The Trace

The trace of an $n \times n$ square matrix $\mathbf{A} = [a_{ij}]$ is denoted as tr$\mathbf{A}$ or trace $\mathbf{A}$ and is defined as the sum of all the elements of the main diagonal, i.e.,

$$\text{tr}\mathbf{A} = \sum_{i=1}^{n} a_{ii} \quad (2.6\text{-}9)$$

b. The Rank

The rank of a matrix is equal to the dimension of its largest non-singular (square) submatrix.

2.6.2 Matrix Operations

1 Matrix Addition

Consider the matrices $\mathbf{A} = [a_{ij}]$ and $\mathbf{B} = [b_{ij}]$, both of whose dimensions are $n \times m$. Then, their addition $\mathbf{A} + \mathbf{B}$ is the $n \times m$ matrix $\mathbf{C} = [c_{ij}]$, whose elements c_{ij} are $c_{ij} = a_{ij} + b_{ij}$.

2 Matrix Multiplication

Consider the matrices $\mathbf{A} = [a_{ij}]$ and $\mathbf{B} = [b_{ij}]$ of dimensions $n \times m$ and $m \times p$, respectively. Then, their product $\mathbf{AB}$ is the $n \times p$ matrix $\mathbf{C} = [c_{ij}]$, whose elements c_{ij} are given by

$$c_{ij} = \sum_{k=1}^{m} a_{ik} b_{kj} = \mathbf{a}_i^{\mathrm{T}} \mathbf{b}_j$$

where $\mathbf{a}_i^{\mathrm{T}}$ is the ith row of matrix $\mathbf{A}$ and $\mathbf{b}_j$ is the jth column of matrix $\mathbf{B}$. Hence, every c_{ij} element is determined by multiplying the ith row of the matrix $\mathbf{A}$ with the jth column of the matrix $\mathbf{B}$.

3 Multiplying a Matrix with a Constant

Consider the matrix $\mathbf{A} = [a_{ij}]$ and the constant k. Then, every element of the matrix $\mathbf{C} = k\mathbf{A}$ is simply $c_{ij} = ka_{ij}$.

4 Transpose of a Matrix Product

Consider the matrix product $\mathbf{A}_1 \mathbf{A}_2 \ldots \mathbf{A}_m$. Then

$$(\mathbf{A}_1 \mathbf{A}_2 \ldots \mathbf{A}_m)^{\mathrm{T}} = \mathbf{A}_m^{\mathrm{T}} \mathbf{A}_{m-1}^{\mathrm{T}} \ldots \mathbf{A}_1^{\mathrm{T}} \qquad (2.6\text{-}10)$$

5 Derivatives of a Matrix

Consider the matrix $\mathbf{A} = [a_{ij}]$, whose elements a_{ij} are functions of time t, and the function f which is a scalar function of time. Then, the following relations hold:

$$\frac{d\mathbf{A}}{dt} = \left[\frac{da_{ij}}{dt}\right] \qquad (2.6\text{-}11)$$

$$\frac{d(f\mathbf{A})}{dt} = \frac{df}{dt}\mathbf{A} + f\frac{d\mathbf{A}}{dt} \qquad (2.6\text{-}12)$$

$$\frac{d(\mathbf{A}+\mathbf{B})}{dt} = \frac{d\mathbf{A}}{dt} + \frac{d\mathbf{B}}{dt} \qquad (2.6\text{-}13)$$

$$\frac{d(\mathbf{AB})}{dt} = \left[\frac{d\mathbf{A}}{dt}\right]\mathbf{B} + \mathbf{A}\left[\frac{d\mathbf{B}}{dt}\right] \qquad (2.6\text{-}14)$$

$$\frac{d(\mathbf{A}^{-1})}{dt} = -\mathbf{A}^{-1}\left[\frac{d\mathbf{A}}{dt}\right]\mathbf{A}^{-1} \qquad (2.6\text{-}15)$$

6 Derivatives of a Matrix with Respect to a Vector

The following relations hold

$$\frac{\partial f}{\partial \mathbf{x}} = \begin{bmatrix} \dfrac{\partial f}{\partial x_1} \\[2mm] \dfrac{\partial f}{\partial x_2} \\[2mm] \vdots \\[2mm] \dfrac{\partial f}{\partial x_n} \end{bmatrix} \quad \text{and} \quad \frac{\partial \mathbf{y}^{\mathrm{T}}}{\partial \mathbf{x}} = \begin{bmatrix} \dfrac{\partial y_1}{\partial x_1} & \dfrac{\partial y_2}{\partial x_1} & \cdots & \dfrac{\partial y_m}{\partial x_1} \\[2mm] \dfrac{\partial y_1}{\partial x_2} & \dfrac{\partial y_2}{\partial x_2} & \cdots & \dfrac{\partial y_m}{\partial x_2} \\[2mm] \vdots & \vdots & & \vdots \\[2mm] \dfrac{\partial y_1}{\partial x_n} & \dfrac{\partial y_2}{\partial x_n} & \cdots & \dfrac{\partial y_m}{\partial x_n} \end{bmatrix} \tag{2.6-16}$$

$$\frac{\partial [\mathbf{Q}(t)\mathbf{y}(t)]}{\partial \mathbf{y}} = \mathbf{Q}^{\mathrm{T}}(t) \tag{2.6-17}$$

$$\frac{1}{2}\frac{\partial [\mathbf{y}^{\mathrm{T}}(t)\mathbf{Q}(t)\mathbf{y}(t)]}{\partial \mathbf{y}} = \mathbf{Q}(t)\mathbf{y}(t) \tag{2.6-18}$$

$$\frac{1}{2}\frac{\partial [\mathbf{y}^{\mathrm{T}}(t)\mathbf{Q}(t)\mathbf{y}(t)]}{\partial \mathbf{x}} = \left[\frac{\partial \mathbf{y}^{\mathrm{T}}(t)}{\partial \mathbf{x}}\right]\mathbf{Q}(t)\mathbf{y}(t) \tag{2.6-19}$$

where $\mathbf{y}^{\mathrm{T}} = (y_1, y_2, \ldots, y_m)$ and $\mathbf{Q}(t)$ is a symmetric matrix with dimensions $m \times m$.

7 Matrix Integration

Here

$$\int \mathbf{A}\, dt = \left[\int a_{ij}\, dt\right] \tag{2.6-20}$$

2.7 DETERMINANT OF A MATRIX

1 Calculation of the Determinant of a Matrix

The determinant of an $n \times n$ matrix $\mathbf{A}$ is denoted by $|\mathbf{A}|$ or $\det \mathbf{A}$, and it is scalar quantity. A popular method to calculate the determinant of a matrix is the Laplace expansion in which the determinant of an $n \times n$ matrix $\mathbf{A} = [a_{ij}]$ is the sum of the elements of a row or a column, where each element is multiplied by the determinant of an appropriate matrix, i.e.,

$$|\mathbf{A}| = \sum_{i=1}^{n} a_{ij}c_{ij}, \qquad j = 1 \text{ or } 2 \text{ or } \cdots \text{ or } n \text{ (column expansion)} \tag{2.7-1}$$

or

$$|\mathbf{A}| = \sum_{j=1}^{n} a_{ij}c_{ij}, \qquad i = 1 \text{ or } 2 \text{ or } \cdots \text{ or } n \text{ (row expansion)} \tag{2.7-2}$$

where c_{ij} is defined as follows:

$$c_{ij} = (-1)^{i+j}|\mathbf{M}_{ij}|$$

where $\mathbf{M}_{ij}$ is the $(n-1) \times (n-1)$ square matrix formed from $\mathbf{A}$ by deleting the ith row and the jth column from the original matrix.

2 The Determinant of a Matrix Product

Consider the square matrices $\mathbf{A}_1, \mathbf{A}_2, \mathbf{A}_3, \ldots, \mathbf{A}_m$ and their product $\mathbf{B} = \mathbf{A}_1 \mathbf{A}_2 \ldots \mathbf{A}_m$. Then,

$$|\mathbf{B}| = |\mathbf{A}_1||\mathbf{A}_2| \cdots |\mathbf{A}_m| \tag{2.7-3}$$

2.8 THE INVERSE OF A MATRIX

1 Calculation of the Inversion of a Matrix

The inverse of an $n \times n$ square matrix $\mathbf{A}$ is denoted by $\mathbf{A}^{-1}$ and has the following property

$$\mathbf{A}^{-1}\mathbf{A} = \mathbf{A}\mathbf{A}^{-1} = \mathbf{I} \tag{2.8-1}$$

The inverse matrix $\mathbf{A}^{-1}$ is determined as follows:

$$\mathbf{A}^{-1} = \frac{\text{Adj}\,\mathbf{A}}{|\mathbf{A}|} \tag{2.8-2}$$

where the matrix adjA, called the adjoint matrix of $\mathbf{A}$, is the matrix whose elements are the adjoint elements of $\mathbf{A}$.

2 The Inverse of a Matrix Product

Consider the $n \times n$ square matrices $\mathbf{A}_1, \mathbf{A}_2, \ldots, \mathbf{A}_m$ and their product $\mathbf{B} = \mathbf{A}_1 \mathbf{A}_2 \cdots \mathbf{A}_m$. Then

$$\mathbf{B}^{-1} = \mathbf{A}_m^{-1} \mathbf{A}_{m-1}^{-1} \cdots \mathbf{A}_1^{-1}$$

2.9 MATRIX EIGENVALUES AND EIGENVECTORS

The eigenvalues of a matrix are of significance in the study of control systems, since their behavior is strongly influenced by their eigenvalues. The issue of the eigenvalues of an $n \times n$ square matrix $\mathbf{A} = [a_{ij}]$ stems from the following problem: consider the n-dimensional vectors $\mathbf{u}^T = (u_1, u_2, \ldots, u_n)$ and $\mathbf{y}^T = (y_1, y_2, \ldots, y_n)$ and the relation

$$\mathbf{y} = \mathbf{A}\mathbf{u} \tag{2.9-1}$$

The foregoing relation is a transformation of the vector $\mathbf{u}$ onto the vector $\mathbf{y}$ through the matrix $\mathbf{A}$. One may ask the following question: are there nonzero vectors $\mathbf{u}$ which maintain their direction after such transformation? If there exists such a vector $\mathbf{u}$, then the vector $\mathbf{y}$ is proportional to the vector $\mathbf{u}$, i.e., the following holds:

$$\mathbf{y} = \mathbf{A}\mathbf{u} = \lambda\mathbf{u} \tag{2.9-2}$$

where λ is a constant. From relation (2.9-2) we get that $\mathbf{A}\mathbf{u} = \lambda\mathbf{u}$ or $\mathbf{A}\mathbf{u} - \lambda\mathbf{u} = \mathbf{0}$, and if we set $\lambda\mathbf{u} = \lambda\mathbf{I}\mathbf{u}$, then we have

$$(\lambda\mathbf{I} - \mathbf{A})\mathbf{u} = \mathbf{0} \tag{2.9-3}$$

The system of equations (2.9-3) has a nonzero solution if the determinant of the matrix $\lambda\mathbf{I} - \mathbf{A}$ is equal to zero, i.e., if

$$|\lambda\mathbf{I} - \mathbf{A}| = 0 \tag{2.9-4}$$

Equation (2.9-4) is a polynomial equation of degree n and has the general form

$$|\lambda \mathbf{I} - \mathbf{A}| = p(\lambda) = \lambda^n + a_1 \lambda^{n-1} + a_2 \lambda^{n-2} + \cdots + a_{n-1}\lambda + a_n$$

$$= \prod_{i=1}^{n}(\lambda - \lambda_i) = 0 \qquad (2.9\text{-}5)$$

The roots $\lambda_1, \lambda_2, \ldots, \lambda_n$ of Eq. (2.9-5) are called the *eigenvalues* of the matrix $\mathbf{A}$. These eigenvalues, if set in Eq. (2.9-3), produce nonzero solutions to the problem of determining the vectors $\mathbf{u}$, which maintain their direction after the transformation $\mathbf{Au}$.

The polynomial $p(\lambda)$ is called the *characteristic polynomial* of the matrix $\mathbf{A}$ and Eq. (2.9-5) is called the *characteristic equation* of the matrix $\mathbf{A}$.

A vector $\mathbf{u}_i$ is an eigenvector of matrix $\mathbf{A}$ and corresponds to the eigenvalue λ_i if $\mathbf{u}_i \neq \mathbf{0}$ and relation (2.9-2) is satisfied, i.e.,

$$\mathbf{Au}_i = \lambda_i \mathbf{u}_i \qquad (2.9\text{-}6)$$

In determining the characteristic polynomial $p(\lambda)$ of $\mathbf{A}$, considerable computational effort may be required in computing the coefficients $a_1, a_2, \ldots, a_n$ of $p(\lambda)$ from the elements of matrix $\mathbf{A}$, particularly as n becomes large. Several numerical methods have been proposed, one of the most popular of which is Bocher's recursive relation:

$$a_k = -\frac{1}{k}(a_{k-1}T_1 + a_{k-2}T_2 + \cdots + a_1 T_{k-1} + T_k), \qquad k = 1, 2, \ldots, n \qquad (2.9\text{-}7)$$

where $T_k = \text{tr}\mathbf{A}^k$, $k = 1, 2, \ldots, n$. The description of this method is simple but has a serious numerical drawback in that it requries the computation up to the nth power of the matrix $\mathbf{A}$.

Certain interesting properties of matrix $\mathbf{A}$, of its eigenvalues, and of the coefficients of $p(\lambda)$, which may immediately be derived from the relations (2.9-5) to (2.9-7), are the following:

1. If the matrix $\mathbf{A}$ is singular, then it has at least one zero eigenvalue and vice versa.

2. $\text{tr}\mathbf{A} = \lambda_1 + \lambda_2 + \cdots + \lambda_n$ (2.9-8)

3. $|\mathbf{A}| = \lambda_1 \lambda_2 \ldots \lambda_n$ (2.9-9)

4. $a_n = (-1)^n|\mathbf{A}| = (-1)^n \lambda_1 \lambda_2 \ldots \lambda_n = p(0)$ (2.9-10)

Example 2.9.1

Find the characteristic polynomial, the eigenvalues, and the eigenvectors of the matrix

$$\mathbf{A} = \begin{bmatrix} -1 & 1 \\ 0 & -2 \end{bmatrix}$$

Solution

The characteristic polynomial $p(\lambda)$ of the matrix $\mathbf{A}$, accoding to definition (2.9-5), is

$$p(\lambda) = |\lambda \mathbf{I} - \mathbf{A}| = \det \begin{bmatrix} \lambda + 1 & -1 \\ 0 & \lambda + 2 \end{bmatrix} = (\lambda + 1)(\lambda + 2) = \lambda^2 + 3\lambda + 2$$

The coefficients of $p(\lambda)$ may be determined from the recursive equation (2.9-7). In this case, we first compute T_1 and T_2. We have: $T_1 = \text{tr}\mathbf{A} = -3$ and $T_2 = \text{tr}\mathbf{A}^2 = 5$. Hence

$$a_1 = -T_1 = 3 \quad \text{and} \quad a_2 = -\tfrac{1}{2}(a_1 T_1 + T_2) = -\tfrac{1}{2}(-9 + 5) = 2$$

Therefore, $p(\lambda) = \lambda^2 + a_1\lambda + a_2 = \lambda^2 + 3\lambda + 2$. We observe that, working in two different ways, we arrive at the same characteristic polynomial, as expected. The eigenvalues of matrix $\mathbf{A}$ are the roots of the characteristic polynomial $p(\lambda) = \lambda^2 + 3\lambda + 2 = (\lambda + 1)(\lambda + 2)$, and hence we immediately have that the eigenvalues of matrix $\mathbf{A}$ are $\lambda_1 = -1$ and $\lambda_2 = -2$. The eigenvectors can be determined from relation (2.9-6). Let the vector $\mathbf{u}_1$, which corresponds to the eigenvalue $\lambda_1 = -1$, have the form $\mathbf{u}_1 = (\omega_1, \omega_2)^T$. Then, relation (2.9-6) becomes

$$\begin{bmatrix} -1 & 1 \\ 0 & -2 \end{bmatrix}\begin{bmatrix} \omega_1 \\ \omega_2 \end{bmatrix} = -\begin{bmatrix} \omega_1 \\ \omega_2 \end{bmatrix} \quad \text{or} \quad \begin{bmatrix} 0 & 1 \\ 0 & -1 \end{bmatrix}\begin{bmatrix} \omega_1 \\ \omega_2 \end{bmatrix} = \begin{bmatrix} 0 \\ 0 \end{bmatrix}$$

Thus, the eigenvector $\mathbf{u}_1$ is $\mathbf{u}_1 = (\omega_1, \omega_2)^T = (d, 0)^T$, $\forall d \in \mathbb{R}$, where $\mathbb{R}$ is the space of real numbers. In the same way, we calculate the eigenvector $\mathbf{u}_2 = (v_1, v_2)^T$, which corresponds to the eigenvalue $\lambda_2 = -2$. From (2.9-6) we have

$$\begin{bmatrix} -1 & 1 \\ 0 & -2 \end{bmatrix}\begin{bmatrix} v_1 \\ v_2 \end{bmatrix} = -2\begin{bmatrix} v_1 \\ v_2 \end{bmatrix} \quad \text{or} \quad \begin{bmatrix} 1 & 1 \\ 0 & 0 \end{bmatrix}\begin{bmatrix} v_1 \\ v_2 \end{bmatrix} = \begin{bmatrix} 0 \\ 0 \end{bmatrix}$$

Hence, the eigenvector $\mathbf{u}_2$ is $\mathbf{u}_2 = (v_1, v_2)^T = (-k, k)^T$, $\forall k \in \mathbb{R}$. Setting $d = k = 1$, we have the following two eigenvectors

$$\mathbf{u}_1 = \begin{bmatrix} 1 \\ 0 \end{bmatrix}, \mathbf{u}_2 = \begin{bmatrix} -1 \\ 1 \end{bmatrix}$$

2.10 SIMILARITY TRANSFORMATIONS

Consider the $n \times n$ square matrices $\mathbf{A}$ and $\mathbf{B}$. The matrix $\mathbf{B}$ is *similar* to the matrix $\mathbf{A}$ if there exists an $n \times n$ nonsingular matrix $\mathbf{T}$ such that:

$$\mathbf{B} = \mathbf{T}^{-1}\mathbf{A}\mathbf{T} \tag{2.10-1}$$

It is obvious that if relation (2.10-1) is true, then it is also true that

$$\mathbf{A} = \mathbf{T}\mathbf{B}\mathbf{T}^{-1}$$

Hence, if $\mathbf{B}$ is similar to $\mathbf{A}$, then $\mathbf{A}$ must be similar to $\mathbf{B}$.

Relation (2.10-1) is called a *similarity transformation*. The similarity transformation usually aims at simplifying the matrix $\mathbf{A}$. Such a simplification facilitates the solution of certain problems. For example, consider the linear system of first-order differential equations

$$\dot{\mathbf{x}} = \mathbf{A}\mathbf{x}, \qquad \mathbf{x}(0) \tag{2.10-2}$$

where $\mathbf{x}^T = (x_1, x_2, \ldots, x_n)$. The vector $\mathbf{x}(0)$ is the initial vector of the system. Setting $\mathbf{x} = \mathbf{T}\mathbf{z}$, where $\mathbf{z}^T = (z_1, z_2, \ldots, z_n)$ is a new vector, the system (2.10-2) becomes

$$\dot{\mathbf{z}} = \mathbf{T}^{-1}\mathbf{A}\mathbf{T}\mathbf{z}, \qquad \mathbf{z}(0) = \mathbf{T}^{-1}\mathbf{x}(0) \tag{2.10-3}$$

If the similarity transformation matrix **T** is chosen so that $\mathbf{T}^{-1}\mathbf{A}\mathbf{T} = \mathbf{\Lambda} = \operatorname{diag}(\lambda_1, \lambda_2, \ldots, \lambda_n)$, where $\lambda_1, \lambda_2, \ldots, \lambda_n$ are the eigenvalues of matrix **A**, then Eq. (2.10-3) becomes

$$\dot{\mathbf{z}} = \mathbf{\Lambda}\mathbf{z} \qquad \text{or} \qquad \dot{z}_i = \lambda_i z_i, \qquad i = 1, 2, \ldots, n \tag{2.10-4}$$

Comparing relations (2.10-2) and (2.10-4), we realize that the similarity transformation has simplified (decoupled) the vector differential system of equations (2.10-2) into n first-order scalar differential equations whose ith solution is given by the simple relation

$$z_i(t) = z_i(0)\, e^{\lambda_i t}$$

In using the similarity transformation (2.10-1), one has to determine the transformation matrix **T**. In the case where the similarity transformation aims at diagonalizing the matrix **A**, then the determination of **T** is directly linked to the problem of eigenvalues and eigenvectors of **A**. In this case, the following holds true:

1. If the eigenvalues $\lambda_1, \lambda_2, \ldots, \lambda_n$ of the matrix **A** are distinct, then the eigenvectors $\mathbf{u}_1, \mathbf{u}_2, \ldots, \mathbf{u}_n$ are linearly independent. In this case, the similarity transformation matrix **T** which diagonalizes **A**, is denoted by **M** and has the form

$$\mathbf{M} = [\mathbf{u}_1 \vdots \mathbf{u}_2 \vdots \cdots \vdots \mathbf{u}_n] \tag{2.10-5}$$

The matrix **M** is called the *eigenvector matrix*.

2. If the eigenvalues $\lambda_1, \lambda_2, \ldots, \lambda_n$ of the matrix **A** are not distinct, then the eigenvectors $\mathbf{u}_1, \mathbf{u}_2, \ldots, \mathbf{u}_n$ are not always linearly independent. Since the diagonalization of **A** has as a necessary and sufficient condition the existence of n linearly independent eigenvectors of the matrix **A**, it follows that in the case where the eigenvalues of **A** are not distinct, the matrix **A** can be diagonalized only if it has n linearly independent eigenvectors. When **A** does not have n linearly independent eigenvectors, the matrix cannot be transformed into a diagonal form. However, it can be transformed to an "almost diagonal" form, known as the *Jordan canonical form* and is denoted by **J**. The general form of the matrix **J** is the following:

$$\mathbf{J} = \begin{bmatrix} \mathbf{J}_{11}(\lambda_1) & & & & & & \\ & \mathbf{J}_{21}(\lambda_1) & & & & \mathbf{0} & \\ & & \ddots & & & & \\ & & & \mathbf{J}_{k1}(\lambda_1) & & & \\ & & & & \mathbf{J}_{12}(\lambda_2) & & \\ & \mathbf{0} & & & & \ddots & \\ & & & & & & \mathbf{J}_{mp}(\lambda_p) \end{bmatrix} \tag{2.10-6}$$

where

$$\mathbf{J}_{ji}(\lambda_i) = \begin{bmatrix} \lambda_i & 1 & 0 & \cdots & 0 & 0 \\ 0 & \lambda_i & 1 & \cdots & 0 & 0 \\ \vdots & \vdots & \vdots & & \vdots & \vdots \\ 0 & 0 & 0 & \cdots & \lambda_i & 1 \\ 0 & 0 & 0 & \cdots & 0 & \lambda_i \end{bmatrix}$$

where $\mathbf{J}_{ji}(\lambda_i)$ is called the *Jordan submatrix*.

Certain useful properties of the similarity transformation are that the characteristic polynomial (and therefore the eigenvalues), the trace, and the determinant of a matrix are invariable under the similarity transformation. However, it should be noted that, in general, the reverse does not hold true. If, for example, two matrices have the same characteristic polynomial, that does not necessarily mean that they are similar.

Moreover, if $\mathbf{B} = \mathbf{T}^{-1}\mathbf{A}\mathbf{T}$ and $\mathbf{u}$ is an eigenvector of $\mathbf{A}$, then the vector $\mathbf{T}^{-1}\mathbf{u}$ is an eigenvector of $\mathbf{B}$.

Example 2.10.1

Diagonalize the matrix $\mathbf{A}$ of Example 2.9.1.

Solution

From Example 2.9.1 we have the two eigenvalues $\mathbf{u}_1 = (1, 0)^{\mathrm{T}}$ and $\mathbf{u}_2 = (-1, 1)^{\mathrm{T}}$. According to the relation (2.10-5), the similarity matrix $\mathbf{M}$ is

$$\mathbf{M} = [\mathbf{u}_1 \vdots \mathbf{u}_2] = \begin{bmatrix} 1 & -1 \\ 0 & 1 \end{bmatrix} \quad \text{and} \quad \mathbf{M}^{-1} = \begin{bmatrix} 1 & 1 \\ 0 & 1 \end{bmatrix}$$

Since $\mathbf{u}_1$ and $\mathbf{u}_2$ are linearly independent, the matrix $\mathbf{M}$ diagonalized the matrix $\mathbf{A}$. Indeed

$$\boldsymbol{\Lambda} = \mathbf{M}^{-1}\mathbf{A}\mathbf{M} = \begin{bmatrix} 1 & 1 \\ 0 & 1 \end{bmatrix}\begin{bmatrix} -1 & 1 \\ 0 & -2 \end{bmatrix}\begin{bmatrix} 1 & -1 \\ 0 & 1 \end{bmatrix} = \begin{bmatrix} -1 & 0 \\ 0 & -2 \end{bmatrix}$$

2.11 THE CAYLEY–HAMILTON THEOREM

The Cayley–Hamilton theorem relates a matrix to its characteristic polynomial, as follows:

Theorem 2.11.1

Consider an $n \times n$ matrix $\mathbf{A}$ with characteristic polynomial $p(\lambda) = |\lambda\mathbf{I} - \mathbf{A}| = \lambda^n + a_1\lambda^{n-1} + \cdots + a_{n-1}\lambda + a_n$. Then, the matrix $\mathbf{A}$ satisfies its characteristic polynomial, i.e.

$$p(\mathbf{A}) = \mathbf{A}^n + a_1\mathbf{A}^{n-1} + \cdots + a_{n-1}\mathbf{A} + a_n\mathbf{I} = \mathbf{0} \qquad (2.11\text{-}1)$$

The Cayley–Hamilton theorem has several interesting and useful applications. Some of them are given below.

a. Calculation of the Inverse of a Matrix

From relation (2.11-1), we have

$$a_n\mathbf{I} = -[\mathbf{A}^n + a_1\mathbf{A}^{n-1} + \cdots + a_{n-1}\mathbf{A}]$$

Let the matrix $\mathbf{A}$ be nonsingular. Then, multiplying both sides of the foregoing relation with $\mathbf{A}^{-1}$, we obtain

$$a_n\mathbf{A}^{-1} = -[\mathbf{A}^{n-1} + a_1\mathbf{A}^{n-2} + \cdots + a_{n-1}\mathbf{I}]$$

or

$$\mathbf{A}^{-1} = -\frac{1}{a_n}[\mathbf{A}^{n-1} + a_1\mathbf{A}^{n-2} + \cdots + a_{n-1}\mathbf{I}] \tag{2.11-2}$$

where $a_n = (-1)^n|\mathbf{A}| \neq 0$. Relation (2.11-2) expresses the matrix $\mathbf{A}^{-1}$ as a matrix polynomial of $\mathbf{A}$ of degree $n-1$ and presents an alternate way of determining $\mathbf{A}^{-1}$, different from that of Section 2.8, which essentially requires the calculation of the matrices $\mathbf{A}^k$, $k = 2, 3, \ldots, n-1$.

Example 2.11.1

Consider the matrix

$$\mathbf{A} = \begin{bmatrix} 0 & 1 \\ -6 & -5 \end{bmatrix}$$

Show that the matrix $\mathbf{A}$ satisfies its characteristic equation and calculate $\mathbf{A}^{-1}$.

Solution

We have

$$p(\lambda) = |\lambda\mathbf{I} - \mathbf{A}| = \lambda^2 + 5\lambda + 6$$

whence

$$p(\mathbf{A}) = \mathbf{A}^2 + 5\mathbf{A} + 6\mathbf{I}$$
$$= \begin{bmatrix} -6 & -5 \\ 30 & 19 \end{bmatrix} + 5\begin{bmatrix} 0 & 1 \\ -6 & -5 \end{bmatrix} + 6\begin{bmatrix} 1 & 0 \\ 0 & 1 \end{bmatrix} = \begin{bmatrix} 0 & 0 \\ 0 & 0 \end{bmatrix}$$

Therefore $p(\mathbf{A}) = \mathbf{0}$. For the calculation of $\mathbf{A}^{-1}$, using relation (2.11-2), we have

$$\mathbf{A}^{-1} = -\frac{1}{6}(\mathbf{A} + 5\mathbf{I}) = -\frac{1}{6}\begin{bmatrix} 5 & 1 \\ -6 & 0 \end{bmatrix}$$

b. Calculation of $\mathbf{A}^k$

From relation (2.11-1) we have

$$\mathbf{A}^n = -[a_1\mathbf{A}^{n-1} + \cdots + a_{n-1}\mathbf{A} + a_n\mathbf{I}] \tag{2.11-3}$$

Multiplying both sides of the above equation with the matrix $\mathbf{A}$ we have

$$\mathbf{A}^{n+1} = -[a_1\mathbf{A}^n + \cdots + a_{n-1}\mathbf{A}^2 + a_n\mathbf{A}] \tag{2.11-4}$$

Substituting Eq. (2.11-3) into Eq. (2.11-4) we have

$$\mathbf{A}^{n+1} = -[-a_1(a_1\mathbf{A}^{n-1} + \cdots + a_{n-1}\mathbf{A} + a_n\mathbf{I}) + \cdots + a_{n-1}\mathbf{A}^2 + a_n\mathbf{A}] \tag{2.11-5}$$

Relation (2.11-5) expresses the matrix $\mathbf{A}^{n+1}$ as a linear combination of the matrices $\mathbf{I}$, $\mathbf{A}, \ldots, \mathbf{A}^{n-1}$. The general case is

$$\mathbf{A}^k = \sum_{i=0}^{n-1}(\beta_k)_i\mathbf{A}^i, \qquad k \geq n \tag{2.11-6}$$

where $(\beta_k)_i$ are constants depending on the coefficients $a_1, a_2, \ldots, a_n$ of the characteristic polynomial $p(\lambda)$.

Example 2.11.2

Calculate the matrix $\mathbf{A}^5$, where

$$\mathbf{A} = \begin{bmatrix} 0 & 1 \\ 2 & 1 \end{bmatrix}$$

Solution

From the Cayley–Hamilton theorem we have $p(\mathbf{A}) = \mathbf{A}^2 - \mathbf{A} - 2\mathbf{I} = \mathbf{0}$ or $\mathbf{A}^2 = \mathbf{A} + 2\mathbf{I}$. From relation (2.11-5) we get $\mathbf{A}^3 = \mathbf{A}^2 + 2\mathbf{A} = (\mathbf{A} + 2\mathbf{I}) + 2\mathbf{A} = 3\mathbf{A} + 2\mathbf{I}$, $\mathbf{A}^4 = 3\mathbf{A}^2 + 2\mathbf{A} = 3(\mathbf{A} + 2\mathbf{I}) + 2\mathbf{A} = 5\mathbf{A} + 6\mathbf{I}$ and $\mathbf{A}^5 = 5\mathbf{A}^2 + 6\mathbf{A} = 5(\mathbf{A} + 2\mathbf{I}) + 6\mathbf{A} = 11\mathbf{A} + 10\mathbf{I}$. Therefore

$$\mathbf{A}^5 = 11\mathbf{A} + 10\mathbf{I} = 11 \begin{bmatrix} 0 & 1 \\ 2 & 1 \end{bmatrix} + 10 \begin{bmatrix} 1 & 0 \\ 0 & 1 \end{bmatrix} = \begin{bmatrix} 10 & 11 \\ 22 & 21 \end{bmatrix}$$

2.12 QUADRATIC FORMS AND SYLVESTER THEOREMS

Consider a second-order polynomial of the form

$$g(\mathbf{x}) = \sum_{j=1}^{n} \sum_{i=1}^{n} a_{ij} x_i x_j \tag{2.12-1}$$

where $\mathbf{x}^T = (x_1, x_2, \ldots, x_n)$ is a vector of real parameters. The polynomial $g(\mathbf{x})$ is called the quadratic form of n variables. Some of the properties and definitions of $g(\mathbf{x})$ are the following:

1. The polynomial $g(\mathbf{x})$ may be written as

    $$g(\mathbf{x}) = \mathbf{x}^T \mathbf{A} \mathbf{x}, \qquad \mathbf{A} = [a_{ij}], \qquad i, j = 1, 2, \ldots, n \tag{2.12-2}$$

    where $\mathbf{A}$ is an $n \times n$ real symmetric matrix (i.e., $\mathbf{A} = \mathbf{A}^T$).
2. The polynomial $g(\mathbf{x})$ may also be written as an inner product as follows:

    $$g(\mathbf{x}) = \langle \mathbf{x}, \mathbf{A}\mathbf{x} \rangle \tag{2.12-3}$$

    where $\langle \mathbf{a}, \mathbf{b} \rangle = \mathbf{a}^T \mathbf{b}$, and $\mathbf{a}$ and $\mathbf{b}$ are vectors of the same dimension.
3. The matrix $\mathbf{A}$ is called the matrix of the quadratic form $g(\mathbf{x})$. The rank of the matrix $\mathbf{A}$ is the order of $g(\mathbf{x})$.
4. The polynomial $g(\mathbf{x})$ is positive (negative) if its value is positive (negative) or zero for every set of real values of $\mathbf{x}$.
5. The polynomial $g(\mathbf{x})$ is positive definite (negative definite) if its value is positive (negative) for all $\mathbf{x}$ and equal to zero only when $\mathbf{x} = \mathbf{0}$.
6. The polynomial $g(\mathbf{x})$ is positive semidefinite (negative semidefinite) if its value is positive (negative) or zero.
7. The polynomial $g(\mathbf{x})$ is undefined if its value can take positive as well as negative values.

A central problem referring to quadratic forms pertains to the determination of whether or not a quadratic form is positive, negative definite, or semidefinite. Certain theorems due to Sylvester, are presented below:

Theorem 2.12.1

Let rank$\mathbf{A} = m$. Then, the polynomial $g(\mathbf{x}) = \mathbf{x}^T\mathbf{A}\mathbf{x}$ may be written in the following form

$$p(\mathbf{y}) = y_1^2 + \cdots + y_\mu^2 - y_{\mu+1}^2 - \cdots - y_{\mu+v}^2 + 0y_{\mu+v+1}^2 + \cdots + 0y_n^2 \qquad (2.12\text{-}4)$$

where $\mathbf{x} = \mathbf{T}\mathbf{y}$. The intergers μ and v correspond to the number of positive and negative eigenvalues of $\mathbf{A}$ and are such that $\mu + v = m$.

Theorem 2.12.2

From Eq. (2.12-4) we obtain $p(\mathbf{y})$ and, hence, the polynomial $g(\mathbf{x})$ is:

1. Positive definite if $\mu = n$ and $v = 0$
2. Positive semidefinite if $\mu < n$ and $v = 0$
3. Negative definite if $\mu = 0$ and $v = n$
4. Negative semidefinite if $\mu = 0$ and $v < n$
5. Undefined if $v\mu \neq 0$

Theorem 2.12.3

The necessary and sufficient conditions for $g(\mathbf{x})$ to be positive definite are

$$\Delta_i > 0, \quad i = 1, 2, \ldots, n \qquad (2.12\text{-}5)$$

where

$$\Delta_1 = a_{11}, \Delta_2 = \begin{vmatrix} a_{11} & a_{21} \\ a_{21} & a_{22} \end{vmatrix}, \ldots, \Delta_n = \begin{vmatrix} a_{11} & \cdots & a_{n1} \\ \vdots & & \vdots \\ a_{n1} & \cdots & a_{nn} \end{vmatrix}$$

Theorem 2.12.4

The necessary and sufficient conditions for $g(\mathbf{x})$ to be negative definite are

$$\Delta_i \begin{cases} > 0 & \text{when } i \text{ is even} \\ < 0 & \text{when } i \text{ is odd} \end{cases} \qquad (2.12\text{-}6)$$

for $i = 1, 2, \ldots, n$.

Theorem 2.12.5

The necessary and sufficient conditions for $g(\mathbf{x})$ to be negative semidefinite are

$$\Delta_i \geq 0, \quad i = 1, 2, \ldots, n - 1 \quad \text{and} \quad \Delta_n = 0 \qquad (2.12\text{-}7)$$

Theorem 2.12.6

The necessary and sufficient conditions for $g(\mathbf{x})$ to be negative semidefinite are

$$\Delta_1 \leq 0, \Delta_2 \geq 0, \Delta_3 \leq 0, \ldots \quad \text{and} \quad \Delta_n = 0 \qquad (2.12\text{-}8)$$

We note that Theorem 2.12.2 has the drawback, over Theorems 2.12.3–2.12.6, in that it requires the calculation of the integers μ and v of the eigenvalues and of the rank m of matrix $\mathbf{A}$. For this reason, Theorems 2.12.3–2.12.6 are used more often.

Example 2.12.1

Find if the quadratic forms

$$g_1(\mathbf{x}) = 4x_1^2 + x_2^2 + 2x_3^2 + 2x_1x_2 + 2x_1x_3 + 2x_2x_3$$
$$g_2(\mathbf{x}) = -2x_1^2 - 4x_2^2 - x_3^2 + 2x_1x_2 + 2x_1x_3$$
$$g_3(\mathbf{x}) = -4x_1^2 - x_2^2 - 12x_3^2 + 2x_1x_2 + 6x_2x_3$$

are positive or negative definite.

Solution

We have

$$g_1(\mathbf{x}) = \mathbf{x}^{\mathsf{T}}\mathbf{A}\mathbf{x}, \qquad \mathbf{x} = \begin{bmatrix} x_1 \\ x_2 \\ x_3 \end{bmatrix}, \qquad \mathbf{A} \begin{bmatrix} 4 & 1 & 1 \\ 1 & 1 & 1 \\ 1 & 1 & 2 \end{bmatrix}$$

and

$$\Delta_1 = 4 > 0, \ \Delta_2 = \begin{vmatrix} 4 & 1 \\ 1 & 1 \end{vmatrix} = 3 > 0, \qquad \Delta_3 = |\mathbf{A}| = 3 > 0$$

Hence, according to Theorem 2.12.3, $g_1(\mathbf{x})$ is positive definite. Also

$$g_2(\mathbf{x}) = \mathbf{x}^{\mathsf{T}}\mathbf{A}\mathbf{X}, \qquad \mathbf{x} = \begin{bmatrix} x_1 \\ x_2 \\ x_3 \end{bmatrix}, \qquad \mathbf{A} = \begin{bmatrix} -2 & 1 & 1 \\ 1 & -4 & 0 \\ 1 & 0 & -1 \end{bmatrix}$$

and

$$\Delta_1 = -2 < 0, \ \Delta_2 = \begin{vmatrix} -2 & 1 \\ 1 & -4 \end{vmatrix} = 7 > 0, \ \Delta_3 = |\mathbf{A}| = -3 < 0$$

Hence, according to Theorem 2.12.4, $g_2(\mathbf{x})$ is negative definite. Finally,

$$g_3(\mathbf{x}) = \mathbf{x}^{\mathsf{T}}\mathbf{A}\mathbf{x}, \qquad \mathbf{x} = \begin{bmatrix} x_1 \\ x_2 \\ x_3 \end{bmatrix}, \qquad \mathbf{A} = \begin{bmatrix} -4 & 1 & 0 \\ 1 & -1 & 3 \\ 0 & 3 & -12 \end{bmatrix}$$

and

$$\Delta_1 = -4 < 0, \ \Delta_2 = \begin{vmatrix} -4 & 1 \\ 1 & -1 \end{vmatrix} = 3 > 0, \ \Delta_3 = |\mathbf{A}| = 0$$

Hence, according to Theorem 2.12.6, $g_3(\mathbf{x})$ is negative semidefinite.

2.13   PROBLEMS

1. Find the Laplace transform of the following functions:

(a) t^3e^{-at}

(b) $\sin(\omega t + \theta)$

(c) $t\cos(\omega t + \theta)$

(d) $Etu(t - T)$

(e) $eT^2u(t - T)$

(f) $(t - 2)u(t - 2)$

(g) $(t + T_1)e^{-at}u(t - T_2)$

(h) $e^{-(t-1)}u(t - 1)$

(i) $(t + T_1)u(t - T_1)$

(j) $(t + T_2)^2u(t - T_2)$

(k) $t\cos at$

(l) $te^{-t}\cos 2t$

(m) $\cos\omega(t - T)$

(n) $t^2\sin\beta t$

(o) $5\sin[\omega(t - \alpha)]u(t - \alpha)$

(p) $\sin(t - T)u(t - T)$

(q) $\sin tu(t - T)$

(r) $\sin(t - T)u(t)$

2. Find the Laplace transform of the periodic waveforms shown in Figure 2.16.
3. Obtain the initial values $f(0^+)$ and $f^{(1)}(0^+)$, given that

$$F(s) = \frac{3}{s+0.2}$$

4. Obtain the final values of the functions having the following Laplace transforms:

(a) $L\{f(t)\} = F(s) = \dfrac{3}{s(s^2+s+2)}$

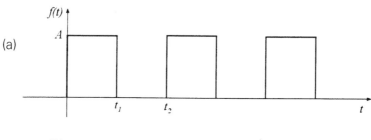

(a)

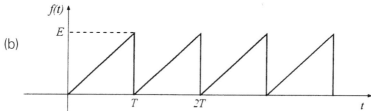

(b)

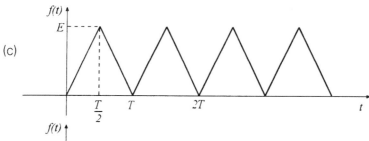

(c)

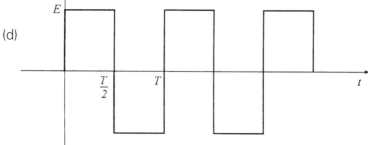

(d)

Figure 2.16

(b) $L\{f(t)\} = F(s) = \dfrac{s+2}{s(s^2+4)}$

5. Find the inverse Laplace transforms of the functions

(a) $\dfrac{1}{(s+1)(s+2)(s+3)}$

(b) $\dfrac{1}{(s+a)(s+b)^2}$

(c) $\dfrac{1}{s^2-2s+9}$

(d) $\dfrac{s+4}{s^2+4s+8}$

(e) $\dfrac{1}{s(s^2+4)}$

(f) $\dfrac{s+1}{s(s^2+4)}$

(g) $\dfrac{s+3}{(s-2)(s+1)}$

(h) $\dfrac{8}{s^3(s^2-s-2)}$

(i) $\dfrac{1}{(s^2+1)(s^2+4s+8)}$

(j) $\dfrac{s^2+1}{s^2+2}$

(k) $\dfrac{e^{-s}}{s^2+1}$

(l) $\dfrac{e^{-Ts}}{s^4+a^4}$

(m) $\dfrac{e^{-Ts}}{s}$

(n) $\dfrac{1-e^{-s}}{s^2}$

6. Find the current $i(t)$ and the voltage $v_L(t)$ across the inductor for the RL circuit shown in Figure 2.17. The switch S is closed at $t = 0$ and the initial condition for the inductor current is zero ($i_L(0) = 0$).
7. Find the current $i(t)$ for the RC circuit shown in Figure 2.18. The switch S is closed at $t = 0$ and the initial condition for the capacitor voltage is zero ($v_c(0) = 0$).
8. Find the current $i(t)$ for the LC circuit shown in Figure 2.19. The switch S is closed at $t = 0$ and the initial conditions for the inductor and the capacitor are $i_L(0) = 0$ and $v_c(0) = 0$.
9. For the RLC circuit shown in Figure 2.20, find the voltage across the resistor R using the Laplace transform method. Assume that the switch S is closed at $t = 0$.
10. Using the Laplace transform method, find the currents of the circuits shown in Figure 2.21.
11. Obtain the output voltage $v_R(t)$ of the highpass filter shown in Figure 2.22.

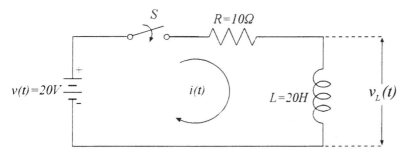

Figure 2.17

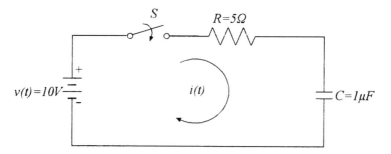

Figure 2.18

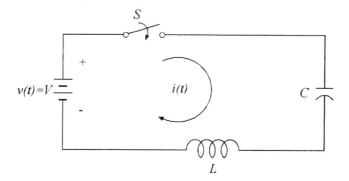

Figure 2.19

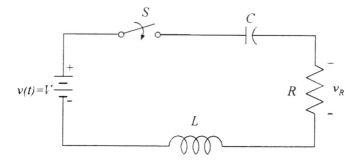

Figure 2.20

12. Find the voltage across the resistor $R = 10\,\Omega$ for the circuit shown in Figure 2.23. Initially, the capacitor is fully charged and the switch S is moved from position a to position b.
13. Consider the frictionless movement of a mass m shown in Figure 2.24, where $f(t)$ is the force applied, $y(t)$ is the displacement of the mass, and $m = 1\,\text{kg}$. Using the Laplace transform, find $y(t)$ for $t > 0$ if $f(t)$ has the form shown in Figure 2.24 and the initial conditions are $y(0) = 1$ and $\dot{y}(0) = 0$.
14. Solve Problem 13 for the system shown in Figure 2.25, where K is the spring constant. Consider the case where $K = 0.5$.

(a)

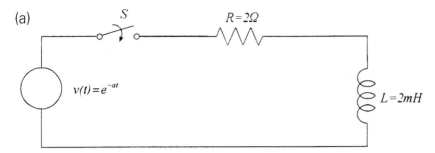

(b)

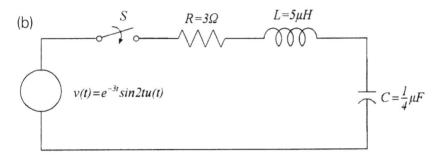

Figure 2.21

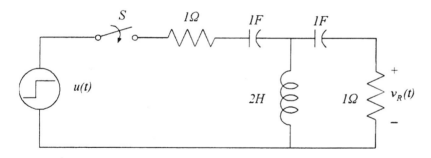

Figure 2.22

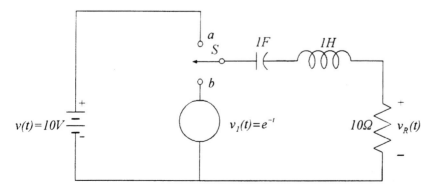

Figure 2.23

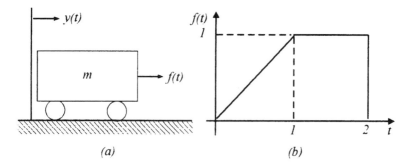

Figure 2.24

15. Determine the eigenvalues and the eigenvectors and diagonalize the following matrices:

$$\mathbf{A} = \begin{bmatrix} 2 & 0 \\ 4 & 3 \end{bmatrix}, \quad \mathbf{B} = \begin{bmatrix} 0 & 1 \\ -2 & -3 \end{bmatrix}, \quad \mathbf{C} = \begin{bmatrix} 0 & 1 & 0 \\ 0 & 0 & 0 \\ 0 & -2 & -3 \end{bmatrix},$$

$$\mathbf{D} = \begin{bmatrix} 3 & 0 & 2 \\ 0 & 3 & -2 \\ 2 & -2 & 1 \end{bmatrix}$$

16. Solve the system of differential equations

$$\dot{\mathbf{x}} = \mathbf{A}\mathbf{x}, \qquad \mathbf{x}(0) = \mathbf{x}_0$$

where

$$\mathbf{x} = \begin{bmatrix} x_1 \\ x_2 \\ x_3 \end{bmatrix}, \qquad \mathbf{A} = \begin{bmatrix} 0 & 1 & 0 \\ 0 & 0 & 1 \\ -6 & -11 & -6 \end{bmatrix}, \qquad \mathbf{x}_0 = \begin{bmatrix} -1 \\ -1 \\ 2 \end{bmatrix}$$

using the diagonalization method for the matrix $\mathbf{A}$.

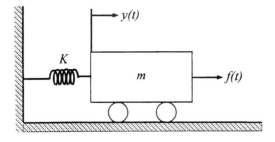

Figure 2.25

17. For the following matrices

$$\mathbf{A} = \begin{bmatrix} 1 & 1 & 0 \\ 1 & -3 & 0 \\ 0 & 0 & 2 \end{bmatrix}, \qquad \mathbf{B} = \begin{bmatrix} 1 & -2 \\ 2 & 3 \end{bmatrix}, \qquad \mathbf{C} = \begin{bmatrix} 1 & 2 & -2 \\ 1 & -3 & 1 \\ 0 & 6 & 3 \end{bmatrix}$$

show that they satisfy their characteristic equations and find $\mathbf{A}^{-1}$, $\mathbf{B}^{-1}$, and $\mathbf{C}^{-1}$. Also obtain $\mathbf{B}^3$ and $\mathbf{C}^8$.

18. Determine whether the following quadratic forms are positive or negative definite.

$g_1(x) = x_1^2 + 2x_1x_2 + 4x_2^2$

$g_2(x) = -x_1^2 - 2x_1x_2 - 2x_2^2$

$g_3(x) = (x_1 - x_2)^2$

$g_4(x) = -(x_1 - x_2)^2$

$g_5(x) = -x_1^2 - x_2^2$

$g_6(x) = 10x_1^2 + x_2^2 + x_3^2 + 2x_1x_2 + 8x_1x_3 + x_2x_3$

$g_7(x) = x_1^2 - 2x_2^2 + 2x_3^2 + 2x_1x_2$

$g_8(x) = x_1^2 - 2x_2^2 + 2x_3^2$

$g_9(x) = 6x_1^2 + 4x_2^2 + 2x_3^2 + 4x_1x_2 - 4x_2x_3 - 2x_1x_3$

$g_{10}(x) = -2x_1^2 - 5x_2^2 - 10x_3^2 + 4x_1x_2 - 2x_1x_3$
$\qquad\qquad - 6x_2x_3$

BIBLIOGRAPHY

Books

1. JJ D'Azzo, CH Houpis. Linear Control System Analysis and Design, Conventional and Modern. New York: McGraw-Hill, 1975.
2. BC Kuo. Automatic Control Systems. London: Prentice Hall, 1995.
3. K Ogata. Modern Control Systems. London: Prentice Hall, 1997.

3

Mathematical Models of Systems

3.1 INTRODUCTION

The general term *system description*, loosely speaking, refers to a mathematical expression that appropriately relates the physical system quantities to the system components. This mathematical relation constitutes the *mathematical model* of the system.

As an example, consider a resistor whose voltage v, current i, and resistance R are related in the well-known equation $v = Ri$. This expression relates the physical quantities (variables) v and i with the component R and constitutes a description; i.e., it constitutes a mathematical model of this simple (one-component) system.

As another example, consider a circuit composed of a resistor and an inductor connected in series. In this case, the voltage v across the circuit, the current i, the resistance R of the resistor, and the self-inductance L of the inductor are related with the well-known relation $v = L(di/dt) + Ri$. This expression relates the physical quantities v and i with the components R and L and constitutes a mathematical model of the system.

As already mentioned in Chap. 1, a system in operation involves the following three elements: the system's input (or excitation), the system itself, and the system's output (or response) (see Figure 3.1). Based on this three-fold element concept (input, system, and output), a more strict definition of the mathematical model of a system can be given.

Definition 3.1.1

The mathematical model of a system is a mathematical relation which relates the input, the system, and the output. This relation must be such as to guarantee that one can determine the system's output for any given input.

From the above definition it follows that the mathematical model is not just any relation, but a very special relation, which offers the capability of system analysis, i.e., the capability to determine the system's response under any excitation. Furthermore, the foregoing definition reveals the basic motive for determining mathematical models. This motive is to have available appropriate tools that will facilitate the system analysis (it is well known that in order to analyze a system, we must have available its

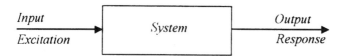

Figure 3.1 A system, its input, and its output.

mathematical model). It should be also noted that the mathematical model is useful for other purposes, as for example to study the system's stability and other properties, to improve the system's performance by applying control techniques, etc.

3.2 GENERAL ASPECTS OF MATHEMATICAL MODELS

The problem of determining a system's mathematical model is essentially a problem of approximating the behavior of a physical system with an ideal mathematical expression. For example, the expression $v = Ri$ is an approximation of the physical relation between v, i, and R. To have greater accuracy, one must take into account additional factors: for example, that the resistance R changes with temperature and, consequently, the relation $v = Ri$ should have the nonlinear form $v = R(i)$, where $R(i)$ denotes that R is a nonlinear function of the current i. However, it is well known that this more accurate model is still an approximation of the physical system. The final conclusion is that, in general, the mathematical model can only give an approximate description of a physical system.

The problem of deriving the mathematical model of a system usually appears as one of the following two cases.

1 Derivation of System's Equations

In this case, the system is considered known: for example, the network shown in Figure 3.2a. Here, we know the components R, L, C_1, and C_2 and their interconnections. To determine a mathematical model of the network, one may apply Kirchhoff's laws and write down that particular system of equations which will constitute the mathematical model. From network theory, it is well known that the system of equations sought are the linearly independent loop or node equations.

2 System Identification

In this case the system is not known. By "not known," we mean that we do not know either the system's components or their interconnections. The system is just a *black box*. In certain cases, we may have available some a priori (in advance) useful information about the system, e.g., that the system is time-invariant or that it has lumped parameters, etc. Based on this limited information about the system, we are asked to determine a mathematical model which describes the given system "satisfactorily." A well-known technique for dealing with this problem is depicted in Figure 3.2b. Here, both the physical system and the mathematical model are excited by the same input $u(t)$. Subsequently, the difference $e(t)$ of the respective responses $y_1(t)$ and $y_2(t)$ is measured. If the error $e(t)$ is within acceptable bounds, then the mathematical model is a satisfactory description of the system. The acceptable bounds depend on the desired degree of accuracy of the model, and they are usually stated in terms of the minimum value of the following cost function:

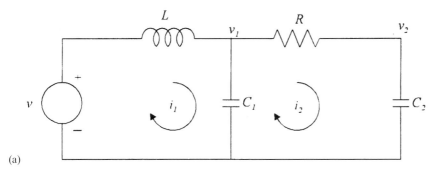

(a)

Figure 3.2a An RLC network.

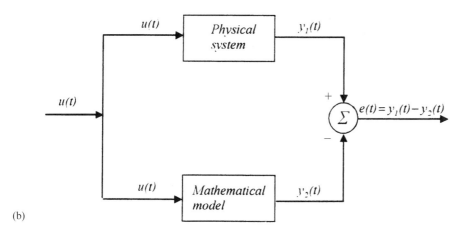

(b)

Figure 3.2b Schematic diagram of system identification.

$$J = \int_{t_0}^{t_1} e^2(t)\, dt$$

The area of system identification is a very interesting but very difficult engineering area. It is for this reason that it is usually taught as a special course or part of a course in the last undergraduate year or first graduate year. In this book, a very useful introduction to the problem of system identification is given in Chap. 13.

3.3 TYPES OF MATHEMATICAL MODELS

Several types of mathematical models have been proposed for the description of systems. The most popular ones, which we will present in this chapter, are the following:

1. The differential equations
2. The transfer function
3. The impulse response

4.   The state equations.

The main reason that all these four models will be studied here is that each model has advantages and disadvantages over the others. Hence, the knowledge of all four models offers the flexibility of using the most appropriate model among the four for a specific system or for a specific application.

The above four mathematical models are based on mathematical relationships and they are presented in Secs 3.4–3.9. Note, however, that there are other ways of describing a system, aiming to give a schematic overview of the system. Two such popular schematic system descriptions are

1.   The block diagrams
2.   The signal-flow graphs

These diagrams and graphs are particularly useful in giving a simplified overall picture of a system and they are presented in Secs 3.10 and 3.11, respectively.

Practical examples, where all the above types of description are applied, are presented in Secs 3.12 and 3.13. In particular, Sec. 3.12 presents mathematical models of components used in control systems, such as motors, error indicators, coupled gears, etc., while Sec. 3.13 presents mathematical models of practical control systems, such as voltage control, position and velocity control, liuqid-level control, temperature control, chemical process control, etc.

3.4   DIFFERENTIAL EQUATIONS

The mathematical model of differential equations (or, more generally, of integrodifferential equations) is the oldest method of system description. This description includes all the linearly independent equations of the system, as well as the appropriate initial conditions.

To illustrate this mathematical model we consider first the case of linear, time-invariant networks. Here, the physical variables v (voltage) and i (current) in any component (e.g., R, L, and C) are related via a linear operator T (see Table 3.1). When many such elements are linked together to form a network, then the variables v and i in each element are constrained according to the following well-known Kirchhoff's laws:

1.   *Kirchhoff's voltage law.* The algebraic sum of the voltages in a loop is equal to zero.
2.   *Kirchhoff's current law.* The algebraic sum of the currents in a node is equal to zero.

Hence, the description of a network using differential equations consists of determining all its linearly independent differential equations. Special methods have been developed for deriving these equations and they are the main subject of a particular branch of network theory, called network topology, which provides a systematic way of choosing the independent loop and node equations.

The linear time-invariant mechanical systems are another area of applications of differential equations, as a mathematical model. Here, the physical variables f (force) and v (velocity) of any component are related via a linear operator T in the

Table 3.1 Physical Variables in Elements of Linear Electrical Networks and Mechanical Systems.

Element	Physical variable	Linear operator $T[\bullet]$	Inverse operator $T^{-1}[\bullet]$
		Electrical Networks	
Resistor R	Voltage $v(t)$	$v(t) = Ri(t)$	$i(t) = \dfrac{1}{R} v(t)$
	Current $i(t)$	$T[\bullet] = R[\bullet]$	$T^{-1}[\bullet] = \dfrac{1}{R}[\bullet]$
Inductor L	Voltage $v(t)$	$v(t) = L\dfrac{d}{dt} i(t)$	$i(t) = \dfrac{1}{L} \displaystyle\int_0^t v(t)\,dt$
	Current $i(t)$	$T[\bullet] = L\dfrac{d}{dt}[\bullet]$	$T^{-1}[\bullet] = \dfrac{1}{L} \displaystyle\int_0^t [\bullet]\,dt$
Capacitor C	Voltage $v(t)$	$v(t) = \dfrac{1}{C} \displaystyle\int_0^t i(t)\,dt$	$i(t) = C\dfrac{d}{dt} v(t)$
	Current $i(t)$	$T[\bullet] = \dfrac{1}{C} \displaystyle\int_0^t [\bullet]\,dt$	$T^{-1}[\bullet] = C\dfrac{d}{dt}[\bullet]$
		Mechanical Systems	
Friction coefficient B	Force $f(t)$	$f(t) = Bv(t)$	$v(t) = \dfrac{1}{B} f(t)$
	Velocity $v(t)$	$T[\bullet] = B[\bullet]$	$T^{-1}[\bullet] = \dfrac{1}{B}[\bullet]$
Mass m	Force $f(t)$	$f(t) = m\dfrac{d}{dt} v(t)$	$v(t) = \dfrac{1}{m} \displaystyle\int_0^t f(t)\,dt$
	Velocity $v(t)$	$T[\bullet] = m\dfrac{d}{dt}[\bullet]$	$T^{-1}[\bullet] = \dfrac{1}{m} \displaystyle\int_0^t [\bullet]\,dt$
Spring constant K	Force $f(t)$	$f(t) = K \displaystyle\int_0^t v(t)\,dt$	$v(t) = \dfrac{1}{K}\dfrac{d}{dt} f(t)$
	Velocity $v(t)$	$T[\bullet] = K \displaystyle\int_0^t [\bullet]\,dt$	$T^{-1}[\bullet] = \dfrac{1}{K}\dfrac{d}{dt}[\bullet]$

same way as in an electrical network (see Table 3.1). Analogous to Kirchhoff's laws for networks is d'Alembert's law for mechanical systems which is stated as follows:

D'Alembert's law of forces: The sum of all forces acting upon a point mass is equal to zero.

Finally, as in the case of electrical networks, in order to derive a description for a mechanical system via differential equations, one must determine all the system's linearly independent differential equations.

The differential equation method is demonstrated by the following examples.

Example 3.4.1

Consider the network shown in Figure 3.3. Derive the network's differential equation mathematical model.

Solution

Applying Kirchhoff's voltage law, we have

$$L\frac{di}{dt} + \frac{1}{C}\int_0^t i\,dt + Ri = v(t)$$

The above integrodifferential equation constitutes a mathematical description of the network. To complete this description, two appropriate initial conditions must be given, since the above mathematical model is essentially a second-order differential equation. As initial conditions, we usually consider the inductor's current $i_L(t)$ and the capacitor's voltage $v_C(t)$ at the moment the switch S closes, which is usually at $t = 0$. Therefore, the initial conditions are

$$i_L(0) = I_0 \qquad \text{and} \qquad v_C(0) = V_0, \qquad \text{where } I_0 \text{ and } V_0 \text{ are given constants}$$

The integrodifferential equation and the two initial conditions constitute a complete description of the network shown in Figure 3.3.

Example 3.4.2

Consider the network shown in Figure 3.4. Derive the network's differential equation mathematical model.

Solution

The differential equations method for describing this network is based on the two differential equations of the first and second loop which arise by applying Kirchhoff's voltage law. These two loop equations are

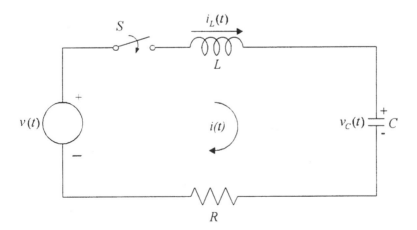

Figure 3.3 RLC network.

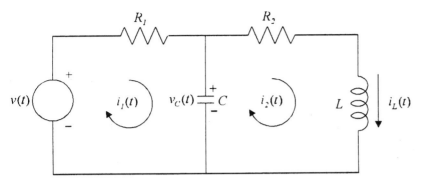

Figure 3.4 A two-loop network.

$$R_1 i_1(t) + \frac{1}{C} \int_0^t i_1(t)\,dt - \frac{1}{C} \int_0^t i_2(t)\,dt = v(t)$$

$$-\frac{1}{C} \int_0^t i_1(t)\,dt + R_2 i_2(t) + L\frac{di_2}{dt} + \frac{1}{C} \int_0^t i_2(t)\,dt = 0$$

with initial conditions $v_c(0) = V_0$ and $i_L(0) = I_0$.

Example 3.4.3

Consider the mechanical system shown in Figure 3.5, where y, K, m, and B are the position of the mass, the spring's constant, the mass, and the friction coefficient, respectively. Derive the system's differential equation mathematical model.

Solution

By using d'Alembert's law of forces, the following differential equation is obtained

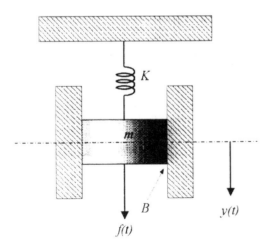

Figure 3.5 A spring and a mass.

$$m\frac{d^2y}{dt^2} + B\frac{dy}{dt} + Ky = f(t)$$

The initial conditions of the above equation are the distance $y(t)$ and the velocity $v(t) = dy/dt$ at the moment $t = 0$, i.e., at the moment where the external force $f(t)$ is applied. Therefore the initial conditions are

$$y(0) = Y_0 \text{ and } v(0) = \left[\frac{dy}{dt}\right]_{t=0} = V_0, \qquad \text{where } Y_0 \text{ and } V_0 \text{ are given constants}$$

The differential equation and the two initial conditions constitute a complete description of the mechanical system shown in Figure 3.5 using the method of differential equations.

Remark 3.4.1

The method of differential equations is a description in the time domain which can be applied to many categories of systems, such as linear and non linear, time-invariant and time-varying, with lumped and distributed parameters, with zero and non-zero initial conditions and many others.

3.5 TRANSFER FUNCTION

In contrast to the differential equation method which is a description in the time domain, the transfer function method is a description in the frequency domain and holds only for a restricted category of systems, i.e., for linear time-invariant systems having zero initial conditions. The transfer function is designated by $H(s)$ and is defined as follows:

Definition 3.5.1

The transfer function $H(s)$ of a linear, time-invariant system with zero initial condtiions is the ratio of the Laplace transform of the output $y(t)$ to the Laplace transform of the input $u(t)$, i.e.,

$$H(s) = \frac{L\{y(t)\}}{L\{u(t)\}} = \frac{Y(s)}{U(s)} \tag{3.5-1}$$

The introductory examples used in Sec. 3.4 will also be used here for the derivation of their transfer function.

Example 3.5.1

Consider the network shown in Figure 3.3. Derive the transfer function $H(s) = I(s)/V(s)$.

Solution

This network, in the frequency domain and with zero initial conditions I_0 and V_0, is as shown in Figure 3.6. To determine the transfer function $H(s) = I(s)/V(s)$, we work as follows: From Kirchhoff's voltage law, we have

$$LsI(s) + RI(s) + \frac{I(s)}{Cs} = V(s)$$

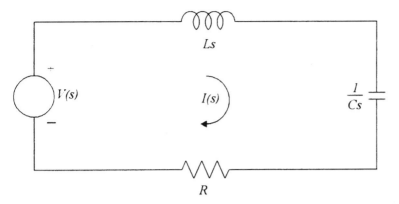

Figure 3.6 RLC circuit.

Therefore, the transfer function sought is given by

$$H(s) = \frac{I(s)}{V(s)} = \frac{I(s)}{\left[Ls + R + \dfrac{1}{Cs}\right]I(s)} = \frac{Cs}{LCs^2 + RCs + 1}$$

Furthermore, consider as circuit output the voltage $V_R(s)$ across the resistor. In this case, the transfer function becomes

$$H(s) = \frac{V_R(s)}{V(s)} = \frac{RI(s)}{V(s)} = \frac{RCs}{LCs^2 + RCs + 1}$$

Example 3.5.2

Consider the electrical network shown in Figure 3.4. Determine the transfer function $H(s) = I_2(s)/V(s)$.

Solution

This network, in the frequency domain and with zero initial conditions, is as shown in Figure 3.7. To determine $H(s) = I_2(s)/V(s)$, we start by writing the two loop equations:

$$\left[R_1 + \frac{1}{Cs}\right]I_1(s) - \frac{1}{Cs}I_2(s) = V(s)$$

$$-\frac{1}{Cs}I_1(s) + \left[R_2 + Ls + \frac{1}{Cs}\right]I_2(s) = 0$$

Next, solve the system for $I_2(s)$. The second equation yields

$$I_1(s) = [LCs^2 + R_2Cs + 1]I_2(s)$$

Substituting this result in the first equation, we have

$$(R_1Cs + 1)(LCs^2 + R_2Cs + 1)I_2(s) - I_2(s) = CsV(s)$$

Hence

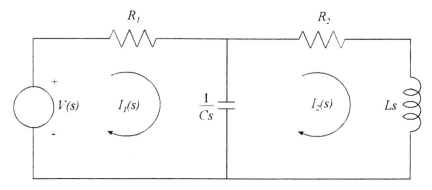

Figure 3.7 A two-loop network.

$$H(s) = \frac{I_2(s)}{V(s)} = \frac{Cs}{(R_1Cs + 1)(LCs^2 + R_2Cs + 1) - 1}$$
$$= \frac{1}{R_1LCs^2 + (R_1R_2C + L)s + R_1 + R_2}$$

Example 3.5.3

Consider the mechanical system shown in Figure 3.5. Determine the transfer function $H(s) = Y(s)/F(s)$.

Solution

This system, in the frequency domain and with zero initial conditions, will be as shown in Figure 3.8. In order to determine the transfer function $H(s) = Y(s)/F(s)$, we first write down the d'Alembert's law of forces. The result is

$$ms^2 Y(s) + BsY(s) + KY(s) = F(s)$$

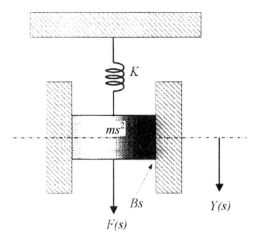

Figure 3.8 A spring and a mass.

Therefore

$$H(s) = \frac{Y(s)}{F(s)} = \frac{1}{ms^2 + Bs + K}$$

Remark 3.5.1

In the above examples, we observe that the transfer function $H(s)$ is the ratio of two polynomials in the freqency domain. In general, $H(s)$ has the following form

$$H(s) = \frac{b(s)}{a(s)} = \frac{b_m s^m + b_{m-1}s^{m-1} + \cdots + b_1 s + b_0}{s^n + a_{n-1}s^{n-1} + \cdots + a_1 s + a_0} = K\frac{\prod_{i=1}^{m}(s + z_i)}{\prod_{i=1}^{n}(s + p_i)}$$

where $-p_i$ are the roots of the denominator's polynomial $a(s)$ of $H(s)$ and are called the *poles* of $H(s)$, and $-z_i$ are the roots of the numerator's polynomial $b(s)$ of $H(s)$ and are called the *zeros* of $H(s)$. Poles and zeros, particularly the poles, play a significant role in the behavior of a system. This fact will most often be demonstrated in many places of this book.

3.6 IMPULSE RESPONSE

Impulse response is a time-domain description and holds only for a limited category of systems; i.e., the linear time-invariant and time-varying systems having zero initial conditions. The impulse response is designated by $h(t)$ and is defined as follows:

Definition 3.6.1

The impulse response $h(t)$ of a linear system with zero initial conditions is the system's output when its input is the unit impulse function $\delta(t)$ (see Eq. (2.2-3) of Chap. 2).

Consider the special case of linear, time-invariant systems where the relation $Y(s) = H(s)U(s)$ holds. Let the input be the unit impulse function, i.e., let $u(t) = \delta(t)$. Then, since $U(s) = 1$, we readily have that $Y(s) = H(s)$. Therefore, the transfer function $H(s)$ and the impulse response $h(t)$ are related as follows:

$$H(s) = L\{h(t)\} \quad \text{or} \quad h(t) = L^{-1}\{H(s)\} \tag{3.6-1}$$

The two relations in Eq. (3.6-1) show that for time-invariant systems, $H(s)$ and $h(t)$ are essentially the same description; whereas $H(s)$ is in the frequency domain, $h(t)$ is in the time domain. Also, Eq. (3.6-1) suggests an easy way of passing from one description to the other. Actually, if $H(s)$ is known, then $h(t) = L^{-1}\{H(s)\}$ and vice versa; if $h(t)$ is known, then $H(s) = L\{h(t)\}$.

Therefore, one way of determining the impulse response of a linear, time-invariant system, provided that its $H(s)$ is known, is to use the relation $h(t) = L^{-1}\{H(s)\}$. Another way of determining the impulse response is through the system's differential equation. In this case, let the system's differential equation have the general form

$$a_n y^{(n)} + a_{n-1}y^{(n-1)} + \cdots + a_1 y^{(1)} + a_0 y = u(t) \tag{3.6-2}$$

Then it can be readily shown that the solution of the homogeneous equation

$$a_n y^{(n)} + a_{n-1} y^{(n-1)} + \cdots + a_1 y^{(1)} + a_0 y = 0 \tag{3.6-3}$$

with initial conditions

$$y(0) = y^{(1)}(0) = \cdots = y^{(n-2)}(0) = 0 \quad \text{and} \quad y^{(n-1)}(0) = 1/a_n \tag{3.6-4}$$

is the system's impulse response.

3.7 STATE EQUATIONS

3.7.1 General Introduction

State-space equations, or simply state equations, is a description in the time domain which may be applied to a very wide category of systems, such as linear and non-linear systems, time-invariant and time-varying systems, systems with nonzero initial conditions, and others. The term *state* of a system refers to the past, present, and future of the system. From the mathematical point of view, the *state* of a system is expressed by its state variables. Usually, a system is described by a finite number of state variables, which are designated by $x_1(t), x_2(t), \ldots, x_n(t)$ and are defined as follows:

Definition 3.7.1

The state variables $x_1(t), x_2(t), \ldots, x_n(t)$ of a system are defined as a (minimum) number of variables such that if we know (a) their values at a certain moment t_0, (b) the input function applied to the system for $t \geq t_0$, and (c) the mathematical model which relates the input, the state variables, and the system itself, then the determination of the system's states for $t > t_0$ is guaranteed.

Consider a system with many inputs and many outputs, as shown in Figure 3.9. The *input vector* is designated by $\mathbf{u}(t)$ and has the form

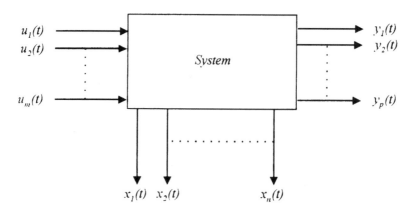

Figure 3.9   System with many inputs and many outputs.

$$\mathbf{u}(t) = \begin{bmatrix} u_1(t) \\ u_2(t) \\ \vdots \\ u_m(t) \end{bmatrix} \tag{3.7-1}$$

where m is the number of inputs. The *output vector* is designated by $\mathbf{y}(t)$ and has the form

$$\mathbf{y}(t) = \begin{bmatrix} y_1(t) \\ y_2(t) \\ \vdots \\ y_p(t) \end{bmatrix} \tag{3.7-2}$$

where p is the number of outputs. The *state vector* is designated by $\mathbf{x}(t)$ and has the form

$$\mathbf{x}(t) = \begin{bmatrix} x_1(t) \\ x_2(t) \\ \vdots \\ x_n(t) \end{bmatrix} \tag{3.7-3}$$

where n is the number of the state variables.

The *state equations* are a set of n first-order differential equations which relate the input vector $\mathbf{u}(t)$ with the state vector $\mathbf{x}(t)$ and has the form

$$\dot{\mathbf{x}}(t) = \mathbf{f}[\mathbf{x}(t), \mathbf{u}(t)] \tag{3.7-4}$$

where $\mathbf{f}$ is a column with n elements. The function $\mathbf{f}$ is, in general, a complex non-linear function of $\mathbf{x}(t)$ and $\mathbf{u}(t)$. Note that Eq. (3.7-4) is a set of *dynamic* equations.

The output vector $\mathbf{y}(t)$ of the system is related to the input vector $\mathbf{u}(t)$ and the state vector $\mathbf{x}(t)$ as follows:

$$\mathbf{y}(t) = \mathbf{g}[\mathbf{x}(t), \mathbf{u}(t)] \tag{3.7-5}$$

where $\mathbf{g}$ is a column with p elements. Relation (3.7-5) is called the *output equation*. The function $\mathbf{g}$ is generally a complex nonlinear function of $\mathbf{x}(t)$ and $\mathbf{u}(t)$. Note that Eq. (3.7-5) is a set of *algebraic* (nondynamic) equations.

The initial conditions of the state-space equations (3.7-4) are the values of the elements of the state vector $\mathbf{x}(t)$ for $t = t_0$ and are denoted as

$$\mathbf{x}(t_0) = \mathbf{x}_0 = \begin{bmatrix} x_1(t_0) \\ x_2(t_0) \\ \vdots \\ x_n(t_0) \end{bmatrix} \tag{3.7-6}$$

The state-space equations (3.7-4), the output equation (3.7-5), and the initial conditions (3.7-6), i.e., the following equations

$$\dot{\mathbf{x}}(t) = \mathbf{f}[\mathbf{x}(t), \mathbf{u}(t)] \tag{3.7-7a}$$

$$\mathbf{y}(t) = \mathbf{g}[\mathbf{x}(t), \mathbf{u}(t)] \tag{3.7-7b}$$

$$\mathbf{x}(t_0) = \mathbf{x}_0 \tag{3.7-7c}$$

constitute the description of a dynamic system in *state space*. Due to the fact that the *dynamic* state equations (3.7-7a) dominate in the three sets of equations (3.7-7a–c), in the sequel, all three equations in (3.7-7a–c), will be called, for simplicity, state equations.

The state equations (3.7-7) are, for the field of automatic control systems, the modern method of system description. This method has special theoretical, computational, and practical importance for the following main reasons:

1. State equations can describe a large category of systems, such as linear and nonlinear systems, time-invariant and time-varying systems, systems with time delays, systems with nonzero initial conditions, and others.
2. Due to the fact that state equations are a set of first-order differential equations, they can be easily programmed and simulated on both digital and analog computers.
3. State equations, by their very nature, greatly facilitate both in formulating and subsequently in investigating a great variety of properties in system theory, such as stability, controllability, and observability. They also facilitate the study of fundamental control problems, such as pole placement, optimal and stochastic control, and adaptive and robust control.
4. State equations provide a more complete description of a system than the other three methods: i.e., differential equations, transfer function, and impulse response. This is because state equations involve very important additional information about the system—namely, the system's state. This information is particularly revealing of the system's structure (e.g., regarding controllability, observability, pole-zero cancellation in the transfer function, etc.).

Overall, state equations is a description which relates the following four elements: input, system, state variables, and output. In contrast, the differential equations, the transfer function, and the impulse response relate three elements: input, system, and output—wherein the input is related to the output via the system directly (i.e., without giving information about the state of the system). It is exactly for this reason that these three system descriptions are called *input–output descriptions*.

3.7.2 Description of Linear Systems via State Equations

If a linear, time-invariant system can be described by a set of ordinary differential equations, then the state equations (3.7-7) take on the following special form:

$$\dot{\mathbf{x}}(t) = \mathbf{A}\mathbf{x}(t) + \mathbf{B}\mathbf{u}(t) \tag{3.7-8a}$$

$$\mathbf{y}(t) = \mathbf{C}\mathbf{x}(t) + \mathbf{D}\mathbf{u}(t) \tag{3.7-8b}$$

$$\mathbf{x}(t_0) = \mathbf{x}(0) = \mathbf{x}_0 \tag{3.7-8c}$$

Matrix $\mathbf{A}$ has dimensions $n \times n$ and it is called the *system matrix*, having the general form

$$\mathbf{A} = \begin{bmatrix} a_{11} & a_{12} & \cdots & a_{1n} \\ a_{21} & a_{22} & \cdots & a_{2n} \\ \vdots & \vdots & & \vdots \\ a_{n1} & a_{n2} & \cdots & a_{nn} \end{bmatrix} \tag{3.7-9}$$

Matrix $\mathbf{B}$ has dimensions $n \times m$ and it is called the *input* matrix, having the general form

$$\mathbf{B} = \begin{bmatrix} b_{11} & b_{12} & \cdots & b_{1m} \\ b_{21} & b_{22} & \cdots & b_{2m} \\ \vdots & \vdots & & \vdots \\ b_{n1} & b_{n2} & \cdots & b_{nm} \end{bmatrix} \tag{3.7-10}$$

Matrix $\mathbf{C}$ has dimensions $p \times n$ and it is called the *output* matrix, having the general form

$$\mathbf{C} = \begin{bmatrix} c_{11} & c_{12} & \cdots & c_{1n} \\ c_{21} & c_{22} & \cdots & c_{2n} \\ \vdots & \vdots & & \vdots \\ c_{p1} & c_{p2} & \cdots & c_{pn} \end{bmatrix} \tag{3.7-11}$$

Matrix $\mathbf{D}$ has dimensions $p \times m$ and it is called the *feedforward* matrix, having the general form

$$\mathbf{D} = \begin{bmatrix} d_{11} & d_{12} & \cdots & d_{1m} \\ d_{21} & d_{22} & \cdots & d_{2m} \\ \vdots & \vdots & & \vdots \\ d_{p1} & d_{p2} & \cdots & d_{pm} \end{bmatrix} \tag{3.7-12}$$

Depending on the dimensions m and p of the input and output vectors, we have the following four categories of systems:

1. *Multi-input–multi-output* systems (MIMO). In this case $m > 1$ and $p > 1$, and the system is described by Eqs (3.7-8).

2. *Multi-input–single-output* systems (MISO). In this case $m > 1$ and $p = 1$, and the system is described by the equations

$$\dot{\mathbf{x}}(t) = \mathbf{A}\mathbf{x}(t) + \mathbf{B}\mathbf{u}(t) \tag{3.7-13a}$$

$$y(t) = \mathbf{c}^T\mathbf{x}(t) + \mathbf{d}^T\mathbf{u}(t) \tag{3.7-13b}$$

$$\mathbf{x}(0) = \mathbf{x}_0 \tag{3.7-13c}$$

where $\mathbf{c}$ and $\mathbf{d}$ are column vectors with n and m elements, respectively.

3. *Single-input–multi-output* systems (SIMO). In this case $m = 1$ and $p > 1$, and the system is described by the equations

$$\dot{\mathbf{x}}(t) = \mathbf{A}\mathbf{x}(t) + \mathbf{b}u(t) \tag{3.7-14a}$$

$$\mathbf{y}(t) = \mathbf{C}\mathbf{x}(t) + \mathbf{d}u(t) \tag{3.7-14b}$$

$$\mathbf{x}(0) = \mathbf{x}_0 \tag{3.7-14c}$$

where $\mathbf{b}$ and $\mathbf{d}$ are column vectors with n and p elements, respectively.

4. *Single-input–single-output* systems (SISO). In this case $m = p = 1$ and the system is described by the equations

$$\dot{\mathbf{x}}(t) = \mathbf{A}\mathbf{x}(t) + \mathbf{b}u(t) \tag{3.7-15a}$$

$$y(t) = \mathbf{c}^T\mathbf{x}(t) + du(t) \tag{3.7-15b}$$

$$\mathbf{x}(0) = \mathbf{x}_0 \tag{3.7-15c}$$

where d is a scalar quantity.

It is noted that very often in the literature, for the sake of brevity, system (3.7-8a,b) is denoted by $[\mathbf{A}, \mathbf{B}, \mathbf{C}, \mathbf{D}]_n$.

If a linear time-varying system can be described by a set of linear differential equations, then the state equations (3.7-7) take on the form

$$\dot{\mathbf{x}}(t) = \mathbf{A}(t)\mathbf{x}(t) + \mathbf{B}(t)\mathbf{u}(t) \tag{3.7-16a}$$

$$\mathbf{y}(t) = \mathbf{C}(t)\mathbf{x}(t) + \mathbf{D}(t)\mathbf{u}(t) \tag{3.7-16b}$$

$$\mathbf{x}(t_0) = \mathbf{x}_0 \tag{3.7-16c}$$

where the time-varying matrices $\mathbf{A}(t)$, $\mathbf{B}(t)$, $\mathbf{C}(t)$, and $\mathbf{D}(t)$ are defined similarly to the constant matrices $\mathbf{A}$, $\mathbf{B}$, $\mathbf{C}$, and $\mathbf{D}$ of system (3.7-8).

3.7.3 Transfer Function and Impulse Response Matrices

1 Transfer Function Matrix

In studying linear, time-invariant systems described by state equations (3.7-8), one may use the Laplace transform. Indeed, applying the Laplace transform to Eqs (3.7-8a,b) yields

$$s\mathbf{X}(s) - \mathbf{x}(0) = \mathbf{A}\mathbf{X}(s) + \mathbf{B}\mathbf{U}(s) \tag{3.7-17a}$$

$$\mathbf{Y}(s) = \mathbf{C}\mathbf{X}(s) + \mathbf{D}\mathbf{U}(s) \tag{3.7-17b}$$

where $\mathbf{X}(s) = L\{\mathbf{x}(t)\}$, $\mathbf{U}(s) = L\{\mathbf{u}(t)\}$ and $\mathbf{Y}(s) = L\{\mathbf{y}(t)\}$. From Eq. (3.7-17a), we have

$$\mathbf{X}(s) = (s\mathbf{I} - \mathbf{A})^{-1}\mathbf{B}\mathbf{U}(s) + (s\mathbf{I} - \mathbf{A})^{-1}\mathbf{x}(0) \tag{3.7-18}$$

If we substitute Eq. (3.7-18) in Eq. (3.7-17b), we have

$$\mathbf{Y}(s) = [\mathbf{C}(s\mathbf{I} - \mathbf{A})^{-1}\mathbf{B} + \mathbf{D}]\mathbf{U}(s) + \mathbf{C}(s\mathbf{I} - \mathbf{A})^{-1}\mathbf{x}(0) \tag{3.7-19}$$

Equation (3.7-19) is an expression of the output vector $\mathbf{Y}(s)$ as a function of the input vector $\mathbf{U}(s)$ and of the initial condition vector $\mathbf{x}(0)$. When the initial condition vector $\mathbf{x}(0) = \mathbf{0}$, then Eq. (3.7-19) reduces to

$$\mathbf{Y}(s) = [\mathbf{C}(s\mathbf{I} - \mathbf{A})^{-1}\mathbf{B} + \mathbf{D}]\mathbf{U}(s) \tag{3.7-20}$$

Obviously, Eq. (3.7-20) is an input–output relation, since it directly relates the input vector $\mathbf{U}(s)$ with the output vector $\mathbf{Y}(s)$ (the state vector $\mathbf{X}(s)$ has been eliminated). Equation (3.7-20) is the same as Eq. (3.5-1), except that Eq. (3.5-1) relates scalars, while Eq. (3.7-20) relates vectors. This difference is because Eq. (3.5-1) holds for SISO systems, while Eq. (3.7-20) holds for MIMO systems. Therefore, Eq. (3.5-1) is a special case of Eq. (3.7-20), where m (number of inputs) $= p$ (number of outputs) $= 1$. The transfer function matrix $\mathbf{H}(s)$ of system (3.7-8) is analogous to the scalar transfer function $H(s)$ of Eq. (3.5-1) and has the following form

$$\mathbf{H}(s) = \mathbf{C}(s\mathbf{I} - \mathbf{A})^{-1}\mathbf{B} + \mathbf{D} \qquad (3.7\text{-}21)$$

The transfer function matrix $\mathbf{H}(s)$ has dimensions $p \times m$ and has the general form

$$\mathbf{H}(s) = \begin{bmatrix} h_{11}(s) & h_{12}(s) & \cdots & h_{1m}(s) \\ h_{21}(s) & h_{22}(s) & \cdots & h_{2m}(s) \\ \vdots & \vdots & & \vdots \\ h_{p1}(s) & h_{p2}(s) & \cdots & h_{pm}(s) \end{bmatrix} \qquad (3.7\text{-}22)$$

Each element $h_{ij}(s)$ of $\mathbf{H}(s)$ is a scalar transfer function which relates the element $y_i(s)$ of the output vector $\mathbf{Y}(s)$ with the element $u_j(s)$ of the input vector $\mathbf{U}(s)$, provided that all the other elements of the ith row of $\mathbf{H}(s)$ are zero. In general, we have

$$y_i(s) = \sum_{j=1}^{m} h_{ij}(s)u_j(s), \qquad i = 1, 2, \ldots, p \qquad (3.7\text{-}23)$$

To give a simplified view of Eq. (3.7-23), consider the special case of a two input–two output system, in which case relation $\mathbf{Y}(s) = \mathbf{H}(s)\mathbf{U}(s)$ has the form

$$y_1(s) = h_{11}(s)u_1(s) + h_{12}(s)u_2(s)$$
$$y_2(s) = h_{21}(s)u_1(s) + h_{22}(s)u_2(s)$$

The block diagram (see Sec. 3.10) of these two equations is shown in Figure 3.10.

2 Impulse Response Matrix

The impulse response matrix $\mathbf{H}(t)$ is analogous to the scalar impulse function $h(t)$ given by Definition 3.6.1. Using Eq. (3.6-1) we readily have

$$L\{\mathbf{H}(t)\} = \mathbf{H}(s) \qquad \text{or} \qquad \mathbf{H}(t) = L^{-1}\{\mathbf{H}(s)\} \qquad (3.7\text{-}24)$$

The impulse response matrix $\mathbf{H}(t)$ has a general form

$$\mathbf{H}(t) = \begin{bmatrix} h_{11}(t) & h_{12}(t) & \cdots & h_{1m}(t) \\ h_{21}(t) & h_{22}(t) & \cdots & h_{2m}(t) \\ \vdots & \vdots & & \vdots \\ h_{p1}(t) & h_{p2}(t) & \cdots & h_{pm}(t) \end{bmatrix} \qquad (3.7\text{-}25)$$

Therefore, the element $h_{ij}(t) = L^{-1}\{h_{ij}(s)\}$ is the scalar impulse response between the element $y_i(t)$ of the output vector $\mathbf{y}(t)$ and the element $u_j(t)$ of the input vector $\mathbf{u}(t)$, provided that all the other vectors of the ith row of $\mathbf{H}(t)$ are zero.

Example 3.7.1

Consider the network shown in Figure 3.11 with initial condition $i_L(0) = I_0$. Determine the state equations of the network.

Solution

The loop equation is

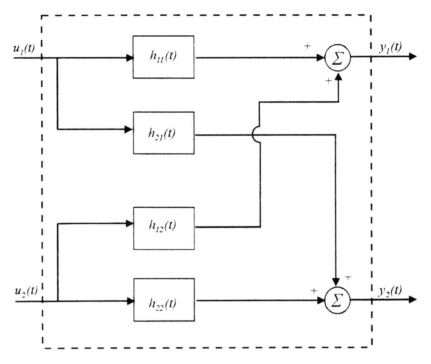

Figure 3.10   Block diagram of a two input–two output system.

$$L\frac{\mathrm{d}i}{\mathrm{d}t} + Ri = v(t) \qquad \text{or} \qquad \frac{\mathrm{d}i}{\mathrm{d}t} = -\frac{R}{L}i + \frac{1}{L}v(t)$$

Define as state variable $x(t)$ of the network the physical quantity $x(t) = i_{\mathrm{L}}(t) =$ the inductor's current. The system output is the voltage across the resistor R. As a result, the network's description via state equations is

$$\dot{\mathbf{x}}(t) = -\frac{R}{L}x(t) + \frac{1}{L}v(t)$$

$$v_{\mathrm{R}}(t) = R\,x(t)$$

$$i_{\mathrm{L}}(0) = i(0) = I_0$$

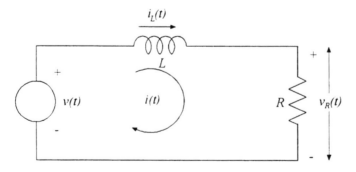

Figure 3.11   RL network.

Since the network is a SISO system, the above state equations have the same form as that of Eqs (3.7-15), with the exception that in this case all values are scalar. This is because the network is a first-order system, and hence there is only one state variable ($n = 1$).

Example 3.7.2

Consider the network shown in Figure 3.3, with initial conditions $i_L(0) = I_0$ and $v_C(0) = V_0$. Determine the state equations of the network.

Solution

The loop equation is

$$L\frac{di}{dt} + Ri + \frac{1}{C}\int_0^t i\,dt = v(t)$$

Define as state variables $x_1(t)$ and $x_2(t)$ of the network the following physical quantities:

$$x_1(t) = i_L(t) = \text{the inductor's current}$$

$$x_2(t) = \int_0^t i\,dt = \text{the capacitor's electrical charge}$$

By differentiating the last equation, the following relation is obtained:

$$\dot{x}_2(t) = i(t) = i_L(t) = x_1(t)$$

If the state variables are inserted in the loop equation, it yields

$$L\dot{x}_1(t) + Rx_1(t) + \frac{1}{C}x_2(t) = v(t)$$

or

$$\dot{x}_1(t) = -\frac{R}{L}x_1(t) - \frac{1}{LC}x_2(t) + \frac{1}{L}v(t)$$

The above relation, together with the relation $\dot{x}_2(t) = x_1(t)$, gives the following system of first-order differential equations:

$$\begin{bmatrix} \dot{x}_1(t) \\ \dot{x}_2(t) \end{bmatrix} = \begin{bmatrix} -\dfrac{R}{L} & -\dfrac{1}{LC} \\ 1 & 0 \end{bmatrix} \begin{bmatrix} x_1(t) \\ x_2(t) \end{bmatrix} + \begin{bmatrix} \dfrac{1}{L} \\ 0 \end{bmatrix} v(t)$$

Let the network's output be the voltage $v_R(t)$ across the resistor R. Then, the network's output equation is

$$y(t) = v_R(t) = Ri(t) = Rx_1(t) = [R \quad 0]\begin{bmatrix} x_1(t) \\ x_2(t) \end{bmatrix}$$

The initial condition vector of the network is

$$\mathbf{x}(0) = \begin{bmatrix} x_1(0) \\ x_2(0) \end{bmatrix} = \begin{bmatrix} i_L(0) \\ Cv_C(0) \end{bmatrix} = \begin{bmatrix} I_0 \\ CV_0 \end{bmatrix}$$

Since the network is a SISO system, the state equations sought have the form of Eqs (3.7-15), where

$$\mathbf{x}(t) = \begin{bmatrix} x_1(0) \\ x_2(0) \end{bmatrix} = \begin{bmatrix} i_L(t) \\ v_C(t) \end{bmatrix}, \qquad u(t) = v(t), \qquad y(t) = v_R(t)$$

$$\mathbf{A} = \begin{bmatrix} -\dfrac{R}{L} & -\dfrac{1}{LC} \\ 1 & 0 \end{bmatrix}, \qquad \mathbf{b} = \begin{bmatrix} \dfrac{1}{L} \\ 0 \end{bmatrix}, \qquad \mathbf{c} = \begin{bmatrix} R \\ 0 \end{bmatrix}, \qquad \mathbf{x}_0 = \begin{bmatrix} I_0 \\ CV_0 \end{bmatrix}$$

Example 3.7.3

Consider the mechanical system shown in Figure 3.5 with initial conditions $y(0) = Y_0$ and $y^{(1)}(0) = U_0$, where Y_0 is the initial position and U_0 is the initial velocity of the mass. Determine the state equations of the system.

Solution

The d'Alembert's law of forces is as follows:

$$m\frac{d^2 y}{dt^2} + B\frac{dy}{dt} + Ky = f(t)$$

Define as state variables: $x_1(t) = y(t) =$ the mass displacement and $x_2(t) = \dot{y}(t) =$ the mass velocity. We notice that $\dot{x}_1(t) = x_2(t)$. By substituting the state variables in d'Alembert's equation, we have

$$m\dot{x}_2(t) + Bx_2(t) + Kx_1(t) = f(t)$$

or

$$\dot{x}_2(t) = -\frac{K}{m}x_1(t) - \frac{B}{m}x_2(t) + \frac{1}{m}f(t)$$

The above equation, together with the relation $\dot{x}_1(t) = x_2(t)$, may be written as the following system of first-order differential equations:

$$\begin{bmatrix} \dot{x}_1(t) \\ \dot{x}_2(t) \end{bmatrix} = \begin{bmatrix} 0 & 1 \\ -\dfrac{K}{m} & -\dfrac{B}{m} \end{bmatrix} \begin{bmatrix} x_1(t) \\ x_2(t) \end{bmatrix} + \begin{bmatrix} 0 \\ \dfrac{1}{m} \end{bmatrix} f(t)$$

Let $y(t)$ be the displacement of the mass m. Then, the system's output equation will be

$$y(t) = \begin{bmatrix} 1 & 0 \end{bmatrix} \begin{bmatrix} x_1(t) \\ x_2(t) \end{bmatrix}$$

The initial condition vector is

$$\mathbf{x}(0) = \begin{bmatrix} x_1(0) \\ x_2(0) \end{bmatrix} = \begin{bmatrix} y(0) \\ \dot{y}(0) \end{bmatrix} = \begin{bmatrix} Y_0 \\ U_0 \end{bmatrix}$$

Hence, the state equations of the mechanical system have the form of Eqs (3.7-15), where

$$\mathbf{x}(t) = \begin{bmatrix} x_1(t) \\ x_2(t) \end{bmatrix} = \begin{bmatrix} y(t) \\ \dot{y}(t) \end{bmatrix}, \qquad u(t) = f(t)$$

$$\mathbf{A} = \begin{bmatrix} 0 & 1 \\ -\dfrac{K}{m} & -\dfrac{B}{m} \end{bmatrix}, \qquad \mathbf{b} = \begin{bmatrix} 0 \\ \dfrac{1}{m} \end{bmatrix}, \qquad \mathbf{c} = \begin{bmatrix} 1 \\ 0 \end{bmatrix}, \qquad \mathbf{x}_0 = \begin{bmatrix} Y_0 \\ U_0 \end{bmatrix}$$

3.8 TRANSITION FROM ONE MATHEMATICAL MODEL TO ANOTHER

As pointed out in Sec. 3.3, every mathematical model has advantages and disadvantages over others: to take advantage of the advantages of all mathematical models, one must have the flexibility of going from one model to another. The issue of transition from one mathematical model to another is obviously of great practical and theoretical importance. In the sequel, we present some of the most interesting such cases.

3.8.1 From Differential Equation to Transfer Function for SISO Systems

Case 1. The Right-Hand Side of the Differential Equation Does Not Involve Derivatives

Consider a SISO system described by the differential equation

$$y^{(n)} + a_{n-1} y^{(n-1)} + \cdots + a_1 y^{(1)} + a_0 y = b_0 u(t) \tag{3.8-1}$$

where all the system's initial conditions are assumed zero, i.e., $y^{(k)}(0) = 0$ for $k = 1, 2, \ldots, n-1$. Applying the Laplace transform to Eq. (3.8-1), we obtain

$$s^n Y(s) + a_{n-1} s^{n-1} Y(s) + \cdots a_1 s Y(s) + a_0 Y(s) = b_0 U(s)$$

Hence, the transfer function sought is given by

$$H(s) = \frac{Y(s)}{U(s)} = \frac{b_0}{s^n + a_{n-1} s^{n-1} + \cdots + a_1 s + a_0} \tag{3.8-2}$$

Case 2. The Right-Hand Side of the Differential Equation Involves Derivatives

Consider a SISO system described by the differential equation

$$y^{(n)} + a_{n-1} y^{(n-1)} + \cdots + a_1 y^{(1)} + a_0 y = b_m u^{(m)} + \cdots + b_1 u^{(1)} + b_0 u \tag{3.8-3}$$

where $m < n$ and where all initial conditions are assumed zero, i.e., $y^{(k)}(0) = 0$, for $k = 0, 1, \ldots, n-1$. We can determine the transfer function of Eq. (3.8-3) as follows: Let $z(t)$ be the solution of Eq. (3.8-1), with $b_0 = 1$. Then, using the superposition principle, the solution $y(t)$ of Eq. (3.8-3) will be

$$y(t) = b_m z^{(m)} + b_{m-1} z^{(m-1)} + \cdots + b_1 z^{(1)} + b_0 z \tag{3.8-4}$$

Applying the Laplace transformation to Eq. (3.8-4) we obtain

$$Y(s) = b_m s^m Z(s) + b_{m-1} s^{m-1} Z(s) + \cdots + b_1 s Z(s) + b_0 Z(s) \tag{3.8-5}$$

Here, we have set $z^{(k)}(0) = 0$, for $k = 0, 1, \ldots, n-1$, since the solution of Eq. (3.8-1), when $b_0 = 1$, is $z(t) = y(t)$, where it was assumed that all initial conditions of $y(t)$ (and hence of $z(t)$) are zero. In Eq. (3.8-2), if $b_0 = 1$, then

$$Z(s) = \left[\frac{1}{s^n + a_{n-1}s^{n-1} + \cdots + a_1 s + a_0} \right] U(s)$$

By substituting the above expression of $Z(s)$ in Eq. (3.8-5), the transfer function $H(s)$ of the differential equation (3.8-3) is obtained; it has the form

$$H(s) = \frac{Y(s)}{U(s)} = \frac{b_m s^m + b_{m-1}s^{m-1} + \cdots + b_1 s + b_0}{s^n + a_{n-1}s^{n-1} + \cdots + a_1 s + a_0} \qquad (3.8\text{-}6)$$

Remark 3.8.1

The transfer function $H(s)$ given by Eq. (3.8-6) can be easily derived from Eq. (3.8-3) if we set s^k in place of the kth derivative and replace $y(t)$ and $u(t)$ with $Y(s)$ and $U(s)$, respectively; i.e., we derive Eq. (3.8-6) by replacing $y^{(k)}(t)$ by $s^k Y(s)$ and $u^{(k)}(t)$ by $s^k U(s)$ in Eq. (3.8-3).

3.8.2   From Transfer Function to Differential Equation for SISO Systems

Let a SISO system be described by Eq. (3.8-6). Then, working backwards the method given in Remark 3.8.1, the differential equation (3.8-3) can be constructed by substituting s^k by the kth derivative and $Y(s)$ and $U(s)$ with $y(t)$ and $u(t)$, respectively.

3.8.3   From $H(s)$ to $H(t)$ and Vice Versa

The matrices $H(s)$ and $H(t)$, according to Eqs (3.6-1) and (3.7-24) are related through the Laplace transform. Therefore, the following transition relations may be used:

$$L\{H(t)\} = H(s) \qquad \text{or} \qquad H(t) = L^{-1}\{H(s)\} \qquad (3.8\text{-}7)$$

3.8.4   From State Equations to Transfer Function Matrix

Consider a system described by the state equations (3.7-8a,b). Then the system's transfer function matrix $H(s)$ is given by (3.7-21), i.e., by the relation

$$H(s) = C(sI - A)^{-1}B + D \qquad (3.8\text{-}8)$$

3.8.5   From Transfer Function Matrix to State Equations

The transition from $H(s)$ to state equations is the well-known problem of state-space *realization*. This is, in general, a difficult problem and has been, and still remains, a topic of research. In the sequel, we will present some introductory results regarding this problem.

Case 1. SISO Systems

Let a system be described by a scalar transfer function of the form

$$H(s) = \frac{Y(s)}{U(s)} = \frac{b_{n-1}s^{n-1} + b_{n-2}s^{n-2} + \cdots + b_1 s + b_0}{s^n + a_{n-1}s^{n-1} + \cdots + a_1 s + a_0} \tag{3.8-9}$$

or, equivalently, by the differential equation

$$y^{(n)} + a_{n-1}y^{(n-1)} + \cdots + a_1 y^{(1)} + a_0 y = b_{n-1}u^{(n-1)} + b_{n-2}u^{(n-2)} + \cdots \\ + b_1 u^{(1)} + b_0 u \tag{3.8-10}$$

Equation (3.8-10) can be expressed in the form of two equations as follows:

$$z^{(n)} + a_{n-1}z^{(n-1)} + \cdots + a_1 z^{(1)} + a_0 z = u \tag{3.8-11}$$

$$y(t) = b_{n-1}z^{(n-1)} + \cdots + b_1 z^{(1)} + b_0 z \tag{3.8-12}$$

Let $z(t)$ be the solution of Eq. (3.8-11). Then, the solution of Eq. (3.8-10) will be given by Eq. (3.8-12), where use is made of the superposition principle.

The state variables $x_1, x_2, \ldots, x_n$ are defined as follows:

$$x_1(t) = z(t)$$
$$x_2(t) = z^{(1)}(t) \quad = x_1^{(1)}(t)$$
$$x_3(t) = z^{(2)}(t) \quad = x_2^{(1)}(t) \tag{3.8-13}$$
$$\vdots \qquad \vdots \qquad \vdots$$
$$x_n(t) = z^{(n-1)}(t) = x_{n-1}^{(1)}(t)$$

If we substitute Eqs (3.8-13) into Eq. (3.8-11), we have

$$\dot{x}_n(t) = -a_{n-1}x_n(t) - \cdots - a_1 x_2(t) - a_0 x_1(t) + u(t) \tag{3.8-14}$$

Also, if we substitute Eqs (3.8-13) into (3.8-12), we have

$$y(t) = b_{n-1}x_n(t) + b_{n-2}x_{n-1}(t) + \cdots + b_1 x_2(t) + b_0 x_1(t) \tag{3.8-15}$$

Equations (3.8-13)–(3.8-15) can be expressed in a matrix form, as follows:

$$\dot{\mathbf{x}}(t) = \mathbf{A}\mathbf{x}(t) + \mathbf{b}u(t) \tag{3.8-16a}$$

$$y(t) = \mathbf{c}^{\mathrm{T}}\mathbf{x}(t) \tag{3.8-16b}$$

where $\mathbf{x}^{\mathrm{T}} = (x_1, x_2, \ldots, x_n)$ and

$$\mathbf{A} = \begin{bmatrix} 0 & 1 & 0 & 0 & \cdots & 0 \\ 0 & 0 & 1 & 0 & \cdots & 0 \\ \vdots & \vdots & \vdots & \vdots & \cdots & \vdots \\ 0 & 0 & 0 & 0 & \cdots & 1 \\ -a_0 & -a_1 & -a_2 & -a_3 & \cdots & -a_{n-1} \end{bmatrix}, \quad \mathbf{b} = \begin{bmatrix} 0 \\ 0 \\ \vdots \\ 0 \\ 1 \end{bmatrix}, \quad \mathbf{c} = \begin{bmatrix} b_0 \\ b_1 \\ \vdots \\ b_{n-2} \\ b_{n-1} \end{bmatrix} \tag{3.8-17}$$

Hence, Eqs (3.8-16) constitute the state equations' description of the transfer function (3.8-9).

Due to the special form of matrix **A** and of vector **b**, wherein there are only ones and zeros, except in the last line in **A** and **b**, we say that the state equations (3.8-16) are in *phase canonical form*, while the state variables are called *phase variables*. Phase variables are, in general, state variables which are defined according to Eqs (3.8-13); i.e., every state variable is the derivative of the previous one. In particular, the special form of matrix **A** and of vector **b** is characterized by as follows: in matrix **A**, if the first column and the last row are deleted, then a $(n - 1) \times (n - 1)$ unit matrix is revealed. Also, the elements of the last row of **A** are the coefficients of the differential equation (3.8-10), placed in reverse order and all having negative signs. The vector **b** has all its elements equal to zero, except for the nth element, which is equal to one. The block diagram of system (3.8-16) is given in Figure 3.12.

Case 2. MIMO Systems

To determine a state-space realization of a MIMO system, where the dimension n of the state vector $\mathbf{x}(t)$ is the smallest possible (i.e., n is minimum), is a rather difficult task and is beyond the scope of this book. However, certain types of nonminimal realizations are easy to attain, as shown in the sequel. Let a system be described by the transfer function matrix $\mathbf{H}(s)$, with dimensions $p \times m$, having the general form

$$\mathbf{H}(s) = \begin{bmatrix} h_{11}(s) & h_{12}(s) & \cdots & h_{1m}(s) \\ h_{21}(s) & h_{22}(s) & \cdots & h_{2m}(s) \\ \vdots & \vdots & & \vdots \\ h_{p1}(s) & h_{p2}(s) & \cdots & h_{pm}(s) \end{bmatrix} \tag{3.8-18}$$

where $h_{ij}(s) = \gamma_{ij}(s)/p_{ij}(s)$, and the degree of the polynomial $\gamma_{ij}(s)$ is smaller than that of $p_{ij}(s)$. Let

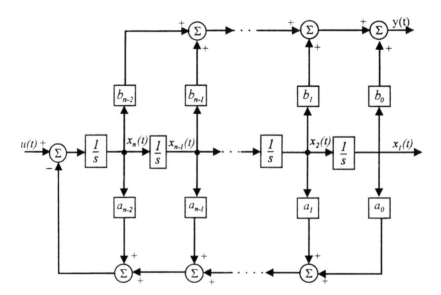

Figure 3.12   Block diagram of the state equations (3.18-16) in phase canonical form of a SISO system.

$$p(s) = s^n + p_{n-1}s^{n-1} + \cdots + p_1 s + p_0$$

be the least common multiplier polynomial of all $p_{ij}(s)$. Then the product $p(s)\mathbf{H}(s)$ is a matrix polynomial which can be expressed as follows:

$$p(s)\mathbf{H}(s) = \mathbf{H}_0 + \mathbf{H}_1 s + \cdots + \mathbf{H}_{n-1}s^{n-1}$$

Let $\mathbf{O}_m$ be a $m \times m$ zero matrix, i.e., a matrix with all its elements equal to zero and $\mathbf{I}_m$ be a $m \times m$ unit matrix. Then, the following state equations constitute a description of Eq. (3.8-18) in state space:

$$\dot{\mathbf{x}} = \mathbf{A}\mathbf{x} + \mathbf{B}\mathbf{u} \tag{3.8-19a}$$

$$\mathbf{y} = \mathbf{C}\mathbf{x} \tag{3.8-19b}$$

where

$$\mathbf{A} = \begin{bmatrix} \mathbf{O}_m & \mathbf{I}_m & \mathbf{O}_m & \cdots & \mathbf{O}_m \\ \mathbf{O}_m & \mathbf{O}_m & \mathbf{I}_m & \cdots & \mathbf{O}_m \\ \vdots & \vdots & \vdots & & \vdots \\ \mathbf{O}_m & \mathbf{O}_m & \mathbf{O}_m & \cdots & \mathbf{I}_m \\ -p_0\mathbf{I}_m & -p_1\mathbf{I}_m & -p_2\mathbf{I}_m & \cdots & -p_{n-1}\mathbf{I}_m \end{bmatrix}, \qquad \mathbf{B} = \begin{bmatrix} \mathbf{O}_m \\ \mathbf{O}_m \\ \vdots \\ \mathbf{O}_m \\ \mathbf{I}_m \end{bmatrix}$$

and

$$\mathbf{C} = [\mathbf{H}_0 \mid \mathbf{H}_1 \mid \cdots \mid \mathbf{H}_{n-2} \mid \mathbf{H}_{n-1}]$$

The dimensions of $\mathbf{x}$, $\mathbf{u}$, $\mathbf{y}$, $\mathbf{A}$, $\mathbf{B}$, and $\mathbf{C}$ are $nm \times 1$, $m \times 1$, $p \times 1$, $nm \times nm$, $nm \times m$, and $p \times nm$, respectively. It is obvious that the results of Case 1 are a special case of the present case, since if we set $m = p = 1$ in Eqs (3.8-19), then we readily derive system (3.8-16).

In system realization, i.e., when going from $\mathbf{H}(s)$ to state equations, several problems arise. We mention the following:

1. The Problem of Realization

This problem examines the existence of a description in state space. Practically, this means that it is not always possible for any given transfer function, or more generally, for any given input–output description, to be able to determine a description in state space of the form (3.7-8).

2. The Problem of Minimum Dimension n

This problem refers to the determination of a state-space realization involving the minimum possible number of state variables. Such a realization is called a minimal realization. For example, realization (3.8-16) is a minimal realization, while (3.8-19), in general, is not.

3. The Problem of Minimum Number of Parameters

This problem refers to the determination of a state-space realization in which the number of parametric elements in the matrices $\mathbf{A}$, $\mathbf{B}$, and $\mathbf{C}$ are as minimum as possible. Realization (3.8-16) is a minimum parameter realization. This is due to the special form of $\mathbf{A}$ and $\mathbf{b}$, where $\mathbf{A}$ has only n parametric elements (the n elements of its last row), while $\mathbf{b}$ has none. Note that the elements 0 and 1 in $\mathbf{A}$ and $\mathbf{b}$ are not parametric, since they do not depend on the parameters of the system.

3.9 EQUIVALENT STATE-SPACE MODELS

The transition from one description to another may or may not be unique. For example, the transition from a state-space description to its input–output description $\mathbf{H}(s)$ or $\mathbf{H}(t)$ is unique. On the contrary, the transition from an input–output description to a state-space model is never unique. That is, the realization problem does not have a unique solution. Hence, the state equations (3.8-16) are one of the many possible descriptions of the transfer function (3.8-9) in state space. The same stands for the state-space realization (3.8-19) of the transfer function matrix (3.8-18). The nonuniqueness in state-space realization generates the problem of description equivalence, examined in the sequel.

Definition 3.9.1

Systems $[\mathbf{A}, \mathbf{B}, \mathbf{C}, \mathbf{D}]_n$ and $[\mathbf{A}^*, \mathbf{B}^*, \mathbf{C}^*, \mathbf{D}^*]_n$ are *strictly equivalent* if and only if there is a constant square matrix $\mathbf{T}$, with $|\mathbf{T}| \neq 0$, such that the following relations hold:

$$\mathbf{A} = \mathbf{T}\mathbf{A}^*\mathbf{T}^{-1} \quad \text{or} \quad \mathbf{A}^* = \mathbf{T}^{-1}\mathbf{A}\mathbf{T} \tag{3.9-1a}$$

$$\mathbf{B} = \mathbf{T}\mathbf{B}^* \quad \text{or} \quad \mathbf{B}^* = \mathbf{T}^{-1}\mathbf{B} \tag{3.9-1b}$$

$$\mathbf{C} = \mathbf{C}^*\mathbf{T}^{-1} \quad \text{or} \quad \mathbf{C}^* = \mathbf{C}\mathbf{T} \tag{3.9-1c}$$

$$\mathbf{D} = \mathbf{D}^* \quad \text{or} \quad \mathbf{D}^* = \mathbf{D} \tag{3.9-1d}$$

The above definition originates from the following: consider the state transformation

$$\mathbf{x} = \mathbf{T}\mathbf{x}^*, \quad \text{where } |\mathbf{T}| \neq 0$$

where $\mathbf{x}$ is the original state vector and $\mathbf{x}^*$ is the new state vector. Substitute this transformation in the state equations of system $[\mathbf{A}, \mathbf{B}, \mathbf{C}, \mathbf{D}]_n$, i.e., of the system

$$\dot{\mathbf{x}} = \mathbf{A}\mathbf{x} + \mathbf{B}\mathbf{u}$$

$$\mathbf{y} = \mathbf{C}\mathbf{x} + \mathbf{D}\mathbf{u}$$

Then, we have

$$\mathbf{T}\dot{\mathbf{x}}^* = \mathbf{A}\mathbf{T}\mathbf{x}^* + \mathbf{B}\mathbf{u}$$

$$\mathbf{y} = \mathbf{C}\mathbf{T}\mathbf{x}^* + \mathbf{D}\mathbf{u}$$

Premuliplying the first equation by $\mathbf{T}^{-1}$ yields the following state-space description

$$\dot{\mathbf{x}}^* = \mathbf{T}^{-1}\mathbf{A}\mathbf{T}\mathbf{x}^* + \mathbf{T}^{-1}\mathbf{B}\mathbf{u} = \mathbf{A}^*\mathbf{x}^* + \mathbf{B}^*\mathbf{u}$$

$$\mathbf{y} = \mathbf{C}\mathbf{T}\mathbf{x}^* + \mathbf{D}\mathbf{u} = \mathbf{C}^*\mathbf{x}^* + \mathbf{D}^*\mathbf{u}$$

where $\mathbf{A}^* = \mathbf{T}^{-1}\mathbf{A}\mathbf{T}$, $\mathbf{B}^* = \mathbf{T}^{-1}\mathbf{B}$, $\mathbf{C}^* = \mathbf{C}\mathbf{T}$, and $\mathbf{D}^* = \mathbf{D}$. These four relations are the relations stated in Eqs (3.9-1).

The input–output description of a linear time-invariant system under strict equivalence is invariant. This property is proven in the following theorem.

Theorem 3.9.1

If the linear time-invariant systems $[\mathbf{A}, \mathbf{B}, \mathbf{C}, \mathbf{D}]_n$ and $[\mathbf{A}^*, \mathbf{B}^*, \mathbf{C}^*, \mathbf{D}^*]_n$ are strictly equivalent, then these two systems have the same input–output description, i.e., their transfer function matrices are equal.

Proof

Let $\mathbf{H}(s) = \mathbf{C}(s\mathbf{I} - \mathbf{A})^{-1}\mathbf{B} + \mathbf{D}$ and $\mathbf{H}^*(s) = \mathbf{C}^*(s\mathbf{I} - \mathbf{A}^*)^{-1}\mathbf{B}^* + \mathbf{D}^*$ be the transfer function matrices of systems $[\mathbf{A}, \mathbf{B}, \mathbf{C}, \mathbf{D}]_n$ and $[\mathbf{A}^*, \mathbf{B}^*, \mathbf{C}^*, \mathbf{D}^*]_n$, respectively. By substituting the relations of strict equivalence (3.9-1) in $\mathbf{H}(s)$, we have

$$\mathbf{H}(s) = \mathbf{C}(s\mathbf{I} - \mathbf{A})^{-1}\mathbf{B} + \mathbf{D} = \mathbf{C}^*\mathbf{T}^{-1}[s\mathbf{I} - \mathbf{T}\mathbf{A}^*\mathbf{T}^{-1}]^{-1}\mathbf{T}\mathbf{B}^* + \mathbf{D}^*$$
$$= \mathbf{C}^*\mathbf{T}^{-1}[\mathbf{T}(s\mathbf{I} - \mathbf{A}^*)\mathbf{T}^{-1}]^{-1}\mathbf{T}\mathbf{B}^* + \mathbf{D}^* = \mathbf{C}^*\mathbf{T}^{-1}\mathbf{T}[s\mathbf{I} - \mathbf{A}^*]^{-1}\mathbf{T}^{-1}\mathbf{T}\mathbf{B}^* + \mathbf{D}^*$$
$$= \mathbf{C}^*[s\mathbf{I} - \mathbf{A}^*]^{-1}\mathbf{B}^* + \mathbf{D}^* = \mathbf{H}^*(s) \tag{3.9-2}$$

All minimal dimension realizations of a linear time-invariant system are strictly equivalent among themselves. This property is given in the following theorem.

Theorem 3.9.2

Let system $[\mathbf{A}, \mathbf{B}, \mathbf{C}, \mathbf{D}]_n$ be a minimum dimension realization of the transfer function matrix $\mathbf{H}(s)$. Then, any other minimum dimension realization $[\mathbf{A}^*, \mathbf{B}^*, \mathbf{C}^*, \mathbf{D}^*]_n$ of $\mathbf{H}(s)$ is given by the relations

$$
\begin{aligned}
\mathbf{A}^* &= \mathbf{T}^{-1}\mathbf{A}\mathbf{T} \\
\mathbf{B}^* &= \mathbf{T}^{-1}\mathbf{B} \\
\mathbf{C}^* &= \mathbf{C}\mathbf{T} \\
\mathbf{D}^* &= \mathbf{D}
\end{aligned}
\tag{3.9-3}
$$

where $\mathbf{T}$ is an arbitrary constant square $n \times n$ matrix, with $|\mathbf{T}| \neq 0$.

From the above, it follows that the transition from state equations to the input–output description is only a computational problem, while the transition from the input–output description to the state equations, is first a theoretical and then a computational problem. Finally, from a practical point of view, the realization problem is essentially the selection of the most appropriate matrices $\mathbf{A}$, $\mathbf{B}$, $\mathbf{C}$, and $\mathbf{D}$ among many equivalent descriptions in state space.

3.10 BLOCK DIAGRAMS

Block diagrams may be considered as a form of system description that provides a simplified overview schematic diagram of a system. Each block is characterized by an input–output description, by its input signal, and by its output signal (see Figure 3.13). The frequency domain appears to facilitate the use of block diagrams and it is for this reason that transfer functions are used to describe each block.

When describing a system by block diagrams, what we actually have is many blocks appropriately linked to form the overall block diagram of the system. To handle complex block diagrams, one needs special rules that specify the significance of each connection and assist in simplifying or modifying the block diagram. The most important of such rules are listed below.

Figure 3.13 Single block system.

3.10.1 Block Diagram Rules

1 Summation Point (Figure 3.14)

The point where two or more signals may be added or subtracted is called a summation point.

2 Blocks in Cascade (Figure 3.15)

Two or more blocks in cascade may be simplified to a single block in the following way. Going from left to right, we have the following relations for each block:

$$X_1(s) = H_1(s)U(s)$$
$$X_2(s) = H_2(s)X_1(s)$$
$$\vdots$$
$$X_{n-1}(s) = H_{n-1}(s)X_{n-2}(s)$$
$$Y(s) = H_n(s)X_{n-1}(s)$$

Eliminating $X_1(s), X_2(s), \ldots, X_{n-1}(s)$ in the above equations yields

$$Y(s) = [H_1(s) \ H_2(s) \ \cdots \ H_n(s)]U(s)$$

Hence, the transfer function $H(s)$ of the equivalent single block diagram is

$$H(s) = H_1(s)H_2(s)\cdots H_n(s) \tag{3.10-1}$$

3 Blocks in Parallel (Figure 3.16)

Two or more blocks in parallel may be simplified to a single block in the following way. Going from top to bottom, the input–output relation for each block is

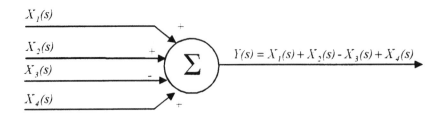

Figure 3.14 Summation point.

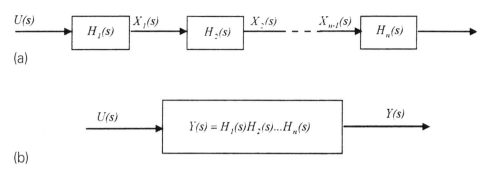

(a)

(b)

Figure 3.15 Block diagram with blocks in cascade. (a) Blocks in cascade; (b) equivalent single block.

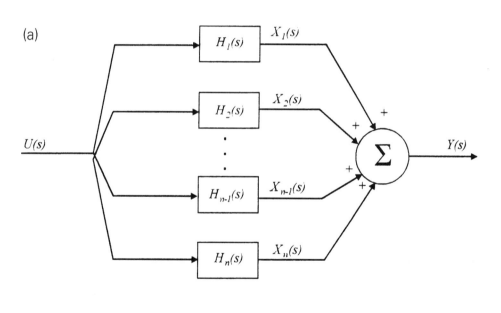

(a)

(b)

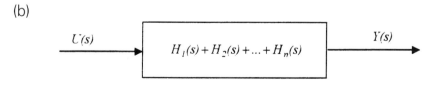

Figure 3.16 Block diagram with blocks in parallel. (a) Blocks in parallel; (b) equivalent block.

$$X_1(s) = H_1(s)U(s)$$
$$X_2(s) = H_2(s)U(s)$$

$$\vdots$$

$$X_{n-1}(s) = H_{n-1}(s)U(s)$$
$$X_n(s) = H_n(s)U(s)$$

Eliminating $X_1(s), X_2(s), \ldots, X_n(s)$ in the above equation yields

$$Y(s) = [H_1(s) + H_2(s) + \cdots + H_n(s)]U(s)$$

Hence, the transfer function $H(s)$ of the equivalent single block diagram is

$$H(s) = H_1(s) + H_2(s) + \cdots + H_n(s) \tag{3.10-2}$$

4 Conversion of a Closed-Loop System to an Open-Loop System (Figure 3.17)

The definitions of closed-loop and open-loop systems have been presented in Sec. 1.3. Consider the closed-loop system of Figure 3.17a. The transfer functions $G(s)$ and $F(s)$ are called the forward-path transfer function and the feedback-path transfer function, respectively.

The closed-loop system of Figure 3.17a may be converted into an open-loop, single-block system as follows. From the closed-loop system, we have

$$U(s) = R(s) \mp F(s)Y(s)$$
$$Y(s) = G(s)U(s)$$

Eliminating $U(s)$ gives

$$Y(s) = \left[\frac{G(s)}{1 \pm G(s)F(s)}\right]R(s)$$

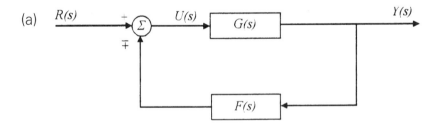

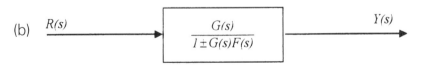

Figure 3.17 Block diagrams of (a) a closed-loop system and (b) its equivalent open-loop system.

Hence, the transfer function $H(s)$ of the equivalent open-loop system is

$$H(s) = \frac{G(s)}{1 \pm G(s)F(s)} \qquad (3.10\text{-}3)$$

Special Case
If the feedback-path transfer function $F(s)$ is *unity*, i.e., if $F(s) = 1$, then Eq. (3.10-3) takes on the form

$$H(s) = \left[\frac{G(s)}{1 \pm G(s)} \right] \qquad (3.10\text{-}4)$$

5 Conversion of an Open-Loop System to a Closed-Loop System (Figure 3.18)

An open-loop system can be converted into a closed-loop system in the following way: Solving Eq. (3.10-4) for $G(s)$ gives

$$G(s) = \frac{H(s)}{1 \pm H(s)} \qquad (3.10\text{-}5)$$

Hence, an open-loop system, having transfer function $H(s)$, is equivalent to a closed-loop system with unity $F(s)$ and with $G(s)$ given by Eq. (3.10-5).

6 Conversion of the Feedback-Path Transfer Function $F(s)$ into Unity

Consider the closed-loop system shown in Figure 3.17a. An equivalent block diagram with unity $F(s)$ is the closed-loop system shown in Figure 3.19. From Figure 3.19 we have

$$U(s) = \frac{R(s)}{F(s)} \mp Y(s) \qquad \text{and} \qquad Y(s) = G(s)F(s)U(s)$$

Eliminating $U(s)$ gives

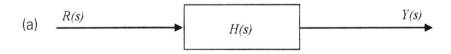

(a)

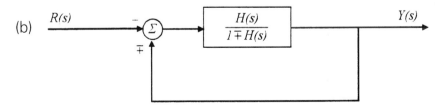

(b)

Figure 3.18 Block diagrams of (a) an open-loop system and (b) its equivalent closed-loop system.

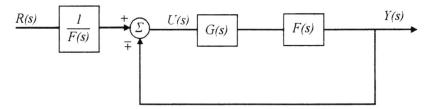

Figure 3.19   Closed-loop system with unity feedback-path transfer function.

$$H(s) = \frac{Y(s)}{R(s)} = \frac{G(s)}{1 \pm G(s)F(s)}$$

Hence, both diagrams have the same transfer function, and thus they are equivalent.

7 Moving a Summation Point

Consider the block diagram shown in Figure 3.20a. An equivalent block diagram with the summation point shifted is as shown in Figure 3.20b. The diagram shown in Figure 3.20a gives

$$Y(s) = H(s)U_1(s) + U_2(s)$$

while the diagram shown in Figure 3.20b gives

$$Y(s) = H(s)\left[U_1(s) + \frac{U_2(s)}{H(s)}\right] = H(s)U_1(s) + U_2(s)$$

Hence, the two block diagrams are equivalent.

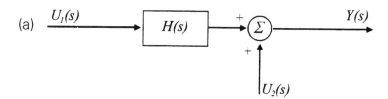

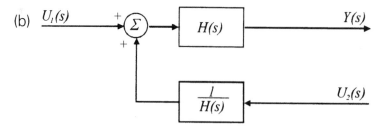

Figure 3.20   Moving a summation point. (a) Summation point following $H(s)$; (b) summation point proceeding $H(s)$.

8 Moving a Pickoff Point

Consider the block diagram shown in Figure 3.21a. An equivalent block diagram is shown in Figure 3.21b. From the diagram shown in Figure 3.21a, we have that $Y(s) = H(s)U(s)$; also, from the diagram shown in Figure 3.21b, we have that $Y(s) = H(s)U(s)$. Hence, both block diagrams are equivalent.

All above cases of equivalent block diagrams are frequently used. Figure 3.22 summarizes these cases, while other additional useful cases of equivalent block diagrams are included.

3.10.2 Simplification of Block Diagrams

In the two examples that follow, we apply the rules of block diagrams of Subsec. 3.10.1 to simplify complex block diagrams.

Example 3.10.1

Simplify the block diagram shown in Figure 3.23 to a single block.

Solution

This can be done step by step as follows. First, simplify the closed-loop system involving the blocks $G_1(s)$, $G_2(s)$, and $F_1(s)$ to an open-loop system, as in Figure 3.24, to yield

$$G_4(s) = \frac{G_1(s)G_2(s)}{1 - G_1(s)G_2(s)F_1(s)}$$

(a)

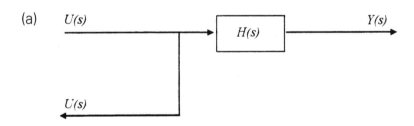

(b)

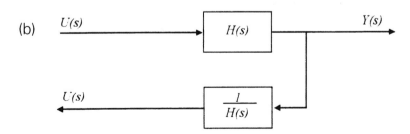

Figure 3.21 Moving a pickoff point. (a) Pickoff point at the input of $H(s)$; (b) pickoff point at the output of $H(s)$.

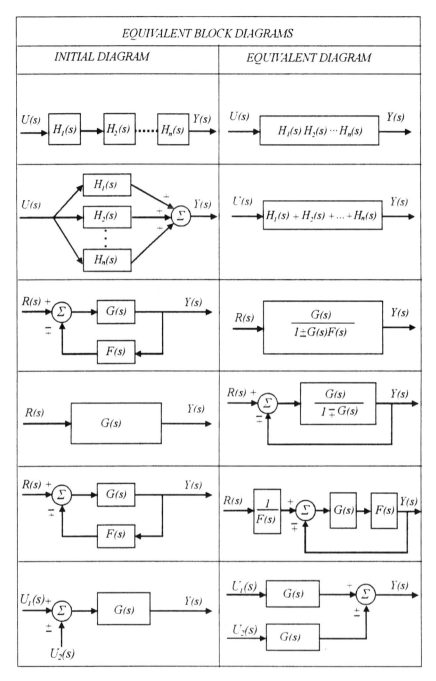

Figure 3.22   Equivalent block diagrams.

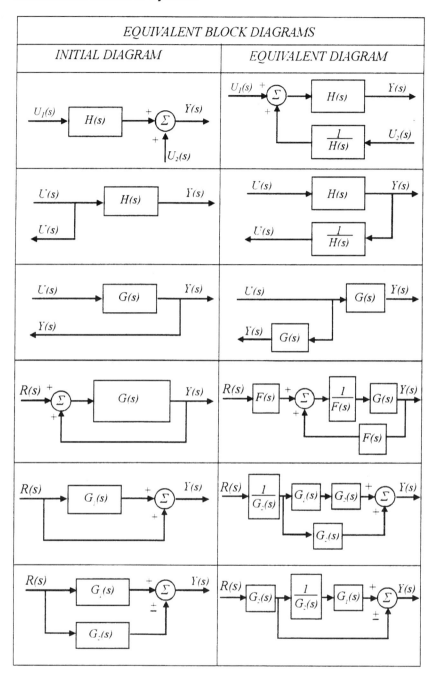

Figure 3.22 (*contd.*)

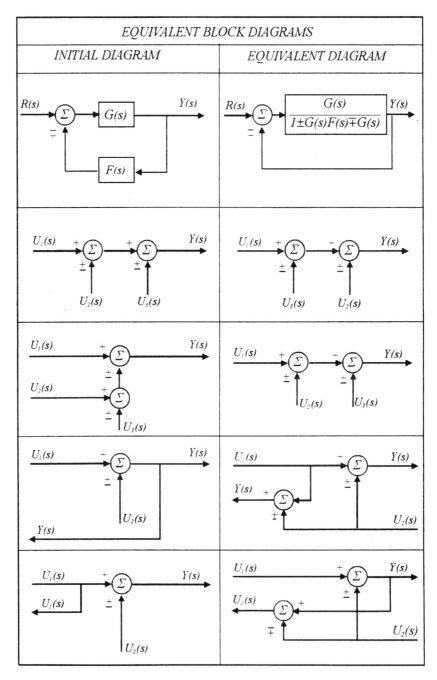

Figure 3.22   (*contd.*)

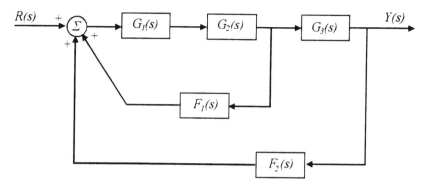

Figure 3.23   Original block diagram.

Then, simplify the two blocks in cascade, as shown in Figure 3.25, where $G_5(s) = G_4(s)G_3(s)$. Finally, simplify the closed-loop system shown in Figure 3.25 to an open-loop system, as in Figure 3.26, to yield

$$H(s) = \frac{G_5(s)}{1 - G_5(s)F_2(s)}$$

Example 3.10.2

Simplify the block diagram shown in Figure 3.27.

Solution

This diagram has two inputs, $R_1(s)$ and $R_2(s)$. To determine the response $Y(s)$ we work as follows. Using the superposition principle, $Y(s)$ can be determined by adding two responses: the response $Y_1(s)$ due to the input $R_1(s)$, where the other input is assumed zero, i.e., when $R_2(s) = 0$; the response $Y_2(s)$ due to the input $R_2(s)$, where the other input is assumed zero, i.e., when $R_1(s) = 0$. Thus, when $R_2(s) = 0$, the block diagram will be as shown in Figure 3.28, and the response $Y_1(s)$, due to the input $R_1(s)$, will be

$$Y_1(s) = \left[\frac{G_1(s)G_2(s)}{1 - G_1(s)G_2(s)F(s)}\right] R_1(s)$$

Similarly, when $R_1(s) = 0$, the block diagram will be as shown in Figure 3.29, and the response $Y_2(s)$, due to the input $R_2(s)$, will be

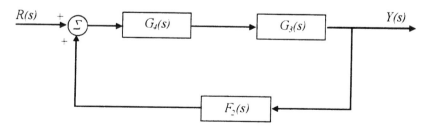

Figure 3.24   First simplification of the block diagram of Figure 3.23.

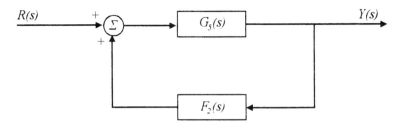

Figure 3.25 Second simplification of the block diagram of Figure 3.23.

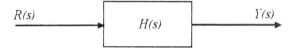

Figure 3.26 Simplification of the block diagram of Figure 3.23 in a single block.

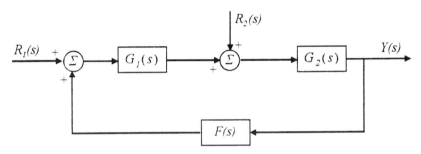

Figure 3.27 Block diagram with two inputs.

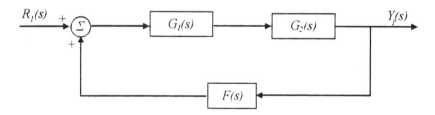

Figure 3.28 The block diagram of Figure 3.27 when $R_2(s) = 0$.

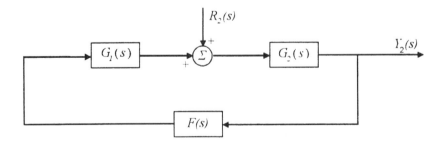

Figure 3.29 Block diagram of Figure 3.27 when $R_1(s) = 0$.

$$Y_2(s) = \left[\frac{G_2(s)}{1 - G_1(s)G_2(s)F(s)}\right]R_2(s)$$

Hence, the total response $Y(s)$ of Figure 3.27 will be

$$Y(s) = Y_1(s) + Y_2(s) = \left[\frac{G_1(s)G_2(s)}{1 - G_1(s)G_2(s)F(s)}\right]R_1(s) + \left[\frac{G_2(s)}{1 - G_1(s)G_2(s)F(s)}\right]R_2(s)$$

Remark 3.10.1

Systems described in state space can also be represented by block diagrams. For example, consider the system (3.7-16a,b), i.e., the system

$$\dot{\mathbf{x}}(t) = \mathbf{A}\mathbf{x}(t) + \mathbf{B}\mathbf{u}(t)$$
$$\mathbf{y}(t) = \mathbf{C}\mathbf{x}(t) + \mathbf{D}\mathbf{u}(t)$$

The block diagram of this system is given in Figure 3.30, where the double lines represent multiple signals.

3.11   SIGNAL-FLOW GRAPHS

3.11.1   Definitions

A signal-flow graph, as in the case of block diagrams, gives a simplified schematic overview of a system. In particular, a signal-flow graph gives this view based on the equations which describe the system.

A signal-flow graph is composed of nodes (junctions) and branches. Each node represents a system's variable. We have three types of nodes:

- *A source or input node* (independent variable) is a node where no branch ends and from which one or more branches begin (Figure 3.31a).
- *A sink or output node* (dependent variable) is a node to which one or more branches end and from which no branch begins (Figure 3.31b).
- *A mixed or generalized node* is a node to which branches both end and begin (Figure 3.31c).

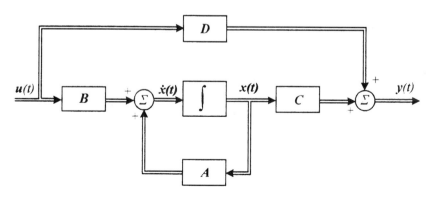

Figure 3.30   State equations block diagram.

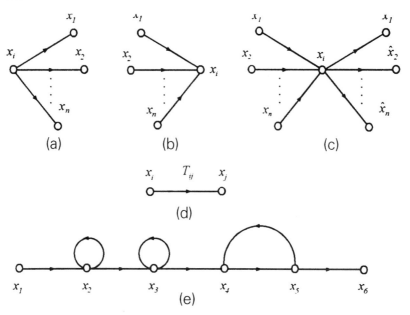

Figure 3.31   Nodes, direction, gain, and paths of signal-flow graphs.

Every branch connects two nodes and has two characteristics: the direction and the gain. *Direction* is simply the direction of the flow of the signal from one node to another (Figure 3.31d). *Gain* or transmittance or transfer function is the quantity T_{ij} which relates the variables (nodes) x_i and x_j in the cause-and-effect sense. Hence, the direction and the gain are related by the equation $x_j = T_{ij}x_i$. If the direction of the branch was opposite, then the equation would be $x_i = T_{ji}x_j$.

A signal-flow graph forms "paths." Depending on the particular form of these paths, the following definitions are given:

- A *path* is a succession of branches which have the same direction. For example, in Figure 3.31e, the combinations $x_1x_2x_3$, $x_1x_2x_3x_4$, and $x_3x_4x_5$ x_6 are paths.
- A *forward path* is the path which starts from the input (source) node and ends at the otput (sink) node. For example, the path $x_1x_2x_3x_4x_5x_6$ in Figure 3.31e is a forward path.
- A *loop* is the path that begins and ends in the same node and along which no other node is encountered more than once. For example, the path $x_4x_5x_4$ in Figure 3.31e is a loop.
- A *single-branch loop* is a loop formed by a single branch. For example, the paths x_2x_2 and x_3x_3 in Figure 3.31e are single-branch loops.

3.11.2   Construction of Signal-Flow Graphs

Consider a system described by the following equations

$$T_{11}x_1 + T_{12}x_2 = x_3$$
$$T_{21}x_1 + T_{22}x_2 = x_4$$

(3.11-1)

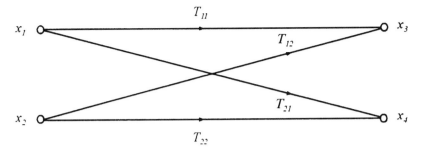

Figure 3.32 Signal-flow graph of system in Eq. (3.11-1).

The signal-flow graph of this system is composed of the nodes x_1, x_2, x_3, and x_4, properly linked with the branch gains T_{11}, T_{12}, T_{21}, and T_{22}, so that the equations of the given system are satisfied. It can be easily shown that the diagram of Figure 3.32 is the signal-flow graph sought of Eq. (3.11-1).

Similarly, for the system described by the equations

$$x_2 = T_{12}x_1 + T_{32}x_3$$
$$x_3 = T_{23}x_2 \tag{3.11-2}$$
$$x_4 = T_{34}x_3$$

one arrives at the signal-flow graph of Figure 3.33.

Because a signal-flow graph is a schematic representation of the equations that describe the system, a good point to start for the construction of a signal-flow graph is to write down the equations of the system. This is done by applying the physical laws that govern the particular system. For example, in electrical networks one applies the Kirchhhoff's laws of currents and voltages, while in mechanical systems one applies d'Alembert's law of forces.

With regard to the network equations, the following are pointed out. We consider as signals the node voltages and the loop currents. For a network with N passive elements (resistors, capacitors, and inductors) we can write $N_k - 1$ node equations and $N - (N_k - 1)$ loop equations, where N_k is the number of nodes. The total number of node and loop equations, and hence the total number of variables (or signals) for the network, is N.

Example 3.11.1

Find the signal-flow graph for the network shown in Figure 3.34.

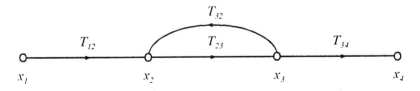

Figure 3.33 Signal-flow graph of system in Eq. (3.11-2).

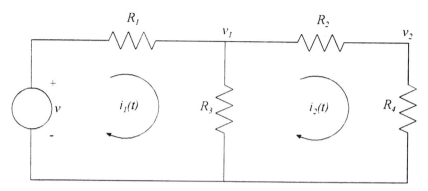

Figure 3.34   A resistive network of two loops.

Solution

The network's variables are v, v_1, v_2, i_1, and i_2. Variable v is known and is the input to the graph. Writing down the network equations, going from left to right, we have the following four equations:

$$i_1 = \frac{v - v_1}{R_1}, \qquad v_1 = R_3(i_1 - i_2), \qquad i_2 = \frac{v_1 - v_2}{R_2}, \qquad v_2 = i_2 R_4$$

The signal-flow graph of the network can be easily constructed and is given in Figure 3.35.

3.11.3   Mason's Rule

Simplification rules, analogous to those for block diagrams presented in Subsec. 3.10.2, have been developed for signal-flow graphs, among which the most popular is Mason's rule. Mason's rule has the attractive characteristic in that it yields the input–output gain T (i.e., the gain from the input to the output node) of a signal-flow graph directly, without intermediate steps. Mason's rule has the form

$$T = \frac{1}{\Delta} \sum_{n=1}^{k} T_n \Delta_n \tag{3.11-3}$$

where

1. T_n is the gain of every forward path.

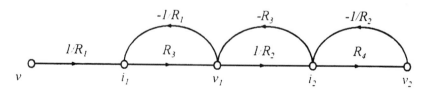

Figure 3.35   The signal-flow graph of the network of Figure 3.34.

2. Δ is the determinant of the signal-flow graph, which is given by

$$\Delta = 1 - \Sigma L_1 + \Sigma L_2 - \Sigma L_3 + \cdots$$

where
a. L_1 is the gain of every loop and ΣL_1 is the sum of the gains of all the loops of the graph.
b. L_2 is the product of the gains of two nontouching loops (two loops are nontouching when they don't have common nodes). ΣL_2 is the sum of the product of the gains of all possible combinations of two nontouching loops.
c. L_3 is the product of the gains of three nontouching loops, and so on.
3. Δ_n is the determinant Δ of the signal-flow graph which remains when the path which produces T_n is removed.
4. k is the number of the forward paths.

Example 3.11.2

Determine the transfer function (the gain) T between v and v_2 of the flow diagram of Figure 3.35.

Solution

The gain T between v and v_2 will be determined using Mason's rule as follows:

1 Forward Path
There is one forward path, as shown in Figure 3.36. The gain T_1 of this path is $T_1 = R_3 R_4 / R_1 R_2$.

2 Evaluation of the Determinant Δ
There are three loops, as shown in Figure 3.37. The gains of the first, second, and third loop are $-R_3/R_1$, $-R_3/R_2$, and $-R_4/R_2$, respectively. Hence

$$\Sigma L_1 = -\left[\frac{R_3}{R_1} + \frac{R_3}{R_2} + \frac{R_4}{R_2}\right]$$

There are two nontouching loops, namely the first and the third loop. Thus, $\Sigma L_2 = R_3 R_4 / R_1 R_2$. Therefore,

$$\Delta = 1 - \Sigma L_1 + \Sigma L_2 = 1 + \left[\frac{R_3}{R_1} + \frac{R_3}{R_2} + \frac{R_4}{R_2}\right] + \frac{R_3 R_4}{R_1 R_2}$$

$$= \frac{R_1 R_2 + R_1 R_3 + R_1 R_4 + R_2 R_3 + R_3 R_4}{R_1 R_2}$$

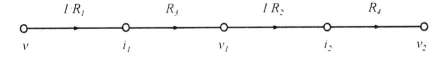

Figure 3.36 Forward path of the signal-flow graph of Figure 3.35.

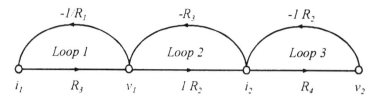

Figure 3.37   Loops in the signal-flow graph of Figure 3.35.

3 Evaluation of Δ_1

Since all branches touch the forward path, it follows that $\Delta_1 = 1$.
Hence

$$T = \frac{T_1 \Delta_1}{\Delta} = \frac{R_3 R_4}{R_1 R_2 + R_1 R_3 + R_1 R_4 + R_2 R_3 + R_3 R_4}$$

3.12   MATHEMATICAL MODELS FOR CONTROL SYSTEMS COMPONENTS

In this section we will derive one or more of the four mathematical models presented in Sections 3.4–3.7 to describe several components which are used in control systems. In particular, we will derive mathematical models for DC and AC motors, tachometers, error detectors, gears, hydraulic actuators, and pneumatic amplifiers. Also, the block diagram and signal-flow graphs of several of these components will be presented. In Sec. 3.13 that follows (as well as in many other places in the book) we present the mathematical descriptions of several practical control systems.

3.12.1   DC Motors

One component which is often used in control systems is the DC motor. There are several types of DC motors. We present here only the separately excited type, because its characteristics present several advantages over others, particularly with regard to linearity. Separately excited DC motors are distinguished in two categories: those that are controlled by the stator, which are usually called field-controlled motors; those that are controlled by the rotor, which are usually called armature-controlled motors.

1 Motors Controlled by the Stator

A simple diagram of a DC motor controlled by its stator (i.e., by its field) is given in Figure 3.38. For simplicity, we make the following approximations:

  a. The rotor's current $i_a(t)$ is constant, i.e., $i_a(t) = I_a$.
  b. The magnetic flux $\varphi(t)$ between the stator and the rotor is given by the linear relation

$$\varphi(t) = K_f i_f(t) \tag{3.12-1}$$

where K_f is a constant and $i_f(t)$ is the stator's current.

  c. The torque $T_m(t)$ that is developed by the motor is given by the relation

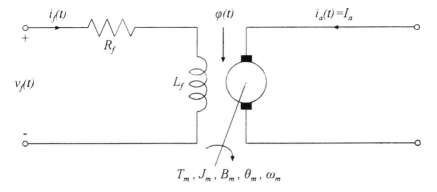

Figure 3.38 A DC motor controlled by the stator.

$$T_m(t) = K_m I_a \varphi(t) \tag{3.12-2}$$

where K_m is a constant.

d. The Kirchhoff's voltage law for the stator network is

$$L_f \frac{di_f}{dt} + R_f i_f = v_f(t) \tag{3.12-3}$$

e. The rotor's rotational motion is described by the differential equation

$$J_m \frac{d\omega_m}{dt} + B_m \omega_m = T_m(t), \qquad \omega_m = \frac{d\theta_m}{dt} \tag{3.12-4}$$

where J_m is the torque inertia, B_m is the coefficient of friction, $\theta_m(t)$ is the angular position or displacement, and $\omega_m(t)$ is the angular velocity of the motor.

Substituting Eq. (3.12-1) into (3.12-2), we have

$$T_m(t) = K_m K_f I_a i_f(t) \tag{3.12-5}$$

Also, if we substitute Eq. (3.12-5) into Eq. (3.12-4), we have

$$J_m \frac{d\omega_m}{dt} + B_m \omega_m = K_m K_f I_a i_f(t) \tag{3.12-6}$$

In the sequel, we will present the four different mathematical models which describe the motor, together with the motor's block diagram and signal-flow graph.

a. Differential Equations

These are the Eqs (3.12-5) and (3.12-4) i.e., the equations

$$L_f \frac{di_f}{dt} + R_f i_f = v_f(t) \tag{3.12-7a}$$

$$J_m \frac{d\omega_m}{dt} + B_m \omega_m = T_m(t) \tag{3.12-7b}$$

b. State Equations

Let $x_1 = \theta_m$, $x_2 = \omega_m$, and $x_3 = i_f$ be the state variables. Then, the system of differential equations (3.12-7), together with the equation $\dot{x}_1 = x_2 = \omega_m$, take on the form

$$\dot{\mathbf{x}} = \mathbf{A}\mathbf{x} + \mathbf{b}u \qquad \text{and} \qquad y = \mathbf{c}^{\mathrm{T}}\mathbf{x} \tag{3.12-8a}$$

where

$$\mathbf{x} = \begin{bmatrix} \theta_{\mathrm{m}} \\ \omega_{\mathrm{m}} \\ i_{\mathrm{f}} \end{bmatrix}, \quad \mathbf{A} = \begin{bmatrix} 0 & 1 & 0 \\ 0 & -T_{\mathrm{m}}^{-1} & J_{\mathrm{m}}^{-1}K_{\mathrm{m}}K_{\mathrm{f}}I_{\mathrm{a}} \\ 0 & 0 & -T_{\mathrm{f}}^{-1} \end{bmatrix}, \quad \mathbf{b} = \begin{bmatrix} 0 \\ 0 \\ L_{\mathrm{f}}^{-1} \end{bmatrix}, \quad \mathbf{c} = \begin{bmatrix} 1 \\ 0 \\ 0 \end{bmatrix}$$

$$\tag{3.12-8b}$$

and where $u = v_{\mathrm{f}}$, $T_{\mathrm{m}} = J_{\mathrm{m}}/B_{\mathrm{m}}$ is the mechanical time constant, and $T_{\mathrm{f}} = L_{\mathrm{f}}/R_{\mathrm{f}}$ is the electrical time constant of the stator.

c. Transfer Function

Applying the Laplace transform to Eqs (3.12-7) or (3.12-8) and after some algebraic manipulations we derive the following transfer function

$$G(s) = \frac{\Theta_{\mathrm{m}}(s)}{V_{\mathrm{f}}(s)} = \frac{K_{\mathrm{m}}K_{\mathrm{f}}I_{\mathrm{a}}}{s(J_{\mathrm{m}}s + B_{\mathrm{m}})(L_{\mathrm{f}}s + R_{\mathrm{f}})} = \frac{B_{\mathrm{m}}^{-1}R_{\mathrm{f}}^{-1}K_{\mathrm{m}}K_{\mathrm{f}}I_{\mathrm{a}}}{s(T_{\mathrm{m}}s + 1)(T_{\mathrm{f}}s + 1)} \tag{3.12-9}$$

where all initial conditions of the motor are assumed zero.

d. Block Diagram

A simple block diagram of the motor is given in Figure 3.39a.

e. Signal-Flow Graph

The signal-flow graph of the state equations (3.12-8) is given in Figure 3.39b.

2 Motors Controlled by the Rotor

A simple diagram of a DC motor controlled by its rotor (i.e., by its armature) is given in Figure 3.40. For simplicity, we assume the following approximations:

a. The stator current $i_{\mathrm{f}}(t)$ is constant, i.e., $i_{\mathrm{f}}(t) = I_{\mathrm{f}}$.

b. The magnetic flux $\varphi(t)$, given by Eq. (3.12-1), will be constant since $i_{\mathrm{f}}(t)$ is constant, i.e., $\varphi(t) = K_{\mathrm{f}}I_{\mathrm{f}} = \phi$.

c. The torque $T_{\mathrm{m}}(t)$, given by Eq. (3.12-2), now has the form

$$T_{\mathrm{m}}(t) = K_{\mathrm{m}}i_{\mathrm{a}}(t)\phi = K_{\mathrm{i}}i_{\mathrm{a}}(t) \tag{3.12-10}$$

where $K_{\mathrm{i}} = K_{\mathrm{m}}\phi$.

d. The voltage $v_{\mathrm{b}}(t)$ is proportional to the angular velocity of the motor, i.e.,

$$v_{\mathrm{b}}(t) = K_{\mathrm{b}}\omega_{\mathrm{m}}(t) \tag{3.12-11}$$

e. The Kirchhoff's law of voltages for the rotor network is

$$L_{\mathrm{a}}\frac{di_{\mathrm{a}}}{dt} + R_{\mathrm{a}}i_{\mathrm{a}} + v_{\mathrm{b}} = v_{\mathrm{a}}$$

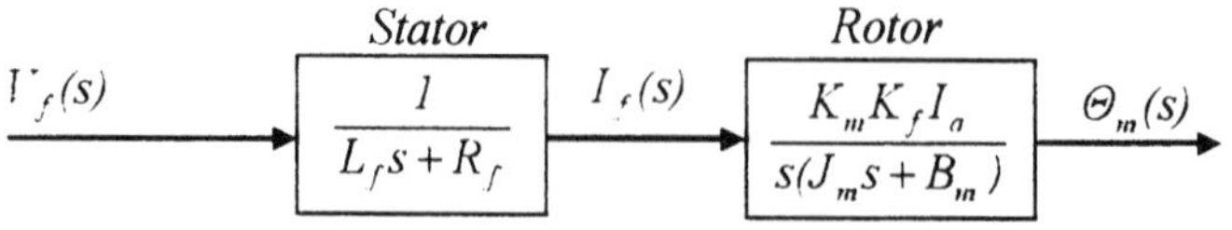

Figure 3.39a Block diagram of a motor controlled by the stator.

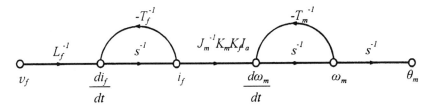

Figure 3.39b Signal-flow graph of a motor controlled by the stator.

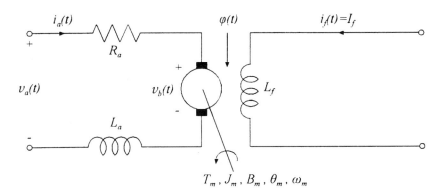

Figure 3.40 A DC motor controlled by the rotor.

or

$$L_a \frac{di_a}{dt} + R_a i_a + K_b \omega_m = v_a \qquad (3.12\text{-}12)$$

where use was made of Eq. (3.12-11).

 f. The rotor's rotational motion is described by the differential equation

$$J_m \frac{d\omega_m}{dt} + B_m \omega_m = T_m(t) = K_i i_a, \qquad \omega_m = \frac{d\theta_m}{dt} \qquad (3.12\text{-}13)$$

In what follows, we present the four mathematical models and the block diagram and signal-flow graph of the motor.

a. Differential Equations

These are the Eqs (3.12-12) and (3.12-13), i.e., the equations

$$L_a \frac{di_a}{dt} + R_a i_a + K_b \omega_m = v_a \qquad (3.12\text{-}14a)$$

$$J_m \frac{d\omega_m}{dt} + B_m \omega_m = K_i i_a \qquad (3.12\text{-}14b)$$

b. State Equations

Let $x_1 = \theta_m$, $x_2 = \omega_m$, and $x_3 = i_a$ be the state variables. Then the system of differential equations (3.12-14), together with the equation $\dot{x}_1 = x_2 = \omega_m$, take on the form

$$\dot{\mathbf{x}} = \mathbf{A}\mathbf{x} + \mathbf{b}u \qquad \text{and} \qquad y = \mathbf{c}^{\mathsf{T}}\mathbf{x} \tag{3.12-15a}$$

where

$$\mathbf{x} = \begin{bmatrix} \theta_m \\ \omega_m \\ i_a \end{bmatrix}, \quad \mathbf{A} = \begin{bmatrix} 0 & 1 & 0 \\ 0 & -T_m^{-1} & J_m^{-1}K_i \\ 0 & -L_a^{-1}K_b & -T_a^{-1} \end{bmatrix}, \quad \mathbf{b} = \begin{bmatrix} 0 \\ 0 \\ L_a^{-1} \end{bmatrix}, \quad \mathbf{c} = \begin{bmatrix} 1 \\ 0 \\ 0 \end{bmatrix}$$

$$\tag{3.12-15b}$$

and where $u = v_a$, and $T_a = L_a/R_a$ is the electrical time constant of the rotor.

c. Transfer Function

Apply the Laplace transform to Eq. (3.12-14) to yield

$$sL_a I_a(s) + R_a I_a(s) + K_b \Omega_m(s) = V_a(s) \tag{3.12-16a}$$

$$J_m s \Omega_m(s) + B_m \Omega_m(s) = K_i I_a(s), \qquad \Omega_m(s) = s\Theta_m(s) \tag{3.12-16b}$$

where all initial conditions are assumed zero. The transfer function will then be

$$G(s) = \frac{\Theta_m(s)}{V_a(s)} = \frac{K_i}{s(sL_a + R_a)(J_m s + B_m) + K_i K_b s}$$

$$= \frac{K_i}{L_a J_m s^3 + (R_a J_m + L_a B_m)s^2 + (R_a B_m + K_i K_b)s} \tag{3.12-17}$$

d. Block Diagram

The equations of system (3.12-16) can be represented by the closed-loop block diagram of Figure 3.41. From this diagram it is clear that the feedback signal $K_b s\Theta_m(s)$ is actually the derivative of the output $\theta_m(t)$. For this reason this type of control is called "derivative" or "speed" feedback control.

e. Signal-Flow Graph

The signal-flow graph of the state equations (3.12-15) is given in Figure 3.42.

3.12.2 AC Motors

In cases where the control requires low power, then two-phase alternating current (AC) motors are usually used. A two-phase AC motor has two coils on the stator which are perpendicular to each other (Figure 3.43). One of the two coils is used as a reference, while the other is used to control the rotor. The transfer function, which

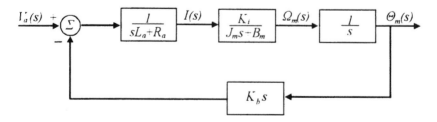

Figure 3.41 Block diagram of a motor controlled by the rotor.

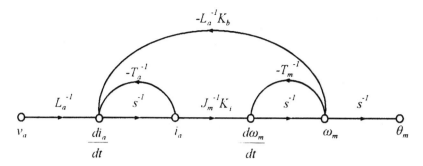

Figure 3.42 Signal-flow graph of a motor controlled by the rotor.

relates the angular position of the rotor $\Theta_m(s)$ to the control voltage $V_2(s)$, is given approximately, by the following relation

$$G(s) = \frac{\Theta_m(s)}{V_2(s)} = \frac{K}{s(Ts+1)} \tag{3.12-18}$$

where K and T are the gain and time constant of the motor, respectively.

3.12.3 Tachometers

The tachometer has the property that its output voltage $v_o(t)$ is proportional to the angular velocity $\omega(t)$ of the motor, i.e.,

$$v_o(t) = K_t \omega(t) \tag{3.12-19}$$

where K_t is the constant of the tachometer.

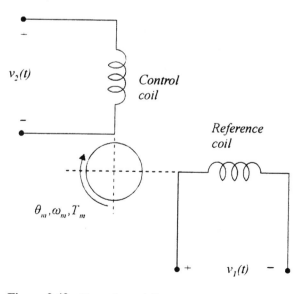

Figure 3.43 Two-phase AC motor.

3.12.4   Error Detectors

Error detectors are special components whose output is the difference of two signals. Such components are very useful in control systems. Two rather well-known types of error detectors are presented below.

1 Operational Amplifier with Resistors

This error detector is given in Figure 3.44. The output voltage $v_o(t)$ of the amplifier is

$$v_o(t) = \frac{R_f}{R_1} v_1(t) - \frac{R_f}{R_2} v_2(t) \tag{3.12-20}$$

Hence, for $R_1 = R_2 = R_f$, the voltage $v_o(t)$ will be

$$v_o(t) = v_1(t) - v_2(t) \tag{3.12-21}$$

2 The Potentiometer

This error detector is given in Figure 3.45. The voltage V_0 is constant. Depending on the position of the points A and B on the resistors, the voltages $v_1(t)$ and $v_2(t)$ are produced, respectively. The voltage $e(t)$ across the points A and B will then be

$$e(t) = K_p[v_1(t) - v_2(t)] \tag{3.12-22}$$

where $e(t)$ is the error voltage and K_p is the potentiometer constant, which depends on the sensitivity of the elements involved.

3 Synchrosystems (Figure 3.46)

A synchrosystem is usually composed of two parts—the transmitter and the receiver. Both the transmitter and the receiver are actually three-phase rotating electrical motors. Let θ_t and θ_r be the angles of the axis of rotation of the transmitter rotor and of the receiver rotor, respectively. It can be shown that if $v_i(t) = \sin \omega t$ is the synchrosystem input voltage, then the amplitude of the output voltage $v_o(t)$ of the synchrosystem will be proportional to the difference of the angles θ_t and θ_r, i.e.,

$$v_o(t) = K(\theta_t - \theta_r) \sin \omega t \tag{3.12-23}$$

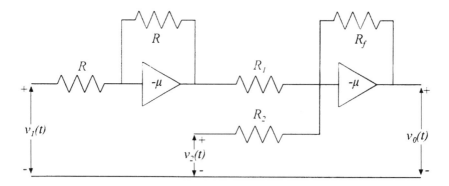

Figure 3.44   Error detector using operational amplifiers and resistors.

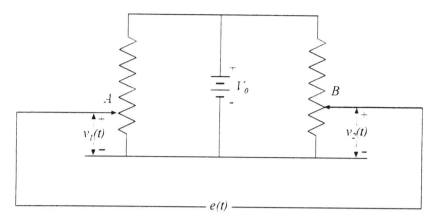

Figure 3.45 Error detector using potentiometers.

where K is the synchrosystem's constant. Relation (3.12-23) suggests that a synchro-system can be used to measure the difference between θ_t and θ_r.

3.12.5 Gears

Gears are used in order to increase or decrease the rotational speed of a motor. Figure 3.47 shows a simple arrangement of gears where

$\quad N_1, N_2 \quad$ = number of teeth of each gear
$\quad R_1, R_2 \quad$ = radius of each gear
$\quad T_1, T_2 \quad$ = torque of each gear
$\quad \theta_1, \theta_2 \quad$ = angular displacement of each gear
$\quad \omega_1, \omega_2 \quad$ = angular velocity of each gear
$\quad J_1, J_2 \quad$ = torque of inertia of each gear
$\quad B_1, B_2 \quad$ = viscous friction coefficient of each gear
$\quad C_1, C_2 \quad$ = Coulomb friction coefficient of each gear

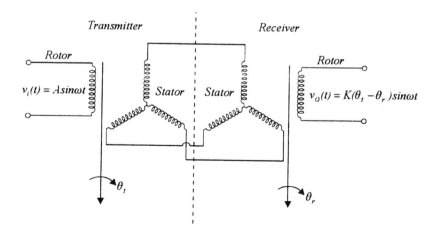

Figure 3.46 A synchrosystem layout.

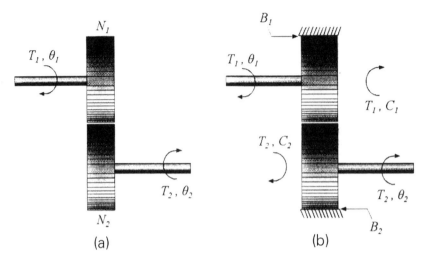

Figure 3.47   (a) Ideal and (b) nonideal gears.

For the ideal system of coupled gears of Figure 3.47a, where the torques of inertia J_1 and J_2 and the friction coefficients B_1, B_2, C_1, and C_2 are considered to be zero, the following relations hold

$$R_1 N_2 = R_2 N_1 \tag{3.12-24a}$$

$$R_1 \theta_1 = R_2 \theta_2 \tag{3.12-24b}$$

$$T_1 \theta_1 = T_2 \theta_2 \tag{3.12-24c}$$

Relation (3.12-24a) expresses the fact that the ratio of the number of teeth is proportional to the radii ratio. Relation (3.12-24b) expresses the fact that the distances run by the two gears are equal. Finally, relation (3.12-24c) expresses the fact that, given that there are no losses, the work done by one gear is equal to the work done by the other. From the above relations, we have that

$$\frac{T_1}{T_2} = \frac{\theta_2}{\theta_1} = \frac{N_1}{N_2} = \frac{\omega_2}{\omega_1} = \frac{R_1}{R_2} \tag{3.12-25}$$

It is well known that, in practice, the torques of inertia J_1 and J_2, as well as the friction coefficients B_1, B_2, C_1, and C_2, are not zero. In this case, the algebraic relation (3.12-24c) is not valid, due to losses. Instead, the following dynamic relations hold:

$$J_1 \frac{d^2\theta_1}{dt^2} + B_1 \frac{d\theta_1}{dt} + C_1 \frac{\omega_1}{|\omega_1|} + T_1 = T \tag{3.12-26}$$

$$J_2 \frac{d^2\theta_2}{dt^2} + B_2 \frac{d\theta_2}{dt} + C_2 \frac{\omega_2}{|\omega_2|} = T_2 \tag{3.12-27}$$

where T is the external torque applied. Relation (3.12-26) is the torque equation of the gear with N_1 number of teeth, while relation (3.12-27) is the torque equation of the gear with N_2 number of teeth.

From relation (3.12-25), we have that $T_2 = N^{-1}T_1$ and $\theta_2 = N\theta_1$, where $N = N_1/N_2$. Substitute these relations in Eq. (3.12-27) to yield

$$NJ_2 \frac{d^2\theta_1}{dt^2} + NB_2 \frac{d\theta_1}{dt} + C_2 \frac{\omega_2}{|\omega_2|} = N^{-1}T_1$$

or

$$N^2J_2 \frac{d^2\theta_1}{dt^2} + N^2B_2 \frac{d\theta_1}{dt} + NC_2 \frac{\omega_2}{|\omega_2|} = T_1 \tag{3.12-28}$$

If we substitute Eq. (3.12-28) into Eq. (3.12-26), we have

$$J_1^* \frac{d^2\theta_1}{dt^2} + B_1^* \frac{d\theta_1}{dt} + T_c = T \tag{3.12-29}$$

where

$$J_1^* = J_1 + N^2J_2, \qquad B_1^* = B_1 + N^2B_2, \qquad \text{and} \qquad T_c = C_1 \frac{\omega_1}{|\omega_1|} + NC_2 \frac{\omega_2}{|\omega_2|}$$

Relation (3.12-29) is the differential equation which describes the dynamic behavior of the gears with reference to the gear with N_1 number of teeth. It is noted that the above analysis of gears is analogous to that of transformers. Indeed, gears and transformers are analogous systems, where the primary transformer coil corresponds to the gear with N_1 number of teeth and the secondary transformer coil corresponds to the gear with N_2 number of teeth.

3.12.6 Hydraulic Actuator or Servomotor

The hydraulic actuator or servomotor is the main component in any hydraulic control system. A typical hydraulic actuator is given in Figure 3.48 and operates as follows: let the actuator input be the position x of the piston valve and the actuator output be the position y of the power cylinder piston. The actuator excitation is caused by pressing the piston valve, which results in moving the valve to the right or to the left from its equilibrium position by Δx. Then the power cylinder piston's position will move correspondingly to the right or to the left from the equilibrium position by Δy. The displacement Δy results as follows: as the piston valve moves to the right, oil will run towards port 1. As this oil is under pressure P_1, the power cylinder piston will move to the right by Δy, because the oil pressure P_2 in port 2 is less than the pressure P_1.

Let q be the oil flow per minute. Then, a first approximation, gives

$$q = K_1\Delta x - K_2\Delta p, \qquad \Delta p = P_2 - P_1 \tag{3.12-30}$$

where K_1 and K_2 are constants. Also, the following relation is valid:

$$A\rho\Delta y = q\Delta t \tag{3.12-31}$$

where A is the area of the power cylinder piston surface and ρ is the oil density. If we eliminate q from relations (3.12-30) and (3.12-31), then we arrive at the relation

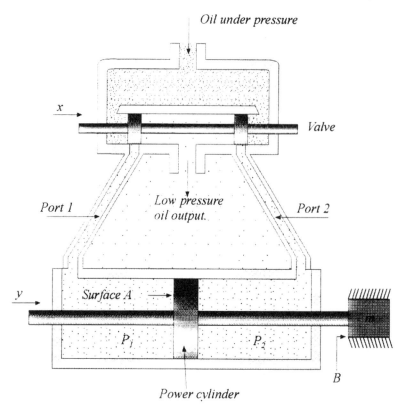

Figure 3.48 The hydraulic actuator.

$$\Delta p = \frac{1}{K^2}\left[K_1 \Delta x - A\rho \frac{\Delta y}{\Delta t}\right] \tag{3.12-32}$$

The force f produced by the power cylinder piston will be

$$f = A\Delta p = \frac{A}{K_2}\left[K_1 \Delta x - A\rho \frac{\Delta y}{\Delta t}\right] \tag{3.12-33}$$

Finally, the differential equation that describes the motion of the load (mass) m will be

$$m\frac{d^2 y}{dt^2} + B\frac{dy}{dt} = f = \frac{AK_1}{K_2}u - \frac{A^2\rho}{K_2}\frac{dy}{dt}$$

or

$$m\frac{d^2 y}{dt^2} + \left[B + \frac{A^2\rho}{K_2}\right]\frac{dy}{dt} = \frac{AK_1}{K_2}u \tag{3.12-34}$$

where we have set $\Delta y/\Delta t = dy/dt$ and $\Delta x = u$.

The transfer function of the hydraulic actuator will then be

$$G(s) = \frac{Y(s)}{U(s)} = \frac{K}{s(Ts+1)} \qquad (3.12\text{-}35)$$

where

$$K = \left[\frac{BK_2}{AK_1} + \frac{A\rho}{K_1}\right]^{-1} \qquad \text{and} \qquad T = \frac{mK_2}{BK_2 + A^2\rho}$$

If we assume that $T \simeq 0$, then $G(s) = K/s$. Therefore, we observe that the hydraulic actuator of Figure 3.48 serves simultaneously as an amplifier and as an integrator.

3.12.7 Pneumatic Amplifier

A typical pneumatic amplifier is given in Figure 3.49a. Let the system input be the distance x between the nozzle and the flapper, and the system output be the pressure P which regulates a control valve. Figure 3.49b shows the relation between x and P, where P_s is the pressure of the forced air and P_a is the atmospheric pressure. This system operates as an amplifier, because the power which is required to move the flapper by a distance x is much less than the power P delivered by the system to the control valve. A practical application of pneumatic amplifiers is in the liquid-level control system given in Figure 1.12.

3.13 MATHEMATICAL MODELS OF PRACTICAL CONTROL SYSTEMS

In this section, the mathematical models of several practical closed-loop control systems are presented. These systems are used to control a specific variable such as voltage, position or speed of a mass, flow or level of a liquid, room temperature, satellite orientation control, etc. Such practical control systems are often used in process industries, as for example in the paper, cement, and food industries, etc.; in controlling airplanes, ships, spacecrafts, etc., and in high-precision control systems such as guns, rockets, radars, radio telescopes, satellites, and others.

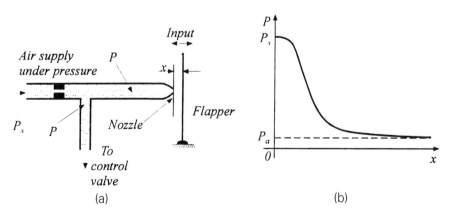

Figure 3.49 (a) Pneumatic amplifier and (b) the effect of distance x upon the pressure P acting upon the control valve.

3.13.1 Voltage Control System

A voltage control system is given in Figure 3.50a. This system is designed so that the load voltage v_L remains constant despite any changes in the load R_L. It operates as follows: The reference voltage v_r and the feedback voltage v_b are constantly compared. Under ideal load conditions, i.e., when $R_L \rightarrow \infty$, the voltage error $v_e = v_r - v_b$ is such that the generator produces the desired voltage v_L. Under nonideal load condtions, the resistance of R_L decreases and the voltage v_b also decreases, resulting in an increase in the error voltage v_e. Increasing the error voltage v_e increases the output voltage v_f of the amplifier, which in turn increases the field current i_f. Further, increasing the current i_f, increases the generator voltage v_g. Finally, increasing the voltage v_g increases the voltage v_L. As a result, the voltage v_L will approach

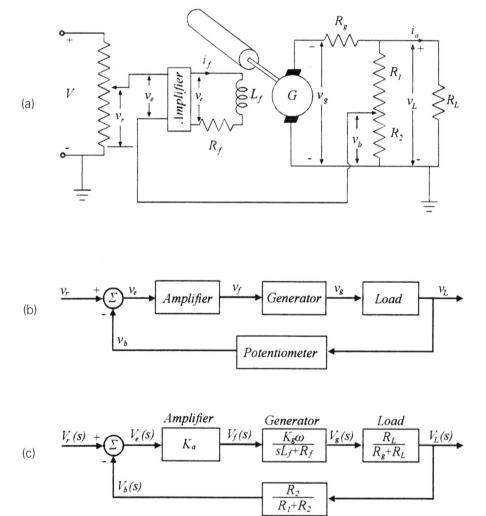

Figure 3.50   Voltage control system. (a) Overall picture of the voltage control system; (b) schematic diagram of the system; (c) block diagram of the system.

its desired initial value, i.e., its value when $R_L \to \infty$. This procedure is shown in the schematic diagram of Figure 3.50b.

The differential equations that describe the voltage control system are

$$v_e = v_r - v_b \tag{3.13-1a}$$

$$v_f = K_a v_e \tag{3.13-1b}$$

$$L \frac{di_f}{dt} + R_f i_f = v_f \tag{3.13-1c}$$

$$v_g = K_g \omega i_f \tag{3.13-1d}$$

$$(R_g + R_L)i_a = v_g \tag{3.13-1e}$$

$$v_b = \frac{R_2}{R_1 + R_2} v_L \tag{3.13-1f}$$

The above relations refer (from top to bottom) to the error, the amplifier, the field of excitation, the generator's voltage (see subsec. 3.12.1), the generator–load network, and the feedback voltage. In relation (3.13-1e), we have assumed that $R_1 + R_1 \gg R_L$.

The block diagram of the system is given in Figure 3.50c. The forward-path transfer function $G(s)$ will be

$$G(s) = \frac{K}{sT_f + 1}, \quad \text{where } K = \left[\frac{R_L R_f}{R_g + R_L} \right] K_a K_g \omega \text{ and } T_f = L_f / R_f \tag{3.13-2}$$

The transfer function $H(s)$ of the closed-loop system is

$$H(s) = \frac{G(s)}{1 + K_p G(s)} = \frac{K}{sT_f + 1 + K_p K} = \frac{K_1}{sT + 1} \tag{3.13-3}$$

where

$$K_p = \frac{R_2}{R_1 + R_2}, \quad T = \frac{T_f}{1 + K_p K}, \quad \text{and} \quad K_1 = \frac{K}{1 + K_p K}$$

3.13.2 Position (or Servomechanism) Control System

A well-known position (or servomechanism) control system is given in Figure 3.51a. This system is designed to control the load angular position. Specifically, the system must be such that any change in the command, i.e., in the input angular position θ_r, is followed as closely as possible by the output angular position θ_y. This is accomplished as follows: the command θ_r is driven manually, rotating the slider of the input potentiometer by θ_r degrees. The input angle θ_r, as well as the output angle θ_y, are converted into electrical signals, which subsequently are subtracted from each other to yield their difference θ_e, which is called the error. The error θ_e is driven into the amplifier. The amplifier output is then used to excite the armature-controlled motor (Subsec. 3.12.1). Finally, the motor rotates the load through the gears. Clearly, if $\theta_e = 0$, then $\theta_r = \theta_y$, meaning that the system has come to a standstill. In the case where $\theta_e \neq 0$, the motor will turn clockwise or counterclockwise, depend-

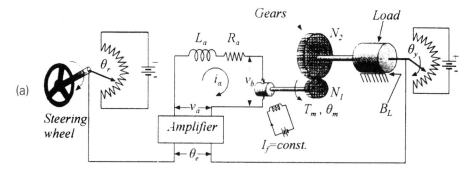

(a)

Steering wheel

Amplifier

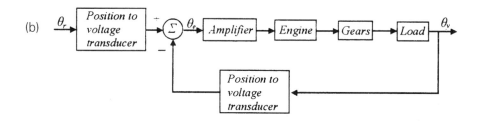

(b)

Position to voltage transducer

Position to voltage transducer

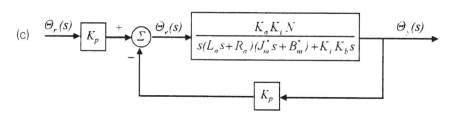

(c)

Figure 3.51 Position (or servomechanism) control system. (a) Overall picture of a position control system; (b) schematic diagram of the system; (c) block diagram of the system.

ing on the sign of the error θ_e, until $\theta_e = 0$. This procedure is depicted in the schematic diagram in Figure 3.51b.

It is remarked that the ideal situation would be, as soon as we apply the command input θ_r, the load rotates instantly and stops at the angle $\theta_y = \theta_r$. But, due to the inertia and other properties of the system, the output θ_y cannot become θ_r instantly (it has the general form shown in Figure 4.2).

The algebraic and differential equations describing each subsystem, going from the input to the output, are

1 The Error Detector
According to relation (3.12.22) we have

$$\theta_e(t) = K_p[\theta_r(t) - \theta_y(t)] \tag{3.13-4}$$

where K_p is the potentiometer detector constant.

2 The DC Amplifier

We have

$$v_a(t) = K_a \theta_e(t) \tag{3.13-5}$$

where K_a is the amplifier gain constant.

3 The Motor

This is an armature-controlled motor with a fixed field current whose dynamic equations are given in Subsec. 3.12.1. To the motor we have added gears plus the load. Hence, the motor equations are modified as follows:

$$L_a \frac{di_a}{dt} + R_a i_a + v_b = v_a, \qquad \text{where } v_b(t) = K_b \omega_m(t) \tag{3.13-6}$$

$$J_m^* \frac{d\omega_m}{dt} + B_m^* \omega_m = T_m, \qquad \text{where } T_m(t) = K_i i_a(t) \tag{3.13-7}$$

where $N = N_1/N_2$, $J_m^* = J_m + N^2 J_L$, and $B_m^* = B_m + N^2 B_L$. The constants J_m^* and B_m^* are determined using the results of Subsec. 3.12.5, under the assumption that the gears are ideal, i.e., that there are no losses. J_L is the moment of inertia and B_L the viscosity coefficient of the load. Likewise, J_m and B_m are the moment of inertia and the viscosity coefficient of the motor and J_m^* and B_m^* of the subsystem, consisting of the motor, the gears, and the load.

4 The Output

We have

$$\frac{d\theta_m}{dt} = \omega_m \qquad \text{and} \qquad \theta_y = N\theta_m \tag{3.13-8}$$

The forward-path transfer function is given by

$$G(s) = \frac{\Theta_y(s)}{\Theta_e(s)} = \frac{K_a K_i N}{s(sL_a + R_a)(J_m^* s + B_m^*) + K_i K_b s} \tag{3.13-9}$$

The above relation is essentially relation (3.12-17), where we have set $\Theta_y(s) = N\Theta_m(s)$ and the constants J_m and B_m have been replaced by J_m^* and B_m^*. The transfer function of the closed-loop system (see Figure 3.51c) will be

$$H(s) = \frac{\Theta_y(s)}{\Theta_r(s)} = K_p \left[\frac{G(s)}{1 + K_p G(s)} \right] = \frac{K_p K_a K_i N}{s(sL_a + R_a)(sJ_m^* + B_m^*) + K_i K_b s + K_p K_a K_i N} \tag{3.13-10}$$

Assuming that $L_a \simeq 0$, then the forward-path transfer function $G(s)$ reduces to

$$G(s) = \frac{K_a K_i N}{s[R_a J_m^* s + R_a B_m^* + K_i K_b]} = \frac{K}{s(As + B)} \tag{3.13-11}$$

where

$$K = \frac{K_a K_i N}{R_a}, \qquad A = J_m^*, \qquad \text{and} \qquad B = B_m^* + \frac{K_i K_b}{R_a} \tag{3.13-12}$$

Relation (3.13-11) shows that for $L_a \simeq 0$ the open-loop system reduces to a second-order system.

3.13.3   Speed Control System

A well-known speed control system is given in Figure 3.52a. This system is designed to control the angular velocity (speed) ω_y of the load, despite changes in the load. For this purpose the Ward–Leonard generator–motor layout is used. The command input v_r is constantly compared with the output of the tachometer v_y. The resulting error v_e is driven into the amplifier, whose output is the voltage v_f. Note that the input to the Ward–Leonard layout is the voltage v_f, which excites the field of the generator G, and that the output is the angular velocity ω_m of the motor M. When the error $v_e = 0$, then the angular velocity ω_y of the load is the desired one. If $v_e \neq 0$, the motor angular velocity ω_m will speed up or slow down, so that v_e reaches zero, in which case ω_y reaches the desired value. It is important to point out that the closed-loop system involves speed feedback action via the tachometer.

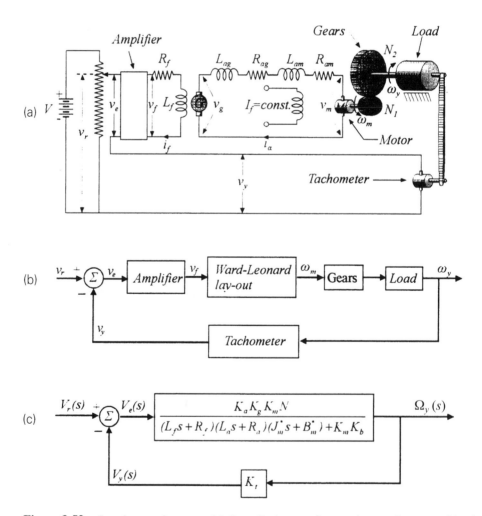

Figure 3.52   Speed control system. (a) Overall picture of a speed control system; (b) schematic diagram of the system; (c) block diagram of the system.

The equations of the Ward–Leonard layout are as follows. The Kirchhoff's law of voltages of the excitation field of the generator G is

$$v_f = R_f i_f + L_f \frac{di_f}{dt} \tag{3.13-13}$$

The voltage v_g of the generator G is proportional to the current i_f, i.e.,

$$v_g = K_g i_f$$

The voltage v_m of the motor M is proportional to the angular velocity ω_m, i.e.,

$$v_m = K_b \omega_m$$

The differential equation for the current i_a is

$$R_a i_a + L_a \frac{di_a}{dt} = v_g - v_m = K_g i_f - K_b \omega_m \tag{3.13-14}$$

where $R_a = R_{ag} + R_{am}$ and $L_a = L_{ag} + L_{am}$. The torque T_m of the motor is proportional to the current i_a, i.e.,

$$T_m = K_m i_a$$

The rotational motion of the rotor is described by

$$J_m^* \frac{d\omega_m}{dt} + B_m^* \omega_m = K_m i_a \tag{3.13-15}$$

where $J_m^* = J_m + N^2 J_L$ and $B_m^* = B_m + N^2 B_L$, where $N = N_1/N_2$. Here, J_m is the moment of inertia and B_m the viscosity coefficient of the motor: likewise, for J_L and B_L of the load. From the above relations, we can determine the transfer function of the Ward–Leonard (WL) layout (including the load):

$$G_{WL}(s) = \frac{\Omega_y(s)}{V_f(s)} = \frac{K_g K_m N}{(L_f s + R_f)[(L_a s + R_a)(J_m^* s + B_m^*) + K_m K_b]} \tag{3.13-16}$$

where use was made of the relation $\omega_y = N\omega_m$. The forward-path transfer function $G(s)$ is given by

$$G(s) = \frac{K_a K_g K_m N}{(L_f s + R_f)[(L_a s + R_a)(J_m^* s + B_m^*) + K_m K_b]}$$

where K_a is the gain constant of the amplifier.

The tachometer (see Subsec. 3.12.3) is mechanically linked to the motor. The tachmotor equation is the following:

$$v_y = K_t \omega_y$$

Finally, the transfer function of the closed-loop system (see Figure 3.52c) is the following third-order system:

$$H(s) = \frac{\Omega_y(s)}{V_r(s)} = \frac{K_a K_g K_m N}{(L_f s + R_f)[(L_a s + R_a)(J_m^* s + B_m^*) + K_m K_b] + K_a K_t K_g K_m N}$$

If we assume that $L_a \simeq 0$, then $H(s)$ becomes the following second-order system:

$$H(s) = \frac{K_a K_g K_m N}{(L_f s + R_f)(R_a J_m^* s + R_a B_m^* + K_m K_b) + K_a K_t K_g K_m N} \tag{3.13-17}$$

3.13.4 Liquid-Level Control System

A simple liquid-level control system is given in Figure 3.53a. This system is designed such that the liquid level (i.e., the height y) remains constant despite the changes in the outflow rate q_o of the liquid. It operates as follows. The load valve regulates the outflow rate q_o of the liquid from the container and the control valve regulates the inflow rate q_i of the liquid in the container. The liquid level y is monitored by the analog measuring device. The output of this device is connected to the control valve in order to control the inflow rate q_i. When q_o increases (decreases), then y decreases (increases) and the analog device opens (closes) the control valve, thus increasing (decreasing) the inflow rate q_i. This procedure keeps the height y of the liquid level constant.

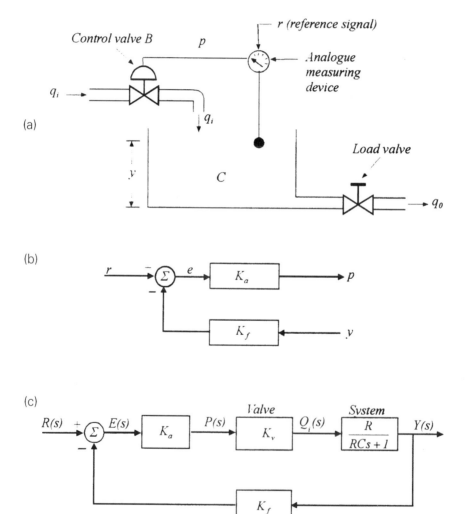

Figure 3.53 Liquid-level control system. (a) Overall picture of a liquid-level control system; (b) block diagram of the liquid-level analog measuring device; (c) block diagram of the closed-loop system.

The differential equation describing the system is determined as follows. The continuity law of fluids states that the amount of liquid which enters the container minus the amount of liquid which leaves the container is equal to the increase of the amount of liquid in the container. Let the load valve be closed (which means that $q_o = 0$). Then, the continuity law of fluids gives

$$q_i dt = A dy \tag{3.13-18}$$

where q_i is the volume of the liquid inflow per unit time, A is the horizontal area of the container, and dt and dy are the differentials of time and height, respectively. If the load valve is open (which means that $q_o \neq 0$), then the continuity law of fluids gives

$$(q_i - q_o) dt = A dy \tag{3.13-19}$$

Let the liquid flow be laminar. A laminar fluid system is characterized by the constants R (resistance of a valve or of a pipe) and C (capacitance of the container), which are defined as follows

$$R = y/q_o \quad \text{and} \quad C = A \tag{3.13-20}$$

Substitute relation (3.13-20) into Eq. (3.13-19) to yield

$$RC \frac{dy}{dt} + y = Rq_i \tag{3.13-21}$$

The differential equation (3.13-21) describes the open-loop system with input as the supply q_i and output as the height y. The transfer function of the open-loop system will then be

$$G(s) = \frac{Y(s)}{Q_i(s)} = \frac{R}{RCs + 1} \tag{3.13-22}$$

The liquid-level analog measuring device operates as shown in Figure 3.53b. The height $Y(s)$ of the liquid level is compared with the reference signal $R(s)$ producing the error $E(s) = R(s) - K_f Y(s)$, where K_f is the constant of the measuring device. The error $E(s)$ is driven into a pneumatic amplifier (see Subsec. 3.12.7) whose output is air with pressure $P(s) = K_a E(s)$, where K_a is the amplification constant of the liquid-level measuring device. This pressure is in turn driven into the control valve. This results in the liquid inflow rate $Q_i(s)$ delivered to the tank, where $Q_i(s) = K_v P(s)$, where K_v is the control valve constant. Hence, the overall block diagram of the closed-loop system will be as shown in Figure 3.53c. The transfer function of the closed-loop system is given by

$$H(s) = \frac{Y(s)}{R(s)} = \frac{K_a K_v R}{RCs + 1 + K_f K_a K_v R} \tag{3.13-23}$$

3.13.5 Temperature Control System

A simple room temperature control system is given in Figure 3.54. This system is designed such that the room temperature remains constant despite changes in the external temperature. The operation of the system is regulated via a temperature analog measurement device. This device compares the room temperature θ_y with the desired temperature θ_r and transforms the error signal θ_e (i.e., the difference $\theta_r - \theta_y$)

(a) *Reference signal* θ_r

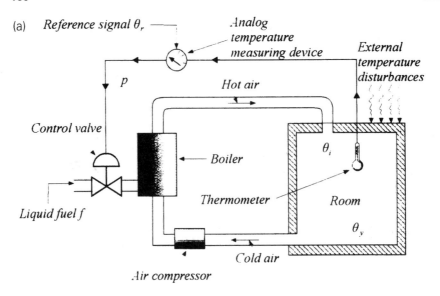

Air compressor

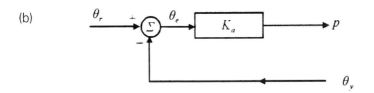

(b)

(c)

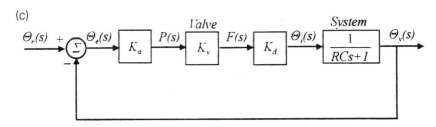

Figure 3.54 Room temperature control system. (a) Overall picture of a room temperature control system; (b) block diagram of the temperature analog measuring device; (c) block diagram of the closed-loop system.

into a pressure signal. This pressure acts on the control valve, which accordingly closes or opens, thus controlling the supply of the liquid fuel. The amount of the liquid fuel which enters the burner, essentially controls the temperature θ_i of the hot air. This procedure results in maintaining the room temperature θ_y constant.

We define the following parameters of the system:

- $C =$ heat capacity of the room's air. If the room's air has mass m and specific heat capacity σ, then $C = \sigma m$.

- R = thermal resistance of the room's air. If A is the horizontal area of the room, h is the height, and k is the thermal conductance coefficient of the air, then we have the approximate relation $R = h/kA$.
- q = flow of the air mass through the room per unit time. If θ_i and θ_y are the input and output temperatures of the room, respectively, then

$$q = (\theta_i - \theta_y)/R \tag{3.13-24}$$

The mathematical model describing the system is determined as follows. Let the room temperature be θ_y. Then, an increase $d\theta_y$ in the temperature θ_y requires $Cd\theta_y$ amount of heat. This amount of heat must be delivered by the hot air flow q in a time dt. Hence, we have

$$Cd\theta_y = qdt = \frac{\theta_i - \theta_y}{R}dt \quad \text{or} \quad RC\frac{d\theta_y}{dt} + \theta_y = \theta_i \tag{3.13-25}$$

Relation (3.13-25) describes an open-loop system with input as the temperature of the hot air and output as the temperature θ_y of the room. The transfer function of this open-loop system is given by

$$G(s) = \frac{\Theta_y(s)}{\Theta_i(s)} = \frac{1}{RCs + 1} \tag{3.13-26}$$

The temperature analog measuring device operates as shown in Figure 3.54b. The block diagram of the closed-loop system is given in Figure 3.54c. Let $F(s) = L\{f(t)\}$ be the flow of the liquid fuel, while K_d is a constant that relates the liquid fuel flow $f(t)$ to the temperature $\theta_i(t)$ of the hot air produced. The transfer function of the closed-loop system is

$$H(s) = \frac{\Theta_y(s)}{\Theta_r(s)} = \frac{K_a K_v K_d}{RCs + 1 + K_a K_v K_d} \tag{3.13-27}$$

where K_v is the valve constant.

3.13.6 Chemical Composition Control System

A simple chemical composition control system is given in Figure 3.55a. This system is designed such that the chemical composition of a certain product, which comes for example from mixing water and a certain liquid, remains the same despite any changes in operating conditions, e.g., in temperature or pressure. The chemical composition of the mixture is compared with the desired composition. The resulting error is converted into pressure, which controls the valve. If the mixture content is smaller than the desired content, then the valve opens, allowing more liquid to enter the container. If the mixture content is greater than the desired content, then the valve closes so that less liquid enters the container. This procedure guarantees that the chemical composition of the product remains constant.

A simplified mathematical model which describes the operation of the chemical composition control system is determined as follows. Let q_v, q_π, and q_y be the flows (m³/sec) of water, liquid, and of produced mixture, where the flow q_v is maintained constant. Also, let x_π and x_y be the contents (kg/m³) of the liquid and of the produced mixture in a certain chemical substance, e.g., calcium. Then, in a time

(a)

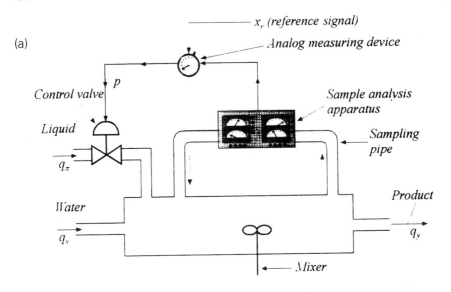

(b)

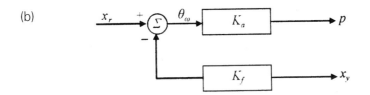

(c)

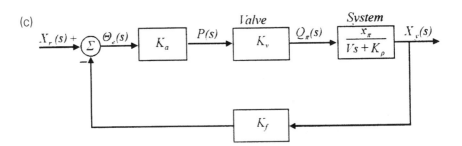

Figure 3.55 Chemical composition control system. (a) Overall picture of a chemical composition control system; (b) block diagram of the chemical composition analog measuring device and sample analysis apparatus; (c) block diagram of the closed-loop system.

interval dt, $q_\pi x_\pi$dt mass of calcium enters the container and $q_y x_y$dt mass of calcium leaves the container. Hence, the change dx_y in the mixture content in calcium will be

$$dx_y = \frac{q_\pi x_\pi dt - q_y x_y dt}{V} \tag{3.13-28}$$

where V is the volume of the container. According to the law of mass preservation we have

$$q_y = q_v + q_\pi \tag{3.13-29}$$

Since $q_v \gg q_\pi$, we have that $q_y \simeq q_v = K_\rho = $ constant. Hence Eq. (3.13-28) becomes

$$\frac{dx_y}{dt} + \frac{K_\rho}{V} x_y = \frac{x_\pi}{V} q_\pi \tag{3.13-30}$$

Relation (3.13-30) describes the open-loop system with input q_π and output x_y. The transfer function of the open-loop system is

$$G(s) = \frac{X_y(s)}{Q_\pi(s)} = \frac{x_\pi}{Vs + K_\rho} \tag{3.13-31}$$

In Figure 3.55b the block diagram of the chemical composition analog measuring device and the sample analysis apparatus are given, where K_f and K_a are the constants of this layout and $X_r(s)$ is the reference signal. From Figure 3.55c we find that the transfer function of the closed-loop system is given by

$$H(s) = \frac{X_y(s)}{X_r(s)} = \frac{K}{Ts + 1} \tag{3.13-32}$$

where

$$K = \frac{K_a K_v x_\pi}{K_\rho + K_f K_a K_v x_\pi} \quad \text{and} \quad T = \frac{V}{K_\rho + K_f K_a K_v x_\pi}$$

3.13.7 Satellite Orientation Control System

This system (see Figure 3.56a) is designed so that the satellite's telescope, which is mounted on the satellite, is constantly in perfect orientation with a star, despite disturbances such as collisions with meteorites, gravity forces, etc. The system operates as follows: the orientation of the telescope is achieved with the help of a reference star, which is much brighter than the star under observation. It is initially assumed that the axis of the telescope is in line with the star under observation. At this position, the axis forms an angle of θ_r with the line which connects the telescope to the reference star. The angle θ_r is known in advance and is the system's excitation. If for any reason the angle θ_y, which is formed by the axis of the telescope with the reference star, is different from the desired angle θ_r, then the control system must align the Z-axis so that $\theta_y = \theta_r$. The block diagram of the closed-loop system is given in Figure 3.56b, where the block designated "thruster" is excited by the controller's output and produces torque forces about the X-axis. These forces will ultimately cause the correction in the angle θ_y, so that $\theta_y = \theta_r$.

The various blocks in Figure 3.56c are described, respectively, by the relations

$$E(s) = K_t[\Theta_r(s) - \Theta_y(s)] \tag{3.13-33a}$$

$$V(s) = G_c(s)E(s) \tag{3.13-33b}$$

$$T(s) = G_b(s)V(s) = K_b V(s) \tag{3.13-33c}$$

$$\Theta_y(s) = G_s(s)T(s) = \frac{1}{Js^2} T(s) \tag{3.13-33d}$$

where K_t is the conversion constant of the position angle to an electrical signal, $G_c(s)$ is the transfer function of the controller, K_b is the constant of the thruster in which

(a)

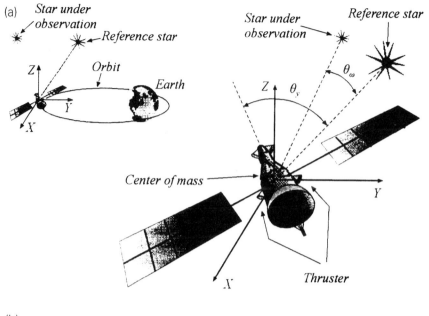

(b)

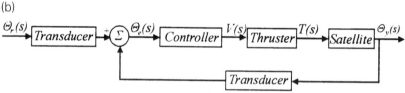

(c)

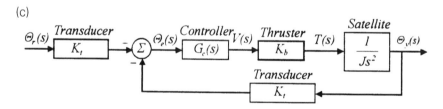

Figure 3.56 Satellite orientation control system. (a) Overall picture of satellite orientation control system; (b) schematic diagram of the closed-loop system; (c) block diagram of the closed-loop system.

the voltage $V(s)$ is converted to the torque $T(s)$, and $G_s(s) = 1/Js^2$ is the satellite's transfer function, where J is the moment of inertia of the satellite about the X-axis. It is noted that relation (3.13-33d) does not involve friction terms, since the satellite is assumed to be in space.

The transfer function of the closed-loop system is

$$H(s) = \frac{\Theta_y(s)}{\Theta_r(s)} = \frac{K_t K_b G_c(s)}{Js^2 + K_t K_b G_c(s)} \tag{3.13-34}$$

From relation (3.13-34) it follows that if the controller is just an amplifier, in which case $G_c(s) = K_a$, where K_a is the amplifier's gain, then $H(s)$ would have the form

$$H(s) = \frac{K_t K_b K_a}{Js^2 + K_t K_b K_a} = \frac{\omega_n^2}{s^2 + \omega_n^2}, \qquad \omega_n^2 = K_t K_b k_a J^{-1} \tag{3.13-35}$$

3.14 PROBLEMS

1. Derive the integrodifferential equations for the two electrical networks shown in Figure 3.57.
2. Obtain the transfer functions and the impulse responses for the circuits shown in Figure 3.58.
3. Derive the transfer functions for the operational amplifier circuits shown in Figure 3.59.
4. Determine the state equations for the systems shown in Figure 3.60. The blocks shown to the right or below the systems define the inputs, the state variables, and the outputs of the system.
5. Find the transfer functions and impulse responses for the systems of Problem 4.
6. Obtain a state-space model for each of the systems having the following transfer functions:

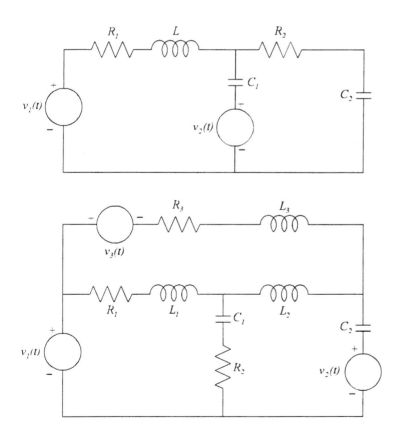

Figure 3.57

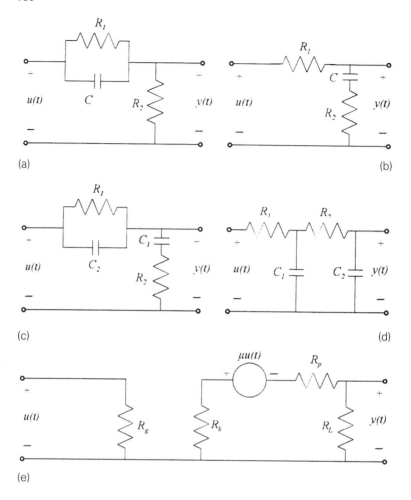

(a)

(b)

(c)

(d)

(e)

Figure 3.58

$$H(s) = \frac{s^2 + 4}{s^3 + 4s + 1}, \qquad H(s) = \frac{2s + 1}{s^4 + 4s^3 + 2s^2 + s + 5}$$

$$\mathbf{H}(s) = \left[\frac{1}{s+1} \ \frac{2}{s+2}\right], \qquad \mathbf{H}(s) = \begin{bmatrix} \dfrac{s+2}{s(s+1)} & \dfrac{2s+4}{s(s+1)} \\ \dfrac{1}{s+1} & \dfrac{2}{s+1} \end{bmatrix}$$

7. A thin glass-walled mercury thermometer has initial temperature θ_1. At time $t = 0$, it is immersed in a bath of temperature θ_0. The thermometer instant temperature is θ_m. Derive the state equations and the transfer function of the thermometer from the two network representations shown in Figure 3.61. The simplified network representation of Figure 3.61a, involves a capacitance C that stores heat and a resistance R that limits the heat flow. The more accurate network representation of Figure 3.61b involves the resistance R_g and the capa-

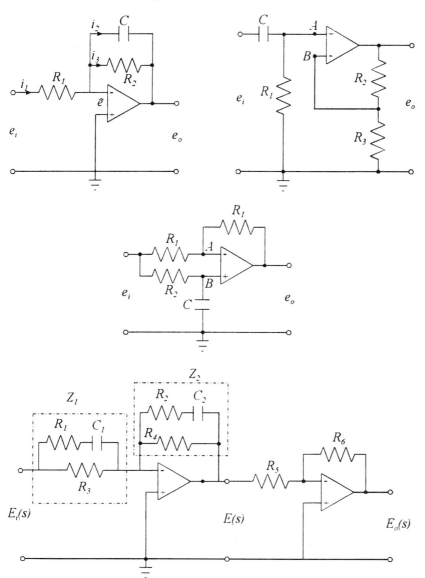

Figure 3.59

citance C_g of the glass, as well as the resistance R_m and the capacitance C_m of the mercury.

8. Find the transfer function and the impulse response of the system shown in Figure 3.62.

9. For the system shown in Figure 3.63, determine (a) the differential equation, (b) the transfer function, and (c) the state equations, considering that the input is the displacement $u(t)$ of the cart and the output is the displacement $y(t)$ of the mass.

10. An inverted pendulum mounted on a motor-driven cart is shown in Figure 3.64. Assume that the input to the system is the control force u applied to the cart and

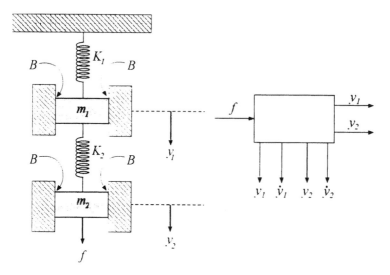

Figure 3.60

the two outputs are the angle θ of the rod from the vertical axis and the position x of the cart. Obtain a state-space representation of the system.

11. A schematic diagram of a pneumatic actuating valve is shown in Figure 3.65. Assume that at $t = 0$, there is a control pressure change p_c from the steady-state value. As a result, there is a change p_v of the valve pressure. The valve pressure variation p_v results in a change x of the valve displacement. Considering that p_c is the input and x is the output of the system, determine the transfer function of the valve. Here, A is the area of the diaphragm and K is the spring constant.

12. A two-tank liquid-level system is shown in Figure 3.66. At steady state, the flow rate is Q and the fluid heights in tank 1 and tank 2 are H_1 and H_2, respectively. Small flow rate changes at the input of tank 1, at the output of tank 1, and at the output of tank 2 and are defined as q, q_1, and q_2, respectively. Small variations of the liquid level of tank 1 and tank 2 from the steady-state values are defined as h_1 and h_2, respectively. Assume that C_1 and C_2 are the capacitances of tanks 1 and 2, respectively, while R_1 is the flow resistance between the two tanks and R_2 is the flow resistance at the output of tank 2. The quantities C and R are defined as follows:

$$C = \frac{\text{Change in liquid stored } (\text{m}^3)}{\text{Change in liquid level}} \quad \text{and}$$

$$R = \frac{\text{Change in level difference } (\text{m})}{\text{Change in flow rate } (\text{m}^3/\text{sec})}$$

Derive a state-space representation of the system when (a) q_2 is considered to be the output and, (b) h_2 is considered to be the output.

13. Find the transfer functions of the block diagrams shown in Figure 3.67.
14. Find the transfer functions of the signal-flow graphs shown in Figure 3.68.
15. For every system described in Section 3.13, find a state-space description.

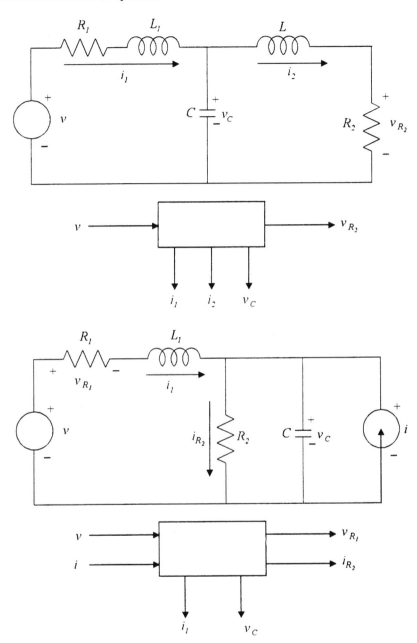

Figure 3.60 *(contd.)*

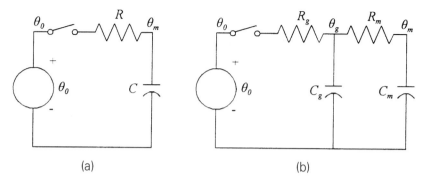

(a)   (b)

Figure 3.61

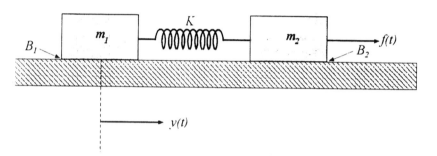

Figure 3.62

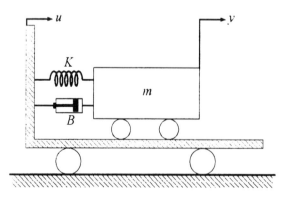

Figure 3.63

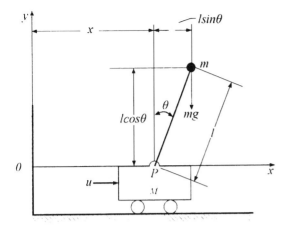

Figure 3.64

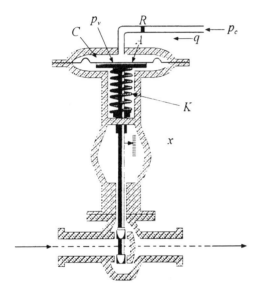

Figure 3.65 [16].

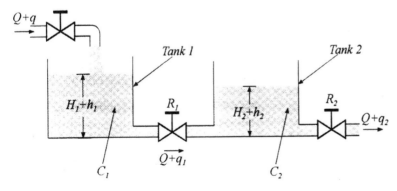

Figure 3.66

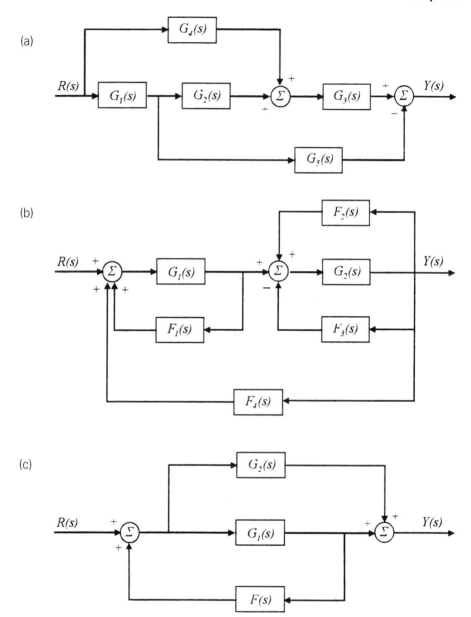

Figure 3.67

(d)

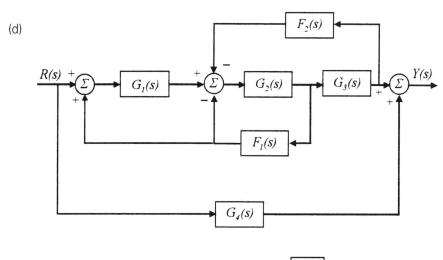

(e)

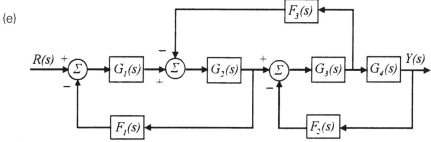

(f)

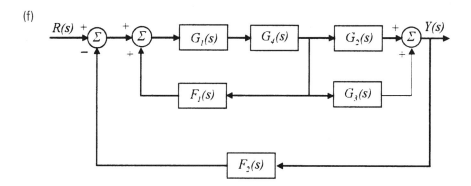

Figure 3.67 (*contd.*)

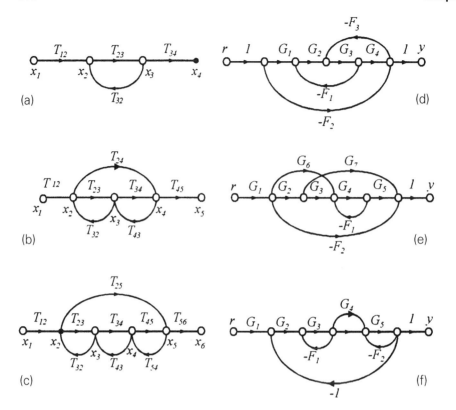

Figure 3.68

BIBLIOGRAPHY

Books

1. PJ Antsaklis, AN Michel. Linear Systems. New York: McGraw-Hill, 1997.
2. RH Cannon Jr. Dynamics of Physical Systems. New York: McGraw-Hill, 1967.
3. JJ D'Azzo, CH Houpis. Linear Control System Analysis and Design, Conventional and Modern. New York: McGraw-Hill, 1975.
4. PM DeRusso, RJ Roy, CM Close. State Variables for Engineers. New York: John Wiley, 1965.
5. JJ DiStefano III, AR Stubberud, IJ Williams. Feedback and Control Systems. Schaum's outline series. New York: McGraw-Hill, 1967.
6. RC Dorf, RE Bishop. Modern Control Analysis. London: Addison-Wesley, 1995.
7. VW Eveleigh. Introduction to Control Systems Design. New York: McGraw-Hill, 1972.
8. TE Fortman, KL Hitz. An Introduction to Linear Control Systems. New York: Marcel Dekker, 1977.
9. GF Franklin, JD Powell, A Emami-Naeini. Feedback Control of Dynamic Systems. Reading MA: Addison-Wesley, 1986.
10. B Friedland. Control System Design. An Introduction to State-Space Methods. New York: McGraw-Hill, 1987.
11. R Johansson. System Modeling and Identification. Englewood Cliffs, New Jersey: Prentice Hall, 1993.

12. T Kailath. Linear Systems. Englewood Cliffs, New Jersey: Prentice Hall, 1980.
13. BC Kuo. Automatic Control Systems. London: Prentice Hall, 1995.
14. AGC MacFarlane. Dynamic System Models. London: George G. Harrap and Co., 1970.
15. NS Nise. Control Systems Engineering. New York: Benjamin and Cummings, 1995.
16. K Ogata. Modern Control Systems. London: Prentice Hall, 1997.
17. FH Raven. Automatic Control Engineering. 4th ed. New York: McGraw-Hill, 1987.

Articles

18. LM Silverman. Realization of linear dynamical systems. IEEE Trans Automatic Control AC-16:554–567, 1971.

4

Classical Time-Domain Analysis of Control Systems

4.1 INTRODUCTION

Chapters 4 and 5 refer to the time-domain analysis of linear time-invariant control systems. In particular, Chap. 4 refers to single-input–single-output (SISO) systems and uses the classical system description techniques: namely, the differential equations and the transfer function. On the other hand, Chap. 5 refers to the general case of systems with many inputs and many outputs (MIMO), and uses the modern system description technique: namely, the state equations. The reason for studying both the classical as well as the modern analysis methods is that both methods have great theoretical and practical importance for control engineers.

The problem of time-domain analysis may be briefly stated as follows: given the system (i.e., given a specific description of the system) and its input, determine the time-domain behavior of the output of the system.

The basic motivation for system analysis is that one can predict (theoretically) the system's behavior.

4.2 SYSTEM TIME RESPONSE

4.2.1 Analytical Expression of Time Response

Consider the SISO linear time-invariant system described by the differential equation

$$y^{(n)} + a_{n-1}y^{(n-1)} + \cdots + a_1 y^{(1)} + a_0 y = b_m u^{(m)} + \cdots + b_1 u^{(1)} + b_0 u \qquad (4.2\text{-}1)$$

with initial conditions $y^{(k)}(0) = c_k$, $k = 0, 1, 2, \ldots, n-1$ and $u^{(k)}(0) = d_k$, $k = 0, 1, 2, \ldots, m-1$ and $m \le n$. To find the general solution of Eq. (4.2-1) we proceed as follows: Apply the Laplace transform to Eq. (4.2-1); after some algebraic simplifications, we arrive at the following:

$$a(s)Y(s) - \sigma(s) = b(s)U(s) - \tau(s) \qquad (4.2\text{-}2)$$

where

$$a(s) = s^n + a_{n-1} + \cdots + a_1 s + a_0, \qquad b(s) = b_m s^m + b_{m-1} s^{m-1} + \cdots + b_1 s + b_0$$

$$\sigma(s) = \sum_{i=0}^{n} \sum_{k=0}^{i-1} a_i c_k s^{i-1-k} = \sigma_{n-1} s^{n-1} + \sigma_{n-2} s^{n-2} + \cdots + \sigma_1 s + \sigma_0$$

$$\tau(s) = \sum_{i=0}^{m} \sum_{k=0}^{i-1} b_i d_k s^{i-1-k} = \tau_{m-1} s^{m-1} + \tau_{m-2} s^{m-2} + \cdots + \tau_1 s + \tau_0$$

It is pointed out that the polynomials $a(s)$ and $b(s)$ are functions of the coefficients a_i and b_i, respectively, while the polynomial $\sigma(s)$ is a function of both the coefficients a_i and the initial conditions c_k and the polynomial $\tau(s)$ is a function of both the coefficients b_i and the initial conditions d_k. The solution of Eq. (4.2-2) is given by

$$Y(s) = \left[\frac{b(s)}{a(s)}\right] U(s) + \frac{\sigma(s)}{a(s)} - \frac{\tau(s)}{a(s)} = \left[\frac{\sigma(s)}{a(s)}\right] + \left[\left[\frac{b(s)}{a(s)}\right] U(s) - \frac{\tau(s)}{a(s)}\right]$$

$$= Y_h(s) + Y_f(s) \tag{4.2-3}$$

where

$$Y_h(s) = \frac{\sigma(s)}{a(s)} \qquad \text{and} \qquad Y_f(s) = \left[\frac{b(s)}{a(s)}\right] U(s) - \frac{\tau(s)}{a(s)}$$

Here, $y_h(t) = L^{-1}\{Y_h(s)\}$ is called the *natural* or *free response* of the system and it is due solely to the initial conditions $c_0, c_1 \cdots c_{k-1}$. In other words $y_h(t)$ is the system's response when there is no external excitation, i.e., when $u(t) = 0$. It is for this reason that $y_h(t)$ is also known as the *homogeneous solution* of the differential equation (4.2-1). On the other hand, $y_f(t) = L^{-1}\{Y_f(s)\}$ is called the *forced response* and it is due solely to the input $u(t)$ and its initial conditions $d_0, d_1, \ldots, d_{m-1}$. The forced response $y_f(t)$ is also known as the *particular solution* of the differential equation (4.2-1).

Clearly, when the initial conditions $c_1, c_2, \ldots, c_k$ are all zero, then $\sigma(s) = 0$, and when the initial conditions $d_1, d_2, \ldots, d_m$ are all zero, then $\tau(s) = 0$. When both $\sigma(s)$ and $\tau(s)$ are zero or $\sigma(s) - \tau(s) = 0$, then $Y(s)$ reduces to

$$Y(s) = \left[\frac{b(s)}{a(s)}\right] U(s) = G(s)U(s), \qquad \text{where} \qquad G(s) = \frac{a(s)}{b(s)} \tag{4.2-4}$$

where $G(s)$ is the system's transfer function.

Let the input $U(s)$ have the general form

$$U(s) = \frac{f(s)}{v(s)} = \frac{f_q s^q + f_{q-1} s^{q-1} + \cdots + f_1 s + f_0}{s^p + v_{p-1} s^{p-1} + \cdots + v_1 s + v_0} \tag{4.2-5}$$

In this case, Eq. (4.2-3) becomes

$$Y(s) = \frac{b(s)f(s)}{a(s)v(s)} + \frac{\sigma(s)}{a(s)} - \frac{\tau(s)}{a(s)}$$

Also, let

$$\frac{b(s)f(s)}{a(s)v(s)} = \frac{\gamma(s)}{a(s)} + \frac{\delta(s)}{v(s)}$$

Then, $Y(s)$ becomes

$$Y(s) = \frac{\gamma(s)}{a(s)} + \frac{\delta(s)}{v(s)} + \frac{\sigma(s)}{a(s)} - \frac{\tau(s)}{a(s)} = \frac{\mu(s)}{a(s)} + \frac{\delta(s)}{v(s)} = Y_1(s) + Y_2(s) \qquad (4.2\text{-}6)$$

where $Y_1(s) = \dfrac{\mu(s)}{a(s)}$, $Y_2(s) = \dfrac{\delta(s)}{v(s)}$, and $\mu(s) = \gamma(s) + \sigma(s) - \tau(s)$. Further, assume that

$$a(s) = \prod_{i=1}^{n}(s - \lambda_i) \qquad \text{and} \qquad v(s) = \prod_{i=1}^{p}(s - \rho_i)$$

To determine $y(t)$, we expand $Y(s)$ in partial fractions. Let

$$Y_1(s) = \frac{\mu(s)}{a(s)} = \sum_{i=1}^{n} \frac{K_i}{s - \lambda_i} \qquad \text{and} \qquad Y_s(s) = \frac{\delta(s)}{v(s)} = \sum_{i=0}^{p} \frac{R_i}{s - \rho_i}$$

Then, $y(t) = L^{-1}\{Y(s)\}$ is given by

$$y(t) = L^{-1}\left\{ \sum_{i=0}^{n} \frac{K_i}{s - \lambda_i} \right\} + L^{-1}\left\{ \sum_{i=0}^{p} \frac{R_i}{s - \rho_i} \right\} = \sum_{i=0}^{n} K_i e^{\lambda_i t} + \sum_{i=0}^{p} R_i e^{\rho_i t} \qquad (4.2\text{-}7)$$

$$= y_1(t) + y_2(t)$$

It is clear from Eq. (4.2-7) that the behavior of $y(t)$ is crucially affected by the poles λ_i of the system transfer function $G(s) = b(s)/a(s)$ and by the poles ρ_i of the input $U(s) = f(s)/v(s)$.

The following definitions are introduced.

Definition 4.2.1

The *transient response* $y_{\text{tr}}(t)$ of a system is that particular part of the response of the system which tends to zero as time increases. This means tht $y_{\text{tr}}(t)$ has the property

$$\lim_{t \to \infty} y_{\text{tr}}(t) = 0$$

Definition 4.2.2

The *steady-state response* $y_{\text{ss}}(t)$ of a system is that particular part of the response of the system which remains after the transient part has reached zero. The steady-state response may be a constant (e.g., $y_{\text{ss}}(t) = c$), a sinusoidal function with constant amplitude (e.g., $y_{\text{ss}}(t) = A \sin \omega t$), a function increasing with time (e.g., $y_{\text{ss}}(t) = ct$ or $y_{\text{ss}}(t) = ct^2$ or $y_{\text{ss}}(t) = At \sin \omega t$), etc.

From the above two definitions it is clear that the total solution $y(t)$ is the sum of the transient response and the steady-state response, i.e.,

$$y(t) = y_{\text{tr}}(t) + y_{\text{ss}}(t)$$

Now, let all poles λ_i of $G(s)$ lie on the left-hand complex plane, or equivalently, let the real part of all poles λ_i be negative ($\text{Re } \lambda_i < 0$). In this case (and this is usually the case) we say that the system is *asumptotically stable* (see Sec. 6.2). It is obvious that if $\text{Re } \lambda_i < 0$, then

$$\lim_{t \to \infty} e^{\lambda_i t} = 0$$

Therefore, for asymptotically stable systems, the part $y_1(t)$ of the response $y(t)$ of relation (4.2-7), where

$$y_1(t) = \sum_{i=0}^{n} K_i e^{\lambda_i t}$$

belongs to the transient response $y_{tr}(t)$ of $y(t)$. This means that the contribution of $\tau(s)$, which is due to the initial conditions of the input $u(t)$, the contribution of $\sigma(s)$, which is due to the initial conditions of the output $y(t)$, and the part of the response $\gamma(s)/a(s)$, all appear only in the system's transient response. Hence, in the steady-state response, only the part $y_2(t)$ of the response $y(t)$ of relation (4.2-7) will appear, where

$$y_2(t) = \sum_{i=0}^{p} R_i e^{\rho_i t}$$

for which $\text{Re}\,\rho_i \geq 0$. The rest of $y_2(t)$, for which $\text{Re}\,\rho_i < 0$, belongs to the transient response of $y(t)$.

Remark 4.2.1

From the aforementioned material, we arrive at the conclusion that the waveform of $y_{ss}(t)$ of asymptotically stable systems solely depends upon the poles ρ_i of the input, for which $\text{Re}\,\rho_i \geq 0$. In control systems, the input is usually the unit step input $u(t) = 1$. In this case, we have

$$Y_2(s) = \frac{\delta(s)}{v(s)} = \frac{\delta_0}{s} = \frac{G(0)}{s} = \frac{b_0}{a_0 s}, \qquad \text{where} \qquad G(s) = \frac{b(s)}{a(s)}$$

For simplicity, assume that $G(0) = b_0/a_0 = 1$. Then, $Y_2(s) = 1/s$, and hence $y_{ss}(t) = 1$, i.e., the output follows exactly the input in the steady state.

To illustrate the material of the present subsection, we present two simple examples. In the first example, in order to simplify the analysis, no use is made of the general procedure presented in this subsection. On the contrary, in the second example, the general procedure is illustrated by applying all steps to yield $y(t)$.

Example 4.2.1

Consider the RL circuit shown in Figure 4.1 with initial conditions $i(0) = I_0$. For simplicity, let the input voltage $v(t) = u(t) = $ the unit step function, $R = 1\,\Omega$ and $L = 1\,H$. Determine the total response of the circuit.

Solution

Kirchhoff's voltage law is

$$L\frac{di}{dt} + Ri = v(t) \qquad \text{or} \qquad \frac{di}{dt} + i = u(t),$$

since $R = 1\,\Omega$, $L = 1\,H$, and $v(t) = u(t)$.

Here, the output $y(t) = v_R(t) = Ri = i$. Substituting this relation in the differential equation yields

$$\frac{dy}{dt} + y = u(t)$$

The above differential equation relates directly the input $u(t)$ with the output $y(t)$. Take the Laplace transform of both sides of the differential equation to yield

(a)

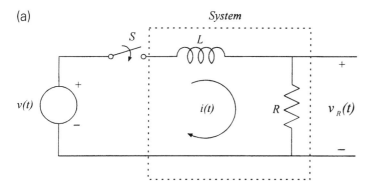

(b)

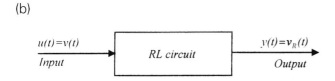

Figure 4.1 RL circuit.

$$sY(s) - y(0) + Y(s) = \frac{1}{s}, \qquad \text{where } y(0) = v_R(0) = Ri(0) = I_0$$

Solving for $Y(s)$ gives

$$Y(s) = \frac{1}{s(s+1)} + \frac{y(0)}{s+1} = \frac{1}{s} - \frac{1}{s+1} + \frac{y(0)}{s+1}$$

Taking the inverse Laplace transform, yields

$$y(t) = 1 - e^{-t} + y(0)e^{-t}$$

The above expression for $y(t)$ is the total response of the circuit. The right-hand side of the equation may be grouped as follows:

$$y(t) = \underbrace{y(0)e^{-t}}_{\text{free response}} + \underbrace{[1 - e^{-t}]}_{\text{forced response}} = y_h(t) + y_f(t)$$

or may be grouped as follows

$$y(t) = \underbrace{[y(0) - 1]e^{-t}}_{\text{transient response}} + \underbrace{1}_{\text{steady-state response}} = y_{tr}(t) + y_{ss}(t)$$

These two ways of writing $y(t)$ clearly demonstrate the four characteristics of the time response of a system:

1. Free response $y_h(t) =$ the part of $y(t)$ which is due to the initial condition $y(0)$.
2. Forced response $y_f(t) =$ the part of $y(t)$ which is due to the external excitation $u(t)$.
3. Transient response $y_{tr}(t) =$ the part of $y(t)$ which goes to zero fast.

4. Steady state $y_{ss}(t)$ = the part of $y(t)$ which remains after the transient response has died out.

Example 4.2.2

Consider a system described by the following differential equation:

$$y^{(2)} + 3y^{(1)} + 2y = 6u^{(1)} + 2u$$

with initial conditions $y(0) = 1$ and $y^{(1)}(0) = 2$. Let $u(t) = e^{5t}$, in which case $u(0) = 1$. To illustrate the general procedure presented in this subsection for determining the total time response of a system, apply this procedure, step by step, to solve the given differential equation.

Solution

Take the Laplace transform of both sides of the differential equation to yield

$$[s^2 Y(s) - sy(0) - y^{(1)}(0)] + 3[sY(s) - y(0)] + 2Y(s) = 6[sU(s) - u(0)] + 2U(s)$$

or

$$(s^2 + 3s + 2)Y(s) - [y(0)s + y^{(1)}(0) + 3y(0)] = (6s + 2)U(s) - 6u(0)$$

Comparing the equation above with Eq. (4.2-2), we have

$$a(s) = s^2 + a_1 s + a_0 = s^2 + 3s + 2 = (s + 1)(s + 2)$$
$$b(s) = b_1 s + b_0 = 6s + 2 = 2(3s + 1)$$
$$\sigma(s) = \sigma_1 s + \sigma_0 = y(0)s + y^{(1)}(0) + 3y(0) = s + 5$$
$$\tau(s) = \tau_0 = 6u(0) = 6$$

The free response of the system is

$$Y_h(s) = \frac{\sigma(s)}{a(s)} = \frac{s+5}{s^2 + 3s + 2} = \frac{s+5}{(s+1)(s+2)} = \frac{4}{s+1} - \frac{3}{s+2}$$

and hence

$$y_h(t) = L^{-1}\{Y_h(s)\} = 4e^{-t} - 3e^{-2t}$$

The force response of the system is

$$Y_f(s) = \left[\frac{b(s)}{a(s)}\right]U(s) - \frac{\tau(s)}{a(s)} = \left[\frac{2(3s+1)}{s^2 + 3s + 2}\right]\left[\frac{1}{s-5}\right] - \frac{6}{s^2 + 3s + 2}$$

$$= \frac{32}{(s+1)(s+2)(s-5)} = \frac{-16/3}{s+1} + \frac{32/7}{s+2} + \frac{16/21}{s-5}$$

and hence

$$y_f(t) = -\frac{16}{3}e^{-t} + \frac{32}{7}e^{-2t} + \frac{16}{21}e^{5t}$$

Therefore, the total time response of the system is

$$y(t) = y_h(t) + y_f(t)$$

$$= [4e^{-t} - 3e^{-2t}] + \left[-\frac{16}{3}e^{-t} + \frac{32}{7}e^{-2t} + \frac{16}{21}e^{5t} \right]$$

$$= -\frac{4}{3}e^{-t} + \frac{11}{7}e^{-2t} + \frac{16}{21}e^{5t}$$

The transient response $y_{tr}(t)$ involves all terms of $y(t)$ which tend to zero as time increases. Therefore

$$y_{tr}(t) = 4e^{-t} - 3e^{-2t} - \frac{16}{3}e^{-t} + \frac{32}{7}e^{-2t} = -\frac{4}{3}e^{-t} + \frac{11}{7}e^{-2t}$$

The steady-state response will then be

$$y_{ss}(t) = \frac{16}{21}e^{5t}$$

Clearly

$$y(t) = y_h(t) + y_f(t) = y_{tr}(t) + y_{ss}(t)$$

4.2.2 Characteristics of the Graphical Representation of Time Response

The total system time response (i.e., both transient and steady state) of an asumptotically stable system with $u(t) = 1$, has the general form shown in Figure 4.2. This waveform gives an overall picture of the system's behavior in the time domain. The basic characteristics of this unit step response waveform are the following.

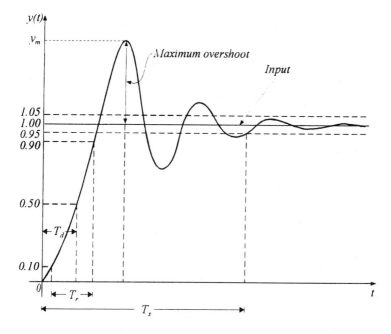

Figure 4.2 Typical unit step response of an asymptotically stable system.

1 Maximum Overshoot

This is the difference between the maximum value y_m and the final value (i.e., the steady state y_{ss}) of $y(t)$. The percentage v of the maximum overshoot is defined as

$$v\% = 100 \left[\frac{y_m - y_{ss}}{y_{ss}} \right] \tag{4.2-7}$$

2 Delay Time T_d

This is the time required for $y(t)$ to reach half of its final value.

3 Rise Time T_r

This is the time required for $y(t)$ to rise from 10% to 90% of its final value.

4 Settling Time T_s

This is the time required for $y(t)$ to reach and remain within a certain range of its final value. This range is usually from 2–5% of the amplitude of the final value.

4.3   TIME RESPONSE OF FIRST- AND SECOND-ORDER SYSTEMS

In Sec. 4.2 we presented a rather general method of analyzing linear systems. In this section we present a detailed analysis of first- and second-order systems with zero initial conditions, since these type of systems are often met in practice. In particular, we derive the analytical expressions of the responses for first- and second-order systems. We also study the effects of the transfer function parameters on the system's output waveform, aiming in waveform standardization. It is shown that the standardization is easy for first- and second-order systems, but becomes more and more difficult as the order of the system increases.

4.3.1   First-Order Systems

The transfer function of a first-order system has the general form

$$G(s) = \frac{b_0}{s + a_0} = \frac{K}{Ts + 1} \tag{4.3-1}$$

where K is the amplification constant and T is the time constant. The system response $Y(s)$, when $U(s) = 1/s$ and $K = 1$, will be $Y(s) = 1/s(Ts + 1)$. If we expand $Y(s)$ in partial fractions, we obtain

$$Y(s) = \frac{1}{s} - \frac{1}{s + 1/T}, \qquad \text{and hence } y(t) = 1 - e^{-(t/T)} \tag{4.3-2}$$

It is clear that first-order systems involve no oscillations and the waveform $y(t)$ is affected only by the time constant T. The waveforms of $y_1(t)$, $y_2(t)$, and $y_3(t)$, which correspond to the time constants $T_1 < T_2 < T_3$, are given in Figure 4.3. The conclusion derived from Figure 4.3 is that the smaller the T, the higher the speed response of the system, where the term *speed response* is defined as the amount of time required for $y(t)$ to reach 63.78% of its final value. For first-order systems, this amount of time coincides with the time constant T, since $y(T) = 1 - e^{-1} = 0.6378$.

4.3.2 Second-Order Systems

Consider a second-order system with transfer function

$$G(s) = \frac{\omega_n^2}{s^2 + 2\zeta\omega_n s + \omega_n^2} \qquad (4.3\text{-}3)$$

where the constants ω_n and ζ are called the *natural undamped frequency* and the *damping ratio* of the system, respectively. The poles of $G(s)$ are

$$s_{1,2} = -\omega_n\left(\zeta \pm \sqrt{\zeta^2 - 1}\right) = -\omega_n\zeta \pm j\omega_n\sqrt{1 - \zeta^2} = -\sigma \pm j\omega_d$$

where $\sigma = \omega_n\zeta$ and $\omega_d = \omega_n\sqrt{1 - \zeta^2}$ are called the *attenuation* or *damping* constant and the *damped natural frequency* of the system, respectively.

The system's response $Y(s)$, when $U(s) = 1/s$, is as follows:

$$Y(s) = \frac{\omega_n^2}{s(s^2 + 2\zeta\omega + \omega_n^2)} \qquad (4.3\text{-}4)$$

Depending on the value, or range of values, of the damping ratio ζ, we distinguish the following four cases:

Case 1 $(\zeta = 0)$

In this case the poles of $G(s)$ are imaginary since $s_{1,2} = \pm j\omega_n$ and relation (4.3-4) becomes

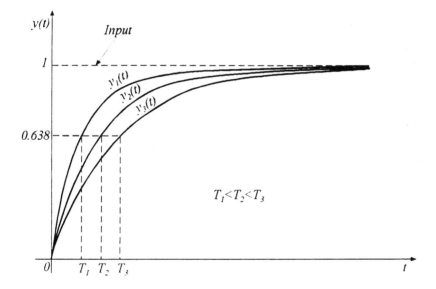

Figure 4.3 Unit step response of a first-order linear system.

$$Y(s) = \frac{\omega_n^2}{s(s^2 + \omega_n^2)}$$

If we expand $Y(s)$ in partial fractions, we have

$$Y(s) = \frac{1}{s} - \frac{s}{s^2 + \omega_n^2}, \qquad \text{and thus } y(t) = 1 - \cos \omega_n t \qquad (4.3\text{-}5)$$

Thus, when $\zeta = 0$ we observe that the response $y(t)$ is a sustained oscillation with constant frequency to ω_n and constant amplitude equal to 1 (see Figure 4.4). In this case, we say that the system is *undamped*.

Case 2 ($0 < \zeta < 1$)

In this case the poles of $G(s)$ are a complex conjugate pair since $s_{1,2} = -\sigma \pm j\omega_d$ and relation (4.3-4) becomes

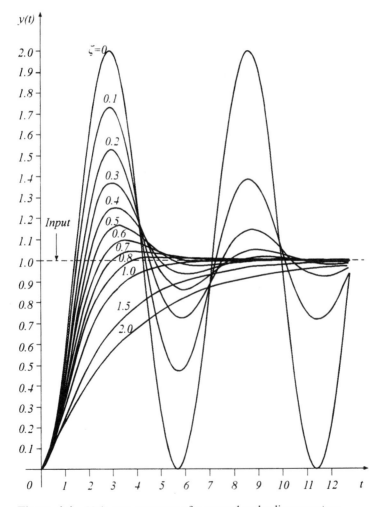

Figure 4.4   Unit step response of a second-order linear system.

$$Y(s) = \frac{\omega_n^2}{s[(s+\sigma)^2 + \omega_d^2]}$$

If we expand $Y(s)$ in partial fractions, we have

$$Y(s) = \frac{1}{s} - \frac{s+2\sigma}{(s+\sigma)^2 + \omega_d^2} = \frac{1}{s} - \frac{s+\sigma}{(s+\sigma)^2 + \omega_d^2} - \begin{bmatrix} \sigma \\ \omega_d \end{bmatrix} \begin{bmatrix} \omega_d \\ (s+\sigma)^2 + \omega_d^2 \end{bmatrix}$$

Taking the inverse Laplace transform, we have

$$y(t) = 1 - e^{-\sigma t}\cos\omega_d t - \frac{\sigma}{\omega_d}e^{-\sigma t}\sin\omega_d t$$

The above expression for $y(t)$ may be written as

$$y(t) = 1 - \frac{e^{-\sigma t}}{\sqrt{1-\zeta^2}}\sin(\omega_d t + \varphi), \qquad \text{where} \qquad \varphi = \tan^{-1}\frac{\sqrt{1-\zeta^2}}{\zeta} \qquad (4.3\text{-}6)$$

Thus, when $0 < \zeta < 1$ we observe that the response $y(t)$ is a damped oscillation which tends to 1 as $t \to \infty$ (see Figure 4.4). In this case, we say that the system is *underdamped*.

Case 3 ($\zeta = 1$)

In this case the poles of G(s) are the real double pole $-\omega_n$, and relation (4.3-4) becomes

$$Y(s) = \frac{\omega_n^2}{s(s+\omega_n)^2}$$

If we expand $Y(s)$ in partial fractions, we have

$$Y(s) = \frac{1}{s} - \frac{1}{s+\omega_n} - \frac{\omega_n^2}{(s+\omega_n)^2}, \qquad \text{and thus } y(t) = 1 - e^{-\omega_n t} - \omega_n t e^{-\omega_n t}$$

$$(4.3\text{-}7)$$

Thus, when $\zeta = 1$ we observe that the waveform of the response $y(t)$ involves no oscillations, and asymptotically tends to 1 as $t \to \infty$ (see Figure 4.4). In this case we say that the system is *critically damped*.

Case 4 ($\zeta > 1$)

In this case the poles of $G(s)$ are both real and negative since $s_{1,2} = -\sigma \pm \omega_n\sqrt{\zeta^2 - 1}$ and relation (4.3-4) becomes

$$Y(s) = \frac{\omega_n^2}{s[(s+\sigma)^2 - a^2]}$$

where $a = \omega_n\sqrt{\zeta^2 - 1}$. If we expand $Y(s)$ in partial fractions, we have

$$Y(s) = \frac{1}{s} - \frac{s+\sigma}{(s+\sigma)^2 - a^2} - \begin{bmatrix} \sigma \\ a \end{bmatrix}\begin{bmatrix} a \\ (s+\sigma)^2 - a^2 \end{bmatrix}$$

Taking the inverse Laplace transform, we have

$$y(t) = 1 - e^{-\sigma t}\cosh at - \frac{\sigma}{a}e^{-\sigma t}\sinh at$$

$$= 1 - e^{-\sigma t}\left[\cosh\left(\omega_n\sqrt{\zeta^2 - 1}\right)t + \frac{\zeta}{\sqrt{\zeta^2 - 1}}\sinh\left(\omega_n\sqrt{\zeta^2 - 1}\right)t\right] \tag{4.3-8}$$

Thus, when > 1 we observe that the response $y(t)$ involves no oscillations and tends asymptotically to 1 as $t \to \infty$, while the speed response decreases as ζ becomes larger (see Figure 4.4). In this case, we say that the system is *overdamped*.

Summarizing the above results we conclude that when $\zeta = 0$, we have sustained (undamped) oscillations. As ζ increases towards 1, these oscillations are damped more and more. When ζ reaches 1, the oscillations stop. Finally, as ζ further increases becoming greater than 1, we have no oscillations and the output requires more and more time to reach asymptotically the value 1.

4.3.3 Special Issues for Second-Order Systems

Here we will study certain issues regarding second-order systems which are of particular interest.

1 The Root Locus

The term *root locus* (see Chap. 7) is defined as the locus of all roots of the characteristic polynomial $p(s)$ of a system in the complex plane. This locus is formed when varying one or more system parameters. Consider a system described by its transfer function $G(s)$, where $p(s)$ is the denominator of $G(s)$. For second-order systems, $p(s)$ has the form

$$p(s) = s^2 + 2\zeta\omega_n s + \omega_n^2 \tag{4.3-9}$$

The roots of $p(s)$ are $s_{1,2} = -\omega_n\zeta \pm \omega_n\sqrt{\zeta^2 - 1}$. In Figure 4.5 the root locus of these two roots is shown, as ζ varies from $-\infty$ to $+\infty$.

2 The Time Response as a Function of the Positions of the Two Poles

In Figure 4.6 several typical positions of the poles of relation (4.3-9) and the corresponding unit-step responses are given.

3 Relation Between the Damping Ratio and the Overshoot

We have found that when $0 < \zeta < 1$ the response $y(t)$ has the analytical form (4.3-6), while its waveform presents an overshoot. To determine this overshoot it suffices to determine the maximum value y_m of $y(t)$. To this end, we take the derivative of relation (4.3-6) to yield

$$y^{(1)}(t) = \frac{e^{-\sigma t}}{\sqrt{1 - \zeta^2}}[\sigma\sin(\omega_d t + \varphi) - \omega_d\cos(\omega_d t + \varphi)] = \frac{\omega_n e^{-\sigma t}}{\sqrt{1 - \zeta^2}}\sin\omega_d t$$

Here $y^{(1)}(t)$ is zero when $\sin\omega_d t = 0$. Furthermore, $\sin\omega_d t$ is zero when $\omega_d t_k = k\pi$, where $k = 1, 2, \ldots$, i.e., when

$$t_k = \frac{k\pi}{\omega_d} = \frac{k\pi}{\omega_n\sqrt{1 - \zeta^2}}, \qquad k = 1, 2, \ldots$$

Hence, the expression for $y(t_k)$ is

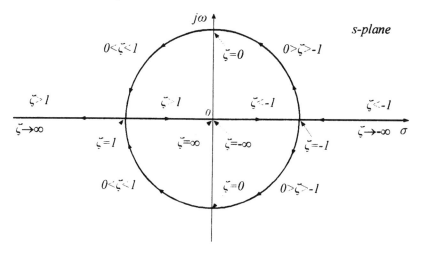

Figure 4.5   The root locus of a second-order system.

$$y(t_k) = 1 - \frac{\exp\left(-\zeta k\pi/\sqrt{1-\zeta^2}\right)}{\sqrt{1-\zeta^2}} \sin(k\pi + \varphi)$$

$$= 1 + (-1)^{k-1} \exp\left(-\zeta k\pi/\sqrt{1-\zeta^2}\right), \qquad k = 1, 2, \ldots$$

Clearly, $y(t_k)$ becomes maximum for $k = 1, 3, 5\ldots$ and minimum for $k = 2, 4, 6, \ldots$ (see Figure 4.7a). Therefore, the maximum value y_m is given by

$$y_m = 1 + \exp\left(-\zeta\pi/\sqrt{1-\zeta^2}\right) \tag{4.3-10}$$

which occurs when $k = 1$, i.e., at the time instant

$$t_1 = \frac{\pi}{\omega_n\sqrt{1-\zeta^2}} \tag{4.3-11}$$

The overshoot percentage defined in Eq. (4.2-7) will be

$$v\% = 100 \exp\left(-\zeta\pi/\sqrt{1-\zeta^2}\right) \tag{4.3-12}$$

The overshoot v as a function of the damping ratio ζ is given in Figure 4.7b.

4.4 MODEL SIMPLICATION

The model simplification or model reduction problem may be stated as follows. We are given a detailed mathematical model of a system, which is usually of very high order and hence very difficult to work with. We wish to find a simpler model, which *approximates* the original model *satisfactorily*. Clearly, the simpler model has the advantage in that it simplifies the system description, but it has the disadvantage in that it is less accurate than the original detailed model.

The motivation for deriving simplified models may be summarized as follows:

    1. It simplifies the description and the analysis of the system

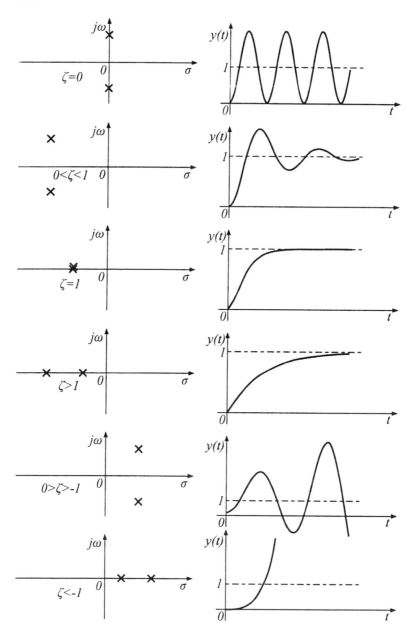

Figure 4.6 Unit step response comparison for various pole locations in the *s*-plane of a second-order system.

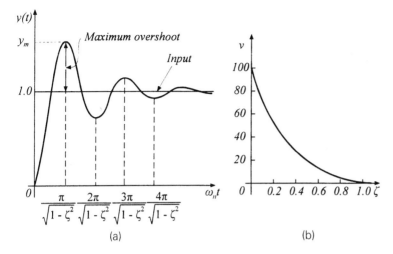

Figure 4.7   (a) The maximum overshoot and (b) its percentage as a function of ζ.

2. It simplifies the computer simulation of the system
3. It facilitates the controller design problem and yields controllers with simpler structures
4. It reduces the computational effort in the analysis and design of control systems.

Obviously, the order of the simplified model is much less than that of the original model. For this reason the model simplification problem is also known as the *order reduction* problem.

Many techniques have been proposed to simplify a model [36–39]. This section aims to introduce the reader to this interesting problem. To this end, we will present here one of the simplest techniques known as the *dominant pole* method.

The dominant pole method is as follows. Let $G(s)$ be the transfer function of the original system. Expand $G(s)$ in partial fractions to yield

$$G(s) = \frac{c_1}{s - \lambda_1} + \frac{c_2}{s - \lambda_2} + \cdots + \frac{c_n}{s - \lambda_n}$$

where $\lambda_1, \lambda_2, \ldots, \lambda_n$ are the poles of $G(s)$. Let $G(s)$ be asymptotically stable, i.e., let $\mathrm{Re}\,\lambda_i < 0$, for $i = 1, 2, \ldots, n$. Then, the closer any pole λ_i is to the imaginary axis, the greater the effect on the system response and vice versa; i.e., the farther away λ_i is from the imaginary axis, the less the effect on the system response. For this reason, the poles that are located close to the imaginary axis are called *dominant poles*.

The dominant pole simplification method yields a simplified model involving only the dominant poles.

To illustrate the dominant pole approach, consider a system with input signal $u(t) = 1$ and transfer function

$$G(s) = \frac{1}{s + \lambda_1} + \frac{1}{s + \lambda_2}, \qquad \text{where } \lambda_1 \text{ and } \lambda_2 \text{ positive and } \lambda_1 \ll \lambda_2.$$

Then

$$Y(s) = G(s)U(s) = \frac{1}{s(s+\lambda_1)} + \frac{1}{s(s+\lambda_1)}$$

$$= \frac{1}{\lambda_1}\left[\frac{1}{s} - \frac{1}{s+\lambda_1}\right] + \frac{1}{\lambda_2}\left[\frac{1}{s} - \frac{1}{s+\lambda_2}\right]$$

Hence

$$y(t) = \frac{1}{\lambda_1}(1 - e^{-\lambda_1 t}) + \frac{1}{\lambda_2}(1 - e^{-\lambda_2 t})$$

Here, $-\lambda_1$ is closer to the imaginary axis than $-\lambda_2$. Hence, the term $e^{-\lambda_2 t}$ goes to zero much faster than $e^{-\lambda_1 t}$. For this reason the dominant pole $-\lambda_1$ is often called *slow mode* and $-\lambda_2$ is called *fast mode*. Let $G_1(s)$ be the satisfactory approximant of $G(s)$ sought. Then, since $e^{-\lambda_2 t}$ goes to zero much faster than $e^{-\lambda_1 t}$, it is clear that $G_1(s)$ should "keep" the slow mode $-\lambda_1$ and "drop" the fast mode $-\lambda_2$. For this reason, we choose as simplified model the transfer function $G_1(s)$, where

$$G_1(s) = \frac{1}{s+\lambda_1}$$

Clearly, $G_1(s)$ is simpler than the original $G(s)$ and involves the most dominant pole $-\lambda_1$. The time response $y_1(t)$ of the simplified model is given by

$$y_1(t) = L^{-1}\{G_1(s)U(s)\} = \frac{1}{\lambda_1}(1 - e^{-\lambda_1 t})$$

Example 4.4.1

Consider the transfer function

$$G(s) = \frac{A_1}{s+1} + \frac{A_2}{s+2} + \frac{A_3}{s+10} + \frac{A_4}{s+100} + \frac{A_5}{s+1000}$$

Find a simplified model of third and second order using the dominant pole method.

Solution

The dominant poles of $G(s)$, in decreasing order, are -1, -2, and -10. Let $G_3(s)$ and $G_2(s)$ be the third- and the second-order approximants of $G(s)$. Then, the simplified models sought are the following:

$$G_3(s) = \frac{A_1}{s+1} + \frac{A_2}{s+2} + \frac{A_3}{s+10} \quad \text{and} \quad G_2(s) = \frac{A_1}{s+1} + \frac{A_2}{s+2}$$

For simplicity, let $A_1 = A_2 = A_3 = A_4 = A_5 = 1$. Also, let the input signal $u(t) = 1$. Then, the output $y(t)$ of the original system $G(s)$ and the output $y_3(t)$ and $y_2(t)$ of the reduced-order systems $G_3(s)$ and $G_2(s)$, respectively, may be determined as follows: For $y(t)$, we have

$$Y(s) = \frac{1}{s}\left[\frac{1}{s+1} + \frac{1}{s+2} + \frac{1}{s+10} + \frac{1}{s+100} + \frac{1}{s+1000}\right]$$

$$= \left[\frac{1}{s} - \frac{1}{s+1}\right] + \frac{1}{2}\left[\frac{1}{s} - \frac{1}{s+2}\right] + \frac{1}{10}\left[\frac{1}{s} - \frac{1}{s+10}\right]$$

$$+ \frac{1}{100}\left[\frac{1}{s} - \frac{1}{100}\right] + \frac{1}{1000}\left[\frac{1}{s} - \frac{1}{1000}\right]$$

and hence

$$y(t) = [1 - e^{-t}] + \tfrac{1}{2}[1 - e^{-2t}] + \tfrac{1}{10}[1 - e^{-10t}] + \tfrac{1}{100}[1 - e^{-100t}] + \tfrac{1}{1000}[1 - e^{-1000t}]$$

For $y_3(t)$ and $y_2(t)$ we keep only the first three and two brackets in the above expression, respectively, i.e.,

$$y_3(t) = [1 - e^{-t}] + \tfrac{1}{2}[1 - e^{-2t}] + \tfrac{1}{10}[1 - e^{-10t}]$$
$$y_2(t) = [1 - e^{-t}] + \tfrac{1}{2}[1 - e^{-2t}]$$

In Figure 4.8 the plots of $y(t)$, $y_3(t)$, and $y_2(t)$ are given. Comparison of these three plots shows that both $y_3(t)$ and $y_2(t)$ are close to $y(t)$. For better accuracy, one may choose $G_3(s)$ as an approximant of $G(s)$. For greater simplicity one may choose $G_2(s)$. Clearly, in choosing either $G_3(s)$ or $G_2(s)$, we reduce the order of the original system from five to three ($G_3(s)$) or to two ($G_2(s)$).

4.5 COMPARISON BETWEEN OPEN- AND CLOSED-LOOP SYSTEMS

There are important differences between open- and closed-loop systems. Three of these differences are of paramount importance and they are described below. From these differences we conclude that closed-loop systems are superior over open-loop systems and, for this reason, they are more often used in practice.

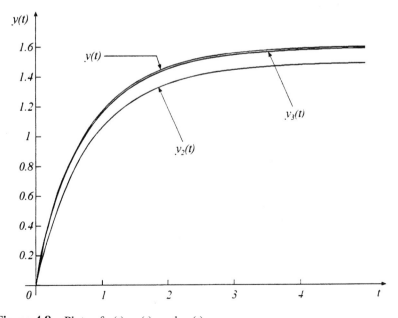

Figure 4.8   Plots of $y(t)$, $y_3(t)$, and $y_2(t)$.

4.5.1 Effect on the Output Due to Parameter Variations in the Open-Loop System

Consider the closed-loop system shown in Figure 4.9, where $G(s)$ is the transfer function of the system under control and $F(s)$ is the feedback transfer function. The open-loop system's output $Y(s)$ (i.e., when $F(s) = 0$) is

$$Y(s) = G(s)R(s) \qquad\qquad\qquad (4.5\text{-}1)$$

The closed-loop system's output $Y_c(s)$ (i.e., when $F(s) \neq 0$) is

$$Y_c(s) = \left[\frac{G(s)}{1 + G(s)F(s)}\right]R(s) \qquad\qquad\qquad (4.5\text{-}2)$$

Now, assume that certain parameters of the transfer function $G(s)$ undergo variations. Let $dG(s)$ be the change in $G(s)$ due to these parameter variations. As a result, the output $Y(s)$ will also vary. In particular, for the open-loop system (4.5-1) the change $dY(s)$ of the output $Y(s)$ will be

$$dY(s) = R(s)dG(s) \qquad\qquad\qquad (4.5\text{-}3)$$

while for the closed-loop system (4.5-2) the change $dY_c(s)$ of the output $Y_c(s)$ will be

$$dY_c(s) = \left[\frac{R(s)}{[1 + G(s)F(s)]^2}\right]dG(s) \qquad\qquad\qquad (4.5\text{-}4)$$

If we divide relations (4.5-3) and (4.5-1), we have

$$\frac{dY(s)}{Y(s)} = \frac{dG(s)}{G(s)} \qquad\qquad\qquad (4.5\text{-}5)$$

Similarly, if we divide relations (4.5-4) and (4.5-2), we have

$$\frac{dY_c(s)}{Y_c(s)} = \left[\frac{1}{1 + G(s)F(s)}\right]\frac{dG(s)}{G(s)} \qquad\qquad\qquad (4.5\text{-}6)$$

Next, consider the magnitudes in Eqs (4.5-5) and (4.5-6). We have

$$\left|\frac{dY(s)}{Y(s)}\right| = \left|\frac{dG(s)}{G(s)}\right| \qquad\qquad\qquad (4.5\text{-}7)$$

$$\left|\frac{dY_c(s)}{Y_c(s)}\right| = \frac{1}{|1 + G(s)F(s)|}\left|\frac{dG(s)}{G(s)}\right| \qquad\qquad\qquad (4.5\text{-}8)$$

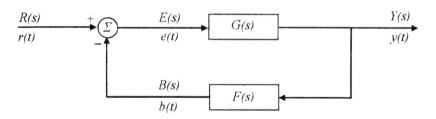

Figure 4.9 Closed-loop system.

In control systems, we usually work in low frequencies, in which case we have that $|G(s)F(s)| \gg 1$. This readily yields

$$|\mathrm{d}Y_c(s)| \ll |\mathrm{d}Y(s)| \qquad (4.5\text{-}9)$$

Releation (4.5-9) reveals that the effects of parameter changes of the transfer function $G(s)$ on the closed-loop system's output $Y_c(s)$ is much smaller than that its effects on the open-loop system's output $Y(s)$. This property is one of the most celebrated basic advantages of closed-loop systems over open-loop systems.

4.5.2 Effect on the Output Due to Parameter Variations in the Feedback Transfer Function

Assume that the parameters of $F(s)$ undergo certain variations. If we differentiate, relation (4.5-2) yields

$$\mathrm{d}Y_c(s) = -\left[\frac{G^2(s)R(s)}{[1 + G(s)F(s)]^2}\right]\mathrm{d}F(s) \qquad (4.5\text{-}10)$$

If we divide relations (4.5-10) and (4.5-2), we have

$$\frac{\mathrm{d}Y_c(s)}{Y_c(s)} = -\left[\frac{G(s)}{1 + G(s)F(s)}\right]\mathrm{d}F(s) \qquad \text{or} \qquad \left|\frac{\mathrm{d}Y_c(s)}{Y_c(s)}\right| = \frac{|G(s)F(s)|}{|1 + G(s)F(s)|}\left|\frac{\mathrm{d}F(s)}{F(s)}\right|$$

$$(4.5\text{-}11)$$

Since $|G(s)F(s)| \gg 1$, it follows that

$$\frac{|G(s)F(s)|}{|1 + G(s)F(s)|} \simeq 1$$

Hence, relation (4.5-11) becomes

$$\left|\frac{\mathrm{d}Y_c(s)}{Y_c(s)}\right| \simeq \left|\frac{\mathrm{d}F(s)}{F(s)}\right| \qquad (4.5\text{-}12)$$

Relation (4.5-12) indicates that the variation $\mathrm{d}F(s)$ crucially affects the system's output. For this reason the feedback transfer function $F(s)$ must be made up of elements which should vary as little as possible.

4.5.3 Effect of Disturbances

Consider the two systems shown in Figure 4.10b, where we assume the presence of the disturbance (or noise) $D(s)$. For the open-loop system of Figure 4.10a, the portion $Y_d(s)$ of the output which is due to the disturbance $D(s)$ will be

$$Y_d(s) = G_2(s)D(s) \qquad (4.5\text{-}13)$$

Similarly for the closed-loop system of Figure 4.10, we have that

$$Y_{cd}(s) = \left[\frac{G_2(s)}{1 + G_1(s)G_2(s)}\right]D(s) \qquad (4.5\text{-}14)$$

where $Y_{cd}(s)$ is the output $Y_c(s)$ of the closed-loop system due to the disturbance $D(s)$. From relations (4.5-13) and 4.5-14) we have that

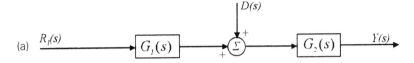

(a)

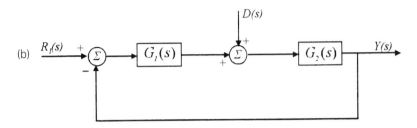

(b)

Figure 4.10   (a) Open- and (b) closed-loop systems with disturbances.

$$Y_{cd}(s) = \left[\frac{1}{1 + G_1(s)G_2(s)}\right] Y_d(s)$$ (4.5-15)

Taking under consideration that $|G_1(s)G_2(s)| \gg 1$, yields

$$|Y_{cd}(s)| \ll |Y_d(s)|$$ (4.5-16)

Relation (4.5-16) reveals another advantage of the feedback action, i.e., of the closed-loop system over the open-loop system. Specifically, it shows that the output of a closed-loop system is much less sensitive to disturbances than that of an open-loop system. This advantage of the closed-loop systems is of paramount importance in practical control applications.

4.6 SENSITIVITY TO PARAMETER VARIATIONS

In the previous section, we dealt with systems or processes which are very often subject to parameter variations, which, in turn have undesirable effects upon the performance of the system. These variations in $G(s)$ are usually due to component aging, changing in the environment, inevitable errors in the system or process model (e.g., in the parameters of its transfer function $G(s)$), and other factors. In Subsec. 4.5.1 it was assessed that in the open-loop case, changes in $G(s)$ have a serious effect on the output $Y(s)$. In the closed-loop case, this effect is considerably reduced, a fact which makes the closed-loop configuration much more attractive in practice than the open-loop configuration.

   In Subsec. 4.5.1, we studied the effect upon the output due to parameter variations. In this section we focus our attention on the determination of the sensitivity of the transfer function and of the poles of the closed-loop system due to parameter variations in $G(s)$. In this section we seek to establish techniques which yield the magnitude of the sensitivity of the transfer function and of the poles of the closed-loop system due to parameter variations.

4.6.1 System Sensitivity to Parameter Variations

Definition 4.6.1

The closed-loop system sensitivity, designated by S_G^H is defined as the ratio of the percentage change in the closed-loop transfer function $H(s)$ to the percentage change in the system transfer function $G(s)$, for a small incremental change. That is

$$S_G^H = \frac{\Delta H(s)/H(s)}{\Delta G(s)/G(s)} \tag{4.6-1}$$

In the limit, as the incremental changes go to zero, Eq. (4.6-1) becomes

$$S = \frac{\partial H/H}{\partial G/G} = \frac{\partial \ln H}{\partial \ln G} \tag{4.6-2}$$

For the open-loop case, we have that $H(s) = G(s)$ and, hence, the sensitivity S_G^G of the open-loop system is

$$S_G^G = \frac{\partial H(s)/H(s)}{\partial Gs)/G(s)} = \frac{\partial G(s)/G(s)}{\partial G(s)/G(s)} = 1 \tag{4.6-3}$$

For the closed-loop case, we will consider the following two problems: the sensitivity of the closed system with respect to the changes in $G(s)$, denoted as S_G^H, and the sensitivity of the closed system with respect to the changes in $F(s)$, denoted as S_F^H. To this end, consider the closed-loop transfer function

$$H(s) = \frac{G(s)}{1 + G(s)F(s)} \tag{4.6-4}$$

Hence,

$$\begin{aligned} S_G^H &= \frac{\partial H(s)/H(s)}{\partial G(s)/G(s)} = \left[\frac{G(s)}{H(s)}\right]\left[\frac{\partial H(s)}{\partial G(s)}\right] = G(s)\left[\frac{1 + G(s)F(s)}{G(s)}\right]\left[\frac{\partial H(s)}{\partial G(s)}\right] \\ &= [1 + G(s)F(s)]\left[\frac{1}{[1 + G(s)F(s)]^2}\right] = \frac{1}{1 + G(s)F(s)} \end{aligned} \tag{4.6-5}$$

Similarly,

$$\begin{aligned} S_F^H &= \frac{\partial H(s)/H(s)}{\partial F(s)/F(s)} = \left[\frac{F(s)}{H(s)}\right]\left[\frac{\partial H(s)}{\partial F(s)}\right] = F(s)\left[\frac{1 + G(s)F(s)}{G(s)}\right]\left[\frac{\partial H(s)}{\partial F(s)}\right] \\ &= -\left[\frac{F(s)[1 + G(s)F(s)]}{G(s)}\right]\left[\frac{G^2(s)}{[1 + G(s)F(s)]^2}\right] = -\frac{G(s)F(s)}{1 + G(s)F(s)} \end{aligned} \tag{4.6-6}$$

Clearly, when $G(s)F(s) \gg 1$, then

$$S_G^H \to 0 \quad \text{and} \quad S_F^H \to -1 \tag{4.6-7}$$

Hence S_G^G and S_F^H indicate that the system is very sensitive to changes in $G(s)$ in the open-loop system case and also very sensitive to changes in $F(s)$ in the closed-loop system case, respectively. On the contrary, S_G^H indicates that the closed-loop system is very insensitive to changes in $G(s)$. These remarks are in complete agreement with the results of Sec. 4.5.

4.6.2 Pole Sensitivity to Parameter Variations

Definition 4.6.2

The closed-loop pole sensitivity, denoted as S_β^s, is the ratio of the change in the position s of the corresponding root β of the characteristic equation of the closed-loop system to the change in the parameter β, for a small increment change, i.e.,

$$S_\beta^s = \frac{ds}{d\beta} \qquad\qquad (4.6\text{-}8)$$

Definition 4.6.2 is very useful in determining the sensitivity of the roots of the characteristic equation (and hence the sensitviity of the poles) of the closed-loop system due to variations in a certain parameter. In computing Eq. (4.6-8), one may find the sensitivity to be, for example, very high, in which case one should take appropriate steps to reduce it.

Example 4.6.1

Consider a typical second-order closed-loop system given in Figure 4.11. Find:

    a. The sensitivity of the transfer function of the closed-loop system with respect to the gain K and to the parameter a (here, $1/a$ is the time constant of the open-loop system).

    b. The sensitivity of the roots r_1 and r_2 with respect to K and a. The nominal values of K and a are $K = 20$ and $a = 4$.

    c. Let $\Delta K = 4$ and $\Delta a = 2$ be the variations in K and a, respectively, in which case the new values of K and a are $K = 20 + \Delta K = 24$ and $a = 4 + \Delta a = 6$. Using the sensitivity approach, find the approximate values of r_1 and r_2 for each variation ΔK and Δa, separately, and compare them with the exact values. Repeat the same step when both variations ΔK and Δa take place simultaneously.

Solution

a. The transfer function $H(s)$ of the closed-loop system is

$$H(s) = \frac{Y(s)}{R(s)} = \frac{G(s)}{1 + G(s)F(s)} = \frac{K}{s(s+a)+K}$$

The sensitivity of $H(s)$ with respect to K is given by

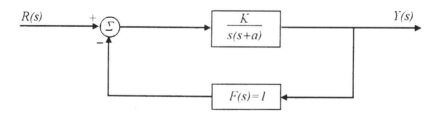

Figure 4.11   Typical second-order system with unity feedback.

$$S_K^H = \left[\frac{K}{H}\right]\left[\frac{dH}{dK}\right] = \left[\frac{K[s(s+a)+K]}{K}\right]\left[\frac{s(s+a)+K]-K}{[s(s+a)+K]^2}\right] = \frac{s(s+a)}{s(s+a)+K}$$

Similarly, the sensitivity of $H(s)$ with respect to a is given by

$$S_a^H = \left[\frac{a}{H}\right]\left[\frac{dH}{da}\right] = \left[\frac{a[s(s+a)+K]}{K}\right]\left[\frac{-sK}{[s(s+a)+K]^2}\right] = \frac{-as}{s(s+a)+K}$$

Clearly, both S_K^H and S_a^H are functions of s.

   b. The characteristic equation of the closed-loop system is

$$s(s+a) + K = (s - r_1)(s - r_2) = 0$$

Taking the derivative with respect to K yields

$$2s\frac{ds}{dK} + a\frac{ds}{dK} + 1 = 0$$

and hence

$$S_K^s = \frac{ds}{dK} = \frac{-1}{a+2s}$$

For the nominal values $K = 20$ and $a = 4$, the roots of the characteristic equation are $r_1 = -2 + j4$ and $r_2 = -2 - 4j$. Hence, the root sensitivity S_K^s with respect to K for the root r_1 (i.e., for $s = r_1$) may be determined as follows:

$$S_K^{r_1} = \frac{ds}{dK}\bigg|_{s=r_1} = \frac{dr_1}{dK} = \frac{-1}{a+2s}\bigg|_{s=r_1} = \frac{-1}{4+2(-2+j4)} = \frac{j}{8}$$

Similarly, for the sensitivity of r_2 (i.e., for $s = r_2$):

$$S_K^{r_2} = \frac{ds}{dK}\bigg|_{s=r_2} = \frac{dr_2}{dK} = \frac{-1}{a+2s}\bigg|_{s=r_2} = \frac{-1}{4+2(-2-j4)} = -\frac{j}{8}$$

Next, the sensitivity S_a^s of the roots r_1 and r_2 of the characteristic polynomial with respect to a will be determined. To this end, take the derivative of the characteristic equation $s(s+a) + K = 0$ with respect to a to yield

$$2s\frac{ds}{da} + \frac{d}{da}[as] = 2s\frac{ds}{da} + \left[a\frac{ds}{da} + s\frac{da}{da}\right] = 0$$

and hence

$$S_a^s = \frac{ds}{da} = \frac{-s}{a+2s}$$

The sensitivity of the root r_1 (i.e., for $s = r_1$) is

$$S_a^{r_1} = \frac{dr_1}{da} = \frac{-(-2+j4)}{4+2(-2+j4)} = -\frac{2+j}{4}$$

Similarly, for the root r_2 (i.e., for $s = r_2$):

$$S_a^{r_2} = \frac{dr_2}{da} = -\frac{2-j}{4}$$

   c. We consider the following three cases:
   1. $\Delta K = 4$ and $\Delta a = 0$. In this case, we make use of the relation

$$dr_1 = \left[\frac{dr_1}{dK}\right] dK = S_K^{r_1} \, dK$$

Using this relation, one may approximately determine r_1, when K varies, as follows:

$$\Delta r_1 \cong S_K^{r_1} \Delta K = \frac{j}{8} \Delta K$$

For $\Delta K = 4$, the above expression gives $\Delta r_1 \cong 0.5j$. Hence, the new value $\tilde{r}_1$ of r_1 due to $\Delta K = 4$ is given by $\tilde{r}_1 \cong r_1 + \Delta r_1 = -2 + j4.5$. To find the exact value $\bar{r}_1$ for r_1, solve the characteristic equation for $K = 20 + 4 = 24$ and $a = 4$ to yield $\bar{r}_1 = -2 + j\sqrt{20} = -2 + j4.47$. Comparing the results, we observe that the approximate value $\tilde{r}_1$ is very close to the exact value $\bar{r}_1$.

2. $\Delta K = 0$ and $\Delta a = 2$. As in case 1, we use the relation

$$dr_1 = \left[\frac{dr_1}{da}\right] da = S_a^{r_1} \, da$$

Using this relation, we determine Δr_1, when a varies, as follows:

$$\Delta r_1 \cong S_a^{r_1} \Delta a = -\frac{2+j}{8} \Delta a$$

For $\Delta a = 2$, the new value $\tilde{r}_1$ of r_1 due to $\Delta a = 2$, is given by

$$\tilde{r}_1 \cong r_1 + \Delta r_1 = (-2 + j4) - \frac{(2+j)}{4} 2 = -3 + j3.5$$

To find the exact value $\bar{r}_1$ of r_1, solve the characteristic equation for $K = 20$ and $a = 4 + 2 = 6$ to yield $\bar{r}_1 = -3 + j\sqrt{11} = -3 + j3.32$. Comparing the results, we observe that the approximate value $\tilde{r}_1$ is very close to its exact value.

3. $\Delta K = 4$ and $\Delta a = 2$. In this case both K and a change simultaneously. For this case, we have

$$ds = \left[\frac{\partial s}{\partial K}\right] dK + \left[\frac{\partial s}{\partial a}\right] da$$

An approximate expression of the above equation is

$$\Delta s \cong \frac{\partial s}{\partial K} \Delta K + \frac{\partial s}{\partial a} \Delta a = S_K^s \Delta K + S_a^s \Delta a$$

When $\Delta K = 4$ and $\Delta a = 2$, we have

$$\Delta r_1 \cong \frac{j}{8} 4 - \frac{2+j}{4} 2 = -1$$

Hence, the new value $\hat{r}_1$ of r_1 due to $\Delta K = 4$ and $\Delta a = 2$ is given by $\hat{r}_1 = (-2 + j4) = -3 + j4$. To find the exact value $\bar{r}_1$ of r_1, solve the characteristic equation for $K = 20 + 4 = 24$ and $a = 4 + 2 = 6$ to yield $\bar{r}_1 = -3 + j\sqrt{15} = -3 + j3.87$. Comparing the results, we observe that the approximate value $\hat{r}_1$ is very close to the exact value $\bar{r}_1$.

Application of the above procedure yields analogous results for the root r_2.

4.7 STEADY-STATE ERRORS

In this section, the behavior of the steady-state performance of closed-loop systems is studied. In the design of a control system the steady-state performance is of special significance, since we seek a system whose output $y(t)$, among other things, has a prespecified desired steady-state value $y_{ss}(t)$. This desired $y_{ss}(t)$ is usually the steady-state value $r_{ss}(t)$ of the input (command) function $r(t)$. That is, control systems are designed in such a way that when they are excited by $r(t)$, they "follow" this input $r(t)$ in the steady state as closedly as possible, which means that in the steady state it is desired to have $y_{ss}(t) = r_{ss}(t)$. If $y_{ss}(t)$ is not exactly equal to $r_{ss}(t)$, then an error appears, which is called the *steady-state error*. The determination of the steady-state error is the subject of this section.

4.7.1 Types of Systems and Error Constants

Consider the unity feedback closed-loop system of Figure 4.12. The general case of nonunity feedback systems is presented in Figure 4.13a, which may readily be reduced to unity feedback as shown in Figure 4.13b. The material of this section is based on the configuration of unity feedback of Figure 4.12. To facilitate the study of nonunity feedback systems, use can be made of its equivalent unity feedback system of Figure 4.13b.

Consider the unity feedback system of Figure 4.12 and assume that $G(s)$ has the form

$$G(s) = K \frac{\prod_{i=1}^{m}(T_i's + 1)}{s^j \prod_{i=1}^{q}(T_is + 1)}, \quad \text{where } j + q = n \leq m \tag{4.7-1}$$

The following definitions are useful.

Definition 4.7.1

A system is called *type j system* when $G(s)$ has j poles at the point $s = 0$, in which case $G(s)$ has the general form (4.7-1).

Definition 4.7.2

The *position (or step) error constant* K_p of a system is defined as $K_p = \lim_{s \to 0} G(s)$. Hence, the cosntant K_p takes on the values

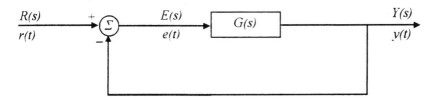

Figure 4.12 Unity feedback system.

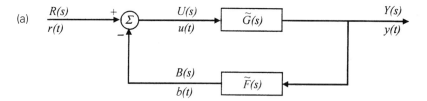

(a)

(a) Non-unity feedback system.

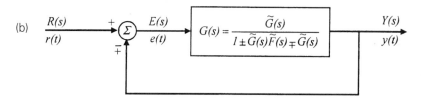

(b)

Figure 4.13   (a) Nonunity feedback system and (b) equivalent unity feedback system.

$$K_{\mathrm{p}} = \lim_{s \to 0} G(s) = \lim_{s \to 0} K \frac{\prod_{i=1}^{m}(T_j's + 1)}{s^j \prod_{i=1}^{q}(T_i s + 1)} = \left[\begin{array}{l} K \text{ when } j = 0 \\ \infty \text{ when } j > 0 \end{array} \right. \qquad (4.7\text{-}2)$$

Definition 4.7.3

The *speed (or velocity, or ramp) error constant* K_{v} of a system is defined as $K_{\mathrm{v}} = \lim_{s \to 0} sG(s)$. Hence, the constant K_{v} takes on the values

$$K_{\mathrm{v}} = \lim_{s \to 0} sG(s) = \lim_{s \to 0} K \frac{\prod_{i=1}^{m}(T_i's + 1)}{s^{j-1} \prod_{i=1}^{q}(T_i s + 1)} = \left[\begin{array}{l} 0 \text{ when } j = 0 \\ K \text{ when } j = 1 \\ \infty \text{ when } j > 1 \end{array} \right. \qquad (4.7\text{-}3)$$

Definition 4.7.4

The *acceleration (or parabolic) error constant* K_{a} of a system is defined as $K_{\mathrm{a}} = \lim_{s \to 0} s^2 G(s)$. Hence, the constant K_{a} takes on the values

$$K_{\mathrm{a}} = \lim_{s \to 0} s^2 G(s) = \lim_{s \to 0} K \frac{\prod_{i=1}^{m}(T_i's + 1)}{s^{j-2} \prod_{i=1}^{q}(T_i s + 1)} = \left[\begin{array}{l} 0 \text{ when } j = 0, 1 \\ K \text{ when } j = 2 \\ \infty \text{ when } j > 2 \end{array} \right. \qquad (4.7\text{-}4)$$

4.7.2 Steady-State Errors with Inputs of Special Forms

Consider the closed-loop system of unity feedback of Figure 4.12. The system error $e(t)$ studied in this section is defined as the difference between the command signal $r(t)$ and the output of the system $y(t)$, i.e.,

$$e(t) = r(t) - y(t) \tag{4.7-5}$$

The steady-state error $e_{ss}(t)$ is given by

$$e_{ss}(t) = r_{ss}(t) - y_{ss}(t) \tag{4.7-6}$$

where

$$e_{ss}(t) = \lim_{t \to \infty} e(t), \qquad r_{ss}(t) = \lim_{t \to \infty} r(t), \qquad \text{and} \qquad y_{ss}(t) = \lim_{t \to \infty} y(t)$$

Clearly, the above definitions may be applied to nonunity feedback systems, as long as their equivalent block diagram of Figure 4.13b is used.

The steady-state error (4.7-6) indicates the deviation of $y_{ss}(t)$ from $r_{ss}(t)$. In practice, we wish $e_{ss}(t) = 0$, i.e., we wish $y_{ss}(t)$ to follow exactly the command signal $r_{ss}(t)$. In cases where $e_{ss}(t) \neq 0$, one seeks ways to reduce or even to zero the steady-state error.

In order to evaluate $e_{ss}(t)$, we work as follows. From Figure 4.12 we have

$$E(s) = \frac{R(s)}{1 + G(s)} \tag{4.7-7}$$

It is noted that for the general case of nonunity feedback systems, using Figure 4.13, relation (4.7-7) becomes

$$E(s) = \left[\frac{1}{1 + G(s)} \right] R(s) = \left[\frac{1 + \tilde{G}(s)\tilde{F}(s) - \tilde{G}(s)}{1 + \tilde{G}(s)\tilde{F}(s)} \right] R(s) \tag{4.7-8}$$

If we apply the final value theorem (see relation (2.3-17)) to Eq. (4.7-7) we have

$$e_{ss}(t) = \lim_{s \to 0} sE(s) = \lim_{s \to 0} \frac{sR(s)}{1 + G(s)} \tag{4.7-9}$$

given that the function $sE(s)$ has all its poles on the left-half complex plane.

We will examine the steady-state error $e_{ss}(t)$ for the following three special forms of the input r(t).

1. $r(t) = P$. In this case the input is a step function with amplitude P. Here, $e_{ss}(t)$ is called the *position error*. We have

$$e_{ss}(t) = \lim_{s \to 0} sE(s) = \lim_{s \to 0} \frac{s[P/s]}{1 + G(s)} = \frac{P}{1 + \lim_{s \to 0} G(s)} = \frac{P}{1 + K_p}$$

$$= \begin{cases} \dfrac{P}{1 + K} & \text{when } j = 0 \\ 0 & \text{when } j > 0 \end{cases} \tag{4.7-10}$$

where use was made of relation (4.7-2). From relation (4.7-10) we observe that for type 0 systems the position error is $P/(1 + K)$, while for type greater than 0 systems the position error is zero.

2. $r(t) = Vt$. In this case the input is a ramp function with slope equal to V. Here $e_{ss}(t)$ is called the *speed* or *velocity error*. We have

$$e_{ss}(t) = \lim_{s \to 0} sE(s) = \lim_{s \to 0} \frac{s[V/s^2]}{1 + G(s)} = \frac{V}{\lim_{s \to 0} sG(s)} = \frac{V}{K_v} = \begin{bmatrix} \infty & \text{when } j = 0 \\ \dfrac{A}{K_v} & \text{when } j = 1 \\ 0 & \text{when } j > 1 \end{bmatrix}$$

(4.7-11)

where use was made of relation (4.7-3). From relation (4.7-11) we observe that for type 0 systems the speed error is infinity, for type 1 systems it is V/K, and for type greater than 1 systems it is zero.

3. $r(t) = (1/2)At^2$. In this case the input is a parabolic function. Here, $e_{ss}(t)$ is called the *acceleration error*. We have

$$e_{ss}(t) = \lim_{s \to 0} sE(s) = \lim_{s \to 0} \frac{s[A/s^3]}{1 + G(s)} = \frac{A}{\lim_{s \to 0} s^2 G(s)} = \frac{A}{K_a} = \begin{bmatrix} \infty & \text{when } j = 0, 1 \\ \dfrac{A}{K} & \text{when } j = 2 \\ 0 & \text{when } j > 2 \end{bmatrix}$$

(4.7-12)

where use was made of relation (4.7-4). From relation (4.7-12) we observe that for type 0 and 1 systems the acceleration error is infinity, for type 2 systems it is A/K, and for type higher than 2 systems it is zero.

In Figure 4.14 we present the error constants and the value of $e_{ss}(t)$ for type 0, 1, and 2 systems.

Example 4.7.1

Consider the liquid-level control system of Subsec. 3.13.4 (Figure 3.53). For simplicity, assume that

$$G(s)F(s) = \left[\frac{K_a K_v R}{RCs + 1}\right] K_f = \frac{K}{RCs + 1}, \qquad K = K_a K_v K_f R$$

Determine $y(t)$ and the steady-state error $e_{ss}(t)$ when the input $r(t)$ is the unit step function.

Solution

We have

$$Y(s) = H(s)R(s) = \left[\frac{K/K_f}{RCs + 1 + K}\right]\left[\frac{1}{s}\right]$$

Thus, $y(t) = L^{-1}\{Y(s)\}$ will be

$$y(t) = \frac{K}{K_f(1 + K)}(1 - e^{-t/T}), \qquad \text{where} \qquad T = \frac{RC}{1 + K}$$

The waveform of $y(t)$ is given in Figure 4.15, from which it is obvious that the liquid-level control system makes an attempt to "follow" the command signal $r(t) = 1$. Unfortuantely, at the steady state it presents an error.

To determine the steady-state error e_{ss} we work as follows. First, we convert the nonunity feedback system to unity feedback, according to Figure 4.13a. Since

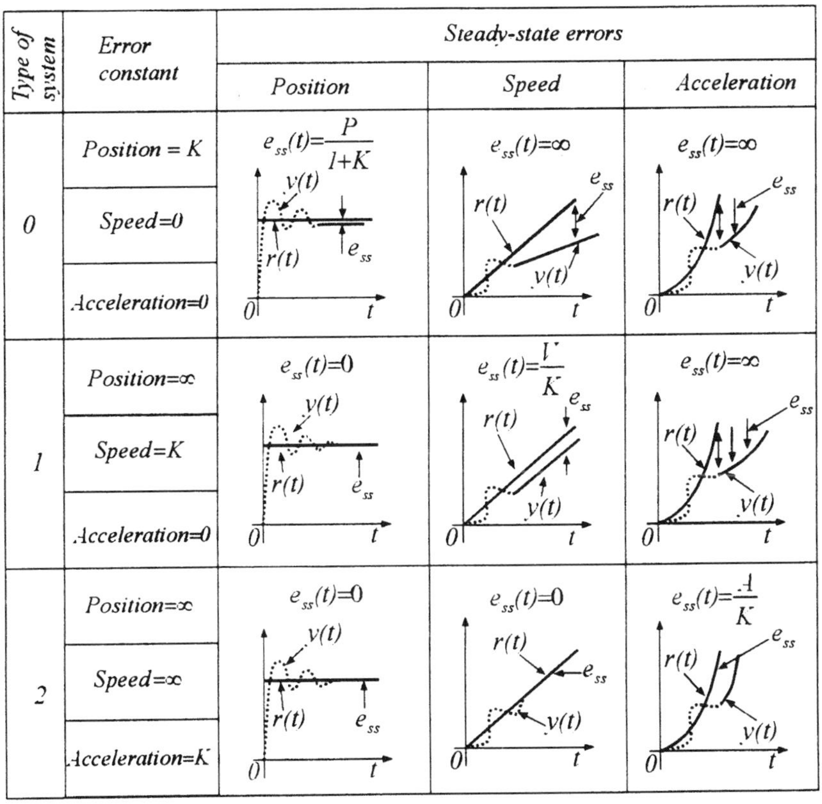

Type of system	Error constant	Steady-state errors		
		Position	Speed	Acceleration
0	Position = K Speed=0 Acceleration=0	$e_{ss}(t)=\dfrac{P}{1+K}$	$e_{ss}(t)=\infty$	$e_{ss}(t)=\infty$
1	Position=∞ Speed=K Acceleration=0	$e_{ss}(t)=0$	$e_{ss}(t)=\dfrac{V}{K}$	$e_{ss}(t)=\infty$
2	Position=∞ Speed=∞ Acceleration=K	$e_{ss}(t)=0$	$e_{ss}(t)=0$	$e_{ss}(t)=\dfrac{A}{K}$

Figure 4.14   Position, speed, and acceleration errors.

$$\tilde{G}(s) = \frac{K_a K_v R}{RCs + 1} \qquad \text{and} \qquad \tilde{F}(s) = K_f$$

it follows that

$$G(s) = \frac{\tilde{G}(s)}{1 + \tilde{G}(s)\tilde{F}(s) - \tilde{G}(s)} = \frac{K/K_f}{RCs + 1 + K - K/K_f}$$

We have

$$\lim_{s \to 0} G(s) = \frac{K/K_f}{1 + K - K/K_f}$$

Now, we are in position to apply relation (4.7-10) to yield

$$e_{ss} = \lim_{s \to 0} sE(s) = \frac{1}{1 + \lim_{s \to 0} G(s)} = \frac{K_f + KK_f - K}{K_f + KK_f}$$

To check the above results, we first determine

$$y_{ss}(t) = \lim_{s \to 0} sY(s) = \frac{K}{K_f(1 + K)}$$

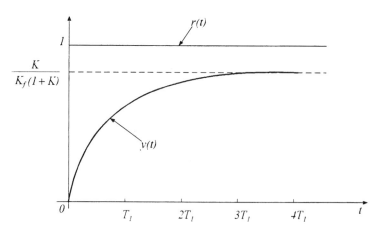

Figure 4.15   Time response of the closed-loop liquid-level control system when $r(t) = 1$.

Therefore, the level $y(t)$ of the liquid will never reach the desired level $y(t) = r(t) = 1$, but it will remain at a level lower than 1. The distance (i.e., the error) of this lower level to the desired level of 1 may be determined as follows:

$$e_{ss} = r_{ss}(t) - y_{ss}(t) = 1 - \frac{K}{K_f(1+K)} = \frac{K_f + K_f K - K}{K_f(1+K)}$$

This error is in complete agreement with the steady-state error e_{ss} found above. It is remarked that this steady-state error may be eliminated if a more complex feedback transfer function $F(s)$ is used, which would include not only the output analog feedback term $K_f y(t)$ but also other terms – for example of the form $K_d y^{(1)}(t)$, i.e., terms involving the derivative of the output. Such feedback controllers, and even more complex ones, are presented in Chap. 9.

Example 4.7.2

Consider the voltage control system of Subsec. 3.13.1 (Figure 3.50). For simplicity, let $T_f = L_f / R_f = 2$ and $K_p = 1$; in which case, $G(s)$ simplifies as follows:

$$G(s) = \frac{K}{2s+1}$$

Investigate the system's steady-state errors.

Solution

If we apply the results of Subsecs 4.7.1 and 4.7.2 we readily have that, since the system is type 0, the error constants will be $K_p = K$, $K_v = 0$, and $K_a = 0$. The steady-state error will be

$$e_{ss}(t) = \begin{bmatrix} \dfrac{P}{1+K} & r(t) = P \\ \infty & r(t) = Vt \\ \infty & r(t) = \frac{1}{2}At^2 \end{bmatrix}$$

Example 4.7.3

Consider the position servomechanism of Subsec. 3.13.2 (Figure 3.51). Assume that $L_a \simeq 0$. Then the open-loop transfer function reduces to

$$G(s) = \frac{K_a K_i N}{s[R_a J_m^* s + R_a B_m^* + K_i K_b]} = \frac{K}{s(As + B)} = \frac{K}{s(s+2)}$$

where we further have chosen $K_p = 1$, $A = 1$, $B = 2$, and where

$$K = \frac{K_a K_i N}{R_a}, \qquad A = J_m^*, \qquad \text{and} \qquad B = B_m^* + K_i K_b / R_a$$

This simplified system is shown in Figure 4.16. Here, $e(t) = \theta_e(t) = \theta_r(t) - \theta_y(t)$. Investigate the steady-state errors of the system.

Solution

If we apply the relations (4.7-2), (4.7-3), and (4.7-4), we readily have that, since the system is of type 1, the error constants will be $K_p = \infty$, $K_v = K/2$, and $K_a = 0$. The steady-state error will be

$$e_{ss}(t) = \begin{cases} 0 & \text{when } \theta_r(t) = P & (4.7\text{-}13\text{a}) \\ \dfrac{2V}{K} & \text{when } \theta_r(t) = Vt & (4.7\text{-}13\text{b}) \\ \infty & \text{when } \theta_r(t) = \frac{1}{2} A t^2 & (4.7\text{-}13\text{c}) \end{cases}$$

4.8 DISTURBANCE REJECTION

In Subsec. 4.5.3, the effect of disturbances upon the output of open- and closed-loop systems is compared. In this section, an approach will be given which aims to reduce or even completely eliminate the effect of disturbances upon the system's output.

Consider the closed-loop system of Figure 4.17, involving the disturbance signal $d(t)$ (or $D(s)$). This disturbance is usually external to systems under control. For example, a sudden change in the wind is an external disturbance for a microwave antenna mounted on the Earth. It may also be internal, e.g., an unexpected variation in the value of the capacitance C of a capacitor which is part of the system under control.

Disturbances appear very often in practice, and they affect the system's output, resulting in a deviation from its normal operating performance. The elimination of

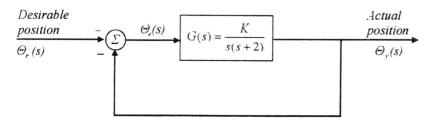

Figure 4.16 Simplified block diagram of the position control system.

the influence of $d(t)$ on $y(t)$ is the well-known problem of *disturbance rejection* and has, for obvious reasons, great practical importance.

Using the general layout of Figure 4.17, we distinguish two cases: the unity feedback and the nonunity feedback systems.

1 Unity Feedback Systems

In this case $F(s) = 1$ and

$$Y(s) = G_c(s)G(s)E(s) + G(s)D(s)$$

Furthermore,

$$Y(s) = R(s) - E(s) \tag{4.8-1}$$

Eliminating $Y(s)$ in the above two equations, we have

$$E(s) = E_r(s) + E_d(s) = \left[\frac{1}{1 + G_c(s)G(s)}\right]R(s) - \left[\frac{G(s)}{1 + G_c(s)G(s)}\right]D(s) \tag{4.8-2}$$

Hence, the steady-state error $e_{ss}(t) = \lim_{s \to 0} sE(s)$ is given by

$$e_{ss}(t) = \lim_{s \to 0}\left[\frac{s}{1 + G_c(s)G(s)}\right]R(s) - \lim_{s \to 0}\left[\frac{sG(s)}{1 + G_c(s)G(s)}\right]D(s) \tag{4.8-3}$$

2 Nonunity Feedback Systems

In this case $F(s) \neq 1$ and one may readily show that

$$E(s) = E_r(s) + E_d(s) = \left[1 - \frac{G_c(s)G(s)}{1 + G_c(s)G(s)F(s)}\right]R(s) - \left[\frac{G(s)}{1 + G_c(s)G(s)F(s)}\right]D(s) \tag{4.8-4}$$

Hence, the steady-state error $e_{ss}(t) = \lim_{s \to 0} sE(s)$ is given by

$$e_{ss}(t) = \lim_{s \to 0}\left[s - \frac{sG_c(s)G(s)}{1 + G_c(s)G(s)F(s)}\right]R(s) - \lim_{s \to 0}\left[\frac{sG(s)}{1 + G_c(s)G(s)F(s)}\right]D(s) \tag{4.8-5}$$

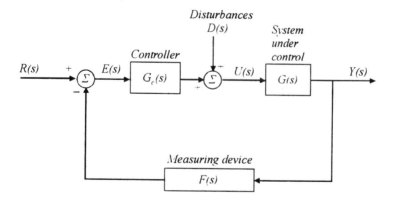

Figure 4.17 Closed-loop system with disturbances.

In the examples that follow, we show how to choose the controller $G_c(s)$ such as as to eliminate the influence of $d(t)$ on $y(t)$ in the steady state.

Example 4.8.1

In this example we study the operation of an industrial company from the point of view of control system theory. In Figure 4.18, a simplified description of an industrial company is given, where it is shown that a company is run (or should be run) as a closed-loop control system, i.e., using the principle of feedback action. The process of producing a particular industrial product requires a certain amount of time. The time constant of the process is $1/a$ and, for simplicity, let $a = 1$. The company's board of directors (which here acts as the "controller" of the company) study in depth the undesirable error $E(s) = R(s) - Y(s)$, where $R(s)$ is the desired productivity and, subsequently, take certain appropriate actions or decisions. These actions may be approximately described by an integral controller $G_c(s) = K_1/s$, which integrates (smooths out) the error $E(s)$. The parameters K_1 and K_2 in the block diagram represent the effort put in by the management and by the production line, respectively. Investigate the steady-state errors for $D(s) = 0$ and $D(s) = 1/s$. The disturbance $D(s) = 1/s$ may be, for example, a sudden increase or decrease in the demand for the product.

Solution

(a) For the case $D(s) = 0$, we have

$$G_c(s)G(s) = \frac{K_1 K_2}{s(s+1)} = \frac{K}{s(s+1)}, \qquad \text{where } K = K_1 K_2$$

$$E(s) = E_r(s) = \left[\frac{1}{1 + G_c(s)G(s)}\right] R(s) = \left[\frac{s(s+1)}{s^2 + s + K}\right] R(s)$$

By applying the results of Subsecs 4.7.1 and 4.7.2, we derive that, since it is a system of type 1, the error constants will be $K_p = \infty$, $K_v = 2K$, and $K_a = 0$. The steady state error is

$$e_{ss}(t) = \begin{bmatrix} 0 & \text{when } r(t) = P \\ V/K & \text{when } r(t) = Vt \\ \infty & \text{when } r(t) = \tfrac{1}{2} At^2 \end{bmatrix}$$

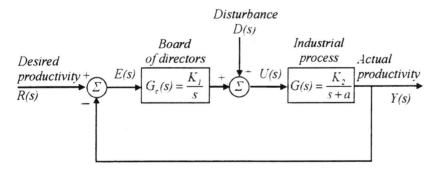

Figure 4.18 Simplified closed-loop block diagram of a company.

(b) For the case $D(s) = 1/s$ the error is given by the relation (4.8-2), i.e., by the relation

$$E(s) = \frac{R(s)}{1 + G_c(s)G(s)} - \frac{G(s)}{1 + G_c(s)G(s)} D(s) = E_r(s) + E_d(s)$$

With regard to $E_r(s)$, the results of case (a) remain the same. With regard to $E_d(s)$, we have

$$E_d(s) = -\left[\frac{K_2 s(s+1)}{s^2 + s + K}\right]\left[\frac{1}{s}\right]$$

Hence, the steady-state error due to the disturbance $D(s)$ will be

$$\lim_{t\to\infty} e_d(t) = \lim_{s\to 0} s E_d(s) = 0$$

Therefore, the effect of the disturbance on the steady-state error is zero. It is noted that for the closed-loop system to be stable there must be $K > 0$.

Example 4.8.2

Consider the block diagram of Figure 4.19, where $d(t) = A\delta(t)$ and $r(t) = 1$ for $t \geq 0$ and $r(t) = 0$ for $t < 0$. Find the range of values of K_1 and K_2 such that the effect of the disturbance $d(t)$ on the system's output is eliminated (rejected) in the steady state, while simultaneously the system's output follows the input signal, i.e., $\lim_{t\to\infty}[(y(t) - r(t)] = 0)$.

Solution

To determine the system's output we will apply the superposition principle. To this end, assume that the disturbance $d(t)$ is the only input, in which case we get

$$Y_d(s) = \left[\frac{K_2(s^2 + s + 1)}{s(s^2 + s + 1) + K_1 K_2(s + 1)}\right] D(s), \qquad \text{where } D(s) = L\{d(t)\} = A$$

Similarly, assuming that $r(t)$ is the only input, we get

$$Y_r(s) = \left[\frac{K_2(s + 1)}{s(s^2 + s + 1) + K_1 K_2(s + 1)}\right] R(s), \qquad \text{where } R(s) = L\{u(t)\} = \frac{1}{s}$$

Using the superposition principle, we have

$$Y(s) = \left[\frac{K_2(s^2 + s + 1)}{s(s^2 + s + 1) + K_1 K_2(s + 1)}\right] A + \left[\frac{K_2(s + 1)}{s(s^2 + s + 1) + K_1 K_{2}(s + 1)}\right] \frac{1}{s}$$

For the disturbance $d(t)$ to be eliminated in the steady-state, the following condition must hold:

$$\lim_{t\to\infty} y_d(t) = 0$$

To investigate the above condition, we use the final value theorem which, as is well known, holds if $sY_d(s)$ is stable. For $sY_d(s)$ to be stable, its denominator must not have any roots in the right-half complex plane. Using the Routh criterion (see Chap. 6) for the characteristic polynomial of the closed-loop system:

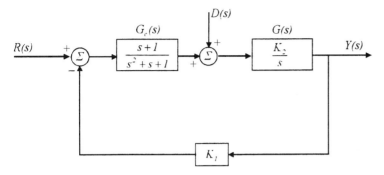

Figure 4.19 System with input $r(t)$ and disturbance $d(t)$.

$$p(s) = s^3 + s^2 + (K_1 K_2 + 1)s + K_1 K_2$$

we form the Routh table

$$
\begin{array}{c|cc}
s^3 & 1 & K_1 K_2 + 1 \\
s^2 & 1 & K_1 K_2 \\
s^1 & 1 & 0 \\
s^0 & K_1 K_2 & 0
\end{array}
$$

Hence, in order for $p(s)$ to be stable, the condition $K_1 K_2 > 0$ must hold. Given that this condition holds and using the final value theorem, we obtain

$$\lim_{t \to \infty} y_d(t) = \lim_{s \to 0} s Y_d(s) = \lim_{s \to 0} \left[\frac{s K_2(s^2 + s + 1)}{s^3 + s^2 + (K_1 K_2 + 1)s + K_1 K_2} \right] A = 0$$

Therefore, the effect of the disturbance in the steady state is eliminated when $K_1 K_2 > 0$.

We will now examine the second condition $\lim_{t \to \infty} [y(t) - r(t)] = 0$. Given that $K_1 K_2 > 0$, we have that $s Y(s)$ is stable. Hence, using the final value theorem and given that $y_d(t) = 0$, in the steady state, we have that

$$\lim_{t \to \infty} [y(t) - r(t)] = \lim_{s \to 0} [s Y(s) - s R(s)] = \lim_{s \to 0} [s Y_r(s) - 1]$$

$$= \lim_{s \to 0} \left[\frac{K_2(s + 1)}{s^3 + s^2 + (K_1 K_2 + 1)s + K_1 K_2} - 1 \right] = 0$$

or

$$\frac{K_2}{K_1 K_2} - 1 = 0$$

The above relation yields $K_1 = 1$ which, in conjunction with the condition $K_1 K_2 > 0$, gives the range of values of K_1 and K_2, which are

$$K_1 = 1 \quad \text{and} \quad K_2 > 0$$

Another approach to solve the problem is to use Eq. (4.8-4), in which case

$$E(s) = E_r(s) + E_d(s)$$

where

$$E_r(s) = \left[1 - \frac{G_c(s)G(s)}{1 + G_c(s)G(s)F(s)}\right]R(s) = \left[1 - \frac{K_2(s+1)}{s(s^2 + s + 1) + K_1K_2(s+1)}\right]R(s)$$

$$E_d(s) = -\left[\frac{G(s)}{1 + G_c(s)G(s)F(s)}\right]D(s) = -\left[\frac{K_2(s+1)}{s(s^2 + s + 1) + K_1K_2(s+1)}\right]D(s)$$

Here, $R(s) = 1/s$ and $D(s) = A$ and, hence,

$$e_{ss} = \lim_{s\to 0} sE_r(s) + \lim_{s\to 0} sE_d(s)$$

Simple calculations yield

$$\lim_{s\to 0} sE_r(s) = \lim_{s\to 0}\left[s - \frac{sK_2(s+1)}{s(s^2 + s + 1) + K_1K_2(s+1)}\right]\frac{1}{s} = 1 - \frac{K_2}{K_1K_2} = \frac{K_1 - 1}{K_1}$$

$$\lim_{s\to 0} sE_d(s) = \lim_{s\to 0}\left[\frac{-sK_2(s+1)}{s(s^2 + s + 1) + K_1K_2(s+1)}\right]A = 0$$

Hence, the problem requirements are satisfied when $K_1 = 1$. However, for the above limits to exist, according to the final value theorem, the characteristic polynomial of the closed-loop system must be stable. This, as shown previously, leads to the condition $K_1K_2 > 0$. Finally, we arrive at the conclusion that K_1 and K_2 must satisfy the conditions $K_1 = 1$ and $K_2 > 0$, which are in perfect agreement with the results of the previous approach.

Example 4.8.3

Consider a telephone network of signal transmission in which noise (disturbance) is introduced as shown in Figure 4.20. In the feedback path introduce a filter $F(s)$ ($F(s)$ is a rational function of s) such that for the closed-loop system in the steady state, the following conditions hold simultaneously:

(a) The effect of the noises $d_1(t)$ and $d_2(t)$ on the output is rejected.
(b) The receiver's signal $y(t)$ is the same as that of the transmitter's signal $r(t)$.

Solution

First, we determine the system's output due to the signals $r(t)$, $d_1(t)$, and $d_2(t)$, one at a time, using the superposition principle. When $r(t)$ is the only input, we have

$$Y_r(s) = \left[\frac{G_1(s)G_2(s)}{1 - G_1(s)G_2(s)F(s)}\right]R(s) = \left[\frac{1}{(s+2)(s^2 + s + 1) - F(s)}\right]R(s)$$

When the disturbance $d_1(t)$ is the only input, we have

$$Y_{d_1}(s) = \left[\frac{G_2(s)}{1 - G_1(s)G_2(s)F(s)}\right]D_1(s) = \frac{s+2}{(s+2)(s^2 + s + 1) - F(s)}$$

When the disturbance $d_2(t)$ is the only input, we have

$$Y_{d_2}(s) = \left[\frac{G_2(s)G_1(s)}{1 - G_1(s)G_2(s)F(s)}\right]D_2(s) = \frac{1}{(s+2)(s^2 + s + 1) - F(s)}$$

According to the problem's requirements, the following must be simultaneously valid

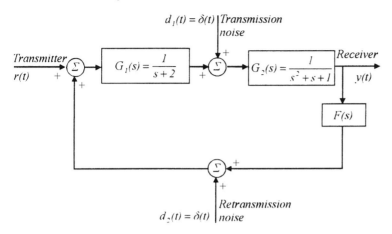

Figure 4.20   Telephone network with transmission noise.

$$\lim_{t\to\infty} y_{d_1}(t) = \lim_{t\to\infty} y_{d_2}(t) = 0$$

$$\lim_{t\to\infty} [y(t) - r(t)] = 0, \qquad \text{for every } r(t)$$

From requirement (a) of the problem, and by choosing $F(s)$ such that the denominator $(s+2)(s^2+s+1) - F(s)$ is stable, we must have

$$\lim_{s\to0} s Y_{d_1}(s) = \lim_{s\to0} s Y_{d_2}(s) = 0$$

From requirement (b) of the problem, we have

$$\lim_{s\to0} s[Y(s) - R(s)] = \lim_{s\to0} s[Y_r(s) - R(s)] = 0, \qquad \text{for every } R(s)$$

under the assumption that the effect of the disturbances has been eliminated. The above relation may be written as follows

$$\lim_{s\to0} s\left[\frac{1}{(s+2)(s^2+s+1) - F(s)} - 1 \right] R(s) = 0.$$

This relation must hold for every $R(s)$. Consequently, the following must hold:

$$\frac{1}{(s+2)(s^2+s+1) - F(s)} = 1$$

The above relation yields $F(s) = (s+1)^3$. By replacing $F(s) = (s+1)^3$ in $Y_{d_1}(s)$ and $Y_{d_2}(s)$ we obtain

$$Y_{d_1}(s) = s+2 \qquad \text{and} \qquad Y_{d_2}(s) = 1$$

Using these values, one may readily prove that the following condition is satisfied:

$$\lim_{s\to0} s Y_{d_1}(s) = \lim_{s\to0} s Y_{d_2}(s) = 0$$

Hence, the function $F(s) = (s+1)^3$ is the transfer function sought.

To realize the function $F(s) = (s+1)^3$, one may use Figure 9.40. From this figure it follows that the gain of the operational amplifier is given by

$$G(s) = -\frac{Z_2(s)}{Z_1(s)} = -(s+1)$$

where use was made of the PD controller case and where the values of R_1, R_2 and C_1 are chosen as follows: $R_1 = R_2 = 1/C_1$. If we have in cascade three such operational amplifiers together with an inverter, then the total transfer function $G_t(s)$ will be

$$G_t(s) = G(s)G(s)G(s)(-1) = (s+1)^3$$

in which case $F(s) = G_t(s) = (s+1)^3$.

PROBLEMS

1. Find the currents $i_1(t)$ and $i_2(t)$ for the network shown in Figure 4.21. Assume that the switch is closed at $t = 0$ and the initial condition for the inductor current is $i_2(0) = 0$.
2. Find and plot the response $y(t)$ of the network shown in Figure 4.22 for $R = 1\,\Omega$, $2\,\Omega$, $4\,\Omega$, and $10\,\Omega$. The input is $u(t) = 1$.
3. Find and plot the response $y(t)$ of the mechanical system shown in Figure 4.23 for $B = 1$, 2, and 4. The input is $u(t) = 1$.
4. The block diagram of the liquid-level control system of Figure 4.24a is shown in Figure 4.24b. The liquid flows into the tank through a valve that controls the input flow Q_1. When the output flow Q_2 through the orifice increases, the liquid-level height H decreases. As a result, the sensor that measures H causes the valve to open in order to increase the input flow Q_1. When the output flow Q_2 decreases, H increases, and the valve closes in order to decrease the input flow Q_1. Determine the parameters K and T, given that for a step input the maximum overshoot is 25.4% and occurs when $t_1 = 3\,\text{sec}$.
5. The block diagram of a system that controls the movement of a robot arm is given in Figure 4.25. For a unit step input, the system has a maximum percent overshoot of 20%, which occurs when $t_1 = 1\,\text{sec}$. Determine:

   (a)   The constants K and K_h
   (b)   The rise time t_r required for the output to reach value 1 for the first time
   (c)   The settling time t_s required for the output to reach and stay within 2% and 5% of its final value

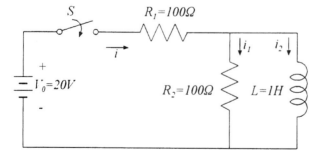

Figure 4.21

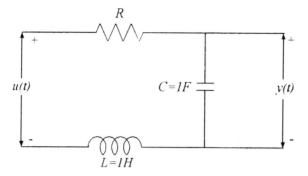

Figure 4.22

6. Consider the closed-loop system shown in Figure 4.12. Find the position, speed, and acceleration error constants when

(a) $G(s) = \dfrac{10}{(s+1)(2s+1)}$ (c) $G(s) = \dfrac{K}{s^2(0.5s+1)(s+1)}$

(b) $G(s) = \dfrac{K}{s(s+1)(2s+1)}$ (d) $G(s) = \dfrac{K(s+4)}{s^2(s^2+6s+2)}$

7. For the closed-loop systems of Problem 6 find the steady-state position, speed, and acceleration errors.

8. The block diagram of an active suspension system for an automobile is shown in Figure 4.26. In this system, the position of the valves of the shock absorber is controller by means of a small electric motor:

(a) Find the position, speed, and acceleration error constants.
(b) Determine the steady-state position, speed, and acceleration errors.

9. The block diagram of a position servomechanism is shown in Figure 4.27. Determine the steady-state error when the input is $r(t) = a_0 + a_1 t + a_2 t^2$.

10. For the Example 4.7.2 find the error $e_{ss}(t)$ when $r(t) = 1 + 2t - t^2$ and when $r(t) = 10t^3$.

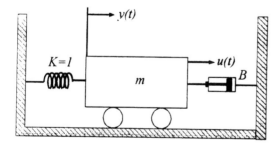

Figure 4.23

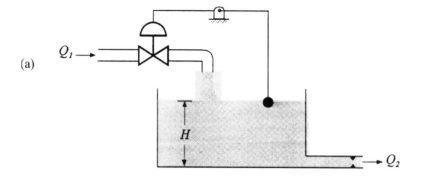

(a)

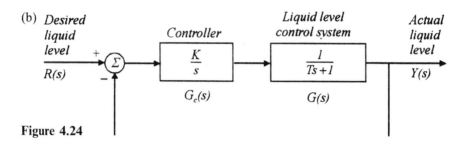

Figure 4.24

11. Find the position, speed, and acceleration error constants of all systems described in Sec. 3.13.
12. Find the steady-state position, speed, and acceleration errors of all systems described in Sec. 3.13.
13. A control system for a human heart with problems related to heart rate is shown in Figure 4.28. The controller used is an electronic pacemaker to keep the heart rate within a desired range. Determine a suitable transfer function for the pacemaker, so that the steady-state error due to a disturbance $d(t) = 1, t > 0$, is zero.

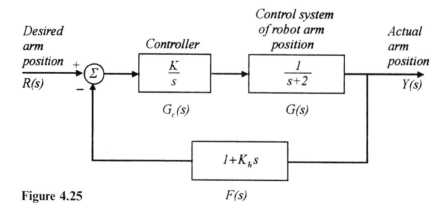

Figure 4.25 $F(s)$

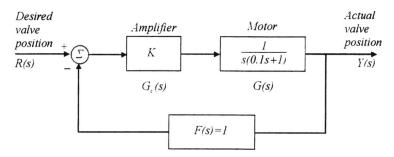

Figure 4.26

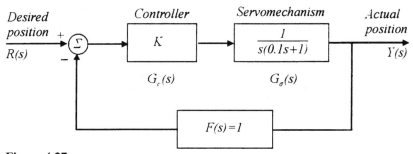

Figure 4.27

14. The block diagram of a position control system of a large microwave antenna is shown in Figure 4.29. To design such a system, we must take into account the disturbance due to large wind gust torques. Determine the range of values of K_1 and K_2, so that the effect of the disturbance $d(t) = \delta(t)$ is minimized in the steady state, while the output follows the input signal $r(t) = 1$.

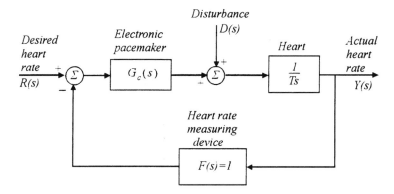

Figure 4.28

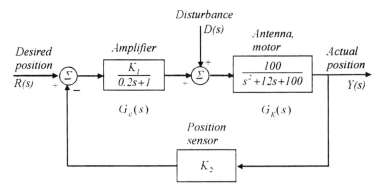

Figure 4.29

15. Consider a remotely controlled vehicle used for reconnaissance missions. The desired speed of the vehicle is transmitted to a receiver mounted on the vehicle. The block diagram of the control system is shown in Figure 4.30, where the disturbance input expresses the transmission noise. Find a transfer function for the feedback controller $F(s)$ so that the effect of noise $d(t) = \delta(t)$ at the output is eliminated as $t \to \infty$, while $\lim_{t \to \infty} [y(t) - r(t)] = 0$ for every input $r(t)$.

16. The block diagram of a system that controls the roll angle of a ship is shown in Figure 4.31. Determine the values of the gain K_p so that the disturbance due to wind is eliminated while the output follows a step input at steady state, when

   (a) $K_i = 0$ (proportional controller)
   (b) $K_i = 1$ (proportional plus integral controller)

17. For the system shown in Figure 4.32, determine K_1 and K_2, so that the effect of noise $d(t) = \delta(t)$ is minimized at steady state.

18. Consider the system shown in Figure 4.33. Determine the transfer function of the controller $G_c(s)$, so that when the disturbance is $d(t) = t$, the steady-state error is zero.

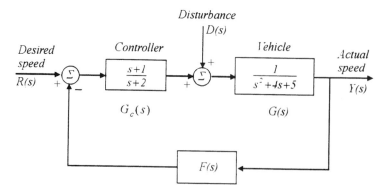

Figure 4.30

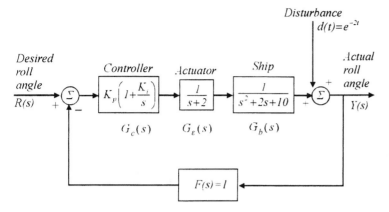

Figure 4.31

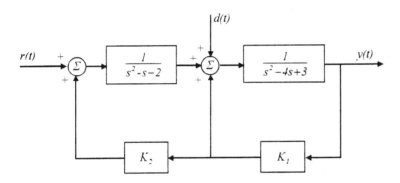

Figure 4.32

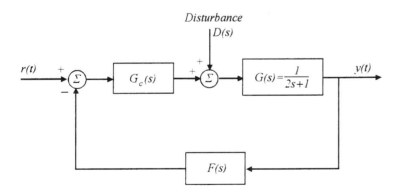

Figure 4.33

BIBLIOGRAPHY

Books

1. DK Anand. Introduction to Control Systems. New York: Pergamon Press, 1974.
2. PJ Antsaklis, AN Michel. Linear Systems. New York: McGraw-Hill, 1997.
3. DM Auslander, Y Takahasi, MJ Rabins. Introducing Systems and Control. New York: McGraw-Hill, 1974.
4. RN Clark. Introduction to Automatic Control Systems. New York: John Wiley, 1962.
5. JB Cruz Jr (ed). System Sensitivity Analysis. Strousdburg, Pennsylvania: Dowden, 1973.
6. JJ D'Azzo, CH Houpis. Linear Control System Analysis and Design, Conventional and Modern. New York: McGraw-Hill, 1975.
7. RA DeCarlo. Linear Systems. Englewood Cliffs, New Jersey: Prentice Hall, 1989.
8. BE DeRoy. Automatic Control Theory. New York: John Wiley, 1966.
9. JJ DiStefano III, AR Stubberud, IJ Williams. Feedback and Control Systems. Schaum's Outline Series. New York: McGraw-Hill, 1967.
10. EO Doebelin. Dynamic Analysis and Feedback Control. New York: McGraw-Hill, 1962.
11. RC Dorf, RE Bishop. Modern Control Analysis. London: Addison-Wesley, 1995.
12. JC Doyle. Feedback Control Theory. New York: Macmillan, 1992.
13. VW Eveleigh. Introduction to Control Systems Design. New York: McGraw-Hill, 1972.
14. TE Fortman, KL Hitz. An Introduction to Linear Control Systems. New York: Marcel Dekker, 1977.
15. GF Franklin, JD Powell, ML Workman. Digital Control of Dynamic Systems. 2nd ed. London: Addison-Wesley, 1990.
16. RA Gabel, RA Roberts. Signals and Linear Systems. 3rd ed. New York: John Wiley, 1987.
17. SC Gupta, L Hasdorff. Automatic Control. New York: John Wiley, 1970.
18. IM Horowitz. Synthesis of Feedback Systems. New York: Academic Press, 1963.
19. BC Kuo. Automatic Control Systems. London: Prentice Hall, 1995.
20. IJ Nagrath, M Gopal. Control Systems Engineering. New Delhi: Wiley Eastern Limited, 1977.
21. GC Newton Jr, LA Gould, JF Kaiser. Analytical Design of Linear Feedback Controls. New York: John Wiley, 1957.
22. NS Nise. Control Systems Engineering. New York: Benjamin and Cummings, 1995.
23. M Noton. Modern Control Engineering. New York: Pergamon Press, 1972.
24. K Ogata. Modern Control Systems. London: Prentice Hall, 1997.
25. CL Phillips, RD Harbor. Feedback Control Systems. 2nd ed. Englewood Cliffs, New Jersey: Prentice Hall, 1991.
26. FH Raven. Automatic Control Engineering. 4th ed. New York: McGraw-Hill, 1987.
27. CE Rohrs, JL Melsa, D Schultz. Linear Control Systems. New York: McGraw-Hill, 1993.
28. IE Rubio. The Theory of Linear Systems. New York: Academic Press, 1971.
29. WJ Rugh. Linear Systems Theory. 2nd ed. Englewood Cliffs, New Jersey: Prentice Hall, 1996.
30. NK Sinha. Control Systems. New York: Holt, Rinehart & Winston, 1986.
31. SM Shinners. Modern Control System Theory and Design. New York: Wiley, 1992.
32. R Tomovic. Sensitivity Analysis of Dynamical Systems. New York: McGraw-Hill, 1963.
33. JG Truxal. Introductory System Engineering. New York: McGraw-Hill, 1972.
34. J Van de Vegte. Feedback Control Systems. Englewood Cliffs, New Jersey: Prentice Hall, 1990.
35. LA Zadeh, CA Desoer. Linear System Theory – the State State Space Approach. New York: McGraw-Hill, 1963.

Articles

36. EJ Davison. A method for simplifying linear dynamic systems. IEE Trans. Automatic Control AC-11:93–101, 1966.
37. TC Hsia. On the simplification of linear systems. IEEE Trans Automatic Control AC-17:372–374, 1972.
38. PN Paraskevopoulos. Techniques in model reduction for large scale systems. Control and Dynamic Systems: Advances in Theory and Applications (CT Leondes, ed.). New York: Academic Press, 20:165–193, 1986.
39. PN Paraskevopoulos, CA Tsonis, SG Tzafestas. Eigenvalue sensitviity of linear time-invariant control systems with repeated eigenvalues. IEEE Trans Automatic Control 19:911–928, 1975.

5

State-Space Analysis of Control Systems

5.1 INTRODUCTION

The classical methods of studying control systems are mostly referred to single-input–single-output (SISO) systems, which are described in the time domain by differential equations or by a scalar transfer function in the frequency domain (see Chap. 4). The modern methods of studying control systems are referred to the general category of multi-input–multi-ouput (MIMO) systems, which are described in the time domain by state equations (i.e., by a set of linear first-order differential equations) or by a transfer function matrix in the frequency domain.

The modern approach of describing a system via the state equations model, compared with the classical models of Chap. 4, has the distinct characteristic of introducing a new concept to the system description—namely, the system's *state variables*. The state variables give information about the *internal structure* of the system, which the classical methods do not. This information is of great significance to the study of the structure and properties of the system, as well as to the solution of high-performance control design problems, such as optimal control, adaptive control, robust control, and pole assignment.

This chapter is devoted to linear, time-invariant systems having the following state-space form:

$$\dot{\mathbf{x}}(t) = \mathbf{A}\mathbf{x}(t) + \mathbf{B}\mathbf{u}(t) \tag{5.1-1a}$$

$$\mathbf{y}(t) = \mathbf{C}\mathbf{x}(t) + \mathbf{D}\mathbf{u}(t) \tag{5.1-1b}$$

$$\mathbf{x}(0) = \mathbf{x}_0 \tag{5.1-1c}$$

where $\mathbf{x}(t)$ is an n-dimensional state vector, $\mathbf{u}(t)$ is an m-dimensional input vector, and $\mathbf{y}(t)$ is a p-dimensional ouput vector. The matrices $\mathbf{A}$, $\mathbf{B}$, $\mathbf{C}$, and $\mathbf{D}$ are time-invariant, and their dimensions are $n \times n$, $n \times m$, $p \times n$, and $p \times m$, respectively. The initial conditions are at $t = 0$ and they are given by Eq. (5.1-1c).

The first objective of this chapter is the solution of Eqs (5.1-1). This will be done in two steps. In Sec. 5.2 the solution of the homogeneous equation $\dot{\mathbf{x}}(t) = \mathbf{A}\mathbf{x}(t)$

will be determined. Subsequently, in Sec. 5.3, the general solution of Eqs (5.1-1) will be derived. The next objective of this chapter is the state vector transformations and special forms of the state equations, presented in Sec. 5.4. Block diagrams and signal-flow graphs are given in Sec. 5.5. The important topics of controllability and observability are presented in Sec. 5.6. Finally, the Kalman decomposition theorem is given in Sec. 5.7.

5.2 SOLUTION OF THE HOMOGENEOUS EQUATION

5.2.1 Determination of the State Transition Matrix

Consider the dynamic part (5.2-1a) of the state equations (5.2-1), i.e., consider the first-order vector differential equation

$$\dot{\mathbf{x}}(t) = \mathbf{A}\mathbf{x}(t) + \mathbf{B}\mathbf{u}(t), \qquad \mathbf{x}(0) = \mathbf{x}_0 \tag{5.2-1}$$

The homogeneous part of Eq. (5.2-1) is

$$\dot{\mathbf{x}}(t) = \mathbf{A}\mathbf{x}(t), \qquad \mathbf{x}(0) = \mathbf{x}_0 \tag{5.2-2}$$

We introduce the following definition.

Definition 5.2.1

The *state transition matrix* of Eq. (5.2-2) is an $n \times n$ matrix, designated by $\boldsymbol{\phi}(t)$, which satisfies the homogeneous equation (5.2-2), i.e.,

$$\dot{\boldsymbol{\phi}}(t) = \mathbf{A}\boldsymbol{\phi}(t) \tag{5.2-3}$$

A rather simple method to solve the homogeneous equation (5.2-2) and simultaneously determine the state transition matrix $\boldsymbol{\phi}(t)$ is to assume the solution of (5.2-2) in a form of Taylor series, i.e., to assume that the state vector $\mathbf{x}(t)$ has the form

$$\mathbf{x}(t) = \mathbf{e}_0 + \mathbf{e}_1 t + \mathbf{e}_2 t^2 + \mathbf{e}_3 t^3 + \cdots \tag{5.2-4}$$

where $\mathbf{e}_0, \mathbf{e}_1, \mathbf{e}_2, \mathbf{e}_3, \ldots$ are n-dimensional constant unknown vectors. To determine these unknown vectors, we successively differentiate Eq. (5.2-4) and then evaluate the derivatives at $t = 0$. Thus, the zero derivative of $\mathbf{x}(t)$ at $t = 0$ is $\mathbf{x}(0) = \mathbf{e}_0$. The first derivative of $\mathbf{x}(t)$ at $t = 0$ is $\mathbf{x}^{(1)}(0) = \mathbf{e}_1$. However, from Eq. (5.2-2) we have that $\mathbf{x}^{(1)}(0) = \mathbf{A}\mathbf{x}(0)$. Hence, $\mathbf{e}_1 = \mathbf{A}\mathbf{x}(0)$. The second derivative of $\mathbf{x}(t)$ at $t = 0$ is $\mathbf{x}^{(2)}(0) = 2\mathbf{e}_2$. However, if we take the second derivative of Eq. (5.2-2), we will have that $\mathbf{x}^{(2)}(t) = \mathbf{A}\mathbf{x}^{(1)}(t) = \mathbf{A}^2\mathbf{x}(t)$ and, therefore, $\mathbf{x}^{(2)}(0) = \mathbf{A}^2\mathbf{x}(0) = 2\mathbf{e}_2$. Further application of this procedure yields the solution of Eq. (5.2-2) sought, having the Taylor series form

$$\mathbf{x}(t) = \left[\mathbf{I} + \mathbf{A}t + \frac{1}{2!}\mathbf{A}^2 t^2 + \frac{1}{3!}\mathbf{A}^3 t^3 + \cdots \right] \mathbf{x}(0) \tag{5.2-5}$$

The power series in the bracket defines the matrix $e^{\mathbf{A}t}$, i.e.,

$$e^{\mathbf{A}t} = \mathbf{I} + \mathbf{A}t + \frac{1}{2!}\mathbf{A}^2 t^2 + \frac{1}{3!}\mathbf{A}^3 t^3 + \cdots \tag{5.2-6}$$

The above series converges for all square matrices $\mathbf{A}$. Therefore, Eq. (5.2-5) takes on the form

$$\mathbf{x}(t) = e^{\mathbf{A}t}\mathbf{x}(0) \tag{5.2-7}$$

Relation (5.2-7) is the solution of the homogeneous equation (5.2-2). From Eq. (5.2-6) one may readily derive that

$$\frac{de^{\mathbf{A}t}}{dt} = \mathbf{A}e^{\mathbf{A}t}$$

that is, the matrix $e^{\mathbf{A}t}$ satisfies Eq. (5.2-3). Hence, it follows that the state transition matrix $\boldsymbol{\phi}(t)$ is given by

$$\boldsymbol{\phi}(t) = e^{\mathbf{A}t} \tag{5.2-8}$$

Another popular method to solve the homogeneous equation (5.2-2) and determine the transition matrix $\boldsymbol{\phi}(t)$ is that of using the Laplace transform. To this end, apply the Laplace transform to (5.2-2) to yield

$$s\mathbf{X}(s) - \mathbf{x}(0) = \mathbf{A}\mathbf{X}(s)$$

Solving for $\mathbf{X}(s)$, we have

$$\mathbf{X}(s) = (s\mathbf{I} - \mathbf{A})^{-1}\mathbf{x}(0) \tag{5.2-9}$$

Using the inverse Laplace transform in (5.2-9), we obtain

$$\mathbf{x}(t) = L^{-1}\{(s\mathbf{I} - \mathbf{A})^{-1}\}\mathbf{x}(0) \tag{5.2-10}$$

Comparing Eqs (5.2-7) and (5.2-10), we immediately have that

$$\boldsymbol{\phi}(t) = e^{\mathbf{A}t} = L^{-1}\{(s\mathbf{I} - \mathbf{A})^{-1}\} \tag{5.2-11}$$

Remark 5.2.1

From the above results, it follows that the state transition matrix $\boldsymbol{\phi}(t)$ depends only upon the matrix $\mathbf{A}$. The state vector $\mathbf{x}(t)$ describes the system's *free response*—namely, the response of the system when it is excited only by its initial condition $\mathbf{x}_0$ (i.e., here $\mathbf{u}(t) = \mathbf{0}$). Furthermore, according to Eq. (5.2-7), $\boldsymbol{\phi}(t)$ completely defines the *transition* of the state vector $\mathbf{x}(t)$, from its initial state $\mathbf{x}(0)$ to any new state $\mathbf{x}(t)$. This is the reason why the matrix $\boldsymbol{\phi}(t)$ is called the state transition matrix.

Remark 5.2.2

For the more general case, where the initial conditions are given for $t = t_0$,

$$\mathbf{x}(t) = \boldsymbol{\phi}(t, t_0)\mathbf{x}(t_0) \tag{5.2-12}$$

where

$$\boldsymbol{\phi}(t, t_0) = \boldsymbol{\phi}(t - t_0) = e^{\mathbf{A}(t - t_0)} \tag{5.2-13}$$

This can easily be proved if the Taylor series (5.2-4) is expanded about the arbitrary point $t = t_0$.

5.2.2 Properties of the State Transition Matrix

The state transition mattrix $\boldsymbol{\phi}(t)$ has various properties. Some of them are useful for the material that follows and are presented in the next theorem.

Theorem 5.2.1

The state transition matrix $\phi(t)$ has the following properties:

$$\phi(0) = \mathbf{I} \qquad\qquad (5.2\text{-}14a)$$

$$\phi^{-1}(t) = \phi(-t) \qquad\qquad (5.2\text{-}14b)$$

$$\phi(t_2 - t_1)\phi(t_1 - t_0) = \phi(t_2 - t_0), \quad \forall\, t_0, t_1, t_2 \qquad\qquad (5.2\text{-}14c)$$

$$[\phi(t)]^k = \phi(kt) \qquad\qquad (5.2\text{-}14d)$$

Proof

If we set $t = 0$ in Eq. (5.2-6) and subsequently use Eq. (5.2-8), we immediately have property (5.2-14a). If we multiply (5.2-8) from the left by $e^{-\mathbf{A}t}$ and from the right by $\phi^{-1}(t)$, we have

$$e^{-\mathbf{A}t}\phi(t)\phi^{-1}(t) = e^{-\mathbf{A}t}e^{\mathbf{A}t}\phi^{-1}(t)$$

Canceling out terms in the above equation, we immediately have property (5.2-14b). Property (5.2-14c) can be proved if we use Eq. (5.2-13), as follows:

$$\phi(t_2 - t_1)\phi(t_1 - t_0) = e^{\mathbf{A}(t_2 - t_1)}e^{\mathbf{A}(t_1 - t_0)} = e^{\mathbf{A}(t_2 - t_0)} = \phi(t_2 - t_0)$$

Finally, property (5.2-14d) can be proven as follows:

$$[\phi(t)]^k = \underbrace{e^{\mathbf{A}t}e^{\mathbf{A}t}\cdots e^{\mathbf{A}t}}_{k-\text{times}} = e^{\mathbf{A}kt} = \phi(kt)$$

5.2.3 Computation of the State Transition Matrix

For the computation of the matrix $e^{\mathbf{A}t}$, many methods have been proposed. We present the three most popular ones.

First Method

This method is based on Eq. (5.2-11), i.e., on the equation

$$\phi(t) = L^{-1}\{(s\mathbf{I} - \mathbf{A})^{-1}\}$$

To apply this method, one must first compute the matrix $(s\mathbf{I} - \mathbf{A})^{-1}$ and subsequently take its inverse Laplace transform. A method for computing the matrix $(s\mathbf{I} - \mathbf{A})^{-1}$ is given by the following theorem.

Theorem 5.2.2 (Leverrier's algorithm)

It holds that

$$\phi(s) = (s\mathbf{I} - \mathbf{A})^{-1} = \frac{s^{n-1}\mathbf{F}_1 + s^{n-2}\mathbf{F}_2 + \cdots + s\mathbf{F}_{n-1} + \mathbf{F}_n}{s^n + a_1 s^{n-1} + \cdots + a_{n-1}s + a_n} \qquad\qquad (5.2\text{-}15a)$$

where $\mathbf{F}_1, \mathbf{F}_2, \ldots, \mathbf{F}_n$ and $a_1, a_2, \ldots, a_n$ are determined by the recursive equations

$$\mathbf{F}_1 = \mathbf{I} \qquad\qquad a_1 = -\mathrm{tr}(\mathbf{AF}_1)$$

$$\mathbf{F}_2 = \mathbf{AF}_1 + a_1\mathbf{I} \qquad a_2 = -\tfrac{1}{2}\mathrm{tr}(\mathbf{AF}_2)$$

$$\vdots \qquad\qquad\qquad \vdots \qquad\qquad\qquad \text{(5.2-15b)}$$

$$\mathbf{F}_n = \mathbf{AF}_{n-1} + a_{n-1}\mathbf{I} \qquad a_n = -\frac{1}{n}\mathrm{tr}(\mathbf{AF}_n)$$

Upon determining the matrix $\boldsymbol{\phi}(s)$ using Leverrier's algorithm, we expand $\boldsymbol{\phi}(s)$ in partial fractions, by extending the results of Sec. 2.4, which hold for scalar functions, to the more general case of matrix functions, to yield

$$\boldsymbol{\phi}(s) = \sum_{i=1}^{n} \frac{1}{s - \lambda_i}\boldsymbol{\phi}_i$$

where $\lambda_1, \lambda_2, \ldots, \lambda_n$ are the roots of the characteristic polynomial $p(s)$ of matrix $\mathbf{A}$, where

$$p(s) = |s\mathbf{I} - \mathbf{A}| = s^n + a_1 s^{n-1} + a_2 s^{n-2} + \cdots + a_{n-1}s + a_n = \prod_{i=1}^{n}(s - \lambda_i)$$

and where $\boldsymbol{\phi}_i$ are constant $n \times n$ matrices. The matrices $\boldsymbol{\phi}_i$ can be computed by extending Eq. (2.4-3) to the matrix case, to yield

$$\boldsymbol{\phi}_i = \lim_{s \to \lambda_i}(s - \lambda_i)\boldsymbol{\phi}(s), \qquad i = 1, 2, \ldots, n$$

If certain roots of the characteristic polynomial $p(s)$ are repeated or complex conjugate pairs, then analogous results to the scalar functions case given in Sec. 2.4 hold for the present case of matrix functions.

Clearly, for the case of distinct roots, we have

$$\boldsymbol{\phi}(t) = L^{-1}\{\boldsymbol{\phi}(s)\} = e^{\mathbf{A}t} = \sum_{i=1}^{n}\boldsymbol{\phi}_i e^{\lambda_i t}$$

Second Method

This method takes place entirely in the time domain and is based on the diagonalization of the matrix $\mathbf{A}$. Indeed, if the eigenvalues of matrix $\mathbf{A}$ are distinct, then the eigenvector matrix $\mathbf{M}$ (see relation (2.10-5)) diagonalizes matrix $\mathbf{A}$, i.e., $\mathbf{A}$ can be transformed to a diagonal matrix $\boldsymbol{\Lambda}$ via the transformation matrix $\mathbf{M}$ as follows: $\boldsymbol{\Lambda} = \mathbf{M}^{-1}\mathbf{AM}$. Matrix $\boldsymbol{\phi}(t)$, under the transformation $\mathbf{M}$, becomes

$$\boldsymbol{\phi}(t) = e^{\mathbf{A}t} = \mathbf{M}e^{\boldsymbol{\Lambda}t}\mathbf{M}^{-1} \qquad\qquad \text{(5.2-16)}$$

Since

$$e^{\boldsymbol{\Lambda}t} = \begin{bmatrix} e^{\lambda_1 t} & & & \mathbf{0} \\ & e^{\lambda_2 t} & & \\ & & \ddots & \\ \mathbf{0} & & & e^{\lambda_n t} \end{bmatrix}, \qquad \boldsymbol{\Lambda} = \begin{bmatrix} \lambda_1 & & & \mathbf{0} \\ & \lambda_2 & & \\ & & \ddots & \\ \mathbf{0} & & & \lambda_n \end{bmatrix}$$

it follows that

$$\phi(t) = e^{\mathbf{A}t} = \mathbf{M} \begin{bmatrix} e^{\lambda_1 t} & & & \mathbf{0} \\ & e^{\lambda_2 t} & & \\ & & \ddots & \\ \mathbf{0} & & & e^{\lambda_n t} \end{bmatrix} \mathbf{M}^{-1} \tag{5.2-17}$$

In the case where the matrix $\mathbf{A}$ has repeated eigenvalues, analogous results may be derived (see Sec. 2.10).

Third Method

This method, as in the case of the second method, takes place entirely in the time domain and is based on the expansion of $e^{\mathbf{A}t}$ in a power series—namely, it is based on Eq. (5.2-6). This particular method does not require the determination of the eigenvalues $\lambda_1, \lambda_2, \ldots, \lambda_n$ of the matrix $\mathbf{A}$, as compared to the first and second methods and is relatively easy to implement on a digital computer.

5.3 GENERAL SOLUTION OF THE STATE EQUATIONS

To determine the state vector $\mathbf{x}(t)$ and the output vector $\mathbf{y}(t)$ of the linear system (5.2-1), we start by solving the first-order vector differential equation (5.2-1a), i.e., the vector differential equation

$$\dot{\mathbf{x}}(t) = \mathbf{A}\mathbf{x}(t) + \mathbf{B}\mathbf{u}(t), \qquad \mathbf{x}(t_0) = \mathbf{x}_0 \tag{5.3-1}$$

In the previous section we found that the *free response* of Eq. (5.3-1), i.e., the solution $\mathbf{x}_h(t)$ of the homogeneous part $\dot{\mathbf{x}}(t) = \mathbf{A}\mathbf{x}(t)$ of Eq. (5.3-1), is given by

$$\mathbf{x}_h(t) = \phi(t - t_0)\mathbf{x}(t_0), \qquad \text{where } \phi(t - t_0) = e^{\mathbf{A}(t - t_0)} \tag{5.3-2}$$

To determine the *forced response* or particular solution $\mathbf{x}_f(t)$ of Eq. (5.3-1), we assume that $\mathbf{x}_f(t)$ has the form

$$\mathbf{x}_f(t) = \phi(t - t_0)\mathbf{q}(t), \qquad \text{with } \mathbf{x}_f(t_0) = \mathbf{0} \tag{5.3-3}$$

where $\mathbf{q}(t)$ is an n-dimensional unknown vector. Substitute Eq. (5.3-3) into Eq. (5.3-1) to yield

$$\dot{\phi}(t - t_0)\mathbf{q}(t) + \phi(t - t_0)\dot{\mathbf{q}}(t) = \mathbf{A}\phi(t - t_0)\mathbf{q}(t) + \mathbf{B}\mathbf{u}(t) \tag{5.3-4}$$

Since $\dot{\phi}(t - t_0) = \mathbf{A}\phi(t - t_0)$, Eq. (5.3-4) takes on the form

$$\mathbf{A}\phi(t - t_0)\mathbf{q}(t) + \phi(t - t_0)\dot{\mathbf{q}}(t) = \mathbf{A}\phi(t - t_0)\mathbf{q}(t) + \mathbf{B}\mathbf{u}(t)$$

or

$$\phi(t - t_0)\dot{\mathbf{q}}(t) = \mathbf{B}\mathbf{u}(t)$$

Hence

$$\dot{\mathbf{q}}(t) = \phi^{-1}(t - t_0)\mathbf{B}\mathbf{u}(t) \tag{5.3-5}$$

If we integrate Eq. (5.3-5) from t_0 to t, we have

$$\mathbf{q}(t) = \int_{t_0}^{t} \phi^{-1}(\lambda - t_0)\mathbf{B}\mathbf{u}(\lambda)d\lambda$$

where $\mathbf{q}(t_0) = \boldsymbol{\phi}^{-1}(t_0 - t_0)\mathbf{x}_f(t_0) = \mathbf{x}_f(t_0) = \mathbf{0}$, where use was made of Eq. (5.3-3). Thus, the forced response of Eq. (5.3-1) will be

$$\mathbf{x}_f(t) = \boldsymbol{\phi}(t - t_0)\mathbf{q}(t) = \boldsymbol{\phi}(t - t_0)\int_{t_0}^{t} \boldsymbol{\phi}^{-1}(\lambda - t_0)\mathbf{B}\mathbf{u}(\lambda)d\lambda$$

$$= \int_{t_0}^{t} \boldsymbol{\phi}(t - t_0)\boldsymbol{\phi}^{-1}(\lambda - t_0)\mathbf{B}\mathbf{u}(\lambda)d\lambda$$

or

$$\mathbf{x}_f(t) = \int_{t_0}^{t} \boldsymbol{\phi}(t - \lambda)\mathbf{B}\mathbf{u}(\lambda)d\lambda$$

where use was made of properties (5.2-14b,c). Hence, the general solution of Eq. (5.3-1) will be

$$\mathbf{x}(t) = \mathbf{x}_h(t) + \mathbf{x}_f(t) = \boldsymbol{\phi}(t - t_0)\mathbf{x}(t_0) + \int_{t_0}^{t} \boldsymbol{\phi}(t - \lambda)\mathbf{B}\mathbf{u}(\lambda)d\lambda$$

The output vector $\mathbf{y}(t)$ of system (5.1-1) can now be easily determined using Eq. (5.3-6) to yield

$$\mathbf{y}(t) = \mathbf{C}\mathbf{x}(t) + \mathbf{D}\mathbf{u}(t) = \mathbf{C}\boldsymbol{\phi}(t - t_0)\mathbf{x}(t_0) + \mathbf{C}\int_{t_0}^{t} \boldsymbol{\phi}(t - \lambda)\mathbf{B}\mathbf{u}(\lambda)d\lambda + \mathbf{D}\mathbf{u}(t)$$

$$(5.3-7)$$

Example 5.3.1

Consider the network of Figure 5.1a with initial conditions $i_L(0) = 1$ and $v_c(0) = 0$. The input, the state variables, and the output of the network are shown in Figure 5.1b. Determine:

(a) A state-space description
(b) The state transition matrix
(c) The solution of the homogeneous equation
(d) The general solution

Solution

(a) The loop equation is

$$\frac{di}{dt} + 3i + 2\int_0^t i\,dt = v(t)$$

The two state variables, which are given in Figure 5.1b, are defined as follows:

$$x_1(t) = \int_0^t i(t)dt \quad \text{and} \quad x_2(t) = \dot{x}_1(t) = i_L(t) = i(t)$$

Using the above definitions, the loop equation can be written in state space as follows: $\dot{x}_2(t) + 3x_2(t) + 2x_1(t) = v(t)$. This equation, combined with the equation $\dot{x}_1(t) = x_2(t)$, yields the following state equations for the given network:

(a)

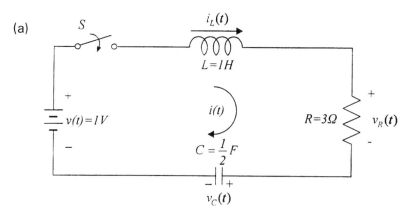

(b)

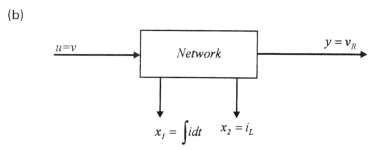

Figure 5.1 (a) RLC network; (b) the input, the state variables, and the output of the net-work.

$$\begin{bmatrix} \dot{x}_1(t) \\ \dot{x}_2(t) \end{bmatrix} = \begin{bmatrix} 0 & 1 \\ -2 & -3 \end{bmatrix} \begin{bmatrix} x_2(t) \\ x_2(t) \end{bmatrix} + \begin{bmatrix} 0 \\ 1 \end{bmatrix} u(t)$$

with initial condition vector $\mathbf{x}(0) = \mathbf{x}_0 = [0, 1]^T$ and $u(t) = v(t)$.

(b) To determine the transition matrix $\boldsymbol{\phi}(t)$, we apply the first method of Sec. 5.2, as follows:

$$\boldsymbol{\phi}(s) = (s\mathbf{I} - \mathbf{A})^{-1} = \begin{bmatrix} s & -1 \\ 2 & s+3 \end{bmatrix}^{-1} = \begin{bmatrix} \dfrac{s+3}{(s+1)(s+2)} & \dfrac{1}{(s+1)(s+2)} \\ \dfrac{-2}{(s+1)(s+2)} & \dfrac{s}{(s+1)(s+2)} \end{bmatrix}$$

Hence

$$\boldsymbol{\phi}(t) = L^{-1}\{\boldsymbol{\phi}(s)\} = \begin{bmatrix} 2e^{-t} - e^{-2t} & e^{-t} - e^{-2t} \\ -2e^{-t} + 2e^{-2t} & -e^{-t} + 2e^{-2t} \end{bmatrix}$$

(c) The solution of the homogeneous equation is

$$\mathbf{x}_h(t) = \boldsymbol{\phi}(t)\mathbf{x}(0)$$

or

$$\mathbf{x}_h(t) = \boldsymbol{\phi}(t) \begin{bmatrix} 0 \\ 1 \end{bmatrix} = \begin{bmatrix} e^{-t} - e^{-2t} \\ -e^{-t} + 2e^{-2t} \end{bmatrix}$$

(d) The general solution of the state vector is

$$\mathbf{x}(t) = \mathbf{x}_h(t) + \mathbf{x}_f(t) = \boldsymbol{\phi}(t)\,\mathbf{x}(0) + \int_0^t \boldsymbol{\phi}(t - \lambda)\mathbf{b}u(\lambda)\,d\lambda$$

The forced response $\mathbf{x}_f(t)$ is determined as follows:

$$\int_0^t \boldsymbol{\phi}(t - \lambda)\mathbf{b}u(\lambda)\,d\lambda = \int_0^t \left[\begin{array}{c} e^{-(t-\lambda)} - e^{-2(t-\lambda)} \\ -e^{-(t-\lambda)} + 2e^{-2(t-\lambda)} \end{array} \right] u(\lambda)\,d\lambda$$

$$= \left[\begin{array}{c} e^{-t}\int_0^t e^{\lambda}d\lambda - e^{-2t}\int_0^t e^{2\lambda}d\lambda \\ -e^{-t}\int_0^t e^{\lambda}d\lambda + 2e^{-2t}\int_0^t e^{2\lambda}d\lambda \end{array} \right]$$

$$= \left[\begin{array}{c} e^{-t}(e^t - 1) - \tfrac{1}{2}e^{-2t}(e^{2t} - 1) \\ -e^{-t}(e^t - 1) + e^{-2t}(e^{2t} - 1) \end{array} \right] = \left[\begin{array}{c} \tfrac{1}{2} - e^{-t} + \tfrac{1}{2}e^{-2t} \\ e^{-t} - e^{-2t} \end{array} \right]$$

Hence, the general solution $\mathbf{x}(t) = \mathbf{x}_h(t) + \mathbf{x}_f(t)$ is as follows:

$$\mathbf{x}(t) = \left[\begin{array}{c} x_1(t) \\ x_2(t) \end{array} \right] = \left[\begin{array}{c} (e^{-t} - e^{-2t}) + (\tfrac{1}{2} - e^{-t} + \tfrac{1}{2}e^{-2t}) \\ (-e^{-t} + 2e^{-2t}) + (e^{-t} - e^{-2t}) \end{array} \right] = \left[\begin{array}{c} \tfrac{1}{2} - \tfrac{1}{2}e^{-2t} \\ e^{-2t} \end{array} \right]$$

The output of the system is

$$y(t) = v_R(t) = Ri(t) = Rx_2(t) = 3e^{-2t}$$

Working in the s-domain, we should arrive at the same result. Indeed we have

$$\mathbf{X}(s) = (s\mathbf{I} - \mathbf{A})^{-1}\mathbf{x}_0 + (s\mathbf{I} - \mathbf{A})^{-1}\mathbf{b}U(s) = (s\mathbf{I} - \mathbf{A})^{-1}[\mathbf{x}_0 + \mathbf{b}U(s)]$$

$$= (s\mathbf{I} - \mathbf{A})^{-1} \left\{ \left[\begin{array}{c} 0 \\ 1 \end{array} \right] + \left[\begin{array}{c} 0 \\ \dfrac{1}{s} \end{array} \right] \right\}$$

$$= \left[\begin{array}{cc} \dfrac{s+3}{(s+1)(s+2)} & \dfrac{1}{(s+1)(s+2)} \\ \dfrac{-2}{(s+1)(s+2)} & \dfrac{s}{(s+1)(s+2)} \end{array} \right] \left[\begin{array}{c} 0 \\ \dfrac{s+1}{s} \end{array} \right] = \left[\begin{array}{c} \dfrac{1}{s(s+2)} \\ \dfrac{1}{s+2} \end{array} \right]$$

and hence

$$\mathbf{x}(t) = L^{-1}\{\mathbf{X}(s)\} = \left[\begin{array}{c} \tfrac{1}{2} - \tfrac{1}{2}e^{-2t} \\ e^{-2t} \end{array} \right]$$

Example 5.3.2

Consider the network of Figure 5.2a with zero initial conditions. The inputs, the state variables, and the outputs of the network are shown in Figure 5.2b. Determine:

(a) The state equations of the network
(b) The transition matrix and the output vector of the network, when $R_1 = 0\,\Omega$, $R_2 = 1.5\,\Omega$, $C = 1\,\text{F}$, $L_1 = L_2 = 1\,\text{H}$, and $v_1(t) = v_2(t) = 1\,\text{V}$.

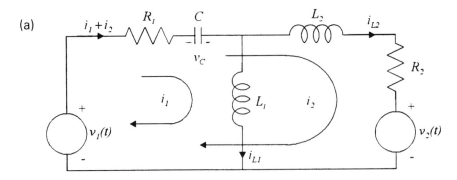

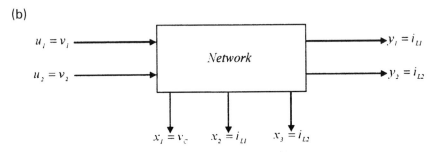

Figure 5.2 (a) A two-loop network; (b) the inputs, the state variables, and the outputs of the network.

Solution

(a) The two loop equations are

$$R_1(i_1 + i_2) + v_c + L_1 \frac{di_1}{dt} = v_1(t)$$

$$R_1(i_1 + i_2) + v_c + L_2 \frac{di_2}{dt} + R_2 i_2 = v_1(t) - v_2(t)$$

Furthermore, we have that

$$C \frac{dv_c}{dt} = i_1 + i_2$$

Hence, the three first-order differential equations that describe the network are

$$\frac{dv_c}{dt} = \frac{1}{C} i_1 + \frac{1}{C} i_2$$

$$\frac{di_1}{dt} = -\frac{1}{L_1} v_c - \frac{R_1}{L_1} i_1 - \frac{R_1}{L_1} i_2 + \frac{1}{L_1} v_1(t)$$

$$\frac{di_2}{dt} = -\frac{1}{L_2} v_c - \frac{R_1}{L_2} i_1 - \frac{R_1 + R_2}{L_2} i_2 + \frac{1}{L_2} v_1(t) - \frac{1}{L_2} v_2(t)$$

Note that $i_1 = i_{L_1}$ and $i_2 = i_{L_2}$. Thus, the state equations of the network are

$$\dot{\mathbf{x}} = \mathbf{A}\mathbf{x} + \mathbf{B}\mathbf{u}$$

$$\mathbf{y} = \mathbf{C}\mathbf{x}$$

where

$$\mathbf{x} = \begin{bmatrix} x_1(t) \\ x_2(t) \\ x_3(t) \end{bmatrix} = \begin{bmatrix} v_c \\ i_{L_1} \\ i_{L_2} \end{bmatrix}, \qquad \mathbf{u} = \begin{bmatrix} v_1(t) \\ v_2(t) \end{bmatrix}, \qquad \mathbf{y} = \begin{bmatrix} i_{L_1} \\ i_{L_2} \end{bmatrix}$$

$$\mathbf{A} = \begin{bmatrix} 0 & \dfrac{1}{C} & \dfrac{1}{C} \\[2mm] -\dfrac{1}{L_1} & -\dfrac{R_1}{L_1} & -\dfrac{R_1}{L_1} \\[2mm] -\dfrac{1}{L_2} & -\dfrac{R_1}{L_2} & -\dfrac{R_1+R_2}{L_2} \end{bmatrix}, \qquad \mathbf{B} = \begin{bmatrix} 0 & 0 \\[2mm] -\dfrac{1}{L_1} & 0 \\[2mm] \dfrac{1}{L_2} & -\dfrac{1}{L_2} \end{bmatrix},$$

$$\mathbf{C} = \begin{bmatrix} 0 & 1 & 0 \\ 0 & 0 & 1 \end{bmatrix}$$

(b) Substituting the given values for each element, we have

$$\mathbf{A} = \begin{bmatrix} 0 & 1 & 1 \\ -1 & 0 & 0 \\ -1 & 0 & -1.5 \end{bmatrix}, \qquad \mathbf{B} = \begin{bmatrix} 0 & 0 \\ 1 & 0 \\ 1 & -1 \end{bmatrix}, \qquad s\mathbf{I} - \mathbf{A} = \begin{bmatrix} s & -1 & -1 \\ 1 & s & 0 \\ 1 & 0 & s+1.5 \end{bmatrix}$$

$$|s\mathbf{I} - \mathbf{A}| = s[s(s+1.5)] + s + 1.5 + s = s^3 + 1.5s^2 + 2s + 1.5$$
$$= (s+1)(s^2 + 0.5s + 1.5)$$

Therefore

$$\boldsymbol{\Phi}(s) = (s\mathbf{I} - \mathbf{A})^{-1} = \dfrac{1}{(s+1)(s^2+0.5s+1.5)}$$

$$\begin{bmatrix} s(s+1.5) & s+1.5 & s \\ -(s+1.5) & s^2+1.5s+1 & -1 \\ -s & -1 & s^2+1 \end{bmatrix}$$

Hence

$$\boldsymbol{\Phi}(t) = L^{-1}\{(s\mathbf{I} - \mathbf{A})^{-1}\} = \begin{bmatrix} \varphi_{11}(t) & \varphi_{12}(t) & \varphi_{13}(t) \\ \varphi_{21}(t) & \varphi_{22}(t) & \varphi_{23}(t) \\ \varphi_{31}(t) & \varphi_{32}(t) & \varphi_{33}(t) \end{bmatrix}$$

where

$$\varphi_{11}(t) = -0.25e^{-t} + e^{-0.25t}(1.25\cos 1.199t + 0.052\sin 1.199t)$$

$$\varphi_{12}(t) = -0.5e^{-t} + e^{-0.25t}(0.5\cos 1.199t + 0.99\sin 1.199t)$$

$$\varphi_{13}(t) = -0.5e^{-t} + e^{-0.25t}(0.5\cos 1.199t + 0.521\sin 1.199t)$$

$$\varphi_{21}(t) = -\varphi_{12}(t)$$

$$\varphi_{22}(t) = 0.25e^{-t} + e^{-0.25t}(0.75\cos 1.199t + 0.365\sin 1.199t)$$

$\varphi_{23}(t) = -0.5e^{-t} + e^{-0.25t}(0.5\cos 1.199t - 0.312\sin 1.199t)$

$\varphi_{31}(t) = -\varphi_{13}(t)$

$\varphi_{32}(t) = \varphi_{23}(t)$

$\varphi_{33}(t) = e^{-t} - 0.417e^{-0.25t}\sin 1.199t$

The output vector is

$$Y(s) = [C(sI - A)^{-1}B]U(s) = \frac{1}{(s+1)(s^2 + 0.5s + 1.5)}\begin{bmatrix} s^2 + 1.5s & 1 \\ s^2 & -s^2 - 1 \end{bmatrix}$$

$$\begin{bmatrix} 1/s \\ 1/s \end{bmatrix}$$

$$= \frac{1}{s(s+1)(s^2 + 0.5s + 1.5)}\begin{bmatrix} s^2 + 1.5s + 1 \\ -1 \end{bmatrix}$$

Therefore

$$y(t) = L^{-1}\{Y(s)\} = \begin{bmatrix} y_1(t) \\ y_2(t) \end{bmatrix}$$

$$= \begin{bmatrix} 0.67 + 0.08e^{-t} + 0.75e^{-0.25t}\cos 1.2t - 0.82e^{-0.25t}\sin 1.199t \\ -0.67 + 0.17e^{-t} + 0.5e^{-0.25t}\cos 1.2t + 0.23e^{-0.25t}\sin 1.199t \end{bmatrix}$$

5.4 STATE VECTOR TRANSFORMATIONS AND SPECIAL FORMS OF STATE EQUATIONS

Consider the linear transformation

$$x = Tz \tag{5.4-1}$$

of the state vector x of system (5.1-1), where T is the transformation matrix $n \times n$ with $|T| \neq 0$ and z is the new n-dimensional state vector. Substitute Eq. (5.4-1) into Eq. (5.1-1) to yield

$$T\dot{z} = ATz + Bu$$

$$y = CTz + Du$$

$$x(0) = Tz(0)$$

or

$$\dot{z} = A^*z + B^*u \tag{5.4-2a}$$

$$y = C^*z + D^*u \tag{5.4-2b}$$

$$z(0) = z_0 \tag{5.4-2c}$$

where

$$A^* = T^{-1}AT \tag{5.4-3a}$$

$$B^* = T^{-1}B \tag{5.4-3b}$$

$$\mathbf{C}^* = \mathbf{CT} \tag{5.4-3c}$$

$$\mathbf{D}^* = \mathbf{D} \tag{5.4-3d}$$

$$\mathbf{z}(0) = \mathbf{T}^{-1}\mathbf{x}(0) \tag{5.4-3e}$$

For brevity, systems (5.1-1) and (5.4-2) are presented as $(\mathbf{A}, \mathbf{B}, \mathbf{C}, \mathbf{D})_n$ and as $(\mathbf{A}^*, \mathbf{B}^*, \mathbf{C}^*, \mathbf{D}^*)_n$, respectively. System $(\mathbf{A}^*, \mathbf{B}^*, \mathbf{C}^*, \mathbf{D}^*)_n$ is called the *transformed state model* of system $(\mathbf{A}, \mathbf{B}, \mathbf{C}, \mathbf{D})_n$. The motivation for transforming the state vector $\mathbf{x}$ to the state vector $\mathbf{z}$ is to select a new coordinate system $z_1, z_2, \ldots, z_n$ which has more advantages than the original coordinate system $x_1, x_2, \ldots, x_n$. These advantages are related to the physical structure and characteristics as well as to the computational and technological aspects of the given system. Therefore, the problem of transforming $\mathbf{x}$ to $\mathbf{z}$ consists in determining a suitable state transformation matrix $\mathbf{T}$ such that the forms of $\mathbf{A}^*, \mathbf{B}^*, \mathbf{C}^*$, and $\mathbf{D}^*$ of the new system (5.4-2) have certain desirable characteristics. Usually, more attention is paid to matrix $\mathbf{A}^*$, so that its form is as simple as possible. The most popular such forms for $\mathbf{A}^*$ are the *diagonal* and the *phase canonical form*.

5.4.1 The Invariance of the Characteristic Polynomial and of the Transfer Function Matrix

Independently of the particular choice of the transformation matrix $\mathbf{T}$, certain characteristics of the initial system $(\mathbf{A}, \mathbf{B}, \mathbf{C}, \mathbf{D})_n$ remain invariant under state transformation. Two such characteirstics are the characteristic polynomial and the transfer function matrix. This is proven in the following theorem.

Theorem 5.4.1

It holds that

$$p(s) = p^*(s) \tag{5.4-4}$$

$$H(s) = H^*(s) \tag{5.4-5}$$

where

$p(s) = $ characteristic polynomial of matrix $\mathbf{A} = |s\mathbf{I} - \mathbf{A}|$
$p^*(s) = $ characteristic polynomial of matrix $\mathbf{A}^* = |s\mathbf{I} - \mathbf{A}^*|$
$H(s) = $ transfer function matrix of system (5.1-1) $= \mathbf{C}(s\mathbf{I} - \mathbf{A})^{-1}\mathbf{B} + \mathbf{D}$
$H^*(s) = $ transfer function matrix of system (5.4-2) $= \mathbf{C}^*(s\mathbf{I} - \mathbf{A}^*)^{-1}\mathbf{B}^* + \mathbf{D}^*$

Proof

We have

$$p^*(s) = |s\mathbf{I} - \mathbf{A}^*| = |s\mathbf{I} - \mathbf{T}^{-1}\mathbf{AT}| = |\mathbf{T}^{-1}(s\mathbf{I} - \mathbf{A})\mathbf{T}|$$
$$= |\mathbf{T}^{-1}||s\mathbf{I} - \mathbf{A}||\mathbf{T}| = |s\mathbf{I} - \mathbf{A}| = p(s)$$

Furthermore,

$$\begin{aligned}
\mathbf{H}^*(s) &= \mathbf{C}^*(s\mathbf{I} - \mathbf{A}^*)^{-1}\mathbf{B}^* + \mathbf{D}^* = \mathbf{CT}(s\mathbf{I} - \mathbf{T}^{-1}\mathbf{A}\mathbf{T})\mathbf{T}^{-1}\mathbf{B} + \mathbf{D} \\
&= \mathbf{CT}[\mathbf{T}^{-1}(s\mathbf{I} - \mathbf{A})\mathbf{T}]^{-1}\mathbf{T}^{-1}\mathbf{B} + \mathbf{D} \\
&= \mathbf{CTT}^{-1}(s\mathbf{I} - \mathbf{A})^{-1}\mathbf{TT}^{-1}\mathbf{B} + \mathbf{D} = \mathbf{C}(s\mathbf{I} - \mathbf{A})^{-1}\mathbf{B} + \mathbf{D} = \mathbf{H}(s)
\end{aligned}$$

where use was made of Eqs (5.4-3). Relations (5.4-5) have been also proved in Theorem 3.9.2.

5.4.2 Special State-Space Forms: The Phase Canonical Form

The main objective of the state vector transformation is to choose an appropriate matrix $\mathbf{T}$ such that the transformed system is as simple as possible. As mentioned earlier, one such simple form is the well-known *phase canonical form*. Another popular simple form is when the matrix $\mathbf{A}^*$ is diagonal. Note that the problem of diagonalizing the matrix $\mathbf{A}$ of the original system is adequately covered in Sec. 2.10. For this reason we will not deal with the subject of diagonalization any further here. It is important to mention that diagonalization is a problem that is strongly related to the eigenvalues and eigenvectors of $\mathbf{A}$, as compared with the phase canonical form, which is a problem strongly related to the controllability of the system (see Sec. 5.6).

With regard to the phase canonical form, we give the following definitions.

Definition 5.4.1

System $(\mathbf{A}^*, \mathbf{B}^*, \mathbf{C}^*, \mathbf{D}^*)_n$ is in phase canonical form when the matrices $\mathbf{A}^*$ and $\mathbf{B}^*$ have the following special forms

$$\mathbf{A}^* = \begin{bmatrix} \mathbf{A}^*_{11} & \mathbf{A}^*_{12} & \cdots & \mathbf{A}^*_{1m} \\ \mathbf{A}^*_{21} & \mathbf{A}^*_{22} & \cdots & \mathbf{A}^*_{2m} \\ \vdots & \vdots & & \vdots \\ \mathbf{A}^*_{m1} & \mathbf{A}^*_{m2} & \cdots & \mathbf{A}^*_{mm} \end{bmatrix} \quad \text{and} \quad \mathbf{B}^* = \begin{bmatrix} \mathbf{B}^*_1 \\ \cdots \\ \mathbf{B}^*_2 \\ \cdots \\ \vdots \\ \cdots \\ \mathbf{B}^*_m \end{bmatrix} \tag{5.4-6a}$$

where

$$\mathbf{A}^*_{ii} = \begin{bmatrix} 0 & 1 & 0 & \cdots & 0 \\ 0 & 0 & 1 & \cdots & 0 \\ \vdots & \vdots & \vdots & \cdots & \vdots \\ -(a^*_{ii})_0 & -(a^*_{ii})_1 & -(a^*_{ii})_2 & \cdots & -(a^*_{ii})_{\sigma_i - 1} \end{bmatrix} \tag{5.4-6b}$$

$$\mathbf{A}^*_{ij} = \begin{bmatrix} & & & 0 & & \\ \cdots & \cdots & \cdots & \cdots & \cdots & \cdots \\ -(a^*_{ij})_0 & -(a^*_{ij})_1 & -(a^*_{ij})_2 & \cdots & -(a^*_{ij})_{\sigma_j - 1} \end{bmatrix} \tag{5.4-6c}$$

$$\mathbf{B}^*_i = \begin{bmatrix} & & & 0 & & & & \\ \cdots & \cdots & \cdots & \cdots & \cdots & \cdots & \cdots & \cdots \\ 0 & 0 & \cdots & 0 & 1 & (b^*_i)_{i+1} & (b^*_i)_{i+2} & \cdots & (b^*_i)_m \end{bmatrix} \tag{5.4-6d}$$

$$\uparrow$$

ith position

where $\sigma_1, \sigma_2, \ldots, \sigma_m$ are positive integer numbers satisfying the relation

$$\sum_{i=1}^{m} \sigma_i = n$$

Definition 5.4.2

Assume that system $(\mathbf{A}^*, \mathbf{B}^*, \mathbf{C}^*, \mathbf{D}^*)_n$ has one input. In this case, the system is in its phase canonical form when the matrix $\mathbf{A}^*$ and the vector $\mathbf{b}^*$ have the special forms

$$\mathbf{A}^* = \begin{bmatrix} 0 & 1 & 0 & & 0 \\ 0 & 0 & 1 & \cdots & 0 \\ \vdots & \vdots & \vdots & & \vdots \\ 0 & 0 & 0 & \cdots & 1 \\ -a_0^* & -a_1^* & -a_2^* & \cdots & -a_{n-1}^* \end{bmatrix} \quad \text{and} \quad \mathbf{b}^* = \begin{bmatrix} 0 \\ 0 \\ \vdots \\ 0 \\ 1 \end{bmatrix} \tag{5.4-7}$$

It is obvious that Definition 5.4.2 constitutes a special (but very important) case of Definition 5.4.1.

With regard to the procedure of transforming a system to its phase canonical form, we present the following theorems.

Theorem 5.4.2

Assume that the system $(\mathbf{A}, \mathbf{B}, \mathbf{C}, \mathbf{D})_n$ has one input and that the matrix

$$\mathbf{S} = [\mathbf{b} \,\vdots\, \mathbf{Ab} \,\vdots\, \mathbf{A}^2\mathbf{b} \,\vdots\, \cdots \,\vdots\, \mathbf{A}^{n-1}\mathbf{b}] \tag{5.4-8}$$

is regular, i.e., $|\mathbf{S}| \neq 0$. In this case, there exists a transformation matrix $\mathbf{T}$ which transforms the given system to its phase canonical form $(\mathbf{A}^*, \mathbf{B}^*, \mathbf{C}^*, \mathbf{D}^*)_n$, where the matrix $\mathbf{T}$ is given by

$$\mathbf{T} = \mathbf{P}^{-1}$$

where

$$\mathbf{P} = \begin{bmatrix} \mathbf{p}_1 \\ \cdots \\ \mathbf{p}_2 \\ \cdots \\ \mathbf{p}_3 \\ \cdots \\ \vdots \\ \cdots \\ \mathbf{p}_n \end{bmatrix} = \begin{bmatrix} \mathbf{q} \\ \cdots \\ \mathbf{qA} \\ \cdots \\ \mathbf{qA}^2 \\ \cdots \\ \vdots \\ \cdots \\ \mathbf{qA}^{n-1} \end{bmatrix} \tag{5.4-9}$$

where $\mathbf{q}$ is the last row of the matrix $\mathbf{S}^{-1}$.

Proof

Since $\mathbf{x} = \mathbf{Tz}$, it follows that $\mathbf{z} = \mathbf{T}^{-1}\mathbf{x} = \mathbf{Px}$, where $\mathbf{P} = \mathbf{T}^{-1}$. We also have $z_1 = \mathbf{p}_1\mathbf{x}$ and $\dot{z}_1 = \mathbf{p}_1\dot{\mathbf{x}}$. If we replace $\dot{\mathbf{x}} = \mathbf{Ax} + \mathbf{b}u$ we have that $\dot{z}_1 = \mathbf{p}_1(\mathbf{Ax} + \mathbf{b}u) = \mathbf{p}_1\mathbf{Ax} + \mathbf{p}_1\mathbf{b}u$. From the structure of the matrix $\mathbf{A}^*$ given in Eq. (5.4-7), it follows that $\dot{z}_1 = z_2, \dot{z}_2 = z_3, \ldots, \dot{z}_{n-1} = z_n$. Hence, the expression for $\dot{z}_1$ takes on the form

$$\dot{z}_1 = z_2 = \mathbf{p}_1\mathbf{A}\mathbf{x} + \mathbf{p}_1\mathbf{b}u$$

Since, according to relation $\mathbf{z} = \mathbf{P}\mathbf{x}$, the elements of $\mathbf{z}$ are functions of the elements of $\mathbf{x}$ only, it follows that $\mathbf{p}_1\mathbf{b} = 0$. Also, since

$$\dot{z}_2 = z_3 = \mathbf{p}_1\mathbf{A}\dot{\mathbf{x}} = \mathbf{p}_1\mathbf{A}(\mathbf{A}\mathbf{x} + \mathbf{b}u) = \mathbf{p}_1\mathbf{A}^2\mathbf{x} + \mathbf{p}_1\mathbf{A}\mathbf{b}u$$

it follows that $\mathbf{p}_1\mathbf{A}\mathbf{b} = 0$. If we repeat this procedure, we arrive at the final equation

$$\dot{z}_{n-1} = z_n = \mathbf{p}_1\mathbf{A}^{n-2}\dot{\mathbf{x}} = \mathbf{p}_1\mathbf{A}^{n-2}(\mathbf{A}\mathbf{x} + \mathbf{b}u) = \mathbf{p}_1\mathbf{A}^{n-1}\mathbf{x} + \mathbf{p}_1\mathbf{A}^{n-2}\mathbf{b}u$$

and, consequently, $\mathbf{p}_1\mathbf{A}^{n-2}\mathbf{b} = 0$. The above results can be summarized as follows:

$$\mathbf{z} = \mathbf{P}\mathbf{x}$$

where

$$\mathbf{P} = \begin{bmatrix} \mathbf{p}_1 \\ \cdots\cdots \\ \mathbf{p}_1\mathbf{A} \\ \cdots\cdots \\ \mathbf{p}_1\mathbf{A}^2 \\ \cdots\cdots \\ \vdots \\ \cdots\cdots \\ \mathbf{p}_1\mathbf{A}^{n-1} \end{bmatrix}$$

where $\mathbf{p}_1$ is, for the time being, an arbitrary row vector which must satisfy the following relations:

$$\mathbf{p}_1\mathbf{b} = \mathbf{p}_1\mathbf{A}\mathbf{b} = \mathbf{p}_1\mathbf{A}^2\mathbf{b} = \cdots = \mathbf{p}_1\mathbf{A}^{n-2}\mathbf{b} = 0$$

or

$$\mathbf{p}_1[\mathbf{b} \,\vdots\, \mathbf{A}\mathbf{b} \,\vdots\, \mathbf{A}^2\mathbf{b} \,\vdots\, \cdots \,\vdots\, \mathbf{A}^{n-2}\mathbf{b}] = 0$$

For the vector $\mathbf{b}^* = \mathbf{P}\mathbf{b}$ to have the form (5.4-7), it must hold that

$$\mathbf{b}^* = \mathbf{P}\mathbf{b} = \begin{bmatrix} \mathbf{p}_1\mathbf{b} \\ \cdots\cdots\cdots \\ \mathbf{p}_1\mathbf{A}\mathbf{b} \\ \cdots\cdots\cdots \\ \vdots \\ \cdots\cdots\cdots \\ \mathbf{p}_1\mathbf{A}^{n-1}\mathbf{b} \end{bmatrix} = \begin{bmatrix} 0 \\ \cdots \\ 0 \\ \cdots \\ \vdots \\ \cdots \\ 1 \end{bmatrix}$$

The above relation holds when $\mathbf{p}_1\mathbf{A}^{n-1}\mathbf{b} = 1$. Hence, the vector $\mathbf{p}_1$ must satisfy the equation

$$\mathbf{p}_1[\mathbf{b} \,\vdots\, \mathbf{A}\mathbf{b} \,\vdots\, \mathbf{A}^2\mathbf{b} \,\vdots\, \cdots \,\vdots\, \mathbf{A}^{n-2}\mathbf{b} \,\vdots\, \mathbf{A}^{n-1}\mathbf{b}] = \mathbf{p}_1\mathbf{S} = [0, 0, \ldots, 0, 1]$$

Hence

$$\mathbf{p}_1 = [0, 0, \cdots, 0, 1]\mathbf{S}^{-1} = \mathbf{q}$$

where $\mathbf{q}$ is the last row of matrix $\mathbf{S}^{-1}$.

Example 5.4.1

Consider a system of the form (5.1-1), where

$$A = \begin{bmatrix} 1 & -1 \\ -1 & 2 \end{bmatrix}, \qquad b = \begin{bmatrix} 1 \\ 1 \end{bmatrix}$$

Transform matrix A and vector b to A^* and b^* of the form (5.4-7).

Solution

We have

$$S = [b \vdots Ab] = \begin{bmatrix} 1 & \vdots & 0 \\ 1 & \vdots & 1 \end{bmatrix}, \qquad S^{-1} = \begin{bmatrix} 1 & 0 \\ -1 & 1 \end{bmatrix}$$

Since S is invertible, matrix A and vector b can be transformed in phase canonical form. Matrix P will then be

$$P = \begin{bmatrix} q_1 \\ \cdots \\ q_1 A \end{bmatrix} = \begin{bmatrix} -1 & 1 \\ \cdots \\ -2 & 3 \end{bmatrix}, \qquad P^{-1} = \begin{bmatrix} -3 & 1 \\ -2 & 1 \end{bmatrix}$$

and hence $T = P^{-1}$ given above. Consequently,

$$A^* = T^{-1}AT = \begin{bmatrix} -1 & 1 \\ -2 & 3 \end{bmatrix}\begin{bmatrix} 1 & -1 \\ -1 & 2 \end{bmatrix}\begin{bmatrix} -3 & 1 \\ -2 & 1 \end{bmatrix} = \begin{bmatrix} 0 & 1 \\ -1 & 3 \end{bmatrix}$$

$$b^* = T^{-1}b = \begin{bmatrix} -1 & 1 \\ -2 & 3 \end{bmatrix}\begin{bmatrix} 1 \\ 1 \end{bmatrix} = \begin{bmatrix} 0 \\ 1 \end{bmatrix}$$

Theorem 5.4.3

Assume that the system $(A, B, C, D)_n$ is a MIMO. Further, assume that there are positive integer numbers $\sigma_1, \sigma_2, \ldots, \sigma_m$ such that the matrix

$$\hat{S} = [b_1 \vdots Ab_1 \vdots \cdots \vdots A^{\sigma_1 - 1}b_1 | b_2 \vdots Ab_2 \vdots \cdots \vdots A^{\sigma_2 - 1}b_2 | \cdots | b_m \vdots Ab_m \vdots \cdots \vdots$$
$$A^{\sigma_m - 1}b_m]$$

$$(5.4\text{-}10)$$

is of full rank and that $\sigma_1 + \sigma_2 + \cdots + \sigma_m = n$, where b_i is the ith column of the matrix B (a systematic way of choosing the integers $\sigma_1, \sigma_2, \ldots, \sigma_m$ is given in [11]). Then, there is a transformation matrix T which transforms the given system to its phase canonical form $(A^*, B^*, C^*, D^*)_n$. Matrix T is given by the relation $T = P^{-1}$, where

$$
\mathbf{P} =
\begin{bmatrix}
\mathbf{P}_1 \\
\cdots\cdots \\
\mathbf{P}_2 \\
\vdots \\
\cdots\cdots \\
\mathbf{P}_m
\end{bmatrix}
=
\begin{bmatrix}
\mathbf{P}_1 \\
\mathbf{P}_2 \\
\vdots \\
\mathbf{P}_{\sigma_1} \\
\cdots\cdots\cdots \\
\mathbf{P}_{\sigma_1+1} \\
\mathbf{P}_{\sigma_2+2} \\
\vdots \\
\mathbf{P}_{\sigma_1+\sigma_2} \\
\cdots\cdots\cdots \\
\vdots \\
\cdots\cdots\cdots \\
\mathbf{P}_{n-\sigma_m} \\
\mathbf{P}_{n-\sigma_m+1} \\
\vdots \\
\mathbf{P}_n
\end{bmatrix}
=
\begin{bmatrix}
\mathbf{q}_1 \\
\mathbf{q}_1\mathbf{A} \\
\vdots \\
\mathbf{q}_1\mathbf{A}^{\sigma_1-1} \\
\cdots\cdots\cdots \\
\mathbf{q}_2 \\
\mathbf{q}_2\mathbf{A} \\
\vdots \\
\mathbf{q}_2\mathbf{A}^{\sigma_2-1} \\
\cdots\cdots\cdots \\
\vdots \\
\cdots\cdots\cdots \\
\mathbf{q}_m \\
\mathbf{q}_m\mathbf{A} \\
\vdots \\
\mathbf{q}_m\mathbf{A}^{\sigma_m-1}
\end{bmatrix}
\tag{5.4-11}
$$

where $\mathbf{q}_k$ is the δ_k row of the matrix $\hat{\mathbf{S}}^{-1}$ and where

$$
\delta_k = \sum_{i=1}^{k} \sigma_i \qquad k = 1, 2, \ldots, m
\tag{5.4-12}
$$

Example 5.4.2

Consider a system of the form (5.1-1), where

$$
\mathbf{A} =
\begin{bmatrix}
-2 & -1 & -1 \\
3 & 1 & 1 \\
5 & 1 & 3
\end{bmatrix},
\qquad
\mathbf{B} =
\begin{bmatrix}
1 & 1 \\
-1 & 0 \\
-1 & -1
\end{bmatrix},
\qquad
\mathbf{C} =
\begin{bmatrix}
2 & 0 & 2 \\
0 & 1 & -1
\end{bmatrix},
$$

$$
\mathbf{D} =
\begin{bmatrix}
0 & 0 \\
0 & 0
\end{bmatrix}
$$

Transform this system to its phase canonical form (5.4-6).

Solution

We have $\hat{\mathbf{S}} = [\mathbf{b}_1 \;\vdots\; \mathbf{A}\mathbf{b}_1 \;\vdots\; \mathbf{b}_2]$. Hence

$$
\hat{\mathbf{S}} =
\begin{bmatrix}
1 & \vdots & 0 & \vdots & 1 \\
-1 & \vdots & 1 & \vdots & 0 \\
-1 & \vdots & 1 & \vdots & -1
\end{bmatrix}
\qquad \text{and} \qquad
\hat{\mathbf{S}}^{-1} =
\begin{bmatrix}
1 & -1 & 1 \\
1 & 0 & 1 \\
0 & 1 & -1
\end{bmatrix}
$$

Here, $\sigma_1 = 2$ and $\sigma_2 = 1$ and thus $\sigma_1 + \sigma_2 = 3 = n$. Also, $\delta_1 = \sigma_1 = 2$ and $\delta_2 = \sigma_1 + \sigma_2 = 3$. The rows $\mathbf{q}_1$ and $\mathbf{q}_2$ are the 2nd and 3rd rows of the matrix $\hat{\mathbf{S}}^{-1}$, i.e., $\mathbf{q}_1 = (1, 0, 1)$ and $\mathbf{q}_2 = (0, 1, -1)$. Therefore, the matrix $\mathbf{P}$ has the form

$$P = \begin{bmatrix} q_1 \\ q_1 A \\ \hdotsfor{1} \\ q_2 \end{bmatrix} = \begin{bmatrix} 1 & 0 & 1 \\ 2 & 0 & 3 \\ \hdotsfor{3} \\ 0 & 1 & -1 \end{bmatrix} \quad \text{and} \quad P^{-1} = \begin{bmatrix} -2 & 1 & 0 \\ 3 & -1 & 1 \\ 3 & -1 & 0 \end{bmatrix}$$

Finally, the matrices of the transformed system $(A^*, B^*, C^*, D^*)_n$ have the form

$$A^* = T^{-1}AT = \begin{bmatrix} 0 & 1 & \vdots & 0 \\ 2 & 2 & \vdots & -1 \\ \hdotsfor{4} \\ -2 & 0 & \vdots & 0 \end{bmatrix} = \begin{bmatrix} A_{11}^* & \vdots & A_{12}^* \\ \hdotsfor{3} \\ A_{21}^* & \vdots & A_{22}^* \end{bmatrix}$$

$$B^* = T^{-1}B = \begin{bmatrix} 0 & 0 \\ 1 & 1 \\ \hdotsfor{2} \\ 0 & 1 \end{bmatrix} = \begin{bmatrix} B_1^* \\ \hdotsfor{1} \\ B_2^* \end{bmatrix}$$

$$C^* = CT = \begin{bmatrix} 2 & 0 & 0 \\ 0 & 0 & 1 \end{bmatrix} \quad \text{and} \quad D^* = D = \begin{bmatrix} 0 & 0 \\ 0 & 0 \end{bmatrix}$$

We observe that the forms of A^* and B^* are in agreement with the forms in Eqs (5.4-6).

5.4.3 Transition from an nth Order Differential Equation to State Equations in Phase Canonical Form

Consider the differential equation

$$y^{(n)} + a_{n-1}y^{(n-1)} + a_{n-2}y^{(n-2)} + \cdots + a_1 y^{(1)} + a_0 y = u(t)$$

Define as state variables $x_1(t) = y(t)$, $x_2(t) = y^{(1)}(t)$, ..., $x_n(t) = y^{(n-1)}(t)$. This particular set of state variables are called *phase variables*. Then, as shown in Subsec. 3.8.5, the above differential equation in state space takes on the following form:

$$\dot{x} = Ax + bu, \qquad y = c^T x$$

where A and b are in phase canonical form of the form (5.4-7). The vector c^T has the form $c^T = (1, 0, \ldots, 0)$.

It is important to observe that the structure of matrix A in phase canonical form is as follows

$$A = \begin{bmatrix} 0 & \vdots & I_{n-1} \\ \hdotsfor{3} \\ & a & \end{bmatrix}$$

where 0 is a zero column of dimension $n - 1$, the matrix I_{n-1} is the unity $(n - 1) \times (n - 1)$ matrix and $a = [-a_0, -a_1, -a_2, \ldots, -a_{n-1}]$ is a row whose elements are the coefficients of the differential equation, in reverse order and with negative sign. Due to this special structure of matrix A, we readily have the following three relations:

$$p(s) = |sI - A| = s^n + a_{n-1}s^{n-1} + a_{n-2}s^{n-2} + \cdots + a_1 s + a_0 \qquad (5.4\text{-}13a)$$

$$(s\mathbf{I} - \mathbf{A})^{-1}\mathbf{b} = \frac{1}{p(s)}\begin{bmatrix} 1 \\ s \\ s^2 \\ \vdots \\ s^{n-1} \end{bmatrix} \qquad (5.4\text{-}13\text{b})$$

$$H(s) = \mathbf{c}^{\mathrm{T}}(s\mathbf{I} - \mathbf{A})^{-1}\mathbf{b} = \frac{1}{p(s)} \qquad (5.4\text{-}13\text{c})$$

5.4.4 Transition from the Phase Canonical Form to the Diagonal Form

Assume that a system is already in phase canonical form and that we want to diagonalize the matrix $\mathbf{A}$ of the system. A general method of diagonalizing any square matrix is given in Sec. 2.10. For the special case, where the matrix $\mathbf{A}$ is already in phase canonical form, the diagonalization is simple and may be carried out as follows: a similarity matrix $\mathbf{T}$, which diagonalizes the matrix $\mathbf{A}$, can be determined by letting

$$\mathbf{\Lambda} = \mathbf{T}^{-1}\mathbf{A}\mathbf{T} \qquad \text{or} \qquad \mathbf{T}\mathbf{\Lambda} = \mathbf{A}\mathbf{T}, \qquad \text{where } \mathbf{\Lambda} = \mathrm{diag}\{\lambda_i\} \qquad (5.4\text{-}14)$$

or

$$\begin{bmatrix} \mathbf{t}_1 \\ \mathbf{t}_2 \\ \mathbf{t}_2 \\ \vdots \\ \mathbf{t}_n \end{bmatrix}\begin{bmatrix} \lambda_1 & 0 & 0 & \cdots & 0 \\ 0 & \lambda_2 & 0 & \cdots & 0 \\ 0 & 0 & \lambda_3 & \cdots & 0 \\ \vdots & \vdots & \vdots & \vdots & \vdots \\ 0 & 0 & 0 & \cdots & \lambda_n \end{bmatrix} = \begin{bmatrix} 0 & 1 & 0 & \cdots & 0 \\ 0 & 0 & 1 & \cdots & 0 \\ 0 & 0 & 0 & \cdots & 0 \\ \vdots & \vdots & \vdots & & \vdots \\ -a_0 & -a_1 & -a_2 & \cdots & -a_{n-1} \end{bmatrix}\begin{bmatrix} \mathbf{t}_1 \\ \mathbf{t}_2 \\ \mathbf{t}_3 \\ \vdots \\ \mathbf{t}_n \end{bmatrix}$$

or

$$\mathbf{t}_1\mathbf{\Lambda} = \mathbf{t}_2$$
$$\mathbf{t}_2\mathbf{\Lambda} = \mathbf{t}_3$$
$$\vdots$$
$$\mathbf{t}_{n-1}\mathbf{\Lambda} = \mathbf{t}_n$$
$$\mathbf{t}_n\mathbf{\Lambda} = \mathbf{a}\mathbf{T}$$

where $\mathbf{a} = [-a_0, -a_1, -a_2, \ldots, -a_{n-1}]$ and $\mathbf{t}_i$ is the ith row of the matrix $\mathbf{T}$. The above relations can be written as

$$\mathbf{t}_2 = \mathbf{t}_1\mathbf{\Lambda}$$
$$\mathbf{t}_3 = \mathbf{t}_2\mathbf{\Lambda} = \mathbf{t}_1\mathbf{\Lambda}^2$$
$$\vdots$$
$$\mathbf{t}_{n-1} = \mathbf{t}_{n-2}\mathbf{\Lambda} = \mathbf{t}_1\mathbf{\Lambda}^{n-2}$$
$$\mathbf{t}_n = \mathbf{t}_{n-1}\mathbf{\Lambda} = \mathbf{t}_1\mathbf{\Lambda}^{n-1}$$
$$\mathbf{t}_n\mathbf{\Lambda} = \mathbf{a}\mathbf{T}$$

These equations yield $\mathbf{t}_1 = [1, 1, \ldots, 1]$. Hence, $\mathbf{T}$ has the following form

$$
\mathbf{T} = \begin{bmatrix} \mathbf{t}_1 \\ \mathbf{t}_2 \\ \vdots \\ \mathbf{t}_{n-1} \\ \mathbf{t}_n \end{bmatrix} = \begin{bmatrix} \mathbf{t}_1 \\ \mathbf{t}_1\boldsymbol{\Lambda} \\ \vdots \\ \mathbf{t}_1\boldsymbol{\Lambda}^{n-2} \\ \mathbf{t}_2\boldsymbol{\Lambda}^{n-1} \end{bmatrix} = \begin{bmatrix} 1 & 1 & \cdots & 1 \\ \lambda_1 & \lambda_2 & \cdots & \lambda_n \\ \vdots & \vdots & & \vdots \\ \lambda_1^{n-2} & \lambda_2^{n-2} & \cdots & \lambda_n^{n-2} \\ \lambda_1^{n-1} & \lambda_2^{n-1} & \cdots & \lambda_n^{n-1} \end{bmatrix}
\tag{5.4-15}
$$

This particular form of $\mathbf{T}$ is known as the Vandermonde matrix, and it is always regular under the condition that the eigenvalues $\lambda_1, \lambda_2, \ldots, \lambda_n$ are distinct.

Example 5.4.3

Diagonalize the following matrix

$$
\mathbf{A} = \begin{bmatrix} 0 & 1 & 0 \\ 0 & 0 & 1 \\ 6 & 11 & 6 \end{bmatrix}
$$

Solution

The characteristic polynomial $p(s)$ of matrix $\mathbf{A}$ is

$$
p(s) = |s\mathbf{I} - \mathbf{A}| = s^3 - 6s^2 - 11s - 6 = (s-1)(s-2)(s-3)
$$

Hence, the eigenvalues of $\mathbf{A}$ are 1, 2, and 3. The transformation matrix $\mathbf{T}$ will be

$$
\mathbf{T} = \begin{bmatrix} 1 & 1 & 1 \\ \lambda_1 & \lambda_2 & \lambda_3 \\ \lambda_1^2 & \lambda_2^2 & \lambda_3^2 \end{bmatrix} = \begin{bmatrix} 1 & 1 & 1 \\ 1 & 2 & 3 \\ 1 & 4 & 9 \end{bmatrix}
$$

Therefore

$$
\boldsymbol{\Lambda} = \mathbf{T}^{-1}\mathbf{A}\mathbf{T} = \begin{bmatrix} 1 & 0 & 0 \\ 0 & 2 & 0 \\ 0 & 0 & 3 \end{bmatrix}
$$

5.5 BLOCK DIAGRAMS AND SIGNAL-FLOW GRAPHS

MIMO systems can be described by block diagrams and signal-flow graphs in a similar way that SISO systems are described (see Secs. 3.10 and 3.11). Consider a MIMO system, which is described in state space by the equations

$$
\dot{\mathbf{x}} = \mathbf{A}\mathbf{x} + \mathbf{B}\mathbf{u}, \qquad \mathbf{x}(0) = \mathbf{x}_0
\tag{5.5-1a}
$$

$$
\mathbf{y} = \mathbf{C}\mathbf{x} + \mathbf{D}\mathbf{u}
\tag{5.5-1b}
$$

This system can also be described in the s-domain as follows:

$$
s\mathbf{X}(s) - \mathbf{x}(0) = \mathbf{A}\mathbf{X}(s) + \mathbf{B}\mathbf{U}(s)
\tag{5.5-2a}
$$

$$
\mathbf{Y}(s) = \mathbf{C}\mathbf{X}(s) + \mathbf{D}\mathbf{U}(s)
\tag{5.5-2b}
$$

The block diagrams of Eqs (5.5-1) and (5.5-2) are given in Figures 5.3 and 5.4, respectively. The signal-flow graph of Eqs (5.5-2) is given in Figure 5.5.

For the special case where the system is described by the nth order differential equation

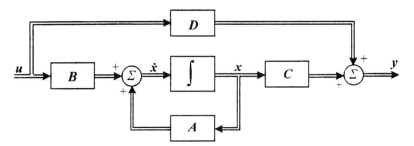

Figure 5.3   Block diagram of state equations in the time domain.

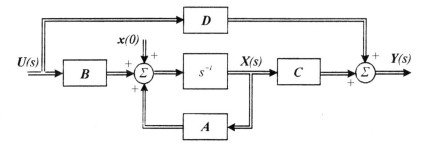

Figure 5.4   Block diagram of state equations in the frequency domain.

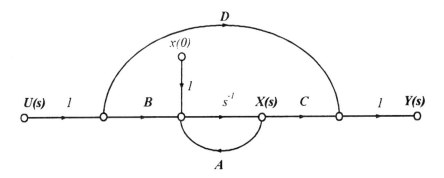

Figure 5.5   Signal-flow graph of state equations.

$$y^{(n)} + a_{n-1}y^{(n-1)} + \cdots + a_1 y^{(1)} + a_0 y = u(t)$$

with zero initial conditions $y^{(k)}(0)$, $k = 0,\ 1, \ldots, n - 1$, the signal-flow graph of the state equations of this differential equation can be constructed as follows. Let $x_1 = y, x_2 = y^{(1)}, \ldots, x_n = y^{(n-1)}$. Then, the differential equation becomes

$$\dot{x}_1 = x_2$$
$$\dot{x}_2 = x_3$$
$$\vdots$$
$$\dot{x}_n = -a_0 x_1 - a_1 x_2 - \cdots - a_{n-1}x_n + u(t)$$

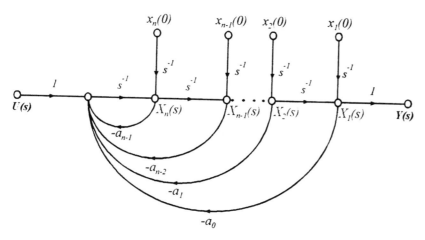

Figure 5.6 Signal-flow graph of state equations of a SISO system.

Take the Laplace transform to yield

$$sX_1(s) - x_1(0) = X_2(s)$$
$$sX_2(s) - x_2(0) = X_3(s)$$

$$\vdots$$

$$sX_n(s) - x_n(0) = -a_0 X_1(s) - a_1 X_2(s) - \cdots - a_{n-1} X_n(s) + U(s)$$

From the above relations we can easily construct the signal-flow graph given in Figure 5.6. The corresponding block diagram has already been presented in Figure 3.12.

5.6 CONTROLLABILITY AND OBSERVABILITY

The concepts of *controllability* and *observability* have been introduced by Kalman [24–30] and are of great theoretical and practical importance in modern control. For example, controllability and observability play an important role in solving several control problems, such as optimal control, adaptive control, pole assignment, etc.

5.6.1 State Vector Controllability

The concept of controllability is related to the state vector as well as to the output vector of a system. Simply speaking, we say that the state (or output) vector is controllable if a control vector $u(t)$ can be found such that the state (or output) vector reaches a preassigned value in a finite period of time. If this is not possible – i.e., even if one state (or output) variable cannot be controlled (in which case we say that this variable is uncontrollable) – it follows that the whole system is uncontrollable.

As an introductionary example, consider the following system:

$$\begin{bmatrix} \dot{x}_1 \\ \dot{x}_2 \end{bmatrix} = \begin{bmatrix} -1 & 0 \\ 0 & -2 \end{bmatrix} \begin{bmatrix} x_1 \\ x_2 \end{bmatrix} + \begin{bmatrix} 0 \\ 1 \end{bmatrix} u$$

From the first equation $\dot{x}_1 = -x_1$, it is obvious that the state variable x_1 is not a function of the input u. Therefore, the behavior of x_1 cannot be affected by the input u, and hence the variable x_1 is uncontrollable. On the contrary, from the second equation $\dot{x}_2 = -2x_2 + u$, it follows that the variable x_2 is controllable since the input u affects x_2, and we can therefore select an input u such that x_2 reaches any preassigned value in a finite period of time.

The strict definition of state controllability of system (5.2-1) is the following.

Definition 5.6.1

The vector $\mathbf{x}(t)$ of system (5.1-1) is *completely controllable* or simply *controllable* if there exists a piecewise continuous control function $\mathbf{u}(t)$ such as to drive $\mathbf{x}(t)$ from its initial condition $\mathbf{x}(t_0)$ to its final value $\mathbf{x}(t_f)$ in a finite period of time $(t_f - t_0) \geq 0$.

In Definition 5.6.1 the expression "a piecewise . . . $\mathbf{u}(t)$" has the meaning that we do not put any limitation on the amplitude or on the energy of $\mathbf{u}(t)$.

The definition of controllability of $\mathbf{x}(t)$ gives a very good insight into the physical meaning of controllability, but it is not easy to apply in order to determine whether or not $\mathbf{x}(t)$ is controllable. To facilitate this problem we give two alternative theorems (criteria) that simplify the determination of the controllability of $\mathbf{x}(t)$.

Theorem 5.6.1

Assume that the matrix $\mathbf{A}$ of system (5.1-1) has distinct eigenvalues. Also assume that the transformation matrix $\mathbf{T}$ diagonalizes matrix $\mathbf{A}$, in which case the diagonalized system will be

$$\dot{z} = \Lambda z + \mathbf{B}^* \mathbf{u}, \qquad \text{where } \Lambda = \text{diag}\{\lambda_i\} = \text{diag}\{\lambda_1, \lambda_2, \ldots, \lambda_n\} \qquad (5.6\text{-}1)$$

where $\mathbf{z} = \mathbf{T}^{-1}\mathbf{x}$, $\Lambda = \mathbf{T}^{-1}\mathbf{A}\mathbf{T}$ and $\mathbf{B}^* = \mathbf{T}^{-1}\mathbf{B}$. Then, $\mathbf{x}(t)$ is controllable if no row of matrix $\mathbf{B}^*$ is a zero row.

Proof

Equation (5.6-1) can be written as

$$\dot{z}_i = \lambda_i z_i + \mathbf{b}^*_i \mathbf{u}, \qquad i = 1, 2, \ldots, n \qquad (5.6\text{-}2)$$

where $\mathbf{b}^*_i$ is the ith row of matrix $\mathbf{B}^*$. From Eq. (5.6-2) it is obvious that the state variable z_i of system (5.6-1) is controllable if at least one of the elements of the row $\mathbf{b}^*_i$ is not zero. Hence, all variables $z_1, z_2, \ldots, z_n$ are controllable if none of the rows of matrix $\mathbf{B}^*$ are zero.

Theorem 5.6.2

The state vector $\mathbf{x}(t)$ of Eq. (5.1-1a) is controllable if and only if

$$\text{Rank}\,\mathbf{S} = n, \qquad \text{where } \mathbf{S} = [\mathbf{B} \,\vdots\, \mathbf{A}\mathbf{B} \,\vdots\, \mathbf{A}^2\mathbf{B} \,\vdots\, \cdots \,\vdots\, \mathbf{A}^{n-1}\mathbf{B}] \qquad (5.6\text{-}3)$$

where $\mathbf{S}$ is called the *controllability matrix* and has dimensions $n \times mn$.

Proof

The solution of Eq. (5.1-1a) is given by Eq. (5.3-6), i.e., by the equation

$$\mathbf{x}(t) = \boldsymbol{\phi}(t - t_0)\mathbf{x}(t_0) + \int_{t_0}^{t} \boldsymbol{\phi}(t - \lambda)\mathbf{Bu}(\lambda)\,d\lambda \tag{5.6-4}$$

To simplify the proof, let $\mathbf{x}(t_f) = \mathbf{0}$. Then, solving Eq. (5.6-4) for $\mathbf{x}(t_0)$ gives

$$\mathbf{x}(t_0) = -\boldsymbol{\phi}^{-1}(t_f - t_0)\int_{t_0}^{t_f} \boldsymbol{\phi}(t_f - \lambda)\mathbf{Bu}(\lambda)\,d\lambda = -\int_{t_0}^{t_f} \boldsymbol{\phi}(-t_f + t_0 + t_f - \lambda)\mathbf{Bu}(\lambda)\,d\lambda$$

$$= -\int_{t_0}^{t_f} \boldsymbol{\phi}(t_0 - \lambda)\mathbf{Bu}(\lambda)\,d\lambda \tag{5.6-5}$$

where use was made of Eq. (5.2-14b). From Cayley–Hamilton's theorem, given in Eq. (2.11-6), we have that

$$\mathbf{A}^k = \sum_{i=0}^{n-1}(\beta_k)_i \mathbf{A}^i, \qquad k \ge n$$

Using this relation, the state transition matrix $\boldsymbol{\phi}(t)$ can be written as

$$\boldsymbol{\phi}(t) = e^{\mathbf{A}t} = \sum_{k=0}^{n-1}\frac{t^k}{k!}\mathbf{A}^k + \sum_{k=n}^{\infty}\frac{t^k}{k!}\sum_{i=0}^{n-1}(\beta_k)_i\mathbf{A}^i = \sum_{i=0}^{n-1}\mathbf{A}^i\sum_{k=0}^{\infty}(\beta_k)_i\frac{t^k}{k!} = \sum_{i=0}^{n-1}\gamma_i(t)\mathbf{A}^i,$$

$$\text{where } \gamma_i(t) = \sum_{k=0}^{\infty}(\beta_k)_i\frac{t^k}{k!}$$

Using the above expression for $\boldsymbol{\phi}(t)$, the matrix $\boldsymbol{\phi}(t_0 - \lambda)$ can be written as

$$\boldsymbol{\phi}(t_0 - \lambda) = \sum_{i=0}^{n-1}\gamma_i(t_0 - \lambda)\mathbf{A}^i \tag{5.6-6}$$

If we substitute Eq. (5.6-6) into Eq. (5.6-5), we have

$$\mathbf{x}(t_0) = -\sum_{i=0}^{n-1}\mathbf{A}^i\mathbf{B}\int_{t_0}^{t_f}\gamma_i(t_0 - \lambda)\mathbf{u}(\lambda)\,d\lambda = -\sum_{i=0}^{n-1}\mathbf{A}^i\mathbf{B}\mathbf{q}_i$$

where the vector $\mathbf{q}_i$ is defined as

$$\mathbf{q}_i = \int_{t_0}^{t_f}\gamma_i(t_0 - \lambda)\mathbf{u}(\lambda)\,d\lambda$$

The above expression for $\mathbf{x}(t_0)$ can be written in compact matrix form, as follows

$$\mathbf{x}(t_0) = -\mathbf{Sq} \tag{5.6-7}$$

where

$$\mathbf{S} = [\mathbf{B} \vdots \mathbf{AB} \vdots \mathbf{A}^2\mathbf{B} \vdots \cdots \vdots \mathbf{A}^{n-1}\mathbf{B}] \quad \text{and} \quad \mathbf{q} = \begin{bmatrix} \mathbf{q}_0 \\ \mathbf{q}_1 \\ \vdots \\ \mathbf{q}_{n-1} \end{bmatrix}$$

Equation (5.6-7) is a system of n equations with $n \times m$ unknowns. The problem at hand, i.e., the determination of an input vector $\mathbf{u}(t)$ such that $\mathbf{x}(t_f) = \mathbf{0}$, has been reduced to that of solving system (5.6-7). From linear algebra, it is well known that for system (5.6-7) to have a solution, the rank of **S** must be equal to n, i.e., condition (5.6-3) must hold.

Example 5.6.1

Determine if the state vector of the system $\dot{\mathbf{x}} = \mathbf{A}\mathbf{x} + \mathbf{b}u$ is controllable, where

$$\mathbf{A} = \begin{bmatrix} 2 & 3 \\ 0 & 5 \end{bmatrix} \quad \text{and} \quad \mathbf{b} = \begin{bmatrix} 1 \\ 0 \end{bmatrix}$$

Solution

Construct the matrix **S**:

$$\mathbf{S} = [\mathbf{b} \vdots \mathbf{A}\mathbf{b}] = \begin{bmatrix} 1 & \vdots & 2 \\ 0 & \vdots & 0 \end{bmatrix}$$

Since $|\mathbf{S}| = 0$, it follows that the rank of **S** is less than $n = 2$. Hence, the state vector is not controllable.

Example 5.6.2

Determine if the state vector of a SISO system in phase canonical form (5.4-7) is controllable.

Solution

Construct the matrix $\mathbf{S}^*$:

$$\mathbf{S}^* = [\mathbf{b}^* \vdots \mathbf{A}^*\mathbf{b}^* \vdots \cdots \vdots (\mathbf{A}^*)^{n-1}\mathbf{b}^*] = \begin{bmatrix} 0 & 0 & \cdots & 1 \\ 0 & 0 & \cdots & \gamma_1 \\ \vdots & \vdots & & \vdots \\ 0 & 1 & \cdots & \gamma_{n-1} \\ 1 & -a_{n-1}^* & \cdots & \gamma_n \end{bmatrix}$$

where $\gamma_1, \gamma_2, \ldots, \gamma_n$ are linear combinations of the coefficients $a_0^*, a_1^*, \ldots, a_{n-1}^*$ of the characteristic polynomial $|s\mathbf{I} - \mathbf{A}^*|$. Due to the lower-diagonal form of matrix $\mathbf{S}^*$, it immediately follows that $|\mathbf{S}^*| = -1$. Hence, the state vector of the SISO system (5.4-7) in phase canonical form is always controllable.

Remark 5.6.1

Example 5.6.2, in combination with Theorem 5.4.2, shows that, for SISO systems, the controllability of the state vector is the necessary and sufficient condition required to transform the system to its phase canonical form. Hence, if a SISO system is already in its phase canonical form, it follows that its state vector is controllable.

5.6.2 Output Vector Controllability

The controllability of the output vector is defined as follows.

Definition 5.6.2

The output vector $\mathbf{y}(t)$ of system (5.1-1) is *completely controllable* or simply *controllable* if there exists a piecewise continuous control function $\mathbf{u}(t)$, which will drive $\mathbf{y}(t)$

from its initial condition $y(t_0)$ to its final value $y(t_f)$, in a finite period of time $(t_f - t_0) \geq 0$.

A simple criterion for determining the controllability of the output vector is given by the following theorem.

Theorem 5.6.3

The output vector $y(t)$ of system (5.1-1) is controllable if and only if

$$\text{Rank} Q = p, \quad \text{where } Q = [D \vdots CB \vdots CAB \vdots CA^2B \vdots \cdots \vdots CA^{n-1}B] \quad (5.6-8)$$

where matrix Q has dimensions $p \times (n+1)m$.

The proof of Theorem 5.6.3 is analogous to the proof of Theorem 5.6.2.

Example 5.6.3

Consider a system of the form (5.1-1), where

$$A = \begin{bmatrix} -1 & 0 \\ 0 & -2 \end{bmatrix}, \quad B = \begin{bmatrix} 1 & 0 \\ 0 & 1 \end{bmatrix}, \quad C = \begin{bmatrix} 1 & 1 \\ 1 & 0 \end{bmatrix}, \quad D = \begin{bmatrix} 1 & 1 \\ 0 & 0 \end{bmatrix}$$

Determine if the state and output vectors are controllable.

Solution

We have

$$S = [B \vdots AB] = \begin{bmatrix} 1 & 0 & \vdots & -1 & 0 \\ 0 & 1 & \vdots & 0 & -2 \end{bmatrix}$$

Since $\text{Rank} S = 2$, it follows that the state vector is controllable. Furthermore, we have

$$Q = [D \vdots CB \vdots CAB] = \begin{bmatrix} 1 & 1 & \vdots & 1 & 1 & \vdots & -1 & -2 \\ 0 & 0 & \vdots & 1 & 0 & \vdots & -1 & 0 \end{bmatrix}$$

Since $\text{Rank} Q = 2$, it follows that the output vector is also controllable.

5.6.3 State Vector Observability

The concept of *observability* is related to the state variables of the system and it is *dual* to the concept of *controllability* (the concept of duality is explained in Remarks 5.6.2 and 5.6.3, which follow). Assume that we have available the input vector $u(t)$ and the corresponding output vector $y(t)$ of system (5.1-1) over a finite period of time. If, on the basis of these measurements of $u(t)$ and $y(t)$, one can determine the vector of initial conditions $x(t_0)$, then we say that the system is observable. In case that this is not possible—i.e., even if one element of the vector of initial conditions $x(t_0)$ cannot be determined—then we say that this element is unobservable, and as a result we say that the whole system is unobservable.

As an introductory example, consider the following system:

$$\begin{bmatrix} \dot{x}_1 \\ \dot{x}_2 \end{bmatrix} = \begin{bmatrix} -1 & 0 \\ 0 & -2 \end{bmatrix} \begin{bmatrix} x_1 \\ x_2 \end{bmatrix} + \begin{bmatrix} 1 \\ 1 \end{bmatrix} u, \quad y = [2 \quad 0] \begin{bmatrix} x_1 \\ x_2 \end{bmatrix}$$

The above description breaks down to the two differential equations $\dot{x}_1 = -x_1 + u$ and $\dot{x}_2 = -2x_2 + u$. Furthermore, the output is given by $y = 2x_1$. Since $\dot{x}_1 = -x_1 + u$ — i.e., x_1 is not a function of x_2 but only a function of u, it follows that the output of the system is affected only by the state x_1 and therefore the output does not involve any information regarding the state x_2. As a result, the determination of the initial condition $x_2(t_0)$ becomes impossible. Hence, the system at hand is unobservable.

The strict definition of observability is as follows.

Definition 5.6.3

The state vector $\mathbf{x}(t)$ of system (5.1-1) is observable in the time interval $[t_0, t_f]$ if, knowing the input $\mathbf{u}(t)$ and the output $\mathbf{y}(t)$ for $t \in [t_0, t_f]$, one can determine the initial condition vector $\mathbf{x}(t_0)$.

In the sequel, we present two alternative theorems (criteria) that simplify the procedure of determining the observability of $\mathbf{x}(t)$.

Theorem 5.6.4

Let system (5.1-1) have distinct eigenvalues. Furthermore, let the transformation matrix $\mathbf{T}$ be given by (5.4-15), in which case $\mathbf{T}$ diagonalizes matrix $\mathbf{A}$. Then the diagonalized system is the following:

$$\dot{\mathbf{z}} = \Lambda \mathbf{z} + \mathbf{B}^* \mathbf{u} \tag{5.6-9a}$$

$$\mathbf{y} = \mathbf{C}^* \mathbf{z} + \mathbf{D} \mathbf{u} \tag{5.6-9b}$$

where $\mathbf{z} = \mathbf{T}^{-1} \mathbf{x}$, $\Lambda = \mathbf{T}^{-1} \mathbf{A} \mathbf{T}$, $\mathbf{B}^* = \mathbf{T}^{-1} \mathbf{B}$, and $\mathbf{C}^* = \mathbf{C} \mathbf{T}$. Then, $\mathbf{x}(t)$ is observable if no column of matrix $\mathbf{C}^*$ is a zero column.

Proof

The output Eq. (5.6-9b) can be written as

$$\mathbf{y} = \mathbf{c}_1^* z_1 + \mathbf{c}_2^* z_2 + \cdots + \mathbf{c}_i^* z_i + \cdots + \mathbf{c}_n^* z_n + \mathbf{D} \mathbf{u} \tag{5.6-10}$$

where $\mathbf{c}_i^*$ is the ith column of matrix $\mathbf{C}^*$. It is obvious that if one column, for example $\mathbf{c}_i$, of matrix $\mathbf{C}^*$, is zero then the corresponding state variable z_i will not appear in the output $\mathbf{y}(t)$. Consequently, we cannot, in this case, determine the initial condition $z_i(t_0)$. As a result, $\mathbf{x}(t)$ is unobservable.

Remark 5.6.2

Theorems 5.6.1 and 5.6.4 are *dual*, in the sense that the role of rows of matrix $\mathbf{B}^*$ for controllability play the columns of matrix $\mathbf{C}^*$ for observability.

Theorem 5.6.5

The state vector $\mathbf{x}(t)$ of system (5.1-1) is observable if and only if

$$\text{rank} \mathbf{R}^T = n, \quad \text{where } \mathbf{R}^T = [\mathbf{C}^T \vdots \mathbf{A}^T \mathbf{C}^T \vdots (\mathbf{A}^T)^2 \mathbf{C}^T \vdots \cdots \vdots (\mathbf{A}^T)^{n-1} \mathbf{C}^T] \tag{5.6-11}$$

where $\mathbf{R}$ is called the *observability matrix* and has dimensions $n \times np$.

Proof

The general solution of Eq. (5.2-1) is given by Eq. (5.3-7), i.e., by the equation

$$\mathbf{y}(t) = \mathbf{C}\boldsymbol{\phi}(t - t_0)\mathbf{x}(t_0) + \mathbf{C}\int_{t_0}^{t} \boldsymbol{\phi}(t - \lambda)\mathbf{B}\mathbf{u}(\lambda)\,d\lambda + \mathbf{D}\mathbf{u}(t) \qquad (5.6\text{-}12)$$

To simplify the proof, let $\mathbf{u}(t) = \mathbf{0}$. Then, Eq. (5.6-12) becomes

$$\mathbf{y}(t) = \mathbf{C}\boldsymbol{\phi}(t - t_0)\mathbf{x}(t_0)$$

If we use Eq. (5.6-6) in the above relation, we have

$$\mathbf{y}(t) = \mathbf{C}\left[\sum_{i=0}^{n-1} \gamma_i(t - t_0)\mathbf{A}^i\right]\mathbf{x}(t_0) = \left[\sum_{i=0}^{n-1} \gamma_i(t - t_0)\mathbf{C}\mathbf{A}^i\right]\mathbf{x}(t_0)$$

The above relation can be written in compact matrix form as follows:

$$\mathbf{y}(t) = \mathbf{E}\mathbf{R}\mathbf{x}(t_0) \qquad (5.6\text{-}13)$$

where

$$\mathbf{E} = [\gamma_0\mathbf{I} \vdots \gamma_1\mathbf{I} \vdots \gamma_2\mathbf{I} \vdots \cdots \vdots \gamma_{n-1}\mathbf{I}] \qquad \text{and} \qquad \mathbf{R} = \begin{bmatrix} \mathbf{C} \\ \mathbf{C}\mathbf{A} \\ \mathbf{C}\mathbf{A}^2 \\ \vdots \\ \mathbf{C}\mathbf{A}^{n-1} \end{bmatrix}$$

From linear algebra it is well known that for system (5.6-13) to have a solution for $\mathbf{x}(t_0)$, the rank of matrix $\mathbf{R}$, or equivalently the rank of its transpose matrix

$$\mathbf{R}^\mathrm{T} = [\mathbf{C}^\mathrm{T} \vdots \mathbf{A}^\mathrm{T}\mathbf{C}^\mathrm{T} \vdots (\mathbf{A}^\mathrm{T})^2\mathbf{C}^\mathrm{T} \vdots \cdots \vdots (\mathbf{A}^\mathrm{T})^{n-1}\mathbf{C}^\mathrm{T}]$$

must be equal to n.

Remark 5.6.3

Theorems 5.6.2 and 5.6.5 are *dual*, in the sense that the role of the matrices $\mathbf{B}$ and $\mathbf{A}$ in the controllability matrix $\mathbf{S}$ play the matrices $\mathbf{C}^\mathrm{T}$ and $\mathbf{A}^\mathrm{T}$ in the observability matrix $\mathbf{R}^\mathrm{T}$. The duality of $\mathbf{S}$ and $\mathbf{R}^\mathrm{T}$ also appears in transforming a system to its phase canonical form. In Subsec. 5.4.2 we presented a method of transforming system $(\mathbf{A}, \mathbf{B}, \mathbf{C}, \mathbf{D})_n$ to its phase canonical form $(\mathbf{A}^*, \mathbf{B}^*, \mathbf{C}^*, \mathbf{D}^*)_n$ based on the controllability matrix $\mathbf{S}$, where the matrices $\mathbf{A}^*$ and $\mathbf{B}^*$ have special forms. In an analogous way, system $(\mathbf{A}, \mathbf{B}, \mathbf{C}, \mathbf{D})_n$ can be transformed to its phase canonical form $(\mathbf{A}^+, \mathbf{B}^+, \mathbf{C}^+, \mathbf{D}^+)_n$ based on the observability matrix $\mathbf{R}$, where the matrices $\mathbf{A}^+$ and $\mathbf{C}^+$ have special forms. The forms of the matrices $\mathbf{A}^*$ and $\mathbf{A}^+$, and of the matrices $\mathbf{B}^*$ and $\mathbf{C}^+$, are dual. In order to distinguish these two cases, we say that system $(\mathbf{A}^*, \mathbf{B}^*, \mathbf{C}^*, \mathbf{D}^*)_n$ is in its *input* phase canonical form, whereas system $(\mathbf{A}^+, \mathbf{B}^+, \mathbf{C}^+, \mathbf{D}^+)_n$ is in its *output* phase canonical form. The example that follows refers to the determination of the output phase canonical form.

Example 5.6.4

Determine if the state vector of a system with matrices

$$A = \begin{bmatrix} -1 & 0 & 0 \\ 0 & -2 & 0 \\ 0 & 0 & -3 \end{bmatrix}, \quad B = \begin{bmatrix} 0 & 0 \\ 1 & 0 \\ 0 & 1 \end{bmatrix}, \quad \text{and} \quad C = [1 \quad 1 \quad 1]$$

is observable. Furthermore, determine the output phase canonical form of the system.

Solution

Construct the matrix $\mathbf{R}^T$:

$$\mathbf{R}^T = [\mathbf{C}^T \vdots \mathbf{A}^T\mathbf{C}^T \vdots (\mathbf{A}^T)^2\mathbf{C}^T] = \begin{bmatrix} 1 & \vdots & -1 & \vdots & 1 \\ 1 & \vdots & -2 & \vdots & 4 \\ 1 & \vdots & -3 & \vdots & 9 \end{bmatrix}$$

Since rank $\mathbf{R} = 3$, it follows that the state vector of the given system is observable. To determine its output phase canonical form, we have

$$(\mathbf{R}^T)^{-1} = \frac{1}{2}\begin{bmatrix} 6 & -6 & 2 \\ 5 & -8 & 3 \\ 1 & -2 & 1 \end{bmatrix}, \quad \mathbf{P} = \begin{bmatrix} \mathbf{q}_3 \\ \mathbf{q}_3\mathbf{A}^T \\ \mathbf{q}_3(\mathbf{A}^T)^2 \end{bmatrix} = \frac{1}{2}\begin{bmatrix} 1 & -2 & 1 \\ -1 & 4 & -3 \\ 1 & -8 & 9 \end{bmatrix},$$

$$\mathbf{T} = \mathbf{P}^{-1} = \begin{bmatrix} 6 & 5 & 1 \\ 3 & 4 & 1 \\ 2 & 3 & 1 \end{bmatrix}$$

where $\mathbf{q}_3$ is the last row of $(\mathbf{R}^T)^{-1}$. The matrices of the given system in output phase canonical form are

$$(\mathbf{A}^+)^T = \mathbf{T}^{-1}\mathbf{A}^T\mathbf{T} = \begin{bmatrix} 0 & 1 & 0 \\ 0 & 0 & 1 \\ -6 & -11 & -6 \end{bmatrix}, \quad (\mathbf{B}^+)^T = \mathbf{B}^T\mathbf{T} = \begin{bmatrix} 3 & 4 & 1 \\ 2 & 3 & 1 \end{bmatrix}$$

$$(\mathbf{C}^+)^T = \mathbf{T}^{-1}\mathbf{C}^T = \begin{bmatrix} 0 \\ 0 \\ 1 \end{bmatrix}$$

5.6.4   The Invariance of Controllability and Observability

The properties of controllability and observability are invariant under state vector similarity transformation. Indeed, the matrices $\mathbf{S}^*$, $\mathbf{Q}^*$, and $\mathbf{R}^*$ of the transformed system $(\mathbf{A}^*, \mathbf{B}^*, \mathbf{C}^*, \mathbf{D}^*)_n$ are related to the matrices $\mathbf{S}$, $\mathbf{Q}$, and $\mathbf{R}$ of the original system $(\mathbf{A}, \mathbf{B}, \mathbf{C}, \mathbf{D})_n$ as follows:

$$\mathbf{S}^* = [\mathbf{B}^* \vdots \mathbf{A}^*\mathbf{B}^* \vdots (\mathbf{A}^*)^2\mathbf{B}^* \vdots \cdots \vdots (\mathbf{A}^*)^{n-1}\mathbf{B}^*]$$

$$= [\mathbf{T}^{-1}\mathbf{B} \vdots \mathbf{T}^{-1}\mathbf{A}\mathbf{T}\mathbf{T}^{-1}\mathbf{B} \vdots (\mathbf{T}^{-1}\mathbf{A}\mathbf{T})^2\mathbf{T}^{-1}\mathbf{B} \vdots \cdots \vdots (\mathbf{T}^{-1}\mathbf{A}\mathbf{T})^{n-1}\mathbf{T}^{-1}\mathbf{B}]$$

$$= [\mathbf{T}^{-1}\mathbf{B} \vdots \mathbf{T}^{-1}\mathbf{A}\mathbf{B} \vdots \mathbf{T}^{-1}\mathbf{A}^2\mathbf{B} \vdots \cdots \vdots \mathbf{T}^{-1}\mathbf{A}^{n-1}\mathbf{B}]$$

$$= \mathbf{T}^{-1}[\mathbf{B} \vdots \mathbf{A}\mathbf{B} \vdots \mathbf{A}^2\mathbf{B} \vdots \cdots \vdots \mathbf{A}^{n-1}\mathbf{B}] = \mathbf{T}^{-1}\mathbf{S} \qquad (5.6\text{-}14)$$

Furthermore

$$\mathbf{Q}^* = [\mathbf{D}^* \vdots \mathbf{C}^* \mathbf{B}^* \vdots \mathbf{C}^* \mathbf{A}^* \mathbf{B}^* \vdots \cdots \vdots \mathbf{C}^* (\mathbf{A}^*)^{n-1} \mathbf{B}^*]$$

$$= [\mathbf{D} \vdots \mathbf{CTT}^{-1} \mathbf{B} \vdots \mathbf{CTT}^{-1} \mathbf{ATT}^{-1} \mathbf{B} \vdots \cdots \vdots \mathbf{CT}(\mathbf{T}^{-1} \mathbf{AT})^{n-1} \mathbf{T}^{-1} \mathbf{B}]$$

$$= [\mathbf{D} \vdots \mathbf{CB} \vdots \mathbf{CAB} \vdots \cdots \vdots \mathbf{CA}^{n-1} \mathbf{B}] = \mathbf{Q} \qquad (5.6\text{-}15)$$

Finally,

$$\mathbf{R}^{*T} = [\mathbf{C}^{*T} \vdots \mathbf{A}^{*T} \mathbf{C}^{*T} \vdots (\mathbf{A}^{*T})^2 \mathbf{C}^{*T} \vdots \cdots \vdots (\mathbf{A}^{*T})^{n-1} \mathbf{C}^{*T}]$$

$$= [\mathbf{T}^T \mathbf{C}^T \vdots \mathbf{T}^T \mathbf{A}^T (\mathbf{T}^{-1})^T \mathbf{T}^T \mathbf{C}^T \vdots [\mathbf{T}^T \mathbf{A}^T (\mathbf{T}^{-1})^T]^2 \mathbf{T}^T \mathbf{C}^T \vdots \cdots \vdots$$
$$[\mathbf{T}^T \mathbf{A}^T (\mathbf{T}^{-1})^T]^{n-1} \mathbf{T}^T \mathbf{C}^T]$$

$$= [\mathbf{T}^T \mathbf{C}^T \vdots \mathbf{T}^T \mathbf{A}^T \mathbf{C}^T \vdots \mathbf{T}^T (\mathbf{A}^T)^2 \mathbf{C}^T \vdots \cdots \vdots \mathbf{T}^T (\mathbf{A}^T)^{n-1} \mathbf{C}^T]$$

$$= \mathbf{T}^T [\mathbf{C}^T \vdots \mathbf{A}^T \mathbf{C}^T \vdots (\mathbf{A}^T)^2 \mathbf{C}^T \vdots \cdots \vdots (\mathbf{A}^T)^{n-1} \mathbf{C}^T] = \mathbf{T}^T \mathbf{R}^T \qquad (5.6\text{-}16)$$

From Eqs (5.6-14)–(5.6-16) it follows that, if the original system is controllable and/
or observable, then the transformed system is also controllable and/or observable.

5.6.5 Relation Among Controllability, Observability, and Transfer Function Matrix

It is clear that the transfer function matrix $\mathbf{H}(s)$ is an input–output description of a
system. That is, it relates the input vector $\mathbf{U}(s)$ to the output vector $\mathbf{Y}(s)$ of the system
without involving the state vector $\mathbf{X}(s)$. At this point, we raise the following question:
Is the transfer function matrix $\mathbf{H}(s)$ affected and how by the properties of controllability and observability of the system? The answer to this question is of great
importance and constitutes one of the basic reasons for preferring the state equations
over transfer function matrices for describing control systems. In the sequel, we will
try to give the answer to this question.

We introduce the following definition.

Definition 5.6.4

Consider the sequences

$$\mathbf{S}_j = [\mathbf{B} \vdots \mathbf{AB} \vdots \cdots \vdots \mathbf{A}^{j-1} \mathbf{B}] \qquad (5.6\text{-}17)$$

$$\mathbf{R}_j^T = [\mathbf{C}^T \vdots \mathbf{A}^T \mathbf{C}^T \vdots \cdots \vdots (\mathbf{A}^T)^{j-1} \mathbf{C}^T] \qquad (5.6\text{-}18)$$

Let α and β be the smallest positive integer numbers such that $\text{rank}\mathbf{S}_\alpha = \text{rank}\mathbf{S}_{\alpha+1}$
and $\text{rank}\mathbf{R}_\beta = \text{rank}\mathbf{R}_{\beta+1}$ (therefore $\text{rank}\mathbf{S}_i = \text{rank}\mathbf{S}_\alpha$, $\forall i > \alpha$ and $\text{rank}\mathbf{R}_i = \text{rank}\mathbf{R}_\beta$,
$\forall i > \beta$). Then, the index α is called the *controllability index* and the index β is called
the *observability index*.

It has been proven that there is a strong relationship among the three system
characteristics: (a) controllability and observability; (b) the matrices $\mathbf{S}_\alpha$ and $\mathbf{R}_\beta$; and
(c) the minimum state-space realization (see Subsec. 3.8.5). This strong relationship
is stated in the following theorem.

Theorem 5.6.6

For system (5.1-1), the following three propositions are equivalent:

(a) The system is observable and controllable—namely, the rank of both matrices $\mathbf{S}$ and $\mathbf{R}$ is n.
(b) Rank$\mathbf{R}_\beta\mathbf{S}_\alpha = n$.
(c) The dimension n of the state-space realization is minimum.

The following theorem relates the classical and the modern control theory, since it relates the classical input–output description $\mathbf{H}(s)$ to the modern description of a system in state space. This relation has come to light from the concepts of controllability and observability.

Theorem 5.6.7

If the transfer function matrix of a system involves pole-zero cancellations, then the system is either uncontrollable or unobservable or both. If the transfer function matrix does not involve any pole-zero cancellation, then the system is both controllable and observable.

Proof

Consider an SISO system which is already in diagonal form, i.e., consider the following diagonal system

$$\dot{\mathbf{x}} = \mathbf{\Lambda}\mathbf{x} = \mathbf{b}u \qquad\qquad (5.6\text{-}19a)$$

$$y = \mathbf{c}^T\mathbf{x} \qquad\qquad (5.6\text{-}19b)$$

For this system we have

$$\dot{x}_i = \lambda_i x_i + b_i u, \qquad i = 1, 2, \ldots, n$$

or

$$X_i(s) = \frac{b_i}{s - \lambda_i} U(s), \qquad i = 1, 2, \ldots, n \qquad\qquad (5.6\text{-}20)$$

where b_i is the ith element of $\mathbf{b}$. Therefore

$$Y(s) = \mathbf{c}^T\mathbf{X}(s) = \sum_{i=1}^{n} c_i X_i(s) = \sum_{i=1}^{n} \frac{c_i b_i}{s - \lambda_i} U(s) \qquad\qquad (5.6\text{-}21)$$

where c_i is the ith element of the vector $\mathbf{c}$. The transfer function $H(s)$ of the system has the general form

$$H(s) = K\frac{(s - z_1)(s - z_2)\cdots(s - z_m)}{(s - \lambda_1)(s - \lambda_2)\cdots(s - \lambda_n)}, \qquad m < n$$

If we expand $H(s)$ in partial functions, we have

$$H(s) = \sum_{i=1}^{n} \frac{h_i}{s - \lambda_i} \qquad\qquad (5.6\text{-}22)$$

Comparing Eqs (5.6-21) and (5.6-22), it follows that

$$h_i = c_i b_i \qquad\qquad (5.6\text{-}23)$$

Consequently, if $H(s)$ involves cancellation of the pole λ_i, then there must be $h_i = 0$. From Eq. (5.6-23) it follows that for $h_i = 0$ to hold, then there must be either $b_i = 0$ or $c_i = 0$, or even $b_i = c_i = 0$. Clearly, if $b_i = 0$, then the system is uncontrollable (because one of the rows of **b** is zero). If $c_i = 0$, then the system is unobservable (because one of the columns of $\mathbf{c}^T$ is zero). Finally, if $b_i = c_i = 0$, then the system is both uncontrollable and unobservable.

If $h_i \neq 0$, $\forall i$, then all c_i's and b_i's are different than zero and hence system (5.6-19) is both controllable and observable.

The following theorem is of great practical importance.

Theorem 5.6.8

The transfer function matrix involves only the controllable and observable part of a system.

The practical importance of the previous theorems, and particularly of Theorem 5.6.8, is demonstrated by the following examples.

Example 5.6.5

Consider the network of Figure 5.7. Determine the transfer function and study the case of pole-zero cancellations.

Solution

The transfer function of the network is

$$H(s) = \frac{Y(s)}{U(s)} = \frac{Z_2(s)}{Z_1(s) + Z_2(s)}$$

where

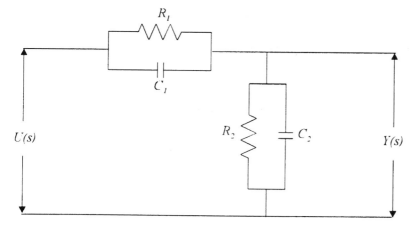

Figure 5.7 RC network.

$$Z_1(s) = \frac{R_1\left[\dfrac{1}{C_1 s}\right]}{R_1 + \dfrac{1}{C_1 s}} = \frac{R_1}{R_1 C_1 s + 1}, \qquad Z_2(s) = \frac{R_2\left[\dfrac{1}{C_2 s}\right]}{R_2 + \dfrac{1}{C_2 s}} = \frac{R_2}{R_2 C_2 s + 1}$$

Furthermore,

$$H(s) = \frac{\dfrac{R_2}{R_c C_2 s + 1}}{\dfrac{R_1}{R_1 C_1 s + 1} + \dfrac{R_2}{R_2 C_2 s + 1}} = \frac{R_2(R_1 C_1 s + 1)}{R_1(R_2 C_2 s + 1) + R_2(R_1 C_1 s + 1)}$$

$$= \left[\frac{C_1}{C_1 + C_2}\right]\left[\frac{s + \alpha}{s + \beta}\right]$$

where

$$\alpha = \frac{1}{R_1 C_1}, \qquad \beta = \frac{R_1 + R_2}{R_1 R_2 (C_1 + C_2)}$$

If we choose R_1, R_2, C_1, and C_2 such that $R_1 C_1 = R_2 C_2$, then $\alpha = \beta$ and

$$H(s) = \frac{C_1}{C_1 + C_2} = \frac{R_2}{R_1 + R_2} = \text{constant}$$

The above results indicate that, in general, $H(s)$ is a function of s. However, in the special case where $R_1 C_1 = R_2 C_2$ the transfer function $H(s)$ reduces to a constant. In this case, $H(s)$ gives misleading information about the network, because one may arrive at the conclusion that the RC network is a pure resistive network, which it is not. Furthermore, when $R_1 C_1 = R_2 C_2$, the network is neither controllable (the voltages of the capacitors C_1 and C_2 cannot be controlled) nor observable (the initial voltages of the capacitors C_1 and C_2 cannot be estimated from input and output data).

Example 5.6.6

Consider the SISO system $\dot{x} = Ax + bu$, $y = c^T x$, where

$$A = \begin{bmatrix} 0 & 1 \\ 2 & -1 \end{bmatrix}, \qquad b = \begin{bmatrix} 0 \\ 1 \end{bmatrix}, \qquad \text{and } c = \begin{bmatrix} -1 \\ 1 \end{bmatrix}$$

Determine if the system is controllable and observable and find its transfer function.

Solution

Since the system is already in its phase canonical form, it immediately follows that it is controllable. To determine if the system is observable, construct the matrix

$$R^T = [c \vdots A^T c] = \begin{bmatrix} -1 & \vdots & 2 \\ 1 & \vdots & -2 \end{bmatrix}$$

Since $|R| = 0$, it follows that rank $R < 2$, and hence the system is not observable. The transfer function of the system is

$$H(s) = c^T(sI - A)^{-1}b = \frac{s - 1}{(s - 1)(s + 2)} = \frac{1}{s + 2}$$

Clearly, since the system is unobservable, it was expected that at least one pole-zero cancellation will take place. The cancellation of the factor $(s - 1)$ results in "concealing" the eigenvalue 1. This can have extremely undesirable results, particularly in cases like the present example, where the cancelled out pole is in the right-half complex plane. In this case, if the system were excited, the states of the system would increase with time, and could even break down or burn out the system. Indeed, if, for simplicity, we let $u(t) = \delta(t)$, then

$$X_1(s) = \frac{1}{(s-1)(s+2)} \qquad \text{and thus} \qquad x_1(t) = \frac{1}{3}(e^t - e^{-2t})$$

$$X_2(s) = \frac{s}{(s-1)(s+2)} \qquad \text{and thus} \qquad x_2(t) = \frac{1}{3}(e^t + 2e^{-2t})$$

Hence, due to the term e^t, both $x_1(t)$ and $x_2(t)$ increase with time. Now, consider determining the output on the bases of the transfer function. We have

$$Y(s) = H(s)\, U(s) = \frac{1}{(s+2)} \qquad \text{and thus} \qquad y(t) = e^{-2t}$$

Clearly, the output gives very misleading results as to the behavior of the system: in reality, the system may break down or burn out and the output leads us to believe that there is "no problem" and that the system's behavior is described by the decaying function $y(t) = e^{-2t}$. This leads to the conclusion that we should not "trust" $H(s)$ unless the following theorem is satisfied.

Theorem 5.6.9

When a system is controllable and observable, then no pole-zero cancellations take place and its transfer function matrix constitutes a *complete* description of the system.

5.7 KALMAN DECOMPOSITION

Kalman showed that it is possible to introduce certain coordinates, using a suitable transformation matrix **T**, such that a system can be decomposed as follows:

$$\dot{\mathbf{x}}(t) = \begin{bmatrix} \mathbf{A}_{11} & \mathbf{A}_{12} & \mathbf{0} & \mathbf{0} \\ \mathbf{0} & \mathbf{A}_{22} & \mathbf{0} & \mathbf{0} \\ \mathbf{A}_{31} & \mathbf{A}_{32} & \mathbf{A}_{33} & \mathbf{A}_{34} \\ \mathbf{0} & \mathbf{A}_{42} & \mathbf{0} & \mathbf{A}_{44} \end{bmatrix} \mathbf{x}(t) + \begin{bmatrix} \mathbf{B}_1 \\ \mathbf{0} \\ \mathbf{B}_3 \\ \mathbf{0} \end{bmatrix} \mathbf{u}(t) \qquad (5.7\text{-}1a)$$

$$\mathbf{y}(t) = [\ \mathbf{C}_1 \quad \mathbf{C}_2 \quad \mathbf{0} \quad \mathbf{0}\ ]\, \mathbf{x}(t) \qquad (5.7\text{-}1b)$$

where $\mathbf{A}_{ij}$, $\mathbf{B}_i$, and $\mathbf{C}_i$ are block matrices of suitable dimensions. The state-space vector $\mathbf{x}(t)$ is accordingly decomposed into four subvectors, each one of which corresponds to one of the following four cases:

- States that are both controllable and observable
- States that are uncontrollable but are observable
- States that are controllable but unobservable
- States that are both uncontrollable and unobservable.

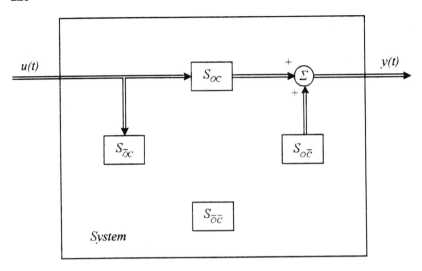

Figure 5.8   Block diagram of the Kalman decomposition.

The following theorem, which is related to the above results, is called the *Kalman decomposition theorem.*

Theorem 5.7.1

A system can be decomposed into four subsystems with the following properties:

- Subsystem S_{OC}: the observable and controllable subsystem (A_{11}, B_1, C_1).
- Subsystem $S_{O\bar{C}}$: the observable but uncontrollable subsystem $(A_{22}, 0, C_2)$.
- Subsystem $S_{\bar{O}C}$: the unobservable but controllable subsystem $(A_{33}, B_3, 0)$.
- Subsystem $S_{\bar{O}\bar{C}}$: the unobservable and uncontrollable subsystem $(A_{44}, 0, 0)$.

The transfer function matrix of system (5.7-1) is unique and can be determined from the subsystem which is both controllable and observable. Indeed, straightforward calculations show that the transfer function matrix of system (5.7-1) is given by

$$H(s) = C_1(sI - A_{11})^{-1}B_1 \qquad (5.7\text{-}2)$$

Relation (5.7-2) contains only the controllable and the observable part of the system.

Figure 5.8 shows the block diagram of the Kalman decomposition, involving all four subsystems and the way in which they are linked to each other. Also, it shows that the input is related to the output only through the subsystem S_{OC}.

5.8   PROBLEMS

1. Consider the frictionless horizontal movement of the mass m shown in Figure 5.9a. A force $f(t)$ is applied to the mass. The displacement and the velocity of the mass are $y(t)$ and $v(t)$, respectively. The input, state, and output variables are defined in Figure 5.9b. Determine (a) the state equations of the system, (b) the transition matrix, (c) the solution of the homogeneous equation, and (d) the general solution, given that $f(t) = f_0 u(t)$, and (e) the output of the system.

(a)

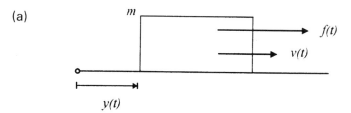

m

$f(t)$

$v(t)$

$y(t)$

(b)

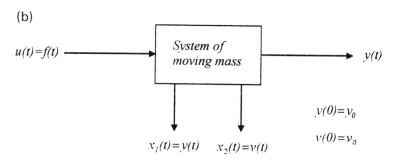

$u(t)=f(t)$ ⟶ System of moving mass ⟶ $y(t)$

$v(0)=v_0$

$v(0)=v_0$

$x_1(t)=y(t)$ $x_2(t)=v(t)$

Figure 5.9

2. For the RLC network shown in Figure 5.10a, determine (a) the state equations, using the definitions of input, state, and output variables shown in Figure 5.10b, (b) the transition matrix, (c) the solution of the homogeneous equation, (d) the general solution, and (e) the output vector of the system.

3. Consider the mechanical system shown in Figure 5.11a, where the two inputs are the forces $f_1(t)$ and $f_2(t)$ applied to the masses m_1 and m_2, respectively. The two outputs are the displacements $y_1(t)$ and $y_2(t)$ of the masses. Using the definitions shown in Figure 5.11b, determine (a) the state equations of the system and (b) the transition matrix and the output vector of the system, when $m_1 = m_2 = 1\,\mathrm{kg}$, $K_1 = 1$, $K_2 = 2$, $B_1 = 1$, $f_1(t) = f_2(t) = 1$, $y_1(0) = 1$, $\dot{y}_1(0) = y_2(0) = \dot{y}_2(0) = 0$.

4. Consider the liquid-level control system shown in Figure 5.12a. Using the definitions of Figure 5.12b, determine (a) the state equations of the system and (b) the transition matrix and the output vector when the horizontal area of both tanks is $A = 1\,m^2$ and the flow resistance at the output of both tanks is $R = 1$.

5. A system of two carts connected by a damper is shown in Figure 5.13a. The input, state, and output variables are defined in Figure 5.13b, where f is the force applied, v_1 and v_2 are the velocities of the two carts, and the difference $v_2 - v_1$ is the system output. (a) Determine the state equations and (b) transform the system into phase canonical form.

6. A control system is described by the following state equations:

$$\begin{bmatrix} \dot{x}_1 \\ \dot{x}_2 \end{bmatrix} = \begin{bmatrix} 0 & 1 \\ -2 & -3 \end{bmatrix} \begin{bmatrix} x_1 \\ x_2 \end{bmatrix} + \begin{bmatrix} 0 \\ 2 \end{bmatrix} u$$

$$y = \begin{bmatrix} 1 & 0 \end{bmatrix} \begin{bmatrix} x_1 \\ x_2 \end{bmatrix}$$

(a)
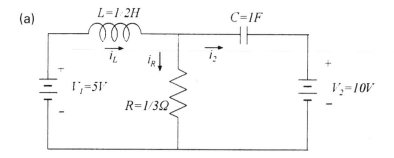

(b)
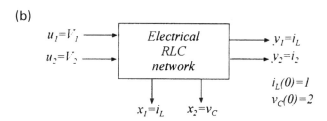

Figure 5.10

(a)
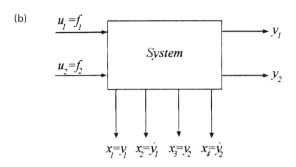

(b)
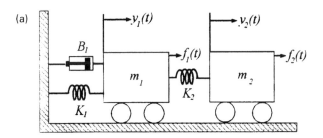

Figure 5.11

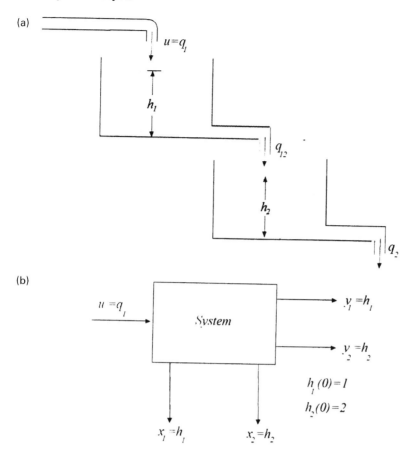

Figure 5.12

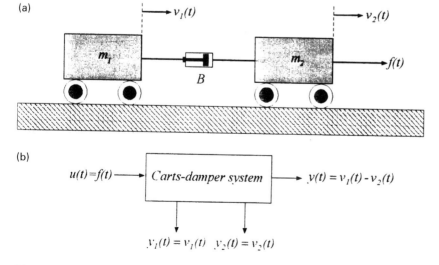

Figure 5.13

(a) Transform the system into diagonal form and (b) determine the general solution for the transformed system, given that $x_1(0) = x_2(0) = 0$ and $u(t) = e^{-t}$, $t \geq 0$.

7. Diagonalize the following matrices

$$\begin{bmatrix} 0 & 1 \\ -2 & -3 \end{bmatrix}, \quad \begin{bmatrix} 0 & 1 & 0 \\ 0 & 0 & 1 \\ 0 & 1 & 0 \end{bmatrix}, \quad \begin{bmatrix} 0 & 1 & 0 & 0 \\ 0 & 0 & 1 & 0 \\ 0 & 0 & 0 & 1 \\ 0 & 2 & 1 & 2 \end{bmatrix}$$

8. Derive the block diagrams and the signal-flow graphs for the following systems

(a) $\mathbf{H}(s) = \begin{bmatrix} \dfrac{1}{s} & 0 \\ 0 & \dfrac{1}{s+1} \end{bmatrix}$ (b) $\mathbf{H}(s) = \begin{bmatrix} 0 & \dfrac{1}{s} & \dfrac{1}{s+2} \\ \dfrac{1}{s+1} & 0 & 0 \\ 0 & 1 & \dfrac{1}{s+3} \end{bmatrix}$

(c) $\dot{\mathbf{x}} = \begin{bmatrix} 0 & 1 & 0 \\ 0 & 0 & 1 \\ 6 & 11 & 6 \end{bmatrix} \mathbf{x} + \begin{bmatrix} 0 \\ 0 \\ 1 \end{bmatrix} u, \qquad \mathbf{x}(0) = \begin{bmatrix} 1 \\ -1 \\ 0 \end{bmatrix}$

9. A simplified picture of a submarine is shown in Figure 5.14a. In Figure 5.14b the input, state, and output variables of the system are defined. The state equations describing the system are the following:

$$\begin{bmatrix} \dot{x}_1 \\ \dot{x}_2 \\ \dot{x}_3 \end{bmatrix} = \begin{bmatrix} 0 & 1 & 0 \\ 1 & -3 & 2 \\ 0 & \beta & -1 \end{bmatrix} \begin{bmatrix} x_1 \\ x_2 \\ x_3 \end{bmatrix} + \begin{bmatrix} 0 \\ 1 \\ -1 \end{bmatrix} u$$

$$y = \begin{bmatrix} 1 & 0 & 0 \end{bmatrix} \begin{bmatrix} x_1 \\ x_2 \\ x_3 \end{bmatrix}$$

Examine the controllability and observability of the system.

10. Consider the electrical network given in Figure 5.15a. Find its state equations using the definitions of input, state, and output variables shown in Figure 5.15b. Investigate the observability of the system in relation to the free values of the network elements.

11. Investigate the controllability and observability of the following systems. Furthermore, transform them into their phase canonical input and output forms:

(a)

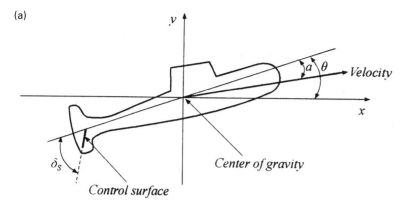

(b)

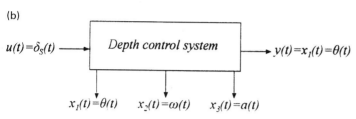

Figure 5.14

(a)

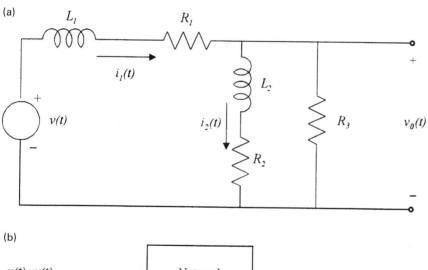

(b)

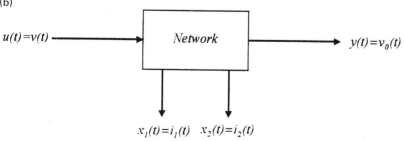

Figure 5.15

(a) $\quad A = \begin{bmatrix} 1 & 0 & 1 \\ 2 & -2 & 0 \\ -1 & 2 & 1 \end{bmatrix}, \quad b = \begin{bmatrix} 1 \\ 1 \\ 0 \end{bmatrix}, \quad c^T = \begin{bmatrix} 1 & 0 & 1 \end{bmatrix}$

(b) $\quad A = \begin{bmatrix} 1 & 2 & 1 & 0 \\ 0 & 1 & 4 & -1 \\ 0 & 0 & 1 & 0 \\ -1 & 0 & 0 & 0 \end{bmatrix}, \quad b = \begin{bmatrix} 0 \\ 1 \\ 1 \\ 1 \end{bmatrix}, \quad C = \begin{bmatrix} 1 & -1 & 1 & 0 \\ 0 & 1 & 0 & 0 \end{bmatrix}$

(c) $\quad A = \begin{bmatrix} 1 & 0 & 0 \\ 0 & -1 & 1 \\ 1 & 0 & 0 \end{bmatrix}, \quad B = \begin{bmatrix} 1 & 1 \\ 0 & 1 \\ 1 & 0 \end{bmatrix}, \quad c^T = \begin{bmatrix} -1 & 1 & -1 \end{bmatrix}$

12. Consider the network of Figure 5.16a, where $R_1 = R_2 = 1\,\Omega$, $C_1 = C_2 = 1\,F$, $L = 2\,H$, $v_{c_1}(0) = 1\,V$, $v_{c_2}(0) = -1\,V$, $i_L(0) = 0\,A$. The input, state, and output variables are defined in Figure 5.16b. Determine (a) the state equations, (b) the state transition matrix, (c) the output of the network, (d) the state and output controllability, (e) the state observability, and (f) a phase canonical form description.

(a)

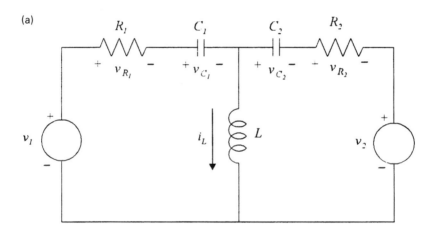

(b)

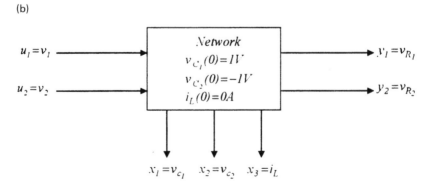

Figure 5.16

13. A system is described by the matrix equations

$$\dot{x} = Ax + Bu, \qquad y = Cx + Du, \qquad x(0) = 0$$

where

$$A = \begin{bmatrix} 0 & 1 & 0 & 0 \\ 0 & 0 & 1 & 0 \\ 0 & 0 & 0 & 1 \\ 0 & 0 & 0 & 1 \end{bmatrix}, \qquad B = \begin{bmatrix} 0 & 0 \\ 0 & 0 \\ 1 & 0 \\ 0 & 1 \end{bmatrix}, \qquad C = \begin{bmatrix} 1 & 0 & 0 & 0 \\ 0 & 0 & 0 & 1 \end{bmatrix},$$

$$D = \begin{bmatrix} 0 & 0 \\ 0 & 1 \end{bmatrix}$$

Determine (a) the matrix $(sI - A)^{-1}$, using the Leverrier's algorithm, (b) the state transition matrix, (c) the transfer function matrix, (d) the output vector $y(t)$, and (e) the state and output controllability and the state observability.

14. Solve Problem 13 for the system with matrices

$$A = \begin{bmatrix} 0 & 1 & 0 \\ 2 & 3 & 0 \\ 1 & 1 & 1 \end{bmatrix}, \qquad B = \begin{bmatrix} 0 & 0 \\ 1 & 0 \\ 0 & 1 \end{bmatrix}, \qquad C = \begin{bmatrix} 1 & 1 & 1 \\ 0 & 0 & 1 \end{bmatrix},$$

$$D = \begin{bmatrix} -1 & 0 \\ 0 & 0 \end{bmatrix}$$

and initial state vector $x^T(0) = [1, -1, 0]$.

BIBLIOGRAPHY

Books

1. PJ Antsaklis, AN Michel. Linear Systems. New York: McGraw-Hill, 1997.
2. RW Brockett. Finite Dimensional Linear Systems. New York: John Wiley, 1970.
3. WL Brogan. Modern Control Theory. 2nd ed. Englewood Cliffs, New Jersey: Prentice Hall, 1985.
4. FM Callier, CA Desoer. Multivariable Feedback Systems. New York: Springer-Verlag, 1982.
5. CT Chen. Linear System Theory and Design. New York: Holt, Rinehart and Winston, 1984.
6. RA DeCarlo. Linear Systems. Englewood Cliffs, New Jersey: Prentice Hall, 1989.
7. PM DeRusso, RJ Roy, CM Close. State Variables for Engineers. New York: John Wiley, 1965.
8. JC Doyle. Feedback Control Theory. New York: Macmillan, 1992.
9. B Friedland. Control System Design. An Introduction to State-Space Methods. New York: McGraw-Hill, 1987.
10. RA Gabel, RA Roberts. Signals and Linear Systems. 3rd ed. New York: John Wiley & Sons, 1987.
11. T Kailath. Linear Systems. Englewood Cliffs, New Jersey: Prentice Hall, 1980.
12. BC Kuo. Automatic Control Systems. London: Prentice Hall, 1995.
13. JM Maciejowski. Multivariable Feedback Design. Reading, MA: Addison-Wesley, 1989.
14. NS Nise. Control Systems Engineering. New York: Benjamin and Cummings, 1995.
15. K Ogata. Modern Control Systems. London: Prentice Hall, 1997.

16. CE Rohrs, JL Melsa, D Schultz. Linear Control Systems. New York: McGraw-Hill, 1993.
17. HH Rosenbrock. State Space and Multivariable Theory. London: Nelson, 1970.
18. HH Rosenbrock. Computer Aided Control System Design. New York: Academic Press, 1974.
19. WJ Rugh. Linear System Theory. 2nd ed. Englewood Cliffs, New Jersey: Prentice Hall, 1996.
20. RJ Schwarz, B Friedland. Linear Systems. New York: McGraw-Hill, 1965.
21. LK Timothy, BE Bona. State Space Analysis: An Introduction. New York: McGraw-Hill, 1968.
22. DM Wiberg. Theory and Problems of State-Space and Linear Systems. Schaums Outline Series. New York: McGraw-Hill, 1971.
23. LA Zadeh, CA Desoer. Linear Systems Theory – The State Space Approach. New York: McGraw-Hill, 1963.

Articles

24. RE Kalman. A new approach to linear filtering and prediction problems. Trans ASME (J Basic Engineering) 82D:35–45, 1960.
25. RE Kalman. On the general theory of control systems. Proceedings First International Congress, IFAC, Moscow, USSR, 1960, pp 481–492.
26. RE Kalman. Contributions to the theory of optimal control. Proceedings 1959 Mexico City Conference on Differential Equations, Mexico City, 1960, pp 102–119.
27. RE Kalman. Canonical structure of linear dynamical systems. Proceedings National Academy of Science 48:596–600, 1962.
28. RE Kalman. Mathematical description of linear dynamic systems. SIAM J Control 1:152–192, 1963.
29. RE Kalman, RS Bucy. New results in linear filtering and prediction theory. Trans ASME (J Basic Engineering) 83D:95–108, 1961.
30. RE Kalman, YC Ho, KS Narenda. Controllability of linear dynamic systems. Contributions to Differential Equations 1:189–213, 1963.
31. BC Moore. Principal component analysis in linear systems: controllability, observability and model reduction. IEEE Trans Automatic Control AC-26:17–32, 1981.

6

Stability

6.1 INTRODUCTION

Systems have several properties—such as controllability, observability, stability, and invertibility—that play a very decisive role in their behavior. From these character-istics, *stability* plays the most important role.

The most basic practical control problem is the design of a closed-loop system such that its output follows its input as closely as possible (see Chap. 4). In the present chapter we will show that unstable systems cannot guarantee such behavior and therefore are not useful in practice. Another serious disadvantage of unstable systems is that the amplitude of at least one of their state and/or output variables tends to infinity as time increases, even though the input of the system is bounded. This usually results in driving the system to saturation and in certain cases the consequences may be even more undesirable: the system may suffer serious damage, such as burn out, break down, explosion, etc. For these and other reasons, in designing an automatic control system, our primary goal is to guarantee stability. As soon as stability is guaranteed, then one seeks to satisfy other design require-ments, such as speed of response, settling time, bandwidth, and steady-state error.

The concept of stability has been studied in depth, and various criteria for testing the stability of a system have been proposed. Among the most celebrated stability criteria are those of Routh, Hurwitz, Nyquist, Bode, Nichols, and Lyapunov. From these criteria, the first five are in the frequency domain, whereas the last one is in the time domain. These criteria are presented in this chapter, except the criteria of Nyquist, Bode, and Nichols, which are presented in Chap. 8. The very popular root locus technique proposed by Evans, which also facilitates the study of stability, is presented in Chap. 7.

6.2 STABILITY DEFINITIONS

In this section, the stability of linear, time-invariant systems is studied in connection with each of the three well-established mathematical models of a system—namely, the state equations, the transfer function matrix, and the impulse response matrix.

1 Stability of Systems Described in State Space

Consider a linear, time-invariant system described in state space as follows

$$\dot{\mathbf{x}} = \mathbf{Ax} + \mathbf{Bu}, \qquad \mathbf{x}(0) = \mathbf{x}_0 \neq \mathbf{0} \tag{6.2-1a}$$

$$\mathbf{y} = \mathbf{Cx} + \mathbf{Du} \tag{6.2-1b}$$

The stability considered here refers to the *zero-input* case, i.e., to the case where the sysem (6.2-1) has zero input ($\mathbf{u}(t) = \mathbf{0}$). This is the case of natural (or free) response of the system, also known as the homogeneous solution (see relation (4.2-3)). The stability defined for $\mathbf{u}(t) = \mathbf{0}$ is called *zero-input stability*.

a   Asymptotic Stability

System (6.2-1) is asymptotically stable if, for $\mathbf{u}(t) = \mathbf{0}$ and for every finite initial state $\mathbf{x}(0) \neq \mathbf{0}$, the following condition is satisfied

$$\lim_{t \to \infty} \|\mathbf{x}(t)\| = 0 \tag{6.2-2}$$

where $\|.\|$ stands for the Euclidean norm of the vector $\mathbf{x}(t)$, i.e.,

$$\|\mathbf{x}(t)\| = (x_1^2 + x_2^2 + \cdots + x_n^2)^{1/2}$$

A simplified picture of the above definition is given in Figure 6.1, where, for simplicity, we consider the case where the state vector has only two variables. In the figure it is shown that the state vector $\mathbf{x}(t)$ of the system (6.2-1) with $\mathbf{u}(t) = \mathbf{0}$ and initial conditions $\mathbf{x}(0)$ moves towards the origin when condition (6.2-2) is satisfied.

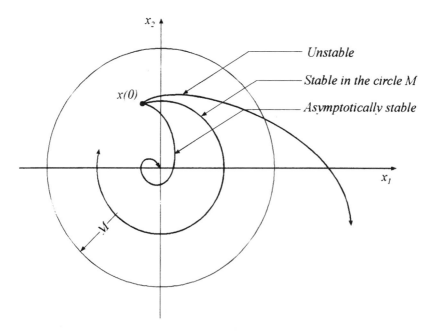

Figure 6.1   A simplified picture of stability of systems described in state space.

b   Stability in the Circle M

System (6.2-1) is stable in the circle M if, for $\mathbf{u}(t) = \mathbf{0}$ and for every finite initial state $\mathbf{x}(0) \neq \mathbf{0}$, the following condition is satisfied:

$$\|\mathbf{x}(t)\| < M, \quad t \geq 0 \tag{6.2-3}$$

In this case, $\mathbf{x}(t)$ remains within a circle of finite radius M, as shown in Figure 6.1.

c   Instability

If $\mathbf{x}(t)$ does not satisfy condition (6.2-3), the system is said to be unstable. In this case, the vector $\mathbf{x}(t)$, for $\mathbf{u}(t) = \mathbf{0}$ and initial state $\mathbf{x}(0) \neq \mathbf{0}$, moves towards infinity as $t \to \infty$, as shown in Figure 6.1.

2   Stability of Systems Described by Their Transfer Function Matrix

The previous definition of stability refers to systems described in state space and relates their stability with the Euclidean norm of the state vector. For systems described by their transfer function matrix $\mathbf{H}(s)$, the asymptotic stability condition follows directly from the asymptotic stability definition (6.2-2). Here, the stability is exclusively related to the poles of $\mathbf{H}(s)$. This marks out the great importance that the poles of a system have in relation to its stability.

Consider a linear, time-invariant system described in the s domain by its transfer function matrix

$$\mathbf{H}(s) = \mathbf{C}[s\mathbf{I} - \mathbf{A}]^{-1}\mathbf{B} + \mathbf{D} \tag{6.2-4}$$

The characteristic polynomial of the system is

$$p(s) = |s\mathbf{I} - \mathbf{A}| = \prod_{i=1}^{n}(s - \lambda_i)$$

where $\lambda_1, \lambda_2, \ldots, \lambda_n$ are the eigenvalues of $\mathbf{A}$, or equivalently, the poles of $\mathbf{H}(s)$.

a   Asymptotic Stability

System (6.2-4) is asymptotically stable if all the poles $\lambda_1, \lambda_2, \ldots, \lambda_n$ of $\mathbf{H}(s)$ lie in the left-half complex plane, i.e., if the following condition holds:

$$\mathrm{Re}\,\lambda_i < 0, \quad i = 1, 2, \ldots, n \tag{6.2-5}$$

In this case, the natural response of the system goes to zero as time goes to infinity.

b   Marginal Stability

System (6.2-4) is marginally stable if there exist poles on the imaginary axis of multiplicity one. The rest of the poles are in the left-half complex plane. In this case the natural response of the system neither grows nor decays. It remains constant or it oscillates with constant amplitude.

It is remarked that stability in the circle M and marginal stability are essentially equivalent.

Remark 6.2.1

In marginal stability, we tacitly assume that none of the poles of the system excitation $\mathbf{u}(t)$ coincides with any of the poles of $\mathbf{H}(s)$ on the imaginary axis. In the opposite case, the system is unstable. For example, let $U(s) = 1/s$ and $H(s) = 1/s(s+1)$. Then $Y(s) = H(s)U(s) = 1/s^2(s+1)$ and, hence, the system is unstable. Similarly, for conjugate poles on the imaginary axis. Let $U(s) = 1/(s^2 + 1)$ (i.e., let $u(t) = \sin t$) and

$H(s) = 1/(s+4)(s^2+1)$; then $Y(s) = H(s)U(s) = 1/(s+4)(s^2+1)^2$, and hence the system is unstable.

Remark 6.2.2

In certain special cases the system is intended to be marginally stable: as, for example, in the case of an integrator ($H(s) = K/s$) or an oscillator ($H(s) = Ks/(s^2 + \omega_n^2)$). In these special cases, we make an exception to the rule and we usually say that the system is stable (even though it is marginally stable).

c Instability

System (6.2-4) is unstable if it has at least one pole in the right-half complex plane or it has poles on the imaginary axis of multiplicity greater than 1. In this case, as time approaches infinity, the natural response of the system approaches infinity.

3 Stability of Systems Described by their Impulse Response Matrix

For systems described by their impulse response matrix $\mathbf{H}(t)$, the asymptotic stability condition follows directly from the asymptotic stability definition (6.2-2). Here, the stability is related to the absolute value of each element or to the integral of the absolute value of each element of $\mathbf{H}(t)$.

Consider a linear, time-invariant system described in the time domain by its impulse response matrix

$$\mathbf{H}(t) = L^{-1}\{\mathbf{C}(s\mathbf{I} - \mathbf{A})^{-1}\mathbf{B} + \mathbf{D}\} = \mathbf{C}e^{\mathbf{A}t}\mathbf{B} + \mathbf{D}\delta(t), \qquad t \geq 0 \tag{6.2-6}$$

a Asymptotic Stability

System (6.2-6) is asymptotically stable if the following condition is satisfied:

$$\int_0^\infty |h_{ij}(t)|\, dt < A, \qquad \forall i, j \tag{6.2-7}$$

where A is a finite positive constant and $h_{ij}(t)$ is the (i,j)th element of the impulse response matrix $\mathbf{H}(t)$.

b Marginal Stability

System (6.2-6) is marginally stable if the following conditon holds

$$|h_{ij}(t)| < B, \qquad \forall i, j \tag{6.2-8}$$

where B is a finite positive number. An analogous remark to Remark 6.2.1 holds for the present case.

c Instability

System (6.2-6) is unstable even if one element $h_{ij}(t)$ of $\mathbf{H}(t)$ does not satisfy Eq. (6.2-8).

At this point, we briefly introduce another definition for stability, called the bounded-input–bounded-output (BIBO) stability. A system is BIBO stable if, for any bounded input, its output is also bounded.

This definition is more general than all previous definitions, because it refers to systems that may not be linear and time invariant, while it gives a simple picture of the concept of stability. For single-input–single-output (SISO) systems, BIBO stability may be interpreted as follows. Consider the bounded input $u(t)$, i.e., assume that $|u(t)| \leq C_1$, for $t \in (0, \infty)$, where C_1 is a finite constant. Also assume that the response of the system to this input is $y(t)$ which is also bounded, i.e., assume that

$|y(t)| \leq C_2$, for $t \in (0, \infty)$, where C_2 is a finite constant. If for all possible bounded inputs the corresponding outputs of the system are also bounded, then the system is said to be BIBO stable.

Finally, it is mentioned that asymptotic stability is the strongest, since it is more stringent than marginal stability and BIBO stability.

Example 6.2.1

Investigate the stability of a system described in state space in the form (6.2-1), where

$$\mathbf{A} = \begin{bmatrix} 0 & 1 \\ -2 & -3 \end{bmatrix}, \quad \mathbf{b} = \begin{bmatrix} 0 \\ 1 \end{bmatrix}, \quad \mathbf{c} = \begin{bmatrix} 1 \\ 0 \end{bmatrix}, \quad \mathbf{D} = 0, \quad \mathbf{x}(0) = \begin{bmatrix} 1 \\ -1 \end{bmatrix}$$

Solution

The state vector, for $u(t) = 0$, will be

$$\mathbf{x}(t) = L^{-1}\{(s\mathbf{I} - \mathbf{A})^{-1}\mathbf{x}(0)\} = L^{-1}\left\{ \begin{bmatrix} \dfrac{s+3}{(s+1)(s+2)} & \dfrac{1}{(s+1)(s+2)} \\ \dfrac{-2}{(s+1)(s+2)} & \dfrac{s}{(s+1)(s+2)} \end{bmatrix} \begin{bmatrix} x_1(0) \\ x_2(0) \end{bmatrix} \right\}$$

$$= \begin{bmatrix} 2e^{-t} - e^{-2t} & e^{-t} - e^{-2t} \\ -2e^{-t} + 2e^{-2t} & -e^{-t} + 2e^{-2t} \end{bmatrix} \begin{bmatrix} 1 \\ -1 \end{bmatrix} = \begin{bmatrix} e^{-t} \\ -e^{-t} \end{bmatrix}$$

The norm of the vector $\mathbf{x}(t)$ is

$$\|\mathbf{x}(t)\| = [x_1^2(t) + x_2^2(t)]^{1/2} = [e^{-2t} + e^{-2t}]^{1/2} = [2e^{-2t}]^{1/2} = \sqrt{2}e^{-t}$$

It is clear that, on the basis of the definition of asymptotic stability, condition (6.2-2) is satisfied, since $\|\mathbf{x}(t)\| < \sqrt{2}$ for $t > 0$ and $\lim_{t\to\infty} \|\mathbf{x}(t)\| \to 0$. Therefore, the system is asymptotically stable.

The characteristic polynomial of the system is

$$p(s) = |s\mathbf{I} - \mathbf{A}| = (s+1)(s+2)$$

The two eigenvalues of the matrix $\mathbf{A}$ are -1 and -2, and they both lie in the left-half complex plane. Therefore, the system, on the basis of condition (6.2-5), is asymptotically stable.

The impulse response of the system is

$$h(t) = \mathbf{c}^T e^{\mathbf{A}t}\mathbf{b} = \mathbf{c}^T \mathbf{M}e^{\mathbf{\Lambda}t}\mathbf{M}^{-1}\mathbf{b}$$

where $\mathbf{M}$ is a transformation matrix which transforms the matrix $\mathbf{A}$ in its diagonal form $\mathbf{\Lambda}$. According to the results of Subsec. 5.4.4, the matrix $\mathbf{M}$ has the form

$$\mathbf{M} = \begin{bmatrix} 1 & 1 \\ \lambda_1 & \lambda_2 \end{bmatrix} = \begin{bmatrix} 1 & 1 \\ -1 & -2 \end{bmatrix}, \quad \mathbf{M}^{-1} = \begin{bmatrix} 2 & 1 \\ -1 & -1 \end{bmatrix}$$

Therefore,

$$h(t) = \mathbf{c}^T \mathbf{M}e^{\mathbf{\Lambda}t}\mathbf{M}^{-1}\mathbf{b} = [1, 0]\begin{bmatrix} 1 & 1 \\ -1 & -2 \end{bmatrix}\begin{bmatrix} e^{-t} & 0 \\ 0 & e^{-2t} \end{bmatrix}\begin{bmatrix} 2 & 1 \\ -1 & -1 \end{bmatrix}\begin{bmatrix} 0 \\ 1 \end{bmatrix}$$

$$= e^{-t} - e^{-2t}$$

If we apply condition (6.2-7), we have

$$\int_0^\infty |h(t)|\, dt = \int_0^\infty |e^{-t} - e^{-2t}|\, dt \le \int_0^\infty e^{-t}\, dt + \int_0^\infty e^{-2t}\, dt = 1 + 0.5 = 1.5 < \infty$$

Hence, the system, on the basis of condition (6.2-7), is asymptotically stable.

Assume that the system is excited by a bounded input $|u(t)| < C < \infty$. Then, the absolute value of the output of the system will be

$$|y(t)| = \left| \int_0^t h(\lambda) u(t - \lambda)\, d\lambda \right| \le \int_0^t |h(\lambda)||u(t - \lambda)|\, d\lambda = C \int_0^t |h(\lambda)|\, d\lambda \le 1.5C < \infty$$

Hence, on the basis of definition of the BIBO stability, it is concluded that the system is BIBO stable.

Example 6.2.2

Investigate the stability of a system described in state-space form (6.2-1), where

$$A = \begin{bmatrix} 0 & 1 \\ 1 & 0 \end{bmatrix}, \qquad b = \begin{bmatrix} 0 \\ 1 \end{bmatrix}, \qquad c = \begin{bmatrix} 0 \\ 1 \end{bmatrix}, \qquad D = 0, \qquad x(0) = \begin{bmatrix} 1 \\ 1 \end{bmatrix}$$

Solution

The state vector, for $u(t) = 0$, will be

$$x(t) = L^{-1}\{(sI - A)^{-1}x(0)\} = L^{-1}\left\{ \begin{bmatrix} \dfrac{s}{s^2 - 1} & \dfrac{1}{s^2 - 1} \\ \dfrac{1}{s^2 - 1} & \dfrac{s}{s^2 - 1} \end{bmatrix} \begin{bmatrix} x_1(0) \\ x_2(0) \end{bmatrix} \right\}$$

$$= \begin{bmatrix} \frac{1}{2}(e^t + e^{-t}) & \frac{1}{2}(e^t - e^{-t}) \\ \frac{1}{2}(e^t - e^{-t}) & \frac{1}{2}(e^t + e^{-t}) \end{bmatrix} \begin{bmatrix} 1 \\ 1 \end{bmatrix} = \begin{bmatrix} e^t \\ e^t \end{bmatrix}$$

The norm of the state vector is

$$\|x(t)\| = [x_1^2(t) + x_2^2(t)]^{1/2} = [2e^{2t}]^{1/2} = \sqrt{2} e^t$$

It is clear that the system is unstable because, as $t \to \infty$, $x(t)$ tends to infinity.

The characteristic polynomial of the system is

$$p(s) = |sI - A| = (s - 1)(s + 1)$$

The two eigenvalues of the matrix A are 1 and -1. From these two eigenvalues, one lies in the right-half complex plane and therefore the system is unstable.

The impulse response of the system is

$$h(t) = L^{-1}\{c^T(sI - A)^{-1}b\} = \frac{1}{2}(e^t + e^{-t})$$

If we apply condition (6.2-7) we have

$$\int_0^\infty |h(t)|\, dt = \frac{1}{2}\int_0^\infty |e^t + e^{-t}|\, dt \le \frac{1}{2}\int_0^\infty e^t\, dt + \frac{1}{2}\int_0^\infty e^{-t}\, dt = \infty$$

Hence the system, is unstable.

Assume that the system is excited by a bounded input $|u(t)| < C < \infty$. Then, the absolute value of the output of the system will be

$$|y(t)| = \left| \int_0^t h(\lambda) u(t - \lambda) \, d\lambda \right| \leq \int_0^t |h(\lambda)| |u(t - \lambda)| \, d\lambda = C \int_0^t |h(\lambda)| \, d\lambda$$

Therefore

$$\lim_{t \to \infty} |y(t)| = C \lim_{t \to \infty} \int_0^t |h(\lambda)| \, d\lambda = \infty$$

Hence the system is not BIBO stable.

A summary of the main points of the present section is given in Figure 6.2.

6.3 STABILITY CRITERIA

Clearly, each of the definitions of Sec. 6.2 may be applied to study the stability of a system. Their application, however, appears to have many difficulties. For example, the definition based on the state-space description requires the determination of the state vector $\mathbf{x}(t)$. This computation is usually quite difficult. The definition based on the transfer function matrix $\mathbf{H}(s)$ requires the computation of the roots of the characteristic polynomial $|s\mathbf{I} - \mathbf{A}|$. This computation becomes more complex as the degree of the characteristic polynomial becomes greater. The definition based on the impulse response matrix $\mathbf{H}(t)$ requires the determination of the impulse response matrix $\mathbf{H}(t)$. This appears to have about the same difficulties as the determination of the transition matrix $\boldsymbol{\phi}(t)$. The BIBO definition appears to be simple and practical to apply, but because of its very nature, it is almost impossible to use. This is because in order to study the stability of a system on the basis of the BIBO stability definition, one must examine all possible bounded inputs, which requires an infinitely long period of time.

From all different definitions mentioned above, the definition based on the transfer function description appears to offer, from the computational point of view, the simplest approach. But even in this case, particularly when the degree of the characteristic polynomial is very high, the determination of the poles could involve numerical difficulties which might make it difficult, if not impossible, to apply.

From the above, one may conclude that in practice it is very difficult to apply the definitions of stability presented in Sec. 6.2 directly in order to study the stability of a system. To circumvent this difficulty, various *stability criteria* have been developed. These criteria give pertinent information regarding the stability of a system without directly applying the definitions for stability and without requiring complicated numerical procedures. The most popular criteria are the following:

1. *The algebraic criteria*: these criteria assume that the analytical expression of the characteristic polynomial of the system is available and give information with regard to the position of the roots of the characteristic polynomial in the left- or the right-half complex plane. Examples of such algebraic criteria are the Routh criterion, the Hurwitz criterion, and the continued fraction expansion criterion. These criteria are simple to apply and, for this reason, they have become most popular in studying the stability of linear systems.

System description	Mathematical description of the system	Conditions		
		Asymptotically stable	Marginal stable or stable in the circle M	Unstable
State space	$\dot{x} = Ax + Bu$ $y = Cx + Du$	$\lim_{t\to\infty}\|x(t)\| \to \infty$	$\|x(t)\| < M,\ t>0$	$\lim_{t\to\infty}\|x(t)\| \to \infty$
Transfer function matrix	$H(s) = C(sI-A)^{-1} + D$	$Re\gamma_i > 0,\ \forall i$	Imaginary axis poles of multiplicity one and the rest in the left half complex plane	$Re\lambda_i > 0$, for at least one i or imaginary axis poles of multiplicity greater than one.
Impulse response matrix	$H(t) = Ce^{At}B + D\delta(t)$	$\int_0^\infty |h_{ij}(t)|dt < A,\ \forall i,j$	$|h_{ij}(t)| < B,\ \forall i,j$	$\int_0^S |h_{ij}(t)|dt = \infty$ for at least one $h_{ij}(t)$
Simple illustration of the concepts of asymptotic stability, marginal stability and instability		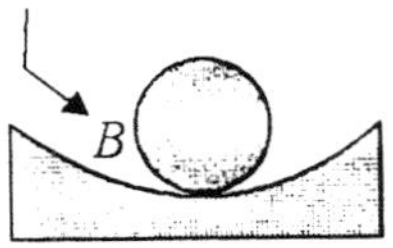 *Friction* · B	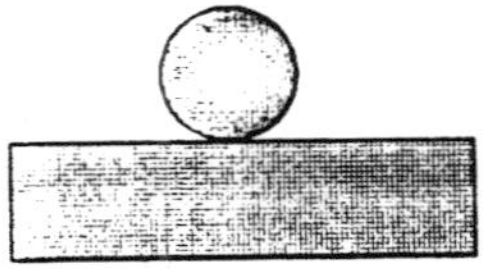	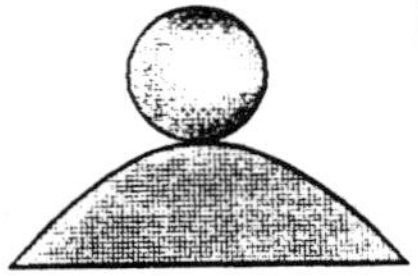

Figure 6.2 Types of system description and their corresponding definitions of asymptotic stability, marginal stability, and instability.

2. *The Nyquist criterion*: this criterion refers to the stability of the closed-loop systems and is based on the Nyquist diagram of the open-loop transfer function.

3. *The Bode criterion*: this criterion is essentialy the Nyquist criterion extended to the Bode diagrams of the open-loop transfer function.

4. *The Nichols criterion*: this criterion, as in the case of the Bode criterion, is essentially an extension of the Nyqist criterion to the Nichols diagrams of the open-loop transfer function.

5. *The root locus*: this method consists of determining the root loci of the characteristic polynomial of the closed-loop system when one or more parameters of the system vary (usually these parameters are gain constants of the system).

6. *The Lyapunov criterion*: this criterion is based on the properties of Lyapunov functions of a system and may be applied to both linear and nonlinear systems.

The algebraic criteria, the Nyquist criterion, the Bode criterion, and the Nichols criterion, as well as the root locus technique, are all criteria in the frequency domain. The Lyapunov criterion is in the time domain.

The algebraic criteria and the Lyapunov criterion are presented in this chapter. The root locus technique is presented in Chap. 7 and the Nyquist, the Bode, and the Nichols criteria are presented in Chap. 8.

6.4   ALGEBRAIC STABILITY CRITERIA

6.4.1   Introductory Remarks

The most popular algebraic criteria are the Routh, Hurwitz, and the continued fraction expansion criteria. The main characteristic of these three algebraic criteria is that they determine whether or not a system is stable by using a very simple numerical procedure, which circumvents the need for determining the roots of the characteristic polynomial.

Consider the characteristic polynomial

$$p(s) = a_n s^n + a_{n-1} s^{n-1} + \cdots + a_1 s + a_0 \tag{6.4-1}$$

where the coefficients $a_n, a_{n-1}, \ldots, a_0$ are real numbers. Here, we assume $a_0 \neq 0$ to avoid having a root at the origin. Next, we state the following well-known theorem of algebra.

Theorem 6.4.1

The polynomial $p(s)$ has one or more roots in the right-half complex plane if at least one of its coefficients is zero and/or all coefficients do not have the same sign.

Theorem 6.4.1 is very useful since it allows one to determine the stability of a system by simply inspecting the characteristic polynomial. However, Theorem 6.4.1 gives only necessary stability conditions. This means that if $p(s)$ satisfies Theorem 6.4.1, then the system with characteristic polynomial $p(s)$ is definitely unstable. For the cases where $p(s)$ does not satisfy Theorem 6.4.1, i.e., none of the coefficients of $p(s)$ is zero and all its coefficients have the same sign, we cannot conclude as to the stability of the system. For these cases, we apply one of the algebraic criteria (Routh, or Hurwitz, or continued fraction expansion), which are presented below.

6.4.2 The Routh Criterion

The Routh criterion determines the number of the roots of the characteristic polynomial $p(s)$ which lie in the right-half complex plane. This criterion is applied by using the Routh array, as shown in Table 6.1. In the Routh array, the elements a_n, $a_{n-1}, a_{n-2}, \ldots, a_1, a_0$ are the coefficients of $p(s)$. The elements $b_1, b_2, b_3, \ldots, c_1, c_2, c_3$, $\ldots$, etc., are computed as follows:

$$b_1 = -\frac{\begin{vmatrix} a_n & a_{n-2} \\ a_{n-1} & a_{n-3} \end{vmatrix}}{a_{n-1}}, \qquad b_2 = -\frac{\begin{vmatrix} a_n & a_{n-4} \\ a_{n-1} & a_{n-5} \end{vmatrix}}{a_{n-1}}, \ldots \qquad (6.4\text{-}2a)$$

$$c_1 = -\frac{\begin{vmatrix} a_{n-1} & a_{n-3} \\ b_1 & b_2 \end{vmatrix}}{b_1}, \qquad c_2 = -\frac{\begin{vmatrix} a_{n-1} & a_{n-5} \\ b_1 & b_3 \end{vmatrix}}{b_1}, \ldots \qquad (6.4\text{-}2b)$$

and so on. The Routh criterion is given by the following theorem.

Theorem 6.4.2

The necessary and sufficient conditions for $\mathrm{Re}\lambda_i < 0$, $i = 1, 2, \ldots, n$, where λ_1, λ_2, $\ldots, \lambda_n$ are the roots of the characteristic polynomial $p(s)$, are that the first column of the Routh array does not involve any sign changes. In cases where it involves sign changes, then the system is unstable and the number of roots of $p(s)$ with positive real part is equal to the number of sign changes.

Example 6.4.1

Investigate the stability of a system with characteristic polynomial $p(s) = s^3 + 10s^2 + 11s + 6$.

Table 6.1 The Routh Array

s^n	a_n	a_{n-2}	a_{n-4}	$\cdots$
s^{n-1}	a_{n-1}	a_{n-3}	a_{n-5}	$\cdots$
s^{n-2}	b_1	b_3	b_3	$\cdots$
s^{n-3}	c_1	c_2	c_3	$\cdots$
$\vdots$	$\vdots$	$\vdots$	$\vdots$	$\vdots$
s^1	$\vdots$			
s^0	$\vdots$			

Solution

Construct the Routh array as follows:

s^3	1	11
s^2	10	6
s^1	52/5	0
s^0	6	0

Since the first column of the Routh array involves no sign changes, it follows that the system is stable.

Example 6.4.2

Investigate the stability of a system with characteristic polynomial $p(s) = s^4 + s^3 + s^2 + 2s + 1$.

Solution

Construct the Routh array as follows:

s^4	1	1	1
s^3	1	2	0
s^2	−1	1	0
s^1	3	0	0
s^0	1	0	0

Since the first column of the Routh array involves two sign changes, it follows that $p(s)$ has two roots in the right-half complex plane and therefore the system is unstable.

It has been proven that we can multiply or divide a column or a row in the Routh array by a constant without influencing the end results of the Routh criterion. We may take advantage of this fact to simplify several operations which are required in constructing the Routh array.

There are two cases in which the Routh criterion, as it has been presented above, cannot be applied. For these two cases certain modifications are necessary so that the above procedure is applicable. These two cases are the following.

1   A Zero Element in the First Column of the Routh Array

In this case the Routh array cannot be completed because the element below the zero element in the first column will become infinite as one applies relation (6.4-2). To circumvent this difficulty, we multiply the characteristic polynomial $p(s)$ with a factor $(s + a)$, where $a > 0$ and $−a$ is not a root of $p(s)$. The conclusions regarding the stability of the new polynomial $\hat{p}(s) = (s + a)p(s)$ are obviously the same as those of the original polynomial $p(s)$.

Example 6.4.3

Investigate the stability of a system with characteristic polynomial $p(s) = s^4 + s^3 + 2s^2 + 2s + 3$.

Solution

Construct the Routh array as follows:

s^4	1	2	3
s^3	1	2	0
s^2	0	3	0
s^1	∞		
s^0			

Since the third element in the first column of the Routh array is zero, it is clear that the Routh array cannot be completed. If we multiply $p(s)$ by the factor $(s+1)$, we have

$$\hat{p}(s) = (s+1)p(s) = s^5 + 2s^4 + 3s^3 + 4s^2 + 5s + 3$$

Next, construct the Routh array of $\hat{p}(s)$:

s^5	1	3	5
s^4	2	4	3
s^3	1	3.5	0
s^2	-3	3	0
s^1	4.5	0	0
s^0	3	0	0

According to the above Routh array, one observes that the polynomials $\hat{p}(s)$ and $p(s)$ have two roots in the right-half complex plane, and therefore the system with characteristic polynomial $p(s)$ is unstable.

2 A Zero Row in the Routh Array

In this case the Routh array cannot be completed, because in computing the rest of the elements that follow the zero row, according to formula (6.4-2), the indeterminate form $0/0$ will appear. To circumvent this difficulty, we proceed as follows:

1. Form the "auxiliary polynomial" $q(s)$ of the row which precedes the zero row.
2. Take the derivative of $q(s)$ and replace the zero row with the coefficients of $q^{(1)}(s)$, where $q^{(1)}(s)$ is the derivative of $q(s)$.
3. Complete the construction of the Routh array in the usual manner.

Example 6.4.4

Investigate the stability of a system with characteristic polynomial $p(s) = s^5 + s^4 + 2s^3 + 2s^2 + 3s + 3$.

Solution

Construct the Routh array as follows:

s^5	1	2	3
s^4	1	2	3
s^3	0	0	0
s^2	?	?	?
s^1			
s^0			

Since the row s^3 of the Routh array involves only zeros, it is clear that the Routh array cannot be completed. At this point we construct the "auxiliary polynomial" $q(s) = s^4 + 2s^2 + 3$ of the row s^4. Taking the derivative of $q(s)$ yields $q^{(1)}(s) = 4s^3 + 4s$. Next, form the new row s^3 of the Routh array using the coefficients of $q^{(1)}(s)$ and, subsequently, complete the Routh array in the usual manner to yield

s^5	1	2	3
s^4	1	2	3
s^3	4	4	0
s^2	1	3	0
s^1	-8	0	0
s^0	3	0	0

Since the first column of the new Routh array appears to have two sign changes, it follows that p(s) has two roots in the right-half complex plane and therefore the system is unstable.

Finally, consider the case where $p(s)$ involves free parameters. Then the Routh criterion can be used to determine the appropriate range of values of these free parameters which guarantee stability of the system. This can be accomplished if one imposes the restriction that all the free parameters appearing in $p(s)$ be such that all the coefficients of the elements of the first column in the Routh array have the same sign. This leads to a system of algebraic inequalities whose solution determines the range of values of the free parameters for which the system is stable.

Example 6.4.5

Determine the range of values of the free parameter K such that the system with characteristic polynomial $p(s) = s^3 + 10s^2 + 11s + K$ is stable.

Solution

Construct the Routh array:

s^3	1	11
s^2	10	K
s^1	$\dfrac{110 - K}{10}$	0
s^0	K	0

For the system to be stable all elements of the first column of the Routh array must have the same sign. Hence, there must be $(110 - K)/10 > 0$ and $K > 0$. The two inequalities are simultaneously satisfied for $0 < K < 110$. Hence the system is stable when $0 < K < 110$.

6.4.3 The Hurwitz Criterion

The Hurwitz criterion determines whether or not the characteristic polynomial has roots in the right-half complex plane. However, compared with the Routh criterion, it does not give any information regarding the number of the roots that the characteristic polynomial has in the right-half complex plane.

The Hurwitz criterion is applied on the basis of the Hurwitz determinants, which are defined as follows:

$$\Delta_0 = a_n$$

$$\Delta_1 = a_{n-1}$$

$$\Delta_2 = \begin{vmatrix} a_{n-1} & a_{n-3} \\ a_n & a_{n-2} \end{vmatrix}$$

$$\Delta_3 = \begin{vmatrix} a_{n-1} & a_{n-3} & a_{n-5} \\ a_n & a_{n-2} & a_{n-4} \\ 0 & a_{n-1} & a_{n-3} \end{vmatrix}$$

$$\vdots$$

$$\Delta_n = \begin{vmatrix} a_{n-1} & a_{n-3} & \cdots & \begin{bmatrix} a_0 \text{ if } n \text{ is odd} \\ a_1 \text{ if } n \text{ is even} \end{bmatrix} & 0 & \cdots & 0 \\ a_n & a_{n-2} & \cdots & \begin{bmatrix} a_1 \text{ if } n \text{ is odd} \\ a_0 \text{ if } n \text{ is even} \end{bmatrix} & 0 & \cdots & 0 \\ 0 & a_{n-1} & & a_{n-3} & \cdots & & 0 \\ 0 & a_n & & a_{n-2} & \cdots & & 0 \\ \vdots & \vdots & & \vdots & \vdots\vdots\vdots & & \vdots \\ 0 & 0 & & 0 & \cdots & & a_n \end{vmatrix}$$

The Hurwitz criterion is given by the following theorem.

Theorem 6.4.3

The necessary and sufficient conditions for $\operatorname{Re} \lambda_i < 0$, $i = 0, 1, 2, \ldots, n$, where $\lambda_1, \lambda_2 , \ldots, \lambda_n$ are the roots of the characteristic polynomial $p(s)$, are that $\Delta_i > 0$, for all $i = 0, 1, 2, \ldots, n$.

Example 6.4.6

Investigate the stability of a system with characteristic polynomial $p(s) = s^3 + 10s^2 + 11s + 6$.

Solution

Compute the Hurtwitz determinants:

$$\Delta_0 = 1, \qquad \Delta_1 = 10, \qquad , \Delta_2 = \begin{vmatrix} 10 & 6 \\ 1 & 11 \end{vmatrix} = 104,$$

$$\Delta_3 = \begin{vmatrix} 10 & 6 & 0 \\ 1 & 11 & 0 \\ 0 & 10 & 6 \end{vmatrix} = 624$$

Since all determinants are positive, it follows that the system is stable.

6.4.4 The Continued Fraction Expansion Criterion

The continued fraction expansion criterion, as in the case of the Hurwitz criterion, determines whether or not the characteristic polynomial has roots in the right-half complex plane. To apply the continued fraction expansion criterion, the characteristic polynomial $p(s)$ is first grouped into two polynomials $p_1(s)$ and $p_2(s)$ as follows:

$$p_1(s) = a_n s^n + a_{n-2} s^{n-2} + a_{n-4} s^{n-4} + \cdots$$

$$p_2(s) = a_{n-1} s^{n-1} + a_{n-3} s^{n-3} + a_{n-5} s^{n-5} + \cdots$$

Next, we examine the ratio of $p_1(s)$ divided by $p_2(s)$ by expanding it as follows:

$$\frac{p_1(s)}{p_2(s)} = h_1 s + \cfrac{1}{h_2 s + \cfrac{1}{h_3 s + \cfrac{1}{\ddots \cfrac{1}{h_n s}}}}$$

The continued fraction expansion criterion is given by the following theorem.

Theorem 6.4.4

If $h_j > 0$, for all $j = 1, 2, \ldots, n$, then $\operatorname{Re} \lambda_j < 0$, $j = 1, 2, \ldots, n$, where $\lambda_1, \lambda_2, \ldots, \lambda_n$ are the roots of the characteristic polynomial $p(s)$ and vice versa.

Example 6.4.7

Investigate the stability of a system with characteristic polynomial $p(s) = s^3 + 10s^2 + 11s + 6$.

Solution

Construct the polynomials $p_1(s)$ and $p_2(s)$:

$$p_1(s) = s^2 + 11s, \qquad p_2(s) = 10s^2 + 6$$

We have

$$\frac{p_1(s)}{p_2(s)} = \frac{s^3 + 11s}{10s^2 + 6} = \frac{1}{10}s + \cfrac{\frac{104}{10}s}{10s^2 + 6} = \frac{1}{10}s + \cfrac{1}{\frac{100}{104}s + \cfrac{1}{\frac{104}{60}s}}$$

Therefore

$$h_1 = \frac{1}{10}, \qquad h_2 = \frac{100}{104}, \qquad h_3 = \frac{104}{60}$$

Since all coefficients of the continued fraction expansion are positive, it follows that the system is stable.

6.4.5 Stability of Practical Control Systems

In what follows, we will present several practical automatic control system examples, investigating their stability using one of the algebraic criteria which we have just presented.

Example 6.4.8

For the closed-loop position control system of Example 4.7.3 presented in Chap. 4, determine the range of values of the parameter K for which the closed-loop system is stable.

Solution

The transfer function of the closed-loop system of the Example 4.7.3 is

$$H(s) = \frac{G(s)}{1 + G(s)} = \frac{\dfrac{K}{s(s+2)}}{1 + \dfrac{K}{s(s+2)}} = \frac{K}{s(s+2) + K}$$

The characteristic polynomial $p(s)$ of the closed-loop system is

$$p(s) = s(s+2) + K = s^2 + 2s + K$$

Construct the Routh array of the characteristic polynomial:

$$
\begin{array}{c|cc}
s^2 & 1 & K \\
s^1 & 2 & 0 \\
s^0 & K & 0
\end{array}
$$

Therefore, for the closed-loop system to be stable, $K > 0$.

Example 6.4.9

Consider the closed-loop speed control system of Example 3.13.3 presented in Chap. 3 and assume that $L_a \cong 0$. The transfer function of the closed-loop system is given by relation (3.13-17). For simplicity, choose all parameters L_f, R_f, J_m^*, B_m^*, K_m, K_b and K_t to be equal to unity and let $K = K_t K_a K_g K_m N$. Determine the range of values of the parameter K for which the closed-loop system is stable.

Solution

The transfer function of the closed-loop system is

$$H(s) = \frac{K_a K_g K_m N}{(L_f s + R_f)(R_a J_m^* s + R_a B_m^* + K_m K_b) + K_t K_a K_g K_m N} = \frac{K_a K_g K_m N}{(s+1)(s+2) + K}$$

Therefore, the characteristic polynomial $p(s)$ of the closed-loop system is

$$p(s) = s^2 + 3s + K + 2$$

Construct the Routh array of the characteristic polynomial:

$$
\begin{array}{c|cc}
s^2 & 1 & K+2 \\
s^1 & 3 & 0 \\
s^0 & K+2 & 0
\end{array}
$$

For the closed-loop system to be stable, $K + 2 > 0$ or $K > -2$.

Example 6.4.10

This example refers to an automatic depth control system for submarines. In Figure 6.3 the block diagram of the closed-loop system is given, where the submarine is

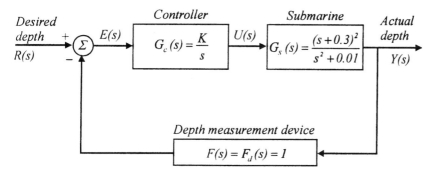

Figure 6.3   Block diagram of the automatic depth control system of a submarine.

approximated by a second-order transfer function. The depth of the submarine is measured by a depth sensor with transfer function $F_d(s)$. It is remarked that, as the value of the gain K of the controller is increased, so does the speed of sinking of the submarine. For simplicity, let $F_d(s) = 1$. Determine the range of values of K for which the closed-loop system is stable.

Solution

The transfer function of the closed-loop system is

$$H(s) = \frac{G_c(s)G_s(s)}{1 + G_c(s)G_s(s)F_d(s)} = \frac{\left[\dfrac{K}{s}\right]\left[\dfrac{(s + 0.3)^2}{(s^2 + 0.01)}\right]}{1 + \left[\dfrac{K}{s}\right]\left[\dfrac{(s + 0.3)^2}{(s^2 + 0.01)}\right]}$$

$$= \frac{K(s + 0.3)^2}{s(s^2 + 0.01) + K(s + 0.3)^2}$$

The characteristic polynomial $p(s)$ of the closed-loop system is

$$p(s) = s(s^2 + 0.01) + K(s + 0.3)^2 = s^3 + Ks^2 + (0.01 + 0.6K)s + 0.09K$$

Construct the Routh array of the characteristic polynomial:

$$
\begin{array}{c|ll}
s^3 & 1 & 0.01 + 0.6K \\
s^2 & K & 0.09K \\
s^1 & 0.6K - 0.08 & 0 \\
s^0 & 0.09K &
\end{array}
$$

For the closed-loop system to be stable, the two inequalities $0.06K - 0.08 > 0$ and $0.09K > 0$ must hold simultaneously. This holds true for $K > 0.1333$.

Example 6.4.11

This example refers to the stabilization of ships due to oscillations resulting from waves and strong winds, presented in paragraph 12 of Sec. 1.4 and shown in Figure 1.20. When a ship exhibits a deviation of θ degrees from the vertical axis, as shown in Figure 1.20, then most ships use fins to generate an opposite torque which restores

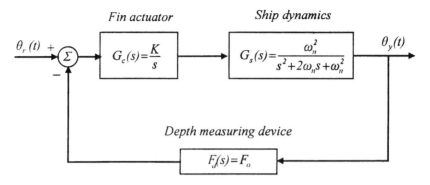

Roll angle measuring device

Figure 6.4   A simplified block diagram of ship stabilization control system.

the ship to the vertical position. In Figure 6.4 the block diagram of the system is given, where, obviously, $\theta_r = 0$ is the desired deviation of the ship. The length of the fins projecting into the water is controlled by an actuator with transfer function $G_c(s) = K/s$. The deviation from the vertical axis is measured by a measuring device with transfer function $F_d(s) = F_0 = $ constant. A simplified mathematical description of the ship is given by the second-order transfer function $G_s(s)$. Typical values of ζ and ω_n in $G_s(s)$ are $\zeta = 0.1$ and $\omega_n = 2$. For simplicity, let $F_0 = 1$. Determine the range of values of K for which the closed-loop system is stable. It is noted that since it is desirable that $\theta_r = 0$, the problem of restoring the ship to its vertical position is a typical regulator problem (see Sec. 11.3).

Solution

The transfer function of the closed-loop system is given by

$$H(s) = \frac{G_c(s)G_s(s)}{1 + G_c(s)G_s(s)F_d(s)} = \frac{\left[\dfrac{K}{s}\right]\left[\dfrac{4}{s^2 + 0.4s + 4}\right]}{1 + \left[\dfrac{K}{s}\right]\left[\dfrac{4}{s^2 + 0.4s + 4}\right]} = \frac{4K}{s^3 + 0.4s^2 + 4s + 4K}$$

The characteristic polynomial $p(s)$ of the transfer function of the closed-loop system is the following:

$$p(s) = s^3 + 0.4s^2 + 4s + 4K$$

Construct the Routh array of the characteristic polynomial:

$$
\begin{array}{c|cc}
s^3 & 1 & 4 \\
s^2 & 0.4 & 4K \\
s^1 & 4 - 10K & 0 \\
s^0 & 4K &
\end{array}
$$

For the closed-loop system to be stable, the inequalities $4 - 10K > 0$ and $K > 0$ must hold simultaneously. This holds true for $0 < K < 0.4$.

Example 6.4.12

This example refers to the problem of controlling the yaw of a fighter jet (Figure 6.5a). A simplified diagram of the closed-loop system is given in Figure 6.5b, where the aircraft is approximated by a fourth-order system. Determine the range of values of K for which the closed-loop system is stable.

Solution

The transfer function of the closed-loop system is given by

$$H(s) = \frac{G_c(s)G_a(s)}{1 + G_c(s)G_a(s)} = \frac{\dfrac{K}{s(s+2)(s^2+s+1)}}{1 + \dfrac{K}{s(s+2)(s^2+s+1)}} = \frac{K}{s(s+2)(s^2+s+1)+K}$$

The characteristic polynomial $p(s)$ of the transfer function of the closed-loop system is the following:

$$p(s) = s^4 + 3s^2 + 3s^2 + 2s + K$$

Construct the Routh array of the characteristic polynomial:

(a)

(b)

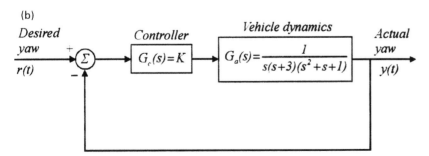

Figure 6.5   Closed-loop system for the control of the yaw of a fighter aircraft. (a) A fighter aircraft; (b) simplified block diagram of the closed-loop system.

$$
\begin{array}{c|ccc}
s^4 & 1 & 3 & K \\
s^3 & 3 & 2 & 0 \\
s^2 & 7/3 & K & \\
s^1 & 2 - 9K/7 & 0 & \\
s^0 & K & &
\end{array}
$$

For the closed-loop system to be stable, the inequalities $2 - 9K/7 > 0$ and $K > 0$ must hold simultaneously. This holds true for $0 < K < 14/9$.

Example 6.4.13

One of the most important applications of industrial robots is that of welding. Many such robots use a vision system to measure the performance of the welding. Figure 6.6 shows a simplified block diagram of such a system. The welding process is approximated by a second-order underdamped system, the vision system by a unity transfer function, and the controller is assumed to be of the integrator type. Determine the range of values of K for which the closed-loop system is stable.

Solution

The transfer function of the closed-loop system is

$$
H(s) = \frac{G_c(s)G_w(s)}{1 + G_c(s)G_w(s)F_v(s)} = \frac{\dfrac{K}{s(s+3)(s+4)}}{1 + \dfrac{K}{s(s+3)(s+4)}} = \frac{K}{s(s+3)(s+4) + K}
$$

The characteristic polynomial $p(s)$ of the transfer function of the closed-loop system is the following:

$$
p(s) = s^3 + 7s^2 + 12s + K
$$

Construct the Routh array of the characteristic polynomial:

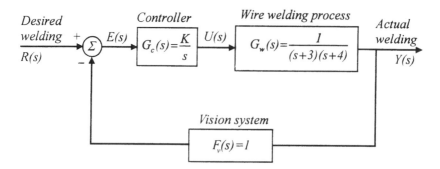

Figure 6.6   Simplified closed-loop block diagram for the control of a welding robot.

s^3	1	12
s^2	7	K
s^1	$\dfrac{84-K}{7}$	0
s^0	K	

For the closed-loop system to be stable, the inequalities $(84-K)/7 > 0$ and $K > 0$ must hold simultaneously. This holds true for $0 < K < 84$.

Example 6.4.14

In digital computers, large disk-storage devices are widely used today. As the disk is spinning, the data head is moved to various positions. This movement must be made very fast and very accurately. A simplified block diagram of the closed-loop head-position control system is given in Figure 6.7. The mathematical model of the head is approximated by a third-order system and the particular controller applied is of the phase lead or lag type (see Chap. 9), depending on the parameter β. Determine:

    (a) For $\beta = 3$, the range of values of K for which the closed-loop system is stable.

    (b) For arbitrary β, the ranges of both β and K for which the closed-loop system is stable.

Solution

(a) For $\beta = 3$, the transfer function of the closed-loop system is given by

$$H(s) = \frac{G_c(s)G_h(s)}{1 + G_c(s)G_h(s)} = \frac{K\left[\dfrac{(s+3)}{(s+1)}\right]\left[\dfrac{1}{s(s+2)(s+5)}\right]}{1 + K\left[\dfrac{(s+3)}{(s+1)}\right]\left[\dfrac{1}{s(s+2)(s+5)}\right]}$$

$$= \frac{K(s+3)}{s(s+1)(s+2)(s+5) + K(s+3)}$$

The characteristic polynomial $p(s)$ of the transfer function of the closed-loop system is the following:

$$p(s) = s^4 + 8s^3 + 17s^2 + (K+10)s + 3K$$

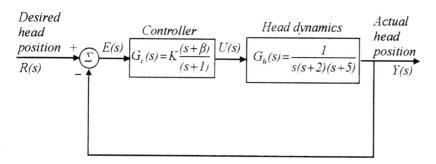

Figure 6.7 Simplified closed-loop block diagram for head-position control.

Construct the Routh array of the characteristic polynomial:

$$
\begin{array}{c|ccc}
s^4 & 1 & 17 & 3K \\
s^3 & 8 & K+10 & 0 \\
s^2 & \dfrac{126-K}{8} & 3K & \\
s^1 & \dfrac{-K^2-76K+1260}{126-K} & 0 & \\
s^0 & 3K & &
\end{array}
$$

For the closed-loop system to be stable, the inequalities $(126-K)/8 > 0$, $-K^2 - 76K + 1260 > 0$ and $3K > 0$ must hold simultaneously, i.e., there must hold $K < 126$, $-90 < K < 14$, and $K > 0$. This holds true for $0 < K < 14$.

(b) For arbitrary β, the transfer function of the closed-loop system is given by

$$
H(s) = \frac{G_c(s)G_h(s)}{1+G_c(s)G_h(s)} = \frac{\dfrac{K(s+\beta)}{s(s+1)(s+2)(s+5)}}{1+\dfrac{K(s+\beta)}{s(s+1)(s+2)(s+5)}}
$$

$$
= \frac{K(s+\beta)}{s(s+1)(s+2)(s+5)+K(s+\beta)}
$$

The characteristic polynomial $p(s)$ of the transfer function of the closed-loop system is the following:

$$
p(s) = s^4 + 8s^3 + 17s^2 + (K+10)s + K\beta
$$

Construct the Routh array of the characteristic polynomial:

$$
\begin{array}{c|ccc}
s^4 & 1 & 17 & K\beta \\
s^3 & 8 & K+10 & 0 \\
s^2 & \dfrac{126-K}{8} & K\beta & \\
s^1 & \dfrac{\dfrac{126-K}{8}(K+10)-8K\beta}{\dfrac{126-K}{8}} & 0 & \\
s^0 & K\beta & &
\end{array}
$$

For the closed-loop system to be stable, the following inequalities must hold simultaneously:

$$
\frac{126-K}{8} > 0
$$

$$
\frac{\dfrac{126-K}{8}(K+10)-8K\beta}{\dfrac{126-K}{8}} > 0
$$

$$
K\beta > 0
$$

The above inequalities may be written as

$$126 > K$$

$$(K + 10)(126 - K) - 64K\beta > 0$$

$$K\beta > 0$$

In Figure 6.8, the hatched area shows the range of values of β and K for which the above inequalities are satisfied and, consequently, the closed-loop system is stable.

Example 6.4.15

This example refers to the human respiratory control system. A simplified block diagram of the system is given in Figure 6.9. Our body has certain special chemo-receptors which measure the percentage of CO_2 in the blood. This percentage is the output $Y(s)$. The ventilation $B(s)$ at the lungs is known to be proportional to the percentage of CO_2 in the blood. That is, our body, by measuring $Y(s)$, indirectly measures the ventilation $B(s)$ at the lungs. Determine the range of values of K for which the closed-loop system is stable.

Solution

The transfer function of the closed-loop system is

$$H(s) = \frac{G_c(s)G_r(s)}{1 + KG_c(s)G_r(s)} = \frac{\dfrac{0.25}{(s+0.5)^2(s+0.1)(s+10)}}{1 + \dfrac{0.25K}{(s+0.5)^2(s+0.1)(s+10)}}$$

$$= \frac{0.25}{s^4 + 11.1s^3 + 11.35s^2 + 3.525s + 0.25(K+1)}$$

Construct the Routh array of the characteristic polynomial of the closed-loop system:

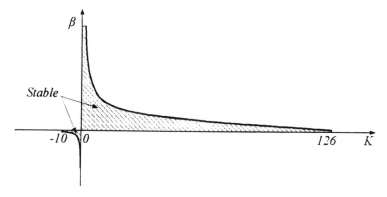

Figure 6.8   Range of values of β and K for which the closed-loop system is stable.

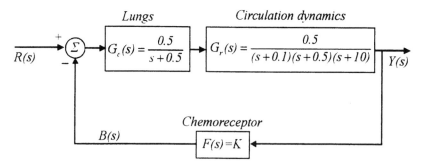

Figure 6.9   Block diagram of the closed-loop humam respiratory control system.

s^4	1	11.35	$0.25(K+1)$
s^3	11.1	3.525	0
s^2	11.032	$0.25(K+1)$	
s^1	$3.273 - 0.252K$	0	
s^0	$0.25(K+1)$		

For the closed-loop system to be stable, the inequalities $3.273 - 0.252K > 0$ and $0.25(K+1) > 0$ must simultaneously be satisfied. This holds true for $-1 < K < 13$.

6.5 STABILITY IN THE SENSE OF LYAPUNOV

6.5.1 Introduction—Definitions

The final objective of this section is to derive the Lyapunov stability criterion for linear, time-invariant systems presented in Subsec. 6.5.4. To this end, we first present some preliminary results from the Lyapunov's stability theory for nonlinear systems which are necessary for the derivation of the results sought in Subsec. 6.5.4. The stability results for nonlinear sytems are, by themselves, of great importance to control engineers.

The Lyapunov approach is based on the differential equations which describe the system and gives information about the stability of the system without requiring the solution of the differential equations. The Lyapunov's results may be grouped in two basic methods: the first method of Lyapunov (or the method of the first approximation) and the second method of Lyapunov (or the direct method).

Before we present the two methods of Lyapunov, we first give some preliminary material and definitions necessary for the results that follow. To this end, consider a system described in state space via the mathematical model

$$\dot{x} = f(x, t), \qquad x(t_0) = x_0 \tag{6.5-1}$$

The solution of Eq. (6.5-1) is denoted by $\phi(t, x_0, t_0)$. This solution depends not only upon x_0 but also upon t_0. Then, the following identity holds:

$$\phi(t_0, x_0, t_0) = x_0 \tag{6.5-2}$$

Definition 6.5.1

The vector x_e is called an equilibrium state of system (6.5-1) if it satisfies the relation

$$\mathbf{f}(\mathbf{x}_e, t) = \mathbf{0}, \qquad \text{for all } t \tag{6.5-3}$$

Obviously, for the determination of the equilibrium states, it is not necessary to solve the dynamic equations (6.5-1) but only the algebraic equations (6.5-3). For example, when the system (6.5-1) is linear time-invariant, i.e., $\mathbf{f}(\mathbf{x}, t) = \mathbf{A}\mathbf{x}$, then there exists only one equilibrium state when $|\mathbf{A}| \neq 0$ and an infinite number of equilibrium states when $|\mathbf{A}| = 0$. When the system (6.5-1) is nonlinear, then one or more equilibrium states may exist. It is noted that each equilibrium state can be shifted to the origin by using an appropriate transformation, where the new equilibrium state will now satisfy the following condition:

$$\mathbf{f}(\mathbf{0}, t) = \mathbf{0}, \qquad \text{for all } t \tag{6.5-4}$$

We give the following definition of stability.

Definition 6.5.2

The equilibrium state $\mathbf{x}_e$ of system (6.5-1) is *stable* if, for every real number $\varepsilon > 0$, there exists a real number $\delta(\varepsilon, t_0) > 0$ such that, if

$$\|\mathbf{x}_0 - \mathbf{x}_e\| \leq \delta \tag{6.5-5}$$

then

$$\|\boldsymbol{\phi}(t, \mathbf{x}_0, t_0) - \mathbf{x}_e\| \leq \varepsilon, \qquad \text{for all } t \tag{6.5-6}$$

If δ does not depend on t_0, then $\mathbf{x}_e$ is *uniformly stable*.

In Figure 6.10, an equilibrium state $\mathbf{x}_e$ of a system with two variables is presented. The regions $S(\varepsilon)$ and $S(\delta)$ are the interiors of two circles with their centers at $\mathbf{x}_e$ and with radii $\varepsilon > 0$ and $\delta > 0$. The region $S(\varepsilon)$ consists of all points which satisfy

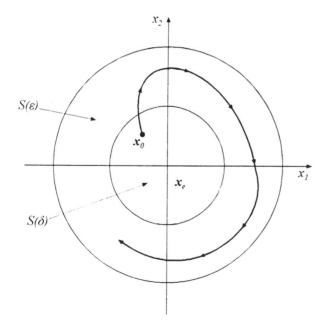

Figure 6.10 A stable equilibrium state.

the condition $\|\mathbf{x} - \mathbf{x}_e\| \leq \varepsilon$. In the figure it is shown that for every $S(\varepsilon)$ there exists an $S(\delta)$ such that, starting with an initial state $\mathbf{x}_0$ which lies inside $S(\delta)$, the trajectory $\boldsymbol{\phi}(t, \mathbf{x}_0, t_0)$ is contained within $S(\varepsilon)$.

Definition 6.5.3

The solution $\boldsymbol{\phi}(t, \mathbf{x}_0, t_0)$ of system (6.5-1) is bounded if for $\delta > 0$ there exists a constant $\varepsilon(\delta, t_0)$ such that, if

$$\|\mathbf{x}_0 - \mathbf{x}_e\| \leq \delta \tag{6.5-7a}$$

then

$$\|\boldsymbol{\phi}(t, \mathbf{x}_0, t_0) - \mathbf{x}_e\| \leq \varepsilon(\delta, t_0), \qquad \text{for all } t \geq t_0 \tag{6.5-7b}$$

If ε does not depend upon t_0, then the solution is uniformly bounded.

Definition 6.5.4

An equilibrium state $\mathbf{x}_e$ of system (6.5-1) is *asymptotically stable* if it is stable and if every solution with $\mathbf{x}_0$ sufficiently close to $\mathbf{x}_e$ converges to $\mathbf{x}_e$ as t increases.

6.5.2 The First Method of Lyapunov

The first method of Lyapunov, or the method of the first approximation, is based on the approximation of the nonlinear differential equation by a linearized differential equation. This approximation is performed for each equilibrium state separately, and conclusions about stability hold only for a small region around the particular equilibrium state. For this reason the first method of Lyapunov is of limited value.

Consider the nonlinear system

$$\dot{\mathbf{x}} = \mathbf{f}(\mathbf{x}) \tag{6.5-8}$$

and let $\mathbf{x}_e$ be an equilibrium state. Expand Eq. (6.5-8) in Taylor series about the point $\mathbf{x} = \mathbf{x}_e$ to yield

$$\dot{\mathbf{x}} = \mathbf{f}(\mathbf{x}) = \mathbf{f}(\mathbf{x}_e) + \left[\frac{\partial \mathbf{f}}{\partial \mathbf{x}}\right]_{\mathbf{x}=\mathbf{x}_e}^{\mathrm{T}} (\mathbf{x} - \mathbf{x}_e) + \frac{1}{2}(\mathbf{x} - \mathbf{x}_e)^{\mathrm{T}} \left[\frac{\partial}{\partial \mathbf{x}} \frac{\partial \mathbf{f}}{\partial \mathbf{x}}\right]_{\mathbf{x}=\mathbf{x}_e}^{\mathrm{T}} (\mathbf{x} - \mathbf{x}_e) + \cdots$$

$$= \mathbf{f}(\mathbf{x}_e) + \mathbf{A}(\mathbf{x} - \mathbf{x}_e) + [\mathbf{B}(\mathbf{x} - \mathbf{x}_e)](\mathbf{x} - \mathbf{x}_e) + \cdots \tag{6.5-9}$$

where

$$\mathbf{f}(\mathbf{x}) = \begin{bmatrix} f_1(\mathbf{x}) \\ f_2(\mathbf{x}) \\ \vdots \\ f_n(\mathbf{x}) \end{bmatrix}, \qquad \mathbf{A} = \left[\frac{\partial \mathbf{f}}{\partial \mathbf{x}}\right]_{\mathbf{x}=\mathbf{x}_e}^{\mathrm{T}} = \begin{bmatrix} \dfrac{\partial f_1}{\partial x_1} & \dfrac{\partial f_1}{\partial x_2} & \cdots & \dfrac{\partial f_1}{\partial x_n} \\[2mm] \dfrac{\partial f_2}{\partial x_1} & \dfrac{\partial f_2}{\partial x_2} & \cdots & \dfrac{\partial f_2}{\partial x_n} \\[2mm] \vdots & \vdots & & \vdots \\[2mm] \dfrac{\partial f_n}{\partial x_1} & \dfrac{\partial f_n}{\partial x_2} & \cdots & \dfrac{\partial f_n}{\partial x_n} \end{bmatrix}_{\mathbf{x}=\mathbf{x}_e}$$

The matrix $\mathbf{B}(\mathbf{x} - \mathbf{x}_e)$ involves higher-order terms. Since $\mathbf{x}_e$ is an equilibrium point, it follows that $\mathbf{f}(\mathbf{x}_e) = \mathbf{0}$. If we let $\mathbf{z} = \mathbf{x} - \mathbf{x}_e$, then Eq. (6.5-9) can be written as follows:

$$\dot{\mathbf{z}} = \mathbf{A}\mathbf{z} + \mathbf{B}(\mathbf{z})\mathbf{z} + \cdots \tag{6.5-10}$$

The first approximation, that is the linear part of (6.5-10), is the following:

$$\dot{z} = Az \tag{6.5-11}$$

The first method of Lyapunov is based on the following theorem.

Theorem 6.5.1

If all the eigenvalues of the matrix A have nonzero real parts, then the conclusions about the stability of the nonlinear system in the neighborhood of x_e may be derived from the study of the stability of the linear system (6.5-11).

Thus, the first method of Lyapunov reduces the problem of studying the stability of nonlinear systems to well-established methods for studying the stability of linear systems.

6.5.3 The Second Method of Lyapunov

The second or direct method of Lyapunov is based on the following idea: if a system has a stable equilibrium state x_e, then the total energy stored in the system decays as time t increases, until this total energy reaches its minimum value in the equilibrium state x_e. The determination of the stability of a linear or nonlinear system via the second method of Lyapunov requires the determination of a special scalar function, which is called the *Lyapunov function*. We give the following definition.

Definition 6.5.5

The time-invariant Lyapunov function, designated by $V(x)$, satisfies the following conditions for all $t_1 > t_0$ and for all x in the neighborhood of $x = 0$, where $x = 0$ is an equilibrium point:

1. $V(x)$ and its partial derivatives are defined and they are continuous
2. $V(0) = 0$
3. $V(x) > 0$, for all $x \neq 0$
4. $\dot{V}(x) < 0$, for all $x \neq 0$, where $\dot{V}(x)$ is the total derivative of $V(x)$, i.e., $\dot{V}(x) = [grad_x V]^T \dot{x}$

The second method of Lyapunov is based on the following theorem.

Theorem 6.5.2

Consider the system

$$\dot{x} = f(x, t), \qquad f(0, t) = 0 \tag{6.5-12}$$

Assume that a Lyapunov function $V(x)$ can be determined for this system. Then, the equilibrium state $x = 0$ is asymptotically stable and the system (6.5-12) is said to be stable in the sense of Lyapunov.

6.5.4 The Special Case of Linear Time-Invariant Systems

From Theorem 6.5.2 it follows that the problem of studying the stability of a system using the second method of Lyapunov is one of determining a Lyapunov function for the particular system. This function may not be unique, while its determination presents great difficulties. It is noted that in cases where we cannot determine even one Lyapunov function for a particular system, this simply means that we cannot conclude about the stability of the system and not that the system is unstable. In what follows we will restrict our presentation to the determination of Lyapunov functions for the special case of linear time-invariant systems. For other types of system, e.g., time-varying, non-linear, etc., see [2–4] and [12].

For the case of time-invariant systems, the following theorem holds.

Theorem 6.5.3

Consider the linear time invariant system $\dot{\mathbf{x}} = \mathbf{Ax}$, with $|\mathbf{A}| \neq 0$ and $\mathbf{x}_e = \mathbf{0}$. Also consider the scalar function $V(\mathbf{x}) = \mathbf{x}^T \mathbf{Px}$, where $\mathbf{P}$ is a positive definite real symmetric matrix. Then, $V(\mathbf{x}) = \mathbf{x}^T \mathbf{Px}$ is a Lyapunov function of the system if, and only if, for any positive definite real symmetric matrix $\mathbf{Q}$ there exists a positive definite real symmetric matrix $\mathbf{P}$ such that the following relation holds:

$$\mathbf{A}^T \mathbf{P} + \mathbf{PA} = -\mathbf{Q} \qquad (6.5\text{-}13)$$

Proof

Taking the derivative of $V(\mathbf{x}) = \mathbf{x}^T \mathbf{Px}$ with respect to time yields

$$\dot{V}(\mathbf{x}) = \dot{\mathbf{x}}^T \mathbf{Px} + \mathbf{x}^T \mathbf{P}\dot{\mathbf{x}} \qquad (6.5\text{-}14)$$

Upon using the relation $\dot{\mathbf{x}} = \mathbf{Ax}$, the expression for $\dot{V}(\mathbf{x})$ becomes

$$\dot{V}(\mathbf{x}) = \mathbf{x}^T \mathbf{A}^T \mathbf{Px} + \mathbf{x}^T \mathbf{PAx} \qquad (6.5\text{-}15)$$

According to Definition 6.5.5, Condition 4, there must be $\dot{V}(\mathbf{x}) < 0$, for all $\mathbf{x} \neq \mathbf{0}$. Hence, Condition 4 is satisfied if we set

$$\dot{V}(\mathbf{x}) = \mathbf{x}^T \mathbf{A}^T \mathbf{Px} + \mathbf{x}^T \mathbf{PAx} = -\mathbf{x}^T \mathbf{Qx} \qquad (6.5\text{-}16)$$

where the right-hand side term $-\mathbf{x}^T \mathbf{Qx} < 0$ due to the choice of $\mathbf{Q}$. For eq. (6.5-16) to hold, the matrices $\mathbf{P}$ and $\mathbf{A}$ must satisfy the following relation:

$$\mathbf{A}^T \mathbf{P} + \mathbf{PA} = -\mathbf{Q} \qquad (6.6\text{-}17)$$

which is the same with relation (6.5-13).

Example 6.5.1

Consider the linear system $\dot{\mathbf{x}} = \mathbf{Ax}$, where

$$\mathbf{A} = \begin{bmatrix} 0 & 1 \\ -2 & -3 \end{bmatrix}$$

Determine the Lyapunov function for the system.

Solution

Consider the relation (6.5-13) where, for simplicity, let $\mathbf{Q} = \mathbf{I}$. Then we have

$$\begin{bmatrix} 0 & -2 \\ 1 & -3 \end{bmatrix} \begin{bmatrix} p_{11} & p_{12} \\ p_{12} & p_{22} \end{bmatrix} + \begin{bmatrix} p_{11} & p_{12} \\ p_{12} & p_{22} \end{bmatrix} \begin{bmatrix} 0 & 1 \\ -2 & -3 \end{bmatrix} = \begin{bmatrix} -1 & 0 \\ 0 & -1 \end{bmatrix}$$

where use was made of the relation $p_{21} = p_{12}$, since we have assumed that the matrix $\mathbf{P}$ is symmetric. The above equation, due to the symmetry in $\mathbf{P}$ and $\mathbf{Q}$, yields the following $n(n+1)/2 = 3$ algebraic equations:

$-4p_{12} = -1$

$p_{11} - 3p_{12} - 2p_{22} = 0$

$2p_{12} - 6p_{22} = -1$

The above equations give the following matrix

$$P = \begin{bmatrix} \frac{5}{4} & \frac{1}{4} \\ \frac{1}{4} & \frac{1}{4} \end{bmatrix}$$

If we apply the Sylvester's criterion (Sec. 2.12), it follows that the matrix P is positive definite. Therefore, the sytem is asymptotically stable. The Lyapunov function is

$$V(\mathbf{x}) = \mathbf{x}^T P \mathbf{x} = [x_1 \quad x_2] \begin{bmatrix} \frac{5}{4} & \frac{1}{4} \\ \frac{1}{4} & \frac{1}{4} \end{bmatrix} \begin{bmatrix} x_1 \\ x_2 \end{bmatrix} = \frac{1}{4}[5x_1^2 + 2x_1 x_2 + x_2^2]$$

To check the results, we investigate $V(\mathbf{x})$ using Definition 6.5.5. It is clear that $V(\mathbf{x})$ satisfies the first three conditions of Definition 6.5.5. For the fourth condition we compute $\dot{V}(\mathbf{x})$ to yield

$$\dot{V}(\mathbf{x}) = [\text{grad}_x \ V]^T \dot{\mathbf{x}} = [\text{grad}_x \ V]^T \mathbf{A}\mathbf{x} = \frac{1}{2}[10x_1 + 2x_2, \ 2x_1 + 2x_2]\mathbf{A}\mathbf{x}$$

$$= \frac{1}{4}[-4x_1 - 4x_2, \ 4x_1 - 4x_2] \begin{bmatrix} x_1 \\ x_2 \end{bmatrix} = -x_1^2 - x_2^2$$

Clearly, $\dot{V}(\mathbf{x})$ also satisfies the fourth condition of Definition 6.5.5. Hence, $V(\mathbf{x})$ is a Lyapunov function and therefore the system is asymptotically stable.

PROBLEMS

1. Investigate the stability of the systems having the following characteristic polynomials:

(a) $s^4 + 2s^3 + 6s^2 + 7s + 5$

(b) $s^3 + s^2 + 2s + 1$

(c) $s^3 + s^2 + 1$

(d) $s^4 + s^3 + s^2 + 2s + 4$

(e) $2s^4 + s^3 + 3s^2 + 5s + 10$

(f) $s^5 + 3s^4 + 2s^3 + 6s^2 + 6s + 9$

(g) $s^4 + 2s^3 + 3s^2 + 4s + 5$

(h) $s^5 + s^4 + 2s^3 + 2s^2 + 3s + 4$

(i) $s^5 + s^4 + s^3 + 2s^2 + 2s + 2$

(j) $s^4 + 3s^3 + 4s^2 + 3s + 3$

2. Find the range of values of the parameter K for which the systems, with the following characteristic polynomials, are stable:

(a) $s^3 + s^2 + Ks + 1$

(b) $s^4 + s^3 + 2s^2 + 3s + K$

(c) $s^4 + (K + 1)s^3 + s^2 + 5s + 2$

3. The block diagram of a system for the speed control of a tape drive is shown in Figure 6.11. Find the range of values of K so that the closed-loop system is stable.

4. Consider a rocket altitude control system having the block diagram shown in Figure 6.12. (a) Given that the transfer function of the controller is

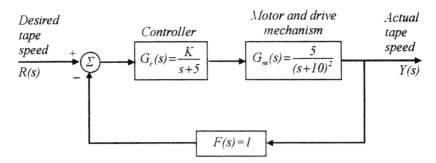

Figure 6.11

$G_c(s) = (s + 3)(s + 2)/s$, determine the range of values of K for which the closed-loop system is stable. (b) Given that the transfer function of the controller is $G_c = s + a$, determine the range of values of K and a for which the closed-loop system is stable.

5. The block diagram of a metal sheet thickness control system (depicted in Figure 1.10) is given in Figure 6.13. Find the range of values of K and a such that the closed-loop system is stable.

6. The block diagram of a feedback control system is shown in Figure 6.14. (a) Determine the range of values of K_2 so that the system is stable for $K_1 = K_3 = 1$, $T_1 = 1$, $T_3 = 1/2$, and $T_4 = 1/3$. (b) Determine the range of values of the parameter T_4 for which the system is stable, given that $K_1 = 1$, $K_2 = 2$, $K_3 = 5$, $T_1 = 1/2$, and $T_3 = 1/3$.

7. Consider a satellite orientation control system shown in Figure 3.56 of Example 3.13.7, where $K_b K_t = 1$ and $J = 1$. Let the transfer function of the controller be $G_s(s) = K_p + (K_i/s) + K_d s$ (PID controller). (a) Find the range of values of the controller parameters so that the closed-loop system is stable. (b) For $K_p = 1$, $K_i = 2$, and $K_d = 1$, determine the number of the closed-loop poles located in the right-half complex plane.

8. The closed-loop control system of an aircraft wing is given in Figure 6.15. Determine the range of values of the parameters K and T of the hydraulic servomotor that guarantee the stability of the closed-loop system.

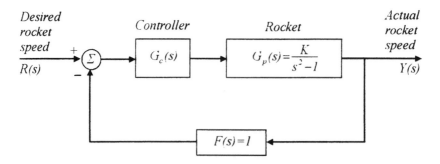

Figure 6.12

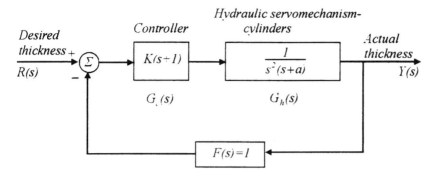

Figure 6.13

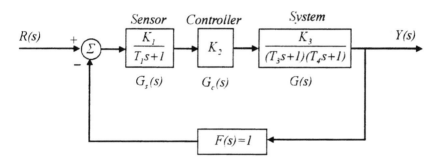

Figure 6.14

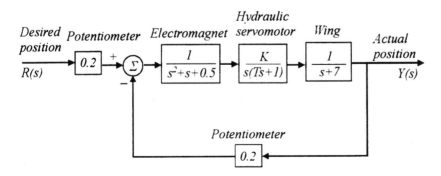

Figure 6.15

9. The differential equation which describes the dynamics of the pendulum shown in Figure 6.16 is the following

$$mR^2\ddot{\theta} + K\dot{\theta} + mgR\sin\theta = 0$$

where K is the friction coefficient, m is the mass at the end of the rod, R is the length of the pendulum, g is the gravitational constant, and θ is the angle of the pendulum from the vertical axis. for this system, (a) find a state-space model and determine the equilibrium states and (b) investigate the stability of the equilibrium states by means of the first method of Lyapunov.

10. Carry out the study of stability of the following nonlinear systems using the corresponding candidate Lyapunov functions:

(a) $\dot{x}_1 = -2x_1 + 2x_2^4, \qquad \dot{x}_2 = -x_2, \qquad$ and
$V(\mathbf{x}) = 6x_1^2 + 12x_2^2 + 4x_1x_2^4 + x_2^8$

(b) $\dot{x}_1 = -x_1 + x_2 + x_1(x_1^2 + x_2^2), \qquad \dot{x}_2 = -x_1 - x_2 + x_2(x_1^2 + x_2^2), \qquad$ and
$V(\mathbf{x}) = x_1^2 + x_2^2$

(c) $\dot{x}_1 = \dfrac{6x_1}{(1 + x_1^2)^2} + 2x_2, \qquad \dot{x}_2 = \dfrac{-2x_1}{(1 + x_1^2)^2} - \dfrac{2x_2}{(1 + x_1^2)^2}, \qquad$ and

$V(\mathbf{x}) = \dfrac{x_1^2}{1 + x_1^2} + x_2^2$

11. For the system described by the equation $\dot{\mathbf{x}} = \mathbf{A}\mathbf{x}$, where

$$\mathbf{A} = \begin{bmatrix} a & 0 \\ 1 & -1 \end{bmatrix}$$

determine the range of values of the parameter a so that the system is asymptotically stable.

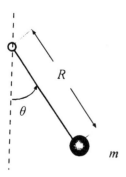

Figure 6.16

12. Determine a Lyapunov function for the systems of the form $\dot{\mathbf{x}} = \mathbf{A}\mathbf{x}$, where

$$\mathbf{A} = \begin{bmatrix} -1 & -2 \\ 1 & -4 \end{bmatrix}, \qquad \mathbf{A} = \begin{bmatrix} 0 & 1 & 0 \\ -\alpha_3 & 0 & 1 \\ 0 & -\alpha_2 & -\alpha_1 \end{bmatrix},$$

$$\mathbf{A} = \begin{bmatrix} 0 & 1 & 0 \\ 0 & 0 & 1 \\ -2 & -5 & -4 \end{bmatrix}, \qquad \mathbf{A} = \begin{bmatrix} 1 & 1 \\ -2 & -3 \end{bmatrix}$$

BIBLIOGRAPHY

Books

1. PJ Antsaklis, AN Michel. Linear Systems. New York: McGraw-Hill, 1997.
2. DP Atherton. Nonlinear Control Engineering. London: Van Nostrand Reinhold, 1975.
3. A Blaguiere. Nonlinear System Analysis. New York: Academic Press, 1966.
4. PA Cook. Nonlinear Dynamical Systems. London: Prentice-Hall, 1986.
5. JJ D'Azzo, CH Houpis. Linear Control System Analysis and Design, Conventional and Modern. New York: McGraw-Hill, 1975.
6. PM DeRusso, RJ Roy, CM Close. State Variables for Engineers. New York: John Wiley, 1965.
7. JJ DiStefano III, AR Stubberud, IJ Williams. Feedback and Control Systems. Schaum's Outline Series. New York: McGraw-Hill, 1967.
8. RC Dorf, RE Bishop. Modern Control Analysis. London: Addison-Wesley, 1995.
9. JC Doyle. Feedback Control Theory. New York: Macmillan, 1992.
10. GF Franklin, JD Powell, A Emami-Naeini. Feedback Control of Dynamic Systems. Reading, MA: Addison-Wesley, 1986.
11. B Friedland. Control System Design. An Introduction to State-space Methods. New York: McGraw-Hill, 1987.
12. JE Gibson. Nonlinear Automatic Control. New York: McGraw-Hill, 1963.
13. MJ Holtzman. Nonlinear System theory, a Functional Analysis Approach. Englewood Cliffs, New Jersey: Prentice Hall, 1970.
14. M Krstic, I Kanellakopoulos, P Kokotovic. Nonlinear and Adaptive Control Design. New York: John Wiley, 1995.
15. BC Kuo. Automatic Control Systems. London: Prentice Hall, 1995.
16. NS Nise. Control Systems Engineering. New York: Benjamin and Cummings, 1995.
17. K Ogata. Modern Control Systems. London: Prentice Hall, 1997.
18. WJ Rugh. Linear System Theory. 2nd ed. Englewood Cliffs, New Jersey: Prentice Hall, 1996.
19. DD Siljak. Nonlinear Systems, the Parameter Analysis and Design. New York: John Wiley, 1969.
20. M Vidyasagar. Nonlinear Systems Analysis. 2nd ed. Englewood Cliffs, New Jersey: Prentice Hall, 1993.
21. LA Zadeh, CA Desoer. Linear System Theory—The State Space Approach. New York: McGraw-Hill, 1963.

7

The Root Locus Method

7.1 INTRODUCTION

The positions of the poles of the transfer function of a system in the complex plane characterize completely the system's stability and play a decisive role in the shape of its time response. For these two basic reasons, the determination of a system's poles is a problem of special importance in control practice.

One of the main problems in control systems is the design of an appropriate controller capable of shifting the poles of the open-loop system to new desired closed-loop pole positions in the complex plane. In its simplest form, such a controller is a gain constant K of an amplifier connected in series with the system's open-loop transfer function. Changing the value of the constant K, from $-\infty$ and $+\infty$, results in shifting the poles of the closed-loop system in the complex plane. Specifically, the locus of the roots of the closed-loop system characteristic polynomial, which is formed in the s-plane as K varies, is the subject of this chapter.

The development of a relatively simple method for constructing the root locus of the closed-loop characteristic polynomial is due to Evans [4, 8, 9]. This method gives an approximate graphical representation of the root locus which is very useful in the design of a closed-loop system since it gives the position of the poles of the closed-loop system in the s-plane for all values of the gain constant K.

7.2 INTRODUCTORY EXAMPLE

To facilitate the understanding of the root locus method, a simple introductory example will first be presented.

Example 7.2.1

Consider the closed-loop position servomechanism system described in Subsec. 3.13.2. Let $L_a \cong 0$, $K_p = 1$, $A = 1$, and $B = 6$. Then, the closed-loop system is simplified, as in Figure 7.1. For this simplified system, draw the root locus of the closed-loop system, for $K \in (-\infty, +\infty)$.

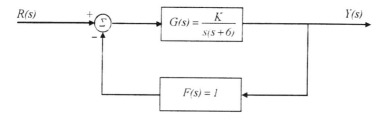

Figure 7.1 Simplified block diagram of the position control system described in Subsec. 3.13.2.

Solution

The closed-loop system transfer function is given by

$$H(s) = \frac{G(s)}{1 + G(s)F(s)} = \frac{K}{s^2 + 6s + K} \qquad (7.2\text{-}1)$$

The characteristic polynomial $p_c(s)$ of the closed-loop system is

$$p_c(s) = s^2 + 6s + K \qquad (7.2\text{-}2)$$

The roots of the characteristic polynomial $p_c(s)$, or equivalently the poles of the closed-loop system transfer function $H(s)$, are

$$s_1 = -3 + \sqrt{9 - K} \qquad \text{and} \qquad s_2 = -3 - \sqrt{9 - K} \qquad (7.2\text{-}3)$$

It is clear that the roots s_1 and s_2 depend upon the parameter K. Therefore, as K varies, the two roots will vary as well. The diagrams of s_1 and s_2 in the complex plane, for $K \in (-\infty, +\infty)$, form the *root locus* of the characteristic polynomial $p_c(s)$.

To draw the root locus of $p_c(s)$, we calculate the roots s_1 and s_2 while K changes from $-\infty$ to $+\infty$. We observe the following:

1. For $-\infty < K < 0$, both roots are real with $s_1 > 0$ and $s_2 < 0$
2. For $K = 0$, $s_1 = 0$ and $s_2 < 0$
3. For $0 < K < 9$, both roots are negative
4. For $K = 9$, we have the double root $s_1 = s_2 = -3$
5. For $9 < K < +\infty$, both roots are complex conjugates with real part -3.

The above remarks for the roots s_1 and s_2 suffice to determine their root locus. In Figures 7.2a and 7.2b, the root locus of s_1 and s_2 are shown, respectively. Usually, the root locus of all roots of a polynomial $p_c(s)$ is given in one single figure. For this example, the root locus of both roots s_1 and s_2 of $p_c(s) = s^2 + 6s + K$ is given in Figure 7.3.

The motivation for constructing the root locus of the characteristic polynomial $p_c(s)$ is that it reveals important information with regard to the behavior of the closed-loop system. The most important information is the following.

1 Stability

From the root locus of Figure 7.3, the stability of the closed-loop system may easily be studied. As already known, the closed-loop system is stable when both roots s_1 and s_2 are in the left-half complex plane, which occurs when $K > 0$. Therefore, the

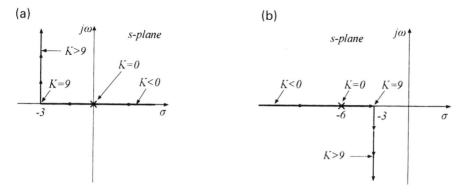

Figure 7.2   The root locus of s_1 and s_2 of the characteristic polynomial $p_c(s) = s^2 + 6s + K$ of the position control system of Figure 7.1. (a) The root locus of $s_1 = -3 + \sqrt{9 - K}$; (b) the root locus of $s_2 = -3 - \sqrt{9 - K}$.

root locus technique can replace the stability criteria for linear time-invariant systems presented in Chap. 6.

2   Transient Response

The closed-loop system's transient response depends mainly on the locations of the roots of $p_c(s)$ in the complex plain (see Secs 4.2 and 4.3). This is demonstrated by the following two cases:

a.   For $0 < K \leq 9$, the system has two negative roots; therefore its response does not invovle any oscillations.

b.   For $K > 9$, the system has two complex roots; therefore its response involves oscillations. Furthermore, the system's damped frequency ω_d increases as K increases.

We must keep in mind that the positions of the zeros of any transfer function also affect the transient response of the system.

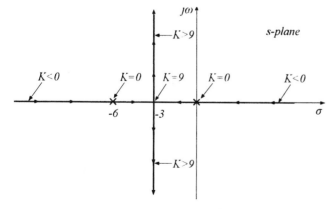

Figure 7.3   The root locus of $p_c(s) = s^2 + 6s + K$ of the position control system of Figure 7.1.

3 Characteristics in the Frequency Domain

Since the bandwidth is proportional to the damped frequency ω_d (see Sec. 8.3), the root locus also gives information about the system's bandwidth. For this example, as K increases, it is clear that the bandwidth also increases.

The above observations are valuable in the study and design of control systems and are presented so as to further motivate the study of the root locus method.

For the general case, where the characteristic polynomial $p_c(s)$ is of higher order, the root locus construction method presented in the above example is difficult, if not impossible, to apply. This is mainly because for very high order polynomials there exists no method for determining the analytical expression of the roots of the polynomial as a function of its coefficients. For this reason, the construction of the root locus for the general case is not done directly, i.e., on the basis of the analytical expressions of the roots of $p_c(s)$, but indirectly. This chapter is devoted to the development of such a method, which, as already mentioned, was first introduced by Evans.

7.3 CONSTRUCTION METHOD OF ROOT LOCUS

7.3.1 Definition of Root Locus

Here, we shall present a more mathematical definition of the root locus. To this end, consider the characteristic equation of any closed-loop system (see for example figure 7.1) given by the following relation:

$$1 + G(s)F(s) = 0 \tag{7.3-1}$$

or equivalently by the relation

$$G(s)F(s) = -1 \tag{7.2-2}$$

The characteristic equation (7.3-2) can also be written as two equations, involving the amplitude and the phase, as follows

$$|G(s)F(s)| = 1 \tag{7.3-3a}$$

and

$$\underline{/G(s)F(s)} = (2\rho + 1)\pi, \qquad \rho = 0, \pm 1, \pm 2, \ldots \tag{7.3-3b}$$

Assume that the open-loop transfer function $G(s)F(s)$ has the following general form:

$$G(s)F(s) = K\frac{(s + z_1)(s + z_2)\cdots(s + z_m)}{(s + p_1)(s + p_2)\cdots(s + p_n)} = K\frac{s^m + d_1 s^{m-1} + \cdots d_{m-1}s + d_m}{s^n + b_1 s^{n-1} + \cdots + b_{n-1}s + b_n} \tag{7.3-4}$$

Hence, Eqs (7.3-3a) and (7.3-3b) will have the form

$$|K| \frac{\prod\limits_{i=1}^{m} |s + z_i|}{\prod\limits_{i=1}^{n} |s + p_i|} = 1, \qquad -\infty < K < +\infty \qquad (7.3\text{-}5a)$$

$$\sum_{i=1}^{m} \underline{/s + z_i} - \sum_{i=1}^{n} \underline{/s + p_i} = \begin{bmatrix} (2\rho + 1)\pi, & K > 0 \\ 2\rho\pi, & K < 0 \end{bmatrix} \qquad \begin{matrix} (7.3\text{-}5b) \\ (7.3\text{-}5c) \end{matrix}$$

for $\rho = 0, \pm1, \pm2, \ldots$

Definition 7.3.1

The root locus of a closed-loop system with characteristic equation (7.3-1) is defined as the locus of all points s which satisfy Eqs (7.3-5a,b) for $K \in (0, +\infty)$. The locus of points s which satisfy Eqs (7.3-5a,c) for $K \in (-\infty, 0)$ is called the *complementary* root locus.

7.3.2 Theorems for Constructing the Root Locus

In this section, several theorems will be presented which greatly facilitate the construction of the root locus. In particular, these theorems are the basic tools in approximately constructing the root locus.

Theorem 7.3.1 (Starting or Departure Points)

The points of the root locus for $K = 0$ are the poles of $G(s)F(s)$. These points are called starting or departure points of the root locus.

Proof

Relation (7.3-5a) may be written as

$$\prod_{i=1}^{n} |s + p_i| = |K| \prod_{i=1}^{m} |s + z_i| \qquad (7.3\text{-}6)$$

For $K = 0$, relation (7.3-6) gives

$$\prod_{i=1}^{n} |s + p_i| = 0 \qquad (7.3\text{-}7)$$

Relation (7.3-7) is satisfied for $s = -p_i$. Therefore, the points of the root locus for $K = 0$ are the poles $-p_1, -p_2, \ldots, -p_n$ of $G(s)F(s)$.

Theorem 7.3.2 (Ending or Arrival Points)

The points of the root locus for $K \to \pm\infty$ are the zeros of $G(s)F(s)$. These points are called ending or arrival points of the root locus.

Proof

Relation (7.3-5a) may be written as

$$\frac{\displaystyle\prod_{i=1}^{m} |s + z_i|}{\displaystyle\prod_{i=1}^{n} |s + p_i|} = \frac{1}{|K|} \tag{7.3-8}$$

For $K = \pm\infty$, relation (7.3-8) gives

$$\prod_{i=1}^{m} |s + z_i| = 0 \tag{7.3-9}$$

Relation (7.3-9) is satisfied for $s = -z_i$. Also, for $K \to \pm\infty$ and for $m < n$, relation (7.3-8) is satisfied for $s \to \infty$. Therefore, the points of the root locus for $K \to \pm\infty$ are the zeros $-z_1, -z_2, \dots, -z_m$ of $G(s)F(s)$ and infinity when $m < n$.

Theorem 7.3.3 (Number of Branches)

The number of branches of the root locus is $\max(m, n)$, where m and n are the number of zeros and poles of $G(s)F(s)$, respectively.

Proof

Introducing Eq. (7.3-4) in Eq. (7.3-1), we obtain

$$1 + G(s)F(s) = 1 + K \frac{\displaystyle\prod_{i=1}^{m}(s + z_i)}{\displaystyle\prod_{i=1}^{n}(s + p_i)} = 0 \tag{7.3-10}$$

The closed-loop system characteristic polynomial $p_c(s)$ is given by

$$p_c(s) = \prod_{i=1}^{n}(s + p_i) + K \prod_{i=1}^{m}(s + z_i) \tag{7.3-11}$$

The degree of $p_c(s)$ will be $\max(m, n)$. Therefore, the number of roots (and hence the number of branches of the root locus) is $\max(m, n)$.

Theorem 7.3.4 (Symmetry About the Real Axis)

The root locus for $K \in (-\infty, +\infty)$ is symmetrical about the real axis.

Proof

Since all complex roots of the $p_c(s)$ appear always in conjugates pairs, it follows that the root locus will be symmetrical about the real axis.

Theorem 7.3.5 (Asymptotes)

For large values of s, the root locus for $K \geq 0$ approaches asymptotically the straight lines having the following angles

$$\theta_\rho = \frac{(2\rho+1)\pi}{n-m}, \qquad \rho = 0, 1, \ldots, |n-m|-1 \tag{7.3-12}$$

The root locus for $K \leq 0$ approaches asymptotically the straight lines having the following angles:

$$\theta_\rho = \frac{2\rho\pi}{n-m}, \qquad \rho = 0, 1, \ldots, |n-m|-1 \tag{7.3-13}$$

Proof

The characteristic equation (7.3-10) can be written as

$$\frac{\displaystyle\prod_{i=1}^{n}(s+p_i)}{\displaystyle\prod_{i=1}^{m}(s+z_i)} = -K$$

Dividing the two polynomials, we obtain

$$s^{n-m} + (b_1 - d_1)s^{n-m-1} + \cdots = -K \tag{7.3-14}$$

where use was made of relation (7.3-4). For large values of s, the left-hand side of relation (7.3-14) may be approximated by its first two terms, as follows:

$$s^{n-m} + (b_1 - d_1)s^{n-m-1} = -K \tag{7.3-15}$$

Relation (7.3-15) can be written as

$$s^{n-m}\left[1 + \frac{b_1 - d_1}{s}\right] = -K$$

Taking the $(n-m)$ root of both parts of the above equation, we obtain

$$s\left[1 + \frac{b_1 - d_1}{s}\right]^{1/(n-m)} = (-K)^{1/(n-m)} \qquad \text{or}$$

$$s\left[1 + \frac{b_1 - d_1}{(n-m)s} + \cdots\right] = (-K)^{1/(n-m)} \tag{7.3-16}$$

where the expansion of $(1+a)^{1/(n-m)}$ was used. For large values of s, this expansion is approximated by its first two terms, in which case Eq. (7.3-16) becomes

$$s + \frac{b_1 - d_1}{n-m} = (-K)^{1/(n-m)} \tag{7.3-17}$$

Since, for $K \geq 0$, it holds that

$$-K = |K|e^{j(2\rho+1)\pi}, \qquad \rho = 0, \pm 1, \pm 2,$$

and for $K \leq 0$, it holds that

$$-K = |K|e^{j2\rho\pi}, \qquad \rho = 0, \pm 1, \pm 2, \ldots$$

it follows that for $K \geq 0$, we have

$$(-K)^{1/(n-m)} = |K|^{1/(n-m)} \exp[j((2\rho+1)\pi/(n-m))], \qquad \rho = 0, \pm 1, \pm 2, \ldots$$

and for $K \leq 0$, we have

$$(-K)^{1/(n-m)} = |K|^{1/(n-m)} \exp[j(2\rho\pi/(n-m))], \qquad \rho = 0, \pm 1, \pm 2, \ldots$$

Therefore, relation (7.3-17) can be written as

$$\sigma + j\omega + \frac{b_1 - d_1}{n - m} = |K|^{1/(n-m)} \exp[j((2\rho + 1)\pi/(n - m))], \qquad \text{for} \qquad K \geq 0$$

(7.3-18a)

$$\sigma + j\omega + \frac{b_1 - d_1}{n - m} = |K|^{1/(n-m)} \exp[j((2\rho\pi)/(n - m))], \qquad \text{for} \qquad K \leq 0$$

(7.3-18b)

for $\rho = 0, \pm 1, \pm 2 \ldots$, where $s = \sigma + j\omega$. Relations (7.3-18a,b) hold if both sides of the two equations have the same phase. That is, if

$$\tan^{-1}\left[\frac{\omega}{\sigma + \dfrac{b_1 - d_1}{n - m}}\right] = \frac{(2\rho + 1)\pi}{n - m}, \qquad \text{for} \qquad K \geq 0$$

and

$$\tan^{-1}\left[\frac{\omega}{\sigma + \dfrac{b_1 - d_1}{n - m}}\right] = \frac{2\rho\pi}{n - m}, \qquad \text{for} \qquad K \geq 0$$

Solving the above equations for ω, we obtain

$$\omega = \left[\tan\frac{2\rho + 1)\pi}{n - m}\right]\left[\sigma + \frac{b_1 - d_1}{n - m}\right], \qquad K \geq 0 \qquad (7.3\text{-}19a)$$

$$\omega = \left[\tan\frac{2\rho\pi}{n - m}\right]\left[\sigma + \frac{b_1 - d_1}{n - m}\right], \qquad K \leq 0 \qquad (7.3\text{-}19b)$$

Relations (7.3-19a,b) are actually straight lines in the s-plane with angles (slopes) as follows:

$$\theta_\rho = \frac{(2\rho + 1)\pi}{n - m}, \qquad \text{for} \qquad K \geq 0$$

$$\theta_\rho = \frac{2\rho\pi}{n - m}, \qquad \text{for} \qquad K \leq 0$$

for $\rho = 0, \pm 1, \pm 2, \ldots$ A careful examination of the values of the angles θ_ρ, as ρ takes on the values $0, \pm 1, \pm 2, \ldots$, shows that there are only $2|n - m|$ asymptotes which correspond to the values of $\rho = 0, 1, \ldots, |n - m| - 1$.

Theorem 7.3.6 (Intersection of Asymptotes)

All the $2|n - m|$ asymptotes of the root locus intersect on the real axis at the point σ_1, where

$$\sigma_1 = -\frac{b_1 - d_1}{n - m} = -\frac{\displaystyle\sum_{i=1}^{n} p_i - \sum_{i=1}^{m} z_i}{n - m} \qquad (7.3\text{-}20)$$

Proof

From relations (7.3-19a,b) it immediately follows that all the asymptotes intersect the real axis at the point σ_1 specified in Eq. (7.3-20).

Theorem 7.3.7 (Real Axis Segments)

A segment of the real axis can be part of the root locus if the following hold:

1. For $K \geq 0$, the number of real poles and zeros of $G(s)F(s)$, which are to the right of the segment, is odd.
2. For $K \leq 0$, the number of real poles and zeros of $G(s)F(s)$, which are to the right of the segment, is even.

Proof

Assume that the point s_1 is located on the real axis. According to relation (7.3-5b), the point s_1 is a point of the root locus for $K \geq 0$, if it satisfies the following equation:

$$\sum_{i=1}^{m} \underline{/s_1 + z_i} - \sum_{i=1}^{n} \underline{/s_1 + p_i} = (2\rho + 1)\pi, \qquad \rho = 0, \pm 1, \pm 2, \ldots \qquad (7.3\text{-}21)$$

It is obvious that the complex conjugate poles and zeros do not contribute in Eq. (7.3-21) because their angles cancel each other. Similarly, with the real poles and zeros in Eq. (7.3-21), which are located to the left of the point s_1. On the other hand, the phase contribution of every real pole and zero which is located to the right of the point s_1, is π. Let p_i, $i = 1, 2, \ldots, q$ and z_i, $i = 1, 2, \ldots, r$ to be the poles and zeros of $G(s)Fs)$ which are located on the real axis and to the right of the point s_1. In this case, Eq. (7.3-21) yields

$$\pi r - \pi q = (r - q)\pi = (2\rho + 1)\pi \qquad (7.3\text{-}22)$$

Relation (7.3-22) holds if $r - q = 2\rho + 1$, i.e., when either r or q is odd (or when $r + q$ is odd).

Working likewise for $K \leq 0$, where relation (7.3-21) has the form

$$\sum_{i=1}^{m} \underline{/s_1 + z_i} - \sum_{i=1}^{n} \underline{/s_1 + p_i} = 2\rho\pi, \qquad \rho = 0, \pm 1, \pm 2, \ldots \qquad (7.3\text{-}23)$$

we conclude that relation (7.3-23) holds if r or q is even (or when $r + q$ is even).

Theorem 7.3.8 (Breakaway Points)

Assume that the characteristic equation

$$1 + G(s)F(s) = 0 \qquad (7.3\text{-}24)$$

has repeated roots. These roots are called breakaway points of the root locus and are also roots of the equation

$$\frac{d}{ds}[G(s)F(s)] = 0 \qquad (7.3\text{-}25)$$

Proof

Let $s = \lambda$ be a repeated root of multiplicity r. In this case, Eq. (7.3-24) can be factored out as follows:

$$1 + G(s)F(s) = (s - \lambda)^r q(s) = 0 \tag{7.3-26}$$

where $q(s)$ is a rational function. If we take the derivative with respect to s of both sides of Eq. (7.3-26), we have

$$\frac{d}{ds}[G(s)F(s)] = r(s - \lambda)^{r-1} q(s) + (s - \lambda)^r \frac{dq(s)}{ds} \tag{7.3-27}$$

Since we have assumed that $r \geq 2$, it follows that the right-hand side of Eq. (7.3-27) becomes zero for $s = \lambda$. Thus, $s = \lambda$ is also a root of Eq. (7.3-25).

Relation (7.3-25) is a necessary but not a sufficient condition. This means that the breakaway points are roots of Eq. (7.3-25), but all roots of Eq. (7.3-25) are not necessarily breakaway points. Therefore, in order to make sure that a certain root of Eq. (7.3-25) is a breakaway point of the root locus, it is sufficient that this root satisfies Eq. (7.3-24) for some real value of K.

7.3.3 Additional Information for Constructing the Root Locus

In addition to the above eight theorems, we also give the following useful information for the construction of a root locus.

1 Angles of Departure and Arrival of the Root Locus

The angle of the root locus at the poles and zeros of $G(s)F(s)$ can be calculated on the basis of Eq. (7.3-5b). As an example, consider calculating the angle at the arbitrary pole $-p_q$. To this end, assume that the point s_1 is very close to the pole $-p_q$. Thus, if we solve Eq. (7.3-5b) for the angle at the pole $-p_q$, we obtain

$$\theta_{-p_q} = \underline{/s_1 + p_q} = -(2\rho + 1)\pi + \sum_{i=1}^{m} \underline{/s_1 + z_i} - \sum_{\substack{i=1 \\ i \neq q}}^{n} \underline{/s_1 + p_i} \tag{7.3-28}$$

The results are similar for the angle of an arbitrary zero $-z_p$.

The angles θ_{-p_q} and θ_{-z_q} are called the departure and arrival angles, respectively. Their graphical representation is given in Figure 7.4. Since the departure angle of the root locus for $K \geq 0$ is equal to the arrival angle for $K \leq 0$, as shown in Figure 7.4, the arrival angle at the point $s_1 \rightarrow -z_q$ can be calculated from the relation (7.3-5c) as follows

$$\theta_{-z_q} = \underline{/s_1 + z_q} = 2\rho\pi - \sum_{\substack{i=1 \\ i \neq q}}^{m} \underline{/s_1 + z_i} + \sum_{i=1}^{n} \underline{/s_1 + p_i} \tag{7.3-29}$$

2 Intersection of the Root Locus with the Imaginary Axis

The intersection of the root locus with the imaginary axis (if there exists such an intersection) is a set of points beyond which the system becomes unstable. A method for determining these points is based on the Routh's criterion and is illustrated in the examples presented in the following subsection.

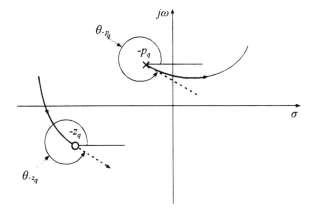

Figure 7.4   Angles of departure and arrival of the root locus.

7.3.4   Root Locus of Practical Control Systems

In this subsection we illustrate the application of the derived theoretical results by constructing the root locus of several practical control systems. To facilitate the understanding of the general procedure for constructing the root locus, we start by applying this general procedure to the introductory Example 7.2.1 in the example that immediately follows.

Example 7.3.1

Determine the root locus of the closed-loop system of Example 7.2.1 using the construction method presented in this section, i.e., the eight theorems of Subsec. 7.3.2 and the additional information of Subsec. 7.3.3.

Solution

Following the root locus construction method step by step, we have:
1. The root locus points for $K = 0$ are the poles of $G(s)F(s)$, e.g., the points $s = 0$ and $s = -6$. These are the root locus starting points for $K \geq 0$.
2. The root locus points for $K \to \pm\infty$ are the zeros of $G(s)F(s)$. Since there are no zeros in $G(s)F(s)$ and since the degree of the numerator is smaller than that of the denominator ($m < n$), the root locus points for $K \to \pm\infty$ are at infinity. These are the ending points of the root locus.
3. The number of branches of the root locus is $\max(m, n) = \max(0, 2) = 2$.
4. The angles of the asymptotes are

$$\theta_\rho = \frac{(2\rho + 1)\pi}{2}, \rho = 0, 1, \quad \text{for} \quad K \geq 0$$

$$\theta_\rho = \frac{2\rho\pi}{2}, \rho = 0, 1, \quad \text{for} \quad K \leq 0$$

Therefore, the asymptotes are straight lines having the following slopes:

$$\theta_0 = \frac{\pi}{2}, \qquad \theta_1 = \frac{3\pi}{2}, \qquad \text{for} \qquad K \geq 0 \qquad \text{and}$$

$$\theta_0 = 0, \qquad \theta_1 = \pi, \qquad \text{for} \qquad K \leq 0$$

5. The point of intersection of the asymptotes is

$$\sigma_1 = -\frac{b_1 - d_1}{n - m} = -\frac{6 - 0}{2} = -3$$

6. The segments of the real axis that can be part of the root locus are

    a.   For $K \geq 0$, the segment from -6 to 0.
    b.   For $K \leq 0$, the segment from $-\infty$ to -6 and from 0 to $+\infty$.

7. The root locus breakaway points are roots of the following equation:

$$\frac{d}{ds}[G(s)F(s)] = \frac{d}{ds}\left[\frac{K}{s(s+6)}\right] = \frac{-K(2s+6)}{s^2(s+6)^2} = 0$$

From the above equation, the point $s = -3$ appears to be a candidate breakaway point. To be a breakaway point, it must satisfy the equation $1 + G(s)F(s)$ for any real value of K. Let $s = -3$. Then

$$1 + G(-3)F(-3) = 1 - \frac{K}{9} = 0$$

Hence, equation $1 + G(-3)F(-3) = 0$ is satisfied for $K = 9$. Thus, the point $s = -3$ is a breakaway point of the root locus.

    8. The departure angles at the two poles of $G(s)F(s)$ are calculated according to the Eq. (7.3-28) as follows:

    a. At the pole $s = 0$: as $s_1 \to 0$, we have

$$\theta_0 = \big/\!\!\underline{s_1 + 0} = -(2\rho + 1)\pi - \big/\!\!\underline{s_1 + 6} = -(2\rho + 1)\pi - 0$$

Taking the smallest angle, e.g., for $\rho = 0$, we have $\theta_0 = -\pi$.

    b. At the pole $s = -6$: as $s_1 \to -6$, we have

$$\theta_{-6} = \big/\!\!\underline{s_1 + 6} = -(2\rho + 1)\pi - \big/\!\!\underline{s_1} = -(2\rho + 1)\pi - \pi$$

Taking the smallest angle, e.g., for $\rho = 0$, we have $\theta_{-6} = -2\pi$ or $\theta_{-6} = 0$.

    9. The root locus intersection with the imaginary axis is determined using Routh's criterion. To this end, construct the Routh's table of the characteristic polynomial $p_c(s) = s(s+6) + K$, as follows:

$$\begin{array}{c|cc} s^2 & 1 & K \\ s^1 & 6 & 0 \\ s^0 & K & \end{array}$$

From the first column of the Routh table we observe that the system changes from stable to unstable (i.e., it intersects the imaginary axis) when $K = 0$. Next, using the row s^2, we form the auxiliary polynomial $A(s) = s^2 + K$. For $K = 0$, the roots of $A(s)$ are $s = \pm j0$. Thus, the point $s = \pm j0$ is the root locus intersection with the imaginary axis.

    Using the above results one can construct the root locus, which, as expected, will be exactly the same as that of Figure 7.3.

Example 7.3.2

This example refers to the control of the nuclear reactor which was presented in Sec. 1.4 (see Figure 1.16). The reactor's block diagram is given in Figure 7.5, where the reactor is described by the simplified model of a first-order system. We assume that the regulator $G_c(s)$ has two terms, an analog and an integral, so that $G_c(s) = K_1 + K_2/s$. Therefore,

$$G(s) = G_c(s)G_r(s) = (K_1 + K_2/s)\left[\frac{1}{Ts + 1}\right] = \frac{K(s + \alpha)}{s(Ts + 1)}$$

where $K = K_1$ and $\alpha = K_2/K_1$. To simplify, let $\alpha = 2$ and $T = 1$, in which case the open-loop transfer function becomes

$$G(s)F(s) = \frac{K(s + 2)}{s(s + 1)}$$

Determine the root locus of the closed-loop system using the construction method presented in this section.

Solution

We have:

1. The root locus points for $K = 0$ of $G(s)F(s)$ are $s = 0$ and $s = -1$. These points are the starting points of the root locus.

2. The root locus points for $K \to +\infty$ are $s = -2$ and infinity. Those points are the ending points of the root locus.

3. The number of branches of the root locus is $\max(m, n) = \max(1, 2) = 2$.

4. The angles of the asymptotes are

$$\theta_\rho = (2\rho + 1)\pi, \qquad \rho = 0, \qquad \text{when } K \geq 0$$

$$\theta_\rho = 2\rho\pi, \qquad \rho = 0, \qquad \text{when } K \leq 0$$

Therefore, the asymptotes are straight lines having the following slopes

$$\theta_0 = \pi, \quad \text{when} \quad K \geq 0 \quad \text{and} \quad \theta_0 = 0, \quad \text{when} \quad K \leq 0$$

5. The point of intersection of the asymptotes is

$$\sigma_1 = -\frac{b_1 - d_1}{n - m} = -(1 - 2) = 1$$

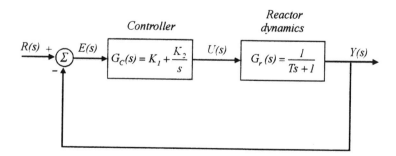

Figure 7.5 Simplified block diagram of a nuclear reactor control system.

It is noted that because $|n - m| - 1 = 0$, we have only one asymptote, which is the real axis. Therefore, there is no point of intersection of the asymptotes, and hence the point $\sigma_1 = 1$ does not have any particular meaning.

6. The segments of the real axis that can be part of the root locus are

    a.   For $K > 0$, the segment from $-\infty$ to -2 and the part from -1 to 0.
    b.   For $K < 0$, the segment from -2 to -1 and from 0 to $+\infty$.

7. The root locus breakaway points are roots of the following equation:

$$\frac{d}{ds}[G(s)F(s)] = \frac{d}{ds}\left[\frac{K(s+2)}{s(s+1)}\right] = K\left[\frac{s(s+1) - (s+2)(2s+1)}{s^2(s+1)^2}\right] = 0$$

Simplifying the above equation yields

$$s^2 + 4s + 2 = 0$$

The roots of the above algebraic equation are $s_1 = -2 + \sqrt{2} = -0.586$ and $s_2 = -2 - \sqrt{2} = -3.414$. For s_1 and s_2 to be breakaway points, they must satisfy the equation $1 + G(s)F(s) = 0$ for any real value of K. For the root s_1 we have

$$1 + G(-0.586)F(-0.586) = 1 + K\frac{1.414}{(-0.586)(0.414)} = 0$$

The above equation is satisfied for $K = 0.1716$. For the root s_2 we have

$$1 + G(-3.414)F(-3.414) = 1 + K\frac{(-1.414)}{(-3.414)(-2.414)} = 0$$

The above equation is satisfied for $K = 5.8274$. Therefore, both points $s_1 = -0.586$ and $s_2 = -3.414$ are breakaway points of the root locus.

8. The root locus departure angles are calculated using Eq. (7.3-28), as follows:
a. At the pole $s = 0$: as $s_1 \to 0$ we have

$$\theta_0 = \underline{/s_1} = -(2\rho + 1)\pi + \underline{/s_1 + 2} - \underline{/s_1 + 1} = -\pi - 0 + 0 = -\pi \text{ or } \pi$$

for $\rho = 0$.
   b. At the pole $s = -1$: as $s_1 \to -1$ we have

$$\theta_{-1} = \underline{/s_1 + 1} = -(2\rho + 1)\pi + \underline{/s_1 + 2} - \underline{/s_1} = -\pi + 0 - \pi = -2\pi \text{ or } 0$$

for $\rho = 0$.

The root locus arrival angles are calculated using Eq. (7.3-29) as follows: here, we have only one zero, namely, the zero $s = -2$. At the zero $s = -2$, as $s_1 \to -2$, we have

$$\theta_{-2} = \underline{/s_1 + 2} = 2\rho\pi + \underline{/s_1} + \underline{/s_1 + 1} = 0 + \pi + \pi = 2\pi \text{ or } 0$$

for $\rho = 0$.

9. The root locus intersection with the imaginary axis is determined using the Routh's criterion. To this end, construct the Routh table of the characteristic polynomial $p_c(s) = s(s + 1) + K(s + 2) = s^2 + (K + 1)s + 2K$, as follows:

$$\begin{array}{c|cc}
s^2 & 1 & 2K \\
s^1 & K+1 & 0 \\
s^0 & 2K &
\end{array}$$

The system is stable if the inequalities $K > -1$ and $K > 0$ hold simultaneously. Thus, for $K > 0$ the system is stable. Next, using the row s^2, we form the auxiliary polynomial $A(s) = s^2 + 2K$. For $K = 0$ the auxiliary polynomial $A(s) = s^2 + 2K$ gives $s = \pm j0$. Therefore, the root locus is intersecting the $j\omega$-axis at the point $s = \pm j0$.

Using the above results one can construct the root locus for $K > 0$, as shown in Figure 7.6.

Example 7.3.3

This example refers to the automatic piloting system for supersonic airplanes (Figure 7.7a), which assists the aerodynamic stability of the plane, thus making the flight more stable and more comfortable. A simplified block diagram of this system is given in Figure 7.7b. The aircraft dynamics are approximated by a second-order system, where K is a parameter which changes according to the flight conditions (e.g., fast landing or take-off, steady flight, etc). Assume that there are no disturbances, i.e., $D(s) = 0$. Determine the closed-loop system root locus for $K > 0$ and the range of values of K such that the closed-loop system is stable.

Solution

From the block diagram of Figure 7.7b we have

$$G(s)F(s) = \frac{K(s+4)}{s(s+6)(s+8)(s^2+2s+2)}$$

Following the root locus construction method step by step, we have:

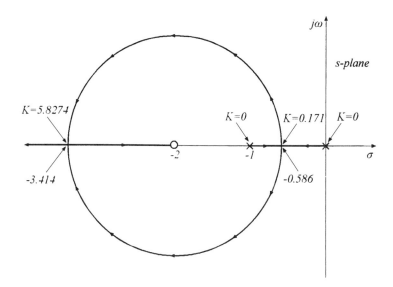

Figure 7.6   The root locus of the nuclear reactor closed-loop system.

(a)

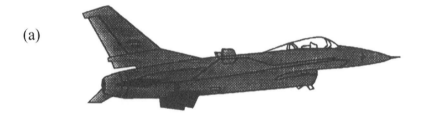

Disturbance
D(s)

(b)

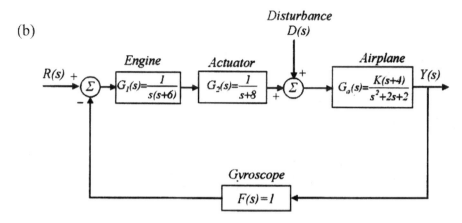

Figure 7.7 Automatic piloting system for supersonic airplanes. (a) Supersonic airplane; (b) simplified block diagram.

1. The root locus points for $K = 0$ are $s = 0$, $s = -6$, $s = -8$, $s = -1 + j$, and $s = -1 - j$. These points are the root locus starting points.

2. The root locus points for $K \to +\infty$ are $s = -4$ and infinity. These points are the root locus ending points.

3. The number of branches of the root locus is $\max(m, n) = \max(1, 5) = 5$.

4. The angles of the asymptotes are

$$\theta_\rho = \frac{(2\rho + 1)\pi}{n - m} = \frac{(2\rho + 1)\pi}{4}, \qquad \rho = 0, 1, 2, 3, \qquad \text{when} \qquad K > 0$$

Therefore, the asymptotes are straight lines having the following slopes

$$\theta_0 = \frac{\pi}{4}, \qquad \theta_1 = \frac{3}{4}\pi, \qquad \theta_2 = \frac{5}{4}\pi, \qquad \text{and} \qquad \theta_3 = \frac{7}{4}\pi$$

5. The point of intersection of the asymptotes is

$$\sigma_1 = -\frac{b_1 - d_1}{n - m} = -\frac{16 - 4}{4} = -3$$

6. The segments of the real axis that can be part of the root locus for $K > 0$ are the segments from 0 to -4 and from -6 to -8.

7. The root locus breakaway points are roots of the equation

$$\frac{d}{ds}[G(s)F(s)] = \frac{d}{ds}\left[\frac{K(s+4)}{s(s+6)(s+8)(s^2+2s+2)}\right] = K\frac{\delta(s)}{\beta(s)}$$

where

$$\delta(s) = s(s+6)(s+8)(s^2+2s+2) - (s+4)[2s(s+7)(s^2+2s+2) + (s+6)(s+8)(3s^2+4s+2)]$$

Because the determination of the roots of the equation $\delta(s) = 0$ is quite difficult, an attempt will be made to come to a conclusion regarding the approximate position of the root locus breakaway points, by circumventing the direct calculation of the roots of the equation $\delta(s) = 0$. To this end, taking advantage of the information that we already have about the root locus, it appears that there is only one breakaway point which lies between the points $(-6, 0)$ and $(-8, 0)$. Indeed, since the points $(-6, 0)$ and $-8, 0)$ are starting points, the root locus which begins from these two points must intersect between the points $(-6, 0)$ and $(-8, 0)$ and then change course, moving away from the real axis.

8. The root locus departure angles are calculated according to Eq. (7.3-28) as follows:

a. At the pole $s = 0$: as $s_1 \to 0$ for $\rho = 0$, we have

$$\theta_0 = \underline{/s_1} = -(2\rho + 1)\pi + \underline{/s+4} - \underline{/s_1+6} - \underline{/s_1+8} - \underline{/s_1+1+j}$$
$$- \underline{/s_1+1-j}$$
$$= -\pi + 0 - 0 - 0 - \frac{\pi}{4} - \left(-\frac{\pi}{4}\right) = -\pi$$

b. At the pole $s = -6$: as $s_1 \to -6$ and for $\rho = 0$, we have

$$\theta_{-6} = \underline{/s_1+6} = -(2\rho + 1)\pi + \underline{/s_1+4} - \underline{/s_1} - \underline{/s_1+8} - \underline{/s_1+1+j}$$
$$- \underline{/s_1+1-j}$$
$$= -\pi + (-\pi) - (-\pi) - 0 - (\alpha) - (-\alpha) = -\pi$$

where $\alpha = \underline{/s_1+1+j}$.

c. At the pole $s = -8$: as $s_1 \to -8$ and for $\rho = 0$, we have

$$\theta_{-8} = \underline{/s_1+8} = -(2\rho + 1)\pi + \underline{/s_1+4} - \underline{/s_1} - \underline{/s_1+6} - \underline{/s_1+1+j}$$
$$- \underline{/s_1+1-j}$$
$$= -\pi + (-\pi) - (-\pi) - (-\pi) - (\alpha) - (-\alpha) = 0$$

d. At the pole $s = -1 - j$: as $s_1 \to -1 - j$ and for $\rho = 0$, we have

$$\theta_{-1-j} = \underline{/s_1 + 1 + j} = -(2\rho + 1)\pi +$$

$$\underline{/s_1 + 4} \quad - \quad \underline{/s_1} \quad - \quad \underline{/s_1 + 6} \quad - \quad \underline{/s_1 + 8} \quad - \quad \underline{/s_1 + 1 - j}$$

$$= -\pi + \underline{/3 - j} \quad - \quad \underline{/-1 - j} \quad - \quad \underline{/5 - j} \quad - \quad \underline{/7 - j} \quad - \quad \underline{/-2j}$$

$$= -180° + (-18.43°) - (-135°) - (-11.30°) - (-8.13°) - (-90°) = 46°$$

e. At the pole $s = -1 + j$: from the root locus symmetry about the real axis we conclude that

$$\theta_{-1+j} = -\theta_{-1-j} = -46°$$

The arrival angle at the zero $s = -4$ is calculated according to Eq. (7.3-29) as follows: as $s_1 \to -4$ and for $\rho = 0$, we have

$$\theta_{-4} = \underline{/s_1 + 4} = 2\rho\pi +$$

$$\underline{/s_1 +} \quad \underline{/s_1 + 6} \quad + \quad \underline{/s_1 + 8} \quad + \quad \underline{/s_1 + 1 + j} \quad + \quad \underline{/s_1 + 1 - j}$$

$$= 0 + \pi + 0 + 0 + \alpha - \alpha = \pi$$

9. The root locus intersection with the imaginary axis is determined as follows: construct the Routh table of the characteristic polynomial

$$p_c(s) = s(s + 6)(s + 8)(s^2 + 2s + 2) + K(s + 4) = s^5 + 16s^4 + 78s^3 + 124s^2$$
$$+ (96 + K)s + 4K$$

as follows:

s^5	1	78	$96 + K$	0
s^4	16	124	$4K$	0
s^3	70.25	$96 + 0.75K$	0	0
s^2	$102.1352 - 0.1708K$	$4K$	0	0
s^1	$\dfrac{9805 - 220.7972K - 0.1281K^2}{102.1352 - 0.1708K}$	0	0	0
s^0	$4K$	0	0	0

From the Routh criterion it is well known that the closed-loop system is stable if all the elements of the first column of the Routh table have the same sign. For this to hold, the inequalities $102.1352 - 0.1708K > 0$ (or $K < 597.89$), $9805 - 220.7972K - 0.1281K^2 > 0$ ($-1767 < K < 43.3$) and $4K > 0$, must be satisfied simultaneously. From these three inequalities it immediately follows that the system is stable when $0 < K < 43.3$. Clearly, for the values of $K = 0$ and $K = 43.3$ the root locus intersects the $j\omega$-axis. The points of intersection are calculated from the auxiliary equation $A(s) = (102.1352 - 0.1708K)s^2 + 4K = 0$, which is formed using the row s^2. For $K = 43.3$ the auxiliary equation becomes $94.739s^2 + 173.2 = 0$ which gives the points of intersection $s = \pm j1.352$. For $K = 0$ the auxiliary equation becomes $102.1352s^2 = 0$, which gives the point of intersection $s = 0$.

Using all the above information we construct the root locus for $K > 0$ as shown in Figure 7.8.

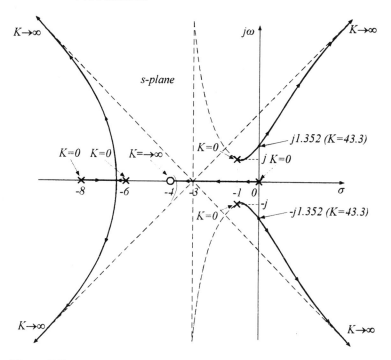

Figure 7.8 The root locus of the supersonic airplane closed-loop system.

Example 7.3.4

Consider the closed-loop control system which controls the thickness of metal sheets, shown in Figure 1.10 of Chap. 1. The system is approximately described as in Figure 7.9. Determine the root locus for the following two cases:

(a) $G_c(s) = K$
(b) $G_c(s) = K(s + 0.5)$

Solution

Case (a)

The open-loop transfer function is

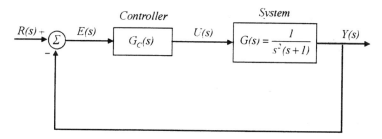

Figure 7.9 Simplified block diagram of the thickness control system.

$$G_c(s)G(s) = \frac{K}{s^2(s+1)}$$

Following the root locus construction method step by step, we have:

1. The points of the root locus for $K = 0$ are $s = 0$ and $s = -1$. These points are the root locus starting points for $K \geq 0$.

2. The points of the root locus for $K \to \infty$ are the root locus ending points, which are at infinity.

3. The number of branches of the root locus is $\max(m, n) = \max(0, 3) = 3$.

4. The angles of the asymptotes are

$$\theta_\rho = \frac{(2\rho + 1)\pi}{3}, \qquad \rho = 0, 1, 2, \qquad \text{for} \qquad K \geq 0$$

$$\theta_\rho = \frac{2\rho\pi}{3}, \qquad \rho = 0, 1, 2, \qquad \text{for} \qquad K \leq 0$$

Therefore, the asymptotes are straight lines having the following slopes:

$$\theta_0 = \frac{\pi}{3}, \qquad \theta_1 = \pi, \qquad \theta_2 = \frac{5\pi}{3}, \qquad \text{when} \qquad K \geq 0$$

$$\theta_0 = 0, \qquad \theta_1 = \frac{2\pi}{3}, \qquad \theta_2 = \frac{4\pi}{3}, \qquad \text{when} \qquad K \leq 0$$

5. The point of intersection of the asymptotes is

$$\sigma_1 = -\frac{\displaystyle\sum_{i=1}^{n} p_i - \sum_{i=1}^{m} z_i}{n - m} = -\frac{1 - 0}{3} = -\frac{1}{3}$$

6. The segments of the real axis that can be part of the root locus are
   a. For $K \geq 0$, the segment from $-\infty$ to -1.
   b. For $K \leq 0$, the segment from -1 to 0 and the segment from 0 to $+\infty$.

7. The root locus breakaway points are roots of the equation

$$\frac{d}{ds}[G_c(s)G(s)] = -K\left[\frac{2s(s+1) + s^2}{s^4(s+1)^2}\right] = 0$$

From the above equation we conclude that candidate breakaway points of the root locus are $s = 0$ and $s = -2/3$. For these points to be breakaway points, they must satisfy the equation

$$1 + G_c(s)G(s) = 0$$

for any real value of K. The point $s = 0$ satisfies the above equation for $K = 0$ and the point $s = 2/3$ for $K = -4/27$. Hence, theya re both breakaway points.

8. The root locus departure angles are
   a. At the double pole $s = 0$: as $s_1 \to 0$ and $\varepsilon \to 0$, we have

$$\theta_0 = \sqrt{s_1 + 0 \mp j\varepsilon} = -(2\rho + 1)\pi - \sqrt{s_1 + 1} - \sqrt{s_1 + 0 \pm j\varepsilon}$$

$$= -(2\rho + 1)\pi \mp \frac{\pi}{2}$$

Choosing the smallest angles, e.g., for $\rho = 0$, we obtain

$$\theta_0 = -\frac{3\pi}{2} \quad \text{and} \quad \theta_0 = -\frac{\pi}{2}.$$

b. At the pole $s = -1$: as $s_1 \to -1$, we have

$$\theta_{-1} = \underline{/s_1 + 1} = -(2\rho + 1)\pi - 2\underline{/s_1} = -(2\rho + 1)\pi - 2\pi$$

Choosing the smallest angles, e.g., for $\rho = 0$, we have $\theta_{-1} = -3\pi$.

The results presented above are adequate to construct the root locus sought, as shown in Figure 7.10a.

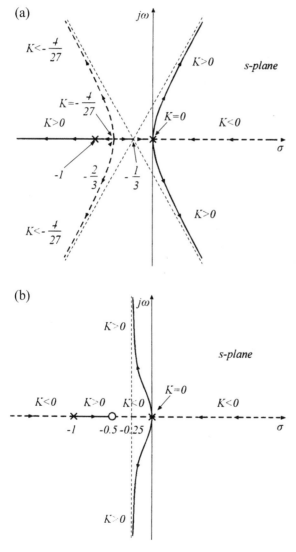

Figure 7.10   The root locus of the closed-loop system of Example 7.4.3: (a) when $G_C(s) = K$; (b) when $G_C(s) = K(s + 0.5)$.

Case (b)

The open-loop transfer function has the form

$$G_c(s)G(s) = \frac{K(s+0.5)}{s^2(s+1)}$$

Following the root locus construction method step by step, we have:

    1. The root locus starting points are at $s = 0$ and $s = -1$ for $K \geq 0$.

    2. The root locus ending points are at $s = -0.5$ and at infinity for $K \geq 0$.

    3. The number of branches of the root locus is $\max(1, 3) = 3$.

    4. The angles of the asymptotes are

    a. For $K \geq 0$, $\theta_0 = \pi/2$, and $\theta_1 = 3\pi/2$

    b. For $K \leq 0$, $\theta_0 = 0$, and $\theta_1 = \pi$.

    5. The point of intersection of the asymptotes is

$$\sigma_1 = -\frac{0.5}{2} = -0.25$$

    6. The segments of the real axis that can be part of the root locus are

    a. For $K \geq 0$, the segment $(-1, 0.5)$

    b. For $K \leq 0$, the segments $(-\infty, -1)$, $(0.5, 0)$, and $(0, +\infty)$.

    7. The root locus breakaway points are roots of the equation

$$\frac{d}{ds}[G_c(s)G(s)] = 0 \qquad \text{or} \qquad s(2s^2 + 2.5s + 1) = 0$$

We have three roots: $s = 0$ and $s = -0.625 \pm j0.33$. From these three roots, only the root $s = 0$ satisfies the equation $1 + G_c(s)G(s) = 0$ for $K = 0$. For the other two roots, there are no real values of K which satisfy the equation $1 + G_c(s)G(s) = 0$.

    8. The root locus departure angles are

    a. At the pole $s = 0$: as $s_1 \to 0$ and $\varepsilon \to 0$, we have

$$\theta_0 = \underline{/s_1 + 0 \mp j\varepsilon} = -2(\rho+1)\pi + \underline{/s_1 + 0.5} - \underline{/s_1 + 0 \pm j\varepsilon}$$
$$= -(2\rho+1)\pi + 0 \mp \frac{\pi}{2}$$

Choosing the smallest angles, i.e., for $\rho = 0$, we have

$$\theta_0 = -\frac{3\pi}{2} \qquad \text{and} \qquad \theta_0 = -\frac{\pi}{2}$$

    b. At the pole $s = -1$: as $s_1 \to -1$, we have

$$\theta_{-1} = \underline{/s_1 + 1} = -(2\rho+1)\pi + \underline{/s_1 + 0.5} - 2\underline{/s_1 + 0} = -(2\rho+1)\pi + \pi - 2\pi$$

Choosing the smallest angles, i.e., for $\rho = 0$, we have

$$\theta_{-1} = -2\pi \qquad \text{or} \qquad 0$$

    9. The root locus arrival angle is at the zero $= -0.5$: as $s_1 \to -0.5$, we have

$$\theta_{-0.5} = \underline{/s_1 + 0.5} = 2\rho\pi + 2\underline{/s_1 + 0} + \underline{/s_1 + 1} = 2\rho\pi + 2\pi + 0$$

or

$\theta_{-0.5} = 2\pi = 0$

Using the above results, one may construct the root locus, as shown in Figure 7.10b. In Fig. 7.10a, where $G_c(s) = K$, the closed-loop system is unstable. In Fig. 7.10b, where $G_c(s) = K(s + 0.5)$, the closed-loop system is stable for $K > 0$. This is obviously because we added the zero $s = -0.5$ in the loop transfer function. It is noted that this zero lies between the poles -1 and 0. The effects of adding poles and/or zeros in the loop transfer function is studied in Sec.7.5.

The following example demonstrates that in several cases it is possible to find an analytical expression for a certain segment of the root locus of a system (another interesting similar problem is stated in Problem 9 of Sec. 7.6).

Example 7.3.5

Consider the closed-loop system of Figure 7.11. Show that the root locus segment which is not located on the real axis is described analytically in polar coordinates as follows: $\rho^3 \cos \theta = -3$, where $s = \rho e^{j\theta}$.

Solution

The closed-loop system transfer function is

$$H(s) = \frac{G(s)}{1 + G(s)} = \frac{\dfrac{Ks}{(s + 1)(s + 2)(s - 3)}}{1 + \dfrac{Ks}{(s + 1)(s + 2)(s - 3)}} = \frac{Ks}{s^3 + (K - 7)s - 6}$$

The characteristic polynomial $p_c(s)$ of the closed-loop system transfer function is $p_c(s) = s^3 + (K - 7)s - 6$. Replacing s with $\rho e^{j\theta}$ in $p_c(s)$, we obtain that the points of the root locus which are not located on the real axis satisfy the equation

$$(\rho e^{j\theta})^3 + (K - 7)\rho e^{j\theta} - 6 = 0 \quad \text{or} \quad \rho^3 e^{3j\theta} + (K - 7)\rho e^{j\theta} - 6 = 0$$

or

$$[\rho^3 \cos 3\theta + (K - 7)\rho \cos \theta - 6] + j[\rho^3 \sin 3\theta + (K - 7)\rho \sin \theta] = 0$$

The points of the root locus must satisfy both the real and the imaginary part in the above equation, i.e., there must be

$$\rho^3 \cos 3\theta + (K - 7)\rho \cos \theta - 6 = 0 \quad \text{and} \quad \rho^3 \sin 3\theta + (K - 7)\rho \sin \theta = 0$$

Solving the second equation for $(K - 7)$ and introducing the result into the first we obtain

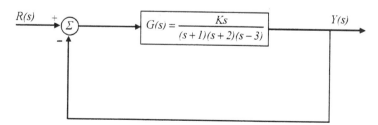

Figure 7.11 The closed-loop system of Example 7.3.5.

$$\rho^3 \cos 3\theta - \frac{\rho^3 \cos \theta \sin 3\theta}{\sin \theta} - 6 = 0$$

Consider the trigonometry relations: $\cos 3\theta = \cos^3 \theta - 3 \sin^2 \theta \cos \theta$ and $\sin 3\theta = 3 \sin \theta \cos^2 \theta - \sin^3 \theta$. Using these two relations, the above equation becomes

$$\rho^3 [\cos^3 \theta + \sin^2 \theta \cos \theta] = -3 \qquad \text{or} \qquad \rho^3 \cos \theta (\cos^2 \theta + \sin^2 \theta) = -3$$

$$\text{or} \qquad \rho^3 \cos \theta = -3$$

Remark 7.3.1

From the examples presented in the present subsection it must be obvious that the form of the root locus changes drastically with the position of the poles and zeros of $G(s)F(s)$. This is demonstrated in Figure 7.12, which gives various pole-zero configurations of $G(s)F(s)$ and their corresponding root loci.

7.4 APPLYING THE ROOT LOCUS METHOD FOR DETERMINING THE ROOTS OF A POLYNOMIAL

The root locus method can be used to determine the roots of a polynomial. This idea can be easily illustrated by the following two simple examples.

Example 7.4.1

Consider the polynomial

$$p(s) = s^2 + 2s + 2 \qquad\qquad\qquad (7.4\text{-}1)$$

Determine the roots of $p(s)$ using the root locus method.

Solution

To determine the roots of $p(s)$ using the root locus method, we find an appropriate form of $1 + G(s)F(s)$ such that the characteristic polynomial of $1 + G(s)F(s)$ is equal to $p(s)$. Such a form is

$$1 + G(s)F(s) = 1 + \frac{K(s+2)}{s(s+1)} \qquad\qquad\qquad (7.4\text{-}2)$$

where, for $K = 1$, the characteristic polynomial of Eq. (7.4-2) is equal to $p(s)$. Therefore, the roots of $p(s)$ are the points of the root locus of Eq. (7.4-2) when $K = 1$. Since the root locus of Eq. (7.4-2) is the root locus of the Example 7.3.2, the roots of $p(s)$ can be found from Figure 7.6 to be $-1 + j$ and $-1 - j$.

Example 7.4.2

Consider the polynomial

$$p(s) = s^3 + 3s^2 + 2s + 6 \qquad\qquad\qquad (7.4\text{-}3)$$

Determine the roots of $p(s)$ using the root locus method.

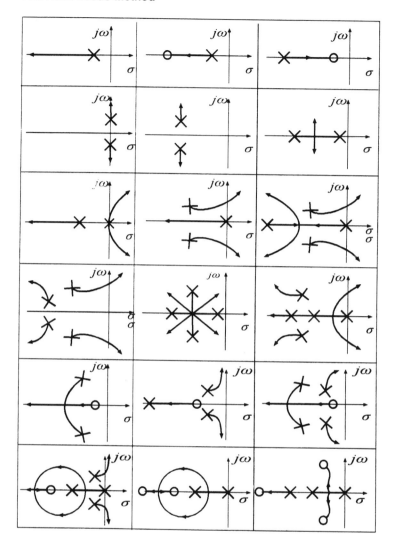

Figure 7.12   Various pole-zero configurations of $G(s)F(s)$ and the corresponding root loci for $K > 0$.

Solution

Repeating the method used in the previous example, we obtain

$$1 + G(s)F(s) = 1 + \frac{K}{s(s+1)(s+2)} \tag{7.4-4}$$

The characteristic polynomial of Eq. (7.4-4), for $K = 6$, is equal to $p(s)$. Therefore, the roots of $p(s)$ are the points of the root locus of Eq. (7.4-4) when $K = 6$. The root locus of Eq. (7.4-4) is given in Figure 7.13, where for $K = 6$ we find that the roots are -3, $j\sqrt{2}$, and $-j\sqrt{2}$. This can also be found from Eq. (7.4-3) if we expand $p(s)$ as follows: $p(s) = s(s^2 + 2) + 3(s^2 + 2) = (s^2 + 2)(s + 3)$.

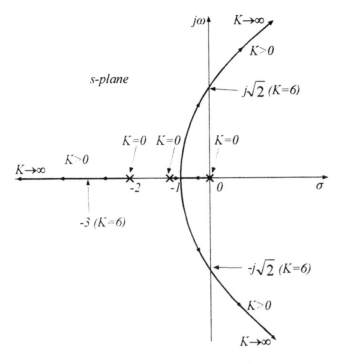

Figure 7.13 The root locus of Example 7.4.2.

7.5 EFFECTS OF ADDITION OF POLES AND ZEROS ON THE ROOT LOCUS

The root locus method is usually used to obtain an overall simple picture of the effect that the gain constant K has on the positions of the poles of a closed-loop system. It is also used to obtain an overall picture of the effect that has on the root locus the addition of poles and/or zeros in the loop transfer function $G(s)F(s)$. The addition of new poles and/or zeros in $G(s)F(s)$ is done as in Figure 7.14b and it aims at the improvement of the closed-loop system behavior. For the closed-loop system 7.14b the characteristic equation takes on the form

$$1 + [G_1(s)F_1(s)]G(s)F(s) = 0 \tag{7.5-1}$$

The transfer functions $G_1(s)$ and $F_1(s)$ are the additional controllers introduced in the closed-loop system. Depending on the particular form of $G_1(s)F_1(s)$ the root locus of Eq. (7.5-1) may change drastically. In the next two subsections, we study this change that the original root locus undergoes when the additional controller $G_1(s)F_1(s)$ is included in the closed-loop system.

7.5.1 Addition of Poles and Its Effect on the Root Locus

Assume that

$$G_1(s)F_1(s) = \frac{1}{(s + \pi_1)(s + \pi_2)\cdots(s + \pi_p)} \tag{7.5-2}$$

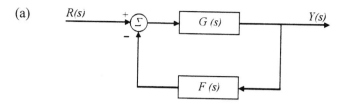

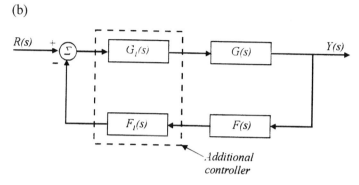

Figure 7.14 Closed-loop system without and with additional controllers. (a) Original closed-loop system; (b) closed-loop system with additional controllers.

Then, Eq. (7.5-1) becomes

$$1 + \frac{G(s)F(s)}{(s + \pi_1) \cdots (s + \pi_p)} = 0 \tag{7.5-3}$$

Here, the root locus of Eqs (7.3-1) and (7.5-3) differ from each other in that the root locus of Eq. (7.5-3) is "moved" or "bended" more to the right of the root locus of Eq. (7.3-1). Thus, the addition of poles to closed-loop systems results in more unstable closed-loop systems. This fact will be illustrated by the following two examples.

Example 7.5.1

Let

$$G(s)F(s) = \frac{K}{s(s + a)}, \qquad a > 0 \tag{7.5-4}$$

and

$$G_1(s)F_1(s) = \frac{1}{s + \pi_1}, \qquad \pi_1 > a$$

Then, Eq. (7.5-3) becomes

$$1 + \frac{K}{s(s + a)(s + \pi_1)} = 0 \tag{7.5-5}$$

Study the stability of Eq. (7.5-5).

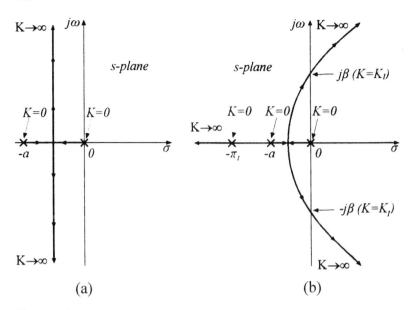

Figure 7.15   The effect of adding a pole to the root locus of Eq. (7.5-4).

Solution

The root locus of Eqs (7.5-4) and (7.5-5) are given in Figure 7.15a and 7.15b, respectively. Figure 7.15b shows that the addition of one pole results in "bending" the root locus of Figure 7.15a more to the right. To be more precise, even though the entire root locus of Eq. (7.5-4) is located in the left-half complex plane for $K \geq 0$, the root locus of Eq. (7.5-7) is partly located in the right-half complex plane. This means that while system (7.5-4) is stable for $K \geq 0$, system (7.5-5) is unstable for large values of K and in particular for $K > K_1$.

Example 7.5.2

Consider the open-loop transfer function (7.5-4). Also, consider the additional controller $G_1(s)F_1(s)$, having the form

$$G_1(s)F_1(s) = \frac{1}{(s + \pi_1)(s + \pi_2)}, \qquad \pi_2 > \pi_1 > a \qquad (7.5\text{-}6)$$

Then, Eq. (7.5-3) becomes

$$1 + \frac{K}{s(s + a)(s + \pi_1)(s + \pi_2)} = 0 \qquad (7.5\text{-}7)$$

Study the stability of Eq. (7.5-7).

Solution

The root locus of Eq. (7.5-7) is given in Figure 7.16. It is clear that the addition of two poles "bends" the root locus of Eq. (7.5-4) even more to the right, a fact which makes the closed-loop system "more" unstable, compared with the case of adding only one pole. Indeed, if one compares Figure 7.15a (case of adding one pole) with

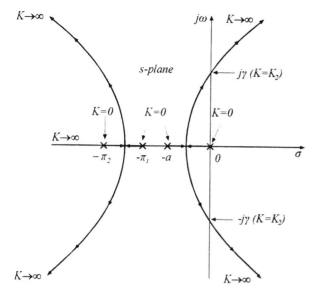

Figure 7.16 The effect of adding two poles in the root locus of Eq. (7.5-4).

Figure 7.16 (case of adding two poles), one observes that the root locus in Figure 7.15a crosses the $j\omega$-axis when $K = K_1$, while in Figure 7.16 it crosses the $j\omega$-axis when $K = K_2$. Since $K_2 < K_1$, it follows that the closed-loop system of Figure 7.16 becomes unstable for smaller values of K compared with Figure 7.15a. Hence, the system with characteristic equation (7.5-7) is more unstable compared with the system with characteristic equation (7.5-5).

7.5.2 Addition of Zeros and Its Effect on the Root Locus

Assume that $G_1(s)F_1(s) = (s + \mu_1)$. Then, Eq. (7.5-1) becomes

$$1 + (s + \mu_1)G(s)F(s) = 0 \tag{7.5-8}$$

Here, the root locus of Eqs (7.3-1) and (7.5-8) differ from each other in that the root locus of Eq. (7.5-8) is "moved" or "bended" more to the left of the root locus of Eq. (7.3-1). Thus, the addition of zeros to closed-loop systems results in more stable closed-loop systems. This fact is illustrated by the following example.

Example 7.5.3

Consider the loop transfer function Eq. (7.5-4). Then for $G_1(s)F_1(s) = s + \mu_1$, Eq. (7.5-8) becomes

$$1 + K\left[\frac{s + \mu_1}{s(s + a)}\right] = 0, \qquad \mu_1 > a \tag{7.5 – 9}$$

Study the stability of Eq. (7.5-9).

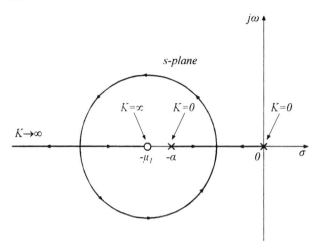

Figure 7.17   The effect of adding a zero in the root locus of Eq. (7.5-4).

Solution

The root locus of Eq. (7.5-9) is given in Figure 7.17, from which we conclude that the root locus of Eq. (7.5-9) is "bended" to the left of the root locus of Eq. (7.5-4) and therefore Eq. (7.5-9) is "more" stable than Eq. (7.5-4).

PROBLEMS

1. Draw the root locus for the closed-loop systems having the following loop transfer functions:

(a) $\dfrac{K}{s^2}$

(b) $\dfrac{K(s^2 + 4s + 8)}{s^2(s + 4)}$

(c) $\dfrac{K(s + 2)}{s(s + 1)(s + 19)}$

(d) $\dfrac{K(s + 2)(s + 6)}{s(s + 4)(s + 3)}$

(e) $\dfrac{K(s + 1)}{s(s + 2)(s + 3)(s + 4)}$

(f) $\dfrac{K}{(s + 1)(s + 2)(s + 3)}$

(g) $\dfrac{K(s + 3)}{(s + 1)(s + 2)(s + 4)}$

(h) $\dfrac{K}{(s + 1)(s + 2)(s + 3)(s + 4)}$

(i) $\dfrac{K}{s(s^2 + 6s + 25)}$

(j) $\dfrac{K(s + 1)}{s(s^2 + 4s + 8)}$

(k) $\dfrac{K(s + 3)}{(s + 1)(s + 2)(s^2 + 4s + 8)}$

(l) $\dfrac{K(s + 3)(s + 5)}{s(s + 1)(s + 4)(s^2 + 4s + 8)}$

2. Figure 7.18 shows the block diagram of the direction control system of an automobile where the controller is a human driver. Draw the root locus of the system.

3. The block diagram of a position control system using a robot is shown in Figure 7.19. Draw the root locus of the system.

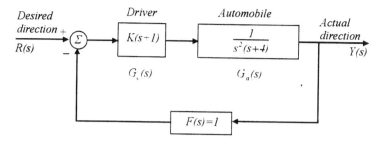

Figure 7.18

4. The block diagram of a speed control system for an aircraft is shown in Figure 7.20. Draw the root locus of the system.

5. Consider the control system of a tape drive shown in Figure 7.21. Determine the root locus of the system for the following two cases:

(a) $G_c(s) = \dfrac{K}{s+4}$

(b) $G_c(s) = \dfrac{K(s+5)}{s+4}$

6. Determine the root locus of the system shown in Figure 7.22, where

$$G_c(s) = \frac{K(s+2)^2}{s} \qquad \text{and} \qquad G(s) = \frac{1}{s(s-1)}$$

7. Determine the root locus of the system in Figure 7.23 for the following three cases:

(a) $G_c(s) = K$
(b) $G_c(s) = K(s+2)$
(c) $G_c(s) = \dfrac{K(s^2 + 2s + 2)}{s}$

8. Draw the root locus of a submarine depth control system shown in Figure 6.3 (Example 6.4.10).

9. Consider the root locus of Example 7.3.2 given in Figure 7.6. Find an analytical expression of the root locus segment which has the shape of a perfect circle.

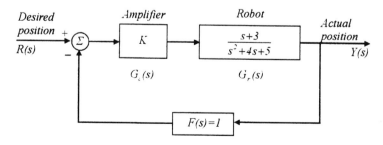

Figure 7.19

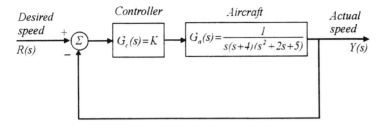

Figure 7.20

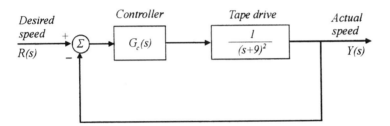

Figure 7.21

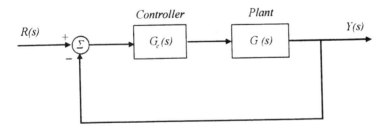

Figure 7.22

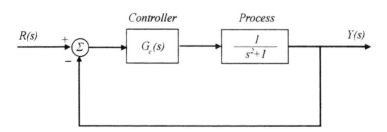

Figure 7.23

10. Find the roots of the following polynomials using the root locus method:
    (a) $s^3 + 4s^2 + 4s + 10$
    (b) $s^3 + s^2 + 10s + 10$

11. Study the effect of variation of the parameter a on the root locus of the loop transfer functions:

    (a) $\dfrac{K}{s(s+a)}$

    (b) $\dfrac{K(s+a)}{s(s+1)}$

    (c) $\dfrac{K}{s(s+1)(s+a)}$

    (d) $\dfrac{K(s+1)}{s(s+2)(s+a)}$

    (e) $\dfrac{K(s+1)}{(s+a)(s+2)(s^2+2s+2)}$

    (f) $\dfrac{K(s+1)}{s^2(s+a)}$

12. Consider the speed control system described in Subsec. 3.13.3. Let $K_g = 1$, $K_m = 1$, $N = 10$, $L_f = 1$, $L_a = 1$, $R_a = 1$, $J_m^* = 1$, and $B_m^* = 3$. Then,

$$G(s) = K_a \frac{10}{(s + R_f)[(s+1)(s+3) + K_b]}$$

For each of the four cases shown in Table 7.1 draw the root locus for $K_a \geq 0$, and compare the results.

13. Consider the position control system described in Subsec. 3.12.2. For simplicity, let $G(s)$ of Eq. (3.13-9) have the form

$$G(s) = K_a \left[\frac{10}{s(s+1)(s+3) + K_b s}\right]$$

For $K_b = 0.1$, 1, and 10, the transfer function $G(s)$ takes on the forms given in Table 7.2. Draw the root locus for the three cases, for $K_a \geq 0$, and compare the results.

Table 7.1

$R_f = 1.5$	$K_b = 1$	$G(s) = G_1(s) = K_a \dfrac{10}{(s+1.5)[(s+1)(s+3)+1]}$
		$= K_a \dfrac{10}{(s+1.5)(s+2)^2}$
	$K_b = 10$	$G(s) = G_2(s) = K_a \dfrac{10}{(s+1.5)[(s+1)(s+3)+10]}$
		$= K_a \dfrac{10}{(s+1.5)(s^2+4s+13)}$
$R_f = 4$	$K_b = 1$	$G(s) = G_3(s) = K_a \dfrac{10}{(s+4)[(s+1)(s+3)+1]}$
		$= K_a \dfrac{10}{(s+4)(s+2)^2}$
	$K_b = 10$	$G(s) = G_4(s) = K_a \dfrac{10}{(s+4)[(s+1)(s+3)+10]}$
		$= K_a \dfrac{10}{(s+4)(s^2+4s+13)}$

Table 7.2

$K_b = 0.1$	$G(s) = G_1(s) = K_a \dfrac{10}{s(s+1)(s+3)+0.1s} = K_a \dfrac{10}{s(s+1.05)(s+2.95)}$
$K_b = 1$	$G(s) = G_2(s) = K_a \dfrac{10}{s(s+1)(s+3)+s} = K_a \dfrac{10}{s(s+2)^2}$
$K_b = 10$	$G(s) = G_3(s) = K_a \dfrac{10}{s(s+1)(s+3)+10s} = K_a \dfrac{10}{s[(s^2+4s+4)+3^2]}$

BIBLIOGRAPHY

Books

1. JJ D'Azzo, and CH Houpis. Linear Control System Analysis and Design, Conventional and Modern. New York: McGraw-Hill, 1975.
2. JJ DiStefano III, AR Stubberud, IJ Williams. Feedback and Control Systems. Schaum's Outline Series. New York: McGraw-Hill, 1967.
3. RC Dorf, RE Bishop. Modern Control Analysis. London: Addison-Wesley, 1995.
4. WR Evans. Control System Dynamics. New York: McGraw-Hill, 1954.
5. GF Franklin, JD Powell, A Emami-Naeini. Feedback Control of Dynamic Systems. Reading, Massachusetts: Addison-Wesley, 1986.
6. NS Nise. Control Systems Engineering. New York: Benjamin and Cummings, 1995.
7. K Ogata. Modern Control Systems. London: Prentice Hall, 1997.

Articles

8. WR Evans. Control system synthesis by root locus method. AIEE Trans, Part II, 69:66–69, 1950.
9. WR Evans. Graphical analysis of control systems. AIEE Trans, Part II, 67:547–551, 1948.

8

Frequency Domain Analysis

8.1 INTRODUCTION

This chapter refers to the behavior of linear time-invariant systems in the frequency domain. Here the important control analysis and design tools of Nyquist, Bode, and Nichols diagrams are presented.

It is important to stress that the control engineer should be able to understand the behavior of a system both in the time and in the frequency domain. Generally speaking, it is easier to understand the time domain behavior of a system compared with its frequency domain behavior. However (particularly with respect to the classical control methods), the time domain has the disadvantage in that it is more difficult to handle (e.g., in designing controllers), compared with the frequency domain. It is therefore important that the control engineer knows both the time- and the frequency-domain system's behavior and is able to correlate the behavior of the system in these two domains. Chapter 8 aims to offer this knowledge, which is particularly necessary for Chap. 9.

8.2 FREQUENCY RESPONSE

Consider a SISO system with transfer function

$$H(s) = \frac{K(s+z_1)(s+z_2)\cdots(s+z_m)}{(s+p_1)(s+p_2)\cdots(s+p_n)}, \qquad m < n \qquad (8.2\text{-}1)$$

Let the input be the sinusoidal function $u(t) = R\sin\omega t$. Then the response $Y(s)$ of the system will be

$$Y(s) = H(s)U(s) = \left[\frac{K(s+z_1)(s+z_2)\cdots(s+z_m)}{(s+p_1)(s+p_2)\cdots(s+p_n)}\right]\left[\frac{R\omega}{s^2+\omega^2}\right]$$

Expand $Y(s)$ in partial fractions as follows:

$$Y(s) = \frac{K}{s+p_1} + \frac{K_2}{s+p_2} + \cdots + \frac{K_n}{s+p_n} + \frac{K_{n+1}}{s+j\omega} + \frac{K_{n+2}}{s-j\omega}$$

The output $y(t) = L^{-1}\{Y(s)\}$ will be

$$y(t) = K_1 e^{-p_1 t} + K_2 e^{-p_2 t} + \cdots + K_n e^{-p_n t} + K_{n+1} e^{-j\omega t} + K_{n+2} e^{j\omega t}$$

Assume that all poles of $H(s)$ lie in the left-half complex plane, i.e., $\mathrm{Re}(-p_i) < 0$, $\forall i$. Then the output of the system in the steady state will be

$$y_{ss}(t) = K_{n+1} e^{-j\omega t} + K_{n+2} e^{j\omega t}$$

Since

$$K_{n+1} = \lim_{s \to -j\omega} (s + j\omega) Y(s) = -\frac{H(-j\omega)R}{2j}$$

and

$$K_{n+2} = \lim_{s \to j\omega} (s - j\omega) Y(s) = \frac{H(j\omega)R}{2j}$$

it follows that

$$y_{ss}(t) = \frac{R}{2j}[-H(-j\omega)e^{-j\omega t} + H(j\omega)e^{j\omega t}] = R|H(j\omega)| \sin[\omega t + \varphi(\omega)] \qquad (8.2\text{-}2)$$

where $|H(j\omega)|$ and $\varphi(\omega)$ are the amplitude and the phase of $H(j\omega)$, respectively.

From relation (8.2-2) it follows that when the transfer function of an SISO system has all its poles in the left-half complex plane, then the output of the system in the steady state, with sinusoidal excitation, is also sinusoidal. In particular, we observe that the amplitude of the output is the amplitude of the input multiplied by $|H(j\omega)|$, while the phase of the output is the phase of the input shifted by $\varphi(\omega)$. The output of the system in steady state, when excited by a sinusoidal function, i.e., $y_{ss}(t)$ of relation (8.2-2), is known as the *frequency response* of the system.

Relation (8.2-2) has a significant characteristic in suggesting a relatively easy laboratory method to determine the transfer function $H(j\omega)$ of a system. Indeed, if a system is excited by a sinusoidal function with amplitude equal to unity and phase equal to zero, then the amplitude of the output steady state will be the amplitude of the transfer function, while the phase of the output steady state will be the phase of the transfer function. Therefore, if we excite Eq. (8.2-1) with $u(t) = \sin \omega t$ and let ω take values for a certain range of frequencies, then we can readily sketch the curves of $|H(j\omega)|$ and $\varphi(j\omega)$ of the transfer function $H(s)$ for this particular range of frequencies.

This graphical representation of $|H(j\omega)|$ and $\varphi(j\omega)$ constitutes an important frequency domain tool for studying stability, specifying the closed-loop desired behavior, developing control design techniques, etc. For this reason, very often, even in cases where we know the analytical expression for the transfer function, we prefer to first determine its graphical representation and subsequently apply the graphical design methods.

The above remarks can also be applied to MIMO systems whose transfer functions are matrices. In this case, the element $h_{ij}(s)$ of the transfer function matrix $H(s)$ relates the ith output $y_i(t)$ to the jth input $u_j(t) = R_j n \mu \omega t$, assuming that all other inputs are zero. In the steady state, the following relation holds

$$[y_i(t)]_{ss} = R_j |h_{ij}(j\omega)| \sin[\omega t + \varphi_{ij}(\omega)] \qquad (8.2\text{-}3)$$

where $|h_{ij}(j\omega)|$ is the amplitude and $\varphi_{ij}(\omega)$ is the phase of the element $h_{ij}(s)$ of $H(s)$.

8.3 CORRELATION BETWEEN FREQUENCY RESPONSE AND TRANSIENT RESPONSE

8.3.1 Characteristics of Frequency Response

The transfer function $H(j\omega)$ of Eq. (8.2-1) can be written as

$$H(j\omega) = |H(j\omega)| \; \underline{/H(j\omega)} = M \; \underline{/\varphi} \tag{8.3-1}$$

where, for simplicity, we have placed $M = |H(j\omega)|$ and $\varphi = \underline{/H(j\omega)}$. Typical graphical representations of M and φ, as functions of the frequency ω, are given in Figure 8.1a and b, respectively.

Certain characteristics of the amplitude curve M have special significance for the frequency response of a system. These characteristics are the following.

1 Resonant Peak M_p

The value M_p is the maximum value of the amplitude M. As we will see in the Subsec. 8.3.4 (see Figure 8.9), large values of M_p usually correspond to large values of overshoot in the time domain, when the excitation is the step function. For most design problems, the value of M_p is chosen to lie between the values 1.1 and 1.5.

2 Resonant Frequency ω_p

The resonant frequency ω_p is defined as the frequency for which $M(\omega_p) = M_p$.

3 Bandwidth (BW)

The bandwidth BW of the amplitude curve of Figure 8.1a is defined as the frequency ω_b for which $M(\omega_b) = 0.707$. The BW in the frequency domain and the rise time T_r of a system's response in the time domain (see Subsec. 4.2.2), are inversely proportional quantities.

In the sequel, we will discuss the correlation between the frequency and time response of a system. In particular, we will study the correlation between the characteristics M_p, ω_p, and BW in the frequency domain with the characteristics T (time constant), y_m (peak value), and ζ (damping ratio) in the time domain.

8.3.2 Correlation for First-Order Systems

The transfer function of a first-order system has the form

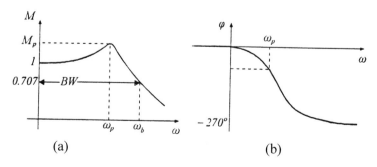

(a)   (b)

Figure 8.1 Typical graphical representations of (a) the amplitude M and (b) the phase φ.

$$H(s) = \frac{K}{Ts + 1}$$

Clearly, $H(s)$ in the sinusoidal steady state will be

$$H(j\omega) = \frac{K}{j\omega T + 1} = \frac{K}{\sqrt{\omega^2 T^2 + 1}} \; \underline{/- \tan^{-1} \omega T} \; = M \; \underline{/\varphi}$$

The waveforms of M and φ are given in Figures 8.2a and 8.2b, respectively. M_p, ω_p, and BW are $M_p = K$, $\omega_p = 0$, and BW $= \omega_b = 1/T$.

The above results in the frequency domain are correlated with those in the time domain (Subsec. 4.3.1) by the relation BW $= 1/T$. This equation shows that BW and T are inversely proportional.

8.3.3 Correlation for Second-Order Systems

Consider a second-order system with transfer function

$$H(s) = \frac{\omega_0^2}{s^2 + 2\zeta\omega_0 s + \omega_0^2} \tag{8.3-2}$$

The transfer function $H(s)$ in the sinsusoidal steady state will be

$$H(j\omega) = \frac{\omega_0^2}{(j\omega)^2 + 2\zeta\omega_0(j\omega) + \omega_0^2} = \frac{1}{1 - u^2 + j2\zeta u} \tag{8.3-3}$$

where, for simplicity, we have set $u = \omega/\omega_0$. The amplitude M and the phase φ of $H(j\omega)$, according to Eq. (8.3-1), will be

$$M = \frac{1}{[(1 - u^2)^2 + (2\zeta u)^2]^{1/2}}, \qquad \varphi = -\tan^{-1}\left[\frac{2\zeta u}{1 - u^2}\right] \tag{8.3-4}$$

Next, we will determine ω_p, M_p, and BW of the system. To this end, to determine ω_p we differentiate M with respect to u and set the derivative equal to zero, i.e.,

$$\frac{dM}{du} = -\frac{[-2(1 - u^2)u + 4\zeta^2 u]}{2[(1 - u^2)^2 + (2\zeta u)^2]^{3/2}} = 0$$

The above relation gives

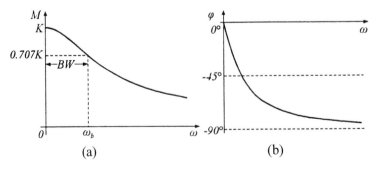

(a)   (b)

Figure 8.2 Diagrams of (a) amplitude M and (b) phase φ of a first-order system.

$$2u^3 - 2u + 4\zeta^2 u = 0 \tag{8.3-5}$$

The roots of Eq. (8.3-5) are $u = 0$, $u = \sqrt{1 - 2\zeta^2}$, and $u = -\sqrt{1 - 2\zeta^2}$. We disregard the third root since it has no physical meaning. Consider the root

$$u = \sqrt{1 - 2\zeta^2}, \qquad \zeta < 0.707 \tag{8.3-6}$$

Its resonant frequency ω_p will be

$$\omega_p = \omega_0\sqrt{1 - 2\zeta^2}, \qquad \zeta < 0.707 \tag{8.3-7}$$

The resonant value M_p is found if we substitute Eq. (8.3-6) into Eq. (8.3-4) to yield

$$M_p = \frac{1}{2\zeta\sqrt{1 - \zeta^2}}, \qquad \zeta < 0.707 \tag{8.3-8}$$

From relations (8.3-7) and (8.3-8) it is clear that ω_p is a function of ω_0 and ζ, while M_p is a function of ζ only. Figure 8.3 presents typical shapes of the amplitude M as a function of the normalized frequency u and with free parameter the damping ratio ζ. From Figure 8.3 it follows that for $\zeta < 0.7$, the peak value M_p increases as ζ decreases. For $\zeta \geq 0.707$, the peak value is $M_p = 1$, which occurs when $u = 0$.

To determine the bandwidth BW, it suffices to specify the frequency ω_b. From Eq. (8.3-4) and the definition of bandwidth, we have

$$M(\omega_b) = \frac{1}{[(1 - u_b^2)^2 + (2\zeta u_b)^2]^{1/2}} = \frac{1}{\sqrt{2}} \tag{8.3-9}$$

where $u_b = \omega_b/\omega_0$. If we solve Eq. (8.3-9) for u_b, we obtain

$$u_b = \left[1 - 2\zeta^2 + \sqrt{4\zeta^4 - 4\zeta^2 + 2}\right]^{1/2} \tag{8.3-10}$$

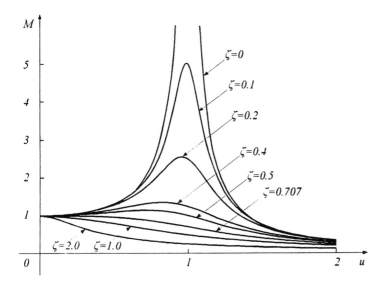

Figure 8.3 The plot of amplitude M of a second-order system versus u.

Thus

$$\mathrm{BW} = \omega_b = \omega_0\left[1 - 2\zeta^2 + \sqrt{4\zeta^4 - 4\zeta^2 + 2}\right]^{1/2} \tag{8.3-11}$$

From relation (8.3-10) we obseve that the normalized bandwidth u_b is a function of the damping ratio ζ only. The dependence of u_b upon ζ is given graphically in Figure 8.4.

To correlate the frequency response with the time response, we make use of the results of Sec. 4.3. First, we compare the maximum value y_m of the response $y(t)$ when the excitation is the unit step function. In this case, y_m is given by Eq. (4.3-10), i.e., by the relation

$$y_m = 1 + \exp\left[-\zeta\pi/\sqrt{1 - \zeta^2}\right], \qquad 0 < \zeta < 1$$

The maximum value M_p of the amplitude of the function $H(j\omega)$ is given by Eq. (8.3-8). We remark that both y_m and M_p are functions of ζ only. In figure 8.5, y_m and M_p are given as functions of ζ.

If we now compare the resonant frequency ω_p (see relation (8.3-7)) with the damped natural frequency ω_d (see Subsec. 4.3.2), where $\omega_d = \omega_n\sqrt{1 - \zeta^2}$, we observe that both frequencies are functions of ω_n and ζ (Figure 8.6). Their ratio is given by

$$\frac{\omega_p}{\omega_d} = \frac{\sqrt{1 - 2\zeta^2}}{\sqrt{1 - \zeta^2}}$$

8.3.4 Correlation for Higher-Order Systems

The analytical correlation between the frequency domain and the time domain response for higher-order systems is, in general, very difficult. This is because the time domain response $y(t)$ and the frequency domain response $Y(j\omega)$ are related by the Fourier integral, as follows:

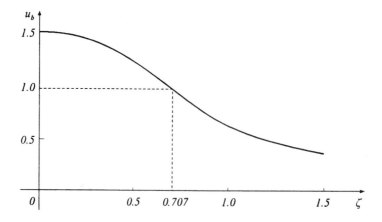

Figure 8.4   The plot of the ratio Bw/ω_0 versus the damping constant ζ.

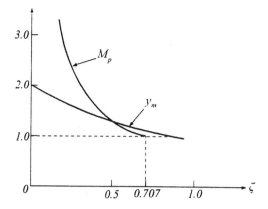

Figure 8.5 The maximum amplitude M_p and the maximum value y_m as a function of ζ.

$$Y(j\omega) = \int_{-\infty}^{\infty} y(t)e^{-j\omega t}dt \qquad (8.3\text{-}12a)$$

$$y(t) = \frac{1}{2\pi}\int_{-\infty}^{\infty} Y(j\omega)e^{j\omega t}d\omega \qquad (8.3\text{-}12b)$$

The above Fourier transform pair requries great computational effort, particularly as the order of the system increases. Another reason which makes it difficult to extend the correlation between the frequency and time domain response to higher-order systems is that there is no analytical expression for the calculation of the poles of a higher-order algebraic equation. To circumvent this difficulty, it is usually proposed to approximate the original high-order transfer function $H(s)$ with a new transfer function $H^*(s)$ of second order. If $H^*(s)$ approximates $H(s)$ satisfactorily, then the study of the system is subsequently based on $H^*(s)$, rather than $H(s)$, which makes the study of the system much simpler. Usually, we choose as poles of the new transfer function $H^*(s)$ the poles of $H(s)$ which are closest to the imaginary

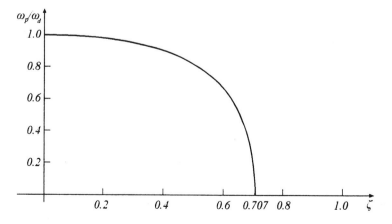

Figure 8.6 The plot of the ratio ω_p/ω_d versus the damping constant ζ.

axis. The issue of the approximation (simplification) of $H(s)$ by $H^*(s)$ has been considered in Sec. 4.4.

We close this sec. with Figure 8.7, 8.8, and 8.9 which refer to second- and higher-order systems and give an overall picture of the correlation between the behavior of a system in the time domain and frequency domain. In Figure 8.7 the correlation between the time domain and frequency domain is given, where the description in the frequency domain is given both with the curves M and φ, as well as with the Nyquist diagrams (see next section). Figure 8.8 gives an overall picture of the correlation between the rise time T_r and the bandwidth BW, from which we reach the conclusion that T_r and BW are inversely proportional. Finally, Figure 8.9 gives the curves related to the overshoot y_m and the resonance value M_p, from which we reach the conclusion that as the resonance value M_p increases, the

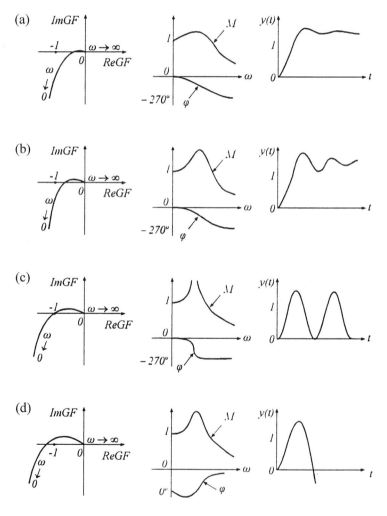

Figure 8.7 Nyquist diagram, amplitude curves of M and φ, and the output $y(t)$ for stable and unstable systems. (a) Stable system without oscillations; (b) stable system with oscillations; (c) stable system with sustained oscillations; (d) unstable system.

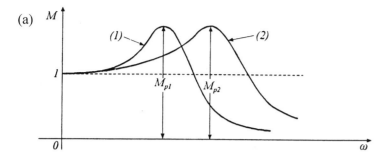

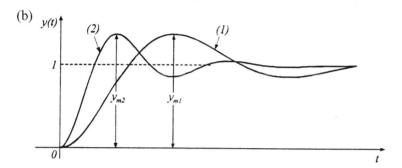

Figure 8.8 (a) Amplitude M and (b) output $y(t)$ curves of two different systems which show the relation between T_r and bandwidth BW.

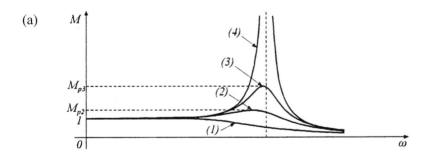

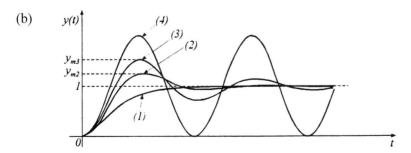

Figure 8.9 (a) The amplitude M and (b) the output $y(t)$ of four different systems, which show the relation between y_m and M_p.

overshoot y_m also increases. In practice, it is widely accepted that a system has a good performance if the value for M_p lies between 1.1 and 1.5, with special preference to the value $M_p = 1.3$.

8.4 THE NYQUIST STABILITY CRITERION

8.4.1 Introduction

The Nyquist stability criterion refers to the stability of closed-loop systems. In cases where a mathematical model is not available and the system transfer function is described graphically, then the use of the Nyquist stability criterion comes in very handy. In cases where a mathematical model is available, then one can apply one of the algebraic criteria to readily determine the stability of the closed-loop system. However, even in this latter case, one can also apply the Nyquist stability criterion to acquire further useful knowledge about the behavior of the closed-loop system. Some important advantages of the Nyquist criterion are given below.

1. It gives information, not only about the stability but also about the *relative stability* of the system. The meaning of the term relative stability may be given on the basis of the roots of the closed-loop characteristic polynomial $p_c(s)$ in the complex plane as follows. Let all roots of $p_c(s)$ lie in the left-half complex plane. Then, as the roots of $p_c(s)$ move to the right and hence closer to the imaginary axis (from left to right), the system becomes *less* stable, and as the root of $p_c(s)$ move to the left and hence away from the imaginary axis, the system becomes *more* stable.
2. It gives information about the behavior (the performance) of the system in the time domain.
3. It can be used for the study of the stability of other categories of systems, such as time-delay systems, nonlinear systems, etc.

Consider the closed-loop system of Figure 8.10. The transfer function $H(s)$ of the closed-loop system is given by

$$H(s) = \frac{G(s)}{1 + G(s)F(s)} = \frac{G(s)}{W(s)} \tag{8.4-1}$$

where

$$W(s) = 1 + G(s)F(s) \tag{8.4-2}$$

Assume that the open-loop transfer function $G(s)F(s)$ has the form

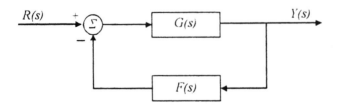

Figure 8.10   Block diagram of a closed-loop system.

$$G(s)F(s) = \frac{K(s+z_1)(s+z_2)\cdots(s+z_m)}{s^j(s+p_1)(s+p_2)\cdots(s+p_q)}, \qquad \text{where } j+q=n \qquad (8.4\text{-}3)$$

Substituting expression (8.4-3) in expression (8.4-2) yields

$$W(s) = 1 + G(s)F(s) \qquad (8.4\text{-}4)$$

or

$$W(s) = \frac{s^j(s+p_1)(s+p_2)\cdots(s+p_q) + K(s+z_1)(s+z_2)\cdots(s+z_m)}{s^j(s+p_1)(s+p_2)\cdots(s+p_q)} = \frac{p_c(s)}{p_\beta(s)}$$

$$(8.4\text{-}5)$$

The polynomials $p_c(s)$ and $p_\beta(s)$ are the characteristic polynomials of the systems with transfer functions $H(s)$ and $G(s)F(s)$, respectively. It is important to note that the stabilities of the closed-loop system and of the system with transfer function $G(s)\,F(s)$ are not correlated. For example, if a system with transfer function $G(s)F(s)$ is unstable, this does not necessarily mean that the closed-loop system is also unstable.

The study of the stability of closed-loop systems through the use of algebraic criteria is based on the characteristic polynomial $p_c(s)$, whereas the Nyquist criterion is based on the graphical representation of the open-loop transfer function $G(s)F(s)$, for $s \in \Gamma_s$, where Γ_s is a specific closed path in the s-plane. This graphical representation of $G(s)F(s)$ gives information not only about the stability but also about the relative stability of the system. In order to formulate the Nyquist criterion, a certain mathematical background on complex function theory is necessary and it is given below.

8.4.2 Background Material on Complex Function Theory for the Formulation of the Nyquist Criterion

Consider a complex function $W(s)$ and let Γ_s be an arbitrary closed path in the s-plane (Figure 8.11a). Then, for $s \in \Gamma_s$, the complex function $W(s)$ (except for some special forms of $W(s)$ which will not be considered here), will also form a closed path Γ_W in the $W(s)$ plane (Figure 8.11b).

If in the closed path Γ_s we give a particular direction, then the path Γ_W will also have a certain direction, not necessarily the same as the direction of Γ_s (see Figure 8.12). Furthermore, if the path Γ_s is a simple closed path, then the path Γ_W will not, in general, be a simple closed path but can be, as for example, in Figure 8.12a and d. Of course, the particular form of the path Γ_W depends not only on the path Γ_s but also on the particular form of the function $W(s)$.

Let the complex function $W(s)$ have the form

$$W(s) = K\frac{(s+z_1)(s+z_2)\cdots(s+z_m)}{(s+p_1)(s+p_2)\cdots(s+p_n)}, \qquad K > 0 \qquad (8.4\text{-}6)$$

Also let the closed path Γ_s enclose P poles and Z zeros of $W(s)$. Then, the numbers P and Z play a very decisive role in the shape of the closed path Γ_W, as stated by the following theorem.

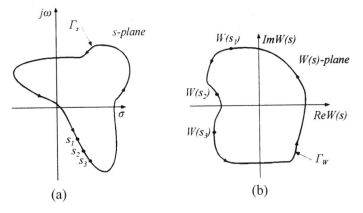

Figure 8.11   The closed path Γ_W of $W(s)$ for $s \in \Gamma_s$.

Theorem 8.4.1

Let the complex function $W(s)$ be a single-valued, rational function of the form (8.4-6) and also assume that this function is analytical (i.e., all its derivatives exist) in all points of a closed path Γ_s. Then, the corresponding path Γ_W will encircle the origin N times, where $N = Z - P$. If $N > 0$ ($N < 0$) then the closed path Γ_W has the same (the opposite) direction as that of the closed path Γ_s.

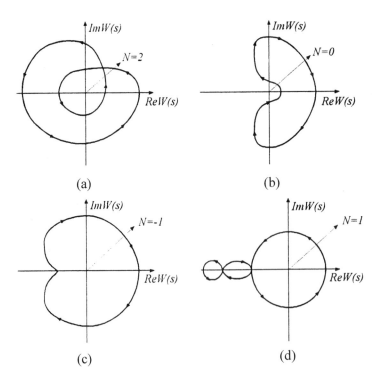

Figure 8.12   Various possible closed paths Γ_W of $W(s)$.

Instead of giving a strict mathematical proof, we will give a rather simple and practical explanation of Theorem 8.4.1. To this end, we write $W(s)$ as follows:

$$W(s) = |W(s)| \underline{/W(s)} \qquad (8.4\text{-}7)$$

where

$$|W(s)| = \text{amplitude of } W(s) = K \frac{|s + z_1||s + z_2| \cdots |s + z_m|}{|s + p_1||s + p_2| \cdots |s + p_n|} \qquad (8.4\text{-}8)$$

$$\underline{/W(s)} = \text{phase of } W(s) = \sum_{i=1}^{m} \varphi_i(\omega) - \sum_{i=1}^{n} \theta_i(\omega) \qquad (8.4\text{-}9)$$

where $\varphi_i(\omega) = \underline{/s + z_i}$ and $\theta_i(\omega) = \underline{/s + p_i}$. The positions of the zeros $-z_1, -z_2, \ldots$ $, -z_m$ and of the poles $-p_1, -p_2, \ldots, -p_n$ of $W(s)$ are depicted in Figure 8.13. Choose the closed path Γ_s such as to encircle Z zeros and P poles of $W(s)$. Also, let the point $s \in \Gamma_s$ move on Γ_s, having the direction of the arrow (i.e., counterclockwise). Then, we observe that as the point s travels around the closed path Γ_s, each factor $s + z_k$ or $s + p_i$ will generate an angle of 360° when $-z_k$ or $-p_i$ lie within the closed path Γ_s, or an angle of 0° when $-z_k$ or $-p_i$ lie outside the closed path Γ_s. Since the phase of $W(s)$, according to the relation (8.4-9), is the sum of the phases of the factors $s + z_k$ of the numerator minus the sum of the phases of the factors $s + p_i$ of the denominator, it follows that the total phase φ which the transfer function $W(s)$ will generate as the point s travels around the closed path Γ_s will be

$$\varphi = \underline{/W(s)} = 2\pi(Z - P) = 2\pi N \qquad (8.4\text{-}10)$$

To determine the number N from the diagram of $W(s)$, we draw an arbitrary radius from the origin to infinity. Then, the diagram of $W(s)$ will intersect this radius Z times in the direction of Γ_s and P times in the opposite direction. Therefore $N = Z - P$. Certain illustrative examples are given in Figure 8.12.

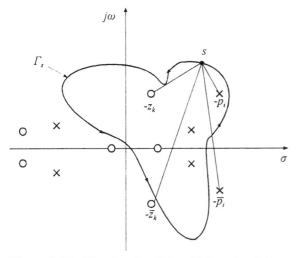

Figure 8.13   The closed path Γ_s which enclosed Z zeros and P poles of $W(s)$.

8.4.3 The Nyquist Stability Criterion

The Nyquist stability criterion is based on Theorem 8.4.1. Since the objective of the Nyquist criterion is the study of the stability of the closed-loop system, the closed path Γ_s is no longer arbitrary. It is that particular path which encloses, in the clockwise direction, the entire right-half complex plane (Figure 8.14). This special path is called the *Nyquist path* and it is designated by Γ_N. If the function $W(s)$ has poles on the imaginary axis, then the Nyquist path circumvents these poles by going around them on a semicircle with infinitesimal radius $\rho \to 0$ (Figure 8.14). This is done in order that the function $W(s)$ remains analytcial on the Nyquist path and, hence, Theorem 8.4.1 holds. The path Γ_W of $W(s)$, which corresponds to the Nyquist path, is called the *Nyquist diagram* of $W(s)$.

Assume that $W(s)$ is the transfer function of a system. Then, it is clear that in the case of a stable system there must be $P = 0$, since there are no poles in the right-half complex plane. This means that $N = Z$, i.e., the Nyquist diagram of a stable system must encircle the origin clockwise as many times as the number of zeros Z of $W(s)$ which lie in the right-half complex plane.

Now let $W(s)$ be given by the expression (8.4-5), i.e., by the expression

$$W(s) = 1 + G(s)F(s) = \frac{p_c(s)}{p_\beta(s)} \qquad (8.4\text{-}11)$$

where $p_c(s)$ is the characteristic polynomial of the closed-loop transfer function $H(s)$ and $p_\beta(s)$ is the characteristic polynomial of the open-loop transfer function $G(s)F(s)$. Also assume that $W(s)$ has Z zeros and P poles within the Nyquist path Γ_N. Apparently, for the closed-loop system to be stable, the roots of $p_c(s)$ must lie in the left-half complex plane, i.e., there must be $Z = 0$. This means that for the closed-loop system to be stable, the Nyquist diagram of $W(s)$, having the same direction as that of the Nyquist path Γ_N, encircles the origin $N = -P$ times. If $P = 0$, then the

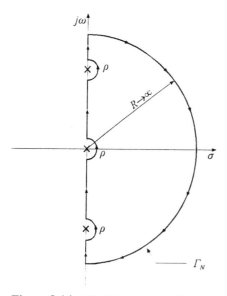

Figure 8.14   The Nyquist path Γ_N.

Nyquist diagram of $W(s)$ does not encircle the origin. Therefore, we have established the following theorem, which is the well-known Nyquist theorem.

Theorem 8.4.2

Assume that the transfer function $W(s) = 1 + G(s)F(s)$ does not have poles in the right-half complex plane. Then, for the closed-loop system (8.4-1) to be stable, the Nyquist diagram of $W(s)$, having the same direction as that of the path Γ_N, must not encircle the origin.

Define

$$W^*(s) = W(s) - 1 = G(s)F(s) \tag{8.4-12}$$

Then, the Nyquist diagrams of $W(s)$ and $W^*(s)$ differ in that the diagram for $W^*(s)$ has been translated by one unit to the left of the diagram for $W(s)$. This means that the conclusions about the stability of $W(s)$ hold for $W^*(s)$ as well, if the origin is now substituted by the point $(-1, j0)$, which is called the *critical point*. Hence, the following theorem holds.

Theorem 8.4.3

Assume that the transfer function $W^*(s) = G(s)F(s)$ does not have poles in the right-half complex plane. Then, for the closed-loop system (8.4-1) to be stable, the Nyquist diagram of $G(s)F(s)$, having the same direction as that of the path Γ_N, does not encircle the critical point $(-1, j0)$.

    As an introductory illustrative example of Theorem 8.4.3, we will study the stability of the two closed-loop systems which have as Nyquist diagrams the Nyquist diagrams of the open-loop transfer functions $G(s)F(s)$ of Figure 8.15. It is assumed that the transfer function $G(s)F(s)$ does not have any poles in the right-half complex plane. In Figure 8.15a, we observe that the Nyquist diagram of $G(s)F(s)$ does not encircle clockwise the critical point $(-1, j0)$, while it does in Figure 8.15b. Hence the

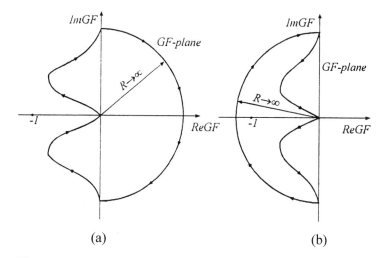

(a)   (b)

Figure 8.15   Nyquist diagrams of (a) a stable and (b) an unstable system.

closed-loop system, having as Nyquist diagram of the open-loop transfer function $G(s)F(s)$ that of Figure 8.15a, is stable, while that of Figure 8.15b is unstable.

We introduce the following definition.

Definition 8.4.1

A transfer function $H(s)$ is called a *minimum phase function* when it does not have any poles or zeros in the right-half complex plane. In the case where it does, then the complex function $H(s)$ is called a *nonminimum phase function*.

Using all the material we have presented up to now regarding the Nyquist criterion, we have essentially proven the following theorem.

Theorem 8.4.4

Assume that the transfer function $W^*(s) = G(s)F(s)$ is a nonminimum phase function and has P poles in the right-half complex plane. Then, for the closed-loop system (8.4-1) to be stable, the Nyquist diagram of $G(s)F(s)$, having the opposite direction to that of the path Γ_N, must encircle P times the critical point $(-1, j0)$.

Theorem 8.4.4 is called the *generalized Nyquist theorem*. It is clear that Theorem 8.4.3 is a special case of Theorem 8.4.4.

As an introductory illustrative example of Theorem 8.4.4 we will study the following open-loop transfer functions:

$$W_1^*(s) = G(s)F(s) = \frac{K}{s(Ts-1)} \quad \text{and} \quad W_2^*(s) = G(s)F(s) = \frac{K(T_1's+1)}{s(T_1s-1)}$$

$$(8.4\text{-}13)$$

The Nyquist diagrams of $W_1^*(s)$ and $W_2^*(s)$ are given in Figure 8.16. For the transfer function $W_1^*(s)$, we have that $P = 1$ (because $W_1^*(s)$ has the pole $1/T$ in the right-half complex plane). According to Theorem 8.4.4, for the closed-loop system to be stable, the Nyquist diagram must encircle the critical point $(-1, j0)$ once in the counterclockwise direction. From Figure 8.16a it follows that the Nyquist diagram encircles the critical point but in the clockwise direction and, hence, the closed-loop system is unstable. In this example we observe that both the open-loop system and the closed-loop system are unstable. For the transfer function $W_2^*(s)$ we also have that $P = 1$. The Nyquist diagram for small values of K does not encircle, in the counterclockwise direction, the point $(-1, j0)$ and, therefore, the closed-loop system is unstable (see Figure 8.16b). For greater values of K, the Nyquist diagram encircles the critical point $(-1, j0)$ once in the counterclockwise direction and, hence, the closed-loop system is stable (see Figure 8.16c). In this example we observe that even though the open-loop system is unstable, the closed-loop system can be stable, provided that K assumes large enough values.

8.4.4 Construction of Nyquist Diagrams

Since the Nyquist diagram of the open-loop transfer function $G(s)F(s)$ constitutes the basis for the application of the Nyquist criterion, it follows that one must be familiar with how to construct such diagrams. To this end, this subsection presents a rather systematic approach on how to construct Nyquist diagrams of various forms of transfer functions.

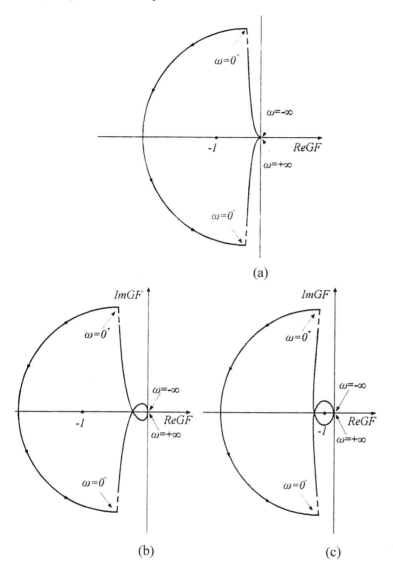

Figure 8.16 Nyquist diagrams of nonminimum phase systems. (a) Plot of $W_1^*(s)$ (unstable system); (b) plot of $W_2^*(s)$ for small values of K (unstable system); (c) plot of $W_2^*(s)$ for large values of K (stable system).

1 Nyquist Diagram for First-Order Systems

Consider the first-order system with transfer function

$$H(s) = \frac{1}{Ts + 1} \tag{8.4-14}$$

where, for simplicity, we use $H(s)$ instead of $G(s)F(s)$. Since $H(s)$ does not have any poles on the imaginary axis, it follows that the Nyquist path Γ_N will cover the entire imaginary axis. Therefore, the Nyquist diagram of $H(s)$ is the function

$$H(j\omega) = \frac{1}{j\omega T + 1}, \qquad \omega \in (-\infty, +\infty) \tag{8.4-15}$$

We have

$$H(j\omega) = |H(j\omega)| \; \underline{/\varphi(\omega)} = \frac{1}{\sqrt{\omega^2 T^2 + 1}} \; \underline{/-\tan^{-1}\omega T}$$

If we let $\omega = 0$, then $|H(j\omega)| = 1$ and $\varphi(\omega) = 0°$. As the frequency ω increases, the amplitude $|H(j\omega|$ becomes smaller while the phase $\varphi(\omega)$ becomes more negative. In the limit where $\omega \to \infty$, the amplitude $|H(j\omega| \to 0$, whereas the phase $\varphi(\omega) \to -90°$. Finally, we observe that $|H(j\omega)| = |H(-j\omega)|$ and $\varphi(\omega) = -\varphi(-\omega)$. Hence, the Nyquist diagram is symmetrical with respect to the $\operatorname{Re} H(j\omega)$ axis and has the form given in Figure 8.17.

It is mentioned that the determination of the analytical expression of the Nyquist diagram with coordinates $\operatorname{Re} H$ and $\operatorname{Im} H$ is difficult, particularly as the order of the system becomes greater. For the case of first-order systems, the determination of the analytical expression of the Nyquist diagram is relatively simple and is derived as follows. We have

$$H(j\omega) = \frac{1}{j\omega T + 1} = \frac{1}{\omega^2 T^2 + 1} - j\frac{\omega T}{\omega^2 T^2 + 1} = x + jy$$

where

$$x = \operatorname{Re} H(j\omega) = \frac{1}{\omega^2 T^2 + 1}, \qquad y = \operatorname{Im} H(j\omega) = -\frac{\omega T}{\omega^2 T^2 + 1}$$

Let $u = \omega T$. then

$$x = \frac{1}{u^2 + 1}, \qquad y = -\frac{u}{u^2 + 1}$$

Since $y = -ux$, it follows that

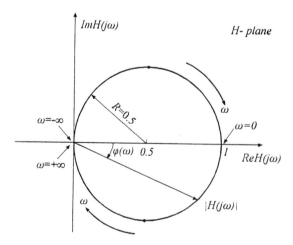

Figure 8.17   The Nyquist diagram of a first-order transfer function.

$$y = -\frac{u}{u^2 + 1} = \frac{\frac{y}{x}}{\left[\frac{-y}{x}\right]^2 + 1} = \frac{xy}{x^2 + y^2} \qquad \text{or} \qquad y^2 + x^2 - x = 0$$

We finally have

$$y^2 + \left[x - \frac{1}{2}\right]^2 = \frac{1}{4} \qquad (8.4\text{-}16)$$

Relation (8.4-16) is the equation of a circle with its center at the point $(1/2, 0)$ and with radius $1/2$ and constitutes the analytical expression of the Nyquist diagram of Eq. (8.4-14). The circle (8.4-16) is shown in Fig. 8.17.

2 Nyquist Diagram for Second-Order Systems

Consider a second-order system with transfer function

$$H(s) = \frac{\omega_n^2}{s^2 + 2\zeta\omega_n s + \omega_n^2}$$

Since $H(s)$ does not have any poles on the imaginary axis, it follows that the Nyquist path will cover the entire imaginary axis. Therefore, the Nyquist diagram of $H(s)$ is

$$H(j\omega) = \frac{\omega_n^2}{(j\omega)^2 + 2\zeta\omega_n(j\omega) + \omega_n^2}, \qquad \omega \in (-\infty, \infty) \qquad (8.4\text{-}17)$$

We have

$$H(j\omega) = |H(j\omega)| \underline{/\varphi(\omega)} = \frac{\omega_n^2}{\sqrt{(\omega_n^2 - \omega^2)^2 + 4\zeta^2\omega_n^2\omega^2}} \underline{/- \tan^{-1}[2\zeta\omega_n\omega/(\omega_n^2 - \omega^2)]}$$

As the frequency ω varies from 0 to ∞, the Nyquist diagram starts with amplitude 1 and zero phase and terminates with amplitude 0 and phase $-180°$. For $\omega = \omega_n$, the amplitude is $(2\zeta)^{-1}$ and the phase is $-90°$. In Figure 8.18 the Nyquist diagram of $H(s)$ is given only for $\omega \in (0, \infty)$. The rest of the diagram for $\omega \in (0, -\infty)$ is omitted since it is symmetrical to the Nyquist diagram for $\omega \in (0, +\infty)$ with respect to the Re H axis.

3 Nyquist Diagrams for Higher-Order Systems

The construction of Nyquist diagrams for systems of third, fourth, etc., order cannot be constructed easily, as opposed to the cases of first- and second-order systems, unless the numerator and the denominator of the transfer function can be factored out. For this last case, very interesting results may be derived, as shown in the sequel.

4 Nyquist Diagrams for Transfer Functions with Poles on the Imaginary Axis

When a transfer function has poles on the imaginary axis, the construction of the Nyquist diagrams must be carried out with extra attention, as shown in the following introductory example.

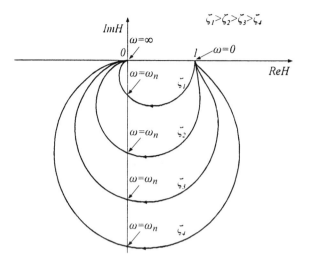

Figure 8.18 The Nyquist diagram of a second-order system.

Example 8.4.1

Consider a system with transfer function

$$H(s) = \frac{K}{s(Ts + 1)} \qquad (8.4\text{-}18)$$

Construct the Nyquist diagram.

Solution

Since $H(s)$ has a pole at $s = 0$ on the imaginary axis, it follows that the path Γ_N will cover the entire imaginary axis, except the point $s = 0$ (see Figure 8.19a). To facil-

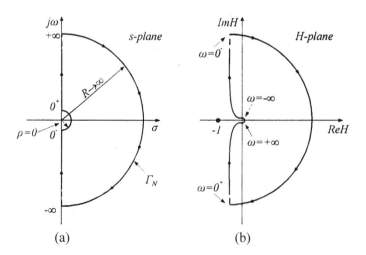

(a)   (b)

Figure 8.19   The Nyquist diagram of a system with a pole on the imaginary axis. (a) The Nyquist path; (b) the Nyquist diagram of Eq. (8.4-18).

itate the construction of the Nyquist diagram of $H(s)$ we divide the Nyquist path Γ_N into four segments and construct the Nyquist diagram for each of the corresponding four segments of Γ_N as follows.

a. The segment for $s \in (j0^+, j\infty)$. This segment of the path Γ_N is the positive part of the imaginary axis not including the origin. For this range of frequencies the transfer function is given by

$$H(j\omega) = \frac{K}{j\omega(j\omega T + 1)}, \qquad \omega \in (0^+, \infty)$$

We have

$$H(j\omega) = |H(j\omega)| \; \underline{/\varphi(\omega)} = \frac{K}{\omega\sqrt{\omega^2 T^2 + 1}} \; \underline{/ -90° - \tan^{-1} \omega T}$$

As the frequency ω varies from 0^+ to $+\infty$, the amplitude $|H(j\omega)|$ varies from ∞ to 0 and the phase $\varphi(\omega)$ from $-90°$ to $-180°$ (see Figure 8.19b).

b. The segment for $s \in (j\infty, -j\infty)$, or $s = Re^{j\theta}$, where $R \to \infty$ and $\theta \in (90°, -90°)$. This segment is the large semicircle of the path Γ_N. In this case, the transfer function becomes

$$\lim_{R \to \infty} H(Re^{j\theta}) = \lim_{R \to \infty} \left[\frac{K}{Re^{j\theta}(TRe^{j\theta} + 1)} \right] = \lim_{R \to \infty} \left[\frac{K}{TR^2 e^{j2\theta}} \right] = \lim_{\rho \to 0} \rho e^{-j2\theta}$$

This means that as the phase θ of the semicircle $Re^{j\theta}$ takes on values in the clockwise direction from $90°$ to $-90°$, the Nyquist diagram of the transfer function $H(s)$ moves counterclockwise, generating a phase change of $360°$ from $-180°$ to $180°$. Therefore, when $s = Re^{j\theta}$, where $R \to \infty$ and $\theta \in (90°, -90°)$, the transfer function $H(s)$ is given by $H(s) = \rho e^{-j2\theta}$, where $\rho \to 0$. Hence, the large semicircle of the path Γ_N maps onto a small circle around the origin in the $H(s)$ plane.

c. The segment for $s \in (-j\infty, j0^-)$. Since this segment is symmetric to the segment $s \in (j0^+, j\infty)$ with respect to the real axis, it follows that the corresponding Nyquist diagram will be symmetric with respect to the $\text{Re } H$ axis.

d. The segment for $s \in (j0^-, j0^+)$, or $s = \rho e^{j\theta}$, where $\rho \to 0$ and $\theta \in (-90°, 90°)$. This segment is the small semicircle, where the path Γ_N moves around the pole $s = 0$. In this case the transfer function becomes

$$\lim_{\rho \to 0} H(\rho e^{j\theta}) = \lim_{\rho \to 0} \left[\frac{K}{\rho e^{j\theta}(T\rho e^{j\theta} + 1)} \right] = \lim_{R \to \infty} Re^{-j\theta}$$

This means that as the phase θ of the semicircle $\rho e^{j\theta}$ takes on values in the counter-clockwise direction from $-90°$ and $90°$, the Nyquist diagram of the transfer function moves clockwise, generating an angle of $180°$ from $90°$ to $-90°$. Therefore, as $s = \rho e^{j\theta}$, where $\rho \to 0$ and $\theta \in (-90°, 90°)$, the transfer function $H(s)$ is given by $H(s) = Re^{-j\theta}$, where $R \to \infty$. Hence, the small interior semicircle of the path Γ_N maps onto the big semicircle in the right-half $H(s)$ plane.

5 The Influence on the Nyquist Diagram of Adding Poles and Zeros to the Transfer Function

Consider the transfer function

$$H_1(s) = \frac{K}{T_1 s + 1}$$

Assume that the transfer function $H_1(s)$ is multiplied by the factor $(T_2 s + 1)^{-1}$, i.e., a pole is added to the transfer function at the point $s = -1/T_2$. Then, the resulting transfer function $H_2(s)$ is given by

$$H_2(s) = \frac{K}{(T_2 s + 1)(T_1 s + 1)} \qquad (8.4\text{-}19)$$

The Nyquist diagrams of $H_1(s)$ and $H_2(s)$ are given in Figure 8.20a and b, respectively. From these two diagrams, the influence of adding a pole to the transfer function $H_1(s)$ is clear. If we add one or two more poles to the original transfer function $H_1(s)$, the Nyquist diagram will be as in Figures 8.20c and d. In cases where the added poles are at the origin, or some are at the origin and some are not, then the Nyquist diagram of $H_1(s)$ is influenced as in Figures 8.20e–h and 8.21a–d, respectively.

On the basis of Figures 8.20 and 8.21a–d, one may derive the following general conclusion: if a new pole is added to a transfer function, then the new Nyquist diagram occupies the next quadrant which lies in the counterclockwise direction. For the particular case where the new pole is at the origin, then the new Nyquist diagram occupies the next quadrant, in the counterclockwise direction, but it no longer occupies the starting quadrant. It is almost like "rotating" the Nyquist diagram by one quadrant in the counterclockwise direction. In the general case, where

$$H(s) = K \frac{(T_1' s + 1)(T_2' s + 1) \cdots}{s^j (T_1 s + 1)(T_2 s + 1) \cdots}$$

the Nyquist diagrams for systems of type 0, 1, 2, and 3, which corresponds to the values of $j = 0$, 1, 2, and 3, are given in Figure 8.22. From the Figures 8.20, 8.21a–d, and 8.22 it becomes clear that adding poles to the transfer function of a system results in a system that is less stable or might even result in an unstable system.

Now, consider multiplying the transfer function $H_1(s)$ with the factor $(T's + 1)$, i.e., a zero is added to the transfer function at the point $s = -1/T'$. Then the new transfer function will be

$$H_2(s) = K \frac{T's + 1}{T_2 s + 1} \qquad (8.4\text{-}20)$$

The Nyquist diagram of $H_2(s)$ is given in the Figure 8.21e, where the influence of adding a zero to the transfer function is clear. Other examples which illustrate the influence of adding zeros to the Nyquist diagrams are given in Figures 8.21f–h. On the basis of these figures one may arrive at the general conclusion that adding zeros results in a more stable system.

It is remarked once again that, for the sake of simplicity, in all material of the present subsection we have been using $H(s)$ instead of $G(s)F(s)$. From this point on, we return to the regular notation $G(s)F(s)$ for the open-loop transfer function.

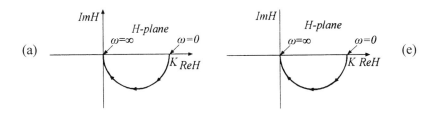

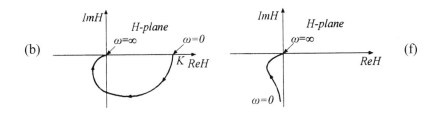

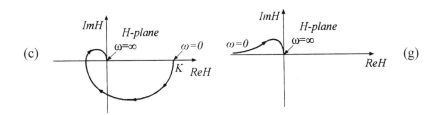

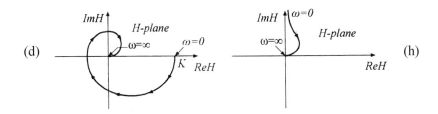

Figure 8.20 Nyquist diagrams. Influence of additional poles.

(a) $\quad H(s) = \dfrac{K}{(T_1 s + 1)}$

(b) $\quad H(s) = \dfrac{K}{(T_2 s + 1)(T_1 s + 1)}$

(c) $\quad H(s) = \dfrac{K}{(T_3 s + 1)(T_2 s + 1)(T_1 s + 1)}$

(d) $\quad H(s) = \dfrac{K}{(T_4 s + 1)(T_3 s + 1)(T_2 s + 1)(T_1 s + 1)}$

(e) $\quad H(s) = \dfrac{K}{(T s + 1)}$

(f) $\quad H(s) = \dfrac{K}{s(T s + 1)}$

(g) $\quad H(s) = \dfrac{K}{s^2(T s + 1)}$

(h) $\quad H(s) = \dfrac{K}{s^3(T s + 1)}$

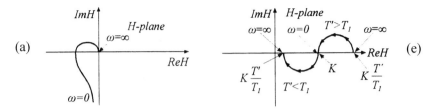

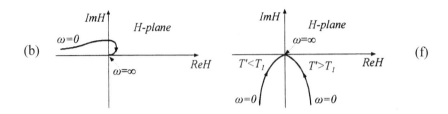

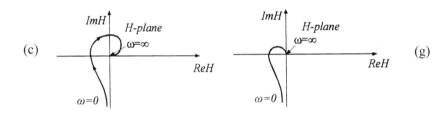

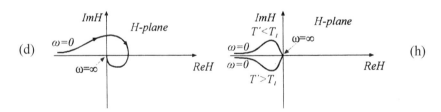

Figure 8.21 Nyquist diagrams. Influence of additional poles and zeros.

(a) $H(s) = \dfrac{K}{s(T_2s + 1)(T_1s + 1)}$

(b) $H(s) = \dfrac{K}{s^2(T_2s + 1)(T_1s + 1)}$

(c) $H(s) = \dfrac{K}{s(T_3s + 1)(T_2s + 1)(T_1s + 1)}$

(d) $H(s) = \dfrac{K}{s^2(T_3s + 1)(T_2s + 1)(T_1s + 1)}$

(e) $H(s) = K\dfrac{T's + 1}{(T_1s + 1)}$

(f) $H(s) = K\dfrac{T's + 1}{s(T_1s + 1)}$

(g) $H(s) = K\dfrac{T's + 1}{s(T_1 + s)(T_2s + 1)}$

(h) $H(s) = K\dfrac{T's + 1}{s^2(T_1s + 1)}$

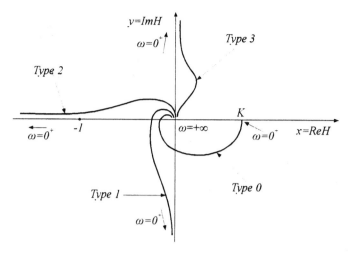

Figure 8.22   Influence of the system's type on the Nyquist diagrams.

In closing the issue of constructing the Nyquist diagram of a system, we point out the following with regard to the influence of the gain K. For first- and second-order systems, the closed-loop system is always stable for all values of K. However, for third- or higher-order systems, the stability is greatly influenced by the gain K. In the general case one may conclude that, as the gain K is increased, the closed-loop system tends to become unstable. This becomes particularly clear for the case of third-order systems for which the open-loop transfer function $G(s)F(s)$ has the form

$$G(s)F(s) = \frac{K}{(T_1 s + 1)(T_2 s + 1)(T_3 s + 1)}$$

From Figure 8.23 it is clear that as K increases the closed-loop system becomes unstable. In particular, when $K = K_1$ the closed-loop system is stable, when $K =$

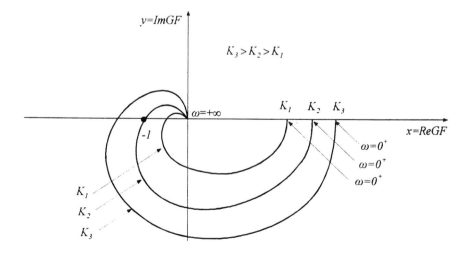

Figure 8.23   The Nyquist diagram of a third-order system as K increases.

K_2 the closed-loop system undergoes sustained oscillations, and for $K = K_3$ the closed-loop system is unstable.

There are, however, cases where the closed-loop system is stable only for certain ranges of values of the gain K. As an example, consider the system having the following open-loop transfer function:

$$G(s)F(s) = \frac{K(T's + 1)^2}{(T_1 s + 1)(T_2 s + 1)(T_3 s + 1)(T_4 s + 1)^2}$$

The Nyquist diagram of $G(s)F(s)$ is given in Figure 8.24. From this figure it follows that the closed-loop system can become unstable not only as K increases but also as K decreases. In this case, the closed-loop system is stable only for certain ranges of values of the gain K.

8.4.5   Gain and Phase Margins

Consider the Nyquist diagrams of Figure 8.25 of a certain open-loop transfer function $G(s)F(s)$ having the form (8.4-3). We introduce the following definitions.

Definitions 8.4.2

Let ω_c be the critical frequency where the Nyquist diagram of $G(s)F(s)$ intersects the Re GF-axis. Then, the *gain margin* K_g of the closed-loop system is given by the following relationship:

$$K_g(\text{dB}) = -20 \log_{10} |G(j\omega_c)F(j\omega_c)| \tag{8.4-21}$$

The physical meaning of the gain margin K_g may be interpreted as follows. The gain margin K_g is the gain in decibels (dB) which the open-loop transfer function is allowed to increase before the closed-loop system becomes unstable. This increase is usually done by increasing the gain K of the open-loop system.

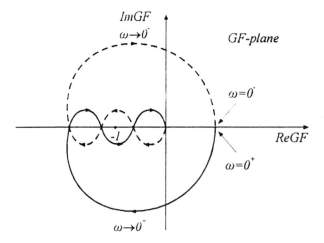

Figure 8.24   The Nyquist diagram of a system which becomes unstable not only as K increases but also as K decreases.

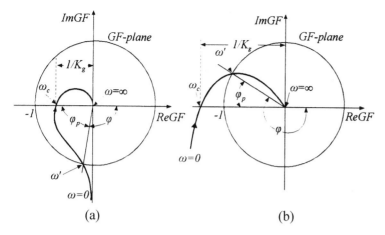

Figure 8.25   Diagrams of the gain and phase margin of (a) a stable and (b) unstable system.

Definition 8.4.3

Let ω' be the critical frequency where the amplitude of the Nyquist diagram of $G(s)F(s)$ is equal to unity, i.e., $|G(j\omega')F(j\omega')| = 1$. Also, let φ be the phase of $G(j\omega')F(j\omega')$. Then, the *phase margin* φ_p of the closed-loop system is given by the following relationship:

$$\varphi_p = 180° + \varphi \tag{8.4-22}$$

The physical meaning of the phase margin φ_p may be interpreted as follows: the phase margin φ_p is the phase in degrees which the Nyquist diagram of $G(j\omega)F(j\omega)$ must rotate about the origin until the point where $|G(j\omega)F(j\omega)| = 1$ passes through the *critical point* $(-1, j0)$. This change in phase is usually done by varying the parameters of the system, as for example the damping ratio, the time constants, etc., and not by varying the gain K of the system.

From Definitions 8.4.2 and 8.4.3 it follows that the phase margin φ_p and the gain margin K_g give an indication of how close the Nyquist diagram of $G(s)F(s)$ is to the critical point $(-1, j0)$. Therefore, φ_p and K_g give pertinent information regarding the *relative stability* of the closed-loop system. For this reason, they constitute important design criteria.

The stability of the closed-loop system may readily be determined by the signs of φ_p and K_g. Indeed, the closed-loop is stable when the margins φ_p and K_g are both positive (Figure 8.25a). On the contrary, a closed-loop system is unstable when one or both margins φ_p and K_g are negative (Figure 8.25b).

The relative stability of a stable closed-loop system is directly related to the values of φ_p and K_g. As the values of φ_p and K_g become greater, the closed-loop systems becomes more stable. That is to say that the greater the margins φ_p and K_g, the less probable it is for the system to become unstable.

All the above results hold for systems whose open-loop transfer function $G(s)F(s)$ is of minimum phase (Definition 8.4.1). However, in cases where $G(s)F(s)$ is of nonminimum phase the above results do not hold because one of the margins may be negative even though the closed-loop system is stable (see Example 8.4.5).

Example 8.4.2

Consider the third-order system with open-loop transfer function

$$G(s)F(s) = \frac{K}{s(T_1 s + 1)(T_2 s + 1)}, \qquad K, T_1, T_2 > 0$$

Construct the Nyquist diagram and study the stability of the closed-loop system.

Solution

For $s \in (j0^+, j\infty)$, we have

$$G(j\omega)F(j\omega) = \frac{K}{j\omega(j\omega T_1 + 1)(j\omega T_2 + 1)} = x + jy$$

where

$$x = -\frac{K(T_1 + T_2)}{1 + \omega^2(T_1^2 + T_2^2) + \omega^4 T_1^2 T_2^2} \qquad \text{and} \qquad y = \frac{K\omega^{-1}(\omega^2 T_1 T_2 - 1)}{1 + \omega^2(T_1^2 + T_n^2) + \omega^4 T_1^2 T_2^2}$$

The point where the Nyquist diagram intersected the real axis is when $y = 0$, i.e., when $K\omega^{-1}(\omega^2 T_1 T_2 - 1) = 0$. The roots of this equation are

$$\omega = \frac{1}{\sqrt{T_1 T_2}}, \qquad \omega = -\frac{1}{\sqrt{T_1 T_2}} \qquad \text{and} \qquad \omega = \infty$$

The negative value of ω is rejected because it has no physical meaning. The transfer function $G(j\omega)F(j\omega)$, for $\omega = \omega_c = 1/\sqrt{T_1 T_2}$, becomes

$$G(j\omega_c)F(j\omega_c) = x = -K\left[\frac{T_1 T_2}{T_1 + T_2}\right]$$

For the closed-loop system to be stable there must be $x > -1$, or equivalently

$$-K\left[\frac{T_1 T_2}{T_1 + T_2}\right] > -1 \qquad \text{or} \qquad 0 < K < \left[\frac{T_1 + T_2}{T_1 T_2}\right]$$

Therefore, the Nyquist diagram is as in Figure 8.26.

Example 8.4.3

Consider the third-order system with open-loop transfer function

$$G(s)F(s) = \frac{K}{s^2(Ts + 1)}, \qquad K, T > 0$$

Construct the Nyquist diagram and study the stability of the closed-loop system.

Solution

For $s \in (j0^+, j\omega)$, we have

$$G(j\omega)F(j\omega) = \frac{K}{-\omega^2(j\omega + 1)} = x + jy$$

where

$$x = -\frac{K}{\omega^2(\omega^2 T^2 + 1)} \qquad \text{and} \qquad y = \frac{KT}{\omega(\omega^2 T^2 + 1)}$$

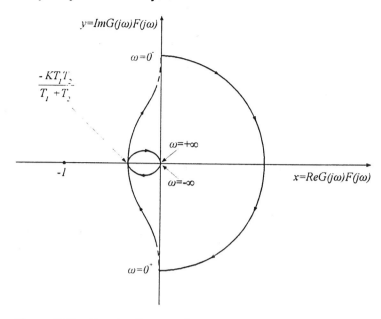

Figure 8.26 Nyquist diagram of Example 8.4.2.

The point where the Nyquist diagram intersects the x-axis is when $y = 0$, i.e., when $\omega \to \infty$. Therefore, the Nyquist diagram has the form of Figure 8.27. Hence there is no value of K for which the closed-loop system is stable, since the Nyquist diagram encircles clockwise the point $(-1, j0)$ permanently.

Example 8.4.4

Consider the numerical control tool machine described in Sec. 1.4 (see Figure 1.19). A simplified block diagram of the closed-loop system is given in Figure 8.28. The

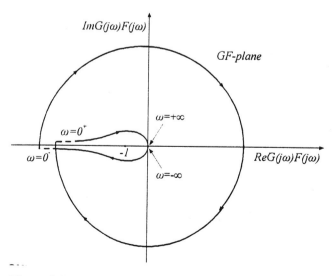

Figure 8.27 Nyquist diagram of Example 8.4.3.

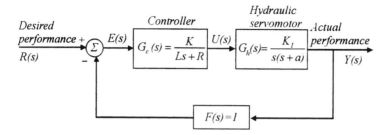

Figure 8.28 Block diagram of the numerical control tool machine.

transfer function $G_h(s)$ of the hydraulic servomotor has been determined in Subsec. 3.12.6. For the servomotor, let $K_1 = 1$ and $a = 1$. For the amplifier controller circuit transfer function $G_c(s)$, let $L = 1$ and $R = 4$. Then, the open-loop transfer function becomes

$$G(s)F(s) = G_c(s)G_h(s)F(s) = \frac{K}{s(s+1)(s+4)}, \qquad K > 0$$

Determine:

(a) The range of values of K for which the closed-loop system is stable
(b) The gain margin K_g when $K = 2$ and when $K = 40$.

Solution

(a) For $s \in (j\omega^+, j\infty)$, we have

$$G(j\omega)F(j\omega) = \frac{K}{j\omega(j\omega+1)(j\omega+4)} = \frac{(K/4)}{j\omega(j\omega+1)(j\omega/4+1)}$$

If we apply the results of Example 8.4.2, we have

$$\omega_c = \frac{1}{\sqrt{T_1 T_2}} = \frac{1}{\sqrt{(1/4)}} = 2$$

Therefore, the closed-loop system is stable when

$$0 < \frac{K}{4} < \frac{T_1 + T_2}{T_1 T_2} \qquad \text{or} \qquad 0 < \frac{K}{4} < \frac{1 + 1/4}{1/4} \qquad \text{or} \qquad 0 < K < 20$$

Indeed, for $\omega = 2$, the Nyquist diagram of $G(j\omega)F(j\omega)$ intersects the real axis and

$$G(j2)F(j2) = -\frac{K}{20}$$

The critical value of K for which $G(j2)F(j2) = -1$, is $K = 20$. Therefore, for $K \in (0, 20)$ the closed-loop system is stable.

(b). For $K = 2$, and upon using definition (8.4-21), we have

$$K_g\,(\mathrm{dB}) = -20\log_{10}\left|-\frac{2}{20}\right| = 20\,(\mathrm{dB})$$

For $K = 40$, we have

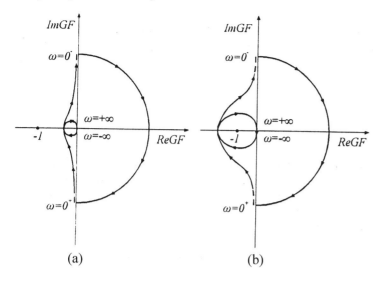

(a)   (b)

Figure 8.29   Nyquist diagram of a machine tool control system. (a) Stable: $K < 20$; (b) unstable: $K > 20$.

$$K_g \,(\mathrm{dB}) = -20\log_{10}\left|-\frac{40}{20}\right| = -20\log 2\,(\mathrm{dB})$$

Therefore, for $K = 2$ the closed-loop system is stable, with gain margin $K_g = 20\,\mathrm{dB}$, whereas when $K = 40$ the closed-loop system is unstable.

In Figure 8.29 the Nyquist diagrams of the open-loop transfer function are given for $K < 20$ and for $K > 20$, where it is clear how the closed-loop system becomes unstable as K increases.

Example 8.4.5

Consider the closed-loop system of Figure 8.30. Determine the range of values of K for which the closed-loop system is stable.

Solution

We have

$$G(s)F(s) = \frac{K}{s-1}, \qquad K > 0$$

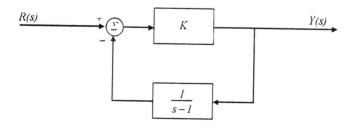

Figure 8.30   The closed-loop block diagram of Example 8.4.5.

Since the open-loop transfer function is of nonminimum phase (it has a pole in the right-half complex plane), it follows that for the closed-loop system to be stable, the Nyquist diagram of $G(s)F(s)$, having the opposite direction than that of the Nyquist path Γ_N, must encircle $N = P = 1$ times the critical point $(-1, j0)$ (Theorem 8.4.4). We next construct the Nyquist diagram. We have

$$G(j\omega)F(j\omega) = \frac{K}{j\omega - 1}$$

If we use the results of case 1 of Subsec. 8.4.4, it follows that the Nyquist diagram will be a circle, as in Figure 8.31. Therefore, since the Nyquist diagram must encircle once the critical point $(-1, j0)$ in the counterclockwise direction, in order that the closed-loop system be stable, it follows that the circle must have a radius greater than $1/2$, i.e., there must be $K > 1$.

Of course, one would arrive at the same results by applying one of the algebraic stability criteria. Indeed, since the denominator of the closed-loop transfer function is given by

$$1 + G(s)F(s) = 1 + \frac{K}{s - 1} = \frac{s - 1 + K}{s - 1}$$

it follows that the characteristic polynomial $p_c(s)$ of the closed-loop system is $p_c(s) = s - 1 + K$. Upon using Routh's criterion, it readily follows that for the closed-loop system to be stable, there must be $K > 1$. Finally, it is worth mentioning that a careful examination of Figure 8.31 shows that for the present example (where the open-loop transfer function $G(s)F(s)$ is nonminimum phase), even though the closed-loop system is stable (i.e., for $K > 1$), the phase margin is positive, whereas the gain margin is negative.

Example 8.4.6

An automatic control system for controlling the thickness of sheet metal is given in Figure 1.10. A simplified block diagram of this system is given in Figure 8.32. Construct the Nyquist diagram and determine the range of values of K for which

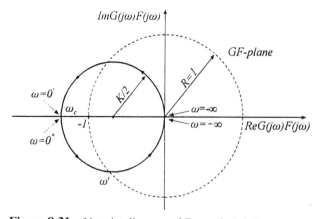

Figure 8.31   Nyquist diagram of Example 8.4.5.

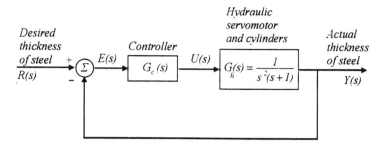

Figure 8.32 The block diagram of the automatic thickness control system.

the closed-loop system is stable. The controller transfer function $G_c(s)$ is specified as follows:

(a) $G_c(s) = K$, i.e., the controller is a gain amplifier
(b) $G_c(s) = K(0.5 + s)$, i.e., when the controller is a PD controller (see Subsec. 9.6.2).

Solution

(a) For this case (i.e., when $G_c(s) = K$), the system is unstable for all values of K (see Example 8.4.3).

(b) For $s \in (j0^+, j\infty)$, we have

$$G(s)F(s) = G_c(j\omega)G_h(j\omega)F(j\omega) = \frac{K(j\omega + 0.5)}{-\omega^2(j\omega + 1)} = x + jy$$

where

$$x = -\frac{K(\omega^2 + 0.5)}{\omega^2(\omega^2 + 1)}, \qquad y = \frac{0.5K}{\omega(\omega^2 + 1)}$$

The point where the Nyquist diagram intersects the x-axis is when $y = 0$, i.e., when $\omega \to \infty$. The Nyquist diagram has the form of Figure 8.33. From this diagram we conclude that the closed-loop system of Figure 8.32 is stable for all values of K in the interval $(0, +\infty)$. We observe that by adding a zero to the open-loop transfer function, which lies to the right of the pole -1 (namely, by adding the zero -0.5 of the controller), the system becomes stable (see also, Figure 8.21h).

8.4.6 Comparison Between Algebraic Criteria and the Nyquist Criterion

We close Sec. 8.4 by making a comparison between the algebraic criteria of Chap. 6 and the Nyquist criterion.

The algebraic criteria have the following characteristics:

a. They require knowledge of the analytical expression of the characteristic polynomial of the system under control
b. The computational effort required to apply the algebraic criteria is extremely small

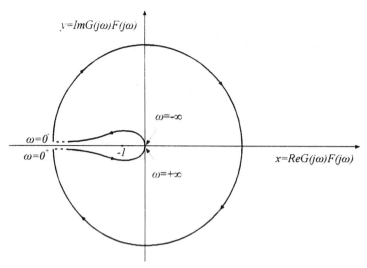

Figure 8.33 Nyquist diagram of Example 8.4.6.

   c.   the algebraic criteria do not give any information on the exact position of
      the poles in the complex plane and, therefore, they do not give any infor-
      mation on the relative stability of the system or its transient response.

The Nyquist criterion has the following characteristics:

   a.   The Nyquist diagram of $G(s)F(s)$ can be determined experimentally and
      relatively easily
   b.   The stability is determined by simply inspecting the Nyquist diagram of the
      open-loop transfer function $G(s)F(s)$
   c.   The relationship between the Nyquist diagram and the amplifier gain K of
      the system is well understood, which allows us to take advantage of the
      influence of the variations of K in order to secure stability of the closed-
      loop system
   d.   The Nyquist diagram gives information on the relative stability and the
      transient response of the system.

8.5 BODE DIAGRAMS

8.5.1 Introduction

Simply speaking, the Bode and the Nyquist diagrams are plots of the transfer function $H(j\omega)$ as a function of the angular frequency ω. The difference between the two
diagrams is that the Nyquist diagram consists of only one curve, while the Bode
diagrams consist of two curves. The two curves of the Bode diagrams are the ampli-
tude M curve of $H(j\omega)$ in decibels, i.e., the curve $A = 20 \log M = 20 \log |H(j\omega)|$ and
the phase φ of $H(j\omega)$, i.e., the curve $\varphi = \angle H(j\omega)$.

     The Bode and Nyquist diagrams essentially offer the same information about
the transfer function $H(j\omega)$. The reason for introducing the Bode diagrams here is

that these diagrams, in comparison with the Nyquist diagrams, can be plotted more easily, a fact that has contributed to the extensive use of Bode diagrams. Both Bode and Nyquist diagrams give information about the stability of the system and its phase and gain margins. They are both rather simple to apply and, for this reason, they have become particularly helpful in control system design.

8.5.2 Bode Diagrams for Various Types of Transfer Function Factors

Usually, the transfer function $H(j\omega)$ involves factors of the form $(j\omega)^{\pm\rho}$, $(j\omega T + 1)^{\pm\rho}$, and $[(j\omega)^2 + 2\zeta\omega_n(j\omega) + \omega_n^2]^{\pm\rho}$. Consider the following transfer function:

$$H(j\omega) = \frac{K(j\omega T_1' + 1)(j\omega T_2' + 1)}{(j\omega)^2(j\omega T_1 + 1)[(j\omega)^2 + 2\zeta\omega_n(j\omega) + \omega_n^2]} \tag{8.5-1}$$

The Bode diagrams of $H(j\omega)$ are the curves of A and φ, where

$$A = 20 \log M = 20 \log |H(j\omega)|$$

$$= 20 \log \frac{|K||j\omega T_1' + 1||j\omega T_2' + 1|}{|(j\omega)^2||j\omega T_1 + 1||(j\omega)^2 + 2\zeta\omega_n(j\omega) + \omega_n^2|}$$

$$= 20 \log |K| + 20 \log |j\omega T_1' + 1| + 20 \log |j\omega T_2' + 1| - 20 \log |(j\omega)^2|$$

$$\quad - 20 \log |j\omega T_1 + 1| - 20 \log |(j\omega)^2 + 2\zeta\omega_n(j\omega) + \omega_n^2| \tag{8.5-2}$$

and

$$\varphi = \underline{/H(j\omega)} = \underline{/K} + \underline{/j\omega T_1' + 1} + \underline{/j\omega T_2' + 1} - \underline{/(j\omega)^2} - \underline{/j\omega T_1 + 1}$$

$$\quad - \underline{/(j\omega)^2 + 2\zeta\omega_n(j\omega) + \omega_n^2} \tag{8.5-3}$$

From Eqs (8.5-2) and (8.5-3) we observe that plotting the curves of A and φ becomes particularly easy because curve A is actually the *sum* of the curves of the individual terms $20 \log |K|$, $20 \log |j\omega T_1' + 1|, \ldots$, etc., and curve φ is actually the *sum* of the curves of the individual terms $\underline{/K}$, $\underline{/j\omega T_1' + 1}, \ldots$, etc. In the sequel we will show that the sketching of each of these individual terms is rather simple.

1 The Constant Term K

In this case we have

$$A = 20 \log |K| \tag{8.5-4a}$$

$$\varphi = \begin{bmatrix} 0°, & \text{when } K > 0 \\ 180°, & \text{when } K > 0 \end{bmatrix} \tag{8.5-4b}$$

The plots of A and φ are given in Figure 8.34.

(a)

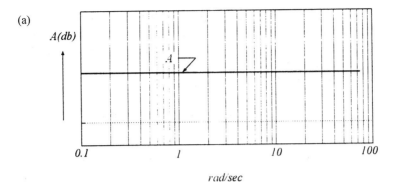

(b)

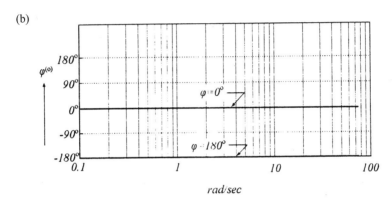

Figure 8.34   The plots of (a) amplitude A and (b) phase φ when $H(s) = K$.

2 Poles or Zeros at the Origin: $(j\omega)^{\pm\rho}$

In this case we have

$$A = 20\log|(j\omega)^{\pm\rho}| = \pm20\rho\log\omega \tag{8.5-5a}$$

$$\varphi = \underline{/(j\omega)^{\pm\rho}} = \pm90°\rho \tag{8.5-5b}$$

Relation (8.5-5a) presents a family of lines on semilogarithmic paper. All these lines meet at the point where $A = 0$ and $\omega = 1$. Their slopes are $\pm20\rho$, in which case we say that these slopes are $\pm20\rho$ dB/decade. This means that if the frequency is increased from ω' to $10\omega'$, the change in amplitude is $\pm20\rho$ dB. Indeed, from relation (8.5-5a), we have

$$\Delta A = A(10\omega') - A(\omega') = 20\log|(j10\omega')^{\pm\rho}| - 20\log|(j\omega')^{\pm\rho}|$$

$$= \pm20\rho\log\omega' \pm 20\rho\log 10 - (\pm20\rho\log\omega') = \pm20\rho$$

The plots of A and φ are given in Figure 8.35.

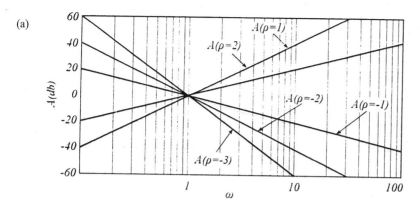

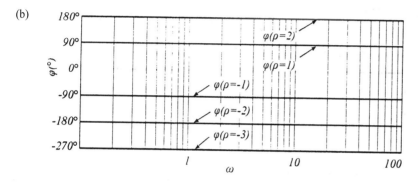

Figure 8.35 The plots of (a) amplitude A and (b) phase φ when $H(s) = s^{\pm\rho}$.

3 Poles or Zeros of the Form $(j\omega T + 1)^{\pm\rho}$

In this case we have

$$A = \pm 20\rho \log |(j\omega T + 1)| = \pm 20\rho \log \sqrt{\omega^2 T^2 + 1} \qquad (8.5\text{-}6a)$$

$$\varphi = \pm \rho \tan^{-1}(\omega T) \qquad (8.5\text{-}6b)$$

Relation (8.5-6a) is a family of curves which may be plotted approximately using the following:

a. When $\omega \ll 1/T$, Eq. (8.5-6a) becomes

$$A = \pm 20\rho \log \sqrt{\omega^2 T^2 + 1} \simeq 20\rho \log 1 = 0 \qquad (8.5\text{-}7a)$$

b. When $\omega = 1/T$, Eq. (8.5-6a) becomes

$$A = \pm 20\rho \log \sqrt{2} \simeq \pm 3\rho \qquad (8.5\text{-}7b)$$

c. When $\omega \gg 1/T$, Eq. (8.5-6a) becomes

$$A \simeq \pm 20\rho \log \omega T \qquad (8.5\text{-}7c)$$

The frequency $\omega = 1/T$ is called the *corner frequency*. Hence, the plot of Eq. (8.5-6a) consists, approximately, of two asymptotes. The first asymptote coincides with the 0 dB axis and holds for $0 \le \omega \le 1/T$. The other asymptote crosses over the ω-axis at the point $\omega = 1/T$, has a slope of $\pm 20\rho$ dB and holds for $1/T \le \omega \le +\infty$. At the corner frequency $\omega = 1/T$, the plot of A, according to relation (8.5-7b), is approximately equal to $\pm 3\rho$ dB.

The plot of phase φ, given by Eq. (8.5-6b), is sketched by assigning several values to ω and calculating the respective values of the phase φ. Some characteristic values of φ are the following:

  a.   When $\omega = 0$, then $\varphi = 0°$

  b.   When $\omega = 1/T$, then $\varphi = \pm 45°\rho$

  c.   When $\omega = \infty$, then $\varphi = \pm 90°\rho$.

Thus, the plot of φ starts from the point zero, passes through the point $\pm 45°\rho$, and terminates asymptotically to the line $\pm 90°\rho$.

The plots of A and φ, when $\rho = \pm 1$, are given in Figure 8.36.

Factors of the Form $[\omega_n^{-2}(j\omega)^2 + 2\zeta\omega_n^{-1}(j\omega) + 1]^{\pm\rho}$

In this case, we have

(a)

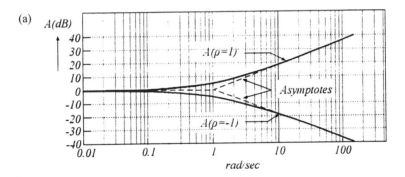

(b)

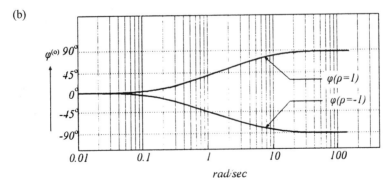

Figure 8.36   The plots of (a) amplitude A and (b) phase φ when $H(s) = (Ts + 1)^{\pm\rho}$.

$$A = \pm 20\rho \log \sqrt{(1-u^2)^2 + 4\zeta^2 u^2} \qquad (8.5\text{-}8a)$$

$$\varphi = \pm \rho \tan^{-1}\left[\frac{2\zeta u}{1-u^2}\right] \qquad (8.5\text{-}8b)$$

where $u = \omega/\omega_n$. The plot of Eq. (8.5-8a) is sketched approximately using the following:

    a.   When $u \ll 1$, then $A \simeq \pm 20\rho \log 1 = 0$
    b.   When $u \gg 1$, then $A \simeq \pm 40\rho \log u$
    c.   When $\zeta = 1$, then $A = \pm 20\rho \log |1 + u^2|$
    d.   When $\zeta = 0$, then $A = \pm 20\rho \log |1 - u^2|$.

Thus, the curve of Eq. (8.5-8a) consists, approximately, of two asymptotes. The first asymptote coincides with the 0 dB-axis and the second has a slope of $\pm 40\rho$ dB and crosses over the u-axis at the point $u = 1$. In the vicinity of the point of intersection of the two asymptotes, the form of the curve of A is decisively influenced by the damping ratio ζ.

The plot of the phase φ, which is given by Eq. (8.5-8b), has the following characteristics:

    a.   When $u = 0$, then $\varphi = 0°$
    b.   When $u = 1$, then $\varphi = \pm 90°\rho$
    c.   When $u = \infty$, then $\varphi = \pm 180°\rho$.

The plots of A and φ, when $\rho = -1$ (which is the most common case), are given in Figure 8.37.

8.5.3 Transfer Function Bode Diagrams

The plots of the amplitude A and the phase φ of a transfer function $H(s)$ start by plotting each factor in A and φ, separately. Subsequently, we add the curves of all factors in A and the curves of all factors in φ, resulting in the curves A and φ sought. This methodology is presented, step by step, in the following example.

Example 8.5.1

Consider the transfer function

$$H(j\omega) = \frac{10(j\omega + 1)}{(j\omega)(10^{-1}j\omega + 1)(3 \times 10^{-3}j\omega + 1)}$$

Plot the Bode diagrams A and φ of $H(j\omega)$.

Solution

With respect to the diagram A, define $A_1 = 20 \log 10 = 20$, $A_2 = 20 \log |j\omega + 1|$, $A_3 = -20 \log |j\omega|$, $A_4 = -20 \log |10^{-1}j\omega + 1|$, and $A_5 = -20 \log |3 \times 10^{-3}j\omega + 1|$. Then, the diagram of the amplitude A is $A = A_1 + A_2 + A_3 + A_4 + A_5$. Thus, in order to plot the curve A, it suffices to plot each curve A_1, A_2, A_3, A_4, and A_5 separately and then add them. To this end, plot the curves A_1, A_2, A_3, A_4, and A_5 by applying the results of Subsec. 8.5.2. The break points of A_2, A_4, and A_5 are 1, 10, and $10^3/3$, respectively. The plots of A_1, A_2, A_3, A_4, and A_5, as well as that of A, are given in Figure 8.38.

(a)

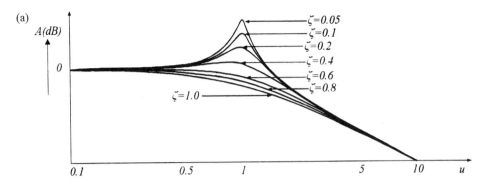

(b)

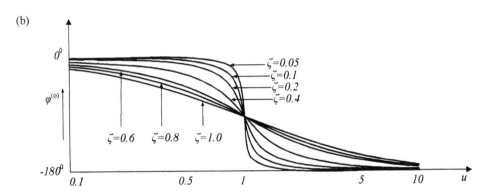

Figure 8.37   The plots of (a) amplitude A and (b) phase φ when $H(s) = [\omega_n^{-2}s^2 + 2\zeta\omega_n^{-1}s + 1]^{-1}$.

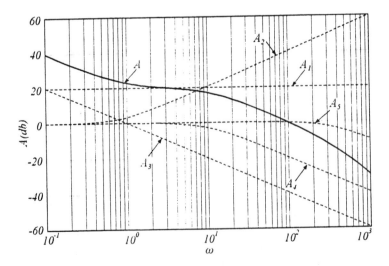

Figure 8.38   The plot of the amplitude A of the transfer function of Example 8.5.1.

With respect to the diagram of φ, define $\varphi_1 = \angle{10} = 0$, $\varphi_2 = \angle{j\omega + 1} = \tan^{-1}\omega$, $\varphi_3 = -\angle{j\omega} = -90°$, $\varphi_4 = -\angle{10^{-1}j\omega + 1} = -\tan^{-1}10^{-1}\omega$, and $\varphi_5 = -\angle{3 \times 10^{-3}j\omega + 1} = -\tan^{-1}3 \times 10^{-3}\omega$. The total phase $\varphi = \varphi_1 + \varphi_2 + \varphi_3 + \varphi_4 + \varphi_5$ is determined following the same steps as for determining A. The plots of φ_1, φ_2, φ_3, φ_4, and φ_5, as well as that of φ, are given in Figure 8.39.

8.5.4 Gain and Phase Margin

Consider the definitions for gain and phase margin given in Eqs (8.4-21) and (8.4-22), respectively. These two definitions are given on the basis of the Nyquist diagram of the open-loop transfer function and are shown in Figure 8.25. Similarly, the gain and phase margins can be defined on the basis of the Bode diagrams. To this end, consider Figure 8.40. Here, the Bode diagrams A and φ of a certain open-loop transfer function $G(j\omega)F(j\omega)$ are depicted. In this figure, when $\omega = \omega'$, then $|G(j\omega')\ F(j\omega')| = 1$ and hence $A(\omega') = 0$, and when $\omega = \omega_c$, then $\varphi(\omega_c) = -180°$. Thus, K_g is the vertical straight line which connects the point ω_c with the curve A, while φ_p is the vertical straight line which connects the point ω' with the curve φ. More specifically, from Figure 8.40, we have

$$K_g = -20 \log |G(j\omega)F(j\omega_c)| = DE \, \text{dB} \tag{8.5-9}$$

$$\varphi_p = 180° + \angle{G(j\omega')F(j\omega')} = 180° + [-180° + (CB)°] = (CB)° \tag{8.5-10}$$

In Figure 8.40, both margins K_g and φ_p are positive. Therefore, for this particular open-loop system, the closed-loop system is stable.

Example 8.5.2

This example refers to the system of automatically adjusting the opening and closing of the pupil of the human eye. A simplified diagram of this system is given in Figure

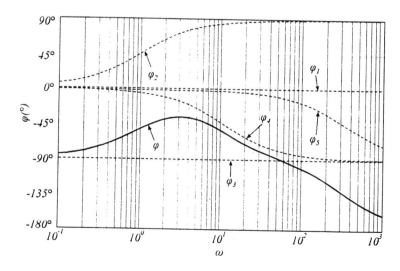

Figure 8.39 The plot of the phase φ of the transfer function of Example 8.5.1.

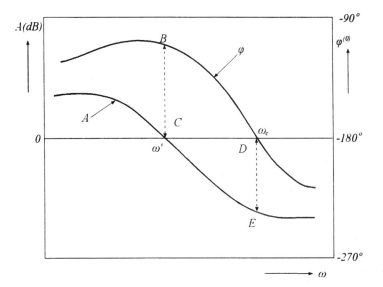

Figure 8.40 Definitions of gain margin and phase margin in Bode diagrams.

8.41. Here, the constant a, which is the inverse of the time constant of the pupil of the eye, is usually 0.5 sec; the constant K is the gain constant of the pupil; and $F(s)$ is the feedback transfer function of the signal from the optic nerve, where T is the time constant, which is usually 0.2 sec. For simplicity, assume that T is zero. Using the Bode diagrams, determine the range of values of K for which the closed-loop system is stable.

Solution

The open-loop transfer function has the form

$$G(j\omega)F(j\omega) = \frac{K}{(j\omega + 0.5)^3} = \frac{8K}{(2j\omega + 1)^3}, \qquad K > 0$$

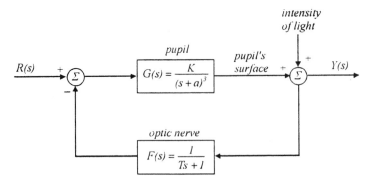

Figure 8.41 Closed-loop system for the automatic control of human vision.

To plot the diagram of the amplitude A, define $A_1 = 20\log(8K)$ and $A_2 = -60\log|2j\omega + 1| = -60\log\sqrt{4\omega^2 + 1}$. The plot of A_2 has a break point at $\omega = 1/2 = 0.5\,\text{rad/sec}$, and is shown in Figure 8.42. The plot of A_1 is a horizontal line which passes through the point $20\log(8K)$. The diagram of the amplitude A is the sum of A_1 and A_2, which means that the diagram A essentially is the plot A_2 moving upwards or downwards depending on the sign of A_1. To plot the diagram of the phase φ, define $\varphi_1 = \underline{/8K} = 0°$ and $\varphi_2 = -3\tan^{-1}(2\omega)$. The diagram of φ is the sum of φ_1 and φ_2, shown in Figure 8.43. The critical frequency ω_c for which the total phase becomes $-180°$ satisfies the following equation

$$\varphi(\omega_c) = -180° \quad \text{or} \quad -3\tan^{-1}(2\omega_c) = -180° \quad \text{or}$$

$$2\omega_c = \tan 60° = \sqrt{3}$$

Hence, the critical frequency is $\omega_c = 0.87\,\text{rad/sec}$. The gain margin K_g is given by

$$K_g = -20\log|G(j\omega_c)F(j\omega_c)| = -20\log(8K) + 60\log\sqrt{4\omega_c^2 + 1}$$

$$= -20\log(8K) + 18$$

For the closed-loop system to be stable, the gain margin K_g and the phase margin φ_p must both be positive. For K_g to be positive, we must have $20\log(8K) < 18$ or $K < 1$. For values of K less than the total amplitude diagram of $G(j\omega)F(j\omega)$ will cross over the 0 dB axis at a frequency whose value is less than ω_c. From Figure 8.43 we conclude that for $\omega < \omega_c$ the phase φ is less (in absolute value) than $-180°$, and hence $\varphi_p > 0$. Therefore, for the closed-loop system to be stable it must hold that $0 < K < 1$.

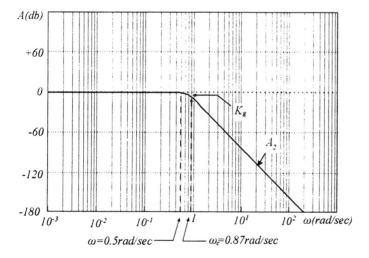

Figure 8.42 The amplitude diagram of a closed-loop system for the automatic control of human vision.

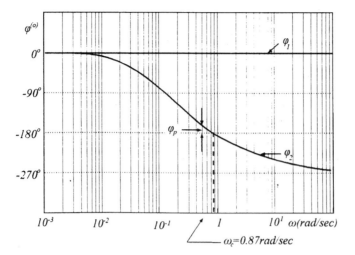

Figure 8.43   The phase diagram of the closed-loop system for the automatic control of human vision.

8.5.5   Bode's Amplitude–Phase Theorem

One of the most important contributions of Bode is the celebrated Bode's amplitude–phase theorem. According to this theorem, for every minimum phase system (e.g., a system without poles or zeros in the right complex plane), the phase $\underline{/G(j\omega)F(j\omega)}$ of the open-loop transfer function $G(j\omega)F(j\omega)$ of the system is related to its amplitude $|G(j\omega)F(j\omega)|$ in a unique manner. The exact expression is

$$\underline{/G(j\omega')F(j\omega')} = \frac{1}{\pi}\int_{-\infty}^{+\infty}\left(\frac{dM}{du}\right)W(u)du \qquad \text{(in radians)}$$

where

$$M = \ln|G(j\omega)F(j\omega)|, \qquad u = \ln\left(\frac{\omega}{\omega'}\right) \qquad \text{and} \qquad W(u) = \ln(\coth|u|/2)$$

Here, ω' is the critical amplitude frequency, where $|G(j\omega')F(j\omega')| = 1$.

The hyperbolic cotangent is defined as $\coth x = (e^x + e^{-x})/(e^x - e^{-x})$. The weighting function $W(u)$ is presented in Figure 8.44. From its form we conclude that the phase of $G(j\omega)F(j\omega)$ depends mainly upon the gradient dM/du at the frequency ω' and to a lesser degree upon the gradient dM/du at neighboring frequencies. If we approximate $W(u)$ with an impulse function at the point ω' and assume that the slope of $G(j\omega)F(j\omega)$ remains constant and is equal to $-20n$ dB/decade for a band of frequencies of about 1 decade above and one below the frequency ω', then we can arrive at the approximate relation $\underline{/G(j\omega')F(j\omega')} \cong -n90°$.

We know that in order to have positive phase margin (stability), $\underline{/G(j\omega')F(j\omega')} > -180°$ at the frequency ω', where $|G(j\omega')F(j\omega')| = 1$. For this reason, if ω' is the desired critical amplitude frequency, it is plausible that the gradient of $|G(j\omega)F(j\omega)|$ is -20 dB/decade ($n = 1$) for 1 decade above and 1 decade below the critical amplitude frequency ω'. Then, according to the approximate

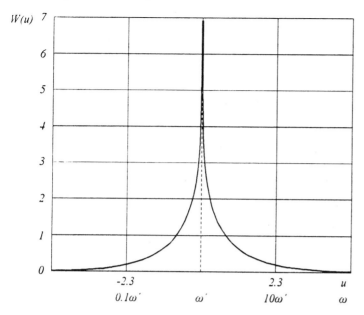

Figure 8.44 The weighting function $W(u) = \ln(\coth |u|/2)$.

relation $\underline{/G(j\omega')}F(j\omega) \cong -90°$, the phase margin is about 90°. To obtain a more desired (smaller) value for the phase margin, it suffices that the gradient of the logarithmic amplitude curve of $G(j\omega)F(j\omega)$ is equal to $-20\,\text{dB/decade}$ for a band of frequencies of 1 decade with center frequency the desired critical frequency ω'.

Example 8.5.3

Using Bode's theorem, design a satisfactory controller for the altitude control of a spaceship, which is described by the transfer function $G_s(s) = 0.9/s^2$. We wish to obtain a satisfactory damping ratio and a bandwidth of about 0.2 rad/sec.

Solution

The block diagram of the compensated closed-loop system is shown in Figure 8.45. The amplitude diagram of the uncompensated system, with transfer function $G_s(s) = 0.9/s^2$, is shown in Figure 8.46, from which it follows that the slope is constant and equal to $-40\,\text{dB/decade}$, due to the double pole at $s = 0$. According

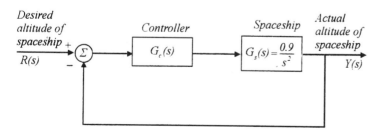

Figure 8.45 The block diagram of the spaceship altitude control system.

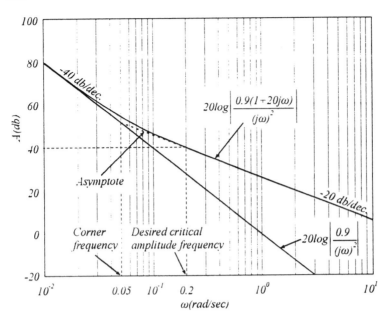

Figure 8.46 Bode amplitude diagrams for $0.9/s^2$ and $0.9(1 + 20s)/s^2$.

to the design requirements, we should obtain a constant slope of -20 dB/decade for a decade of frequencies near the desired amplitude frequency.

Thus, it is obvious that the controller must add, in a band of frequencies near the desired critical amplitude frequency, a slope of $+20$ dB/decade. This means that the controller $G_c(s)$ must be of the form $G_c(s) = K(Ts + 1)$, which is a PD controller (see Subsec. 9.6.2). This controller adds a zero, which must yield a gradient -20 dB/ decade near the critical amplitude frequency, as well as an amplification K, which must yield the desired bandwidth. At first, we assume that the criical frequency and the bandwidth are the same for the system, a fact which will be checked later. Since we wish a bandwidth (and hence a critical amplitude frequency) of 0.2 rad/sec, we choose the corner frequency $1/T$ to be four times lower than the desired critical frequency, e.g., we choose $T = 20$. This is done in order to keep the slope -20 dB/ decade for frequencies lower than 0.2 rad/sec. Hence, the open-loop transfer function has the form

$$G(s)F(s) = G_c(s)G_s(s) = (20s + 1)\left[\frac{0.9}{s^2}\right]$$

The amplitude curve of $G(s)F(s)$ is given in Figure 8.46. From this figure, it follows that the gradient is -20 dB/decade in the band of frequencies from 0.1 to 1 rad/sec (one decade of frequencies near the desired critical frequency 0.2 rad/sec). However, for $\omega = 0.2$ rad/sec, the amplitude of $G(s)F(s)$ is

$$20 \log \left|\frac{0.9(1 + 20j\omega)}{(j\omega)^2}\right|_{\omega=0.2} \cong 39.3 \text{ dB}$$

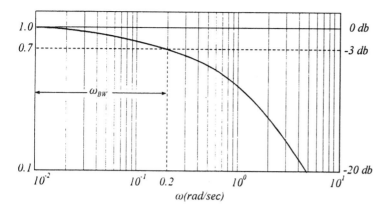

Figure 8.47   Frequency response of the closed-loop system.

Thus, if we choose $20 \log K = -39.3 \, \text{dB}$, which yields $K = 0.0108$, it follows that the frequency $\omega = 0.2 \, \text{rad/sec}$ is the critical frequency of the compensated system. At this point, our design is completed. Checking the results, from Figure 8.47 it follows that it is true that the critical frequency and the bandwidth are the same. If we further draw the phase curve of the open-loop system transfer function $G(s)F(s)$, we will have that the phase margin is $75°$, which is quite satisfactory.

8.6   NICHOLS DIAGRAMS

8.6.1   Consant Amplitude Loci

Consider a closed-loop system with unity feedback transfer function, i.e., with $F(s) = 1$. The transfer function of the closed-loop system is given by

$$H(s) = \frac{G(s)}{1 + G(s)} \tag{8.6-1}$$

In the sinusoidal steady state, expression (8.6-1) becomes

$$H(j\omega) = \frac{G(j\omega)}{1 + G(j\omega)} = |H(j\omega)| \ \underline{/H(j\omega)} = M \ \underline{/\varphi} \tag{8.6-2}$$

Let

$$G(j\omega) = \text{Re}\, G(j\omega) + j\text{Im}\, G(j\omega) = x + jy \tag{8.6-3}$$

Then, the amplitude M may be written as

$$M = \frac{|G(j\omega)|}{|1 + G(j\omega)|} = \frac{\sqrt{x^2 + y^2}}{\sqrt{(1 + x)^2 + y^2}} \tag{8.6-4}$$

Equation (8.6-4) is the mathematical expression which defines the locus of constant amplitude M in the $G(j\omega)$-plane. This locus is the circumference of a circle. Indeed, Eq. (8.6-4) can be written in the well-known form of a circle, as follows

$$\left[x - \frac{M^2}{1 - M^2}\right]^2 + y^2 = \left[\frac{M}{1 - M^2}\right]^2 \tag{8.6-5}$$

Consequently, for any consant M, relation (8.6-5) represents the circumference of a circle with at center the point (x_c, y_c) and radius R, where

$$x_c = \frac{M^2}{1 - M^2}, \qquad y_c = 0, \qquad \text{and} \qquad R = \left|\frac{M}{1 - M^2}\right|$$

For the particular value of $M = 1$, Eq. (8.6-5) is not defined. However, from Eq. (8.6-4) we can obtain that for $M = 1$, the constant M locus is the straight line.

$$x = -\frac{1}{2} \tag{8.6-6}$$

Typical curves of the loci (8.6-5) and (8.6-6) are given in Figure 8.48, where we can observe that the circles of constant amplitude M are symmetrical to the lines $y = 0$ and $x = -1/2$. To the left of the line $x = -1/2$ are the circles with $M > 1$ and to the right are the circles with $M < 1$. At the boundary values, i.e., when $M \to \infty$ and $M \to 0$, the radii tend to zero, which means that the circles degenerate to the points $(-1, 0)$ and $(0, 0)$, respectively.

The above results are very useful for correlating the curve of the open-loop transfer function $G(j\omega)$ in the $G(j\omega)$-plane with the curve of the amplitude $M(\omega)$ of $H(j\omega)$. This correlation is given in Figure 8.49, where it is shown that the tangent point of the two curves is also the resonant point, which means that the curve $M(\omega)$ reaches its maximum value M_p for the frequency for which the curve $G(j\omega)$ is tangent to the circumference of the circle with constant M and amplitude equal to M_p.

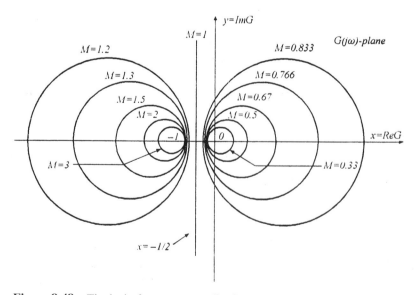

Figure 8.48 The loci of constant amplitude M.

(a)

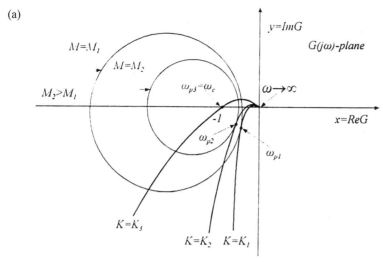

(b)

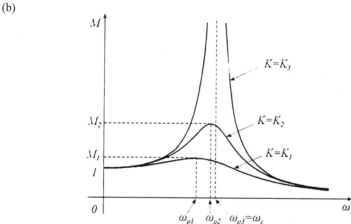

Figure 8.49   The correlation between the amplitude M curves and the corresponding $G(j\omega)$ curves. (a) Constant amplitude M circles and the curves of three open-loop transfer functions; (b) amplitude curves of three open-loop transfer functions.

8.6.2   Constant Phase Loci

Using relation (8.6-3), the phase φ of $H(j\omega)$ may be written as

$$\varphi = \underline{/H(j\omega)} = \tan^{-1}\left[\frac{y}{x}\right] - \tan^{-1}\left[\frac{y}{1+x}\right] \tag{8.6-7}$$

Next, take the tangent of both sides of Eq. (8.6-7) to yield

$$\tan\varphi = \tan\left\{\tan^{-1}\left[\frac{y}{x}\right] - \tan^{-1}\left[\frac{y}{1+x}\right]\right\} = \frac{y}{x^2 + x + y^2} \tag{8.6-8}$$

If we set $N = \tan\varphi$ in Eq. (8.6-8), we have

$$x^2 + x + y^2 - \frac{y}{N} = 0$$

The above relation may be written as

$$\left[x + \frac{1}{2}\right]^2 + \left[y - \frac{1}{2N}\right]^2 = \frac{N^2 + 1}{4N^2} \qquad (8.6\text{-}9)$$

Equation (8.6-9) represents a circle with center at the point (x_c, y_c) and radius R, where

$$x_c = -\frac{1}{2}, \qquad y_c = \frac{1}{2N}, \qquad \text{and} \qquad R = \sqrt{\frac{N^2 + 1}{4N^2}}$$

Figure 8.50 shows the circles of Eq. (8.6-9). As in the case of Figure 8.48, the circles of constant phase $N = \tan \varphi$ are symmetrical to the lines $y = 0$ and $x = -1/2$.

8.6.3 Constant Amplitude and Phase Loci: Nichols Charts

In Subsec. 8.6.1 and 8.6.2 we studied the constant amplitude and phase curves, respectively, of a closed-loop system with unity feedback in the $G(j\omega)$-plane. The Nichols diagrams, which are presented in the sequel, are curves with the following coordinates: the y-axis is the amplitude $|G(j\omega)|$, in dB, and the x-axis is the phase $\theta = \angle G(j\omega)$, in degrees.

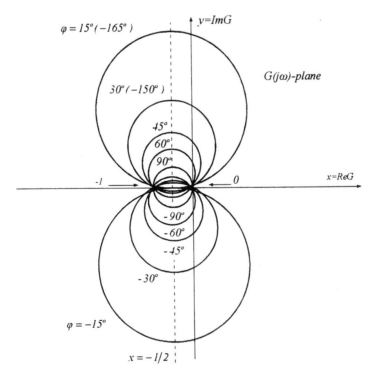

Figure 8.50   The loci of constant phase φ.

1 Curves with Constant Amplitude M

From Relation (8.6-2), we have

$$M = \frac{|G(j\omega)|}{|1 + |G(j\omega)|e^{j\theta}|}, \qquad \theta = \underline{/G(j\omega)}$$

or

$$M^2 = \frac{|G(j\omega)|^2}{1 + 2|G(j\omega)| \cos\theta + |G(j\omega)|^2}$$

The above relation may also be written as

$$|G(j\omega)|^2 + \left[\frac{2M^2}{M^2 - 1}\right]|G(j\omega)| \cos\theta + \frac{2M^2}{M^2 - 1} = 0 \qquad (8.6\text{-}10)$$

For every value of the amplitude M, relation (8.6-10) is a curve whose coordinates are $|G(j\omega)|$ (in dB) and θ (in degrees), as shown in Figure 8.51. These curves are called curves of constant M of the closed-loop system.

2 Curves with Constant Phase φ

From relation (8.6-2), we have

$$\varphi = \underline{/G(j\omega)} - \underline{/1 + G(j\omega)} = \underline{/G(j\omega)} - \underline{/1 + |G(j\omega)| \cos\theta + j|G(j\omega)| \sin\theta}$$

$$= \theta - \tan^{-1}\left\{\frac{|G(j\omega)| \sin\theta}{1 + |G(j\omega)| \cos\theta}\right\} = \theta - \psi$$

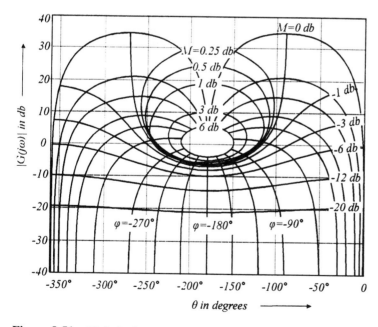

Figure 8.51   Nichols charts: curves of constant amplitude M and phase φ.

Examining the relation $N = \tan \varphi$, we have

$$N = \tan \varphi = \tan(\theta - \psi) = \frac{\tan \theta - \tan \psi}{1 + \tan \theta \tan \psi} = \frac{\tan \theta - \dfrac{|G(j\omega)| \sin \theta}{1 + |G(j\omega)| \cos \theta}}{1 + \dfrac{|G(j\omega)| \sin \theta \tan \theta}{1 + |G(j\omega)| \cos \theta}}$$

$$= \frac{\sin \theta[1 + |G(j\omega)| \cos \theta] - |G(j\omega)| \sin \theta \cos \theta}{\cos \theta + |G(j\omega)| \cos^2 \theta + |G(j\omega)| \sin^2 \theta} = \frac{\sin \theta}{\cos \theta + |G(j\omega)|}$$

Finally, we arrive at the relation

$$|G(j\omega)| + \cos \theta - \frac{1}{N} \sin \theta = 0 \qquad\qquad (8.6\text{-}11)$$

For every value of the phase φ, relation (8.6-11) is a curve whose coordinates are $|G(j\omega)|$ (in dB) and θ (in degrees), as shown in Figure 8.51. These curves are called curves of constant N (or φ) of the closed-loop system.

3 Closed-Loop System Response Curves

The curves of constant M and φ of Figure 8.51 are essentially the same curves as those of constant M and φ of Figures 8.48 and 8.50, except for the fact that Figure 8.51 has coordinates the amplitude $|G(j\omega)|$ and phase θ of $G(j\omega)$, while Figures 8.48 and 8.50 have coordinates $\mathrm{Re}\, G(j\omega)$ and $\mathrm{Im}\, G(j\omega)$. The basic advantage of Nichols charts, i.e., of the diagrams of Figure 8.51, is that for every change of the gain constant K, the response curve of the closed-loop system moves upwards or downwards without affecting the shape of the response curve. The gain margin K_g and phase margin φ_p in Nichols charts are defined in Figure 8.52. In this figure both K_g and φ_p are positive. A comparison among Nyquist, Bode, and Nichols diagrams, as far as the gain and phase margins are concerned, is given in Figure 8.53. Finally, an example, where there is correlation between Nichols charts and the response curve of M and φ is given in Figure 8.54. In particular, in Figure 8.54a the Nichols curve of a transfer function $G(j\omega)$ is given. This curve is plotted by calculating the value for each coordinate $|G(j\omega)|$ in dB and $\theta = \underline{/G(j\omega)}$ for different values of the frequency

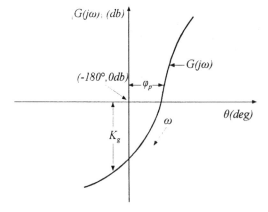

Figure 8.52   The gain margin K_g and the phase margin φ_p.

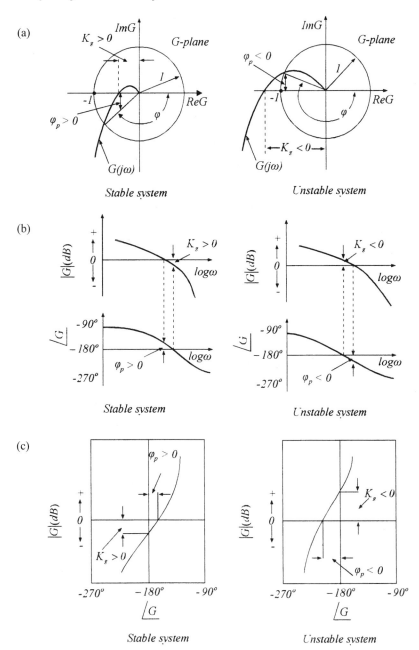

Figure 8.53 Comparison among (a) Nyquist, (b) Bode, and (c) Nicholas diagrams for the gain and phase margins.

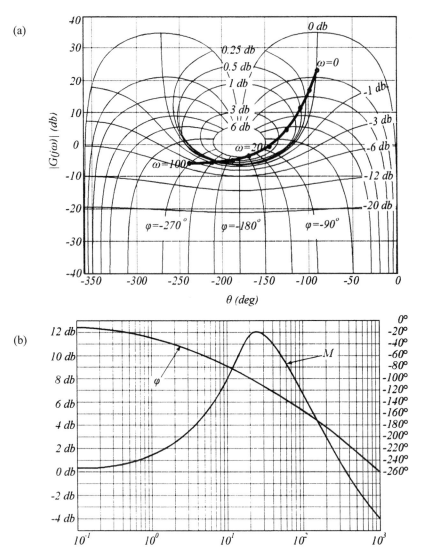

Figure 8.54 Curves of constant amplitude M and constant phase φ. (a) The Nichols curve of the transfer function $G(j\omega)$; (b) the curves M and φ of the transfer function of the closed-loop system.

ω. In Figure 8.54b, the Bode curves of a closed-loop system are given. These curves are plotted as follows. For each value of the frequency ω, the Nichols crosses over the curves of M and φ. By using several values of ω, we find the respective values of M and φ in Figure 8.54a and, subsequently, we transfer these values to Figure 8.54b. By joining these different values of M and φ, we obtain the curves of Figure 8.54b.

Remark 8.6.1

The Nyquist and Nichols diagrams of $G(j\omega)$ have the common characteristic in that they both consist of only one curve, with the frequency ω as a free parameter.

However, they differ in that the Nyquist diagrams have coordinates $\mathrm{Re}\,G(j\omega)$ and $\mathrm{Im}\,G(j\omega)$, while the Nichols diagrams have coordinates $|G(j\omega)|$ and $\angle G(j\omega)$. On the contrary, Bode diagrams consists of two separate curves: the amplitude curve and the phase curve. Both of these curves are functions of the frequency ω.

In Figure 8.55 we present, for some typical transfer functions, the root loci and the Nyquist diagrams. These diagrams are worth studying because they facilitate the comparison of the basic concepts developed in Chaps 7 and 8, in relation to the very popular classical control design tools in the frequency domain: namely, the root

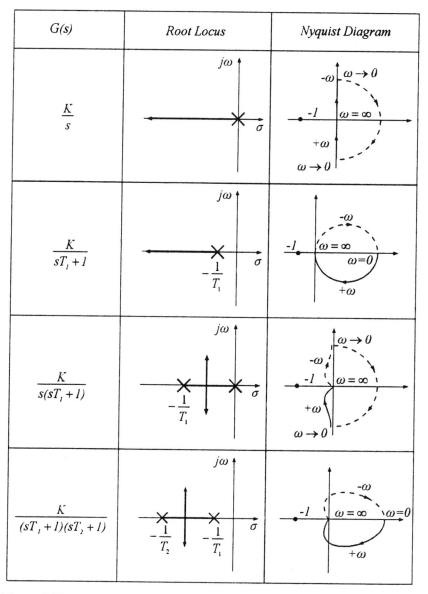

$G(s)$	Root Locus	Nyquist Diagram
$\dfrac{K}{s}$		
$\dfrac{K}{sT_1 + 1}$		
$\dfrac{K}{s(sT_1 + 1)}$		
$\dfrac{K}{(sT_1 + 1)(sT_2 + 1)}$		

Figure 8.55 Root loci and Nyquist diagrams for typical transfer functions (*continued*).

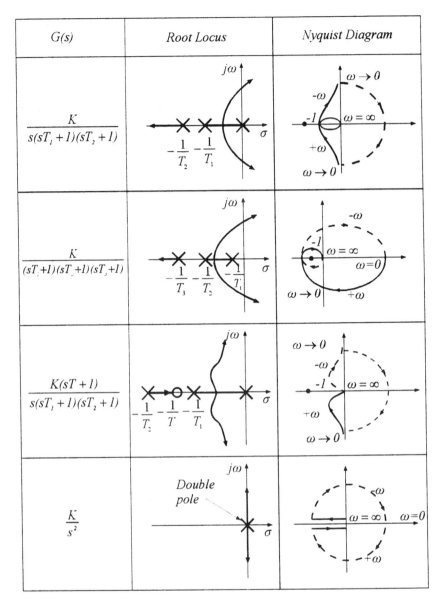

Figure 8.55 *(contd.)*

locus and the Nyquist diagrams (the Bode diagrams may readily be derived from the Nyquist diagrams).

Remark 8.6.2

For simplicity, in the presentation of Nichols charts it was assumed that $F(s) = 1$, in which case the open-loop transfer function is $G(s)$. When $F(s) \neq 1$, we have the more general case where

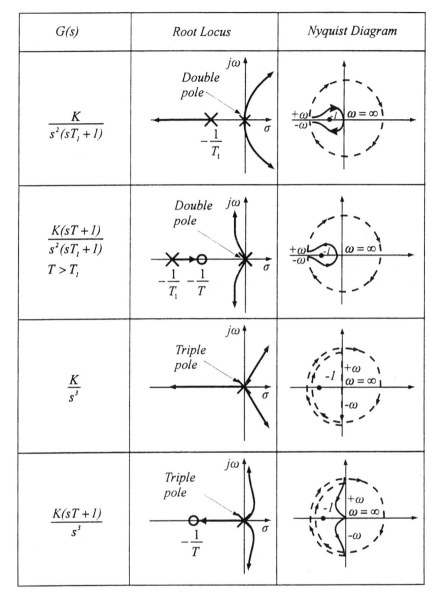

$G(s)$	Root Locus	Nyquist Diagram
$\dfrac{K}{s^2(sT_1+1)}$		
$\dfrac{K(sT+1)}{s^2(sT_1+1)}$ $T > T_1$		
$\dfrac{K}{s^3}$		
$\dfrac{K(sT+1)}{s^3}$		

Figure 8.55 (*contd.*)

$$H(s) = \frac{G(s)}{1+G(s)F(s)} = \left[\frac{1}{F(s)}\right]\left[\frac{G(s)F(s)}{1+G(s)F(s)}\right] = \left[\frac{1}{F(s)}\right]\left[\frac{G^*(s)}{1+G^*(s)}\right] = \frac{H^*(s)}{F(s)}$$

From the above relation we conclude that when $F(s) \neq 1$, we can directly apply the results of the present section for $H^*(s)$, as long as, instead of $G(s)$, we set $G^*(s) = G(s)F(s)$. To determine $H(s) = H^*(s)F^{-1}(s)$, we must multiply the two transfer functions $H^*(s)$ and $1/F(s)$. This can be done easily by using Bode diagrams.

PROBLEMS

1. Plot the Nyquist diagrams and investigate the stability of the closed-loop systems for the following open-loop transfer functions:

(a) $G(s)F(s) = \dfrac{K(s-1)}{s(s+1)}$

(d) $G(s)F(s) = \dfrac{K(s+3)}{s(1+s)(1+2s)}$

(b) $G(s)F(s) = \dfrac{10K(s+2)}{s^3 + 3s^2 + 10}$

(e) $G(s)F(s) = \dfrac{K(s-3)}{s(1+s)(1+2s)}$

(c) $G(s)F(s) = \dfrac{Ks}{1-0.5s}$

(f) $G(s)F(s) = \dfrac{K}{s^2(1+s)(1+2s)}$

2. Find the gain and phase margins of the systems of Problem 1.
3. The block diagram of a laser beam control system used for metal processing is shown in Figure 8.56. Plot the Nyquist diagram for $K > 0$ and investigate the stability of the system.
4. The block diagram of a position control system for a space robot arm is given in Figure 8.57. Determine the stability of the system using the Nyquist diagram, for $K > 0$.
5. Let $G(s)F(s) = K(Ts+1)/s^2$. Plot the Nyquist diagram and determine the value of T so that the phase margin is 45°.
6. Consider a field-controlled DC motor represented by the block diagram in Figure 8.58. Draw the Nyquist diagram. Furthermore:

(a) Determine the gain K so that the gain margin is 20 dB
(b) Determine the value of K so that the phase margin is 60°.

7. Plot the Bode diagrams and determine the gain and phase margins of the systems having the following open-loop transfer functions:

(a) $G(s)F(s) = \dfrac{s+1}{0.1s+1}$

(d) $G(s)F(s) = \dfrac{0.1s+1}{s^2(s+1)(0.2s+1)^2}$

(b) $G(s)F(s) = \dfrac{10}{s^2(s+1)}$

(e) $G(s)F(s) = \dfrac{100}{s^2(s^2+2s+1)}$

(c) $G(s)F(s) = \dfrac{s+1}{s(0.1s+1)}$

(f) $G(s)F(s) = \dfrac{(s+1)^2}{s(0.1s+1)^3(0.01s+1)}$

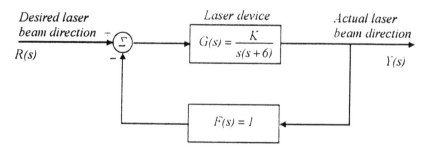

Figure 8.56

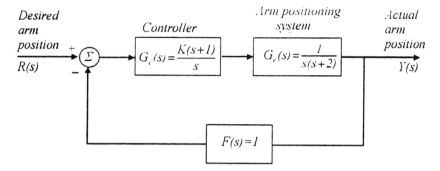

Figure 8.57

8. The block diagram of the orientation control system of a space telescope is shown in Figure 8.59. Determine the value of the gain K for which the phase margin is 50°. Find the gain margin for this case.

9. Consider the field-controlled DC motor system of Problem 6.

   (a) For $K = 4$, plot the Bode diagram of the system, find the phase-crossover and gain-crossover frequencies and determine the gain margin and the phase margin. Is the system stable?

   (b) Determine the value of K for which the phase margin is 50°.

   (c) Find the value of K so that the gain margin is 16 dB.

10. The Bode diagram of a system is given in Figure 8.60. Determine the transfer function of the system.

11. Determine the transfer function $G(s)$ of a system, based on the measurement data shown in Table 8.1.

12. For $K = 1$, plot the Nichols diagram for the unity feedback systems having

   (a) $G(s) = \dfrac{K(s + 1)}{s(0.1s + 1)(0.01s + 1)}$,   (b) $G(s) = \dfrac{K}{s^2(s + 1)}$

   Plot the response of the closed-loop systems and find the values of K for which $M_p = 1.3$.

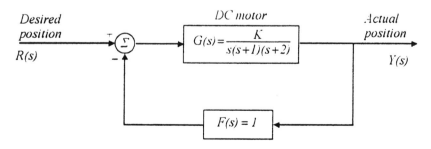

Figure 8.58

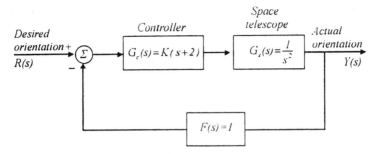

Figure 8.59

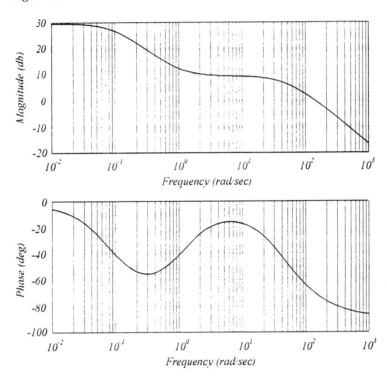

Figure 8.60

Table 8.1

ω	$|G(j\omega)|$	φ°
0.1000	0.0481	−0.2204
0.2154	0.0481	−0.4750
0.4642	0.0482	−1.0249
1.0000	0.0485	−2.2240
2.1544	0.0501	−4.9571
4.6416	0.0592	−12.6895
10.0000	0.1244	−84.2894
21.5443	0.0135	−166.5435
46.4159	0.0024	−174.8261
100.0000	0.0005	−177.6853

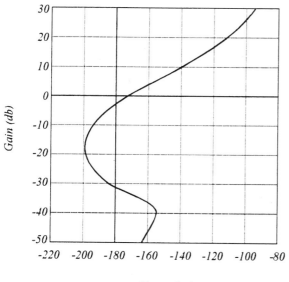

Phase (deg)

Figure 8.61

13. Consider the open-loop transfer function

$$G(s)F(s) = \frac{K(0.25s^2 + 0.5s + 1)}{s(1 + 2s)^2(1 + 0.25s)}$$

The Nichols diagram is shown in Figure 8.61, for $K = 1$. Determine the gain K so that the gain margin is at least 10 dB and the phase margin is at least 45°.

14. The Nichols chart of a system is shown in Figure 8.62. Using the data given below determine: (a) the resonance peak M_p in dB, (b) the resonant frequency, (c) the bandwidth, (d) the phase margin, and (e) the gain margin of the system.

Angular frequency	ω_1	ω_2	ω_3	ω_4
rad/sec	1	3	6	10

15. For the control system shown in Figure 8.63, plot the Nyquist, Bode, and Nichols diagrams, the constant M and φ loci, and the curves $M(\omega)$ and $\varphi(\omega)$, for $K = 1$, 10, and 100. Comment on the results by comparing the diagrams.

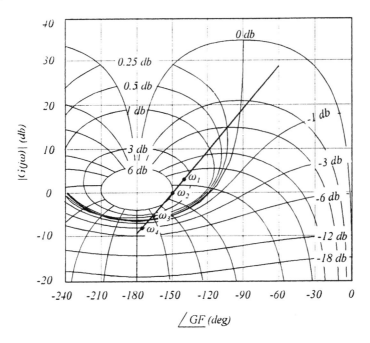

Figure 8.62

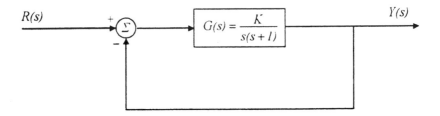

Figure 8.63

BIBLIOGRAPHY

Books

1. JJ D'Azzo, CH Houpis. Linear Control System Analysis and Design, Conventional and Modern. New York: McGraw-Hill, 1975.
2. JJ DiStefano III, AR Stubberud, IJ Williams. Feedback and Control Systems. Schaum's Outline Series. New York: McGraw-Hill, 1967.
3. RC Dorf, RE Bishop. Modern Control Analysis. London: Addison-Wesley, 1995.
4. GF Franklin, JD Powell, A Emami-Naeini. Feedback Control of Dynamic Systems. Reading, MA: Addison-Wesley, 1986.
5. NS Nise. Control Systems Engineering. New York: Benjamin and Cummings, 1995.
6. K Ogata. Modern Control Systems. London: Prentice Hall, 1997.

9

Classical Control Design Methods

9.1 GENERAL ASPECTS OF THE CLOSED-LOOP CONTROL DESIGN PROBLEM

The problem of designing an open- or closed-loop control system can be stated using Figure 9.1 as follows (see also Subsec. 1.3): given the system G under control and its desirable behavior $y(t)$, find an appropriate controller G_c such that the composite system (i.e., the system composed of the controller G_c plus the system G) yields the desired output $y(t)$.

A general form of the controller G_c in closed-loop systems is given in Figure 9.2, where G_1 and G_2 are the controllers in "series" and F is the controller in "parallel" or the feedback loop controller. In practice, in most case, various combinations of G_1, G_2, and F are used, as for example G_2 alone, F alone, G_1 and F in pair, G_2 and F in pair, etc.

For the design of the controller G_c, many methods have been developed, which may be distinguished in two categories: the *classical* and the *modern*. The classical methods are based mainly on the root locus techniques and the Nyquist, Bode, and Nichols diagrams. These methods are graphical and they are developed in the frequency domain. The advantage of the classical methods is that they are rather simple to apply. However, they have certain disadvantages. One disadvantage is that classical methods can be applied to SISO systems. In recent years, major efforts have been made to extend many of the SISO classical methods to MIMO systems. Another disadvantage is that in many cases, due to their graphical nature, these methods do not give the necessary and sufficient conditions which must be satisfied for the design problem to have a solution. This means that in situations where the design requirements cannot be satisfied, the designer will be searching in vain for the solutions of the problem.

In contrast to classical methods, modern control design methods can be characterized as analytical and are mostly developed in the time domain. Necessary and sufficient conditions are established for the design problem to have a solution. Many of these methods are based upon the idea of minimizing a cost function (or performance index). In particular, one of the major problems of modern control theory can be formulated on the basis of Figure 9.3, as follows: we are given the system G whose

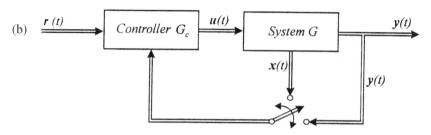

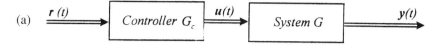

Figure 9.1   Block diagrams of (a) open- and (b) closed-loop control systems.

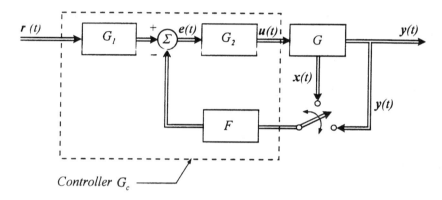

Figure 9.2   Typical structure of a closed-loop system.

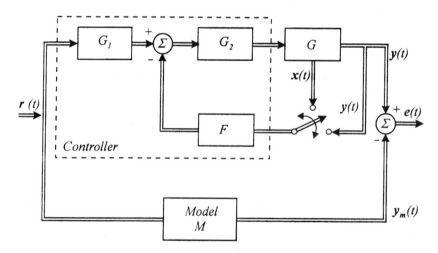

Figure 9.3   Block diagram of a closed-loop system with a reference model.

behavior is considered unsatisfactory. We are also given a second system, system M, whose behavior is considered ideal. This system M is called the reference model or simply the model. Find an appropriate controller G_c (i.e., find G_1, G_2, and F) which minimizes a specific cost function. The cost function is designated by the letter J and its typical form is given by

$$J = \lim_{T \to \infty} \frac{1}{T} \int_0^T e^T(t) e(t \, dt \qquad (9.1\text{-}1)$$

where $e(t) = y(t) - y_m(t)$ is the error between the desired behavior (output) $y_m(t)$ of the reference model and the actual behavior (output) $y(t)$ of the given system. It is clear that the solution of Eq. (9.1-1) is the optimal solution to the problem. The field of modern control engineering which is based upon the minimization of cost functions is called optimal control. Chapter 11 constitutes an extensive introduction to this very important approach of optimal control.

   To give a simple comparison between classical and optimal design methods, the following example is presented.

Example 9.1.1.

Consider the closed-loop system of Figure 9.4. Let $r(t) = 1$. Find the appropriate value of the parameter K which minimizes the cost function

$$J = \int_0^\infty |e(t)| \, dt \qquad (9.1\text{-}2)$$

Solution

To facilitate the understanding of the main idea of optimal control, no strict mathematical proof will be given here, but rather a simple and practical explanation of the solution of the problem. In Figure 9.5a the output $y(t)$ is given for various values of K. In Figures 9.5b and 9.5c the waveforms of $|e(t)|$ and $J(K)$ are given, where $J(K)$ is the cost function (9.1-2) with the amplification constant K as a parameter. From these figures we conclude that the optimal control approach guarantees the optimal solution K_3. If a classical method were applied, the resulting solution for K would be, in general, different from the optimal solution K_3.

   This chapter is devoted to the classical control design methods. In particular, we will present control design techniques using proportional controllers and PID controllers, i.e., controllers consisting of three terms: P(proportional), I(integral), and D(derivative). Also, use of special types of circuits, such as phase-lead, phase-

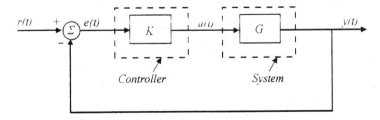

Figure 9.4   A closed-loop system with a proportional controller.

(a)

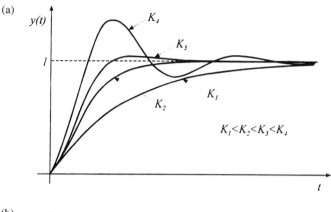

(b)

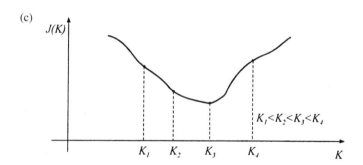

(c)

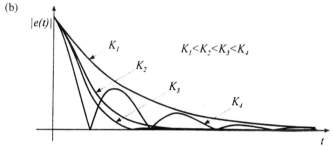

Figure 9.5 Waveforms of (a) $y(t)$, (b) $|e(t)|$, and (c) $J(K)$ of Example 9.1.1.

lag, and phase lag-lead circuits will be used for closed-loop system compensation. At the end of the chapter we give a brief description of certain quite useful classical methods of optimal cotnrol, which preceded the modern advanced techniques of optimal control presented in Chap. 11.

Chapters 10 and 11 give an introduction to modern state-space control techniques. Specifically, in Chap. 10 the following very important algebraic control design techniques are presented: eigenvalue assignment, input–output decoupling, exact model-matching, and state observers. Chapter 11 gives an introduction to optimal control covering the well-known problems of optimal regulator and optimal servomechanism. Further material on even more recent results on modern control design techniques are presented in the remaining chapters (Chaps 12–16).

9.2 GENERAL REMARKS ON CLASSICAL CONTROL DESIGN METHODS

As already mentioned in Sec. 9.1, the classical control design methods are mainly graphical and as a result they are mostly based upon experience. A typical example of system design with classical control methods is the closed-loop system of Figure 9.6a. Assume that the performance of the given system $G(s)$ is not satisfactory, e.g., assume that the output $y(t)$ is slower than expected. To improve its performance, we introduce the controller $G_c(s)$. Suppose that the transfer function $G(j\omega)$ of the system under control is tangent to the circumference of the circle of constant M at the frequency ω_1 (Figure 9.6b). To improve the performance, we choose the dynamic

(a)

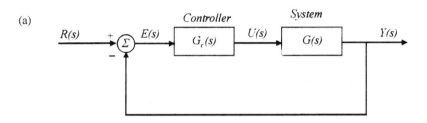

(b)

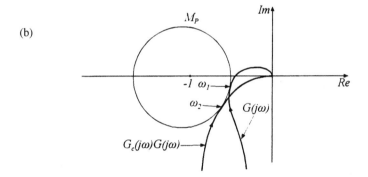

(c)

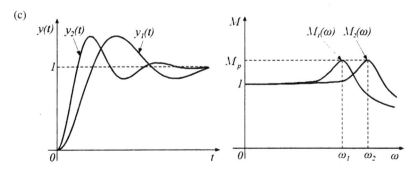

Figure 9.6 Performance improvement of a closed-loop system using a dynamic controller. (a) Unity feedback closed-loop system; (b) Nyquist diagrams of $G(j\omega)$ and $G_c(j\omega)G(j\omega)$; (c) time response of an open- and closed-loop system; (d) frequency response of an open- and closed-loop system.

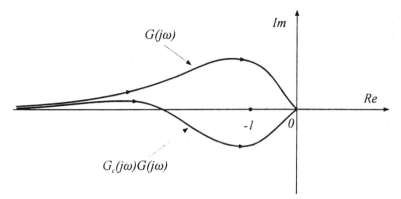

Figure 9.7   Nyquist diagrams of $G(j\omega)$ and $G_c(j\omega)G(j\omega)$.

controller $G_c(j\omega)$ such that $G_c(j\omega)G(j\omega)$ is tangent to the same circle M but at the frequency ω_2, where $\omega_2 > \omega_1$. In Figures 9.6c and d, the waveforms $M_1(\omega)$ and $y_1(t)$ correspond to the open-loop system and the waveforms $M_2(\omega)$ and $y_2(t)$ correspond to the closed-loop system. From these waveforms we conclude that the closed-loop system has a wider bandwidth than the open-loop system and as a result it is a faster system than the open-loop system. The maximum value $M = M_p$ for both systems is the same. Thus, as a result of the introduction of the controller $G_c(j\omega)$, the speed of response of the closed-loop system is greatly improved since it is much faster than that of the open-loop system.

Another typical design example is the case of making an unstable system stable. To this end, consider the unstable system with transfer function

$$G(s) = \frac{1}{s^2(sT + 1)}$$

Its Nyquist diagram is given in Figure 9.7. Here, the control design problem is to find an appropriate $G_c(s)$ so that the closed-loop system becomes stable. This can be done if $G_c(s)$ is chosen such that the Nyquist diagram of $G_c(j\omega)G(j\omega)$ takes the particular form shown in Figure 9.7. Clearly, such a choice of $G_c(s)$ makes the closed-loop stable.

9.3 CLOSED-LOOP SYSTEM SPECIFICATIONS

The desired improvement of a system's behavior can be specified either in the time domain, the frequency domain, or in both domains. In the time domain the requirements are specified on the basis of the output function $y(t)$, and they refer mainly to the transient and the steady-state response of $y(t)$. In the case of the transient response it is desired that the system responds as fast as possible (and rarely more slowly), i.e., we want a short rise time, while the overshoot should be kept small. In the case of the steady-state response, it is desired that the error in the steady state (see Sec. 4.7) be zero, and if this is not possible, made as small as possible.

In the frequency domain the specifications are given on the basis of the Nyquist, Bode, or Nichols diagrams of the transfer function $G_c(s)G(s)$ and they mainly refer to the gain and phase margins and to the bandwidth. In the case of gain and phase margins, it is desirable to have large margins to guarantee sufficient relative stability. In the case of the bandwidth, we seek to make it as wide as possible to reduce the rise time.

Certain of the aforementioned specifications are equivalent or conflicting (opposing). Equivalent requirements are, for example, the short rise time and wide bandwidth, since wide bandwidth results in short rise time and vice versa (see Secs 4.3 and 8.3). Conflicting specifications are the cases where as one tries to improve one requirement one does damage to the other and vice versa. Such specifications are, for example, the small steady-state error and the large gain and phase margins. Here, in order to obtain a small steady-state error, the open-loop transfer function $G_c(s)G(s)$ must have a big amplification factor or many integrations or both, as opposed to obtaining large gain and phase margins, which require small amplification and no integrations (see Sec. 4.7 and Subsec. 8.4.5). Other examples of conflicting specifications are the steady-state and the transient response of $y(t)$, because as the steady-state error improves, i.e., as the steady-state error decreases, the closed-loop system tends to become unstable with its transient response becoming oscillatory.

In our effort to improve the behavior of a system we are often confronted with conflicting desired specifications, a common problem in all branches of engineering. In this situation the classical control theory deals with the problem by appropriately comprising the conflicting specifications. Modern control theory uses, for the same purpose, the minimization of a cost function which refers to one or more requirements—for example, the minimization of time and/or energy (see chap. 11).

Classical control theory compromises the conflicting specifications most often by using controllers, which are composed of an amplifier with an amplifications constant K in series with electric circuits (as well as other types of circuits such as hydraulic and pneumatic) connected in such a way that the transfer function $G_c(s)$ of the controller has the general form

$$G_c(s) = K \frac{\prod_{i=1}^{d}(1 + T_i's)}{\prod_{i=1}^{r}(1 + T_i s)} \tag{9.3-1}$$

The circuits used to realize $G_c(s)$ are called controller circuits. The most common controller circuits used include the phase-lead, the phase-lag, the phase lag-lead, the bridged T, the proportional (P), the proportional plus derivative (PD), the proportional plus integral (PI), and the proportional plus integral plus derivative (PID). Since later in this chapter we will use these circuits for the realization of controllers, we next give a short description of these circuits. Note that these controller circuits are also known in the literature as *compensating networks*, since they are inserted in the closed-loop system to compensate for certain undesirable performances appearing in the uncompensating closed-loop system.

9.4 CONTROLLER CIRCUITS

9.4.1 Phase-Lead Circuit

The most common phase-lead circuit is the simple circuit shown in Figure 9.8. When this circuit is excited by a sinusoidal signal, the phase of the output signal leads the phase of the input signal, i.e., the circuit introduces a "positive" phase. For this reason it is called a phase-lead circuit.

Using Figure 9.8 one may readily determine the transfer function $G_c(s)$ of the phase-lead circuit as follows

$$G_c(s) = \frac{Y(s)}{U(s)} = \frac{R_2 + R_1 R_2 Cs}{R_1 + R_2 + R_1 R_2 Cs} = a^{-1}\left[\frac{1 + aTs}{1 + Ts}\right] = a^{-1}G_c^*(s) \qquad (9.4\text{-}1)$$

where

$$G_c^*(s) = \frac{1 + aTs}{1 + Ts}, \qquad a = \frac{R_1 + R_2}{R_2} > 1, \qquad \text{and} \qquad T = \frac{R_1 R_2}{R_1 + R_2}C$$

Hence $G_c(s)$ has a real zero at $s = -1/aT$ and a real pole at $s = -1/T$. Since $a > 1$, the pole is always to the left of the zero.

The $G_c(s)$ diagram for $s = j\omega$, i.e., the Nyquist diagram of $G_c(s)$, is a semicircle. Indeed, if we set $s = j\omega$ and $u = T\omega$, $G_c^*(s)$ becomes

$$G_c^*(j\omega) = aG_c(j\omega) = \frac{1 + jaT\omega}{1 + jT\omega} = \frac{1 + jau}{1 + ju} = \operatorname{Re} G_c^*(j\omega) + j\operatorname{Im} G_c^*(j\omega) = x + jy$$

$$(9.4\text{-}2)$$

Equation (9.4-2) is further written as

$$x + jy = \frac{(1 + jau)(1 - ju)}{1 + u^2} = \left[\frac{1 + au^2}{1 + u^2}\right] + j\left[\frac{(a-1)u}{1 + u^2}\right]$$

Equating the real and the imaginary parts of both sides in the above equation yields

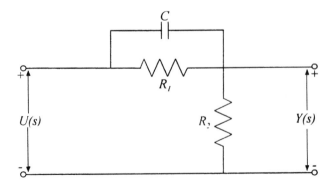

Figure 9.8 Typical phase-lead circuit.

$$x = \frac{1 + au^2}{1 + u^2} \tag{9.4-3a}$$

$$y = \frac{(a - 1)u}{1 + u^2} \tag{9.4-3b}$$

From Eq. (9.4-3a), we have

$$u^2 = \frac{1 - x}{x - a} \tag{9.4-4}$$

If we square both sides of Eq. (9.4-3b) and use Eq. (9.4-4), we have

$$y^2 = \frac{(a - 1)^2 u^2}{(1 + u^2)^2} = (a - 1)^2 \frac{\dfrac{1 - x}{x - a}}{\left[1 + \dfrac{1 - x}{x - a}\right]^2} = (a - 1)^2 \frac{(1 - x)(x - a)}{(a - 1)^2}$$

$$= x - a - x^2 + ax$$

or

$$x^2 - (a + 1)x + y^2 = -a$$

or

$$\left[x - \frac{1 + a}{2}\right]^2 + y^2 = \left[\frac{a - 1}{2}\right]^2 \tag{9.4-5}$$

Equation (9.4-5) represents a circle with center at the point $((1 + a)/2, 0)$ and radius $(a - 1)/2$. However, due to Eqs (9.4-3a and b), x and y are always positive, and hence Eq. (9.4-5) refers only to positive x and y. Such semicircles are given in Figure 9.9. Note that Figure 9.9 is the same as that of the right semicircle of the diagram of Figure 8.21e, with $K = 1$ and $T' = aT > T = T_1$, where $a > 1$.

For large values of the parameter a, the denominator of $G_c^*(s)$ will be smaller than the numerator. Therefore, as $a \to +\infty$ then $G_c^*(s) \to 1 + aTs$ and, hence, $G_c^*(s)$ becomes a straight line, as can be seen in figure 9.9. In this case $G_c^*(s)$ has two terms: an analog and a differential term. For this reason this controller is called a proportional plus derivative (PD) controller.

The phase φ of $G_c^*(s)$ and, consequently, of $G_c(s)$ is given by

$$\varphi = \tan^{-1} au - \tan^{-1} u = \tan^{-1} \left[\frac{(a - 1)u}{1 + au^2}\right] \tag{9.4-6}$$

The angle φ_m will have its maximum value when

$$\frac{d\varphi}{du} = \frac{a}{1 + a^2 u^2} - \frac{1}{1 + u^2} = 0 \tag{9.4-7}$$

Equation (9.4-7) is satisfied when

$$u = u_m = \frac{1}{\sqrt{a}} \quad \text{or} \quad \omega = \omega_m = \frac{1}{T\sqrt{a}} \tag{9.4-8}$$

and the maximum angle is

$$\varphi_m = \tan^{-1} \left[\frac{a - 1}{2\sqrt{a}}\right] \tag{9.4-9}$$

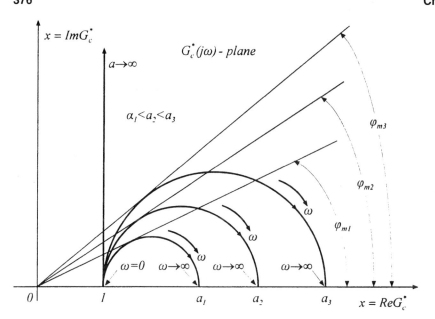

Figure 9.9 Nyquist diagram of the phase-lead circuit transfer function $G_c^*(j\omega)$.

From Eq. (9.4-9), we have

$$\sin \varphi_m = \frac{a-1}{a+1}$$ (9.4-10)

Equation (9.4-10) is very useful for the calculation of the suitable value of a for the maximum leading phase. The relation between φ_m and a is given in Figure 9.10.

The Bode diagrams of the transfer function $G_c^*(s)$ are obtained in the usual way and are presented in Figure 9.11.

9.4.2 Phase-Lag Circuit

The most common phase-lag circuit is the simple circuit shown in Figure 9.12. When this circuit is excited by a sinusoidal signal, the phase of the output signal is lagging

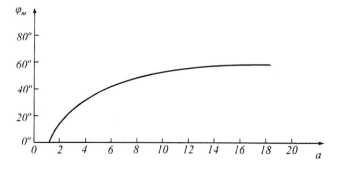

Figure 9.10 The maximum angle φ_m as a function of parameter a for the phase-lead circuit.

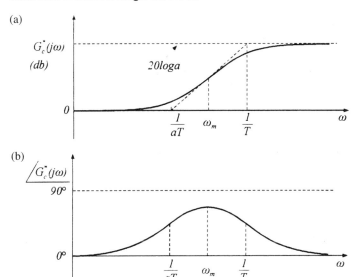

Figure 9.11 Bode diagrams of the phase-lead circuit transfer function $G_c^*(j\omega)$. (a) Magnitude Bode diagram of $G_c^*(j\omega)$; (b) phase Bode diagram of $G_c^*(j\omega)$.

the phase of the input signal, i.e., the circuit introduces a "negative" phase. For this reason it is called a phase-lag circuit.

Using Figure 9.12 one may readily determine the transfer function $G_c(s)$ of the phase-lag circuit as follows:

$$G_c(s) = \frac{Y(s)}{U(s)} = \frac{1 + R_2 Cs}{1 + (R_2 + R_2)Cs} = \frac{1 + aTs}{1 + Ts} \tag{9.4-11}$$

where

$$a = \frac{R_2}{R_1 + R_2} < 1 \quad \text{and} \quad T = (R_1 + R_2)C$$

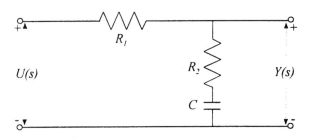

Figure 9.12 Typical phase-lag circuit.

The transfer function $G_c(s)$ has a real zero at $s = -1/aT$ and a real pole at $s = -1/T$. Since $a < 1$, the pole is always to the right of the zero.

The Nyquist diagram of $G_c(s)$ is similar to the Nyquist diagram of $G_c^*(s)$, and is given in Figure 9.13. In the present case, the circles are described by Eq. (9.4-5), with the only difference that x is positive (Eq. (9.4-3a)) but y is negative ((Eq. (9.4-3b)). Such semicircles are given in Figure 9.13. Note that Figure 9.13 is the same as that of the left semicircle of the diagram of Figure 8.21e, with $K = 1$ and $T' = aT < T = T_1$, where $a < 1$.

The Bode diagrams of the transfer function $G_c(s)$ are obtained in the usual way and are presented in Figure 9.14.

9.4.3 Phase Lag-Lead Circuit

The most common phase lag-lead circuit is given in Figure 9.15. When this circuit is excited by a sinusoidal signal, the phase of the output signal shows a phase lag in the low frequencies and a phase lead in the high frequencies. This circuit combines the characteristics of the phase-lag and the phase-lead circuits studied previously. The transfer function $G_c(s)$ of the phase lag-lead network is given by

$$G_c(s) = \frac{(1 + R_1 C_1 s)(1 + R_2 C_2 s)}{1 + (R_1 C_1 + R_1 C_2 + R_2 C_2)s + R_1 R_2 C_1 C_2 s^2}$$

In the special case where $ab = 1$, $G_c(s)$ can be expressed as the product of $G_1(s)$ and $G_2(s)$, as follows:

$$G_c(s) = \left[\frac{1 + bT_2 s}{1 + T_2 s}\right]\left[\frac{1 + aT_1 s}{1 + T_1 s}\right] = G_1(s)G_2(s) \tag{9.4-12}$$

where $G_1(s)$ and $G_2(s)$ are the phase-lag and phase-lead circuit transfer functions, respectively, and where $aT_1 = R_1 C_1$, $bT_2 = R_2 C_2$, $T_1 + T_2 = R_1 C_1 + R_1 C_2 + R_2 C_2$, and $T_1 T_2 = R_1 R_2 C_1 C_2$. To verify this result, multiply the first two equations $aT_1 =$

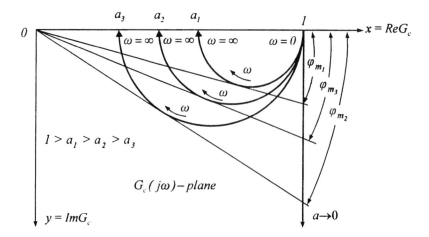

Figure 9.13 Nyquist diagram of the phase-lag circuit transfer function $G_c^*(j\omega)$.

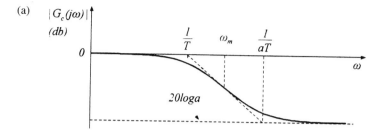

(a) $|G_c(j\omega)|$ (db)

(b) $\angle G_c(j\omega)$

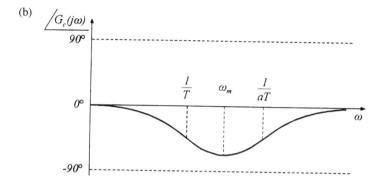

Figure 9.14 Bode diagrams of the phase-lag circuit transfer function $G_c(s)$. (a) Magnitude Bode diagram of $G_c(j\omega)$; (b) phase Bode diagram of $G_c(j\omega)$.

$R_1 C_1$ and $bT_2 = R_2 C_2$ to yield $abT_1 T_2 = R_1 R_2 C_1 C_2$. Upon using the last of the four above equations, $T_1 T_2 = R_1 R_2 C_1 C_2$, we readily have that $ab = 1$. Clearly, in order to have the convenient property of Eq. (9.4-12), a and b cannot take any independent aribtrary values, but they are constrained by the equation $ab = 1$.

The Nyquist and Bode diagrams of $G_c(s)$ are shown in Figures 9.16 and 9.17, respectively. From Eq. (9.4-12) and from Figures 9.16 and 9.17, it follows that $G_c(s)$ behaves in the low frequencies as a phase-lag circuit and in the high frequencies as a phase-lead circuit, thus combining the advantages of both phase-lag and phase-lead circuits.

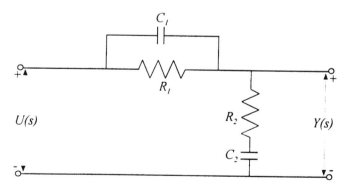

Figure 9.15 Typical phase lag-lead circuit.

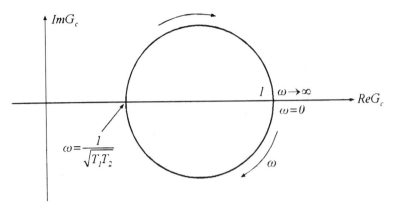

Figure 9.16   Nyquist diagram of the phase lag-lead circuit transfer function $G_c(j\omega)$.

Remark 9.4.1

A close examination of the three types of controller circuits presented thus far shows that these circuits have, in general, the following design capabilities:

    1.   Phase-lead circuits, in the frequency domain, can improve the open-loop transfer function in the high frequencies. In the time domain, they can improve the transient response by decreasing the rise time and the over-shoot and to some extent can reduce the steady-state error.

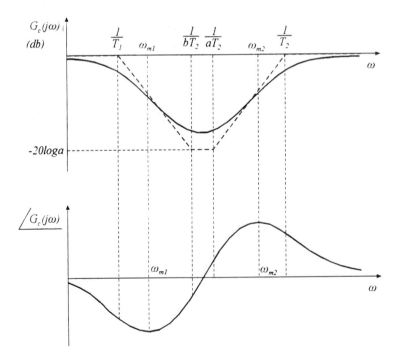

Figure 9.17   Bode diagrams of the phase lead-lag circuit transfer function $G_c(s)$.

2. Phase-lag circuits, in the frequency domain, can improve the open-loop transfer function in the low frequencies. In the time domain, they can improve the steady-state error at the expense of the transient response, because the rise time increases.

3. Phase lag-lead circuits combine the characteristics of phase-lag and phase-lead circuits. However, the phase lag-lead circuits increase the order of the system by two compared with phase-lead or phase-lag circuits, which increases the order of the system only by one. This means that although phase lag-lead circuits give better design flexibility than phase-lag and phase-lead circuits, they are not often used because the increase in the order of the system by two complicates the system analysis and design.

9.4.4 Bridged T Circuit

The most common bridged T circuit may have one of the two versions given in Figure 9.18. The circuit transfer function of Figure 9.18a is given by

$$G_1(s) = \frac{1 + 2RC_2s + R^2C_1C_2s^2}{1 + R(C_1 + 2C_2)s + R^2C_1C_2s^2} \qquad (9.4\text{-}13)$$

and that of Figure 9.18b is given by

$$G_2(s) = \frac{1 + 2R_1Cs + C^2R_1R_2s^2}{1 + C(2R_1 + R_2)s + C^2R_1R_2s^2} \qquad (9.4\text{-}14)$$

Note that $G_1(s)$ and $G_2(s)$ have essentially the same form. These circuits are different from the phase-lead, phase-lag, and phase lag-lead circuits, because their transfer functions can have complex conjugate zeros which may prove useful in realizing a controller.

9.4.5 Other Circuits

Besides the above four types of circuits, many other types of circuits can be used for controller realization. In Figure 9.19 we give a variety of such circuits.

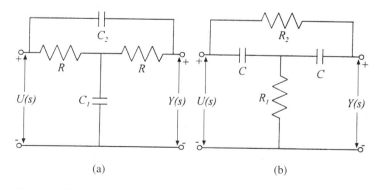

(a)   (b)

Figure 9.18 Typical T-bridged circuits.

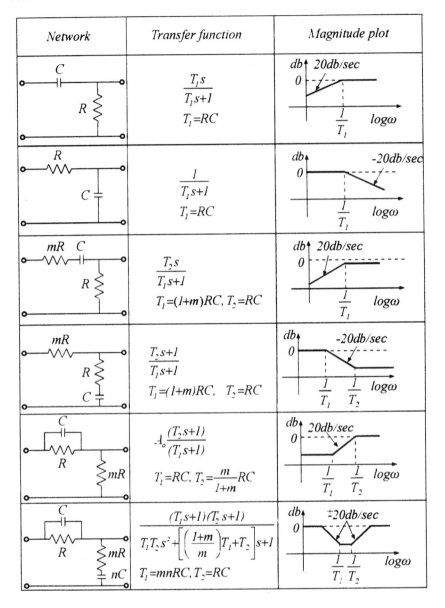

Figure 9.19 Controller circuits (*continued*).

9.5 DESIGN WITH PROPORTIONAL CONTROLLERS

In this case the controller is an amplifier of gain K and, hence, $G_c(s) = K$ (Figure 9.20a). Since $G_c(s)$ is constant, one should expect that this type of controller has limited capabilities for improving the system's performance. Indeed, when $G_c(s) = K$ the only thing that one can do is to increase or decrease K. When increasing K, the steady state of the closed-loop system of Figure 9.20a decreases. This is demonstrated in Figure 9.20b for various types of systems, as well as for various types of

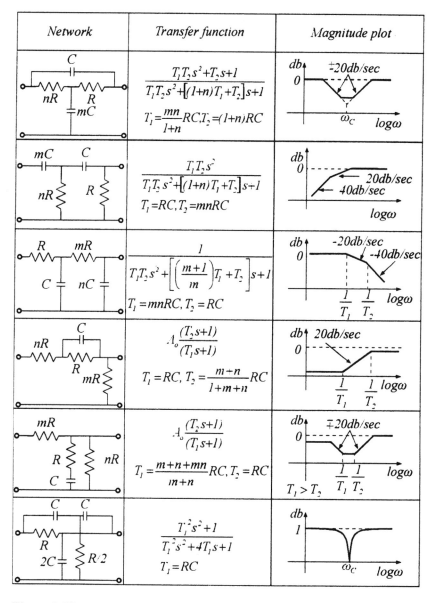

Network	Transfer function	Magnitude plot
	$\dfrac{T_1T_2s^2+T_2s+1}{T_1T_2s^2+[(1+n)T_1+T_2]s+1}$ $T_1=\dfrac{mn}{1+n}RC, T_2=(1+n)RC$	
	$\dfrac{T_1T_2s^2}{T_1T_2s^2+[(1+n)T_1+T_2]s+1}$ $T_1=RC, T_2=mnRC$	
	$\dfrac{1}{T_1T_2s^2+\left[\left(\dfrac{m+1}{m}\right)T_1+T_2\right]s+1}$ $T_1=mnRC, T_2=RC$	
	$A_0\dfrac{(T_2s+1)}{(T_1s+1)}$ $T_1=RC, T_2=\dfrac{m+n}{1+m+n}RC$	
	$A_0\dfrac{(T_2s+1)}{(T_1s+1)}$ $T_1=\dfrac{m+n+mn}{m+n}RC, T_2=RC$	
	$\dfrac{T_1^2s^2+1}{T_1^2s^2+4T_1s+1}$ $T_1=RC$	

Figure 9.19   (*contd.*)

excitation signals. Furthermore, as K increases, the gain margin decreases, and for larger values of K the closed-loop system becomes unstable, as shown in Figure 9.20c. In practice, it is desirable to have a small steady-state error and a large gain margin. These specifications are apparently conflicting, since in improving one we damage the other. In this case, a compromise is usually made, where the steady-state error is small enough, while the gain margin is large enough. A satisfactory such choice for K is $K=K_1$ shown in Figure 9.20c.

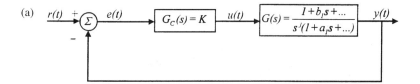

(a)

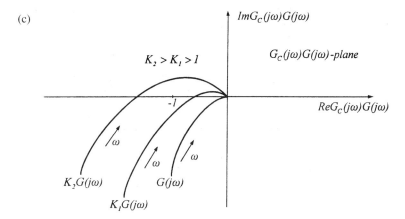

(b)

Type of system	Type of excitation		
	$r(t) = P$	$r(t) = Vt$	$r(t) = \frac{1}{2}At^2$
$j=0$	$\dfrac{P}{1+K}$	∞	∞
$j=1$	0	$\dfrac{V}{K}$	∞
$j=2$	0	0	$\dfrac{A}{K}$

(c)

Figure 9.20 Proportional controller and its influence upon the behavior of a closed-loop system. (a) Proportional controller in a closed-loop system; (b) steady-state error as a function of K; (c) influence of the parameter K upon the stability of the closed-loop system.

9.6 DESIGN WITH PID CONTROLLERS

9.6.1 Introduction to PID Controllers

A PID controller involves three terms: the proportional term designated as K_p, the integral term designated as K_i/s, and the derivative term designated as sK_d. Thus, the transfer function of a PID controller has the general form

$$G_c(s) = K_p + \frac{K_i}{s} + K_d s = \frac{K_p s + K_i + K_d s^2}{s} = \frac{K_d\left[s^2 + \dfrac{K_p}{K_d}s + \dfrac{K_i}{K_d}\right]}{s} \qquad (9.6\text{-}1a)$$

where K_p, K_i, and K_d are the proportional, integral, and derivative gains, respectively. PID controllers are also expressed as follows:

$$G_c(s) = K_p\left[1 + \frac{1}{T_i s} + T_d s\right], \qquad \text{where} \qquad T_i = \frac{K_p}{K_i} \qquad \text{and} \qquad T_d = \frac{K_d}{K_p}$$

$$(9.6\text{-}1b)$$

where K_p is the proportional gain, T_i is called the integration time constant, and T_d is called the derivative or rate time constant. The block diagram of the PID controller (Eq. (9.6-1b)) is given in Figure 9.21. Clearly, the transfer function of a PID controller involves one pole at the origin and two zeros whose position depends upon the parameters K_p, K_i, and K_d or K_p, T_i, and T_d. The overall problem of PID controllers is how to select (or tune) the arbitrary parameters involved in $G_c(s)$ such as to satisfy the design requirements as best as possible [1].

Special cases of PID controllers are the PI and PD controllers. To facilitate the study of PID controllers, we will first examine the PD and PI controllers.

9.6.2 PD Controllers

The transfer function of a PD controller is given by

$$G_c(s) = K_p + K_d s = K_d\left[s + \frac{K_p}{K_d}\right] = K_p(1 + T_d s)$$

$$(9.6\text{-}2)$$

Consider the closed-loop system of Figure 9.22, where the controller is of the PD type and, for simplicity, the system's transfer function $G_p(s)$ is of second order. Then, the forward-path transfer function $G(s)$ of Figure 9.22 is given by

$$G(s) = \frac{Y(s)}{E(s)} = G_c(s)G_p(s) = \frac{\beta(K_p + K_d s)}{s(s + \alpha)}$$

$$(9.6\text{-}3)$$

Thus, in this case, we add to the original system a zero at $s = -K_p/K_d = -1/T_d$, but the order of the closed-loop system remains the same. As a result, the closed-loop

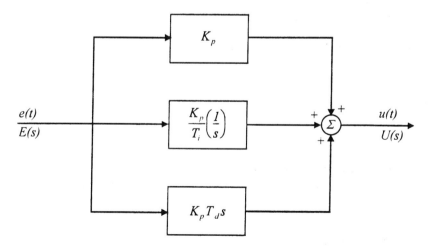

Figure 9.21 Block diagram of the PID controller.

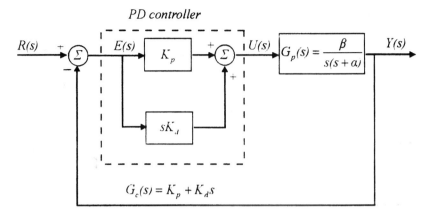

Figure 9.22 Closed-loop system with PD controller.

system becomes more stable (see Subsec. 7.5.2). The influence of $G_c(s)$ upon the behavior of the closed-loop system may be interpreted using Figure 9.23. All three waveforms concern a closed-loop system, where the input signal $r(t)$ is the step function and the controller is just a proportional controller. We observe that we have a rather high overshoot and the system is quite oscillatory. The waveform of the derivative $\dot{e}(t)$ of the error signal $e(t)$ gives information about the expected overshoot increase or decrease of $y(t)$. Indeed, in linear systems, if the slope of $e(t)$ or of $y(t)$ is large, then the overshoot will also be large. Now, if a PD controller were used, then its term sK_d predicts this fact and tries to reduce the overshoot. That is, the derivative term of the PD controller acts as an "anticipatory" controller, wherein, by knowing the slope of $e(t)$, the derivative term can anticipate the direction of the error and use it to improve the performance of the closed-loop system.

Return to Eq. (9.6-3). For simplicity, set $\alpha = \beta = 1$, in which case the closed-loop system transfer function becomes

$$H(s) = \frac{G(s)}{1 + G(s)} = \frac{K_p + K_d s}{s^2 + (1 + K_d)s + K_p} \tag{9.6-4}$$

To further simplify our study, let $K_p = 4$ and $K_d = 0$ and 1. Then, the closed-loop transfer functions $H_1(s)$ and $H_2(s)$ for $K_d = 0$ and $K_d = 1$, respectively, become

$$H_1(s) = \frac{4}{s^2 + s + 4} \qquad \text{for} \qquad K_d = 0 \qquad \text{and}$$

$$H_2(s) = \frac{4 + s}{s^2 + 2s + 4} \qquad \text{for} \qquad K_d = 1$$

The response of the closed-loop system for $r(t) = 1$ is given in Figure 9.24. The figure shows that the derivative term reduces the overshoot and damps the oscillations.

Example 9.6.1

Consider the orientation control system of a satellite which is described in Subsec. 3.13.7. Examine the behavior of the satellite when the controller $G_c(s)$ is P and PD,

(a)

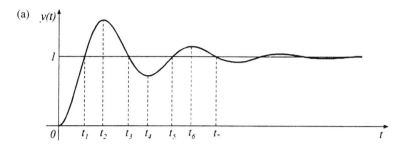

(b)

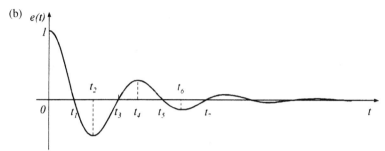

(c)

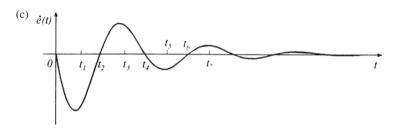

Figure 9.23 The waveforms of (a) $y(t)$, (b) $e(t)$, and (c) $\dot{e}(t)$.

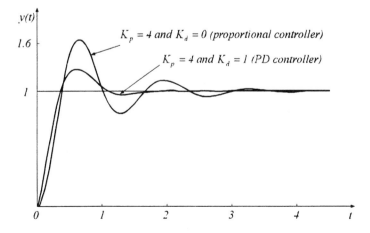

Figure 9.24 Closed-loop system response of system of Figure 9.22 for $\alpha = \beta = 1$, $K_p = 4$, and $K_d = 0$ and 1.

i.e., when $G_c(s) = K_p$ and $G_c(s) = K_p(1 + T_d s)$. Assume that the input of the closed-loop system is $\theta_r(t) = 1$.

Solution

The angular position of the satellite, when $G_c(s) = K_p$, is given by (see Eq. (3.13-35))

$$\Theta_y(s) = H(s)\Theta_r(s) = \frac{\omega_n^2}{s(s^2 + \omega_n^2)} = \frac{1}{s} - \frac{s}{s^2 + \omega_n^2}, \qquad \omega_n^2 = K_t K_b K_p J^{-1}$$

Taking the inverse Laplace transform, we have $\theta_y(t) = 1 - \cos \omega_n t$. This expression for $\theta_y(t)$ shows that the output of the system is an undamped oscillation of the satellite about the x-axis. This is because there is no friction in space (i.e., the damping ratio is $\zeta = 0$). Therefore, the system with $G_c(s) = K_p$, is not behaving satisfactory at all! To improve its behavior, we introduce the PD controller

$$G_c(s) = K_p(1 + T_d s)$$

For this case, the transfer function of the closed-loop system is given by

$$H(s) = \frac{K_t K_b K_p(1 + T_d s)}{Js^2 + K_t K_b K_p(1 + T_d s)} = \frac{\omega_n^2(1 + T_d s)}{s^2 + 2\zeta\omega_n s + \omega_n^2}, \qquad \omega_n^2 = K_t K_b K_p J^{-1}$$

where $\zeta = T_d \omega_n / 2$. Let $\Theta_r(s) = 1/s$. Then the ouput of the system is given by

$$\Theta_y(s) = H(s)\Theta_r(s) = \frac{H(s)}{s} = \frac{\omega_n^2(1 + T_d s)}{s(s^2 + 2\zeta\omega_n s + \omega_n^2)}$$

If we select a value for the constant T_d such that $0 < \zeta < 1$, then according to case 2 of Subsec. 4.3.2 and Eq. (2.4-7), we have

$$\Theta_y(s) = \frac{1}{s} - \frac{s + 2\sigma + \omega_n^2 T_d}{(s + \sigma)^2 + \omega_d^2}, \qquad \sigma = \omega_n \zeta, \qquad \text{and} \qquad \omega_d = \omega_n(1 - \zeta^2)^{1/2}$$

or

$$\Theta_y(s) = \frac{1}{s} - \frac{s + \sigma}{(s + \sigma)^2 + \omega_d^2} - \left[\frac{\sigma + \omega_n^2 T_d}{\omega_d}\right]\left[\frac{\omega_d}{(s + \sigma)^2 + \omega_d^2}\right]$$

The response in the time domain will be

$$\theta_y(t) = 1 - e^{-\sigma t}\left[\cos \omega_d t + \frac{\sigma + \omega_n^2 T_d}{\omega_d} \sin \omega_d t\right] = 1 - C e^{-\sigma t} \sin(\omega_d t + \varphi)$$

where

$$C = \frac{\sqrt{\omega_d^2 + (\sigma + \omega_n^2 T_d)^2}}{\omega_d} \qquad \text{and} \qquad \varphi = \tan^{-1}\left[\frac{\omega_d}{\sigma + \omega_n^2 T_d}\right]$$

From the above expression for $\theta_y(t)$ we see that, when a PD controller is used, the response of the system will be as in Figure 9.25. In this case, the system will exhibit some damped oscillations and will settle in the position $\theta_y = \theta_r = 1$. The amplitude of these oscillations is influenced by the damping ratio (Figure 4.4) and, therefore, it can be adjusted from the constants T_d and ω_n, since $\zeta = T_d\omega_n/2$, where $0 < \zeta < 1$ and $\omega_n^2 = K_t K_b K_p J^{-1}$.

The present example reveals in a very clear way the influence of the PD controller on the behavior of a closed-loop system. Here, when $G_c(s) = K_p$ the system contantly oscillates, whereas when $G_c(s) = K_p(1 + T_d s)$ the system quickly comes to a standstill. This is solely due to the differential term, which by anticipating the direction of the error, acts accordingly, and reduces the overshoot and the oscillations and brings the system to the desired standstill position $\theta_y = \theta_r = 1$.

9.6.3 PI Controllers

The transfer function of a PI controller is given by

$$G_c(s) = K_p + \frac{K_i}{s} = \frac{K_p\left(s + \dfrac{K_i}{K_p}\right)}{s} = K_p\left(1 + \frac{1}{T_i s}\right) \tag{9.6-5}$$

Consider the closed-loop system of Figure 9.26, where the controller is of the PI type and, for simplicity, the system's transfer function $G_p(s)$ is of second order. Then, the forward-path transfer function $G(s)$ of Figure 9.26 is

$$G(s) = \frac{Y(s)}{E(s)} = G_c(s)G_p(s) = \frac{\beta(K_i + K_p s)}{s^2(s + \alpha)} \tag{9.6-6}$$

Thus, the PI controller adds to the original system a zero at $s = -K_i/K_p = -1/T_i$ and a pole at $s = 0$. Here, as compared with the PD controller, the order of the system increases by one. We also have an increase by one in the system type (see subsection 4.7.1). As a result, the PI controller has a beneficiary influence on the

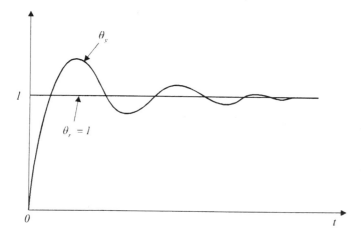

Figure 9.25   Time response of the closed-loop system for the control of a satellite orientation system when $G_c(s) = K_p(1 + T_d s)$.

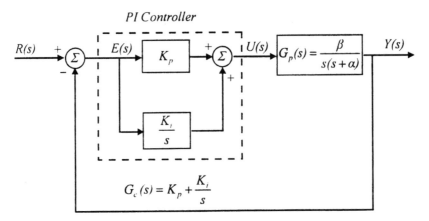

Figure 9.26 Closed-loop system with PI controller.

steady-state error, since it increases the type of the system by one (see Sec. 4.7). On the contrary, the relative stability decreases due to the pole $s = 0$ (see Subsec.7.5.1). In particular, the closed-loop response of the system for $\alpha = \beta = 1$, and $(K_p)_1 > (K_p)_2 > (K_p)_3$ and for $(K_i)_1 > (K_i)_2 > (K_i)_3$ is shown in Figure 9.27. This figure shows that as the gains K_p and K_i decrease, the oscillations decrease and the system becomes more stable. Appropriate choice of K_p and K_i may yield the desired behavior, which is usually an overshoot of about 25%.

Overall, the PI controllers behave as low-frequency filters that improve the steady-state error, but they decrease the relative stability.

Example 9.6.2

Consider a system under control, with transfer function

$$G(s) = \frac{K}{1 + T_1 s}$$

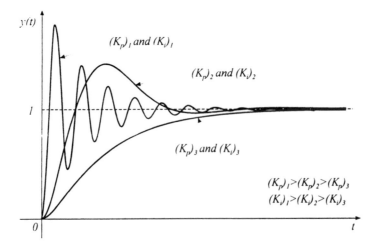

Figure 9.27 Closed-loop system response of system of Figure 9.26.

Show that by using the PI controller (Eq. (9.6-5)), one can achieve arbitrary pole placement and simultaneously drive the position steady-state error to zero.

Solution

Clearly, by choosing a PI controller, the type of the system becomes one, which guarantees zero position steady-state error (see Sec. 4.7). With regard to pole placement, the transfer function $H(s)$ of the closed-loop system is given by

$$H(s) = \frac{G_c(s)G(s)}{1 + G_c(s)G(s)}$$

We have

$$1 + G_c(s)G(s) = 1 + \left[\frac{K}{1 + T_1 s}\right]\left[K_p\left(1 + \frac{1}{T_i s}\right)\right] = \frac{b(s)}{a(s)}$$

The characteristic equation $b(s) = 0$ is as follows:

$$s^2 + \left[\frac{1}{T_1} + \frac{KK_p}{T_1}\right]s + \frac{KK_p}{T_i T_1} = 0$$

Hence, the closed-loop system is a second-order system. The general form of the characteristic equation of a second-order system is

$$s^2 + 2\zeta\omega_n s + \omega_n^2 = 0$$

Equating coefficients of like powers of s in the last two equations and solving for the two PI controller parameters – namely, for the parameters K_p and T_i – we get

$$K_p = \frac{2\zeta\omega_n T_1 - 1}{K} \quad \text{and} \quad T_i = \frac{2\zeta\omega_n T_1 - 1}{\omega_n^2 T_1}$$

Hence, for any desired values of ζ and ω_n we can always find K_p and T_i such that the closed-loop system has the desired characteristic polynomial $s^2 + 2\zeta\omega_n s + \omega_n^2$. Since the roots of this polynomial are the poles of the closed-loop system, it follows that we can achieve arbitrary pole placement of the given first-order system with a PI controller. This result is of paramount importance, since controlling the poles one controls completely the stability and may greatly influence the time response of the closed-loop system.

From the above expressions for K_p and T_i, we observe the following:

1. The gain K_p is positive for $\omega_n > 2\zeta T_1$.
2. For large values of ω_n, $T_i \cong 2\zeta/\omega_n$. In this case, T_i does not depend on the system's time constant T_1.

9.6.4 PID Controllers

A typical closed-loop system involving a PID controller is given in Figure 9.28. The PID controller transfer function $G_c(s)$ defined in Eq. (9.6-1b) may further be written as

$$G_c(s) = K_p\left[1 + \frac{1}{T_i s} + T_d s\right] = K_p\frac{(as + 1)(bs + 1)}{s} \tag{9.6-7a}$$

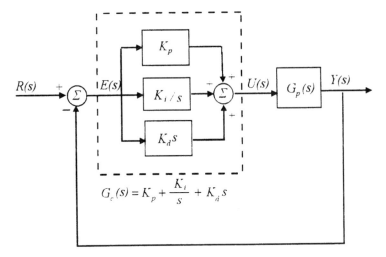

Figure 9.28   Closed-loop system with PID controller.

where

$$a + b = T_i \quad \text{and} \quad ab = T_i T_d \tag{9.6-7b}$$

Hence, the PID controller increases the number of zeros by two and the number of poles by one, where the two zeros are located at $s = -1/a$ and $s = -1/b$ and the pole is located at $s = 0$.

   The PID controller is designed by properly choosing K_p, K_i, and K_d, or K_p, T_i, and T_d such as to control the system with all the advantages of the PD and PI controllers combined. The resulting desired closed-loop system's behavior is shown in Figure 9.29. Here, $y(t)$ has a small rise time, a small overshoot, a small settling time, and a zero steady-state error. Such a response is, of course, close to the ideal. The difficulty in achieving such a response is the selection (or tuning) of the appropriate K_p, K_i, and K_d for any specific system under control [1]. Two practical methods for selecting the appropriate K_p, K_i, and K_d are presented in Subsec. 9.6.5 that follows.

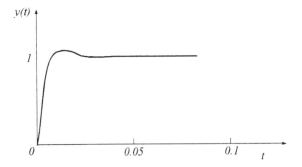

Figure 9.29   Response of a closed-loop system with PID controller when $r(t) = 1$.

Example 9.6.3

Consider a system under control, with transfer function

$$G(s) = \frac{K}{(1 + sT_1)(1 + sT_2)}$$

Show that using the PID controller (Eq. (9.6-1)), one can achieve arbitrary pole placement and simultaneously drive the position steady-state error to zero.

Solution

Clearly, by choosing a PID controller, the type of the system becomes one, which guarantees zero position steady-state error (see Sec. 4.7). With regard to pole placement, we work as in Example 9.6.2, and find that the characteristic equation of the closed-loop system is as follows:

$$s^3 + \left[\frac{1}{T_1} + \frac{1}{T_2} + \frac{KK_p T_d}{T_1 T_2}\right]s^2 + \left[\frac{1}{T_1 T_2} + \frac{KK_p}{T_1 T_2}\right]s + \frac{KK_p}{T_1 T_2 T_i} = 0$$

Hence, the closed-loop system is a third-order system. The general form of the characteristic equation of a third-order system is

$$(s + \tau\omega)(s^2 + 2\zeta\omega_n s + \omega_n^2) = 0$$

Equating coefficients of like powers of s in the last two characteristic equations, and after some algebraic manipulations, we arrive at the following values for the PID controller parameters K_p, T_i, and T_d:

$$K_p = \frac{T_1 T_2 \omega_n^2 (1 + 2\zeta\tau) - 1}{K}, \qquad T_i = \frac{T_1 T_2 \omega_n^2 (1 + 2\zeta\tau) - 1}{T_1 T_2 \tau \omega_n^3},$$

$$T_d = \frac{T_1 T_2 \omega_n (\tau + 2\zeta) - T_1 - T_2}{T_1 T_2 \omega (1 + 2\tau) - 1}$$

Hence, for any desired values of ζ, ω_n, and τ we can always find K_p, T_i, and T_d such that the closed-loop characteristic polynomial has the desired form $(s + \tau\omega_n)(s^2 + 2\zeta\omega_n s + \omega_n^2)$. Since the roots of this polynomial are the poles of the closed-loop system, it follows that we can achieve arbitrary pole placement of the given second-order system with a PID controller. This result is of paramount importance, since controlling the poles one controls completely the stability and may greatly influence the time response of the closed-loop system.

Example 9.6.4

Consider the closed-loop system shown in Figure 9.30, where a PID controller is used to control the direction of a ship. The disturbance (e.g., due to the wind) affects the system, as shown in the block diagram. The input reference signal $r(t)$ is usually constant. Design a control system such that the closed-loop response to a unit step disturbance decays fast (e.g., the settling time T_s is 2 or 3 sec with 2% final value tolerance) and the damping of the system is satisfactory. Choose the position of the poles so that the closed-loop system has a pair of dominant poles. Find the time response of the system to a unit step disturbance and to a unit step input reference signal.

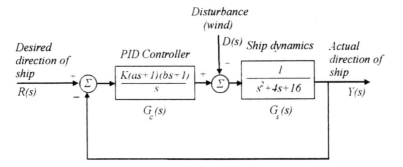

Figure 9.30 Control system which uses a PID controller to control the direction of a ship.

Solution

According to relation (9.6-7), the transfer function of the PID controller can be written as

$$G_c(s) = \frac{K(as+1)(bs+1)}{s} = K_p\left[1 + \frac{1}{T_i s} + T_d s\right]$$

where $K = K_p$, $a + b = T_i$, and $ab = T_i T_d$. When $R(s) = 0$ and $D(s) \neq 0$, the closed-loop transfer function due to the disturbance $D(s)$ is given by

$$\frac{Y_d(s)}{D(s)} = \frac{G_c(s)G_s(s)}{1 + G_c(s)G_s(s)} = \frac{s}{s(s^2 + 4s + 16) + K(as+1)(bs+1)}$$

$$= \frac{s}{s^3 + (4 + Kab)s^2 + (16 + Ka + Kb)s + K}$$

where $Y_d(s)$ is the output due to the disturbance $D(s)$.

The closed-loop specifications require that the response to a unit step disturbance be such that the settling time T_s is 2–3 sec with 2% final value tolerance, and the system has satisfactory damping. However, we have that

$$T_s = \frac{4}{\zeta\omega_0} = 2\,\text{sec} \qquad \text{and therefore we have } \zeta\omega_0 = 2$$

We can choose $\zeta = 0.4$ and $\omega_n = 5\,\text{rad/sec}$ for the dominant poles of the closed-loop system. We choose the third pole at $s = -12$, so that it has negligible effect on the response. Then, the desired characteristic equation can be written as

$$(s + 12)[s^2 + 2(0.4)5s + 5^2] = (s + 12)(s^2 + 4s + 25) = s^3 + 16s^2 + 73s + 300 = 0$$

The characteristic equation of the closed-loop transfer function (having as input the disturbance $D(s)$), is given by

$$s^3 + (4 + Kab)s^2 + (16 + Ka + Kb)s + K = 0$$

Comparing the coefficients of the two characteristic polynomials, it follows that $4 + Kab = 16$, $16 + Ka + Kb = 73$, and $K = 300$. Hence, $ab = 0.04$ and $a + b = 0.19$. Therefore, the PID controller becomes

$$G_c(s) = \frac{K[abs^2 + (a+b)s + 1]}{s} = \frac{12(s^2 + 4.75s + 25)}{s}$$

Using this PID controller, the response $Y_d(s)$ to the disturbance $D(s)$ is

$$Y_d(s) = \left[\frac{s}{(s+12)(s^2+4s+25)}\right]D(s)$$

For $D(s) = 1/s$, the output in the steady state is zero, since, according to the final value theorem, we have

$$\lim_{t\to\infty} y_d(t) = \lim_{s\to 0} s Y_d(s) = \lim_{s\to 0}\left[\frac{s^2}{(s+12)(s^2+4s+25)}\right]\left[\frac{1}{s}\right] = 0$$

More specifically, we have

$$Y_d(s) = \left[\frac{s^2}{(s+12)(s^2+4s+25)}\right]\left[\frac{1}{s}\right] = \frac{0.099174}{s+12} + \left[\frac{-0.099174s + 0.206612}{s^2+4s+25}\right]$$

$$= \frac{0.099174}{s+12} - \frac{0.099174(s+2)}{(s+2)^2+(\sqrt{21})^2} + \frac{0.08837\sqrt{21}}{(s+2)^2+(\sqrt{21})^2}$$

Taking the inverse Laplace transform, we have

$$y_d(t) = 0.099174e^{-12t} - 0.099174e^{-2t}\cos\left[\sqrt{21}\,t\right] + 0.08837e^{-2t}\sin\left[\sqrt{21}\,t\right]$$

The unit step response is shown in Figure 9.31, from which we get a settling time around 0.72 sec and a satisfactory damping. Therefore, the design with regard to the disturbance is satisfactory.

When $R(s) \neq 0$ and $D(s) = 0$, the closed-loop transfer function due to the input reference signal $R(s)$ is given by

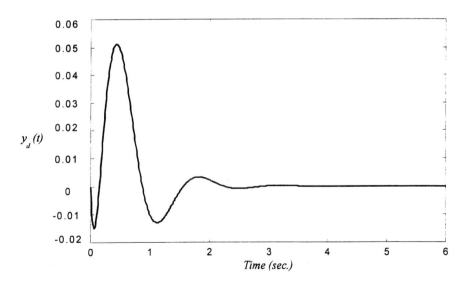

Figure 9.31 Closed-loop response to unit step disturbance.

$$\frac{Y_r(s)}{R(s)} = \frac{G_c(s)G_s(s)}{1 + G_c(s)G_s(s)} = \frac{12(s^2 + 4.75s + 25)}{s^3 + 16s^2 + 73s + 300}$$

For $R(s) = 1/s$, the output $Y_r(s)$ becomes

$$Y_r(s) = \left[\frac{12(s^2 + 4.75s + 25)}{s^3 + 16s^2 + 73s + 300}\right]\left[\frac{1}{s}\right]$$

$$= \frac{1}{s} - \frac{0.92562}{s + 12} - \frac{0.07438(s + 2)}{(s + 2)^2 + (\sqrt{21})^2} + \frac{0.16231\sqrt{21}}{(s + 2)^2 + (\sqrt{21})^2}$$

Taking the inverse Laplace transform, we have

$$y_r(t) = 1 - 0.92562\,e^{-12t} - 0.07438e^{-2t}\cos\left[\sqrt{21}\,t\right] + 0.1623\,e^{-2t}\sin\left[\sqrt{21}\,t\right]$$

The unit step response is shown in Figure 9.32, from which we find that the maximum overshoot is 7.15% and the settling time is 1.1 sec, which are very satisfactory.

9.6.5 Design of PID Controllers Using the Ziegler–Nichols Methods

The PID controller has the flexibility of simultaneously tuning three parameters: namely, the parameters K_p, T_i, and T_d. This allows a PID controller to satisfy the design requirements in many practical cases, a fact which makes the PID controller the most frequently met controller in practice. The appropriate values of the parameters K_p, T_i, and T_d of the PID controller may be chosen by trial and error. This is usually a formidable task, even in cases where the design engineer has great experience on the subject. To facilitate the determination of the appropriate values of the parameters K_p, T_i, and T_d, even for cases where a mathematical model for the system under control is not available, Ziegler and Nichols [11] have suggested the following two rather simple and practically useful methods.

1 The Transient Response Method

In this case, the system under control is excited with the unit step function (Figure 9.33a). The shape of the transient response of the open-loop system may have the

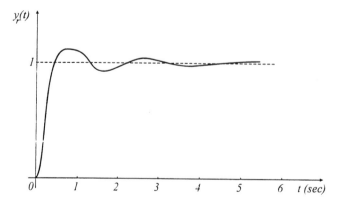

Figure 9.32 Closed-loop response to unit step reference signal.

(a)

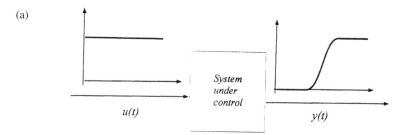

(b)

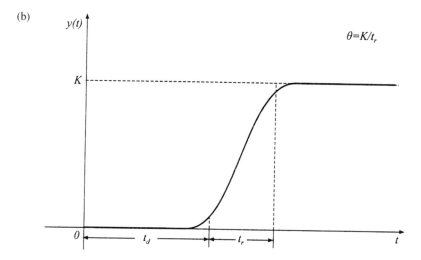

Figure 9.33 The transient response method. (a) Experimental step response; (b) detailed step response.

general form shown in Figure 9.33b. In this case, we introduce the parameters t_d = delay time and t_r = rise time. Aiming in achieving a damping ratio ζ of about 0.2 (which corresponds to an overshoot of about 25%), the values of the parameters K_p, T_i, and T_d of the PID controller are chosen according to Table 9.1.

It is useful to mention that here the transfer function of the system under control may be approximated as follows

$$G(s) = K\left[\frac{e^{-t_d s}}{1 + t_r s}\right]$$ (9.6-8)

Furthermore, upon using Table 9.1, the PID controller transfer function $G_c(s)$ becomes

$$G_c(s) = K_p\left[1 + \frac{1}{T_i s} + T_d s\right] = 1.2\frac{t_r}{t_d}\left[1 + \frac{1}{2t_d s} + 0.5t_d s\right]$$

$$= (0.6t_r)\left[\frac{(s + 1/t_d)^2}{s}\right]$$ (9.6-9)

That is, the PID controller has a pole at the origin and a double zero at $s = -1/t_d$.

Table 9.1 The Values of the Parameters K_p, T_i, and T_d Using the Ziegler–Nichols Transient Response Method

Controller		K_p	T_i	T_d
Proportional	P	$\dfrac{t_r}{t_d}$	∞	0
Proportional–integral	PI	$0.9\dfrac{t_r}{t_d}$	$\dfrac{t_d}{0.3}$	0
Proportional–integral–derivative	PID	$1.2\dfrac{t_r}{t_d}$	$2t_d$	$0.5t_d$

Example 9.6.5

In Figure 9.34, the unit step response of a plant is given. Determine K_p, K_i, and K_d of a PID controller using the transient response method.

Solution

From the figure we have that $t_d = 150\,\text{sec}$ and $t_r = 75\,\text{sec}$. Using Table 9.1, we readily have

$$K_p = 1.2\frac{t_r}{t_d} = (1.2)\frac{75}{150} = 0.6$$

$$T_i = 2t_d = 2(150) = 300\,\text{sec}$$

$$T_d = 0.5t_d = (0.5)(150) = 75\,\text{sec}$$

2 The Stability Limit Method

Here, we start by controlling the system only with a proportional controller (Figure 9.35a). The gain K_p is slowly increased until a persistent oscillation is reached (Figure 9.35b). At this point, we mark down the value of the parameter K_p, denoted as $\tilde{K}_p$, as

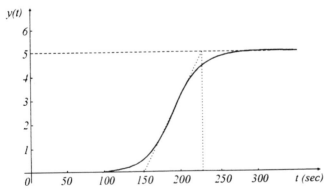

Figure 9.34 The unit step response of a plant of Example 9.6.5.

(a)

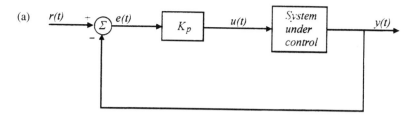

(b)

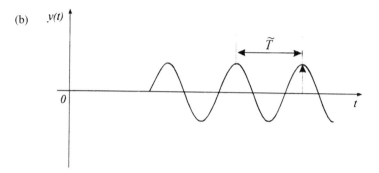

Figure 9.35 The stability limit method. (a) Closed-loop system with proportional controller; (b) sustained oscillations with period $\tilde{T}$.

well as the value of the respective oscillation period, denoted as $\tilde{T}$. Then, the parameters K_p, T_i, and T_p of the PID controller are chosen according to Table 9.2.

For the present case, and upon using Table 9.2, the transfer function of the PID controller becomes

$$G_c(s) = K_p\left(1 + \frac{1}{T_i s} + T_d s\right) = 0.6\tilde{K}_p\left(1 + \frac{1}{0.5\tilde{T}} + \frac{\tilde{T}}{8}\right)$$

$$= (0.075\tilde{K}_p\tilde{T})\left[\frac{(s + 4/\tilde{T})^2}{s}\right] \tag{9.6-10}$$

That is, the PID controller has a pole at the origin and a double zero at $s = -4/\tilde{T}$.

Example 9.6.6

The position control system of an object holder robot is shown in Figure 9.36. The block diagram is given in Figure 9.37. Use the stability limit method of Ziegler-Nichols to determine the parameters of the PID controller. Use the Routh stability criterion to calculate the values of $\tilde{K}_p$ and $\tilde{T}$.

Table 9.2 The Values of the Parameters K_p, T_i, and T_d Using the Ziegler–Nichols Stability Limit Method

Controller		K_p	T_i	T_d
Proportional	P	$0.5\tilde{K}_p$	∞	0
Proportional–integral	PI	$0.45\tilde{K}_p$	$\tilde{T}/1.2$	0
Proportional–integral–derivative	PID	$0.6\tilde{K}_p$	$\tilde{T}/2$	$\tilde{T}/8$

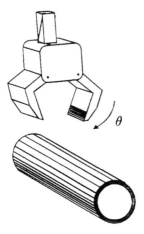

Figure 9.36 Object holder robot.

Solution

Using only a proportional controller, as in Figure 9.35a, yields the following closed-loop transfer function

$$\frac{Y(s)}{R(s)} = \frac{K_p}{s(s+1)(s+4) + K_p}$$

The value of K_p that makes the system marginally stable, in which case sustained oscillations occur, can be obtained using Routh's stability criterion. The characteristic equation of the closed-loop system is given by

$$s^3 + 5s^2 + 4s + K_p = 0$$

The Routh array is as follows:

$$
\begin{array}{c|cc}
s^3 & 1 & 4 \\
s^2 & 5 & K_p \\
s^1 & \dfrac{20 - K_p}{5} & \\
s^0 & K_p &
\end{array}
$$

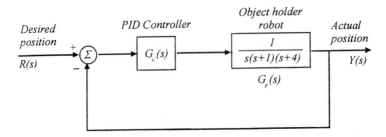

Figure 9.37 Position control system with PID controller of an object holder robot.

Examining the coefficients of the first column of the Routh table, we find that sustained oscillation will occur when $K_p = 20$. Thus, the critical gain $\tilde{K}_p$ is 20. With the gain K_p set equal to $\tilde{K}_p$, the characteristic equation becomes

$$s^3 + 5s^2 + 4s + 20 = 0$$

Substituting $s = j\omega$ into the characteristic equation, we can find the frequency of the sustained oscillations. We have

$$(j\omega)^3 + 5(j\omega)^2 + 4j\omega + 20 = 0$$

or

$$5(4 - \omega^2) + j\omega(4 - \omega^2) = 0 \quad \text{and hence} \quad \omega^2 = 4 \quad \text{or} \quad \omega = 2$$

The period $\tilde{T}$ of the sustained oscillations is

$$\tilde{T} = \frac{2\pi}{\omega} = \frac{2\pi}{2} = \pi = 3.14$$

Referring to Table 9.2, we determine K_p, T_i, and T_d as follows:

$$K_p = 0.6\tilde{K}_p = (0.6)(20) = 12$$

$$T_i = 0.5\tilde{T} = (0.5)(3.14) = 1.57$$

$$T_d = 0.125\tilde{T} = (0.125)(3.14) = 0.3925$$

Hence, the transfer function of the PID controller is the following:

$$G_c(s) = K_p\left[1 + \frac{1}{T_i s} + T_d s\right] = 12\left[1 + \frac{1}{1.57s} + 0.3925s\right]$$

$$= \frac{4.71(s + 1.27389)^2}{s}$$

The PID controller has a pole at the origin and double zero at $s = -1.27389$. The transfer function $H(s)$ of the closed-loop system is given by

$$H(s) = \frac{G_c(s)G_r(s)}{1 + G_c(s)G_r(s)} = \frac{\left[\dfrac{4.71(s + 1.27389)^2}{s}\right]\left[\dfrac{1}{s(s + 1)(s + 4)}\right]}{1 + \left[\dfrac{4.71(s + 1.27389)^2}{s}\right]\left[\dfrac{1}{s(s + 1)(s + 4)}\right]}$$

$$= \frac{4.71s^2 + 12s + 7.643}{s^4 + 5s^3 + 8.71s^2 + 12s + 7.643}$$

The unit step response $Y(s)$ of the closed-loop system is given by

$$Y(s) = H(s)R(s) = H(s)\left[\frac{1}{s}\right] = \left[\frac{4.71s^2 + 12s + 7.643}{s^4 + 5s^3 + 8.71s^2 + 12s + 7.643}\right]\left[\frac{1}{s}\right]$$

The unit step response of the closed-loop system is shown in Figure 9.38. If the maximum overshoot is excessive, it can be reduced by fine tuning the controller parameters.

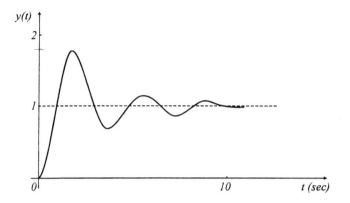

Figure 9.38   Closed-loop response to unit step reference signal.

9.6.6   Active Circuit Realization for PID Controllers

In Sec. 9.4, several circuits were presented, which are the circuit realizations of many popular controllers, such as phase-lead, phase-lag, phase lag-lead, etc. These realizations involve the passive elements R, L, and C and for this reason they are called passive circuit controller realizations.

Here, we will present another approach to controller realization by using an active element, namely, the operational amplifier. A typical such circuit involving an operational amplifier is given in Figure 9.39. Since the operational amplifier is an active element, it is for this reason that such realization is called *active*-circuit realization of a controller, as compared with *passive*-circuit realizations presented in Sec. 9.4.

The active circuit of Figure 9.39 appears to have great flexibility in realizing many types of controllers, by making the proper choice of $Z_1(s)$ and $Z_2(s)$. Note that here the input voltage $V_i(s)$ is related to the output voltage $V_0(s)$ via the transfer function

$$G_c(s) = \frac{V_0(s)}{V_i(s)} = -\frac{Z_2(s)}{Z_1(s)} \tag{9.6-11}$$

Figure 9.40 presents several configurations which lead to active realizations for various types of controllers, including PI, PD, and PID.

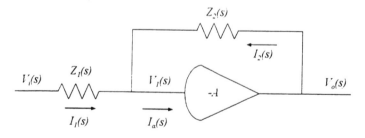

Figure 9.39   Active-circuit realization of a controller.

Function	$Z_1(s)$	$Z_2(s)$	$G_c(s) = -\dfrac{Z_1(s)}{Z_2(s)}$
Gain	R_2 (resistor)	R_2 (resistor)	$-\dfrac{R_2}{R_1}$
Integration	R_2 (resistor)	C (capacitor)	$-\dfrac{1/RC}{s}$
Differentiation	C (capacitor)	R_2 (resistor)	$-RCs$
Lag compensation	C_1 and R_1 (parallel)	C_1 and R_1 (parallel)	$-\dfrac{C_1}{C_2}\dfrac{\left(s+\dfrac{1}{R_1C_1}\right)}{\left(s+\dfrac{1}{R_2C_2}\right)}$ and $R_2C_2 > R_1C_1$
Lead compensation	C_1 and R_1 (parallel)	C_1 and R_1 (parallel)	$-\dfrac{C_1}{C_2}\dfrac{\left(s+\dfrac{1}{R_1C_1}\right)}{\left(s+\dfrac{1}{R_2C_2}\right)}$ and $R_1C_1 > R_2C_2$
PI controller	R_2 (resistor)	R_2, C_2 (series)	$-\dfrac{R_2}{R_1}\dfrac{(s+1/R_1C)}{s}$
PD controller	C_1 and R_1 (parallel)	R_2 (resistor)	$-R_2C\left(s+\dfrac{1}{R_1C}\right)$
PID controller	C_1 and R_1 (parallel)	R_2, C_2 (series)	$-\left[K+R_2C_1s+\dfrac{\dfrac{1}{R_1C_2}}{s}\right]$ where $K=\dfrac{R_2}{R_1}+\dfrac{C_1}{C_2}$

Figure 9.40 Active-circuit realization of controllers.

9.7 DESIGN WITH PHASE-LEAD CONTROLLERS

Phase-lead controllers are used to introduce a positive phase in the closed-loop transfer function, aimed at improving the transition response of the system in the time domain (i.e., decreasing the rise time and overshoot) and in the frequency domain at improving the gain and phase margins and the bandwidth (i.e., increasing K_g, φ_p, and BW). The transfer function $G_c^*(s)$ defined in Eq. (9.4-1) of the phase-lead circuit influences the open-loop transfer function at the high frequencies. This is easily seen if we consider $G_c^*(s)$ to have the special form

$$G_c^*(s) = 1 + sK_d \qquad (9.7\text{-}1)$$

Equation (9.7-1) is the special case of $G_c^*(s)$ defined by Eq. (9.4-1) as $a \to +\infty$ (see Figure 9.9). In this case we observe that the controller $G_c^*(s) = 1 + sK_d$ involves two terms: the proportional term 1 and the differential term sK_d. Hence, $G_c^*(s)$ is a PD controller. For this reason we say that the phase-lead controller, as $a \to \infty$, behaves like a PD controller. A simple way of realizing $G_c^*(s) = 1 + sK_d$ is given in Figure 9.41a. The influence of $G_c^*(j\omega) = 1 + j\omega K_d$ upon the transfer function $G(j\omega)$ of a given system under control is shown in Figure 9.41b, wherein the closed-loop sytem becomes stable.

When the $G_c^*(j\omega)$ does not have the special form of $1 + j\omega K_d$, but instead has the general form $G_c^*(j\omega) = [(1 + ja\omega T)/(1 + j\omega T)]$, whose Nyquist diagram is a semi-circle as shown in Figure 9.9, then $G_c^*(j\omega)$ is actually an approximation of $1 + j\omega K_d$ and, as a result, it influences the diagram of $G_c^*(j)G(j\omega)$ in approximately the same way as in Figure 9.41b.

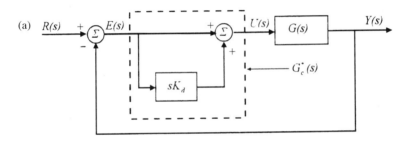

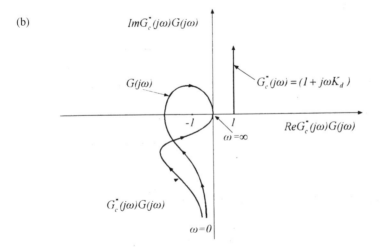

Figure 9.41 Closed-loop system with a phase-lead controller and the influence of this controller on the Nyquist diagram of the closed-loop system. (a) Closed-loop system with controller $G_c^*(s) = 1 + sK_d$; (b) Nyquist diagram of $G(j\omega)$, $G_c^*(j\omega)$, and $G_c^*(j\omega)G(j\omega)$.

The design of a phase-lead controller is done preferably with the use of Bode diagrams, since the influence of the controller is calculated by simply adding the amplitude and phase diagrams of the controller to the corresponding diagrams of the given system. Let the design specifications refer only to the steady-state error and the gain and phase margins. Also let the open-loop transfer function involve an amplification factor K. Then, the main steps in determining the amplification constant K and the transfer function $G_c(s) = a^{-1}G_c^*(s)$ of the controller are

1. The amplification constant K is chosen so as to compensate for the parameter a^{-1} and to satisfy the specifications of the steady-state error.
2. Using the bode diagrams of the transfer function $G(j\omega)$ of the system under control, the angle φ_m of $G_c^*(j\omega)$ is determined, which must be added to the system in order to satisfy the phase margin specifications.
3. On the basis φ_m, the parameter a of $G_c^*(j\omega)$ is determined by using Eq. (9.4-10) or from Figure 9.10.
4. The parameter T is calculated in such a way that, for $\omega = \omega_m$, $20 \log |G(j\omega_m)| = -0.5[20 \log a]$. Since ω_m is the geometric mean of the corner frequencies $1/aT$ and $1/T$, we can conclude that the parameter T is calculated from the equation

$$T = \frac{1}{\omega_m \sqrt{a}} \tag{9.7-2}$$

5. Finally, we draw the compensated open-loop diagram $G_c(s)G(s)$. If the value of the parameter T does not give satisfactory results we repeat the above steps by giving new (usually bigger) values to the parameter φ_m until we get satisfactory results.

The above procedure for determining K and $G_c(s)$, is illustrated in the example that follows.

Example 9.7.1

Consider the sun-seeker control system shown in Figure 1.21 in Chap. 1. The block diagram of this system is given in Figure 9.42. The transfer function of each block is as follows:

$$G_1(s) = I_s, \qquad G_2(s) = K_a, \qquad G_3(s) = \frac{K_i}{s(R_a J_m s + R_a B_m + K_i K_b)}, \qquad \text{and}$$

$$G_4(s) = K_4$$

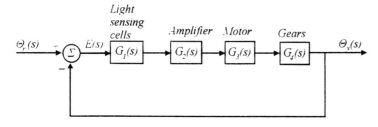

Figure 9.42 Block diagram of the sun-seeker orientation control system.

Note that the transfer function $G_3(s)$ of the motor is given by Eq. (3.12-17) when $L_a \to 0$. The open-loop transfer function $G(s)$ will then be

$$G(s) = G_1(s)G_2(s)G_3(s)G_4(s) = \frac{I_s K_a K_i K_4}{s(R_a J_m s + R_a B_m + K_i K_b)}$$

Typical values for the parameters of $G(s)$ are approximately $R_a J_m = 1$, $R_a B_m + K_i K_b = 10$, and $I_s K_i K_4 = 10^3$. For simplicity let $K = K_a$. Then $G(s)$ takes on the form

$$G(s) = \frac{10^3 K}{s(s + 10)}$$

The transfer function $H(s)$ of the closed-loop system will be

$$H(s) = \frac{10^3 K}{s^2 + 10s + 10^3 K}$$

Let the angular position of $\theta_r(t)$ of the sun be given by

$$\theta_r(t) = t$$

Since the Laplace transform of t is $1/s^2$, the output $\Theta_y(s)$ becomes

$$\Theta_y(s) = H(s)\Theta_r(s) = \left[\frac{10^3 K}{s^2 + 10s + 10^3 K}\right]\left[\frac{1}{s^2}\right]$$

The inverse Laplace transformation of $\Theta_y(s)$ has the general form

$$\theta_y(t) = C_0 + C_1 t + C_2 e^{-\sigma t} \sin(\omega t + \varphi)$$

where $C_1 = 1$. The constant C_0 is the velocity steady error which $\theta_y(t)$ presents in comparison to $\theta_r(t) = t$ (Figure 9.43). The constant C_0 is calculated according to Eq. (4.7-11), as follows:

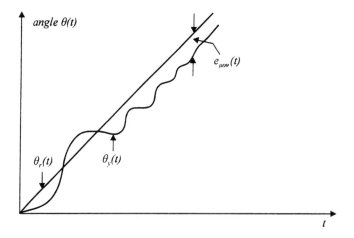

Figure 9.43   The steady-state error of the system of Figure 9.42 when $\theta_r(t) = t$.

$$C_0 = e_{ss}(t) = \lim_{s \to 0} sE(s) = \frac{1}{\lim_{s \to 0} sG(s)} = \frac{1}{100K}$$

The design problem here is the following. We are given the closed-loop system of Figure 9.44a. In this system we introduce a phase-lead controller with transfer function $G_c(s)$ (Figure 9.44b). Determine the parameters of $G_c(s)$, as well as the system amplification constant K, such that the compensated closed-loop system of Figure 9.44b satisfies the following:

(a) The velocity steady-state error, when $\theta_r(t) = t$, is equal to or less than 0.01.

(b) The phase margin is larger than 40°.

Solution

According to Eq. (4.7-11), the velocity error of the compensated closed-loop system of Figure 9.44b is as follows

$$e_{ss}(t) = \lim_{s \to 0} \left[\frac{s\Theta_r(s)}{1 + G_c(s)G(s)} \right], \qquad G_c(s) = a^{-1} \left[\frac{1 + aTs}{1 + Ts} \right]$$

Therefore

$$e_{ss}(t) = \lim_{s \to 0} \left\{ \frac{s \left[\frac{1}{s^2} \right]}{1 + \left[a^{-1} \frac{1 + aTs}{1 + Ts} \right] \left[\frac{1000K}{s(s + 10)} \right]} \right\} = \frac{a}{100K} \leq 10^{-2}$$

Thus, we must have $K \geq a$. For simplicity, let $K = a$. Then, the open-loop transfer function becomes

$$G_c(s)G(s) = \left[a^{-1} \frac{1 + aTs}{1 + Ts} \right] \left[\frac{1000a}{s(s + 10)} \right] = \left[\frac{1 + aTs}{1 + Ts} \right] \left[\frac{1000}{s(s + 10)} \right] = G_c^*(s)G^*(s)$$

where

(a)
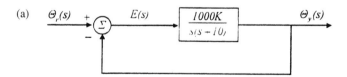

(b)
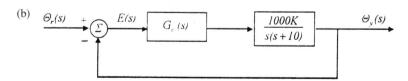

Figure 9.44 The sun-seeker orientation control system (a) without a controller (uncompensated system) and (b) with a controller (compensated system).

$$G_c^*(s) = \left[\frac{1 + aTs}{1 + Ts}\right] \quad \text{and} \quad G^*(s) = \left[\frac{1000}{s(s + 10)}\right]$$

From the above equation we can conclude that when the amplification constant K is selected to be equal to a, then the amplifier compensates the amplitude damping which is introduced by the factor a^{-1}. Hence, the selection of the parameters a and T is facilitated by working with the transfer functions $G_c^*(s)$ and $G^*(s)$, rather than the transfer functions $G_c(s)$ and $G(s)$. This is the approach used in the sequel.

The critical frequency ω' (see Sec. 8.5.4 and Figure 8.40) for which the curve $20 \log |G^*(j\omega)|$ becomes 0 dB satisfies the relation $|G^*(j\omega')| = 1$, i.e., satisfies the relation

$$\frac{10^2}{\omega'\sqrt{(0.01)(\omega')^2 + 1}} = 1 \quad \text{or} \quad (\omega')^2[(0.01)(\omega')^2 + 1] = 10^4$$

From the above equation and Figure 9.45, we find that $\omega' \simeq 31$ rad/sec. For this frequency we calculate the phase margin φ_p for the uncompensated system. We have

$$\varphi_p = 180° + \angle G(j\omega') = 180° - 90° - \varepsilon\varphi^{-1}(0.1\omega') = 90° - \varepsilon\varphi^{-1}3.1$$
$$= 90° - 72° = 18°$$

Since this phase margin is too small, the controller $G_c^*(s)$ must introduce a positive phase in the high frequencies. The phase-lead circuit achieves just that, as we have

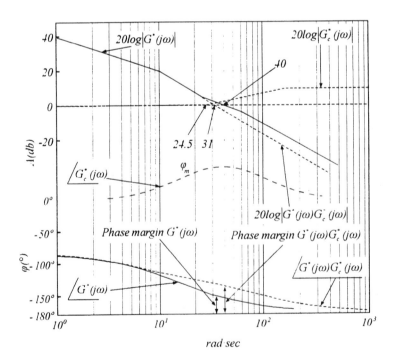

Figure 9.45 Phase and amplitude Bode diagrams of the uncompensated system $G^*(j\omega)$ and the compensated $G^*(j\omega)G_c^*(j\omega)$ system of Example 9.7.1.

already explained in Remark 9.4.1 of Sec. 9.4. To satisfy the specification for a phase margin of 40°, the phase of the controller transfer function

$$G_c^*(j\omega) = \frac{1 + j\omega aT}{1 + j\omega T}$$

should be 22°.

Remark 9.7.1

Using the relation $20\log|G^*(j\omega_m) = -0.5[20\log a]$, as mentioned in step 4 of the phase-lead controller design approach presented at the beginning of this section, we seek to secure that when $\omega = \omega_m$, the amplitude of $G_c^*(j\omega)G^*(j\omega)$ is zero and its phase is φ_m (given by Eq. (9.4-9)). This, unfortunately, is not possible to achieve exactly for the following reason: since $\omega_m > \omega'$, the phase of $G^*(j\omega)$ at the frequency ω_m is different (smaller) than the phase at the frequency ω'. This means that the correction of 22° that we expect from $G_c^*(j\omega)$ to have in order to secure a phase margin of 40° at the frequency ω_m, is no longer valid. To overcome this obstacle, we increase the angle of 22° by an amount analogous to the slope of $\underline{/G(j\omega)}$.

Using Remark 9.7.1, we increase the phase by 5°, in which case the maximum angle φ_m of $G_c^*(j\omega)$ becomes $\varphi_m = 27°$. Then, using Eq. (9.4-10) we can determine the constant a. We have

$$\sin\varphi_m = \sin 27° = \frac{a-1}{a+1} = 0.454$$

thus, $a = 2.663$. Of course, we would have arrived at the same value for a if we had used Figure 9.10.

The frequency ω_m is calculated from the equation $20\log|G^*(j\omega_m)| = -0.5[20\log a] = -4.25$. From the plot of $20\log|G^*(j\omega)|$ of figure 9.45 we obtain $\omega_m = 40$ rad/sec.

Using the values of a and ω_m we can specify the time constant T. The corner frequencies $1/aT$ and $1/T$ of the Bode curve of $G^*(j\omega)$ (Figure 9.11) are selected such that the maximum angle φ_m is equal to $\varphi_m = \underline{/G_c^*(j\omega_m)}$, where ω_m is the geometrical mean of the two corner frequencies. In other words, we choose

$$\omega_m = \frac{1}{T\sqrt{a}}$$

Hence

$$T = \frac{1}{\omega_m\sqrt{a}} = 0.0153$$

Consequently, the transfer function $G_c(s)$ of the phase-lead controller is given by

$$G_c(s) = \left[\frac{1}{a}\right]\left[\frac{1+aTs}{1+Ts}\right] = \left[\frac{1}{2.66}\right]\left[\frac{1+0.04s}{1+0.0153s}\right]$$

Hence, the open-loop compensated transfer function becomes

$$G_c(s)G(s) = \left\{\left[\frac{1}{2.663}\right]\left[\frac{1+0.045s}{1+0.0153s}\right]\right\}\left\{\frac{(1000)(2.663)}{s(s+10)}\right\} = \frac{2663(s+24.5)}{s(s+10)(s+65.2)}$$

The phase margin for the compensated system (see Figure 9.45) is about 43°.

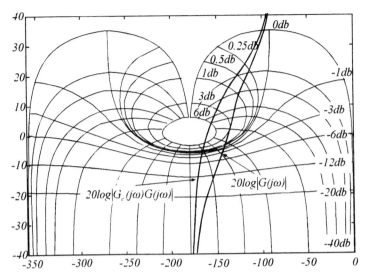

Figure 9.46 Nichols diagrams of the uncompensated and the compensated system of Example 9.7.1.

In Figure 9.45 we show the phase and magnitude Bode plots of the compensated and uncompensated system. In Figure 9.46, the same plots are shown using Nichols diagrams. In Figure 9.47, we give the time response of those two systems when the input is the unit step function. One can observe the improvement of the transient response of the compensated system (shorter rise time and smaller overshoot). In Figure 9.48 the amplitude M of the two systems is given as a function of

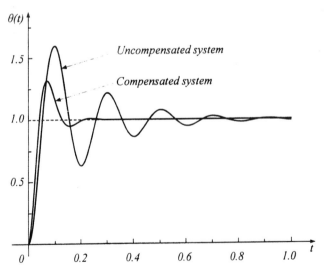

Figure 9.47 Time response of the uncompensated and the compensated system of Example 9.7.1.

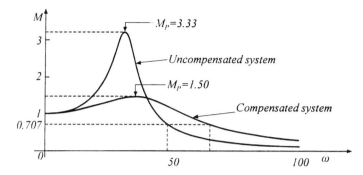

Figure 9.48   The amplitude M of the uncompensated and the compensated system of Example 9.7.1.

the frequency ω, where the effect of the controller on the bandwidth is shown. Finally, in Figure 9.49 the root locus of both systems is given, where the effect of the phase-lead controller upon the root locus of the closed-loop system is clearly shown.

9.8   DESIGN WITH PHASE-LAG CONTROLLERS

The phase-lag controllers are used to introduce a negative phase in the closed-loop transfer function aimed at improving the overshoot and the relative stability (note that, the rise time usually increases). The transfer function $G_c(s)$ of the phase-lag circuit affects the closed-loop transfer function in the low frequencies. This is easily seen if one considers the special case where $G_c(s)$ has the form $G_c(s) = 1 + K_i/s$. This form of $G_c(s)$ is the phase-lag controller (Eq. (9.4-11)) as $a \to 0$. This special case is shown in Figure 9.13. Since this special case involves a proportional and an integral term, we say that the phase-lag controller, as $a \to 0$, behaves like a PI controller. A simple way to realize $G_c(s) = 1 + K_i/s$ is as in Figure 9.50a. Note that the phase of $G_c(j\omega)$ is constant and equal to $-90°$. The influence of $G_c(s) = 1 + K_i/s$ on the closed-loop system performance is shown in Figure 9.50b. The influence of the phase-lag controller $G_c(s) = (1 + aTs)/(1 + Ts)$ on the closed-loop system performance is quite similar to the influence of $G_c(s) = 1 + K_i/s$.

The main steps in determining the system amplification constant K of the open-loop transfer function and of the parameter of the controller transfer function $G_c(s)$ of the phase-lag controller, using the Bode diagrams, are the following:

1. The amplification constant K is chosen such as to satisfy the specifications of the steady-state error.
2. From the Bode diagrams of $G(j\omega)$ we determine the phase and gain margins. Next, we find the frequency ω' corresponding to the specified phase margin. By knowing the frequency ω' we can determine the value of the parameter a of $G_c(j\omega)$ such that the magnitude diagram of the open-loop transfer function $G_c(j\omega)G(j\omega)$, for $\omega = \omega'$, becomes $0\,dB$, i.e., at the frequency ω', we have $20\log|G_c(j\omega')G(j\omega')| = 0$. We choose $G_c(j\omega)$ so that the frequency ω' is bigger than the corner frequency $1/aT$ (see Figure

(a)

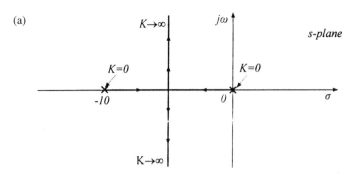

(b)

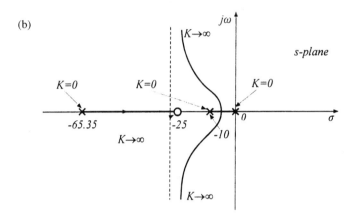

Figure 9.49 The root locus of (a) the uncompensated and (b) the compensated system of Example 9.7.1.

9.14), in which case $20 \log |G_c(j\omega')| = -20 \log a$. Thus, the parameter a may be calculated according to the relation

$$20 \log |G_c(j\omega')| = -20 \log a = -20 \log a |G(j\omega')|$$

or according to the relation

$$a = 10^{-\frac{20 \log |G(j\omega')|}{20}} \qquad (9.8\text{-}1)$$

3. The selection of the parameter T is done by approximation: usually, it is selected so that the highest corner frequency $1/aT$ is 10% of the new critical frequency ω', i.e., we choose

$$\frac{1}{aT} = \frac{\omega'}{10} \qquad (9.8\text{-}2)$$

The choice of T is made so that the phase of $G_c(s)$ does not affect the phase of $G(j\omega)$ at the frequency ω'.

4. Finally, we construct the diagram of the compensated open-loop transfer function $G_c(s)G(s)$. If the values of the parameters a and T do not give satisfactory results, we repeat step 2 by giving new (usually smaller) values to the critical frequency ω' until we get satisfactory results.

(a)

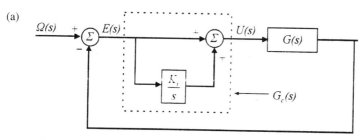

(b)

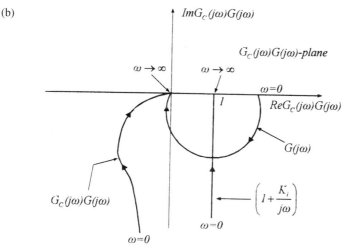

Figure 9.50 Closed-loop system with phase-lag controller and the effect of the controller on the Nyquist diagram of the closed-loop system. (a) Closed-loop system with controller $G_c(s) = 1 + (K_i/s)$; (b) Nyquist diagram of $G(j\omega)$, $G_c(j\omega)$, and $G_c(j\omega)G(s)$.

Example 9.8.1

Design a phase-lag controller that it satisfies the design specifications of the system of Example 9.7.1.

Solution

Consider Figure 9.44b. For the present example, the controller $G_c(s)$ is the phase-lag controller, whose transfer function has the form

$$G_c(s) = \frac{1 + aTs}{1 + Ts}, \qquad a < 1$$

The steady-state velocity error requirement is met by appropriately choosing the amplification constant K using Eq. (4.7-11) as follows:

$$e_{ss}(t) = \lim_{s \to 0}\left\{\frac{s\Theta_i(s)}{1 + G_c(s)G(s)}\right\} = \lim_{s \to 0}\left\{\frac{s\left[\dfrac{1}{s^2}\right]}{1 + \left[\dfrac{1 + aTs}{1 + Ts}\right]\left[\dfrac{1000K}{s(s+10)}\right]}\right\} = \frac{1}{100K} \le 10^{-2}$$

Hence, $K \ge 1$.

The Bode diagrams of $G(j\omega)$ are given in Figure 9.51. From these diagrams we can conclude that at the frequency $\omega = 14\,\text{rad/sec}$ the phase of $G(j\omega)$ is 40°. The frequency ω' is usually chosen to be smaller than $\omega = 14\,\text{rad/sec}$ for the reasons given in the following remark.

Remark 9.8.1

For the case of the phase-lead circuits (as mentioned in Remark 9.7.1) there are certain difficulties in selecting ω_m for the appropriate maximum angle φ_m. Similar difficulties arise for the case of phase-lag circuits in selecting ω' which will lead to the appropriate parameter a. This difficulty is handled in the same way as in Remark 9.7.1. Specifically, in place of the frequency ω', we choose a smaller frequency, analogous to the slope of the magnitude diagram of $G(j\omega)$, because the critical frequency of $G_c(j)G(j\omega)$ will be smaller than the critical frequency ω' of $G(j\omega)$. This is because in $G_c(j\omega)G(j\omega)$ the factor $G_c(j\omega)$ shifts the magnitude plot of $G(j\omega)$ downwards.

Using Remark 9.8.1 we select ω' a bit smaller than the value $14\,\text{rad/sec}$. For example, let $\omega' = 10\,\text{rad/sec}$. Then, the magnitude of $G(j\omega)$ at the frequency $\omega' = 10\,\text{rad/sec}$ is about 20 dB. Hence the constant a, according to Eq. (9.8-1), is given by

$$a = 10^{-\frac{20}{20}} = 0.1$$

The time constant T is calculated according to Eq. (9.8-2). We have

$$\frac{1}{aT} = \frac{\omega'}{10} = \frac{10}{10} = 1$$

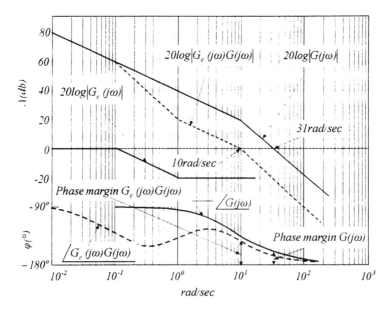

Figure 9.51 Bode diagrams of phase and magnitude plots of the uncompensated and the compensated system of Example 9.8.1.

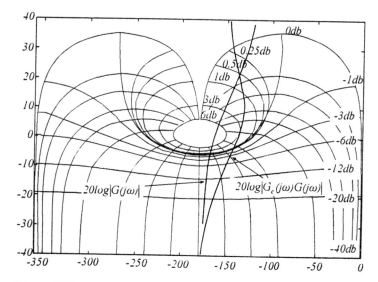

Figure 9.52 Nichols diagrams of the uncompensated and the compensated system of Example 9.8.1.

Hence, $T = 10$.

The transfer function $G_c(s)$ of the phase-lag controller has the form

$$G_c(s) = \frac{1 + aTs}{1 + Ts} = \frac{1 + s}{1 + 10s}$$

The open-loop transfer function of the compensated system becomes

$$G_c(s)G(s) = \frac{100(s + 1)}{s(s + 0.1)(s + 10)}$$

The phase margin of the compensated system (see Figure 9.51) is close to 50°. This margin is about 10° bigger than the required margin of 40°, and it is therefore very satisfactory.

In Figure 9.52 the Nichols diagrams of $G(j\omega)$ and $G_c(j\omega)G(j\omega)$ are given. In Figure 9.53 we present the time response of these two systems when the input is the unit step function, from which it is clear that the compensated system has a smaller overshoot but higher rise time. The increase in the rise time is because the bandwidth of the compensated system has decreased.

9.9 DESIGN WITH PHASE LAG-LEAD CONTROLLERS

The phase lag-lead controllers are used in cases where a phase-lead or a phase-lag controller alone cannot satisfy the design specifications. For the selection of the appropriate phase lag-lead controller there is no systematic method. For this reason, it is usually done by successive approximations. Note that a special form of the transfer function $G_c(s)$ of the phase lag-lead networks is the PID controller, which has been presented in Sec. 9.6.

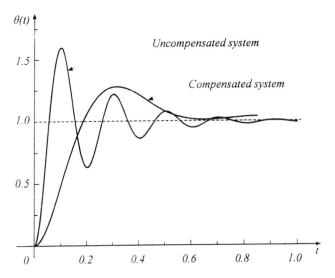

Figure 9.53 Time response of the uncompensated and the compensated system of Example 9.8.1.

Example 9.9.1

Consider a system with transfer function

$$G(s) = \frac{K}{s(1+0.1s)(1+0.4s)}$$

Find a phase lag-lead controller such as to satisfy the following closed-loop specifications:

(a) Velocity constant $K_v = 100 \sec^{-1}$
(b) Phase margin $\varphi_p \geq 45°$.

Solution

Using the definition (4.7-3) of the velocity constant K_v yields

$$K_v = \lim_{s \to 0}[sG(s)] = K$$

Hence, $K = 100 \sec^{-1}$.

The transfer function $G_c(s)$ of the phase lag-lead controller is given by Eq. (9.4-12), i.e., by the equation

$$G_c(s) = \left[\frac{1+bT_2s}{1+T_2s}\right]\left[\frac{1+aT_1s}{1+T_1s}\right] = G_1(s)G_2(s), \qquad \text{with} \qquad ab = 1$$

The determination of $G_c(s)$ will be done in two steps. First, we determine the parameters of $G_1(s)$ and, secondly, the parameters of $G_2(s)$, as follows.

Step 1

Determination of $G_1(s)$. We draw the Bode diagrams of $G(j\omega)$ (Figure 9.54). The critical frequency ω' is $\omega' = 14\,\text{rad/sec}$. Assume that we want to move ω' from

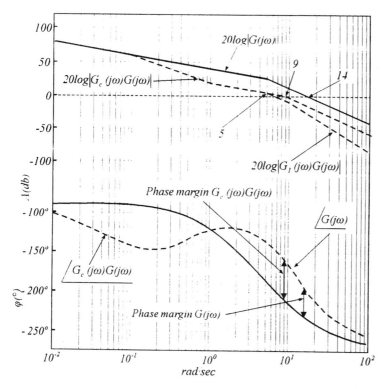

Figure 9.54 The Bode diagrams of the magnitude and phase plots of the uncompensated and the compensated system of Example 9.9.1.

$\omega' = 14 \, \text{rad/sec}$ to the new position $\omega' = 5 \, \text{rad/sec}$ using only $G_1(s)$. Since $20 \log |G(j5)| = 20 \, \text{dB}$ and using Eq. (9.8-1), the constant b is given by

$$b = 10^{-\frac{20}{20}} = 0.1$$

To find T_2, we use Eq. (9.8-2), in which case we have

$$\frac{1}{bT_2} = \frac{5}{10}$$

and, thus, $T_2 = 20$. Hence $G_1(s)$ has the form

$$G_1(s) = \frac{1 + 2s}{1 + 20s}$$

Step 2

Determination of $G_2(s)$. The parameter a is calculated by the constraint relation $ab = 1$, which yields $a = 10$. The maximum angle which corresponds to $a = 10$ is found by the relation

$$\sin \varphi_m = \frac{a - 1}{a + 1} = \frac{9}{11}$$

which yields $\varphi_m = 54.9°$. The frequency ω_m is calculated by the relation $20 \log |G(j\omega_m) = -0.5[20 \log a] = -10\,\text{dB}$. From the diagram $20 \log |G(j\omega)|$ of Figure 9.54 we get that $\omega_m = 9\,\text{rad/sec}$. Finally, from relation (9.7-2), the parameter T_1 may be determined as follows:

$$T_1 = \frac{1}{\omega_m \sqrt{a}} = \frac{1}{9\sqrt{10}} = \frac{1}{28.46}$$

Therefore, the transfer function $G_2(s)$ has the form

$$G_2(s) = \frac{1 + 0.35s}{1 + 0.035s}$$

Finally, $G_c(s)$ has the form

$$G_c(s) = G_1(s)G_2(s) = \left[\frac{1 + 0.35s}{1 + 0.035s}\right]\left[\frac{1 + 2s}{1 + 20s}\right]$$

The open-loop transfer function of the compensated system becomes

$$G_c(s)G(s) = \frac{2500(s + 2.86)(s + 0.5)}{s(s + 10)(s + 2.5)(s + 28.6)(s + 0.05)}$$

In Figure 9.54 one can observe that the phase margin of the compensated system is about 50°. Hence, both specifications of the problem are satisfied.

9.10 DESIGN WITH CLASSICAL OPTIMAL CONTROL METHODS

Generally speaking, the "classical" optimal control approach aims to determine the parameters of the controller such as to minimize a specific cost function for a particular type of input signal. More specifically, the classical optimal control problem is formulated as follows. Given a linear time-invariant SISO system, described by the transfer function $G(s)$, apply output feedback as shown in Figure 9.55. The cost function J may have several forms. In this section we consider the two cost functions J_e and J_u, where

$$J_e = \int_0^\infty e^2(t)\,dt \qquad\qquad (9.10\text{-}1)$$

$$J_u = \int_0^\infty u^2(t)\,dt \qquad\qquad (9.10\text{-}2)$$

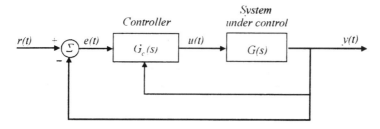

Figure 9.55   Block diagram of classical optimal control system.

The cost function J_e is called the *integral square error* (ISE) and expresses the specifications of the closed-loop system that refer to features such as overshoot, rise time, and relative stability. The cost function J_u is called the *integral square effort* and expresses the energy that is consumed by the control signal in performing the specified control action.

The control design problem considered in this chapter is to find the appropriate parameters of the controller $G_c(s)$ in Figure 9.55 such that one of the two cost functions, J_e or J_u, or a combination of both, is a minimum.

There are two general categories of classical optimal control problems: the case where the structure of the controller is free and the case where the structure of the controller is fixed. Both categories are studied in the material that follows.

It is noted that in Chap. 11 we present an introduction to "modern" optimal control approach. In this case, the system under control is described in state space and the derivation of the optimal controller is based on very advanced mathematical techniques, such as the calculus of variations, the maximum principle, and the principle of optimality. Before we present these modern techniques, we thought that it is worthwhile to present in this chapter, in conjunction with other classical control techniques that we have already presented, the "classical" optimal control techniques that were founded before the modern theories of optimal and stochastic control appeared. The classical control methods that we are about to present are useful in practice and are helpful in understanding the modern optimal control methods that follow in Chap. 11.

9.10.1 Free Structure Controllers

In this case there are no restrictions on the form of the type of the controller. More specifically, the design problem for the present case is the following: find a controller which minimizes J_e (or J_u), with $J_u = K$ (or $J_e = K$), where K is a constant.

Assume that we wish to minimize J_e, while $J_u = K$. To solve the problem, we make use of the Lagrange multiplier method. This method begins by expressing J_e and J_u as a single cost criterion, as follows

$$J = J_e + \lambda J_u \tag{9.10-3}$$

where λ is the Lagrange multiplier. The minimization of J will be done in the s-domain by using Parseval's theorem. Parseval's theorem relates a function $f(t)$ described in the time domain with its complex frequency counterpart $F(s)$, as follows:

$$\int_0^\infty f^2(t)\,dt = \frac{1}{2\pi j}\int_{-j\infty}^{j\infty} F(s)F(-s)\,ds \tag{9.10-4}$$

where $F(s)$ is the Laplace transformation of $f(t)$. The theorem is valid under the condition that $F(s)$ has all its poles in the left-hand side of the complex plane. Using Eqs (9.10-1) and (9.10-2), the cost function (9.10-3) may be written as

$$J = J_e + \lambda J_u = \int_0^\infty [e^2(t) + \lambda u^2(t)]\,dt \tag{9.10-5}$$

If we apply Parseval's theorem (9.10-4) and (9.10-5), the cost function J becomes

$$J = \frac{1}{2\pi j} \int_{-j\infty}^{j\infty} [E(s)E(-s) + \lambda U(s)U(-s)]\, ds \tag{9.10-6}$$

From Figure 9.55, we have

$$E(s) = R(s) - G(s)U(s) \tag{9.10-7}$$

Hence, the cost function J takes on the form

$$J = \frac{1}{2\pi j} \int_{-j\infty}^{j\infty} \left[[R(s) - G(s)U(s)][R(-s) - G(-s)U(-s)] + \lambda U(s)U(-s) \right] ds \tag{9.10-8}$$

The cost function J is a function of $U(s)$ and λ. To study the maxima and minima of J we apply the method of calculus of variations (see also Subsec. 11.2.1). To this end, assume that $U(s)$ is a rational function of s, which is given by the equation

$$U(s) = \hat{U}(s) + \varepsilon R_1(s) = \hat{U}(s) + \delta U(s) \tag{9.10-9}$$

where $\hat{U}(s)$ is the optimal control signal sought, ε is a constant, $R_1(s)$ is any rational function whose poles lie in the left-half complex plane, and $\delta U(s)$ is the change of $U(s)$ about the optimal control signal $\hat{U}(s)$. If we substitute Eq. (9.10-9) into Eq. (9.10-8), and after some appropriate grouping, we have

$$J = J_e + \lambda J_u = J_1 + J_2 + J_3 + J_4 \tag{9.10-10}$$

where

$$J_1 = \frac{1}{2\pi j} \int_{-j\infty}^{j\infty} [R\overline{R} - R\overline{G}\hat{\overline{U}} - \overline{R}G\hat{U} + \lambda \hat{U}\hat{\overline{U}} + G\overline{G}\hat{U}\hat{\overline{U}}]\, ds \tag{9.10-11a}$$

$$J_2 = \frac{1}{2\pi j} \int_{-j\infty}^{j\infty} [\lambda\hat{\overline{U}} - G\overline{R} + G\overline{G}\hat{\overline{U}}]\varepsilon R_1\, ds \tag{9.10-11b}$$

$$J_3 = \frac{1}{2\pi j} \int_{-j\infty}^{j\infty} [\lambda\hat{U} - \overline{G}R + G\overline{G}\hat{U}]\varepsilon \overline{R}_1\, ds \tag{9.10-11c}$$

$$J_4 = \frac{1}{2\pi j} \int_{-j\infty}^{j\infty} [\lambda + G\overline{G}]\varepsilon^2 R_1 \overline{R}_1\, ds \tag{9.10-11d}$$

where, for simplicity, we use G instead of $G(s)$, $\overline{G}$ instead of $G(-s)$, U instead of $U(s)$, $\overline{U}$ instead of $U(-s)$, etc.

Next, we calculate the linear part δJ of the first differential of J. We observe the following with regard to the factor εR_1: the factor εR_1 does not appear in J_1, it appears to the first power in J_2 and J_3, and it appears to higher (second) power in J_4. Furthermore, from Eqs (9.10-11b and c) we have that $J_2 = J_3$, which can be easily proven if we substitute s by $-s$ and vice versa. Consequently, δJ becomes

$$\delta J = J_2 + J_3 = \frac{2}{2\pi j} \int_{-j\infty}^{j\infty} [\lambda\hat{U} - \overline{G}R + G\overline{G}\hat{U}]\varepsilon \overline{R}_1\, ds \tag{9.10-12}$$

A necessary condition for J to be a minimum for $U(s) = \hat{U}(s)$ is that

$$\delta J = 0 \tag{9.10-13}$$

For Eq. (9.10-13) to be valid for every $\varepsilon \overline{R}_1$, the function $X(s)$, where

$$X(s) = \lambda \hat{U} - \overline{G}R + G\overline{G}\hat{U} = [\lambda + G\overline{G}]\hat{U} - \overline{G}R \qquad (9.10\text{-}14)$$

must satisfy the following condition:

$$\delta J = \frac{2}{2\pi j} \int_{-j\infty}^{j\infty} X(s)\varepsilon \overline{R}_1 \, ds = 0 \qquad (9.10\text{-}15)$$

We assume that the control signal $u(t)$ is bounded. Then, it follows that the function $R_1(s)$ has all its poles in the left-half complex plane. This means that $\overline{R}_1 = R_1(-s)$ will have all its poles in the right-half complex plane. A sufficient condition for $\delta J = 0$, independently of $\overline{R}_1$, is that the function $X(s)$ has all its poles in the right-half complex plane and that the integration should be performed around the left-half complex plane. Indeed, in this case, if we integrate going from $-j\infty$ to $j\infty$ and passing only through the left-half complex plane, the integral in Eq. (9.10-15) will become zero. Next, define

$$Y(s)Y(-s) = Y\overline{Y} = \lambda + G\overline{G} \qquad (9.10\text{-}16)$$

Since the function $\lambda + G\overline{G}$ is symmetrical about the $j\omega$-axis, it follows that the function $Y(s)$ has poles and zeros only in the left-half complex plane, while $Y(-s)$ has poles and zeros only on the right-half complex plane. We can express these remarks by the following definitions:

$$Y = [\lambda + G\overline{G}]^+ \qquad (9.10\text{-}17a)$$

$$\overline{Y} = [\lambda + G\overline{G}]^- \qquad (9.10\text{-}17b)$$

The factorization of the function $\lambda + G\overline{G}$ in the sense of definitions (9.10-17a and b) is called *spectral factorization*. Hence, $X(s)$ may be written as

$$X = Y\overline{Y}\hat{U} - \overline{G}R \qquad \text{or} \qquad \frac{X}{Y} = Y\hat{U} - \frac{\overline{G}R}{Y}$$

or

$$\frac{X}{Y} + \left[\frac{\overline{G}R}{Y}\right]_- = Y\hat{U} - \left[\frac{\overline{G}R}{Y}\right]_+ \qquad (9.10\text{-}18)$$

where

$$\left[\frac{\overline{G}R}{Y}\right]_+ = \begin{array}{l} \text{the part of the partial fraction expansion of } \overline{G}R/Y \\ \text{whose poles lie in left-half complex plane} \end{array}$$

$$\left[\frac{\overline{G}R}{Y}\right]_- = \begin{array}{l} \text{the part of the partial fraction expansion of } \overline{G}R/Y \\ \text{whose poles lie in right-half complex plane} \end{array}$$

The left-hand side of Eq. (9.10-18) involves terms whose poles lie only in the right-half complex plane, while the right-hand side involves terms whose poles lie only in the left-half complex plane. Hence, in order for Eq. (9.10-18) to hold, both sides must be equal to zero. This yields

$$\hat{U} = \frac{1}{Y}\left[\frac{\overline{G}R}{Y}\right]_+ \qquad (9.10\text{-}19)$$

Equation (9.10-19) is the optimal control signal. However, this signal is a function of the parameter λ. The value of the parameter λ can be found from the constraint $J_u = K$, i.e., from the equation

$$J_u = \frac{1}{2\pi j} \int_{-j\infty}^{j\infty} \hat{U}(s)\hat{U}(-s)\,ds = K \tag{9.10-20}$$

To facilitate the calculations of the integral (9.10-20), define

$$I_n = \frac{1}{2\pi j} \int_{-j\infty}^{j\infty} M(s)M(-s)\,ds = \frac{1}{2\pi j} \int_{-j\infty}^{j\infty} \left[\frac{c(s)c(-s)}{d(s)d(-s)}\right]ds$$

where

$$c(s) = c_{n-1}s^{n-1} + \cdots + c_1 s + c_0 \quad \text{and}$$
$$d(s) = d_n s^n + d_{n-1}s^{n-1} + \cdots + d_1 s + d_0$$

The integrals I_1, I_2, I_3, and I_4, are given in Table 9.3. To derive the general expression of I_n is a formidable task.

The transfer function $H(s)$ of the optimal closed-loop system is given by

$$H(s) = \frac{Y(s)}{R(s)} = \frac{G(s)}{R(s)}\hat{U}(s) \tag{9.10-21}$$

Using Eq. (9.10-21) we can determine the transfer function $G_c(s)$ of the optimal controller. Indeed, if for example the closed-loop system is as shown in Figure 9.56, the optimal controller has the form

$$G_c(s) = \frac{1}{G(s)}\left[\frac{H(s)}{1 - H(s)}\right] \tag{9.10-22}$$

Table 9.3 The Integrals I_1, I_2, I_3, and I_4

$$I_1 = \frac{c_0^2}{2d_0 d_1}$$

$$I_2 = \frac{c_1^2 d_0 + c_0^2 d_2}{2d_0 d_1 d_2}$$

$$I_3 = \frac{c_1^2 d_0 d_1 + (c_1^2 - 2c_0 c_2)d_0 d_3 + c_1^2 d_2 d_3}{2d_0 d_3(-d_0 d_3 + d_1 d_2)}$$

$$I_4 = \frac{c_3^2(-d_0^2 d_3 + d_0 d_1 d_2) + (c_2^2 - 2c_1 c_3)d_0 d_1 d_4}{2d_0 d_3(-d_0 d_3^2 - d_1^2 d_4 + d_1 d_2 d_3)}$$
$$+ \frac{(c_1^2 - 2c_0 c_2)d_0 d_3 d_4 + c_0^2(-d_1 d_4^2 + d_2 d_3 d_4)}{2d_0 d_3(-d_0 d_3^2 - d_1^2 d_4 + d_1 d_2 d_3)}$$

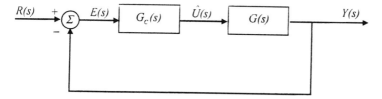

Figure 9.56 Block diagram with optimal controller.

Remark 9.10.1

The results of the present subsection may also be applied for the case of MIMO systems. However, this extension involves great difficulties, particularly in dealing with the problem of the spectral factorization.

Example 9.10.1

Consider the closed-loop system of Figure 9.56, where $G(s) = 1/s^2$ and $R(s) = 1/s$. Find the transfer function $G_c(s)$ of the optimal controller such that J_e = minimum and $J_u \leq 1$.

Solution

We have

$$\lambda + G\overline{G} = \lambda + \left[\frac{1}{s^2}\right]\left[\frac{1}{(-s)^2}\right] = \frac{\lambda s^4 + 1}{s^4}$$

To factorize the function $\lambda + G\overline{G}$, we assume that the output $Y(s)$ has the form $Y(s) = [a_2 s^2 + a_1 s + a_0]/s^2$. Then

$$Y(s)Y(-s) = \left[\frac{a_2 s^2 + a_1 s + a_0}{s^2}\right]\left[\frac{a_2 s^2 - a_1 s + a_0}{s^2}\right] = \frac{a_2^2 s^4 - (a_1^2 - 2a_0 a_2)s^2 + a_0^2}{s^4}$$

From Eq. (9.10-16), we have

$$\lambda s^4 + 1 = a_2^2 s^4 - (a_1^2 - 2a_0 a_2)s^2 + a_0^2$$

Equating the coefficients of like powers of s of both sides in the above equation, we obtain $\lambda = a_2^2$, $a_1^2 - 2a_0 a_2 = 0$ and $a_0^2 = 1$. We finally obtain $a_0 = 1$, $a_1 = \sqrt{2}\rho$, and $a_2 = \rho^2$, where $\lambda = \rho^4$. Hence,

$$Y(s) = \frac{\rho^2 s^2 + \sqrt{2}\rho s + 1}{s^2} \quad \text{and} \quad Y(-s) = \frac{\rho^2 s^2 - \sqrt{2}\rho s + 1}{s^2}$$

We also have

$$\frac{\overline{G}R}{\overline{Y}} = \frac{\left[\frac{1}{s^2}\right]\left[\frac{1}{s}\right]}{\frac{\rho^2 s^2 - \sqrt{2}\rho s + 1}{s^2}} = \frac{1}{s[\rho^2 s^2 - \sqrt{2}\rho s + 1]}$$

Therefore

$$\left[\frac{\overline{GR}}{\overline{Y}}\right]_+ = \frac{1}{s}$$

where we have considered that the pole $s = 0$ lies in the left-half complex plane. Using Eq. (9.10-19), we obtain

$$\hat{U}(s) = \frac{s}{\rho^2 s^2 + \sqrt{2}\rho s + 1}$$

Therefore, we have determined the optimal control signal $\hat{U}(s)$, as a function of $\lambda = \rho^4$. To find λ, we use Eq. (9.10-20) with $K = 1$. This yields

$$J_u = \frac{1}{2\pi j} \int_{-j\infty}^{j\infty} \left[\frac{s}{\rho^2 s^2 + \sqrt{2}\rho s + 1}\right]\left[\frac{-s}{\rho^2 s^2 - \sqrt{2}\rho s + 1}\right] ds$$

$$= \frac{c_1^2 d_0 + c_0^2 d_2}{2 d_0 d_1 d_2} = \frac{1}{2\sqrt{2}\rho^3} \leq 1$$

where

$$\rho = \left[\frac{1}{2\sqrt{2}}\right]^{1/3}, \qquad \lambda = \left[\frac{1}{2\sqrt{2}}\right]^{4/3}, \qquad \text{and} \qquad U(s) = \frac{2}{s^2 + 2s + 2}$$

where for the calculation of J_u use was made of Table 9.3. Finally, the transfer function of $G_c(s)$ of the optimal controller has the form

$$G_c(s) = \frac{1}{G(s)}\left[\frac{H(s)}{1 - H(s)}\right] = s^2\left[\frac{\frac{1}{s}\hat{U}(s)}{1 - \frac{1}{s}\hat{U}(s)}\right] = \frac{s}{\rho^2 s + \sqrt{2}\rho} = \frac{2s}{s + 2}$$

9.10.2 Fixed Structure Controllers

In this case, the transfer function $G_c(s)$ of the controller has a preassigned fixed structure. For example, for the closed-loop system of Figure 9.56, the following specific form for $G_c(s)$ may be assigned:

$$G_c(s) = \frac{b_m s^m + b_{m-1} s^{m-1} + \cdots + b_1 s + b_0}{s^n + a_{n-1} s^{n-1} + \cdots + a_1 s + a_0}, \qquad m \leq n \tag{9.10-23}$$

The problem here is to find the appropriate values of the parameters $a_0, a_1, \ldots, a_{n-1}$, $b_0, b_1, \ldots, b_m$ of $G_c(s)$ which minimize a cost function J. This method is called the parameter optimization method and is actually a minimization problem of a function involving many variables. To illustrate the method, three examples are presented which show the procedure involved. The last example is of practical interest because it refers to the optimal control of a position control system.

Example 9.10.2

Consider the closed-loop system of Figure 9.57. Find the value of K such that the cost function $J = J_e + \lambda J_u$ is a minimum when $r(t) = 1$.

Solution

The error $E(s)$ and the signal $U(s)$ are given by

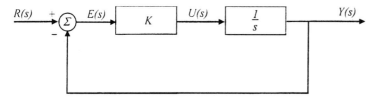

Figure 9.57 Block diagram of Example 9.10.2.

$$E(s) = \frac{1}{s+K} \quad \text{and} \quad U(s) = KE(s) = \frac{K}{s+K}$$

Therefore

$$J = J_e + \lambda J_u = \frac{1}{2\pi j}\int_{-j\infty}^{j\infty} E(s)E(-s)\,ds + \lambda\frac{K^2}{2\pi j}\int_{-j\infty}^{j\infty} E(s)E(-s)\,ds$$

$$= \frac{1}{2K} + \lambda\frac{K^2}{2K} = \frac{1}{2K} + \frac{\lambda K}{2}$$

The value of K which minimizes J can be found using well-known techniques. For example, take the partial derivative of J with respect to K to yield

$$\frac{\partial J}{\partial K} = -\frac{1}{2K^2} + \frac{\lambda}{2} = 0$$

Solving the above equation yields that J is minimum when $K = \hat{K} = \lambda^{-1/2}$.

Example 9.10.3

Consider the block diagram of Figure 9.58. The transfer functions $G(s)$ and $G_c(s)$ are given by

$$G(s) = \frac{K}{s^2(Ts+1)} \quad \text{and} \quad G_c(s) = 1 + K_d s$$

Find the value of the constant K_d which minimizes J_e when $r(t) = 1$.

Solution

The error $E(s)$ is given by

$$E(s) = R(s) - Y(s)$$

Also, we have that

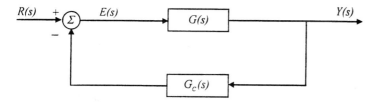

Figure 9.58 Block diagram of Example 9.10.3.

$$Y(s) = \left[\frac{G(s)}{1 + G(s)G_c(s)}\right] R(s)$$

Therefore

$$E(s) = R(s) - Y(s) = \left[\frac{1 + G(s)G_c(s) - G(s)}{1 + G(s)G_c(s)}\right] R(s)$$

Substitute the expressions of $G(s)$ and of $G_c(s)$ in the above relation to yield

$$E(s) = \frac{Ts^2 + s + KK_d}{Ts^3 + s^2 + KK_d s + K}$$

Using definition (9.10-1) and the Parseval's theorem, the cost function J_e becomes

$$J_e = \int_0^\infty e^2(t)\,dt = \frac{1}{2\pi j}\int_{-j\infty}^{j\infty} E(s)E(-s)\,ds = \frac{1}{2}\left[K_d + \frac{1}{K(K_d - T)}\right]$$

where use was made of Table 9.3. The value K_d for which J_e is minimum can be found using well-known techniques. For example, take the partial derivative of J_e with respect to K_d to yield

$$\frac{\partial J_e}{\partial K_d} = \frac{1}{2}\left[1 - \frac{1}{K(K_d - T)^2}\right] = 0$$

The above equation gives

$$K_d = \hat{K}_d = T \pm K^{-1/2}$$

It is noted that the optimal value $\hat{K}_d$ of K_d, as $K \to \infty$, becomes $\hat{K}_d = T$. In this case $G_c(s)$ becomes $G_c(s) = 1 + Ts$. In other words, the zero of $G_c(s)$ coincides with one of the poles of $G(s)$. As a result, the closed-loop system is of second order.

Example 9.10.4

Consider the position control system described in Subsec. 3.13.2 (Figure 3.51). Find the values of the unspecified parameters of the closed-loop system such as to minimize the cost function

$$J = \int_0^\infty [\theta_e(t)]^2\,dt$$

For simplicity, let $L_a \simeq 0$ and $K_p = 1$.

Solution

As we have already shown in subsec. 3.13.2, when $L_a \simeq 0$ and $K_p = 1$, then the block diagram 3.51c of the closed-loop system is simplified as shown in Figure 9.59. The forward-path transfer function $G(s)$ reduces to

$$G(s) = \frac{K}{As^2 + Bs}, \qquad \text{where} \qquad K = \frac{K_a K_i N}{R_a}, \qquad A = J_m^*, \qquad \text{and}$$

$$B = B_m^* + \frac{K_i K_b}{R_a}$$

Let the input $\theta_r(t)$ of the system be the unit step function, i.e., let $\theta_r(t) = 1$. In $G(s)$, all the unspecified parameters K, A, and B of the system are to be chosen so as to

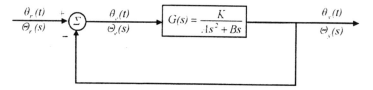

Figure 9.59   Simplified block diagram of the servomechanism.

minimize J. To determine the values of K, A, and B which minimize J, use the Parseval's theorem given by relation (9.10-4). For this example, we have

$$\int_0^\infty [\theta_e(t)]^2 \, dt = \frac{1}{2\pi j} \int_{-j\infty}^{j\infty} \Theta_e(s)\Theta_e(-s) \, ds$$

It can be easily shown that

$$\Theta_e(s) = \left[\frac{As^2 + Bs}{As^2 + Bs + K}\right]\Theta_r(s)$$

Since $\theta_r(t) = 1$, or $\Theta_r(s) = 1/s$, the above equation becomes

$$\Theta_e(s) = \frac{As + B}{As^2 + Bs + K}$$

Since $\Theta_e(s)$ is of the form $c(s)/d(s)$, where $c(s) = c_0 + c_1 s$ and $d(s) = d_0 + d_1 s + d_2 s^2$, the calculation of the Parseval's integral can be done by using Table 9.3, where $c_0 = B$, $c_1 = A$, $d_0 = K$, $d_1 = B$ and $d_2 = A$. using Table 9.3 yields

$$I_2 = \frac{c_1^2 d_0 + c_0^2 d_2}{2 d_0 d_1 d_2} = \frac{A^2 K + B^2 A}{2KBA} = \frac{AK + B^2}{2KB}$$

Hence

$$J = \int_0^\infty [\theta_e(t)]^2 \, dt = \frac{AK + B^2}{2KB}$$

Clearly, the above cost function J is an analytical expression of the cost function J in terms of the parameters K, A, and B of the closed-loop system.

We will further investigate the above expression for J in terms of K, A, and B, wherein we distinguish the following four interesting cases:

Case 1

Let K and B be constants. Then, for J to be minimum we must have $A = 0$, in which case $J = B/2K$.

Case 2

A more realistic approach is to assume that K and A are constant. Then, J becomes maximum with respect to B when $\partial J/\partial B = 0$, which gives $B = \sqrt{KA}$. Returning to Figure 9.59, we may write the differential equation of the closed-loop system as follows:

$$\frac{d^2\theta_y(t)}{dt^2} + 2\zeta\omega_n\frac{d\theta_y(t)}{dt} + \omega_n^2\theta_y(t) = \omega_n^2\theta_r(t)$$

where

$$\zeta = \frac{B}{2\sqrt{KA}} \qquad \text{and} \qquad \omega_n = \sqrt{\frac{K}{A}}$$

Clearly, in the present case where $B = \sqrt{KA}$, we have $\zeta = 0.5$. That is, we have the very interesting result that the value of the damping ratio $\zeta = 0.5$.

Case 3

Let A and B be constants. The parameter K is strongly influenced by the amplification constant K_a of the amplifier. If we differentiate J with respect to K, then the derivative tends to zero as $K \to \infty$. For $K \to \infty$, we obtain

$$\lim_{K\to\infty} J = \frac{A}{2B}$$

Case 4

One more useful case is to limit the values of K and B such that $KB = C$, where C is a cosntant. Then, since $K = C/B$, the cost function J becomes

$$J = \frac{ACB^{-1} + B^2}{2C}$$

The partial derivative $\partial J/\partial B$ is zero when $2B^3 = AC = AKB$. Hence

$$B = \sqrt{\frac{KA}{2}}$$

In this case the damping ratio $\zeta = [2\sqrt{2}]^{-1} \cong 0.353$.

Finally, it is noted that by using the above results, one may study other combinations of K, A, and B.

9.11 PROBLEMS

1. For the control system shown in Figure 9.60, solve the design problems given in Table 9.4. Furthermore:

   (a) draw the Bode and the Nichols diagrams
   (b) plot the amplitudes M and the step responses of the systems.

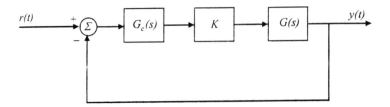

Figure 9.60

Table 9.4

$G(s)$		Compensator type	Design requirements
1	$\dfrac{1}{s(1+0.2s)}$	Phase-lead controller	$K_v = 4\,\mathrm{sec}^{-1}$ $\varphi_p \geq 40°$
2	$\dfrac{1}{s(1+0.1s)}$	Phase-lag controller	$K_v = 10\,\mathrm{sec}^{-1}$ $\varphi_p \geq 40°$
3	$\dfrac{1}{s^2(1+0.2s)}$	Phase-lead controller	$K_a = 4\,\mathrm{sec}^{-2}$ $M_p \leq 2$
4	$\dfrac{1}{(s+0.5)(s+0.1)(s+0.2)}$	Phase-lag controller	$K_p = 1$ $M_p \leq 0.7$

2. The closed-loop block diagram for controlling the altitude of a space vehicle is given in Figure 9.61. Determine a phase-lead controller so that for the closed-loop system the settling time (2%) is $T_s \leq 4\,\mathrm{sec}$ and the maximum percent over-shoot is less than 20%.

3. Consider the case of controlling the angle θ of the robot arm shown in Figure 9.62a. Determine a phase-lag network so that $K_v = 20\,\mathrm{sec}^{-1}$ and $\zeta = 0.707$ for the compensated closed-loop system shown in Figure 9.62b.

4. The open-loop transfer function of a position control servomechanism is given by

$$G(s)F(s) = \frac{K}{s(0.1s+1)(0.2s+1)}$$

Design a phase lag-lead compensator such that for the compensated closed-loop system the velocity error constant is $K_v = 30\,\mathrm{sec}^{-1}$, the phase margin is $\varphi_p \cong 50°$, and the bandwidth BW $\cong 12\,\mathrm{rad/sec}$.

5. Consider the orientation control system of a satellite described in Subsec. 3.13.7 and Example 9.6.1, where the controller is a PD controller and $K_t = K_b = J = 1$. Determine the parameters of the PD controller such that for the closed-loop system $\zeta = 0.7$ and $\omega_n = 2$.

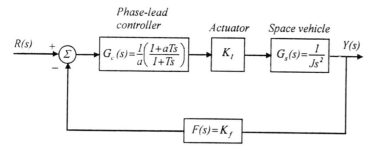

Figure 9.61

(a)

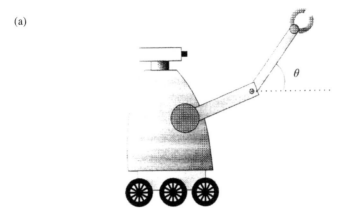

(b)

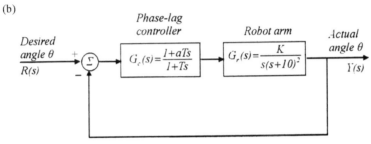

Figure 9.62

6. Consider the position servomechanism shown in Figure 9.63, where a PI controller is used. Determine the parameters of the PI controller for a 25% overshoot.

7. Consider the system shown in Figure 9.64. Determine the parameters of the PI controller, such that the poles of the closed-loop system are -2 and -3.

8. Consider the system shown in Figure 9.65. Determine the parameters of the PID controller, such that the poles of the closed-loop system are $-2+j$, $-2-j$, and -5.

9. Consider the system shown in Figure 9.66. Using the Ziegler–Nichols stability limit method, determine the parameters of the PID controller in order to achieve an overshoot of 25%.

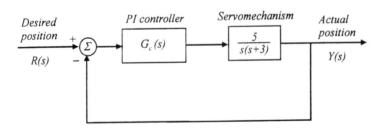

Figure 9.63

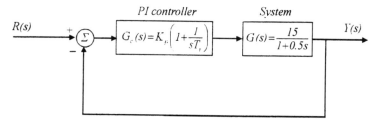

Figure 9.64

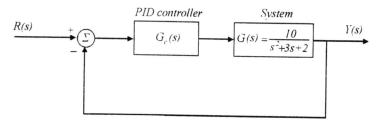

Figure 9.65

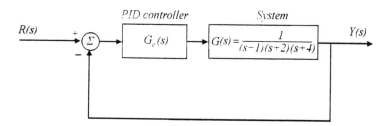

Figure 9.66

10. Consider the system with transfer function

$$G(s) = \frac{1}{(s+1)(0.2s+1)(0.05s+1)(0.01s+1)}$$

Draw the step response of the system and determine the parameters of the PID controller using the Ziegler–Nichols method.

11. Find an active-circuit realization for each of the controllers found in Problems 4, 5, 6, and 7.

12. Consider the system shown in Figure 9.67, where $r(t) = 1$. Find the transfer functions of the controllers $G_c(s)$ and $F(s)$ so that

$$J_e = \int_0^\infty e^2(t)\, dt \text{ is minimized, while } J_u = \int_0^\infty u^2(t)\, dt \le 2.$$

13. The orientation control system of a space telescope is shown in Figure 9.68. Given that $R(s) = 0.5/s$, determine the optimal control signal $U(s)$, the optimal

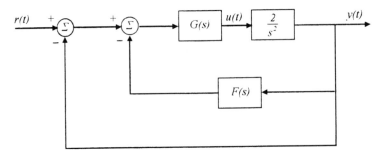

Figure 9.67

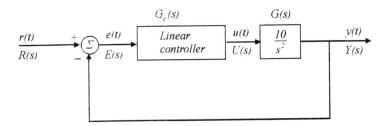

Figure 9.68

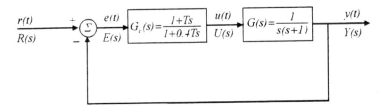

Figure 9.69

linear controller, and the optimal closed-loop transfer function $H(s)$, so that $J_e(t) = \int_0^\infty e^2(t)\,dt$ is minimized, while $J_u = \int_0^\infty u^2(t) \le 2.5$.

14. For the system shown in Figure 9.69, find the value of the parameter T for which the cos $J_e = \int_0^\infty e^2(t)\,dt$ is minimized for $r(t) = 1$.

BIBLIOGRAPHY

Books

1. K Astrom, T Hagglund. PID Controllers: Theory, Design, and Tuning. Research Triangle Park, North Carolina: Instrument Society of America, 1995.
2. JJ D'Azzo, CH Houpis. Linear Control System Analysis and Design, Conventional and Modern. New York: McGraw-Hill, 1975.
3. JJ DiStefano III, AR Stubberud, IJ Williams. Feedback and Control Systems. Schaum's Outline Series. New York: McGraw-Hill, 1967.
4. RC Dorf, RE Bishop. Modern Control Analysis. London: Addison-Wesley, 1995.

5. AF D'Souza. Design of Controls Systems. Englewood Cliffs, New Jersey: Prentice hall, 1988.
6. GF Franklin, JD Powell, A Emami-Naeini. Feedback Control of Dynamic Systems. Reading, MA: Addison-Wesley, 1986.
7. GH Hostetter, CJ Savant Jr, RT Stefani. Design of Feedback Control Systems. 2nd ed. New York: Saunders College Publishing, 1989.
8. NS Nise. Control Systems Engineering. New York: Benjamin and Cummings, 1995.
9. K Ogata. Modern Control Systems. London: Prentice Hall, 1997.
10. FG Shinskey. Process Control Systems. New York: McGraw-Hill, 1979.

Articles

11. JG Ziegler, NB Nichols. Optimum settings for automatic controllers. Trans ASME 64:759–768, 1942.

10

State-Space Design Methods

10.1 INTRODUCTION

This chapter presents an introduction to certain modern state-space control design methods. The specific methods presented are distinguished into two categories: the *algebraic control methods* and the *observer design methods*. There are many other interesting modern control design methods presented in the remainder of this book: optimal control (Chap. 11), digital control (Chap. 12), system identification (Chap. 13), adaptive control (Chap. 14), robust control (Chap. 15), and fuzzy control (Chap. 16). All these modern control methods are of paramount theoretical and practical importance to the control engineer. It should be mentioned that there are several control design methods—such as geometrical control, hierarchical control, and neural control—that are not presented here, since they are beyond the scope of this book.

Algebraic control refers to a particular category of modern control design problems wherein the controller has a prespecified structure. In this case, the design problem reduces to that of determining the controller parameters such that certain closed-loop requirements are met. This is not achieved via minimization of some cost functions (as is done, for example, in optimal control in Chap. 11), but via the solution of algebraic equations. It is for this reason that these techniques are called *algebraic* control design techniques. These algebraic techniques are used to solve many interesting practical control problems, such as pole placement, input–output decoupling, and exact model matching. These three problems are studied in Secs 10.3, 10.4, and 10.5, respectively. In Sec. 10.2 an overview of the structure of state and output feedback laws is given, which are subsequently used for the study of the three aforementioned algebraic control problems.

State observers are used in order to produce a good estimate of the state vector $\mathbf{x}(t)$. It is well known that, in practice, most often not all state variables of a system are accessible to measurement. This obstacle can be circumvented by the use of state observers which yield a good estimate $\hat{\mathbf{x}}(t)$ of the real state vector $\mathbf{x}(t)$, provided that a mathematical model of the system is available. Estimating $\hat{\mathbf{x}}(t)$ makes it possible to use state feedback techniques to solve many important control problems, such as pole assignment, input–output decoupling, and model matching, presented in Secs

10.3–10.5; optimal regulator and optimal servomechanism, presented in Chap. 11; and many others.

10.2 LINEAR STATE AND OUTPUT FEEDBACK LAWS

In designing control systems using algebraic techniques, we usually apply linear state or output feedback.

1 State Feedback Controllers

Consider the linear time-invariant system

$$\dot{\mathbf{x}} = \mathbf{A}\mathbf{x} + \mathbf{B}\mathbf{u} \tag{10.2-1a}$$

$$\mathbf{y} = \mathbf{C}\mathbf{x} \tag{10.2-1b}$$

where $\mathbf{x} \in \mathbf{R}^n$, $\mathbf{u} \in \mathbf{R}^m$, $\mathbf{y} \in \mathbf{R}^p$ and the matrices $\mathbf{A}$, $\mathbf{B}$, and $\mathbf{C}$ are of appropriate dimensions. Let the controller have the linear state feedback form

$$\mathbf{u} = \mathbf{F}\mathbf{x} + \mathbf{G}\mathbf{r} \tag{10.2-2}$$

where $\mathbf{r} \in \mathbf{R}^{m^*}$ is a new vector with m^* inputs and $\mathbf{F}$ and $\mathbf{G}$ are the unknown controller matrices with dimensions $m \times n$ and $m \times m^*$, respectively (Figure 10.1). Substituting Eq. (10.2-2) in Eq. (10.2-1) yields the closed-loop system

$$\dot{\mathbf{x}} = (\mathbf{A} + \mathbf{B}\mathbf{F})\mathbf{x} + \mathbf{B}\mathbf{G}\mathbf{r} \tag{10.2-3a}$$

$$\mathbf{y} = \mathbf{C}\mathbf{x} \tag{10.2-3b}$$

The control problem here is to determine the control law (10.2-2), i.e., to determine the controller matrices $\mathbf{F}$ and $\mathbf{G}$, such that the closed-loop system has the desired prespecified characteristics.

2 Output Feedback Controllers

Consider the system (10.2-1) and the linear output feedback controller

$$\mathbf{u} = \mathbf{K}\mathbf{y} + \mathbf{N}\mathbf{r} \tag{10.2-4}$$

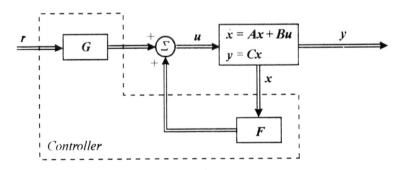

Figure 10.1 Closed-loop system with state feedback.

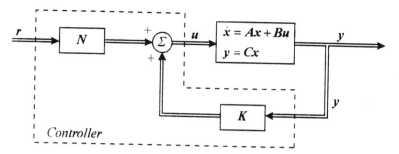

Figure 10.2 Closed-loop system with output feedback.

where **K** and **N** are the unknown controller matrices with dimensions $m \times p$ and $m \times m^*$, respectively (Figure 10.2). Substituting Eq. (10.2-4) in Eq. (10.2-1) yields the closed-loop system

$$\dot{x} = (A + BKC)x + BNr \tag{10.2-5a}$$

$$y = Cx \tag{10.2-5b}$$

The problem here is to determine the control law (10.2-4), i.e., to determine the controller matrices **K** and **N**, such that the closed-loop system (10.2-5) has the desired prespecified characteristics.

By inspection, we observe that the controller matrices (**F**, **G**) and (**K**, **N**) of the foregoing controller design problems via state or output feedback, respectively, are related via the following equations

$$F = KC \tag{10.2-6a}$$

$$G = N \tag{10.2-6b}$$

In fact, if Eqs (10.2-6a and b) hold true, the closed-loop systems (10.2-4) and (10.2-5) are identical. This shows that if the problem via state feedback has a solution, Eqs (10.2-6a and b) may facilitate the solution of the problem via output feedback. In this latter case, the solution procedure will be simple since Eqs (10.2-6a and b) are linear in **K** and **N**.

The main differences between the state and the output feedback methods are the following. The state feedback method has the advantage over the output feedback method in that it has greater degrees of freedom in the controller parameters. This is true since the matrix **F** has nm arbitrary elements, while **K** has $mp < mn$ arbitrary elements. However, the output feedback method is superior to the state feedback method from the practical point of view, because the output vector $y(t)$ is known and measurable. On the contrary, it is almost always difficult, if not impossible, to measure the entire state vector $x(t)$, in which case we are forced to use a special type of system, called the *state observer*, for the estimation of the vector $x(t)$ (see Sec. 10.6).

In the sequel, the problem of determining the matrices **F** and **G** (or **K** and **N**) is considered for the following three specific problems: pole placement, input–output decoupling, and exact model matching. These three problems have been chosen because they are very useful in practice.

10.3 POLE PLACEMENT

10.3.1 Pole Placement via State Feedback

Consider the linear, time-invariant system

$$\dot{\mathbf{x}}(t) = \mathbf{A}\mathbf{x}(t) + \mathbf{B}\mathbf{u}(t) \qquad\qquad (10.3\text{-}1)$$

where we assume that all states are accessible and known. To this system we apply a linear state feedback control law of the form

$$\mathbf{u}(t) = -\mathbf{F}\mathbf{x}(t) \qquad\qquad (10.3\text{-}2)$$

Then, the closed-loop system (see Figure 10.3) is given by the homogeneous equation

$$\dot{\mathbf{x}}(t) = (\mathbf{A} - \mathbf{B}\mathbf{F})\mathbf{x}(t) \qquad\qquad (10.3\text{-}3)$$

It is remarked that the feedback law $\mathbf{u}(t) = -\mathbf{F}\mathbf{x}(t)$ is used rather than the feedback law $\mathbf{u}(t) = \mathbf{F}\mathbf{x}(t)$. This difference in sign is chosen to facilitate the observer design problem presented in Sec. 10.6.

Here, the design problem is to find the appropriate controller matrix $\mathbf{F}$ so as to improve the performance of the closed-loop system (10.3-3). One such method of improving the performance of (10.3-3) is that of pole placement. The pole-placement method consists in finding a particular matrix $\mathbf{F}$, such that the poles of the closed-loop system (10.3-3) take on desirable preassigned values. Using this method, the behavior of the open-loop system may be improved significantly. For example, the method can stabilize an unstable system, increase or decrease the speed of response, widen or narrow the system's bandwidth, increase or decrease the steady-state error, etc. For these reasons, improving the system performance via the pole-placement method is widely used in practice.

The pole placement or eigenvalue assignment problem can be defined as follows: let $\lambda_1, \lambda_2, \ldots, \lambda_n$ be the eigenvalues of the matrix $\mathbf{A}$ of the open-loop system (10.3-1) and $\hat{\lambda}_1, \hat{\lambda}_2, \ldots, \hat{\lambda}_n$ be the desired eigenvalues of the matrix $\mathbf{A} - \mathbf{B}\mathbf{F}$ of the closed-loop system (10.3-3), where all complex eigenvalues exist in complex conjugate pairs. Also, let $p(s)$ and $\hat{p}(s)$ be the respective characteirstic polynomials, i.e., let

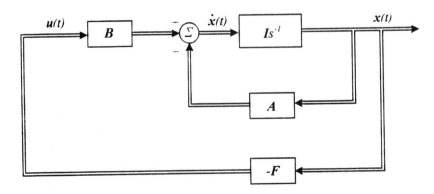

Figure 10.3 Closed-loop system with a linear state feedback law.

$$p(s) = \prod_{i=1}^{n} (s - \lambda_i) = |s\mathbf{I} - \mathbf{A}| = s^n + a_1 s^{n-1} + \cdots + a_{n-1} s + a_n \qquad (10.3\text{-}4)$$

$$\hat{p}(s) = \prod_{i=1}^{n} (s - \hat{\lambda}_i) = |s\mathbf{I} - \mathbf{A} + \mathbf{B}\mathbf{F}| = s^n + \hat{a}_1 s^{n-1} + \cdots + \hat{a}_{n-1} s + \hat{a}_n \qquad (10.3\text{-}5)$$

Find a matrix $\mathbf{F}$ so that Eq. (10.3-5) is satisfied.

The pole-placement problem has attracted considerable attention for many years. The first significant results were established by Wonham in the late 1960s and are given by the following theorem [25].

Theorem 10.3.1

There exists a state feedback matrix $\mathbf{F}$ which assigns to the matrix $\mathbf{A} - \mathbf{B}\mathbf{F}$ of the closed-loop system any arbitrary eigenvalues $\hat{\lambda}_1, \hat{\lambda}_2, \ldots, \hat{\lambda}_n$, if and only if the state vector of the open-loop system (10.3-1) is controllable, i.e., if and only if

$$\text{rank}\,\mathbf{S} = n, \qquad \text{where} \qquad \mathbf{S} = [\mathbf{B} \,\vdots\, \mathbf{A}\mathbf{B} \,\vdots\, \mathbf{A}^2\mathbf{B} \,\vdots\, \cdots \,\vdots\, \mathbf{A}^{n-1}\mathbf{B}] \qquad (10.3\text{-}6)$$

where all complex eigenvalues of the set $\{\hat{\lambda}_1, \ldots, \hat{\lambda}_n\}$ appear in conjugate pairs.

According to this theorem, in cases where the open-loop system (10.3-1) is not controllable, at least one eigenvalue of the matrix $\mathbf{A}$ remains invariant under the state feedback law (10.3-2). In such cases, in order to assign all eigenvalues, one must search for an appropriate dynamic controller wherein the feedback law (10.3-2) may involve, not only propontial, but also derivative, integral and other terms (a special category of dynamic controllers are the PID controllers presented in Sec. 9.6). Dynamic controllers have the disadvantage in that they increase the order of the system.

Now, consider the case where the system $(\mathbf{A}, \mathbf{B})$ is controllable, a fact which guarantees that there exists an $\mathbf{F}$ which satisfies the pole-placement problem. Next, we will deal with the problem of determining such a feedback matrix $\mathbf{F}$. For simplicity, we will first study the case of single-input systems, in which case the matrix $\mathbf{B}$ reduces to a column vector $\mathbf{b}$ and the matrix $\mathbf{F}$ reduces to a row vector $\mathbf{f}^T$. Equation (10.3-5) then becomes

$$\hat{p}(s) = \prod_{i=1}^{n} (s - \hat{\lambda}_i) = |s\mathbf{I} - \mathbf{A} + \mathbf{b}\mathbf{f}^T| = s^n + \hat{a}_1 s^{n-1} + \cdots + \hat{a}_{n-1} s + \hat{a}_n \qquad (10.3\text{-}7)$$

It is remarked that the solution of Eq. (10.3-7) for $\mathbf{f}$ is unique.

Several methods have been proposed for determining $\mathbf{f}$. We present three well-known such methods.

Method 1. The Base–Gura Formula. One of the most popular pole-placement methods, due to Bass & Gura [3], gives the following simple solution:

$$\mathbf{f} = [\mathbf{W}^T \mathbf{S}^T]^{-1} (\hat{\mathbf{a}} - \mathbf{a}) \qquad (10.3\text{-}8)$$

where $\mathbf{S}$ is the controllability matrix defined in Eq. (10.3-6) and

$$\mathbf{W} = \begin{bmatrix} 1 & a_1 & \cdots & a_{n-1} \\ 0 & 1 & \cdots & a_{n-2} \\ \vdots & \vdots & & \vdots \\ 0 & 0 & \cdots & 1 \end{bmatrix}, \qquad \hat{\mathbf{a}} = \begin{bmatrix} \hat{a}_1 \\ \hat{a}_2 \\ \vdots \\ \hat{a}_n \end{bmatrix}, \qquad \mathbf{a} = \begin{bmatrix} a_1 \\ a_2 \\ \vdots \\ a_n \end{bmatrix} \qquad (10.3\text{-}9)$$

Method 2. The Phase Canonical Form Formula. Consider the special case where the system under control is described in its phase-variable canonical form, i.e., $\mathbf{A}$ and $\mathbf{b}$ have the special forms $\mathbf{A}^*$ and $\mathbf{b}^*$, where (see Subsec. 5.4.2)

$$\mathbf{A}^* = \begin{bmatrix} 0 & 1 & 0 & \cdots & 0 \\ 0 & 0 & 1 & \cdots & 0 \\ 0 & 0 & 0 & \cdots & 0 \\ \vdots & \vdots & \vdots & & \vdots \\ 0 & 0 & 0 & \cdots & 1 \\ -a_n^* & -a_{n-1}^* & -a_{n-2}^* & \cdots & -a_1^* \end{bmatrix}, \qquad \mathbf{b}^* = \begin{bmatrix} 0 \\ 0 \\ 0 \\ \vdots \\ 0 \\ 1 \end{bmatrix} \qquad (10.3\text{-}10)$$

Then, it can be easily shown that the matrix $\mathbf{S}^* = [\mathbf{b}^* \vdots \mathbf{A}^*\mathbf{b}^* \vdots \mathbf{A}^{*2}\mathbf{b}^* \vdots \cdots \vdots \mathbf{A}^{*n-1}\mathbf{b}^*]$ is such that the product $\mathbf{W}^T\mathbf{S}^{*T}$ reduces to the simple form

$$\mathbf{W}^T\mathbf{S}^{*T} = \tilde{\mathbf{I}} = \begin{bmatrix} 0 & 0 & \cdots & 0 & 1 \\ 0 & 0 & \cdots & 1 & 0 \\ \vdots & \vdots & & \vdots & \vdots \\ 1 & 0 & \cdots & 0 & 0 \end{bmatrix} \qquad (10.3\text{-}11)$$

In this case, the vector $\mathbf{f}^*$ in expression (10.3-8) reduces to $\mathbf{f}^* = \tilde{\mathbf{I}}(\hat{\mathbf{a}} - \mathbf{a})$, i.e., it reduces to the following form [22]:

$$\mathbf{f}^* = \tilde{\mathbf{I}}(\hat{\mathbf{a}} - \mathbf{a}) = \begin{bmatrix} \hat{a}_n - a_n \\ \hat{a}_{n-1} - a_{n-1} \\ \vdots \\ \hat{a}_1 - a_1 \end{bmatrix} \qquad (10.3\text{-}12)$$

where use is made of the property $(\tilde{\mathbf{I}})^{-1} = \tilde{\mathbf{I}}$. It is evident that expression (10.3-12) is extremely simple to apply, provided that the matrix $\mathbf{A}$ and the vector $\mathbf{b}$ of the system under control are in the phase-variable canonical form (10.3-10).

Method 3. The Ackermann's Formula. Another approach for computing $\mathbf{f}$ has been proposed by Ackermann, leading to the following expression [5]:

$$\mathbf{f}^T = \mathbf{e}^T\mathbf{S}^{-1}\hat{\mathbf{p}}(\mathbf{A}) \qquad (10.3\text{-}13)$$

where the matrix $\mathbf{S}$ is given in Eq. (10.3-6) and $\mathbf{e}^T = (0, 0, \ldots, 0, 1)$. The matrix polynomial $\hat{\mathbf{p}}(\mathbf{A})$ is given by Eq. (10.3-5), wherein the variable s is substituted by the matrix $\mathbf{A}$, i.e.,

$$\hat{\mathbf{p}}(\mathbf{A}) = \mathbf{A}^n + \hat{a}_1\mathbf{A}^{n-1} + \cdots + \hat{a}_{n-1}\mathbf{A} + \hat{a}_n\mathbf{I} \qquad (10.3\text{-}14)$$

In the general case of multi-input systems, the determination of the matrix $\mathbf{F}$ is somewhat complicated. A simple approach to the problem is to assume that $\mathbf{F}$ has the following outer product form:

$$\mathbf{F} = \mathbf{q}\mathbf{p}^T \qquad (10.3\text{-}15)$$

where $\mathbf{q}$ and $\mathbf{p}$ are n-dimensional vectors. Then, the matrix $\mathbf{A} - \mathbf{BF}$ becomes

$$\mathbf{A} - \mathbf{BF} = \mathbf{A} - \mathbf{Bqp}^T = \mathbf{A} - \boldsymbol{\beta}\mathbf{p}^T, \qquad \text{where} \qquad \boldsymbol{\beta} = \mathbf{Bq} \tag{10.3-16}$$

Therefore, assuming that $\mathbf{F}$ has the form (10.3-15), then the multi-input system case is reduced to the single-input case studied previously. In other words, the solution for the vector $\mathbf{p}$ is Eq. (10.3-8) or Eq. (10.3-13) and differs only in that the matrix $\mathbf{S}$ is now the matrix $\tilde{\mathbf{S}}$, having the form

$$\tilde{\mathbf{S}} = [\boldsymbol{\beta} \,\vdots\, \mathbf{A}\boldsymbol{\beta} \,\vdots\, \mathbf{A}^2\boldsymbol{\beta} \,\vdots\, \cdots \,\vdots\, \mathbf{A}^{n-1}\boldsymbol{\beta}] \tag{10.3-17}$$

The vector $\boldsymbol{\beta} = \mathbf{Bq}$ involves arbitrary parameters, which are the elements of the arbitrary vector $\mathbf{q}$. These arbitrary parameters can have any value, provided that rank $\tilde{\mathbf{S}} = n$. In cases where this condition cannot be satisfied, other approaches for determining $\mathbf{F}$ may be found in the literature [22].

Example 10.3.1

Consider a system in the form (10.3-1), where

$$\mathbf{A} = \begin{bmatrix} 0 & 1 \\ -1 & 0 \end{bmatrix} \qquad \text{and} \qquad \mathbf{b} = \begin{bmatrix} 0 \\ 1 \end{bmatrix}$$

Find a vector $\mathbf{f}$ such that the closed-loop system has eigenvalues $\hat{\lambda}_1 = -1$ and $\hat{\lambda}_2 = -1.5$.

Solution

We have

$$p(s) = |s\mathbf{I} - \mathbf{A}| = s^2 + 1 \qquad \text{and} \qquad \hat{p}(s) = (s - \hat{\lambda}_1)(s - \hat{\lambda}_2) = s^2 + 2.5s + 1.5$$

Method 1. Here we use Eq. (10.3-8). Equations (10.3-9) and (10.3-6) give

$$\mathbf{W} = \begin{bmatrix} 1 & a_1 \\ 0 & 1 \end{bmatrix} = \begin{bmatrix} 1 & 0 \\ 0 & 1 \end{bmatrix} \qquad \text{and} \qquad \mathbf{S} = [\mathbf{b} \,\vdots\, \mathbf{Ab}] = \begin{bmatrix} 0 & 1 \\ 1 & 0 \end{bmatrix}$$

Therefore

$$\mathbf{W}^T\mathbf{S}^T = \begin{bmatrix} 1 & 0 \\ 0 & 1 \end{bmatrix}\begin{bmatrix} 0 & 1 \\ 1 & 0 \end{bmatrix} = \begin{bmatrix} 0 & 1 \\ 1 & 0 \end{bmatrix} \qquad \text{and} \qquad (\mathbf{W}^T\mathbf{S}^T)^{-1} = \begin{bmatrix} 0 & 1 \\ 1 & 0 \end{bmatrix}$$

Hence

$$\mathbf{f} = (\mathbf{W}^T\mathbf{S}^T)^{-1}(\hat{a} - a) = \begin{bmatrix} 0 & 1 \\ 1 & 0 \end{bmatrix}\left\{\begin{bmatrix} 2.5 \\ 1.5 \end{bmatrix} - \begin{bmatrix} 0 \\ 1 \end{bmatrix}\right\} = \begin{bmatrix} 0 & 1 \\ 1 & 0 \end{bmatrix}\begin{bmatrix} 2.5 \\ 0.5 \end{bmatrix} = \begin{bmatrix} 0.5 \\ 2.5 \end{bmatrix}$$

Method 2. Since the system is in phase-variable canonical form, the vector $\mathbf{f}$ can readily be determined by Eq. (10.3-12), as follows:

$$\mathbf{f} = \mathbf{f}^* = \begin{bmatrix} \hat{a}_2 - a_2 \\ \hat{a}_1 - a_1 \end{bmatrix} = \begin{bmatrix} 1.5 - 1 \\ 2.5 - 0 \end{bmatrix} = \begin{bmatrix} 0.5 \\ 2.5 \end{bmatrix}$$

Method 3. Here we apply Eq. (10.3-13). We have

$$\hat{p}(\mathbf{A}) = \mathbf{A}^2 + \hat{a}_1\mathbf{A} + \hat{a}_2\mathbf{I} = \mathbf{a}^2 + 2.5\mathbf{A} + 1.5\mathbf{I}$$

$$= \begin{bmatrix} 0 & 1 \\ -1 & 0 \end{bmatrix}^2 + 2.5\begin{bmatrix} 0 & 1 \\ -1 & 0 \end{bmatrix} + 1.5\begin{bmatrix} 1 & 0 \\ 0 & 1 \end{bmatrix}$$

$$= \begin{bmatrix} -1 & 0 \\ 0 & -1 \end{bmatrix} + \begin{bmatrix} 0 & 2.5 \\ -2.5 & 0 \end{bmatrix} + \begin{bmatrix} 1.5 & 0 \\ 0 & 1.5 \end{bmatrix} = \begin{bmatrix} 0.5 & 2.5 \\ -2.5 & 0.5 \end{bmatrix}$$

$$\mathbf{S}^{-1} = [\mathbf{b} \vdots \mathbf{Ab}]^{-1} = \begin{bmatrix} 0 & 1 \\ 1 & 0 \end{bmatrix}$$

Therefore

$$\mathbf{f}^T = \mathbf{e}^T\mathbf{S}^{-1}\hat{p}(\mathbf{A}) = \begin{bmatrix} 0 & 1 \end{bmatrix}\begin{bmatrix} 0 & 1 \\ 1 & 0 \end{bmatrix}\begin{bmatrix} 0.5 & 2.5 \\ -2.5 & 0.5 \end{bmatrix} = \begin{bmatrix} 0.5 & 2.5 \end{bmatrix}$$

Clearly, the resulting three controller vectors derived by the three methods are identical. This is due to the fact that for single-input systems, $\mathbf{f}$ is unique.

Example 10.3.2

Consider a system in the form (10.3-1), where

$$\mathbf{A} = \begin{bmatrix} 0 & 1 & 0 \\ 0 & 0 & 1 \\ 1 & 0 & 0 \end{bmatrix} \quad \text{and} \quad \mathbf{b} = \begin{bmatrix} 0 \\ 0 \\ 1 \end{bmatrix}$$

Find a vector $\mathbf{f}$ such that the closed-loop system has eigenvalues $\hat{\lambda}_1 = -1$, $\hat{\lambda}_2 = -2$, and $\hat{\lambda}_3 = -2$.

Solution

We have

$$p(s) = |s\mathbf{I} - \mathbf{A}| = s^3 - 1 \quad \text{and} \quad \hat{p}(s) = (s - \hat{\lambda}_1)(s - \hat{\lambda}_2)(s - \hat{\lambda}_3)$$
$$= s^3 + 5s^2 + 8s + 4$$

Method 1. Here, we make use of Eq. (10.3-8). Equations (10.3-9) and (10.3-6) give

$$\mathbf{W} = \begin{bmatrix} 1 & a_1 & a_2 \\ 0 & 1 & a_1 \\ 0 & 0 & 1 \end{bmatrix} = \begin{bmatrix} 1 & 0 & 0 \\ 0 & 1 & 0 \\ 0 & 0 & 1 \end{bmatrix}, \quad \mathbf{S} = [\mathbf{b} \vdots \mathbf{Ab} \vdots \mathbf{A}^2\mathbf{b}] = \begin{bmatrix} 0 & 0 & 1 \\ 0 & 1 & 0 \\ 1 & 0 & 0 \end{bmatrix}$$

Therefore,

$$\mathbf{W}^T\mathbf{S}^T = \begin{bmatrix} 1 & 0 & 0 \\ 0 & 1 & 0 \\ 0 & 0 & 1 \end{bmatrix}\begin{bmatrix} 0 & 0 & 1 \\ 0 & 1 & 0 \\ 1 & 0 & 0 \end{bmatrix} = \begin{bmatrix} 0 & 0 & 1 \\ 0 & 1 & 0 \\ 1 & 0 & 0 \end{bmatrix} \quad \text{and}$$

$$(\mathbf{W}^T\mathbf{S}^T)^{-1} = \begin{bmatrix} 0 & 0 & 1 \\ 0 & 1 & 0 \\ 1 & 0 & 0 \end{bmatrix}$$

Hence

$$\mathbf{f} = (\mathbf{W}^T \mathbf{S}^T)^{-1} (\hat{\mathbf{a}} - \mathbf{a}) = \begin{bmatrix} 0 & 0 & 1 \\ 0 & 1 & 0 \\ 1 & 0 & 0 \end{bmatrix} \left\{ \begin{bmatrix} 5 \\ 8 \\ 4 \end{bmatrix} - \begin{bmatrix} 0 \\ 0 \\ -1 \end{bmatrix} \right\} = \begin{bmatrix} 0 & 0 & 1 \\ 0 & 1 & 0 \\ 1 & 0 & 0 \end{bmatrix} \begin{bmatrix} 5 \\ 8 \\ 5 \end{bmatrix}$$

$$= \begin{bmatrix} 5 \\ 8 \\ 5 \end{bmatrix}$$

Method 2. Since the system is in phase-variable canonical form, the vector $\mathbf{f}$ can readily be determined by Eq. (10.3-12), as follows:

$$\mathbf{f} = \mathbf{f}^* = \begin{bmatrix} \hat{a}_3 - a_3 \\ \hat{a}_2 - a_2 \\ \hat{a}_1 - a_1 \end{bmatrix} = \begin{bmatrix} 4+1 \\ 8+0 \\ 5+0 \end{bmatrix} = \begin{bmatrix} 5 \\ 8 \\ 5 \end{bmatrix}$$

Method 3. Here, we make use of Eq. (10.3-13). We have

$$\hat{\mathbf{p}}(\mathbf{A}) = \mathbf{A}^3 + \hat{a}_1 \mathbf{A}^2 + \hat{a}_2 \mathbf{A} + \hat{a}_3 \mathbf{I} = \mathbf{A}^3 + 5\mathbf{A}^2 + 8\mathbf{A} + 4\mathbf{I}$$

$$= \begin{bmatrix} 0 & 1 & 0 \\ 0 & 0 & 1 \\ 1 & 0 & 0 \end{bmatrix}^3 + 5 \begin{bmatrix} 0 & 1 & 0 \\ 0 & 0 & 1 \\ 1 & 0 & 0 \end{bmatrix}^2 + 8 \begin{bmatrix} 0 & 1 & 0 \\ 0 & 0 & 1 \\ 1 & 0 & 0 \end{bmatrix} + 4 \begin{bmatrix} 1 & 0 & 0 \\ 0 & 1 & 0 \\ 0 & 0 & 1 \end{bmatrix}$$

$$= \begin{bmatrix} 1 & 0 & 0 \\ 0 & 1 & 0 \\ 0 & 0 & 1 \end{bmatrix} + \begin{bmatrix} 0 & 0 & 5 \\ 5 & 0 & 0 \\ 0 & 5 & 0 \end{bmatrix} + \begin{bmatrix} 0 & 8 & 0 \\ 0 & 0 & 8 \\ 8 & 0 & 0 \end{bmatrix} + \begin{bmatrix} 4 & 0 & 0 \\ 0 & 4 & 0 \\ 0 & 0 & 4 \end{bmatrix}$$

$$= \begin{bmatrix} 5 & 8 & 5 \\ 5 & 5 & 8 \\ 8 & 5 & 5 \end{bmatrix}$$

$$\mathbf{S}^{-1} = [\mathbf{b} \vdots \mathbf{Ab} \vdots \mathbf{A}^2\mathbf{b}]^{-1} = \begin{bmatrix} 0 & 0 & 1 \\ 0 & 1 & 0 \\ 1 & 0 & 0 \end{bmatrix}$$

Therefore

$$\mathbf{f}^T = \mathbf{e}^T \mathbf{S}^{-1} \hat{\mathbf{p}}(\mathbf{A}) = [0 \quad 0 \quad 1] \begin{bmatrix} 0 & 0 & 1 \\ 0 & 1 & 0 \\ 1 & 0 & 0 \end{bmatrix} \begin{bmatrix} 5 & 8 & 5 \\ 5 & 5 & 8 \\ 8 & 5 & 5 \end{bmatrix} = [5 \quad 8 \quad 5]$$

The resulting three controller vectors derived by the three methods are identical. As mentioned in the previous example, this is becuase, for single-input systems, $\mathbf{f}$ is unique.

Example 10.3.3

Consider the position control system shown in Figure 10.4. The state variables of the system are as follows. State $x_1 = y = \theta_m$ is the angular position of the motor axis which is converted into an electrical signal through the use of a potentiometer. State $x_2 = \dot{\theta}_m = \omega_m$ is the angular velocity of the motor which is mea-

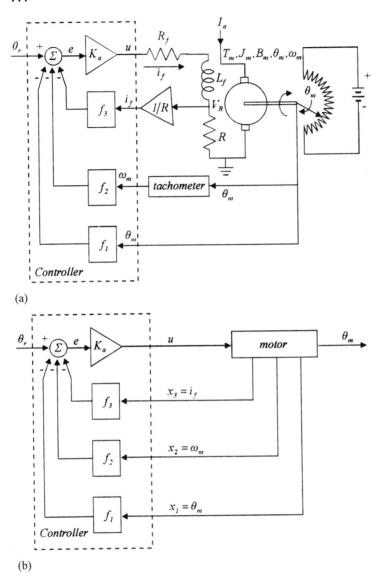

(a)

(b)

Figure 10.4 Position control system of Example 10.3.3. (a) Overall picture of a position control system, with a motor controlled by the stator; (b) schematic diagram of the position control system.

sured by the tachometer. State $x_3 = i_f$ is the current of the stator. To measure i_f, we insert a small resistor R in series with the inductor. The voltage $v_R(t) = Ri_f$ is fed into an amplifier with gain $1/R$, which produces an output $i_f = x_3$. Using the state equations (3.12-8) and Figure 3.39 of Chap. 3, we can construct the block diagram for the closed-loop system as shown in Figure 10.5. It is noted that the amplifier with gain K_a, which is inserted between $e(t)$ and $u(t)$, is used to amplify the signal $e(t)$, which is usually small. The control of the angular position θ_m is achieved in the following way. The external control signal θ_r is the desired angu-

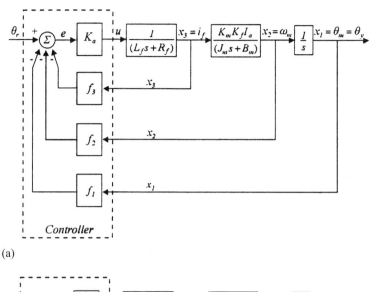

(a)

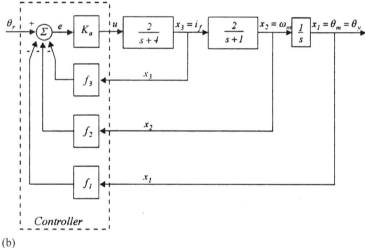

(b)

Figure 10.5 Block diagram of the closed-loop system in Example 10.3.3. (a) Block diagram of the position control system; (b) simplified block diagram for $L_f = 0.5$, $R_f = 2$, $J_m = 1$, $B_m = 1$, and $K_m K_f I_a = 2$.

lar position $\theta_y = \theta_m$ of the motor axis. If $\theta_y \neq \theta_r$, then part of the error $e(t)$ is due to the difference $\theta_y - \theta_r$. This difference is amplified by the amplifier, which subsequently drives the motor, resulting in a rotation of the axis so that the error $\theta_y - \theta_r$ reduces to zero. The problem here is to study the pole placement problem of the closed-loop system via state feedback.

Solution

The state equations of the closed-loop system in Figure 10.5b are (compare with Eq. (3.12-8))

$$\dot{\mathbf{x}} = \mathbf{A}\mathbf{x} + \mathbf{b}u$$

$$y = \mathbf{c}^T \mathbf{x}$$

$$u = K_a[-\mathbf{f}^T \mathbf{x} + \theta_r]$$

where

$$\mathbf{x} = \begin{bmatrix} x_1 \\ x_2 \\ x_3 \end{bmatrix} = \begin{bmatrix} \theta_y \\ \omega_m \\ i_f \end{bmatrix}, \qquad \mathbf{f} = \begin{bmatrix} f_1 \\ f_2 \\ f_3 \end{bmatrix}, \qquad \mathbf{A} = \begin{bmatrix} 0 & 1 & 0 \\ 0 & -1 & 2 \\ 0 & 0 & -4 \end{bmatrix},$$

$$\mathbf{b} = \begin{bmatrix} 0 \\ 0 \\ 2 \end{bmatrix}, \qquad \mathbf{c} = \begin{bmatrix} 1 \\ 0 \\ 0 \end{bmatrix}$$

where use was made of the definitions $T_f^{-1} = R_f/L_f = 4$ and $T_m^{-1} = B_m/J_m = 1$. The controllability matrix of the open-loop system is

$$\mathbf{S} = [\mathbf{b} \vdots \mathbf{A}\mathbf{b} \vdots \mathbf{A}^2\mathbf{b}] = \begin{bmatrix} 0 & 0 & 4 \\ 0 & 4 & -20 \\ 2 & -8 & 32 \end{bmatrix}$$

The determinant of the matrix $\mathbf{S}$ is $|\mathbf{S}| = -32 \neq 0$. Consequently, we may arbitrarily shift all poles of the closed-loop system of Figure 10.5b via state feedback. The characteristic polynomials $p(s)$ and $\hat{p}(s)$ of the open-loop and closed-loop systems are

$$p(s) = |s\mathbf{I} - \mathbf{A}| = s^3 + 5s^2 + 4s = s(s+1)(s+4)$$

$$\hat{p}(s) = |\mathbf{I} - \mathbf{A} + K_a\mathbf{b}\mathbf{f}^T| = s^3 + \gamma_2 s^2 + \gamma_2 s + \gamma_0 = (s - \hat{\lambda}_1)(s - \hat{\lambda}_2)(s - \hat{\lambda}_3)$$

where $\hat{\lambda}_1$, $\hat{\lambda}_2$, and $\hat{\lambda}_3$ are the desired poles of the closed-loop system. To determine the vector $\mathbf{f}$ we use formula (10.3-8). The matrix $\mathbf{W}$ has the form

$$\mathbf{W} = \begin{bmatrix} 1 & 5 & 4 \\ 0 & 1 & 5 \\ 0 & 0 & 1 \end{bmatrix} \qquad \text{and} \qquad [\mathbf{W}^T\mathbf{S}^T]^{-1} = \begin{bmatrix} 0 & 0 & \frac{1}{4} \\ -\frac{1}{4} & \frac{1}{4} & 0 \\ \frac{1}{2} & 0 & 0 \end{bmatrix}$$

Hence

$$K_a\mathbf{f} = [\mathbf{W}^T\mathbf{S}^T]^{-1}(\hat{\mathbf{a}} - \mathbf{a}) \qquad \text{or} \qquad \mathbf{f} = \frac{1}{K_a}[\mathbf{W}^T\mathbf{S}^T]^{-1}(\hat{\mathbf{a}} - \mathbf{a})$$

Finally

$$\mathbf{f} = \begin{bmatrix} f_1 \\ f_2 \\ f_3 \end{bmatrix} = \begin{bmatrix} -\dfrac{1}{4K_a}\gamma_0 \\[2mm] -\dfrac{1}{4K_a}(5 - \gamma_2) + \dfrac{1}{4K_a}(4 - \gamma_1) \\[2mm] \dfrac{1}{2K_a}(5 - \gamma_2) \end{bmatrix}$$

10.3.2 Pole Placement via Output Feedback

For the case of pole placement via output feedback wherein $\mathbf{u} = -\mathbf{Ky}$, a theorem similar to the Theorem 10.3.1 has not yet been proven. The determination of the output feedback matrix $\mathbf{K}$ is, in general, a very difficult task. A method for determining the matrix $\mathbf{K}$, which is closely related to the method of determining the matrix $\mathbf{F}$ presented earlier, is based on Eq. (10.2-6a), namely on the equation

$$\mathbf{F} = \mathbf{KC} \tag{10.3-18}$$

This method starts with the determination of the matrix $\mathbf{F}$ and in the sequel the matrix $\mathbf{K}$ is determined by using Eq. (10.3-18). It is fairly easy to determine the matrix $\mathbf{K}$ from Eq. (10.3-18) since this equation is linear in $\mathbf{K}$. A more general method to determine matrix $\mathbf{K}$ is given in [16]. Note that Eq. (10.3-18) is only a sufficient condition. That is, if Eq. (10.3-18) does not have a solution for $\mathbf{K}$, it does not follow that pole placement by output feedback is impossible.

Example 10.3.4

Consider the multi-input–multi-output (MIMO) system of the form

$$\dot{\mathbf{x}} = \mathbf{Ax} + \mathbf{Bu}, \qquad \mathbf{y} = \mathbf{Cx}$$

where

$$\mathbf{A} = \begin{bmatrix} 0 & 1 & 0 \\ -2 & 3 & 0 \\ 5 & 1 & 3 \end{bmatrix}, \qquad \mathbf{B} = \begin{bmatrix} 0 & 0 \\ 1 & 3 \\ 0 & 1 \end{bmatrix}, \qquad \mathbf{C} = \begin{bmatrix} 0 & 0 & 7 \\ 7 & 9 & 0 \end{bmatrix}$$

Find an output feedback matrix $\mathbf{K}$ such that the poles of the closed-loop system are -3, -3, and -4.

Solution

First, we determine a state feedback matrix $\mathbf{F}$ which satisifes the problem. Using the techniques of Subsec. 10.3.1, we obtain the following matrix $\mathbf{F}$:

$$\mathbf{F} = \begin{bmatrix} 7 & 9 & -21 \\ 0 & 0 & 7 \end{bmatrix}$$

Now, consider the equation $\mathbf{F} = \mathbf{KC}$ and investigate if the above matrix $\mathbf{F}$ is sufficient for the determination of the matrix $\mathbf{K}$. Simple algebraic calculations lead to the conclusion that $\mathbf{F} = \mathbf{KC}$ has a solution for $\mathbf{K}$ having the following form:

$$\mathbf{K} = \begin{bmatrix} -3 & 1 \\ 1 & 0 \end{bmatrix}$$

Checking the results, we have

$$\mathbf{A} - \mathbf{BKC} = \begin{bmatrix} 0 & 1 & 0 \\ -2 & 3 & 0 \\ 5 & 1 & 3 \end{bmatrix} + \begin{bmatrix} 0 & 0 \\ 1 & 3 \\ 0 & 1 \end{bmatrix} \begin{bmatrix} -3 & 1 \\ 1 & 0 \end{bmatrix} \begin{bmatrix} 0 & 0 & 7 \\ 7 & 9 & 0 \end{bmatrix}$$

$$= \begin{bmatrix} 0 & 1 & 0 \\ -9 & -6 & 0 \\ 5 & 1 & -4 \end{bmatrix}$$

and hence

$$|s\mathbf{I} - \mathbf{A} + \mathbf{BKC}| = \begin{vmatrix} s & -1 & 0 \\ 9 & s+6 & 0 \\ -5 & -1 & s+4 \end{vmatrix} = (s+4)(s+3)^2$$

Example 10.3.5

Consider the position control system of Example 10.3.3. To this system apply output feedback for pole placement.

Solution

In practice, the output variable y is usually the output position θ_y. Thus $y = \theta_y = \theta_m = x_1 = \mathbf{c}^T\mathbf{x}$, where $\mathbf{c}^T = (1 \quad 0 \quad 0)$. Hence, the output feedback law here is $u = K_a[-ky + \theta_r] = K_a[-k\mathbf{c}^T\mathbf{x} + \theta_r]$, where k is the output feedback controller or gain. The characteristic polynomial $\hat{p}(s)$ of the closed-loop system then becomes

$$\hat{p}(s) = |s\mathbf{I} - \mathbf{A} + K_a\mathbf{b}k\mathbf{c}^T| = s^3 + 5s^2 + 4s + 4K_ak$$

By using one of the algebraic stability criteria of Chap. 6 we conclude that the closed-loop system is stable when $0 < K_ak < 5$. By using the material of Chap. 7, one may draw the root-locus diagram for $\hat{p}(s)$, thus revealing the regions of the root locus where the closed-loop system is stable. It must be clear that for single-output systems, using output feedback, we may be able to make the closed-loop system stable, but we cannot shift the poles to any arbitrary positions, as in the case of state feedback.

10.4 INPUT–OUTPUT DECOUPLING

The problem of input–output decoupling of a system may be stated as follows. Consider the system (10.2-1) and assume that it has the same number of inputs and outputs, i.e., assume that $p = m$. Determine a pair of matrices $\mathbf{F}$ and $\mathbf{G}$ of the state feedback law (10.2-2) (or a pair of matrices $\mathbf{K}$ and $\mathbf{N}$ of the output feedback law (10.2-4)) such that every input of the closed-loop system (10.2-3) (or of the closed-loop system (10.2-5)) influences only one of the systems outputs, and vice-versa, every output of the closed-loop system is influenced by only one of its inputs. More precisely, in an input–output decoupled system the following relation must hold

$$y_i = f(r_i), \qquad i = 1, 2, \ldots, m \tag{10.4-1}$$

The transfer function matrix $\mathbf{H}(s)$ of the closed-loop system (10.2-3) is given by

$$H(s) = C(sI - A - BF)^{-1}BG \tag{10.4-2}$$

and the transfer function matrix $\hat{H}(s)$ of the closed-loop system (10.2-5) is given by

$$\hat{H}(s) = C(sI - A - BKC)^{-1}BN \tag{10.4-3}$$

Since $Y(s) = H(s)R(s)$ (or $Y(s) = \hat{H}(s)R(s)$) it follows that a definition, equivalent to the foregoing definition of the input–output decoupling problem, is the following: determine a pair of matrices F and G (or a pair of matrices K and N) such that the transfer function matrix $H(s)$ (or $\hat{H}(s)$) is regular and diagonal. In fact, if $H(s)$ is regular and diagonal, that is if $H(s)$ has the form

$$H(s) = \begin{bmatrix} h_{11}(s) & 0 & \cdots & 0 \\ 0 & h_{22}(s) & \cdots & 0 \\ \vdots & \vdots & \ddots & \vdots \\ 0 & 0 & \cdots & h_{mm}(s) \end{bmatrix} \tag{10.4-4}$$

with $|H(s)| \neq 0$, then equation $Y(s) = H(s)R(s)$ may be written as follows:

$$y_i(s) = h_{ii}(s)r_i(s), \qquad i = 1, 2, \ldots, m \tag{10.4-5}$$

Equation (10.4-5) is equivalent to Eq. (10.4-1). A similar definition may be given for the matrix transfer function $\hat{H}(s)$.

The basic motivation for input–output decoupling of a system is that by making each output of the system depend only upon one input and vice versa, we convert a MIMO system to m single-input–single-output (SISO) systems. This fact significantly simplifies and facilitates the control of the closed-loop system, since one has to deal with m scalar systems rather than a MIMO system. For these reasons the problem of input–output decoupling is of great practical importance. A block diagram representation of input–output decoupling via state feedback is given in Figure 10.6.

10.4.1 Decoupling via State Feedback

For the case of input–output decoupling via state feedback the following theorem holds which was first proven by Falb and Wolovich [9].

Theorem 10.4.1

System (10.2-1) can be decoupling using the state-variable feedback law (10.2-2), if and only if the matrix B^+, where

$$B^+ = \begin{bmatrix} c_1 A^{d_1} B \\ \cdots\cdots\cdots \\ c_2 A^{d_2} B \\ \cdots\cdots\cdots \\ \vdots \\ \cdots\cdots\cdots \\ c_m A^{d_m} B \end{bmatrix} \tag{10.4-6}$$

is regular, i.e., $|B^+| \neq 0$, where c_i is the ith row of matrix C and $d_1, d_2, \ldots, d_m$ are integers, which are defined as follows:

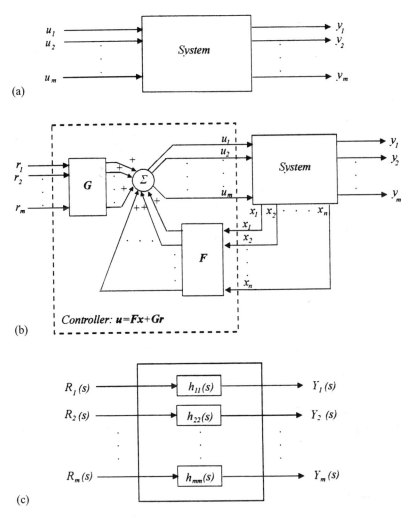

Figure 10.6 Input–output decoupling via state feedback. (a) Open-loop system: $\dot{\mathbf{x}} = \mathbf{Ax} + \mathbf{Bu}$, $\mathbf{y} = \mathbf{Cx}$; (b) closed-loop system: $\dot{\mathbf{x}} = (\mathbf{A} + \mathbf{BF})\mathbf{x} + \mathbf{BGr}$, $\mathbf{y} = \mathbf{Cx}$; (c) closed-loop system transfer function: $\mathbf{H}(s) = \mathbf{C}(s\mathbf{I} - \mathbf{A} - \mathbf{BF})\mathbf{BG} = \mathrm{diag}\{h_{11}(s), h_{22}(s), \ldots, h_{mm}(s)\}$.

$$d_i = \begin{bmatrix} \min j : \mathbf{c}_i \mathbf{A}^j \mathbf{B} \neq \mathbf{0}, & j = 0, 1, \ldots, n-1 \\ n-1 & \text{if} \quad \mathbf{c}_i \mathbf{A}^j \mathbf{B} = \mathbf{0} \quad \text{for all } j \end{bmatrix} \qquad (10.4\text{-}7$$

A pair of matrices $\mathbf{F}$ and $\mathbf{G}$ which satisfy the problem of decoupling is the following

$$\mathbf{F} = -(\mathbf{B}^+)^{-1}\mathbf{A}^+ \qquad (10.4\text{-}8a)$$

$$\mathbf{G} = (\mathbf{B}^+)^{-1} \qquad (10.4\text{-}8b)$$

where matrix $\mathbf{A}^+$ and the transfer function matrix $\mathbf{H}(s) = \mathbf{C}(s\mathbf{I} - \mathbf{A} - \mathbf{BF})^{-1}\mathbf{BG}$ of the closed-loop system have the following forms:

$$
\mathbf{A}^{+} = \begin{bmatrix} \mathbf{c}_1 \mathbf{A}^{d_1+1} \\ \\ \mathbf{c}_2 \mathbf{A}^{d_2+1} \\ \\ \vdots \\ \\ \mathbf{c}_m \mathbf{A}^{d_m+1} \end{bmatrix} \quad \text{and} \quad \mathbf{H}(s) = \begin{bmatrix} \dfrac{1}{s^{d_1+1}} & 0 & \cdots & 0 \\ \\ 0 & \dfrac{1}{s^{d_1+1}} & \cdots & 0 \\ \\ \vdots & \vdots & \ddots & \vdots \\ \\ 0 & 0 & \cdots & \dfrac{1}{s^{d_m+1}} \end{bmatrix} \tag{10.4-9}
$$

From Theorem 10.4.1 we conclude that in order to solve the input–output decoupling problem one must first construct the matrix $\mathbf{B}^{+}$ and then calculate its determinant. If $|\mathbf{B}^{+}| = 0$ it follows that decoupling is not possible via feedback of the form (10.2-2), i.e., no matrices $\mathbf{F}$ and $\mathbf{G}$ exist such that the closed-loop transfer function matrix $\mathbf{H}(s)$ is diagonal and regular. In this case and provided that the open-loop system is invertible, i.e., $\det[\mathbf{C}(s\mathbf{I} - \mathbf{A})^{-1}\mathbf{B}] \neq 0$, one seeks a "dynamic" form of state feedback (not considered in this book) to solve the problem. But if $|\mathbf{B}^{+}| \neq 0$, decoupling is possible using Eq. (10.2-2), and a simple form of the matrices $\mathbf{F}$ and $\mathbf{G}$ that make $\mathbf{H}(s)$ regular and diagonal is given by relation (10.4-8). For the general form of $\mathbf{F}$ and $\mathbf{G}$, involving arbitrary parameters, see [19].

Example 10.4.1

Consider the system of the form (10.2-1), where

$$
\mathbf{A} = \begin{bmatrix} 1 & 2 \\ 2 & 3 \end{bmatrix}, \qquad \mathbf{B} = \begin{bmatrix} 1 & 4 \\ 1 & 4 \end{bmatrix}, \qquad \mathbf{C} = \begin{bmatrix} 1 & 0 \\ -1 & 1 \end{bmatrix}
$$

Find matrices $\mathbf{F}$ and $\mathbf{G}$ such that the closed-loop system is input–output decoupled.

Solution

Determine the integers d_1 and d_2 according to definition (10.4-7). We have

$$
\mathbf{c}_1 \mathbf{A}^0 \mathbf{B} = \begin{bmatrix} 1 & 0 \end{bmatrix} \begin{bmatrix} 1 & 4 \\ 1 & 4 \end{bmatrix} = \begin{bmatrix} 1 & 4 \end{bmatrix} \neq \mathbf{0}
$$

$$
\mathbf{c}_2 \mathbf{A}^0 \mathbf{B} = \begin{bmatrix} -1 & 1 \end{bmatrix} \begin{bmatrix} 1 & 4 \\ 1 & 4 \end{bmatrix} = \begin{bmatrix} 0 & 0 \end{bmatrix} = \mathbf{0}
$$

$$
\mathbf{c}_2 \mathbf{A}^1 \mathbf{B} = \begin{bmatrix} -1 & 1 \end{bmatrix} \begin{bmatrix} 1 & 2 \\ 2 & 3 \end{bmatrix} \begin{bmatrix} 1 & 4 \\ 1 & 4 \end{bmatrix} = \begin{bmatrix} 2 & 8 \end{bmatrix} \neq \mathbf{0}
$$

Therefore $d_1 = 0$ and $d_2 = 1$. Consequently, matrix $\mathbf{B}^{+}$ has the form

$$
\mathbf{B}^{+} = \begin{bmatrix} \mathbf{c}_1 \mathbf{A}^{d_1} \mathbf{B} \\ \cdots\cdots \\ \mathbf{c}_2 \mathbf{A}^{d_2} \mathbf{B} \end{bmatrix} = \begin{bmatrix} \mathbf{c}_1 \mathbf{B} \\ \cdots \\ \mathbf{c}_2 \mathbf{A} \mathbf{B} \end{bmatrix} = \begin{bmatrix} 1 & 4 \\ \cdots \\ 2 & 8 \end{bmatrix}
$$

Examining the determinant of the matrix $\mathbf{B}^{+}$ shows that $|\mathbf{B}^{+}| = 0$. Therefore, we conclude that the system under control cannot be decoupled using the linear state feedback law (10.2-2).

Example 10.4.2

Consider a system of the form (10.2-1), where

$$
\mathbf{A} = \begin{bmatrix} 1 & 0 & 2 \\ 0 & 1 & 1 \\ -1 & 2 & 0 \end{bmatrix}, \qquad \mathbf{B} = \begin{bmatrix} 1 & 0 \\ 1 & -2 \\ -1 & 2 \end{bmatrix}, \qquad \mathbf{C} = \begin{bmatrix} 1 & 0 & 0 \\ 0 & 1 & 1 \end{bmatrix}
$$

Find matrices $\mathbf{F}$ and $\mathbf{G}$ such that the closed-loop system is input–output decoupled.

Solution

Determine the integers d_1 and d_2 using definition (10.4-7). We have

$$
\mathbf{c}_1 \mathbf{A}^0 \mathbf{B} = \begin{bmatrix} 1 & 0 & 0 \end{bmatrix} \begin{bmatrix} 1 & 0 \\ 1 & -2 \\ -1 & 2 \end{bmatrix} = \begin{bmatrix} 1 & 0 \end{bmatrix} \neq \mathbf{0}
$$

$$
\mathbf{c}_2 \mathbf{A}^0 \mathbf{B} = \begin{bmatrix} 0 & 1 & 1 \end{bmatrix} \begin{bmatrix} 1 & 0 \\ 1 & -2 \\ -1 & 2 \end{bmatrix} = \begin{bmatrix} 0 & 0 \end{bmatrix} = \mathbf{0}
$$

$$
\mathbf{c}_2 \mathbf{A}^1 \mathbf{B} = \begin{bmatrix} 0 & 1 & 1 \end{bmatrix} \begin{bmatrix} 1 & 0 & 2 \\ 0 & 1 & 1 \\ -1 & 2 & 0 \end{bmatrix} \begin{bmatrix} 1 & 0 \\ 1 & -2 \\ -1 & 2 \end{bmatrix} = \begin{bmatrix} 1 & -4 \end{bmatrix} \neq \mathbf{0}
$$

Therefore, $d_1 = 0$ and $d_2 = 1$. Consequently, the matrix $\mathbf{B}^+$ has the form

$$
\mathbf{B}^+ = \begin{bmatrix} \mathbf{c}_1 \mathbf{A}^{d_1} \mathbf{B} \\ \cdots\cdots\cdots \\ \mathbf{c}_2 \mathbf{A}^{d_2} \mathbf{B} \end{bmatrix} = \begin{bmatrix} \mathbf{c}_1 \mathbf{B} \\ \cdots\cdots \\ \mathbf{c}_2 \mathbf{A}\mathbf{B} \end{bmatrix} = \begin{bmatrix} 1 & 0 \\ \cdots\cdots \\ 1 & -4 \end{bmatrix}
$$

Examining the determinant of the matrix $\mathbf{B}^+$ shows that $|\mathbf{B}^+| = -4$. Consequently, the system under control can be decoupled using the feedback law (10.2-2). To determine the matrices $\mathbf{F}$ and $\mathbf{G}$ according to relation (10.4-8), we must first compute the matrix $\mathbf{A}^+$. We have

$$
\mathbf{A}^+ = \begin{bmatrix} \mathbf{c}_1 \mathbf{A}^{d_1+1} \\ \cdots\cdots\cdots \\ \mathbf{c}_2 \mathbf{A}^{d_2+1} \end{bmatrix} = \begin{bmatrix} \mathbf{c}_1 \mathbf{A} \\ \cdots\cdots \\ \mathbf{c}_2 \mathbf{A}^2 \end{bmatrix} = \begin{bmatrix} 1 & 0 & 2 \\ \cdots\cdots\cdots \\ -2 & 5 & 1 \end{bmatrix}
$$

Hence, the matrices $\mathbf{G}$ and $\mathbf{F}$ are given by

$$
\mathbf{G} = (\mathbf{B}^+)^{-1} = \frac{1}{4}\begin{bmatrix} 4 & 0 \\ 1 & -1 \end{bmatrix} \qquad \text{and} \qquad \mathbf{F} = -(\mathbf{B}^+)^{-1}\mathbf{A}^+ = -\frac{1}{4}\begin{bmatrix} 4 & 0 & 8 \\ 3 & -5 & 1 \end{bmatrix}
$$

The decoupled closed-loop system is the following

$$
\dot{\mathbf{x}} = (\mathbf{A} + \mathbf{B}\mathbf{F})\mathbf{x} + \mathbf{B}\mathbf{G}\mathbf{r} \qquad \text{and} \qquad \mathbf{y} = \mathbf{C}\mathbf{x}
$$

If we substitute the matrices $\mathbf{A}$, $\mathbf{B}$, $\mathbf{C}$, $\mathbf{F}$, and $\mathbf{G}$ in the above state equations we obtain

$$
\dot{\mathbf{x}} = \frac{1}{2}\begin{bmatrix} 0 & 0 & 0 \\ 1 & -3 & -1 \\ -3 & 9 & 3 \end{bmatrix}\mathbf{x} + \frac{1}{2}\begin{bmatrix} 2 & 0 \\ 1 & 1 \\ -1 & -1 \end{bmatrix}\mathbf{r} \qquad \text{and} \qquad \mathbf{y} = \begin{bmatrix} 1 & 0 & 0 \\ 0 & 1 & 1 \end{bmatrix}\mathbf{x}
$$

The transfer function matrix $\mathbf{H}(s)$ of the closed-loop system is given by

$$\mathbf{H}(s) = \mathbf{C}(s\mathbf{I} - \mathbf{A} - \mathbf{BF})^{-1}\mathbf{BG} = \begin{bmatrix} 1/s^{d_1+1} & 0 \\ 0 & 1/s^{d_2+1} \end{bmatrix} = \begin{bmatrix} 1/s & 0 \\ 0 & 1/s^2 \end{bmatrix}$$

As expected, the matrix $\mathbf{H}(s)$ is diagonal and regular. Using the relation $\mathbf{Y}(s) = \mathbf{H}(s)\mathbf{R}(s)$ we obtain

$$Y_1(s) = \frac{1}{s} R_1(s) \qquad \text{and} \qquad Y_2(s) = \frac{1}{s^2} R_2(s)$$

Hence, it is clear that $y_1(s)$ is only a function of $r_1(s)$ and that $y_2(s)$ is only a function of $r_2(s)$. The corresponding differential equations of the decoupled closed-loop system are

$$\frac{dy_1}{dt} = r_1 \qquad \text{and} \qquad \frac{d^2 y_2}{dt^2} = r_2$$

Example 10.4.3

Consider the following system:

$$\begin{bmatrix} \dot{x}_1(t) \\ \dot{x}_2(t) \end{bmatrix} = \mathbf{A} \begin{bmatrix} x_1(t) \\ x_2(t) \end{bmatrix} + \mathbf{B} \begin{bmatrix} u_1(t) \\ u_2(t) \end{bmatrix}, \qquad \begin{bmatrix} y_1(t) \\ y_2(t) \end{bmatrix} = \mathbf{C} \begin{bmatrix} x_1(t) \\ x_2(t) \end{bmatrix}$$

which it is assumed can be decoupled and for which $\mathbf{c}_1\mathbf{B} \neq \mathbf{0}$ and $\mathbf{c}_2\mathbf{B} \neq \mathbf{0}$, where $\mathbf{c}_1$ and $\mathbf{c}_2$ are the rows of the matrix $\mathbf{C}$. (a) Find matrices $\mathbf{F}$ and $\mathbf{G}$ such that the closed-loop system is decoupled and (b) determine the transfer function matrix of the decoupled closed-loop system.

Solution

(a) Since $\mathbf{c}_1\mathbf{B} \neq \mathbf{0}$ and $\mathbf{c}_2\mathbf{B} \neq \mathbf{0}$, it follows that $d_1 = d_2 = 0$. Thus,

$$\mathbf{B}^+ = \begin{bmatrix} \mathbf{c}_1\mathbf{B} \\ \cdots\cdots \\ \mathbf{c}_2\mathbf{B} \end{bmatrix} = \mathbf{CB} \qquad \text{and} \qquad \mathbf{A}^+ = \begin{bmatrix} \mathbf{c}_1\mathbf{A} \\ \cdots\cdots \\ \mathbf{c}_2\mathbf{A} \end{bmatrix} = \mathbf{CA}$$

Consequently, the matrices $\mathbf{F}$ and $\mathbf{G}$ are given by the following relations:

$$\mathbf{G} = (\mathbf{B}^+)^{-1} = (\mathbf{CB})^{-1} \qquad \text{and} \qquad \mathbf{F} = -(\mathbf{B}^+)^{-1}\mathbf{A}^+ = -(\mathbf{CB})^{-1}\mathbf{CA}$$

(b) The transfer function of the closed-loop system is $\mathbf{H}(s) = \mathbf{C}(s\mathbf{I} - \mathbf{A} - \mathbf{BF})^{-1}\mathbf{BG}$. If we expand $\mathbf{H}(s)$ in negative power series of s, we obtain

$$\mathbf{H}(s) = \mathbf{C}\left[\frac{\mathbf{I}}{s} + \frac{\mathbf{A} + \mathbf{BF}}{s^2} + \frac{(\mathbf{A} + \mathbf{BF})^2}{s^3} + \cdots\right]\mathbf{BG}$$

Using the foregoing expressions for $\mathbf{F}$ and $\mathbf{G}$, the matrix $\mathbf{C}(\mathbf{A} + \mathbf{BF})$ takes on the form

$$\mathbf{C}(\mathbf{A} + \mathbf{BF}) = \mathbf{C}[\mathbf{A} - \mathbf{B}(\mathbf{CB})^{-1}\mathbf{CA}] = \mathbf{CA} - \mathbf{CB}(\mathbf{CB})^{-1}\mathbf{CA} = \mathbf{0}$$

Consequently $\mathbf{C}(\mathbf{A} + \mathbf{BF})^k = \mathbf{0}$, for $k \geq 1$ and therefore the transfer function of the closed-loop system reduces to

$$\mathbf{H}(s) = \frac{\mathbf{CBG}}{s} = \frac{\mathbf{CB(CB)}^{-1}}{s} = \frac{\mathbf{I}}{s}, \qquad \text{where } \mathbf{I} \text{ the unit matrix}$$

We can observe that the closed-loop system has been decoupled into two subsystems, each of which is a simple integrator.

10.4.2 Decoupling via Output Feedback

A simple approach to solve the problem of decoupling via output feedback is to use relation (10.2-6). This method requires prior knowledge of the matrices $\mathbf{F}$ and $\mathbf{G}$. For any given pair of matrices $\mathbf{F}$ and $\mathbf{G}$, which decouples the system under control, relation (10.2-6) may have a solution for $\mathbf{K}$ and $\mathbf{N}$. Since $\mathbf{N} = \mathbf{G}$, it follows that the problem of decoupling via output feedback has a solution provided that equation $\mathbf{KC} = \mathbf{F}$ can be solved for $\mathbf{K}$. Note that the solution of the equation $\mathbf{KC} = \mathbf{F}$ is only a sufficient decoupling condition. This means that in cases where the equation $\mathbf{KC} = \mathbf{F}$ does not have a solution for $\mathbf{K}$, it does not follow that the decoupling problem via output feedback does not have a solution.

Results analogous to Subsec. 10.4.1 are difficult to derive for the case of output feedback and for this reason they are omitted here. For a complete treatment of this problem see [19].

10.5 EXACT MODEL MATCHING

The problem of exact model matching is defined as follows. Consider a system whose behavior is not satisfactory and a model whose behavior is the ideal one. Determine a control law such that the behavior of the closed-loop system follows exactly the behavior of the model.

It is obvious that the solution of such a problem is of great practical importance, since it makes it possible to modify the behavior of a system so as to match an ideal one.

In Sec. 9.1, the criterion (9.1-1) expresses the basic idea behind the problem of exact model matching, wherein we seek a controller such that the behavior of the closed-loop system follows, as closely as possible, the behavior of a model. Of course, this matching may become *exact* when the cost criterion (9.1-1) reduces to zero. The present section is devoted to this latter case—namely, to the *exact* model matching problem.

The problem of exact model matching of linear time-invariant systems, from the algebraic point of view adopted in this chapter, is as follows. Consider a system under control described in state space by Eqs (10.2-1a and b) and a model described by its transfer function matrix $\mathbf{H}_m(s)$. Determine the controller matrices $\mathbf{F}$ and $\mathbf{G}$ of the feedback law (10.2-2) [or the matrices $\mathbf{K}$ and $\mathbf{N}$ of the feedback law (10.2-4)] so that the transfer function $\mathbf{H}(s)$ of the closed-loop system (10.2-3) (or the transfer function matrix $\hat{\mathbf{H}}(s)$ of the closed-loop system (10.2-5)) is equal to the transfer function matrix of the model, i.e., such that

$$\mathbf{H}(s) \text{ (or } \hat{\mathbf{H}}(s)) = \mathbf{H}_m(s) \tag{10.5-1}$$

where

$$\mathbf{H}(s) = \mathbf{C}(s\mathbf{I} - \mathbf{A} - \mathbf{BF})^{-1}\mathbf{BG} \tag{10.5-2a}$$

$$\hat{\mathbf{H}}(s) = \mathbf{C}(s\mathbf{I} - \mathbf{A} - \mathbf{BKC})^{-1}\mathbf{BN} \tag{10.5-2b}$$

The solution of Eq. (10.5-1) for $\mathbf{F}$ and $\mathbf{G}$ is, in the general case, quite difficult. However, the solution of Eq. (10.5-1) for $\mathbf{K}$ and $\mathbf{N}$ is rather simple [15]. For this reason we will examine here this latter case.

To this end, consider Figure 10.7. This figure is a closed-loop system with output feedback, where $\mathbf{G}(s) = \mathbf{C}(s\mathbf{I} - \mathbf{A})^{-1}\mathbf{B}$ is the transfer function matrix of the open-loop system under control (10.2-1). On the basis of this figure we obtain $\mathbf{U}(s) = \mathbf{KY}(s) + \mathbf{NR}(s)$, $\mathbf{Y}(s) = \mathbf{G}(s)\mathbf{U}(s)$ and, consequently, $\mathbf{Y}(s) = \hat{\mathbf{H}}(s)\mathbf{R}(s)$, where $\hat{\mathbf{H}}(s)$ is the transfer function matrix of the closed-loop system of Figure 10.7 and it is given by

$$\hat{\mathbf{H}}(s) = [\mathbf{I}_p - \mathbf{G}(s)\mathbf{K}]^{-1}\mathbf{G}(s)\mathbf{N} \tag{10.5-3}$$

where $\mathbf{I}_p$ is the $p \times p$ identity matrix. Substituting Eq. (10.5-3), into Eq. (10.5-1) gives

$$[\mathbf{I}_p - \mathbf{G}(s)\mathbf{K}]^{-1}\mathbf{G}(s)\mathbf{N} = \mathbf{H}_{\mathrm{m}}(s) \tag{10.5-4}$$

By premultiplying Eq. (10.5-4) by the matrix $[\mathbf{I}_p - \mathbf{G}(s)\mathbf{K}]$, we obtain

$$\mathbf{G}(s)\mathbf{N} = \mathbf{H}_{\mathrm{m}}(s) - \mathbf{G}(s)\mathbf{KH}_{\mathrm{m}}(s)$$

or

$$\mathbf{G}(s)\mathbf{N} + \mathbf{G}(s)\mathbf{KH}_{\mathrm{m}}(s) = \mathbf{H}_{\mathrm{m}}(s) \tag{10.5-5}$$

Relation (10.5-5) is a polynomial equation in s, whose coefficients are matrices. Therefore, Eq. (10.5-5) has a solution only when a pair of matrices $\mathbf{K}$ and $\mathbf{N}$ exists such that the coefficients of all like powers of s in both sides of Eq. (10.5-5) are equal. If we carry out the matrix multiplications involved in Eq. (10.5-5) and appropriately group the results such that each side is a matrix polynomial in s and subsequently equate the coefficients of like powers in s, we obtain a linear algebraic system of equations having the general form

$$\mathbf{P}\theta = \mathbf{h} \tag{10.5-6}$$

where $\mathbf{P}$ and $\mathbf{h}$ are known matrices, while θ is the unknown vector, whose elements are the elements of the matrices $\mathbf{K}$ and $\mathbf{N}$.

Relation (10.5-6) is, in general, a system of equations with more equations than unknowns. Solving Eq. (10.5-6) using the least-squares method we obtain

$$\theta^* = (\mathbf{P}^T\mathbf{P})^{-1}\mathbf{P}^T\mathbf{h} \tag{10.5-7}$$

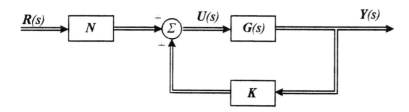

Figure 10.7   Closed-loop system with output feedback.

Expression (10.5-7) is the exact solution of system (10.5-6) when the error $J = \mathbf{e}^T\mathbf{e}$ equals zero, where $\mathbf{e} = \mathbf{P}(\mathbf{P}^T\mathbf{P})^{-1}\mathbf{P}\mathbf{h} - \mathbf{h}$. In this case, the solution θ^* yields the matrices $\mathbf{K}$ and $\mathbf{N}$ which satisfy Eq. (10.5-5) (and, consequently, the relation of exact model matching (10.5-4)), exactly, provided that the matrix $[\mathbf{I}_p - \mathbf{G}(s)\mathbf{K}]$ is regular. This procedure is demonstrated by the example that follows.

Example 10.5.1

Consider an unstable system under control with transfer function matrix $\mathbf{G}(s)$ and a stable model with transfer function matrix $\mathbf{H}_m(s)$, where

$$\mathbf{G}(s) = \begin{bmatrix} \dfrac{s}{s^2+s-1} \\ \dfrac{s+1}{s^2+s-1} \end{bmatrix} = \frac{1}{\alpha(s)}\begin{bmatrix} s \\ s+1 \end{bmatrix}, \qquad \mathbf{H}_m(s) = \begin{bmatrix} \dfrac{2}{s+1} \\ \dfrac{2}{s} \end{bmatrix} = \frac{1}{\beta(s)}\begin{bmatrix} 2s \\ s(s+1) \end{bmatrix}$$

where $\alpha(s) = s^2 + s - 1$ and $\beta(s) = s(s+1)$. Determine the output controller matrices $\mathbf{K}$ and $\mathbf{N}$ such that the transfer function matrix $\hat{\mathbf{H}}(s)$ of the closed-loop system is equal to the transfer function matrix $\mathbf{H}_m(s)$ of the model.

Solution

In this example $\mathbf{K} = [k_1, k_2]$ and $\mathbf{N} = n$. Hence relation (10.5-5) becomes

$$\frac{1}{\alpha(s)}\begin{bmatrix} s \\ \cdots \\ s+1 \end{bmatrix}n + \frac{1}{\alpha(s)\beta(s)}\begin{bmatrix} s \\ \cdots \\ s+1 \end{bmatrix}[k_1, k_2]\begin{bmatrix} 2s \\ \cdots \\ 2(s+1) \end{bmatrix} = \frac{1}{\beta(s)}\begin{bmatrix} 2s \\ \cdots \\ 2(s+1) \end{bmatrix}$$

or

$$\begin{bmatrix} ns\beta(s) \\ \cdots\cdots \\ n(s+1)\beta(s) \end{bmatrix} + \begin{bmatrix} 2s^2k_1 + 2s(s+1)k_2 \\ \cdots\cdots\cdots\cdots\cdots \\ 2(s+1)sk_1 + 2(s+1)^2k_2 \end{bmatrix} = \begin{bmatrix} 2s\alpha(s) \\ \cdots\cdots \\ 2(s+1)\alpha(s) \end{bmatrix}$$

The above relation can be written more compactly as follows:

$$\mathbf{Q}(s)\theta = \mathbf{d}(s)$$

where

$$\mathbf{Q}(s) = \begin{bmatrix} s\beta(s) & 2s^2 & 2s(s+1) \\ (s+1)\beta(s) & 2s(s+1) & 2(s+1)^2 \end{bmatrix}$$

$$\theta = \begin{bmatrix} n \\ k_1 \\ k_2 \end{bmatrix}, \qquad \mathbf{d}(s) = \begin{bmatrix} 2s\alpha(s) \\ 2(s+1)\alpha(s) \end{bmatrix}$$

If we substitute the polynomials $\alpha(s)$ and $\beta(s)$ in the matrix $\mathbf{Q}(s)$ and in the vector $\mathbf{d}(s)$ we obtain

$$\mathbf{Q}(s) = \begin{bmatrix} s^3+s^2 & 2s^2 & 2s^2+2s \\ s^3+2s^2+s & 2s^2+2s & 2s^2+4s+2 \end{bmatrix} = \mathbf{Q}_0 + \mathbf{Q}_1s + \mathbf{Q}_2s^2 + \mathbf{Q}_3s^3$$

where

$$\mathbf{Q}_0 = \begin{bmatrix} 0 & 0 & 0 \\ 0 & 0 & 2 \end{bmatrix}, \qquad \mathbf{Q}_1 = \begin{bmatrix} 0 & 0 & 2 \\ 1 & 2 & 4 \end{bmatrix}, \qquad \mathbf{Q}_2 = \begin{bmatrix} 1 & 2 & 2 \\ 2 & 2 & 2 \end{bmatrix},$$

$$\mathbf{Q}_3 = \begin{bmatrix} 1 & 0 & 0 \\ 1 & 0 & 0 \end{bmatrix}$$

and

$$\mathbf{d}(s) = \begin{bmatrix} 2s^3 + 2s^2 - 2s \\ 2s^3 + 4s^2 - 2 \end{bmatrix} = \mathbf{d}_0 + \mathbf{d}_1 s + \mathbf{d}_2 s^2 + \mathbf{d}_3 s^3$$

where

$$\mathbf{d}_0 = \begin{bmatrix} 0 \\ -2 \end{bmatrix}, \qquad \mathbf{d}_1 = \begin{bmatrix} -2 \\ 0 \end{bmatrix}, \qquad \mathbf{d}_2 = \begin{bmatrix} 2 \\ 4 \end{bmatrix}, \qquad \mathbf{d}_3 = \begin{bmatrix} 2 \\ 2 \end{bmatrix}$$

Consequently, relation $\mathbf{Q}(s)\boldsymbol{\theta} = \mathbf{d}(s)$ can be rewritten as

$$[\mathbf{Q}_0 + \mathbf{Q}_1 s + \mathbf{Q}_2 s^2 + \mathbf{Q}_3 s^3]\boldsymbol{\theta} = \mathbf{d}_0 + \mathbf{d}_1 s + \mathbf{d}_2 s^2 + \mathbf{d}_3 s^3$$

For the above relation to hold, the vector $\boldsymbol{\theta}$ must be such that the coefficients of the like powers of s in both sides of the equation are equal, i.e.,

$$\mathbf{Q}_i \boldsymbol{\theta} = \mathbf{d}_i, \qquad i = 0, 1, 2, 3$$

This relation can be rewritten in the compact form of Eq. 10.5-6), i.e., in the form

$$\mathbf{P}\boldsymbol{\theta} = \mathbf{h}$$

where

$$\mathbf{P} = \begin{bmatrix} \mathbf{Q}_0 \\ \mathbf{Q}_1 \\ \mathbf{Q}_2 \\ \mathbf{Q}_3 \end{bmatrix}, \qquad \mathbf{h} = \begin{bmatrix} \mathbf{d}_0 \\ \mathbf{d}_1 \\ \mathbf{d}_2 \\ \mathbf{d}_3 \end{bmatrix}$$

If we substitute the values of $\mathbf{Q}_i$ and $\mathbf{d}_i$ in the equation $\mathbf{P}\boldsymbol{\theta} = \mathbf{h}$ and solve for $\boldsymbol{\theta}$ we obtain the exact solution $\boldsymbol{\theta} = [2, 1, -1]^T$. Thus, the matrices $\mathbf{N}$ and $\mathbf{K}$ of the compensator are $\mathbf{N} = 2$ and $\mathbf{K} = [k_1, k_2] = [1, -1]$. To check the results, we substitute $\mathbf{K}$ and $\mathbf{N}$ in Eq. (10.5-3), which yields $\hat{\mathbf{H}}(s) = \mathbf{H}_m(s)$. Hence, an exact model matching has been achieved.

10.6 STATE OBSERVERS

10.6.1 Introduction

In designing a closed-loop system using modern control techniques, the control strategy applied is usually a feedback loop involving feedback of the system state vector $\mathbf{x}(t)$. Examples of such strategies is the application of the state feedback law $\mathbf{u} = \mathbf{Fx} + \mathbf{Gr}$ for pole assignment, decoupling, and model matching presented in the previous sections. This means that for this type of feedback law to be applicable, the entire state vector $\mathbf{x}$ must be available (measurable).

In practice, however, it happens very often that not all state variables of a system are accessible to measurement. This obstacle can be circumvented if a mathematical model for the system is available, in which case it is possible to estimate the state vector.

A widely known method for state-vector estimation or reconstruction is that of using an observer. This technique was first proposed by Luenberger [11–13] and is presented in the sequel.

10.6.2 State-Vector Reconstruction Using a Luenberger Observer

Consider the system

$$\dot{\mathbf{x}}(t) = \mathbf{A}\mathbf{x}(t) + \mathbf{B}\mathbf{u}(t) \tag{10.6-1a}$$

$$\mathbf{y}(t) = \mathbf{C}\mathbf{x}(t) \tag{10.6-1b}$$

Assume that state vector $\mathbf{x}(t)$ is given approximately by the state vector $\hat{\mathbf{x}}(t)$ of the following system

$$\dot{\hat{\mathbf{x}}}(t) = \hat{\mathbf{A}}\hat{\mathbf{x}}(t) + \hat{\mathbf{B}}\mathbf{u}(t) + \mathbf{K}\mathbf{y}(t) \tag{10.6-2}$$

where $\hat{\mathbf{x}}$ is an n-dimensional vector and the matrices $\hat{\mathbf{A}}$, $\hat{\mathbf{B}}$, and $\mathbf{K}$ are unknown. System (10.6-2) is called the *state observer* of system (10.6-1). A closer examination of Eq. (10.6-2) shows that the observer is a dynamic system having two inputs, the input vector $\mathbf{u}(t)$ and the output vector $\mathbf{y}(t)$ of the initial system (10.6-1) (see Figure 10.8). Clearly, the observer matrices $\hat{\mathbf{A}}$, $\hat{\mathbf{B}}$, and $\mathbf{K}$ should be chosen such that $\hat{\mathbf{x}}(t)$ is as close as possible to $\mathbf{x}(t)$. In cases where $\hat{\mathbf{x}}(t)$ and $\mathbf{x}(t)$ are of equal dimension, then the observer is referred to as a *full-order* observer. This case is studied in the present section. When the dimension of $\hat{\mathbf{x}}(t)$ is smaller than that of $\mathbf{x}(t)$, then the observer is referred to as a *reduced-order* observer, which is studied in Subsec. 10.6.3.

Define the state error

$$\mathbf{e}(t) = \mathbf{x}(t) - \hat{\mathbf{x}}(t) \tag{10.6-3}$$

The formal definition of the problem of designing the observer (10.6-2) is the following: determine appropriate matrices $\hat{\mathbf{A}}$, $\hat{\mathbf{B}}$, and $\mathbf{K}$, such that the error $\mathbf{e}(t)$ tends to zero as fast as possible.

To solve the problem, we proceed as follows. Using Eqs (10.6-1) and (10.6-2), it can be shown that the error $\mathbf{e}(t)$ satisfies the differential equation

$$\dot{\mathbf{e}}(t) = \dot{\mathbf{x}}(t) - \dot{\hat{\mathbf{x}}}(t) = \mathbf{A}\mathbf{x}(t) + \mathbf{B}\mathbf{u}(t) - \hat{\mathbf{A}}[\mathbf{x}(t) - \mathbf{e}(t)] - \hat{\mathbf{B}}\mathbf{u}(t) - \mathbf{K}\mathbf{C}\mathbf{x}(t)$$

or

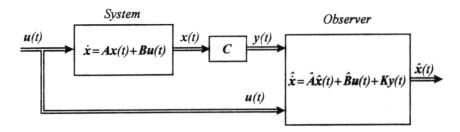

Figure 10.8 Simplified presentation of the system (10.6-1) and the observer (10.6-2).

$$\dot{e}(t) = \hat{A}e(t) + [A - KC - \hat{A}]x(t) + [B - \hat{B}]u(t)$$

For the error $e(t)$ to tend to zero, independently of $x(t)$ and $u(t)$, the following three conditions must be satisfied simultaneously:

1. $\hat{A} = A - KC$
2. $\hat{B} = B$
3. matrix $\hat{A}$ is stable.

From the above we conclude that the error $e(t)$ satisfies the differential equation

$$\dot{e}(t) = \hat{A}e(t) = [A - KC]e(t)$$

while the state observer (10.6-2) takes on the form

$$\hat{x}(t) = [A - KC]\hat{x}(t) + Bu(t) + Ky(t) \tag{10.6-4a}$$

or

$$\hat{x}(t) = A\hat{x}(t) + Bu(t) + K[y(t) - C\hat{x}(t)] \tag{10.6-4b}$$

According to Eq. (10.6-4a), the observer can be considered as a system involving the matrices A, B, and C of the original system together with an arbitrary matrix K. This matrix K must be chosen so that the eigenvalues of the matrix $\hat{A} = A - KC$ effectively force the error $e(t)$ to zero as fast as possible. According to Eq. (10.6-4b), the observer appears to be exactly the original system plus an additional term $K[y(t) - C\hat{x}(t)]$. The term $\omega(t) = y(t) - \hat{y}(t) = y(t) - C\hat{x}(t)$ can be considered as a corrective term, often called a *residual*. Of course, if $\hat{x}(t) = x(t)$, then $\omega(t) = 0$. Therefore, a residual exists if the system output vector $y(t)$ and the observer vector $\hat{y}(t) = C\hat{x}(t)$ are different.

Remark 10.6.1

To construct the observer it is necessary to construct the model of the original system itself, plus the corrective term $K[y(t) - C\hat{x}(t)]$. One may then ask: Why not build the model $\hat{x}(t) = A\hat{x}(t) + Bu(t)$ of the original system with initial condition $\hat{x}(t_0)$, and on the basis of this model estimate the state vector $\hat{x}(t)$? Such an approach is not used in practice because it presents certain serious drawbacks. The most important drawback is the following. Since $\hat{x}(t_0)$ is only an estimate of $x(t_0)$, the initial condition $x(t_0)$ of the system and the initial condition $\hat{x}(t_0)$ of the model differ in most cases. As a result, $\hat{x}(t)$ may not converge fast enough to $x(t)$. To secure rapid convergence of $\hat{x}(t)$ to $x(t)$, we add the term $K[y(t) - C\hat{x}(t)]$ to the model $\hat{x}(t) = A\hat{x}(t) + Bu(t)$, resulting in an observer of the form (10.6-4b). Under the assumption that the system (A, C) is observable, the matrix K provides adequate design flexibility, as shown below, so that $\hat{x}(t)$ converges to $x(t)$ very fast.

The block diagram for the state observer (10.6-4) is presented in Figure 10.9.

The state observer design problem therefore reduces to one of determining an appropriate matrix K, such that all eigenvalues of the matrix $\hat{A} = A - KC$ lie in the left-half complex plane. A closer look at the problem reveals that it comes down to one of solving a pole-placement problem for the matrix $\hat{A} = A - KC$. As a matter of fact, this problem is *dual* to the pole-placement problem discussed earlier in Sec. 10.3. In what follows, we will use the results of Sec. 10.3 to solve the observer design problem.

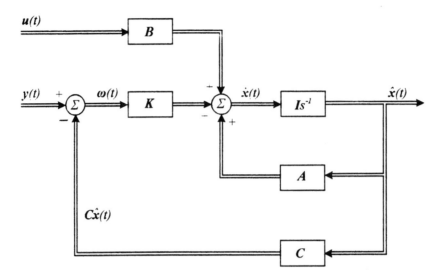

Figure 10.9 Block diagram of the observer (10.6-4).

As already mentioned in Sec. 10.3, the necessary and sufficient condition for a matrix $\mathbf{F}$ to exist, such that the matrix $\mathbf{A} - \mathbf{BF}$ may have any desired eigenvalues, is that the system $(\mathbf{A}, \mathbf{B})$ is controllable, i.e.,

$$\text{rank}\,\mathbf{S} = n, \qquad \text{where } \mathbf{S} = [\mathbf{B} \vdots \mathbf{AB} \vdots \cdots \vdots \mathbf{A}^{n-1}\mathbf{B}] \qquad (10.6\text{-}5)$$

In the case of the observer, the necessary and sufficient condition for a matrix $\mathbf{K}$ to exist, so that the matrix $\hat{\mathbf{A}} = \mathbf{A} - \mathbf{KC}$ or, equivalently, the matrix $\hat{\mathbf{A}}^T = \mathbf{A}^T - \mathbf{C}^T\mathbf{K}^T$ has any desired eigenvalues, is that the system $(\mathbf{A}^T, \mathbf{C}^T)$ is controllable or, equivalently, that the system $(\mathbf{A}, \mathbf{C})$ is observable, i.e.,

$$\text{rank}\,\mathbf{R} = n, \qquad \text{where } \mathbf{R}^T = [\mathbf{C}^T \vdots \mathbf{A}^T\mathbf{C}^T \vdots \cdots \vdots (\mathbf{A}^T)^{n-1}\mathbf{C}^T] \qquad (10.6\text{-}6)$$

Hence, the following theorem holds.

Theorem 10.6.1

The necessary and sufficient conditions for the reconstruction of the state of system (10.6-1) is that the system is completely observable.

For the system $(\mathbf{A}, \mathbf{B}, \mathbf{C})$, we say that the conditions (10.6-5) and (10.6-6) are *dual*.

We will first consider the single-output case for system (10.6-1). For this case, the matrix $\mathbf{C}$ reduces to a row vector $\mathbf{c}^T$, thus reducing $\mathbf{R}$ to the $n \times n$ matrix:

$$\mathbf{R}^T = [\mathbf{c} \vdots \mathbf{A}^T\mathbf{c} \vdots \cdots \vdots (\mathbf{A}^T)^{n-1}\mathbf{c}]$$

The matrix $\hat{\mathbf{A}}$ becomes $\hat{\mathbf{A}} = \mathbf{A} - \mathbf{k}\mathbf{c}^T$, where $\mathbf{k}^T = [k_1, k_2, \ldots, k_n]$. Define

$$p(s) = |s\mathbf{I} - \mathbf{A}| = s^n + a_1 s^{n-1} + \cdots + a_n = \prod_{i=1}^{n}(s - \lambda_i)$$

$$\hat{p}(s) = |s\mathbf{I} - \hat{\mathbf{A}}| = s^n + \hat{a}_1 s^{n-1} + \cdots + \hat{a}_n = \prod_{i=1}^{n}(s - \hat{\lambda}_i)$$

where λ_i are the eigenvalues of the system (10.6-1) and $\hat{\lambda}_i$ are the desired eigenvalues of the observer (10.6-4). Hence, the problem here is to find $\mathbf{k}$ so that the observer has the desired eigenvalues $\hat{\lambda}_1, \hat{\lambda}_2, \ldots, \hat{\lambda}_n$. The vector $\mathbf{k}$ sought is uniquely defined. In Sec. 10.3, three alternative methods were presented to solve for $\mathbf{k}$. Applying the Buss–Gura formula (10.3-8) yields the following solution:

$$\mathbf{k} = [\mathbf{W}^T \mathbf{R}]^{-1}(\hat{\mathbf{a}} - \mathbf{a}) \tag{10.6-7}$$

where

$$\mathbf{W} = \begin{bmatrix} 1 & a_1 & \cdots & a_{n-1} \\ 0 & 1 & \cdots & a_{n-2} \\ \vdots & \vdots & & \vdots \\ 0 & 0 & \cdots & 1 \end{bmatrix}, \qquad \mathbf{R} = \begin{bmatrix} \mathbf{c}^T \\ \mathbf{c}^T \mathbf{A} \\ \vdots \\ \mathbf{c}^T \mathbf{A}^{n-1} \end{bmatrix}, \qquad \hat{\mathbf{a}} = \begin{bmatrix} \hat{a}_1 \\ \hat{a}_2 \\ \vdots \\ \hat{a}_n \end{bmatrix}, \qquad \text{and}$$

$$\mathbf{a} = \begin{bmatrix} a_1 \\ a_2 \\ \vdots \\ a_n \end{bmatrix}$$

Clearly, the solution (10.6-7) corresponds to the solution (10.3-8).

For the multi-output case, determining the matrix $\mathbf{K}$, as discussed in Sec. 10.3, is usually a complicated task. A simple approach to the problem is to assume that $\mathbf{K}$ has the following outer product form:

$$\mathbf{K} = \mathbf{q}\mathbf{p}^T \tag{10.6-8}$$

where $\mathbf{q}$ and $\mathbf{p}$ are n-dimensional vectors. Then $\hat{\mathbf{A}} = \mathbf{A} - \mathbf{KC} = \mathbf{A} - \mathbf{q}\mathbf{p}^T \mathbf{C} = \mathbf{A} - \mathbf{q}\gamma^T$, where $\gamma^T = \mathbf{p}^T \mathbf{C}$. Therefore, assuming $\mathbf{K}$ to be of the form (10.6-8), the multi-output case reduces to the single-output case studied previously. Hence, the solution for $\mathbf{q}$ is given by Eq. (10.6-7), where the matrix $\mathbf{R}$ must be replaced by the matrix $\tilde{\mathbf{R}}$, where $\tilde{\mathbf{R}}^T = [\gamma \vdots \mathbf{A}^T \gamma \vdots \cdots \vdots (\mathbf{A}^T)^{n-1}\gamma]$. It is noted that the vector $\gamma = \mathbf{C}^T \mathbf{p}$ involves arbitrary parameters, which are the elements of the arbitrary vector $\mathbf{p}$. These arbitrary parameters may take any values as long as rank$\tilde{\mathbf{R}} = n$. If the condition rank$\tilde{\mathbf{R}} = n$ cannot be satisfied, other methods for determining $\mathbf{K}$ may be found in the literature.

10.6.3 Reduced-Order Observers

Suppose that the matrix $\mathbf{C}$ in the output equation $\mathbf{y}(t) = \mathbf{Cx}(t)$ is square and nonsingular. Then $\mathbf{x}(t) = \mathbf{C}^{-1}\mathbf{y}(t)$, thus eliminating the need for an observer.

Now, assume that only one of the state variables is not accessible to measurement. Then, it is reasonable to expect that the required state observer will not be of order n, but of lower order. This is in fact true, and can be stated as a theorem.

Theorem 10.6.2

If system (10.6-1) is observable, then the smallest possible order of the state observer is $n - p$.

We will next present some useful results regarding the design of reduced-order observers. To this end, we assume that the vector $\mathbf{x}(t)$ and the matrices $\mathbf{A}$ and $\mathbf{B}$ may be decomposed as follows:

$$\mathbf{x}(t) = \begin{bmatrix} \mathbf{q}_1(t) \\ \mathbf{q}_2(t) \end{bmatrix}, \qquad \mathbf{A} = \begin{bmatrix} \mathbf{A}_{11} & \mathbf{A}_{12} \\ \mathbf{A}_{21} & \mathbf{A}_{22} \end{bmatrix}, \qquad \mathbf{B} = \begin{bmatrix} \mathbf{B}_1 \\ \mathbf{B}_2 \end{bmatrix}$$

Thus

$$\dot{\mathbf{q}}_1(t) = \mathbf{A}_{11}\mathbf{q}_1(t) + \mathbf{A}_{12}\mathbf{q}_2(t) + \mathbf{B}_1\mathbf{u}(t)$$
$$\dot{\mathbf{q}}_2(t) = \mathbf{A}_{21}\mathbf{q}_1(t) + \mathbf{A}_{22}\mathbf{q}_2(t) + \mathbf{B}_2\mathbf{u}(t) \tag{10.6-9}$$

where $\mathbf{q}_1(t)$ is a vector whose elements are all the measurable state variables of $\mathbf{x}(t)$, i.e.,

$$\mathbf{y}(t) = \mathbf{C}_1\mathbf{q}_1(t), \qquad \text{with} \qquad |\mathbf{C}_1| \neq 0 \tag{10.6-10}$$

In cases where the system is not in the form of Eqs (10.6-9) and (10.6-10), it can easily be converted to this form by using an appropriate transformation matrix. The observer of the form (10.6-4b) for the system (10.6-9) and (10.6-10) will then become

$$\dot{\hat{\mathbf{q}}}_1(t) = \mathbf{A}_{11}\hat{\mathbf{q}}_1(t) + \mathbf{A}_{12}\hat{\mathbf{q}}_2(t) + \mathbf{B}_1\mathbf{u}(t) + \mathbf{K}_1[\mathbf{y}(t) - \mathbf{C}_1\hat{\mathbf{q}}_1(t)] \tag{10.6-11a}$$

$$\dot{\hat{\mathbf{q}}}_2(t) = \mathbf{A}_{21}\hat{\mathbf{q}}_1(t) + \mathbf{A}_{22}\hat{\mathbf{q}}_2(t) + \mathbf{B}_2\mathbf{u}(t) + \mathbf{K}_2[\mathbf{y}(t) - \mathbf{C}_1\hat{\mathbf{q}}_1(t)] \tag{10.6-11b}$$

According to Eq. (10.6-10), we have

$$\mathbf{q}_1(t) = \hat{\mathbf{q}}_1(t) = \mathbf{C}_1^{-1}\mathbf{y}(t) \tag{10.6-12}$$

Therefore, there is no need for the observer (10.6-11a), while the observer (10.6-11b) becomes

$$\dot{\hat{\mathbf{q}}}_2(t) = \mathbf{A}_{22}\hat{\mathbf{q}}_2(t) + \mathbf{B}_2\mathbf{u}(t) + \mathbf{A}_{21}\mathbf{C}_1^{-1}\mathbf{y}(t) \tag{10.6-13}$$

where use was made of Eq. (10.6-12). The observer (10.6-13) is a dynamic system of order equal to the number of the state variables which are not accessible to measurement.

It is obvious that for the observer (10.6-13), the submatrix $\mathbf{A}_{22}$ plays an important role. If $\mathbf{A}_{22}$ has by luck satisfactory eigenvalues, then system (10.6-13) suffices for the estimation of $\mathbf{q}_2(t)$. On the other hand, if the eigenvalues of $\mathbf{A}_{22}$ are not satisfactory, then the following observer is proposed for estimating $\mathbf{q}_2(t)$:

$$\hat{\mathbf{q}}_2(t) = \boldsymbol{\phi}\mathbf{y}(t) + \mathbf{v}(t) \tag{10.6-14}$$

where $\mathbf{v}(t)$ is an $(n - p)$ vector governed by the vector difference equation

$$\dot{\mathbf{v}}(t) = \mathbf{F}\mathbf{v}(t) + \mathbf{H}\mathbf{u}(t) + \mathbf{G}\mathbf{y}(t) \tag{10.6-15}$$

Define the error as before, i.e., let

$$\mathbf{e}(t) = \mathbf{x}(t) - \hat{\mathbf{x}}(t) = \begin{bmatrix} \mathbf{q}_1(t) - \hat{\mathbf{q}}_1(t) \\ \mathbf{q}_2(t) - \hat{\mathbf{q}}_2(t) \end{bmatrix} = \begin{bmatrix} \mathbf{e}_1(t) \\ \mathbf{e}_2(t) \end{bmatrix} = \begin{bmatrix} \mathbf{0} \\ \mathbf{e}_2(t) \end{bmatrix}$$

The differential equation for $\mathbf{e}_2(t)$ is the following:

$$\dot{\mathbf{e}}_2(t) = \dot{\mathbf{q}}_2(t) - \dot{\hat{\mathbf{q}}}_2(t) = \mathbf{A}_{21}\mathbf{q}_1(t) + \mathbf{A}_{22}\mathbf{q}_2(t) + \mathbf{B}_2\mathbf{u}(t) - \boldsymbol{\phi}\dot{\mathbf{y}}(t) - \dot{\mathbf{v}}(t)$$

After some algebraic manipulations and simplifications, we have

$$\dot{\mathbf{e}}_2(t) = \mathbf{F}\mathbf{e}_2(t) + (\mathbf{A}_{21} - \boldsymbol{\phi}\mathbf{C}_1\mathbf{A}_{11} - \mathbf{G}\mathbf{C}_2 + \mathbf{F}\boldsymbol{\phi}\mathbf{C}_1)\mathbf{q}_1(t)$$
$$+ (\mathbf{A}_{22} - \boldsymbol{\phi}\mathbf{C}_1\mathbf{A}_{12} - \mathbf{F})\mathbf{q}_2(t) + (\mathbf{B}_2 - \boldsymbol{\phi}\mathbf{C}_1\mathbf{B}_1 - \mathbf{H})\mathbf{u}(t) \tag{10.6-16}$$

In order for $\mathbf{e}_2(t)$ to be independent of $\mathbf{q}_1(t)$, $\mathbf{q}_2(t)$ and $\mathbf{u}(t)$, as well as to tend rapidly to zero, the following conditions must hold:

1. $\mathbf{G}\mathbf{C}_2 = \mathbf{A}_{21} - \boldsymbol{\phi}\mathbf{C}_1\mathbf{A}_{11} + \mathbf{F}\boldsymbol{\phi}\mathbf{C}_1$ or $\mathbf{G} = (\mathbf{A}_{21} - \boldsymbol{\phi}\mathbf{C}_1\mathbf{A}_{11})\mathbf{C}_1^{-1} + \mathbf{F}\boldsymbol{\phi}$

$$\tag{10.6-17a}$$

2. $\mathbf{F} = \mathbf{A}_{22} - \boldsymbol{\phi}\mathbf{C}_1\mathbf{A}_{12}$ $\tag{10.6-17b}$
3. $\mathbf{H} = \mathbf{B}_2 - \boldsymbol{\phi}\mathbf{C}_1\mathbf{B}_1$ $\tag{10.6-17c}$
4. Matrix $\mathbf{F}$ is stable

If the foregoing conditions are met, then Eq. (10.6-16) becomes

$$\dot{\mathbf{e}}_2(t) = \mathbf{F}\mathbf{e}_2(t)$$

The matrix $\boldsymbol{\phi}$ may be chosen such that the matrix $\mathbf{F} = \mathbf{A}_{22} - \boldsymbol{\phi}\mathbf{C}_1\mathbf{A}_{12}$ has any desired eigenvalues, as long as the system $(\mathbf{A}_{22}, \mathbf{C}_1\mathbf{A}_{12})$ is observable, i.e., as long as

$$\text{rank}\mathbf{R}_1 = n - p, \quad \text{where}$$

$$\mathbf{R}_1^T = \left[[\mathbf{C}_1\mathbf{A}_{12}]^T \ \vdots \ \mathbf{A}_{22}^T[\mathbf{C}_1\mathbf{A}_{12}]^T \ \vdots \ \cdots \ \vdots \ [\mathbf{A}_{22}^T]^{n-p-1}[\mathbf{C}_1\mathbf{A}_{12}]^T \right]$$

The following useful theorem has been proven [6].

Theorem 10.6.3

The pair $(\mathbf{A}_{22}, \mathbf{C}_1\mathbf{A}_{12})$ is observable, if and only if the pair $(\mathbf{A}, \mathbf{C})$ is observable. The final form of the observer (10.6-15) is

$$\dot{\mathbf{v}}(t) = \mathbf{F}\mathbf{v}(t) + \mathbf{H}\mathbf{u}(t) + [(\mathbf{A}_{21} - \boldsymbol{\phi}\mathbf{C}_1\mathbf{A}_{11})\mathbf{C}_1^{-1} + \mathbf{F}\boldsymbol{\phi}]\mathbf{y}(t)$$

or

$$\dot{\mathbf{v}}(t) = \mathbf{F}\hat{\mathbf{q}}_2(t) + \mathbf{H}\mathbf{u}(t) + (\mathbf{A}_{21} - \boldsymbol{\phi}\mathbf{C}_1\mathbf{A}_{11})\mathbf{C}_1^{-1}\mathbf{y}(t) \tag{10.6-18}$$

The block diagram of the observer (10.6-18) is presented in Figure 10.10.

10.6.4 Closed-Loop System Design Using State Observers

Consider the system

$$\dot{\mathbf{x}}(t) = \mathbf{A}\mathbf{x}(t) + \mathbf{B}\mathbf{u}(t), \quad \mathbf{y}(t) = \mathbf{C}\mathbf{x}(t) \tag{10.6-19}$$

with the state observer

$$\dot{\hat{\mathbf{x}}}(t) = \mathbf{A}\hat{\mathbf{x}}(t) + \mathbf{B}\mathbf{u}(t) + \mathbf{K}_2[\mathbf{y}(t) - \mathbf{C}\hat{\mathbf{x}}(t)] \tag{10.6-20}$$

Apply the control law

$$\mathbf{u}(t) = -\mathbf{K}_1\hat{\mathbf{x}}(t) \tag{10.6-21}$$

Then, system (10.6-19) becomes

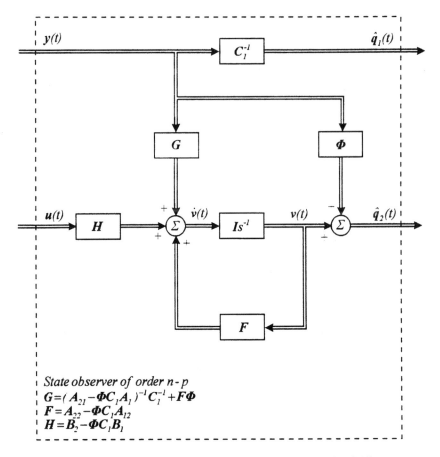

Figure 10.10   Block diagram of the reduced-order observer (10.6-18).

$$\dot{\mathbf{x}}(t) = \mathbf{A}\mathbf{x}(t) - \mathbf{B}\mathbf{K}_1\hat{\mathbf{x}}(t) \tag{10.6-22}$$

and the observer (10.6-20) takes on the form

$$\dot{\hat{\mathbf{x}}}(t) = \mathbf{A}\hat{\mathbf{x}}(t) - \mathbf{B}\mathbf{K}_1\hat{\mathbf{x}}(t) + \mathbf{K}_2[\mathbf{C}\mathbf{x}(t) - \mathbf{C}\hat{\mathbf{x}}(t)] \tag{10.6-23}$$

If we use the definition $\mathbf{e}(t) = \mathbf{x}(t) - \hat{\mathbf{x}}(t)$, then Eq. (10.6-22) becomes

$$\dot{\mathbf{x}}(t) = \mathbf{A}\mathbf{x}(t) - \mathbf{B}\mathbf{K}_1[\mathbf{x}(t) - \mathbf{e}(t)]$$

or

$$\dot{\mathbf{x}}(t) = (\mathbf{A} - \mathbf{B}\mathbf{K}_1)\mathbf{x}(t) + \mathbf{B}\mathbf{K}_1\mathbf{e}(t) \tag{10.6-24}$$

Subtracting Eq. (10.6-23) from Eq. (10.6-22), we have

$$\dot{\mathbf{e}}(t) = (\mathbf{A} - \mathbf{K}_2\mathbf{C})\mathbf{e}(t) \tag{10.6-25}$$

The foregoing results are very interesting, because they illustrate the fact that the matrix $\mathbf{K}_1$ of the closed-loop system (10.6-24) and the matrix $\mathbf{K}_2$ of the error equation (10.6-25) can be designed independently of each other. Indeed, if system $(\mathbf{A}, \mathbf{B})$ is controllable, then the matrix $\mathbf{K}_1$ of the state feedback law (10.6-21) can be chosen so that the poles of the closed-loop system (10.6-24) have any desired arbitrary values. The same applies to the error equation (10.6-25), where, if the system $(\mathbf{A}, \mathbf{C})$ is observable, the matrix $\mathbf{K}_2$ of the observer (10.6-20) can be chosen so as to force the error to go rapidly to zero. This property, where the two design problems (the observer and the closed-loop system) can be handled independently, is called the *separation principle*. This principle is clearly a very important design feature, since it reduces a rather difficult design task to two *separate* simpler design problems.

Figure 10.11 presents the closed-loop system (10.6-24) and the error equation (10.6-25). Figure 10.12 gives the block diagram representation of the closed-loop system with state observer.

Finally, the transfer function $\mathbf{G}_c(s)$ of the compensator defined by the equation $\mathbf{U}(s) = \mathbf{G}_c(s)\mathbf{Y}(s)$ will be

$$\mathbf{G}_c(s) = -\mathbf{K}_1[s\mathbf{I} - \mathbf{A} + \mathbf{B}\mathbf{K}_1 + \mathbf{K}_2\mathbf{C}]^{-1}\mathbf{K}_2 \qquad (10.6\text{-}26)$$

The results above cover the case of the full-order observer (order n). In the case of a reduced-order observer, e.g., of an observer of order $n - p$, similar results can be derived relatively easily.

Remark 10.6.2

Consider the pole placement and the observer design problems. The pole-placement problem is called the *control problem* and it is rather a simple control design tool for improving the closed-loop system performance. The observer design problem is called the *estimation problem*, since it produces a good estimate of $\mathbf{x}(t)$ in cases where $\mathbf{x}(t)$ is not measurable. The solution of the estimation problem reduces to that of solving a pole-placement problem. In cases where an estimate of $\mathbf{x}(t)$ is used in the control problem, one faces the problem of *simultaneous* solving the *estimation* and the *control* problem. At first sight this appears to be a formidable task. However, thanks to the separation theorem, the solution of the combined problem of estimation and control breaks down to separately solving the estimation and the control problem. Since the solution of these two problems is essentially the same, we conclude that the solution of the combined problem of estimation and control requires twice the solution of the pole placement problem. These results are usually referred to as *algebraic* techniques and cover the case of deterministic environment (i.e., deterministic systems and signals).

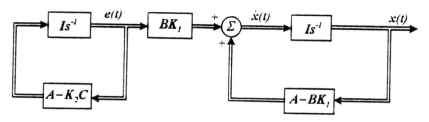

Figure 10.11 Representation of closed-loop system (10.6-24) and the error equation (10.6-25).

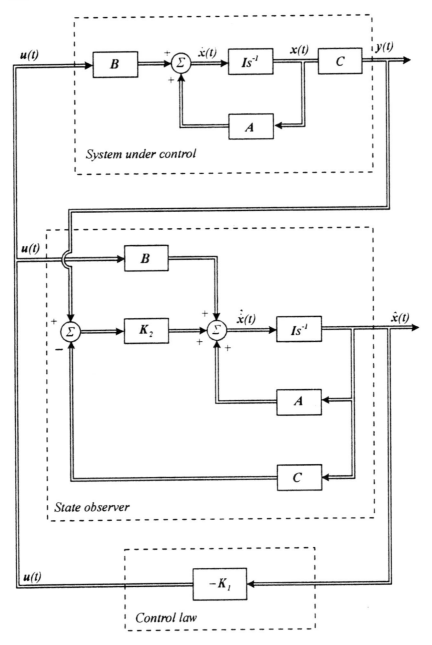

Figure 10.12   Block diagram of closed-loop system with state observer.

Remark 10.6.3

Going from the *algebraic* design techniques pointed out in Remark 10.6.2 to optimal design techniques (see Chap. 11), one realizes that there is a striking analogy between the two approaches: the optimal control problem reduces to that of solving a first-order matrix differential equation, known as the Ricatti equation. When we are in a stochastic environment, the problem of estimating $\mathbf{x}(t)$, known as Kalman filtering, also reduces to that of solving a Ricatti equation. Now, consider the case where Kalman filtering is needed to estimate $\mathbf{x}(t)$, which subsequently is to be applied for optimal control (this is the well-known linear quadratic gaussian or LQG problem). Then, thanks again to the separation theorem, the combined estimation and control problem, i.e., the LQG problem, breaks down to solving two separate Ricatti equations, i.e., solving *twice* a Ricatti equation.

Remark 6.10.4

Clearly, Remark 10.6.2 summarizes the *crux of the algebraic design approach results*, whereas Remark 10.6.3 summarizes the *crux of the optimal design approach results*. It is most impressive that in both cases the separation theorem holds, a fact which greatly facilitates the solution of the combined estimation and control problem. As one would expect, the difficulty in solving estimation and control problems increases as one goes from algebraic to optimal techniques and as one goes from deterministic (Luenberger observer) to stochastic environment (Kalman filtering and LQG problem).

10.6.5   Observer Examples

Example 10.6.1

Consider the system

$$\dot{\mathbf{x}}(t) = \mathbf{A}\mathbf{x}(t) + \mathbf{B}\mathbf{u}(t), \qquad y(t) = \mathbf{c}^T\mathbf{x}(t)$$

where

$$\mathbf{A} = \begin{bmatrix} 6 & 1 & 0 \\ -11 & 0 & 1 \\ 6 & 0 & 0 \end{bmatrix}, \qquad \mathbf{B} = \begin{bmatrix} 1 & 0 \\ 0 & 1 \\ 1 & 0 \end{bmatrix}, \qquad \text{and} \qquad \mathbf{c} = \begin{bmatrix} 1 \\ 0 \\ 0 \end{bmatrix}$$

Design:

    (a)   A full-order state observer, i.e., of order $n = 3$
    (b)   A reduced-order state observer, i.e., of order $n - p = 3 - 1 = 2$
    (c)   The closed-loop system for both cases

Solution

(a) Examine the system's observability. We have

$$\mathbf{R}^T = [\mathbf{c} \,\vdots\, \mathbf{A}^T\mathbf{c} \,\vdots\, (\mathbf{A}^T)^2\mathbf{c}] = \begin{bmatrix} 1 & 6 & 25 \\ 0 & 1 & 6 \\ 0 & 0 & 1 \end{bmatrix}$$

Since rank$\mathbf{R} = 3$, there exists a full-order state observer having the form

$$\dot{\hat{\mathbf{x}}}(t) = [\mathbf{A} - \mathbf{k}\mathbf{c}^T]\hat{\mathbf{x}}(t) + \mathbf{B}\mathbf{u}(t) + \mathbf{k}y(t)$$

The characteristic polynomial of the open-loop system is $p(s) = s^3 - 6s^2 + 11s - 6$. Suppose that the desired observer characteristic polynomial is chosen as $\hat{p}(s) = (s+1)(s+3)(s+4) = s^3 + 8s^2 + 19s + 12$. From Eq. (10.6-7), we have

$$W = \begin{bmatrix} 1 & -6 & 11 \\ 0 & 1 & -6 \\ 0 & 0 & 1 \end{bmatrix}, \qquad \hat{a} = \begin{bmatrix} \hat{a}_1 \\ \hat{a}_2 \\ \hat{a}_3 \end{bmatrix} = \begin{bmatrix} 8 \\ 19 \\ 12 \end{bmatrix}, \qquad \text{and}$$

$$a = \begin{bmatrix} a_1 \\ a_2 \\ a_3 \end{bmatrix} = \begin{bmatrix} -6 \\ 11 \\ -6 \end{bmatrix}$$

and hence

$$k = \begin{bmatrix} k_1 \\ k_2 \\ k_3 \end{bmatrix} = [W^T R]^{-1}(\hat{a} - a) = \begin{bmatrix} 14 \\ 7 \\ 18 \end{bmatrix}$$

(b) The system of the present example is in the form (10.6-9), where

$$A_{11} = 6, \qquad A_{12} = [1 \quad 0], \qquad A_{21} = \begin{bmatrix} -11 \\ 6 \end{bmatrix}, \qquad \text{and} \qquad A_{22} = \begin{bmatrix} 0 & 1 \\ 0 & 0 \end{bmatrix}$$

$$B_1 = [1 \quad 0], \qquad B_2 = \begin{bmatrix} 0 & 1 \\ 1 & 0 \end{bmatrix}, \qquad \text{and} \qquad c_1 = 1$$

and where

$$q_1(t) = x_1(t) \qquad \text{and} \qquad q_2(t) = \begin{bmatrix} x_2(t) \\ x_3(t) \end{bmatrix}$$

Here, $q_1(t) = x_1(t) = y(t)$. The proposed observer for the estimation of the vector $q_2(t)$ is

$$q_2(t) = \varphi y(t) + v(t)$$

where $v(t)$ is a two-dimensional vector described by the vector difference equation

$$\dot{v}(t) = Fv(t) + Hu(t) + gy(t)$$

and where

$$F = A_{22} - \varphi c_1 A_{12} = \begin{bmatrix} 0 & 1 \\ 0 & 0 \end{bmatrix} - \begin{bmatrix} \varphi_1 \\ \varphi_2 \end{bmatrix} [1 \quad 0] = \begin{bmatrix} -\varphi_1 & 1 \\ -\varphi_2 & 0 \end{bmatrix}$$

$$g = (A_{21} - \varphi c_1 A_{11}) c_1^{-1} + F\varphi$$

$$= \begin{bmatrix} -11 \\ 6 \end{bmatrix} - \begin{bmatrix} \varphi_1 \\ \varphi_2 \end{bmatrix} 6 + \begin{bmatrix} -\varphi_1 & 1 \\ -\varphi_2 & 0 \end{bmatrix} \begin{bmatrix} \varphi_1 \\ \varphi_2 \end{bmatrix} = \begin{bmatrix} -11 - 6\varphi_1 - \varphi_1^2 + \varphi_2 \\ 6 - 6\varphi_2 - \varphi_1\varphi_2 \end{bmatrix}$$

$$H = B_2 - \varphi c_1 B_1 = \begin{bmatrix} 0 & 1 \\ 1 & 0 \end{bmatrix} - \begin{bmatrix} \varphi_1 \\ \varphi_2 \end{bmatrix} [1 \quad 0] = \begin{bmatrix} -\varphi_1 & 1 \\ 1 - \varphi_2 & 0 \end{bmatrix}$$

Since

$$\text{rank} R_1^T = \text{rank} \left[[c_1 A_{12}]^T \vdots A_{22}^T [c_1 A_{12}]^T \right] = \text{rank} \begin{bmatrix} 1 & 0 \\ 0 & 1 \end{bmatrix} = 2$$

we can find a vector φ such that the matrix F has the desired eigenvalues. The characteristic polynomial of A_{22} is $p_2(s) = s^2$. Let $\hat{p}_2(s) = (s + 1)(s + 2) = s^2 + 3s + 2$ be the desired characteristic polynomial of matrix F. From Eq. (10.6-7), we have

$$\mathbf{W} = \begin{bmatrix} 1 & 0 \\ 0 & 1 \end{bmatrix}, \qquad \hat{\mathbf{a}} = \begin{bmatrix} 3 \\ 2 \end{bmatrix}, \qquad \mathbf{a} = \begin{bmatrix} 0 \\ 0 \end{bmatrix}$$

and therefore

$$\varphi = \begin{bmatrix} \varphi_1 \\ \varphi_2 \end{bmatrix} = [\mathbf{W}^T \mathbf{R}_1]^{-1}(\hat{\mathbf{a}} - \mathbf{a}) = \begin{bmatrix} 3 \\ 2 \end{bmatrix}$$

Introducing the value of φ into **g**, **F**, and **H** yields

$$\mathbf{g} = \begin{bmatrix} -36 \\ -12 \end{bmatrix}, \qquad \mathbf{F} = \begin{bmatrix} -3 & 1 \\ -2 & 0 \end{bmatrix}, \qquad \mathbf{H} = \begin{bmatrix} -3 & 1 \\ -1 & 0 \end{bmatrix}$$

(c) Let $p_c(s) = (s+1)(s+2)(s+3) = s^3 + 6s^2 + 11s + 6$ be the desired characteristic polynomial of the closed-loop system. The system is controllable because rank$\mathbf{S}$ = rank $[\mathbf{B} \vdots \mathbf{AB} \vdots \mathbf{A}^2\mathbf{B}] = 3$. Consequently, a feedback matrix $\mathbf{K}_1$ exists such that the closed-loop system poles are the roots of $p_c(s) = s^3 + 6s^2 + 11s + 6$. Using Eqs (10.3-15) and (10.3-16), the following matrix may be determined:

$$\mathbf{K}_1 = \begin{bmatrix} 12 & 0 & 0 \\ 0 & 0 & 0 \end{bmatrix}$$

Checking, we have that $|s\mathbf{I} - (\mathbf{A} - \mathbf{B}\mathbf{K}_1)| = s^3 + 6s^2 + 11s + 6 = p_c(s)$. Of course, in the case of a full-order observer, $\mathbf{k}_2 = \mathbf{k}$, where $\mathbf{k}$ was determined in part (a) above. In the case of a reduced-order observer, $\mathbf{k}_2 = \varphi$, where φ was determined in part (b) above.

Example 10.6.2

Consider the system

$$\dot{\mathbf{x}}(t) = \mathbf{A}\mathbf{x}(t) + \mathbf{b}u(t), \qquad y(t) = \mathbf{c}^T\mathbf{x}(t)$$

where

$$\mathbf{A} = \begin{bmatrix} 0 & 1 \\ 0 & -\gamma \end{bmatrix}, \qquad \mathbf{b} = \begin{bmatrix} 0 \\ \beta \end{bmatrix}, \qquad \mathbf{c} = \begin{bmatrix} 1 \\ 0 \end{bmatrix}$$

In this example we suppose that only the state $x_1(t) = y(t)$ can be directly measured. Design:

(a) A full-order state observer, i.e., of order $n = 2$
(b) A reduced-order state observer, i.e., of order $n - p = 2 - 1 = 1$. In other words, find an observer to estimate only the state $x_2(t)$, which we assume it not accessible to measurement. Note that $x_1(t)$ is measurable since $x_1(t) = y(t)$.
(c) The closed-loop system for both cases.

Solution

(a) Examine the system's observability. We have

$$\mathbf{R}^T = [\mathbf{c} \vdots \mathbf{A}^T\mathbf{c}] = \begin{bmatrix} 1 & 0 \\ 0 & 1 \end{bmatrix}$$

Since rank$\mathbf{R} = n = 2$, there exists a full-order state observer having the form

$$\dot{\hat{\mathbf{x}}}(t) = [\mathbf{A} - \mathbf{kc}^T]\hat{\mathbf{x}}(t) + \mathbf{B}u(t) + \mathbf{k}y(t)$$

The characteristic polynomials $p(s)$ and $\hat{p}(s)$ of the open-loop system and of the observer are $p(s) = |s\mathbf{I} - \mathbf{A}| = s^2 + \gamma s$ and $\hat{p}(s) = |s\mathbf{I} - (\mathbf{A} - \mathbf{kc}^T)| = s^2 + \hat{a}_1 s + \hat{a}_2$, respectively. From Eq. (10.6-7) we have

$$\mathbf{W} = \begin{bmatrix} 1 & \gamma \\ 0 & 1 \end{bmatrix}, \qquad \hat{\mathbf{a}} = \begin{bmatrix} \hat{a}_1 \\ \hat{a}_2 \end{bmatrix}, \qquad \text{and} \qquad \mathbf{a} = \begin{bmatrix} \gamma \\ 0 \end{bmatrix}$$

and thus

$$\mathbf{k} = \begin{bmatrix} k_1 \\ k_2 \end{bmatrix} = [\mathbf{W}^T\mathbf{R}]^{-1}(\hat{\mathbf{a}} - \mathbf{a}) = \begin{bmatrix} \hat{a}_1 - \gamma \\ \hat{a}_2 - \gamma(\hat{a}_1 - \gamma) \end{bmatrix}$$

From a practical point of view, we choose $\hat{a}_1$ and $\hat{a}_2$ in $\hat{p}(s)$ so that the error $\mathbf{e}(t) = \mathbf{x}(t) - \hat{\mathbf{x}}(t)$ tends rapidly to zero. Of course, both roots of $\hat{p}(s)$ must lie in the left-hand complex plane.

(b) The system of the present example is in the form (10.6-9), where $A_{11} = 0$, $A_{12} = 1$, $A_{21} = 0$, $A_{22} = -\gamma$, $b_1 = 0$, $b_2 = \beta$, and $c_1 = 1$. Here $q_1(t) = x_1(t)$ and $q_2(t) = x_2(t)$. Moreover $\hat{q}_1(t) = x_1(t) = y(t)$. For the estimation of $x_2(t)$ the proposed observer is

$$q_2(t) = \hat{x}_2(t) = \varphi y(t) + \mathbf{v}(t)$$

where $\mathbf{v}(t)$ is a scalar function governed by the differential equation

$$\dot{\mathbf{v}}(t) = f\mathbf{v}(t) + hu(t) + gy(t)$$

and where

$$f = A_{22} - \varphi c_1 A_{12} = -\gamma - \varphi$$
$$g = (A_{21} - \varphi c_1 A_{11})c_1^{-1} + f\varphi = (-\gamma - \varphi)\varphi = -\gamma\varphi - \varphi^2$$
$$h = B_2 - \varphi c_1 B_1 = \beta$$

Since rank $R_1^T = \text{rank}\,[(c_1 A_{12})^T] = \text{rank}\,(1) = 1$ we can find a φ such that f has the desired eigenvalue. Let $-\rho$ be the desired eigenvalue of f. Then, $\varphi = \rho - \gamma$. Introducing the value of φ into g and f we have

$$g = -\rho^2 + \rho\gamma \qquad \text{and} \qquad f = -\rho$$

(c) Let $p_c(s) = s^2 + \gamma_1 s + \gamma_2$ be the desired characteristic polynomial of the closed-loop ystem. The parameters γ_1 and γ_2 are arbitrary, but they will ultimately be specified in order to meet closed-loop system requirements. The system is controllable since rank $\mathbf{S} = \text{rank}\,[\mathbf{b} \; \vdots \; \mathbf{Ab}] = 2$. Therefore, we can choose a feedback vector $\mathbf{k}_1$ such that the closed-loop system poles are the roots of $p_c(s) = s^2 + \gamma_1 s + \gamma_2$. Using the results of Sec. 10.3, the following vector is determined:

$$\mathbf{k}_1^T = \begin{bmatrix} \dfrac{\gamma_2}{\beta} & -\dfrac{\gamma - \gamma_1}{\beta} \end{bmatrix}$$

Checking the results, we have $|s\mathbf{I} - (\mathbf{A} - \mathbf{bk}_1^T)| = s^2 + \gamma_1 s + \gamma_2 = p_c(s)$. Of course, in the case of a full-order observer, $\mathbf{k}_1 = \mathbf{k}$, where $\mathbf{k}$ has been determined in part (a) above. In the case of a reduced-order observer, $k_2 = \varphi$, where φ has been determined in part (b) above. The block diagram representations of the closed-loop systems in both cases are given in Figures 10.13 and 10.14.

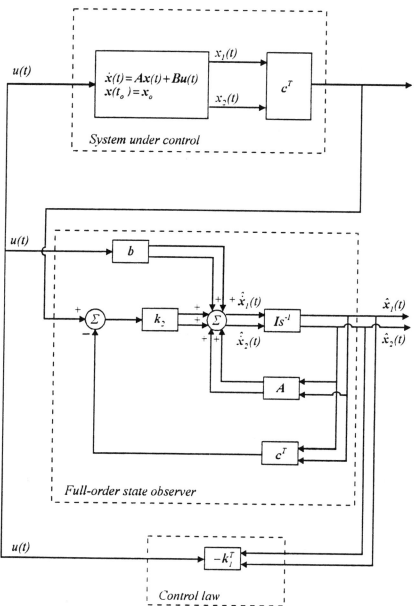

Block diagram of the closed-loop system of example 10.6.2.
with a full order state observer.

Figure 10.13 Block diagram of the closed-loop system of Example 10.6.2 with a full-order state observer.

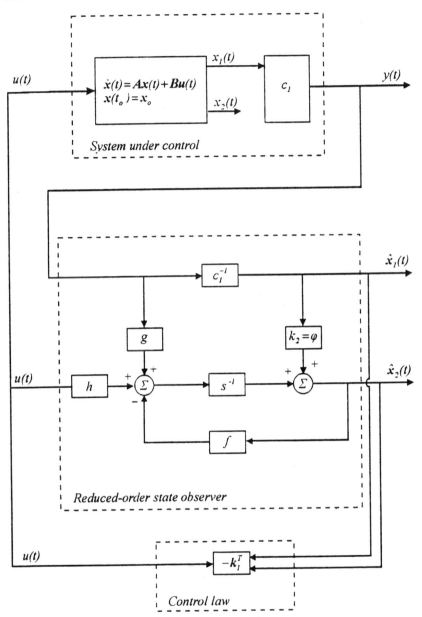

Figure 10.14 Block diagram of the closed-loop system of Example 10.6.2 with a reduced-order state observer.

10.7 PROBLEMS

1. Consider the linear system with

$$
\mathbf{A} = \begin{bmatrix} 1 & 1 & 0 \\ 0 & 1 & 1 \\ -1 & 2 & 1 \end{bmatrix}, \qquad \mathbf{b} = \begin{bmatrix} 1 \\ 0 \\ 1 \end{bmatrix}, \qquad \mathbf{c} = \begin{bmatrix} 0 \\ 0 \\ 1 \end{bmatrix}
$$

Find a state feedback control law of the form $u(t) = -\mathbf{f}^T\mathbf{x}(t)$ and an output feedback control law of the form $u(t) = -\mathbf{k}^T\mathbf{y}(t)$, such that the closed-loop eigenvalues are -1, -2, and -3.

2. Consider the linear system with

$$
\mathbf{A} = \begin{bmatrix} -4 & -1 & -3 \\ 3 & 1 & 1 \\ 5 & 1 & 3 \end{bmatrix}, \qquad \mathbf{B} = \begin{bmatrix} 1 & 1 \\ -1 & 0 \\ -1 & -1 \end{bmatrix}, \qquad \mathbf{C} = \begin{bmatrix} 4 & -1 & 4 \\ 0 & 0 & 0 \end{bmatrix},
$$

$$
\mathbf{D} = \begin{bmatrix} 0 & 0 \\ 0 & 1 \end{bmatrix}
$$

Find a state feedback control law of the form $\mathbf{u}(t) = -\mathbf{F}\mathbf{x}(t)$ and an output feedback control law of the form $\mathbf{u}(t) = -\mathbf{K}\mathbf{y}(t)$, such that the closed-loop eigenvalues are 0, -2, and -2.

3. The lateral motion of a helicopter can be approximately described by the following third-order linear state-space model [2]:

$$
\begin{bmatrix} \dot{q}(t) \\ \dot{\theta}(t) \\ \dot{v}(t) \end{bmatrix} = \begin{bmatrix} -0.4 & 0 & -0.01 \\ 1 & 0 & 0 \\ -1.4 & 9.8 & -0.02 \end{bmatrix} \begin{bmatrix} q(t) \\ \theta(t) \\ v(t) \end{bmatrix} + \begin{bmatrix} 6.3 \\ 0 \\ 0.8 \end{bmatrix} \delta(t)
$$

or

$$
\dot{\mathbf{x}}(t) = \mathbf{A}\mathbf{x}(t) + \mathbf{B}u(t)
$$

where $q(t)$ is the pitch rate, $\theta(t)$ is the pitch angle of the fuselage, $v(t)$ the horizontal velocity of the helicopter, and $\delta(t)$ is the rotor inclination angle. Determine a state feedback control law of the form $u = -\mathbf{f}^T\mathbf{x}$ so that the closed-loop system eigenvalues are -2, $-1+j$, $-1-j$.

4. Consider the system of the inverted pendulum on a cart (Chap. 3, Sec. 3.14, Problem 10), where $M = 3\,\text{kg}$, $m = 200\,\text{g}$, $I = 60\,\text{cm}$.

   (a) Find a state feedback control law $u = -\mathbf{f}^T\mathbf{x}$, such that the eigenvalues of the closed-loop system are $-2+j$, $-2-j$, and -5.
   (b) Find an output feedback control law of the form $u = -\mathbf{k}^T\mathbf{y}$ which assigns the same eigenvalues.

5. Decouple via both state and output feedback the linear systems:

   (a) $\mathbf{A} = \begin{bmatrix} -1 & 0 \\ 0 & -2 \end{bmatrix}$, $\mathbf{B} = \begin{bmatrix} 1 & 0 \\ 0 & 1 \end{bmatrix}$, $\mathbf{C} = \begin{bmatrix} 1 & 1 \\ 1 & 1 \end{bmatrix}$

   (b) $\mathbf{A} = \begin{bmatrix} 0 & 1 & 0 \\ 2 & 3 & 0 \\ 1 & 1 & 1 \end{bmatrix}$, $\mathbf{B} = \begin{bmatrix} 0 & 0 \\ 1 & 0 \\ 0 & 1 \end{bmatrix}$, $\mathbf{C}^T = \begin{bmatrix} 1 & 0 \\ 1 & 0 \\ 0 & 1 \end{bmatrix}$

(c) $\mathbf{A} = \begin{bmatrix} 0 & 1 & 0 & 0 \\ 0 & 0 & 1 & 0 \\ 0 & 0 & 0 & 0 \\ 0 & 0 & 0 & 1 \end{bmatrix}$, $\mathbf{B} = \begin{bmatrix} 0 & 0 \\ 0 & 0 \\ 1 & 0 \\ 0 & 1 \end{bmatrix}$, $\mathbf{C}^T = \begin{bmatrix} 1 & 0 \\ 0 & 0 \\ 0 & 0 \\ 0 & 1 \end{bmatrix}$

6. Consider controlling the lateral motion of the helicopter of Problem 3. If the output is the horizontal velocity v of the helicopter, determine the gains k and n of the output feedback control law of the form

$$\delta = kv + n\delta_{\text{ref}}$$

where δ_{ref} is a reference input, such that the transfer function of the closed-loop system has the form

$$H(s) = \frac{9}{s^2 + 3s^2 + 9s}$$

7. Consider the linear system with

$$\mathbf{A} = \begin{bmatrix} -1 & 0 \\ 0 & -2 \end{bmatrix}, \quad \mathbf{B} = \begin{bmatrix} 1 & 0 \\ 0 & 1 \end{bmatrix}, \quad \mathbf{C} = \begin{bmatrix} 1 & 1 \\ 1 & 0 \end{bmatrix}, \quad \mathbf{D} = \begin{bmatrix} 1 & 1 \\ 0 & 0 \end{bmatrix}$$

Show that if the state feedback law (10.2-2) is applied to this system with

$$\mathbf{F} = \begin{bmatrix} -1 & -2 \\ 0 & -2 \end{bmatrix} \quad \text{and} \quad \mathbf{G} = \begin{bmatrix} 0 \\ 1 \end{bmatrix}$$

then the transfer function matrix of the closed-loop system becomes

$$\mathbf{H}(s) = \frac{1}{(s+2)(s+4)} \begin{bmatrix} (s+1)(s+2) \\ -2 \end{bmatrix}$$

Now assume that $\mathbf{H}_m(s) = \mathbf{H}(s)$ and $\mathbf{F}$ and $\mathbf{G}$ are unknown matrices. Solve the exact model matching problem.

8. Show that if we apply the law (10.2-2) to the system of Problem 2 with

$$\mathbf{F} = \begin{bmatrix} -13 & 4 & -11 \\ 12 & -3 & 12 \end{bmatrix}, \quad \mathbf{G} = \begin{bmatrix} 1 & 0 \\ -1 & 1 \end{bmatrix}$$

then the transfer function matrix of the closed-loop system becomes

$$\mathbf{H}(s) = \begin{bmatrix} \dfrac{1}{s+3} & 0 \\ \dfrac{-s}{s+3} & 1 \end{bmatrix}$$

Now assume that $\mathbf{H}_m(s) = \mathbf{H}(s)$ and $\mathbf{F}$ and $\mathbf{G}$ are unknown matrices. Solve the exact model matching problem.

9. Solve the exact model matching problem via output feedback for the following two cases:

System under control   Model

a. $\quad \mathbf{G}(s) = \begin{bmatrix} \dfrac{s}{(s+1)(s+2)} \\[2mm] \dfrac{-1}{s+1} \end{bmatrix}$ and $\mathbf{H}_{\mathrm{m}}(s) = \begin{bmatrix} \dfrac{1}{(s+2)} \\[2mm] \dfrac{-1}{s+1} \end{bmatrix}$

b. $\quad \mathbf{G}(s) = \begin{bmatrix} \dfrac{1}{s(s+1)} & 1 \\[2mm] -2/(s+2) & 1/s \end{bmatrix}$ and $\mathbf{H}_{\mathrm{m}}(s) = \begin{bmatrix} \dfrac{1}{s+1} & \dfrac{1}{s+2} \\[2mm] -1 & \dfrac{s}{s+1} \end{bmatrix}$

10. Consider the helicopter of Problem 3. Let the pitch rate $q(t)$ be the output of the system. Find

   (a) A full-order state observer
   (b) A reduced-order observer

11. The state-space model of a satellite position control system is as follows:

$$\dot{\mathbf{x}}(t) = \mathbf{A}\mathbf{x}(t) + \mathbf{b}u(t)$$

$$y = \mathbf{c}^T \mathbf{x}(t)$$

where $\mathbf{x}^T = [x_1, x_2] = [\theta_y, \omega]$ with θ_y the angular position and ω the angular velocity, and

$$\mathbf{A} = \begin{bmatrix} 0 & 1 \\ 0 & 0 \end{bmatrix}, \quad \mathbf{b} = \begin{bmatrix} 0 \\ \gamma \end{bmatrix}, \quad \mathbf{c} = \begin{bmatrix} 1 \\ 0 \end{bmatrix}$$

   (a) Find a full-order state observer.
   (b) Find a reduced-order observer.
   (c) Draw the closed-loop system diagram for (a) and (b).

12. Consider the speed control system described in Subsec. 3.13.3. A description of the system in state space is as follows

$$\dot{\mathbf{x}}(t) = \mathbf{A}\mathbf{x}(t) + \mathbf{b}u(t)$$

$$y = \mathbf{c}^T \mathbf{x}(t)$$

where $\mathbf{x}^T = [x_1, x_2]$, where x_1 is the angular speed ω_{m} of the motor, x_2 is the current i_f, u is the input voltage v_f, and y is the angular speed ω_y of the load. The system matrices are

$$\mathbf{A} = \begin{bmatrix} \gamma_1 & \gamma_2 \\ 0 & \gamma_3 \end{bmatrix}, \quad \mathbf{b} = \begin{bmatrix} 0 \\ L_f^{-1} \end{bmatrix}, \quad \mathbf{c} = \begin{bmatrix} N \\ 0 \end{bmatrix}$$

where

$$\gamma_1 = -\left(\frac{B_{\mathrm{m}}^*}{J_{\mathrm{m}}^*} + \frac{K_{\mathrm{m}}^2}{J_{\mathrm{m}}^* R_{\mathrm{a}}} \right), \quad \gamma_2 = \frac{K_{\mathrm{m}} K_{\mathrm{g}}}{J_{\mathrm{m}}^* R_{\mathrm{a}}}, \quad \text{and} \quad \gamma_3 = -\frac{R_f}{K_f}$$

Assume that only the state variable $x_1(t)$ can be measured. Then

(a) Find a full-order state observer.
(b) Find a reduced-order state observer for the estimation of the current i_f, which is not measurable.
(c) Draw the block diagram of the closed-loop system for both cases (a) and (b).

13. Consider the system

$$\dot{\mathbf{x}}(t) = \begin{bmatrix} 0 & 1 & 0 \\ 0 & 0 & 1 \\ 1 & 3 & 3 \end{bmatrix} \mathbf{x}(t) + \begin{bmatrix} 0 \\ 0 \\ 1 \end{bmatrix} u(t)$$

$$y(t) = [\, 1 \quad 1 \quad 0\,]\mathbf{x}(t)$$

(a) Determine the state feedback control law of the form $u(t) = -\mathbf{f}^T\mathbf{x}(t)$, so that the closed loop eigenvalues are $-1, -1.5, -2$.
(b) Find a full-order observer with characteristic polynomial $\hat{p}(s) = s^3 + 12s^2 + 47s + 60$.
(c) Draw the block diagram of the system incorporating the controller and the observer found in (a) and (b).

BIBLIOGRAPHY

Books

1. PJ Antsaklis, AN Michel. Linear Systems. New York: McGraw-Hill, 1997.
2. GJ Franklin, JD Powell, A Emani-Naeini. Feedback Control of Dynamic Systems. 2nd ed. London: Addison-Wesley, 1991.
3. T Kailath. Linear Systems. Englewood Cliffs, New Jersey: Prentice-Hall, 1980.
4. J O'Reilly. Observers for Linear Systems. New York: Academic Press, 1983.

Articles

5. J Ackermann. Parameter space design of robust control systems. IEEE Trans Automatic Control AC-25:1058–1072, 1980.
6. AT Alexandridis, PN Paraskevopoulos. A new approach for eigenstructure assignment by output feedback. IEEE Trans Automatic Control AC-41:1046–1050, 1996.
7. RW Brockett. Poles, zeros and feedback: state space interpretation. Trans Automatic Control AC-10:129–135, 1965.
8. JC Doyle, G Stein. Robustness with observers. IEEE Trans Automatic Control AC-24:607–611, 1979.
9. PL Falb, WA Wolovich. Decoupling in the design and synthesis of multivariable control systems. IEEE Trans Automatic Control AC-12:651–659, 1967.
10. EG Gilbert. Controllability and observability in multivariable control systems. J Soc Ind Appl Math Control Ser A, 1(2):128–151, 1963.
11. DG Luenberger. Observing the state of a linear system. IEEE Trans Military Electronics MIL-8:74–80, 1964.
12. DG Luenberger. Observers for multivariable systems. IEEE Trans Automatic Control AC-11:190–197, 1966.
13. DG Luenberg. An introduction to observers. IEEE Trans Automatic Control AC-16:596–602, 1971.

14. BC Moore. Principal component analysis in linear systems: controllability, observability and model reduction. IEEE Trans Automatic Control AC-26:17–32, 1981.
15. PN Parskevopoulos. Exact transfer-function design using output feedback. Proc IEE 123:831–834, 1976.
16. PN Parskevopoulos. A general solution to the output feedback eigenvalue assignment problem. Int J Control 24:509–528, 1976.
17. PN Parskevopoulos, FN Koumboulis. Decoupling and pole assignment in generalized state space systems. IEE Proc, Part D, Control Theory and Applications 138:547–560, 1991.
18. PN Paraskevopoulos, FN Koumboulis. A unifying approach to observers for regular and singular systems. IEE Proc, Part D, Control Theory and Applications 138:561–572, 1991.
19. PN Paraskevopoulos, FN Koumboulis. A new approach to the decoupling problem of linear time-invariant systems. J Franklin Institute 329:347–369, 1992.
20. PN Paraskevopoulos, FN Koumboulis. The decoupling of generalized state space systems via state feedback. IEEE Trans Automatic Control AC-37:148–152, 1992.
21. PN Paraskevopoulos, FN Koumboulis. Observers for singular systems. IEEE Trans Automatic Control AC-37:1211–1215, 1992.
22. PN Paraskevopoulos, SG Tzafestas. New results on feedback modal-controller design. Int J Control 24:209–216, 1976.
23. PN Paraskevopoulos, FN Koumboulis, DF Anastasakis. Exact model matching of generalized state space systems. J Optimization Theory and Applications (JOTA) 76:57–85, 1993.
24. JC Willems, SK Mitter. Controllability, observability, pole allocation, and state reconstruction. IEEE Trans Automatic Control AC-16:582–595, 1971.
25. WM Wonham. On pole assignment in multi-input controllable linear systems. IEEE Trans Automatic Control AC-12:660–665, 1967.

11

Optimal Control

11.1 GENERAL REMARKS ABOUT OPTIMAL CONTROL

Optimal control deals with the solution of one of the most celebrated problems of modern control theory. Generally speaking, this problem is defined as the determination of the best possible control strategy (usually of the optimum control vector $\mathbf{u}(t)$), which minimizes a certain *cost function* or *performance index*.

In this chapter we consider the optimal control of linear systems (the nonlinear case is far too difficult for the level of this book). The system under control is described in state space by the equations

$$\dot{\mathbf{x}}(t) = \mathbf{A}(t)\mathbf{x}(t) + \mathbf{B}(t)\mathbf{u}(t) \tag{11.1-1a}$$

$$\mathbf{y}(t) = \mathbf{C}(t)\mathbf{x}(t) + \mathbf{D}(t)\mathbf{u}(t) \tag{11.1-1b}$$

$$\mathbf{x}(t_0) = \mathbf{x}_0 \tag{11.1-1c}$$

where the matrices $\mathbf{A}(t)$, $\mathbf{B}(t)$, $\mathbf{C}(t)$, and $\mathbf{D}(t)$ have dimensions $n \times n$, $n \times m$, $p \times n$, and $p \times m$, respectively. The objective of the optimal control problem is to determine a control vector $\mathbf{u}(t)$ which will "force" the behavior of the system under control to minimize some type of cost function, while at the same time satisfying the physical constraints of the system—namely, the state equations (11.1-1). The cost criterion is usually formulated so as to express some physical quantity. This way, the very idea of minimization of a cost criterion has a practical meaning, as for example the minimization of the control effort energy, the energy dissipated by the system, etc.

A particular form of the cost function, which in itself is very general, is the following:

$$J = \theta[\mathbf{x}(t), t] \bigg|_{t=t_0}^{t=t_f} + \int_{t_0}^{t_f} \varphi[\mathbf{x}(t), \mathbf{u}(t), t]\, dt \tag{11.1-2}$$

The first term in Eq. (11.1-2) refers to the cost on the boundaries of the optimization time interval $[t_0, t_f]$. More precisely, $\theta[\mathbf{x}(t_0), t_0]$ is the cost at the beginning, while $\theta[\mathbf{x}(t_f), t_f]$ is the cost at the end of the interval. The second term in Eq. (11.1-2) is an integral which refers to the cost in the entire optimization interval.

479

Depending on the requirements of the particular optimization problem, the functions $\theta[\mathbf{x}(t), t]$ and $\varphi[\mathbf{x}(t), \mathbf{u}(t), t]$ take on special forms. In the sequel, some of the most well-known special forms of Eq. (11.1-2) are given, together with a description of the corresponding optimal control problem.

1 The Minimum Time Control Problem

In this case the cost function has the form

$$J = \int_{t_0}^{t_f} dt = t_f - t_0 \tag{11.1-3}$$

It is obvious that this criterion refers only and exclusively to the time duration $t_f - t_0$. The control problem here is to find a control vector $\mathbf{u}(t)$ such that the time required for $\mathbf{x}(t)$ to go from its initial state $\mathbf{x}(t_0)$ to its final state $\mathbf{x}(t_f)$, is a minimum. Examples of "final states" are the crossing of the finish line by a car or a sprinter in a race, the time needed to complete a certain task, etc.

2 The Terminal Control Problem

In this case the cost function has the form

$$J = [\mathbf{x}(t_f) - \xi(t_f)]^T \mathbf{S}[\mathbf{x}(t_f) - \xi(t_f)] \tag{11.1-4}$$

where $\xi(t_f)$ is the desired final value of the vector $\mathbf{x}(t)$ and $\mathbf{S}$ is an $n \times n$ real, symmetric, positive semidefinite weighting matrix (see Sec. 2.12). This form of J shows clearly that here our attention has been exclusively concentrated on the final value $\mathbf{x}(t_f)$ of $\mathbf{x}(t)$. Here, we want to determine a control vector $\mathbf{u}(t)$ so that the error $\mathbf{x}(t_f) - \xi(t_f)$ is minimal. An example of such a problem is the best possible aiming at a point on earth, in the air, on the moon, or elsewhere.

3 The Minimum Control Effort Problem

In this case the cost function has the form

$$J = \int_{t_0}^{t_f} \mathbf{u}^T(t)\mathbf{R}(t)\mathbf{u}(t) \, dt \tag{11.1-5}$$

where $\mathbf{R}(t)$ is an $m \times m$ real, symmetric, positive definite weighting matrix for $t \in (t_0, t_f)$. This expression for J represents the energy consumed by the control vector $\mathbf{u}(t)$ in controlling the system. We want this energy to be the least possible. As an example, consider the gasoline and break pedals of a car. The driver must use each pedal in such a way as to reach his destination by consuming the least possible fuel. Note that the present cost criterion is essentially the same with the cost criterion J_u of relation (9.10-2) which is used in the classical optimal control techniques (Sec. 9.10).

4 The Optimal Servomechanism or Tracking Problem

In this case the cost function has the form

$$J = \int_{t_0}^{t_f} [\mathbf{x}(t) - \xi(t)]^T \mathbf{Q}(t)[\mathbf{x}(t) - \xi(t)] \, dt = \int_{t_0}^{t_f} \mathbf{e}^T(t)\mathbf{Q}(t)\mathbf{e}(t) \, dt \tag{11.1-6}$$

where $\mathbf{Q}(t)$ is an $n \times n$ real, symmetric, positive semidefinite weighting matrix for $t \in (t_0, t_f)$ and $\xi(t)$ is the prespecified desired path of the state vector $\mathbf{x}(t)$. The vector $\mathbf{e}(t) = \mathbf{x}(t) - \xi(t)$ is the error, which we want to minimize. This may be accomplished by determining an appropriate $\mathbf{u}(t)$ so that J in Eq. (11.-6) becomes minimal. The track of a space shuttle, the desired course of a missile, of a ship, of a car or even of a pedestrian, are optimal control problems of the form (11.1-6). If we desire to incorporate the least-effort problem (11.1-5), then Eq. (11.1-6) takes on the more general form

$$J = \int_{t_0}^{t_f} \left[[\mathbf{x}(t) - \xi(t)]^\mathrm{T} \mathbf{Q}(t)[\mathbf{x}(t) - \xi(t)] + \mathbf{u}^\mathrm{T}(t)\mathbf{R}(t)\mathbf{u}(t) \right] dt \qquad (11.1\text{-}7)$$

We often also want to include the terminal control problem, in which case Eq. (11.1-7) takes on the even more general form

$$J = [\mathbf{x}(t_f) - \xi(t_f)]^\mathrm{T} \mathbf{S}[\mathbf{x}(t_f) - \xi(t_f)]$$
$$+ \int_{t_0}^{t_f} \left[[\mathbf{x}(t) - \xi(t)]^\mathrm{T} \mathbf{Q}(t)[\mathbf{x}(t) - \xi(t)] + \mathbf{u}^\mathrm{T}(t)\mathbf{R}(t)\mathbf{u}(t) \right] dt \qquad (11.1\text{-}8)$$

5 The Optimal Regulator Problem

In this case the cost function is a special case of Eq. (11.1-8), where $\xi(t) = \mathbf{0}$. Hence, in this case we have

$$J = \mathbf{x}^\mathrm{T}(t_f)\mathbf{S}\mathbf{x}(t_f) + \int_{t_0}^{t_f} \left[\mathbf{x}(t)^\mathrm{T}\mathbf{Q}(t)\mathbf{x}(t) + \mathbf{u}^\mathrm{T}(t)\mathbf{R}(t)\mathbf{u}(t) \right] dt \qquad (11.1\text{-}9)$$

A well-known optimal regulator example is the restoring of a system to its equilibrium position after it has been disturbed.

It is noted that the weighting matrices $\mathbf{S}$, $\mathbf{Q}(t)$, and $\mathbf{R}(t)$ are chosen according to the "weight," i.e., to the importance we want to assign to each element of the error vector $\mathbf{e}(t) = \mathbf{x}(t) - \xi(t)$ and the input vector $\mathbf{u}(t)$. The choice of suitable $\mathbf{S}$, $\mathbf{Q}(t)$, and $\mathbf{R}(t)$ for a specific problem is usually difficult and requires experience and engineering insight.

Among the many problems that have been solved thus far using the modern optimal control techniques, only two are presented in this book due to space limitations. These problems are the optimal linear regulator (Sec. 11.3) and the optimal linear servomechanism (Sec. 11.4). These two problems are of great theoretical as well as of practical interest. For more information on these problems, as well as on other problems of optimal control (for example bang-bang control, stochastic control, adaptive control, etc.) see [1–20].

To facilitate the study of the optimal linear regulator and servomechanism control problems, we present the necessary mathematical background in Sec. 11.2 that follows. This mathematical background covers two very basic topics: (i) maxima and minima using the calculus of variations and (ii) the maximum principle.

11.2 MATHEMATICAL BACKGROUND

11.2.1 Maxima and Minima Using the Method of Calculus of Variations

In what follows, we will use the method of calculus of variations to study the following two problems: maxima and minima of a functional without constraints and maxima and minima of a functional with constraints.

1 Maxima and Minima of a Functional Without Constraints

Consider the cost function or performance index

$$J(x) = \int_{t_0}^{t_f} \varphi[x(t),\ \dot{x}(t),\ t]\,dt \tag{11.2-1}$$

This performance index $J(x)$ is a *functional*, i.e., $J(x)$ is a function of another function, namely, of the function $x(t)$. We are asked to find a function $x(t)$ in the interval $[t_0, t_f]$ such that $J(x)$ is a minimum. A convenient method to solve this problem is to apply the method of *calculus of variations* presented in the sequel.

Let $x(t)$ and $\dot{x}(t)$ be presented as follows:

$$x(t) = \hat{x}(t) + \varepsilon\eta(t) = \hat{x}(t) + \delta x \tag{11.2-2a}$$

$$\dot{x}(t) = \hat{\dot{x}}(t) + \varepsilon\dot{\eta}(t) = \hat{\dot{x}}(t) + \delta\dot{x} \tag{11.2-2b}$$

where $\hat{x}(t)$ is an admissable optimal trajectory, i.e., $\hat{x}(t)$ minimizes J, $\eta(t)$ is a deviation of $x(t)$ and ε is small number. Substitute Eq. (11.2-2) in Eq. (11.2-1). Next, expand $\varphi(x, \dot{x}, t)$ in Taylor series about the point $\varepsilon = 0$, to yield

$$\varphi\big[\hat{x}(t) + \varepsilon\eta(t),\ \hat{\dot{x}}(t) + \varepsilon\dot{\eta}(t),\ t\big] = \varphi(\hat{x}, \hat{\dot{x}}, t) + \frac{\partial\varphi}{\partial\hat{x}}\varepsilon\eta(t) + \frac{\partial\varphi}{\partial\hat{\dot{x}}}\varepsilon\dot{\eta}(t) + \text{hot} \tag{11.2-3}$$

where hot stands for higher-order terms and includes all the Taylor series terms which involve ε raised to a power equal or greater than two. Let ΔJ be a small deviation of J from its optimal value, i.e., let

$$\Delta J = J[\hat{x} + \varepsilon\eta(t),\ \hat{\dot{x}} + \varepsilon\dot{\eta}(t),\ t] - J(\hat{x}, \hat{\dot{x}}, t) \tag{11.2-4}$$

Substitute Eq. (11.2-3) in Eq. (11.2-4) and, using Eq. (11.2-1), we have

$$\begin{aligned}
\Delta J &= \int_{t_0}^{t_f}\Big[\varphi[\hat{x} + \varepsilon\eta(t),\ \hat{\dot{x}} + \varepsilon\dot{\eta}(t),\ t] - \varphi(\hat{x}, \hat{\dot{x}}, t)\Big]\,dt \\
&= \int_{t_0}^{t_f}\left[\frac{\partial\varphi}{\partial\hat{x}}\varepsilon\eta(t) + \frac{\partial\varphi}{\partial\hat{\dot{x}}}\varepsilon\dot{\eta}(t) + \text{hot}\right]dt \\
&= \int_{t_0}^{t_f}\left[\frac{\partial\varphi}{\partial\hat{x}}\delta x + \frac{\partial\varphi}{\partial\hat{\dot{x}}}\delta\dot{x} + \text{hot}\right]dt
\end{aligned} \tag{11.2-5}$$

where $\delta x = \varepsilon\eta(t)$ and $\delta\dot{x} = \varepsilon\dot{\eta}(t)$. Let δJ be the linear part of ΔJ with respect to δx and $\delta\dot{x}$. Then δJ takes on the form

$$\delta J = \int_{t_0}^{t_f}\left[\frac{\partial\varphi}{\partial\hat{x}}\delta x + \frac{\partial\varphi}{\partial\hat{\dot{x}}}\delta\dot{x}\right]dt \tag{11.2-6}$$

The following theorem holds.

Theorem 11.2.1

A necessary condition for J to be maximum or minimum when $x(t) = \hat{x}(t)$ is that $\delta J = 0$.

If we apply Theorem 11.2.1 to Eq. (11.2-6) and if, for simplicity, we drop the symbol " $\hat{}$ " from the optimal trajectory $\hat{x}(t)$, then we readily have

$$\int_{t_0}^{t_f} \left[\frac{\partial \varphi}{\partial x} \delta x + \frac{\partial \varphi}{\partial \dot{x}} \delta \dot{x} \right] dt = 0 \tag{11.2-7}$$

The above integral may be simplified by using the "integration by parts" method, as follows. Let

$$\int_a^b u \, dv = uv \Big|_a^b - \int_a^b v \, du$$

where

$$u = \frac{\partial \varphi}{\partial \dot{x}} \qquad \text{and} \qquad dv = \delta \dot{x} \, dt = d(\delta x)$$

Then

$$du = d \left[\frac{\partial \varphi}{\partial \dot{x}} \right] = \frac{d}{dt} \left[\frac{\partial \varphi}{\partial \dot{x}} \right] dt \qquad \text{and} \qquad v = \delta x$$

Hence, the second term in Eq. (11.2-7) becomes

$$\int_{t_0}^{t_f} \frac{\partial \varphi}{\partial \dot{x}} \delta \dot{x} \, dt = \int_{t_0}^{t_f} \frac{\partial \varphi}{\partial \dot{x}} \, d(\delta x) = \frac{\partial \varphi}{\partial \dot{x}} \delta x \Big|_{t_0}^{t_f} - \int_{t_0}^{t_f} (\delta x) \frac{d}{dt} \left[\frac{\partial \varphi}{\partial \dot{x}} \right] dt$$

Thus the integral (11.2-7) may be written as

$$\int_{t_0}^{t_f} \left[\frac{\partial \varphi}{\partial x} - \frac{d}{dt} \left[\frac{\partial \varphi}{\partial \dot{x}} \right] \right] \delta x \, dt + \frac{\partial \varphi}{\partial \dot{x}} \delta x \Big|_{t_0}^{t_f} \tag{11.2-8}$$

For Eq. (11.2-8) to be equal to zero, independently of the variation δx, the following two conditions must hold simultaneously:

$$\frac{\partial \varphi}{\partial x} - \frac{d}{dt} \left[\frac{\partial \varphi}{\partial \dot{x}} \right] = 0 \tag{11.2-9}$$

$$\frac{\partial \varphi}{\partial \dot{x}} \delta x = 0, \qquad \text{for } t = t_0 \text{ and } t_f \tag{11.2-10}$$

Equation (11.2-9) is the *Euler–Lagrange equation* and Eq. (11.2-10) represents the *boundary conditions* of the problem.

The linear portion $\delta^2 L$ of the second differential $\Delta^2 L$ may be determined in a similar way to that of determining δJ, to yield

$$\delta^2 J = \frac{1}{2} \int_{t_0}^{t_f} \left[\left[\frac{\partial^2 \varphi}{\partial x^2} - \frac{d}{dt} \left[\frac{\partial^2 \varphi}{\partial x \partial \dot{x}} \right] \right] (\delta x)^2 + \frac{\partial^2 \varphi}{\partial \dot{x}^2} (\delta \dot{x})^2 \right] dt \tag{11.2-11}$$

For J to be maximum (minimum), there must be $\delta^2 J \leq 0$ ($\delta^2 J \geq 0$). For example, for J to be minimum, from Eq. (11.20-11) it follows that the following relations must hold:

$$\frac{\partial^2 \varphi}{\partial x^2} - \frac{d}{dt}\left[\frac{\partial^2 \varphi}{\partial x \partial \dot{x}}\right] \geq 0 \quad \text{and} \quad \frac{\partial^2 \varphi}{\partial \dot{x}^2} \geq 0 \tag{11.2-12}$$

Remark 11.2.1

With regard to the boundary conditions (11.2-10), we distinguish the following four cases:

Case 1. The trajectory $x(t)$ is fixed at t_0 and t_f, in which case

$$x(t_0) = C_1 \quad \text{and} \quad x(t_f) = C_2$$

where C_1 and C_2 are the given constants. In this case, no restriction is placed upon $\partial \varphi / \partial \dot{x}$.

Case 2. The trajectory $x(t)$ is fixed at t_0 and free at t_f, in which case

$$x(t_0) = C_1 \quad \text{and} \quad \frac{\partial \varphi}{\partial \dot{x}} = 0, \quad \text{for} \quad t = t_f$$

The condition $\partial \varphi / \partial \dot{x} = 0$ for $t = t_f$ is because since we don't know the value of $x(t)$ at t_f, it follows that we don't know δx for $t = t_f$. Hence, to satisfy the boundary condition (11.2-10) at $t = t_f$ we must have $\partial \varphi / \partial \dot{x} = 0$ for $t = t_f$.

Case 3. The trajectory $x(t)$ is free at t_0 and fixed at t_f, in which case

$$\frac{\partial \varphi}{\partial \dot{x}} = 0 \quad \text{for} \quad t = t_0 \quad \text{and} \quad x(t_f) = C_2$$

Case 4. The trajectory $x(t)$ is free at both t_0 and t_f, in which case

$$\frac{\partial \varphi}{\partial \dot{x}} = 0 \quad \text{for} \quad t = t_0 \quad \text{and} \quad t = t_f$$

Remark 11.2.2

The results of this section can readily be expanded to cover the more general case where $x(t)$ is no longer a scalar function but rather a vector function, i.e., when

$$J = \int_{t_0}^{t_f} \varphi(\mathbf{x}, \dot{\mathbf{x}}, t)\, dt, \quad \text{where} \quad \mathbf{x}^{\mathrm{T}} = (x_1 x_2, \ldots, x_n) \tag{11.2-13}$$

Here, the Euler–Lagrange equation is

$$\frac{\partial \varphi}{\partial \mathbf{x}} - \frac{d}{dt}\left[\frac{\partial \varphi}{\partial \dot{\mathbf{x}}}\right] = 0 \tag{11.2-14}$$

and the boundary conditions are

$$(\delta \mathbf{x})^{\mathrm{T}} \frac{\partial \varphi}{\partial \dot{\mathbf{x}}} = 0, \quad \text{for} \quad t = t_0 \quad \text{and} \quad t_f \tag{11.2-15}$$

Remark 11.2.3

From the foregoing material it follows that the problem of determining the maxima and minima of a functional using the calculus of variations reduces to that of solving a *two-point boundary value problem* (TPVBP).

Example 11.2.1

Determine the optimum $x(t)$ which minimizes the cost function

$$J = \int_{t_0}^{t_f} \varphi[x(t), \dot{x}(t), t]\, dt = \int_0^{\pi/2} [x^2(t) - \dot{x}^2(t)]\, dt$$

with boundary conditions $x(t_0) = x(0) = 0$ and $x(t_f) = x(\pi/2) = 1$.

Solution

The Euler–Lagrange equation is

$$\frac{\partial \varphi}{\partial x} - \frac{d}{dt}\left[\frac{\partial \varphi}{\partial \dot{x}}\right] = 2x - \frac{d}{dt}(-2\dot{x}) = 2x + 2x^{(2)} = 0 \qquad \text{or} \qquad x^{(2)} + x = 0$$

where $x^{(2)}$ is the second derivative of x with respect to t. The general solution of the Euler–Lagrange equation is

$$x(t) = A_1 \sin t + A_2 \cos t$$

The constants A_1 and A_2 are determined using the boundary conditions (11.2-10). For the present example, we have (see Case 1 of Remark 11.2.1)

$$x(t_0) = x(0) = 0 \qquad \text{and} \qquad x(t_f) = x\left(\frac{\pi}{2}\right) = 1$$

Thus

$$x(0) = A_2 = 0 \qquad \text{and} \qquad x\left(\frac{\pi}{2}\right) = A_1 = 1$$

Hence, the optimum $x(t)$ sought is $x(t) = \sin t$. The graphical representation of the optimum $x(t)$ is given in Figure 11.1.

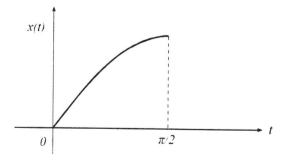

Figure 11.1 Graphical representation of the solution of Example 11.2.1.

Example 11.2.2

Determine a trajectory $x(t)$ such as to minimize the distance between the point $x(t_0) = x(0) = 1$ and the straight line $t = t_f = 2$.

Solution

In Figure 11.2 we present a few possible trajectories which may satisfy the problem specifications, since they all start from the point $x(t_o) = 1$ and end on the straight line $t = t_f = 2$. However, the optimum trajectory sought is the one which will minimize the cost function

$$J = \int_{t_)}^{t_f} \varphi[x(t), \dot{x}(t), t] dt = \int_0^2 ds$$

where $(ds)^2 = (dx)^2 + (dt)^2$ and hence $ds = (1 + \dot{x}^2)^{1/2} dt$. Therefore

$$J = \int_0^2 (1 + \dot{x}^2)^{1/2} dt, \qquad \text{where} \qquad \varphi(x, \dot{x}, t) = (1 + \dot{x}^2)^{1/2}$$

The Euler–Lagrange equation is

$$\frac{\partial \varphi}{\partial x} - \frac{d}{dt}\left[\frac{\partial \varphi}{\partial \dot{x}}\right] = -\frac{d}{dt}\left[\frac{\dot{x}}{(1 + \dot{x}^2)^{1/2}}\right] = \frac{x^{(2)}(1 + \dot{x}^2) - \dot{x}^2 x^{(2)}}{(1 + \dot{x}^2)^{3/2}} = 0$$

The above equation reduces to the differential equation $x^{(2)}(t) = 0$. The general solution of the Euler–Lagrange equation will then be

$$x(t) = A_1 t + A_2$$

The constants A_1 and A_2 are determined by using the boundary conditions (11.2-10). For the present example, the boundary conditions are fixed at $t_0 = 0$ but free at $t_f = 2$ (see Case 2 of Remark 11.2.1). Thus

$$x(0) = A_2 = 1 \qquad \text{and} \qquad \frac{\partial \varphi}{\partial \dot{x}}\bigg|_{t=2} = \frac{\dot{x}}{(1 + \dot{x}^2)^{1/2}}\bigg|_{t=2} = \frac{A_1}{(1 + A_1^2)^{1/2}} = 0$$

Hence, the optimum trajectory $x(t)$ is the straight line $x(t) = 1$.

Example 11.2.3

Determine the optimum trajectory which minimizes the cost function

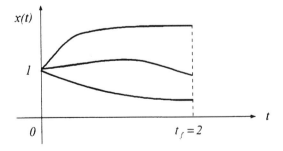

Figure 11.2 Several possible trajectories for Example 11.2.2.

$$J = \int_{t_0}^{t_f} \varphi[x(t), \dot{x}(t), t] \, dt = \int_0^2 \left[\frac{1}{2} \dot{x}^2 + x\dot{x} + \dot{x} + x \right] dt$$

where no restrictions are placed upon the optimum $x(t)$ at the boundaries $t_0 = 0$ and $t_f = 2$.

Solution

The Euler–Lagrange equation is given by

$$\frac{\partial \varphi}{\partial x} - \frac{d}{dt}\left[\frac{\partial \varphi}{\partial \dot{x}} \right] = \dot{x} + 1 - \frac{d}{dt}(\dot{x} + x + 1) = 1 - x^{(2)} = 0$$

The general solution of the above differential equation is

$$x(t) = \frac{1}{2} t^2 + A_1 t + A_2$$

The constants A_1 and A_2 are determined by using the boundary conditions (11.2-10). For the present example, the boundary conditions are free at both $t_0 = 0$ and at $t_f = 2$ (see Case 4 of Remark 11.2.1). We thus have

$$\frac{\partial \varphi}{\partial \dot{x}} = \dot{x} + x + 1 = 0, \quad \text{for } t = 0 \text{ and } 2$$

This leads to the following two algebraic system of equations:

$$\left. \frac{\partial \varphi}{\partial \dot{x}} \right|_{t=0} = \left[(t + A_1) + \left(\frac{1}{2} t^2 + A_1 t + A_2 \right) + 1 \right]_{t=0} = A_1 + A_2 + 1 = 0$$

$$\left. \frac{\partial \varphi}{\partial \dot{x}} \right|_{t=2} = \left[(t + A_1) + \left(\frac{1}{2} t^2 + A_1 t + A_2 \right) + 1 \right]_{t=2} = 3A_1 + A_2 + 5 = 0$$

The solution of these two algebraic equations yields $A_1 = -2$ and $A_2 = 1$. Hence, the optimal trajectory is $x(t) = \frac{1}{2} t^2 - 2t + 1$.

2 Maxima and Minima of Functionals with Constraints

Here, we will extend the results of maxima and minima of functionals without constraints to the more general case where constraint equations are imposed upon the problem of optimization. More specifically, we will study the case where the cost function has the form

$$J = \int_{t_0}^{t_f} \varphi(\mathbf{x}, \dot{\mathbf{x}}, t) \, dt \tag{11.2-16}$$

where $\mathbf{x}(t)$ is constrained by the following set of equations:

$$\mathbf{f}(\mathbf{x}, \dot{\mathbf{x}}, t) = \mathbf{0} \quad \text{for} \quad t \in [t_0, t_f] \tag{11.2-17}$$

To determine the maxima and minima of J under the constraint (11.12-17) we will apply the method of Lagrange multipliers. To this end, define a new cost function J' as follows:

$$J' = \int_{t_0}^{t_f} [\varphi(\mathbf{x}, \dot{\mathbf{x}}, t) + \boldsymbol{\lambda}^T(t)\mathbf{f}(\mathbf{x}, \dot{\mathbf{x}}, t)] \, dt = \int_{t_0}^{t_f} \psi(\mathbf{x}, \dot{\mathbf{x}}, \boldsymbol{\lambda}, t) \, dt \tag{11.2-18}$$

where $\lambda = (\lambda_1, \lambda_2, \ldots, \lambda_n)^{\mathrm{T}}$ is the vector of Language multipliers and

$$\psi(\mathbf{x}, \dot{\mathbf{x}}, \lambda, t) = \varphi(\mathbf{x}, \dot{\mathbf{x}}, t) + \lambda^{\mathrm{T}}(t)\mathbf{f}(\mathbf{x}, \dot{\mathbf{x}}, t)$$

Since $\mathbf{f}(\mathbf{x}, \dot{\mathbf{x}}, t) = \mathbf{0}$, it follows that $\lambda^{\mathrm{T}}(t)\mathbf{f}(\mathbf{x}, \dot{\mathbf{x}}, t) = 0$ and hence $J = J'$. If we extend the results of Remark 11.2.2 to the present case where constraints are involved, we will arrive at the following Euler–Lagrange equation:

$$\frac{\partial \psi}{\partial \mathbf{x}} - \frac{\mathrm{d}}{\mathrm{d}t}\left[\frac{\partial \psi}{\partial \dot{\mathbf{x}}}\right] = 0 \tag{11.2-19}$$

Clearly, when no constraints are involved, then Eq. (11.2-19) reduces to Eq. (11.2-14) of Remark 11.2.2.

Example 11.2.4

Determine the optimum trajectory $\mathbf{x}(t)$ which minimizes the cost function

$$J = \int_{t_0}^{t_f} \varphi[u(t), t]\,\mathrm{d}t = \int_0^1 \frac{1}{2}u^2(t)\,\mathrm{d}t$$

where $\mathbf{x}(t)$ is subject to the constraint

$$\dot{\mathbf{x}} = \mathbf{A}\mathbf{x} + \mathbf{b}u \qquad \text{or} \qquad \mathbf{f}(\mathbf{x}, \dot{\mathbf{x}}, t) = \mathbf{A}\mathbf{x} + \mathbf{b}u - \dot{\mathbf{x}} = \mathbf{0}$$

where

$$\mathbf{x} = \begin{bmatrix} x_1 \\ x_2 \end{bmatrix}, \qquad \mathbf{A} = \begin{bmatrix} 0 & 1 \\ 0 & 0 \end{bmatrix}, \qquad \text{and} \qquad \mathbf{b} = \begin{bmatrix} 0 \\ 1 \end{bmatrix}$$

with boundary conditions

$$\mathbf{x}(0) = \begin{bmatrix} x_1(0) \\ x_2(0) \end{bmatrix} = \begin{bmatrix} 1 \\ 1 \end{bmatrix} \qquad \text{and} \qquad \mathbf{x}(1) = \begin{bmatrix} x_1(1) \\ x_2(1) \end{bmatrix} = \begin{bmatrix} 0 \\ 0 \end{bmatrix}$$

Solution

The present problem is the minimum effort problem defined in Eq. (11.1-5), subject to the constraints of the system model, namely, the state-space equations $\dot{\mathbf{x}} = \mathbf{A}\mathbf{x} + \mathbf{b}u$. Apply the method of Lagrange multipliers to yield

$$J' = \int_0^1 [\varphi(u, t) + \lambda^{\mathrm{T}}(t)\mathbf{f}(\mathbf{x}, \dot{\mathbf{x}}, t)]\,\mathrm{d}t = \int_0^1 \left[\frac{1}{2}u^2 + \lambda^{\mathrm{T}}(\mathbf{A}\mathbf{x} + \mathbf{b}u - \dot{\mathbf{x}})\right]\mathrm{d}t$$

$$= \int_0^1 \left[\frac{1}{2}u^2 + \lambda_1(x_2 - \dot{x}_1) + \lambda_2(u - \dot{x}_2)\right]\mathrm{d}t = \int_0^1 \psi(u, \mathbf{x}, \dot{\mathbf{x}}, \lambda, t)\,\mathrm{d}t$$

where $\lambda^{\mathrm{T}}(t) = [\lambda_1(t), \lambda_2(t)]$ is the Lagrange multiplier vector and

$$\psi(u, \mathbf{x}, \dot{\mathbf{x}}, \lambda, t) = \frac{1}{2}u^2 + \lambda_1(x_2 - \dot{x}_1) + \lambda_2(u - \dot{x}_2)$$

In what follows, we will determine simultaneously both the optimum $\mathbf{x}(t)$ and the optimum $u(t)$. This means that we must determine the three functions $x_1(t)$, $x_2(t)$, and $u(t)$. The Euler–Lagrange equation (11.2-19) is actually the following three differential equations:

$$\frac{\partial \psi}{\partial x_1} - \frac{d}{dt}\left[\frac{\partial \psi}{\partial \dot{x}_1}\right] = \dot{\lambda}_1 = 0$$

$$\frac{\partial \psi}{\partial x_2} - \frac{d}{dt}\left[\frac{\partial \psi}{\partial \dot{x}_2}\right] = \lambda_1 + \dot{\lambda}_2 = 0$$

$$\frac{\partial \psi}{\partial u} - \frac{d}{dt}\left[\frac{\partial \psi}{\partial \dot{u}}\right] = u + \lambda_2 = 0$$

The general solution of these three equations is

$$\lambda_1(t) = A_1, \qquad \lambda_2(t) = -A_1 t + A_2, \qquad \text{and} \qquad u(t) = -\lambda_2(t) = A_1 t - A_2$$

Note that from the state-space system model we have

$$\dot{x}_1 = x_2$$
$$\dot{x}_2 = u(t)$$

Therefore $\dot{x}_2 = A_1 t - A_2$. Hence

$$x_2(t) = \int_0^t u(t)\, dt = \int_0^t (A_1 t - A_2)\, dt = \frac{1}{2}A_1 t^2 - A_2 t + A_3$$

and

$$x_1(t) = \int_0^t x_2(t)\, dt = \frac{1}{6}A_1 t^3 - \frac{1}{2}A_2 t^2 + A_3 t + A_4$$

The constants A_1, A_2, A_3, and A_4 will be determined using the boundary conditions. We have

$$x_1(0) = A_4 = 1$$
$$x_2(0) = A_3 = 1$$
$$x_1(1) = \frac{1}{6}A_1 - \frac{1}{2}A_2 + A_3 + A_4 = 0$$
$$x_2(1) = \frac{1}{2}A_1 - A_1 + A_3 = 0$$

This system of four algebraic equations has the following solution: $A_1 = 18$, $A_2 = 10$, $A_3 = 1$, and $A_4 = 1$. Hence, the optimum $x(t)$ is given by

$$\mathbf{x}(t) = \begin{bmatrix} x_1(t) \\ x_2(t) \end{bmatrix} = \begin{bmatrix} 3t^3 - 5t^2 + t + 1 \\ 9t^2 - 10t + 1 \end{bmatrix}$$

while the optimum $u(t)$ is given by

$$u(t) = 18t - 10$$

Note that here the final system is an open-loop system.

11.2.2 The Maximum Principle

The method of calculus of variations, presented in Subsec. 11.2.1, constitutes a general methodology for the study of maxima and minima of a functional. Here, we will restrict our interest to specialized optimization methods which facilitate the solution of optimal control design problems. Such a method is the maximum principle which was initially proposed by Pontryagin [17]. This method is based on the calculus of variations and yields a general solution to optimal control problems.

More specifically, the following general control problem will be studied. Consider the cost function

$$J = \theta(\mathbf{x}, t)\bigg|_{t_0}^{t_f} + \int_{t_0}^{t_f} \varphi(\mathbf{x}, \mathbf{u}, t)\, dt \tag{11.2-20}$$

Determine the optimum control vector $\mathbf{u}(t)$ which minimizes the cost function J, where the system under control is described by a mathematical model in state space having the general form

$$\dot{\mathbf{x}} = \mathbf{f}(\mathbf{x}, \mathbf{u}, t) \tag{11.2-21}$$

To solve the problem, apply the Lagrange multipliers method. To this end, define the new cost function J' as follows:

$$J' = \theta(\mathbf{x}, t)\bigg|_{t_0}^{t_f} + \int_{t_0}^{t_f} \left[\varphi(\mathbf{x}, \mathbf{u}, t) + \boldsymbol{\lambda}^{\mathrm{T}}(t)[\mathbf{f}(\mathbf{x}, \mathbf{u}, t) - \dot{\mathbf{x}}]\right] dt \tag{11.2-22}$$

Clearly $J = J'$. To facilitate the study of the new cost function J', we introduce the *Hamiltonian function*, defined as

$$H(\mathbf{x}, \mathbf{u}, \boldsymbol{\lambda}, t) = \varphi(\mathbf{x}, \mathbf{u}, t) + \boldsymbol{\lambda}^{\mathrm{T}}\dot{\mathbf{x}} = \varphi(\mathbf{x}, \mathbf{u}, t) + \boldsymbol{\lambda}^{\mathrm{T}}\mathbf{f}(\mathbf{x}, \mathbf{u}, t) \tag{11.2-23}$$

where $\boldsymbol{\lambda}$ is the vector of Lagrange multipliers. If we substitute Eq. (11.2-23) in Eq. (11.2-22), we have

$$J' = \theta(\mathbf{x}, t)\bigg|_{t_0}^{t_f} + \int_{t_0}^{t_f} [H(\mathbf{x}, \mathbf{u}, \boldsymbol{\lambda}, t) - \boldsymbol{\lambda}^{\mathrm{T}}\dot{\mathbf{x}}]\, dt \tag{11.2-24}$$

If we apply the integration by parts method, Eq. (11.2-24) becomes

$$J' = [\theta(\mathbf{x}, t) - \boldsymbol{\lambda}^{\mathrm{T}}\mathbf{x}]\bigg|_{t_0}^{t_f} + \int_{t_0}^{t_f} [H(\mathbf{x}, \mathbf{u}, \boldsymbol{\lambda}, t) - \dot{\boldsymbol{\lambda}}^{\mathrm{T}}\mathbf{x}]\, dt \tag{11.2-25}$$

The first differential $\delta J'$ with respect to the vectors $\mathbf{x}$ and $\mathbf{u}$ is given by

$$\delta J' = \left[\delta\mathbf{x}^{\mathrm{T}}\left[\frac{\partial\theta}{\partial\mathbf{x}} - \boldsymbol{\lambda}\right]\right]_{t_0}^{t_f} + \int_{t_0}^{t_f} \left[\delta\mathbf{x}^{\mathrm{T}}\left[\frac{\partial H}{\partial\mathbf{x}} + \dot{\boldsymbol{\lambda}}\right] + \delta\mathbf{u}^{\mathrm{T}}\frac{\partial H}{\partial\mathbf{u}}\right] dt \tag{11.2-26}$$

Using Theorem 11.2.1, it follows that a necessary condition for J' to be maximum or minimum is that $\delta J' = 0$. Application of this theorem in Eq. (11.2-26) yields that for $\delta J'$ to be zero, for every $\delta\mathbf{x}$ and $\delta\mathbf{u}$, the vectors $\mathbf{x}$ and $\mathbf{u}$ must satisfy the equations

$$\frac{\partial H}{\partial\mathbf{x}} = -\dot{\boldsymbol{\lambda}} \tag{11.2-27a}$$

$$\frac{\partial H}{\partial\mathbf{u}} = \mathbf{0} \tag{11.2-27b}$$

$$\frac{\partial H}{\partial\boldsymbol{\lambda}} = \dot{\mathbf{x}} = \mathbf{f}(\mathbf{x}, \mathbf{u}, t) \tag{11.2-27c}$$

with boundary conditions

$$\delta \mathbf{x}^T \left[\frac{\partial \theta}{\partial \mathbf{x}} - \lambda \right] = \mathbf{0}, \qquad \text{for } t = t_0 \text{ and } t_f \tag{11.2-27d}$$

The first three equations (Eqs (11.2-27a,b,c)) are of paramount importance in control engineering and they are called *canonical Hamiltonian equations*.

Now, consider the second differential $\delta^2 J'$ of the cost function J'. We have

$$\delta^2 J' = \frac{1}{2} \left[\delta \mathbf{x}^T \frac{\partial^2 \theta}{\partial \mathbf{x}^2} \delta \mathbf{x} \right]_{t_0}^{t_f} + \frac{1}{2} \int_{t_0}^{t_f} [\delta \mathbf{x}^T \vdots \delta \mathbf{u}^T] \mathbf{P} \begin{bmatrix} \delta \mathbf{x} \\ \cdots \\ \delta \mathbf{u} \end{bmatrix} dt \tag{11.2-28}$$

where $\mathbf{P}$ is an $(n+m) \times (n+m)$ square matrix having the form

$$\mathbf{P} = \begin{bmatrix} \dfrac{\partial^2 H}{\partial \mathbf{x}^2} & \dfrac{\partial}{\partial \mathbf{u}} \dfrac{\partial H}{\partial \mathbf{x}} \\ \left[\dfrac{\partial}{\partial \mathbf{u}} \dfrac{\partial H}{\partial \mathbf{x}} \right]^T & \dfrac{\partial^2 H}{\partial \mathbf{u}^2} \end{bmatrix}$$

where use was made of the first differential of Eq. (11.2-21), i.e., use was made of the relation

$$\delta \dot{\mathbf{x}} = \frac{\partial f}{\partial \mathbf{x}} \delta \mathbf{x} + \frac{\partial f}{\partial \mathbf{u}} \delta \mathbf{u} \tag{11.2-29}$$

For J' (and hence for J, since $J' = J$) to be minimum, the matrices $\mathbf{P}$ and $\partial^2 \theta / \partial \mathbf{x}^2$ must be negative definite (see Sec. 2.12).

Remark 11.2.4

It has been shown that the control signal $\mathbf{u}(t)$ which minimizes the cost function J, necessarily minimizes the Hamiltonian function, i.e., it holds that

$$H(\mathbf{x}, \mathbf{u}, \lambda, t) \leq H(\mathbf{x}, \tilde{\mathbf{u}}, \lambda, t)$$

where $\tilde{\mathbf{u}}$ is any control signal, different from the optimum control signal $\mathbf{u}$. For this reason, the present method is known as the minimum principle method. However, because of a sign difference in the Hamiltonian function, the method has become known as the *maximum principle*.

Example 11.2.5

Minimize the cost function

$$J = \int_0^1 \frac{1}{2} u^2(t) \, dt$$

when the system under control is the following:

$$\begin{bmatrix} \dot{x}_1 \\ \dot{x}_2 \end{bmatrix} = \begin{bmatrix} 0 & 1 \\ 0 & -1 \end{bmatrix} \begin{bmatrix} x_1 \\ x_2 \end{bmatrix} + \begin{bmatrix} 0 \\ 1 \end{bmatrix} u$$

with boundary conditions

$$\mathbf{x}(0) = 0 \qquad \text{and} \qquad \mathbf{x}(1) = [4 \quad 2]^T$$

Solution

For the present problem the Hamiltonian function has the form

$$H = \tfrac{1}{2}u^2 + \boldsymbol{\lambda}^T(\mathbf{Ax} + \mathbf{b}u) = \tfrac{1}{2}u^2 + \lambda_1 x_2 - \lambda_2 x_2 + \lambda_2 u$$

The canonical Hamiltonian equations are

$$\frac{\partial H}{\partial \mathbf{x}} = \begin{bmatrix} \dfrac{\partial H}{\partial x_1} \\[2mm] \dfrac{\partial H}{\partial x_2} \end{bmatrix} = \begin{bmatrix} -\dot{\lambda}_1 \\[1mm] -\dot{\lambda}_2 \end{bmatrix} = \begin{bmatrix} 0 \\[1mm] \lambda_1 - \lambda_2 \end{bmatrix}$$

$$\frac{\partial H}{\partial \mathbf{u}} = u + \lambda_2 = 0$$

$$\frac{\partial H}{\partial \boldsymbol{\lambda}} = \begin{bmatrix} \dot{x}_1 \\[1mm] \dot{x}_2 \end{bmatrix} = \begin{bmatrix} x_2 \\[1mm] -x_2 + u \end{bmatrix}$$

The general solution of the above equations is given by

$$\lambda_1(t) = C_3$$
$$\lambda_2(t) = C_3[1 - e^{-t}] + C_4 e^t$$
$$x_1(t) = C_1 + C_2[1 - e^{-t}] + C_3\left[-t - \tfrac{1}{2}e^{-t} + \tfrac{1}{2}e^t\right] + C_4\left[1 - \tfrac{1}{2}e^{-t} - \tfrac{1}{2}e^t\right]$$
$$x_2(t) = C_2 e^{-t} + C_3\left[-1 + \tfrac{1}{2}e^{-t} + \tfrac{1}{2}e^t\right] + C_4\left[\tfrac{1}{2}e^{-t} - \tfrac{1}{2}e^t\right]$$

The parameters C_1, C_2, C_3, and C_4 are determined using the boundary conditions. We have $x_1(0) = C_1 = 0$, $x_2(0) = C_2 = 0$, $x_1(1) = 4$, and $x_2(1) = 2$. From these algebraic equations we readily have that $C_1 = 0$, $C_2 = 0$, $C_3 = -40.5$, and $C_4 = -20.42$. Hence

$$x_1(t) = 40.5t - 20.42 + 30.46 e^{-t} - 10.04 e^t$$

$$x_2(t) = 40.5 - 30.46 e^{-t} - 10.04 e^t$$

and the optimum control signal $u(t)$ is given by

$$u(t) = -\lambda_2(t) = 40.5 - 20.08 e^t$$

Note that here the final system is an open-loop system.

11.3 OPTIMAL LINEAR REGULATOR

11.3.1 General Remarks

The optimal linear regulator problem is a special, but very important, optimal control problem. Simply speaking, the regulator problem can be stated as follows. Consider a linear homogeneous system with zero input and nonzero initial state vector conditions $\mathbf{x}(t_0)$. Here, the vector $\mathbf{x}(t_0)$ is the only excitation to the system. An optimal control signal $\mathbf{u}(t)$ is to be determined such as to restore the state vector to its equilibrium point, i.e., such that $\mathbf{x}(t_f) \simeq \mathbf{0}$, while minimizing a certain cost function.

As a practical example of an optimal regulator, consider a ground antenna having a fixed orientation. Assume that the antenna undergoes a disturbance, e.g., due to a sudden strong wind. As a result, the antenna will be forced to a new position $\mathbf{x}(t_0)$. It is obvious that in the present situation it is desirable to implement a control strategy which will restore the antenna to its equilibrium position. Furthermore, this restoration must take place in the time interval $[t_0, t_f]$, while minimizing a certain cost function. This cost function normally includes the following three characteristics:

a. The amplitude of the optimal control signal $\mathbf{u}(t)$ should be as small as possible, making the required control effort (control energy) for restoring the antenna to its equilibrium position as small as possible.
b. The amplitude of $\mathbf{x}(t)$ should be small enough to avoid saturations or even damage (i.e., from overheating) to the system under control.
c. The final value $\mathbf{x}(t_f)$ of $\mathbf{x}(t)$ should be as close as possible to the equilibrium point of the system, i.e., $\mathbf{x}(t_f) \simeq \mathbf{0}$.

Another practical example of an optimal regulator is the problem of ship stabilization presented in Figure 1.20 of Chapter 1 and in Example 6.4.11 of Chapter 6.

From a mathematical point of view, the optimal regulator problem may be formulated as follows. Consider the linear, time-varying system described in state space by the vector differential equation

$$\dot{\mathbf{x}}(t) = \mathbf{A}(t)\mathbf{x}(t) + \mathbf{B}(t)\mathbf{u}(t), \qquad \mathbf{x}(t_0) = \mathbf{x}_0 \tag{11.3-1}$$

Find a control signal $\mathbf{u}(t)$ which minimizes the cost function

$$J = \frac{1}{2}\mathbf{x}^{\mathrm{T}}(t_f)\mathbf{S}\mathbf{x}(t_f) + \frac{1}{2}\int_{t_0}^{t_f} [\mathbf{x}^{\mathrm{T}}(t)\mathbf{Q}(t)\mathbf{x}(t) + \mathbf{u}^{\mathrm{T}}(t)\mathbf{R}(t)\mathbf{u}(t)]\,\mathrm{d}t \tag{11.3-2}$$

The foregoing cost function J is identical to the cost function (11.1-9). This criterion is a sum of inner products of the vectors $\mathbf{x}(t)$ and $\mathbf{u}(t)$, and for this reason it is called the *quadratic cost function*. The matrices $\mathbf{S}$, $\mathbf{Q}(t)$, and $\mathbf{R}(t)$ are weighting matrices and are chosen to be symmetric. Here, we stress again that the main reason for including the energy-like quadratic terms $\mathbf{x}^{\mathrm{T}}(t)\mathbf{Q}(t)\mathbf{x}(t)$ and $\mathbf{u}^{\mathrm{T}}(t)\mathbf{R}(t)\mathbf{u}(t)$ in the cost function J is to minimize the dissipated energy in the system and the required input energy (control effort), respectively. The quadratic term $\mathbf{x}^{\mathrm{T}}(t_f)\mathbf{S}\mathbf{x}(t_f)$ is included in J to force the final value $\mathbf{x}(t_f)$ of $\mathbf{x}(t)$ to be as close as possible to the equilibrium point of the system. Note that $\mathbf{x}(t_f)$ is unspecified.

11.3.2 Solution of the Optimal Linear Regulator Problem

The minimization of the cost function J will be done using the method of maximum principle. To this end, define the Hamiltonian

$$H(\mathbf{x}, \mathbf{u}, \lambda, t) = \tfrac{1}{2}\mathbf{x}^{\mathrm{T}}(t)\mathbf{Q}(t)\mathbf{x}(t) + \tfrac{1}{2}\mathbf{u}^{\mathrm{T}}(t)\mathbf{R}(t)\mathbf{u}(t) + \lambda^{\mathrm{T}}(t)[\mathbf{A}(t)\mathbf{x}(t) + \mathbf{B}(t)\mathbf{u}(t)]$$

$$\tag{11.3-3}$$

where $\lambda(t)$ is the vector of the Lagrange multipliers. Next, define the new cost criterion $J' = J$ by adding the zero term $\lambda^{\mathrm{T}}(t)[\mathbf{A}(t)\mathbf{x}(t) + \mathbf{B}(t)\mathbf{u}(t) - \dot{\mathbf{x}}(t)]$ to the initial cost function J as follows:

$$J' = \frac{1}{2}\mathbf{x}^{\mathrm{T}}(t_{\mathrm{f}})\mathbf{S}\mathbf{x}(t_{\mathrm{f}}) + \frac{1}{2}\int_{t_0}^{t_{\mathrm{f}}} \{[\mathbf{x}^{\mathrm{T}}(t)\mathbf{Q}(t)\mathbf{x}(t) + \mathbf{u}^{\mathrm{T}}(t)\mathbf{R}(t)\mathbf{u}(t)]$$
$$+ \boldsymbol{\lambda}^{\mathrm{T}}(t)[\mathbf{A}(t)\mathbf{x}(t) + \mathbf{B}(t)\mathbf{u}(t) - \dot{\mathbf{x}}(t)]\} \, dt$$

or

$$J' = \frac{1}{2}\mathbf{x}^{\mathrm{T}}(t_{\mathrm{f}})\mathbf{S}\mathbf{x}(t_{\mathrm{f}}) + \int_{t_0}^{t_{\mathrm{f}}} [H(\mathbf{x}, \mathbf{u}, \boldsymbol{\lambda}, t) - \boldsymbol{\lambda}^{\mathrm{T}}(t)\dot{\mathbf{x}}(t)] dt$$

where use was made of the Hamiltonian defined in Eq. (11.3-3). Using the method of integration by parts, the cost criterion J' becomes

$$J' = \frac{1}{2}\mathbf{x}^{\mathrm{T}}(t_{\mathrm{f}})\mathbf{S}\mathbf{x}(t_{\mathrm{f}}) - [\boldsymbol{\lambda}^{\mathrm{T}}(t)\mathbf{x}(t)]_{t_0}^{t_{\mathrm{f}}} + \int_{t_0}^{t_{\mathrm{f}}} [H(\mathbf{x}, \mathbf{u}, \boldsymbol{\lambda}, t) - \dot{\boldsymbol{\lambda}}^{\mathrm{T}}(t)\mathbf{x}(t)] dt$$

The first differential $\delta J'$ with respect to the vectors $\mathbf{x}$ and $\mathbf{u}$ is given by

$$\delta \mathbf{J}' = \mathbf{S}\mathbf{x}(t_{\mathrm{f}}) - \boldsymbol{\lambda}(t_{\mathrm{f}}) + \int_{t_0}^{t_{\mathrm{f}}} \left\{\delta\mathbf{x}^{\mathrm{T}}\left[\frac{\partial H}{\partial \mathbf{x}} + \dot{\boldsymbol{\lambda}}\right] + \delta\mathbf{u}^{\mathrm{T}}\left[\frac{\partial H}{\partial \mathbf{u}}\right]\right\} dt$$

where use was made of the fact that $\mathbf{x}(t_0)$ is fixed and that $\mathbf{x}(t_{\mathrm{f}})$ is unspecified. It has been proven that a necessary condition for J to be maximum or minimum is that $\delta \mathbf{J}' = \mathbf{0}$ (see Theorem 11.2.1). Consequently, the vectors $\mathbf{x}$ and $\mathbf{u}$ should satisfy the equation $\delta \mathbf{J}' = \mathbf{0}$, in which case the following relations should hold:

$$\frac{\partial H}{\partial \mathbf{x}} = -\dot{\boldsymbol{\lambda}}(t) = \mathbf{Q}(t)\mathbf{x}(t) + \mathbf{A}^{\mathrm{T}}(t)\boldsymbol{\lambda}(t) \tag{11.3-4a}$$

$$\frac{\partial H}{\partial \mathbf{u}} = \mathbf{0} = \mathbf{R}(t)\mathbf{u}(t) + \mathbf{B}^{\mathrm{T}}(t)\boldsymbol{\lambda}(t) \tag{11.3-4b}$$

$$\frac{\partial H}{\partial \boldsymbol{\lambda}} = \mathbf{A}(t)\mathbf{x}(t) + \mathbf{B}(t)\mathbf{u}(t) \tag{11.3-4c}$$

$$\frac{\partial \theta}{\partial \mathbf{x}}\bigg|_{t=t_{\mathrm{f}}} = \mathbf{S}\mathbf{x}(t_{\mathrm{f}}) = \boldsymbol{\lambda}(t_{\mathrm{f}}) \tag{11.3-4d}$$

where use was made of the following vector and matrix properties (see relations (2.6-17) and (2.6-18) of Chapter 2):

$$\frac{\partial}{\partial \mathbf{x}}[\mathbf{q}^{\mathrm{T}}(t)\mathbf{x}(t)] = \mathbf{q}(t) \qquad \text{and} \qquad \frac{1}{2}\frac{\partial}{\partial \mathbf{x}}[\mathbf{x}^{\mathrm{T}}(t)\mathbf{Q}(t)\mathbf{x}(t)] = \mathbf{Q}(t)\mathbf{x}(t)$$

where $\mathbf{Q}(t)$ is a symmetric matrix. As first pointed out in Subsec. 11.2.2, Eqs. (11.3-4a,b,c) are called *canonical Hamiltonian equations* and relation (11.3-4d) represents the *boundary conditions* of the problem. Note that for the present case Eq. (11.3-4d) refers only to the final condition, i.e., for $t = t_{\mathrm{f}}$.

In the sequel, we will solve the canonical Hamiltonian equations (11.3-a,b,c) with respect to $\mathbf{u}(t)$. This solution must satisfy the boundary condition (11.3-4d). For simplicity, assume that $\mathbf{R}(t)$ is invertible, i.e., $|\mathbf{R}(t)| \neq 0$ for every $t \in [t_0, t_{\mathrm{f}}]$. Thus, from relation (11.3-4b), we obtain

$$\mathbf{u}(t) = -\mathbf{R}^{-1}(t)\mathbf{B}^{\mathrm{T}}(t)\boldsymbol{\lambda}(t) \tag{11.3-5}$$

At this point we make the assumption that the solution of Eq. (11.3-5) can be expressed as a linear state feedback law, i.e., we assume that

$$\mathbf{u}(t) = \mathbf{K}(t)\mathbf{x}(t) \tag{11.3-6}$$

where $\mathbf{K}(t)$ is called the state feedback matrix. We also assume that the vector of Lagrange multipliers $\lambda(t)$ is linear in $\mathbf{x}(t)$, i.e., we assume that

$$\lambda(t) = \mathbf{P}(t)\mathbf{x}(t) \tag{11.3-7}$$

Note that the vector of Lagrange multipliers $\lambda(t)$ is called the costate vector. If we substitute Eq. (11.3-5) in Eq. (11.3-1), we have

$$\dot{\mathbf{x}}(t) = \mathbf{A}(t)\mathbf{x}(t) - \mathbf{B}(t)\mathbf{R}^{-1}(t)\mathbf{B}^{\mathrm{T}}(t)\lambda(t) \tag{11.3-8a}$$

If we substitute Eq. (11.3-7) in Eq. (11.3-8a), we obtain

$$\dot{\mathbf{x}}(t) = \mathbf{A}(t)\mathbf{x}(t) - \mathbf{B}(t)\mathbf{R}^{-1}(t)\mathbf{B}^{\mathrm{T}}(t)\mathbf{P}(t)\mathbf{x}(t) \tag{11.3-8b}$$

If we differentiate Eq. (11.3-7), we have

$$\dot{\lambda}(t) = \dot{\mathbf{P}}(t)\mathbf{x}(t) + \mathbf{P}(t)\dot{\mathbf{x}}(t) = -\mathbf{Q}(t)\mathbf{x}(t) - \mathbf{A}^{\mathrm{T}}(t)\lambda(t) \tag{11.3-9}$$

where use was made of Eq. (11.3-4a). Finally, if we substitute Eq. (11.3-8b) in Eq. (11.3-9) and use Eq. (11.3-7), we arrive at the relation

$$[\dot{\mathbf{P}}(t) - \mathbf{P}(t)\mathbf{A}(t) + \mathbf{A}^{\mathrm{T}}(t)\mathbf{P}(t) + \mathbf{Q}(t) - \mathbf{P}(t)\mathbf{B}(t)\mathbf{R}^{-1}(t)\mathbf{B}^{\mathrm{T}}(t)\mathbf{P}(t)]\mathbf{x}(t) = \mathbf{0} \tag{11.3-10}$$

Relation (11.3-10) must hold for all vectors $\mathbf{x}(t) \neq 0$. For this to be valid the coefficient of $\mathbf{x}(t)$ must be equal to zero, i.e.,

$$\dot{\mathbf{P}}(t) + \mathbf{P}(t)\mathbf{A}(t) + \mathbf{A}^{\mathrm{T}}(t)\mathbf{P}(t) - \mathbf{P}(t)\mathbf{B}(t)\mathbf{R}^{-1}(t)\mathbf{B}^{\mathrm{T}}(t)\mathbf{P}(t) = -\mathbf{Q}(t) \tag{11.3-11}$$

Relation (11.3-11) is known as the *matrix Riccati differential equation* where the matrix $\mathbf{P}(t)$ is unknown. The final condition of matrix $\mathbf{P}(t)$, according to relation (11.3-4d) and definition (11.3-7), will be

$$\mathbf{S}\mathbf{x}(t_{\mathrm{f}}) = \lambda(t_{\mathrm{f}}) = \mathbf{P}(t_{\mathrm{f}})\mathbf{x}(t_{\mathrm{f}})$$

Consequently

$$\mathbf{P}(t_{\mathrm{f}}) = \mathbf{S} \tag{11.3-12}$$

If we substitute relation (11.3-7) in relation (11.3-5), we obtain

$$\mathbf{u}(t) = -\mathbf{R}^{-1}(t)\mathbf{B}^{\mathrm{T}}(t)\mathbf{P}(t)\mathbf{x}(t) \tag{11.3-13}$$

By comparing relations (11.3-13) and (11.3-6), we have

$$\mathbf{K}(t) = -\mathbf{R}^{-1}(t)\mathbf{B}^{\mathrm{T}}(t)\mathbf{P}(t) \tag{11.3-14}$$

Henceforth, the optimal solution of the linear optimal regulator problem is of the form (11.3-6), where the matrix $\mathbf{K}(t)$ is given by relation (11.3-14). To determine the feedback matrix $\mathbf{K}(t)$ one has to solve the Ricatti equation (11.3-11) for $\mathbf{P}(t)$. The solution (11.3-13) was first determined by Kalman, and it is for this reason that matrix $\mathbf{K}(t)$ is called the *Kalman matrix* [8].

The second differential $\delta^2 J$ of the cost function J is given by

$$\delta^2 J = \frac{1}{2}\delta\mathbf{x}^{\mathrm{T}}(t_f)\mathbf{S}\delta\mathbf{x}(t_f) + \frac{1}{2}\int_{t_0}^{t_f}\left[\delta\mathbf{x}^{\mathrm{T}}(t)\mathbf{Q}(t)\delta\mathbf{x}(t) + \delta\mathbf{u}^{\mathrm{T}}(t)\mathbf{R}(t)\delta\mathbf{u}(t)\right]\mathrm{d}t \qquad (11.3\text{-}15)$$

For the cost function J to be minimal it must hold that $\delta^2 J \geq 0$. We observe that $\delta^2 J$ is a sum of three terms which are in quadratic form. Consequently for $\delta^2 J \geq 0$ to hold true, every term in Eq. (11.3-15) must be positive definite. Thus, for Eq. (11.3-13) to be the solution of the problem, the matrices $\mathbf{S}$, $\mathbf{Q}(t)$, and $\mathbf{R}(t)$ should be at least positive semidefinite matrices.

The following theorems hold true for the Ricatti equation (11.3-11).

Theorem 11.3.1

If $\mathbf{S}$ is positive definite and $\mathbf{Q}(t)$ is at least nonnegative definite, or vice versa, and $\mathbf{R}(t)$ is positive definite, then a minimum J exists if and only if the solution $\mathbf{P}(t)$ of the Riccati equation (11.3-11) exists, is bounded, and is positive definite for all $t < t_f$. Under these conditions the minimum cost function J becomes

$$J = \frac{1}{2}\mathbf{x}^{\mathrm{T}}(t_0)\mathbf{P}(t_0)\mathbf{x}(t_0)$$

Theorem 11.3.2

If $\mathbf{S}$, $\mathbf{Q}(t)$, and $\mathbf{R}(t)$ are symmetric, then the solution of the Riccati equation (11.3-11) is also a symmetric matrix. This means that in this case the $n \times n$ matrix $\mathbf{P}(t)$ has $n(n+1)/2$ unknown elements and, consequently, the solution of eq. (11.3-11) requires only the solution of $n(n+1)/2$ equations.

Two block diagrams referring to the problem of the optimal regulator and its solution are given in Figures 11.3 and 11.4.

The Ricatti equation (11.3-11) is usually solved using a digital computer rather than analytically. Since the final condition $\mathbf{P}(t_f)$ is given, the solution using a digital computer is carried out starting from the final point $t = t_f$ and going backwards until we reach the starting part $t = t_0$. However, this can be avoided if we change variables in the following way. Let $\tau = t_f - t$. Then the Ricatti equation becomes

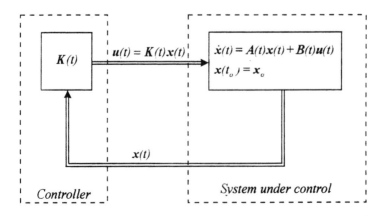

Figure 11.3   A simplified block diagram of the optimal linear regulator.

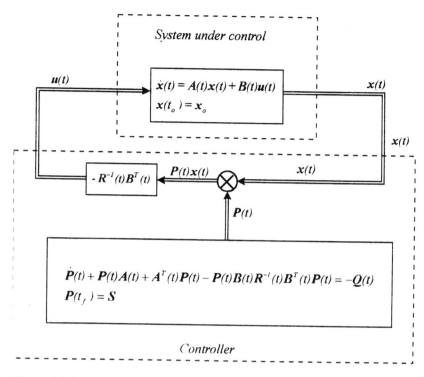

Figure 11.4 Block diagram of the optimal linear regulator, using the Riccati equation.

$$\frac{d\mathbf{P}(t_f - \tau)}{d\tau} - \mathbf{P}(t_f - \tau)\mathbf{A}(t_f - \tau) - \mathbf{A}^T(t_f - \tau)\mathbf{P}(t_f - \tau)$$

$$+ \mathbf{P}(t_f - \tau)\mathbf{B}(t_f - \tau)\mathbf{R}^{-1}(t_f - \tau)\mathbf{B}^T(t_f - \tau)\mathbf{P}(t_f - \tau) = \mathbf{Q}(t_f - \tau)$$

$$(11.3\text{-}16)$$

with initial condition $\mathbf{P}(0) = \mathbf{S}$. Equation (11.3-16) is solved using a digital computer by starting at point $\tau = 0$ and ending at point $\tau = t_f - t_0$.

Remark 11.3.1

Another method of determining the optimal control vector $\mathbf{u}(t)$ is the following. Rewrite relations (11.3-8a) and (11.3-4a) in the form

$$\begin{bmatrix} \dot{\mathbf{x}}(t) \\ \dot{\lambda}(t) \end{bmatrix} = \begin{bmatrix} \mathbf{A}(t) & -\mathbf{B}(t)\mathbf{R}^{-1}(t)\mathbf{B}^T(t) \\ -\mathbf{Q}(t) & -\mathbf{A}^T(t) \end{bmatrix} \begin{bmatrix} \mathbf{x}(t) \\ \lambda(t) \end{bmatrix} \qquad (11.3\text{-}17)$$

with boundary conditions $\mathbf{x}(t_0) = \mathbf{x}_0$ and $\lambda(t_f) = \mathbf{S}\mathbf{x}(t_f)$. The solution of Eq. (11.3-17) yields the vector $\lambda(t)$, on the basis of which the optimal control vector $\mathbf{u}(t)$ may be calculated using relation (11.3-5). To determine $\lambda(t)$, we work as follows. The solution of Eq. (11.3-17) has the general form

$$\begin{bmatrix} \mathbf{x}(t_f) \\ \lambda(t_f) \end{bmatrix} = \phi(t_f, t) \begin{bmatrix} \mathbf{x}(t) \\ \lambda(t) \end{bmatrix} \qquad (11.3\text{-}18)$$

where the $2n \times 2n$ matrix $\phi(t_f, t)$ is the transition matrix of Eq. (11.3-17). Partition Eq. (11.3-18) as follows:

$$\begin{bmatrix} \mathbf{x}(t_f) \\ \lambda(t_f) \end{bmatrix} = \begin{bmatrix} \phi_{11}(t_f, t) & \phi_{12}(t_f, t) \\ \phi_{21}(t_f, t) & \phi_{22}(t_f, t) \end{bmatrix} \begin{bmatrix} \mathbf{x}(t) \\ \lambda(t) \end{bmatrix} \tag{11.3-19}$$

where all four submatrices ϕ_{11}, ϕ_{12}, ϕ_{21}, and ϕ_{22} have dimensions $n \times n$. Since, according to Eq. (11.3-4d), we have that $\lambda(t_f) = \mathbf{S}\mathbf{x}(t_f)$, relation (11.3-19) is written as

$$\mathbf{x}(t_f) = \phi_{11}(t_f, t)\mathbf{x}(t) + \phi_{12}(t_f, t)\lambda(t)$$
$$\mathbf{S}\mathbf{x}(t_f) = \phi_{21}(t_f, t)\mathbf{x}(t) + \phi_{22}(t_f, t)\lambda(t)$$

These two equations yield

$$\lambda(t) = \mathbf{P}(t)\mathbf{x}(t) \tag{11.3-20a}$$

where

$$\mathbf{P}(t) = [\phi_{22}(t_f, t) - \mathbf{S}\phi_{12}(t_f, t)]^{-1}[\mathbf{S}\phi_{11}(t_f, t) - \phi_{21}(t_f, t)] \tag{11.320b}$$

Finally

$$\mathbf{u}(t) = -\mathbf{R}^{-1}(t)\mathbf{B}^{\mathrm{T}}(t)\mathbf{P}(t)\mathbf{x}(t) = \mathbf{K}(t)\mathbf{x}(t) \tag{11.3-21a}$$

where

$$\mathbf{K}(t) = -\mathbf{R}^{-1}(t)\mathbf{B}^{\mathrm{T}}(t)[\phi_{22}(t_f, t) - \mathbf{S}\phi_{12}(t_f, t)]^{-1}[\mathbf{S}\phi_{11}(t_f, t) - \phi_{21}(t_f, t)] \tag{11.3-21b}$$

The present method is easily applied when the matrices $\mathbf{A}$, $\mathbf{B}$, $\mathbf{Q}$, and $\mathbf{R}$ are time invariant. In this case the transition matrix $\phi(t)$ is the inverse Laplace transform of the matrix

$$\phi(s) = \left[s\mathbf{I} - \begin{bmatrix} \mathbf{A} & -\mathbf{B}\mathbf{R}^{-1}\mathbf{B}^{\mathrm{T}} \\ -\mathbf{Q} & -\mathbf{A}^{\mathrm{T}} \end{bmatrix} \right]^{-1} \tag{11.3-22}$$

where t has been replaced by $t_f - t$. When the matrices $\mathbf{A}$, $\mathbf{B}$, $\mathbf{Q}$, and $\mathbf{R}$ are time varying, the transition matrix is usually computed numerically.

11.3.3 The Special Case of Linear Time-Invariant Systems

Consider the special case of linear time-invariant systems. Furthermore, assume that the weighting matrix $\mathbf{S} = \mathbf{0}$ and that $t_f \to +\infty$. Then the matrix $\mathbf{P}(t)$ and, consequently, the matrix $\mathbf{K}(t)$, are time invariant and the Riccati equation reduces to the nonlinear algebraic equation

$$\mathbf{P}\mathbf{A} + \mathbf{A}^{\mathrm{T}}\mathbf{P} - \mathbf{P}\mathbf{B}\mathbf{R}^{-1}\mathbf{B}^{\mathrm{T}}\mathbf{P} = -\mathbf{Q} \tag{11.3-23}$$

The following interesting and very useful results have been proven concerning the solution of Eq. (11.3-23):

1. If there exists a matrix $\mathbf{P}$, it is positive definite and unique.
2. If there exists a matrix $\mathbf{P}$, the closed-loop system is asymptotically stable.
3. If λ_i is an eigenvalue of matrix $\mathbf{M}$, then $-\bar{\lambda}_i$ is also an eigenvalue of matrix $\mathbf{M}$, where

$$\mathbf{M} = \begin{bmatrix} \mathbf{A} & -\mathbf{B}\mathbf{R}^{-1}\mathbf{B}^{\mathrm{T}} \\ -\mathbf{Q} & -\mathbf{A}^{\mathrm{T}} \end{bmatrix} \tag{11.3-24}$$

4. If matrix $\mathbf{M}$ has n distinct eigenvalues $\lambda_1, \lambda_2, \ldots, \lambda_n$ with $\mathrm{Re}\, \lambda_i < 0$, then

$$\mathbf{P} = \mathbf{G}\mathbf{T}^{-1} \qquad\qquad (11.3\text{-}25)$$

where

$$\mathbf{G} = [\mathbf{g}_1 \ \vdots\ \mathbf{g}_2 \ \vdots\ \cdots\ \vdots\ \mathbf{g}_n] \qquad \text{and} \qquad \mathbf{T} = [\mathbf{t}_1 \ \vdots\ \mathbf{t}_2 \ \vdots\ \cdots\ \vdots\ \mathbf{t}_n]$$

where the vectors $\mathbf{g}_i$ and $\mathbf{t}_i$ are defined as

$$\lambda_i \boldsymbol{\omega}_i = \mathbf{M}\boldsymbol{\omega}_i, \qquad \boldsymbol{\omega}_i = \begin{bmatrix} \mathbf{t}_i \\ \mathbf{g}_i \end{bmatrix}, \qquad i = 1, 2, \ldots, n$$

That is, the vector $\boldsymbol{\omega}_i$ is an eigenvector of the matrix $\mathbf{M}$.
5. The state vector $\mathbf{x}(t)$ of the closed-loop system is given by

$$\mathbf{x}(t) = \mathbf{T}e^{\boldsymbol{\Lambda}(t-t_0)}\mathbf{T}^{-1}\mathbf{x}_0, \qquad \boldsymbol{\Lambda} = \mathrm{diag}(\lambda_1, \lambda_2, \ldots, \lambda_n) \qquad (11.3\text{-}26)$$

6. The matrix $\mathbf{A} - \mathbf{B}\mathbf{R}^{-1}\mathbf{B}^T\mathbf{P}$ of the closed-loop system has eigenvalues λ_1, $\lambda_2, \ldots, \lambda_n$ and eigenvectors $\mathbf{t}_1, \mathbf{t}_2, \ldots, \mathbf{t}_n$.

Example 11.3.1

Consider the scalar system

$$\dot{x}(t) = u(t), \qquad x(0) = 1$$

with cost function

$$J = \int_0^\infty [x^2(t) + u^2(t)]\, dt$$

Thus, here we have $S = 0$, $Q(t) = 2$, and $R(t) = 2$. Find the optimal $u(t)$ and $x(t)$, both in the form an open-loop system as well as in the form of a closed-loop system.

Solution

First we study the case of the open-loop system using relation (11.3-17) of Remark 11.3.1. For the present example, we have

$$\begin{bmatrix} \dot{x}(t) \\ \dot{\lambda}(t) \end{bmatrix} = \begin{bmatrix} 0 & -0.5 \\ -2 & 0 \end{bmatrix}\begin{bmatrix} x(t) \\ \lambda(t) \end{bmatrix}, \qquad x(0) = 1, \qquad \text{and} \qquad \lambda(\infty) = Sx(\infty) = 0$$

From this vector differential system, we have that $\dot{x} = 0.5\lambda$ and $\dot{\lambda} = -2x$. Thus $x^{(2)} = x$, where $x^{(2)}$ is the second derivative of $x(t)$ with respect to t. The solution of this last differential equation is $x(t) = Ae^{-t} + Be^{t}$ and, hence, $\lambda(t) = 2Ae^{-t} - 2Be^{t}$. From the boundary conditions $x(0) = 1$ and $\lambda(+\infty) = 0$, we have that $A = 1$ and $B = 0$. Therefore $\lambda(t) = 2e^{-t}$ and, consequently, the optimal $u(t)$ and $x(t)$ are

$$u(t) = -e^{-t} \qquad \text{and} \qquad x(t) = e^{-t}$$

Consequently, for the cost function to be minimal we must excite the system with the input $u(t) = -e^{-t}$. This input can be produced, for example, by a waveform generator. In this case, of course, we have an open-loop system.

For the case of the closed-loop system we solve the algebraic Riccati equation, which for the present example is $-0.5p^2 = -2$. Consequently, $p = \pm 2$. Since p must be positive definite we keep only the value $p = 2$. Thus, the optimal $u(t)$ and $x(t)$ are

$$u(t) = -x(t) \quad \text{and} \quad x(t) = e^{-t}$$

We observe that the optimal $x(t)$ is, as expected, the same with that of the open-loop system. We also observe that the optimal $u(t)$ has not a specific waveform, as in the case of the open-loop system, but it is a function of $x(t)$.

Example 11.3.2

Consider the scalar system

$$\dot{x}(t) = u(t), \qquad x(t_0) = x_0$$

with cost function

$$J = \frac{1}{2}sx^2(t_f) + \frac{1}{2}\int_{t_0}^{t_f} u^2(t)\,dt$$

Find the optimal $u(t)$ as a function of $x(t)$.

Solution

The Hamiltonian is given by

$$H = \tfrac{1}{2}u^2 + \lambda u$$

Hence, the canonical equations are

$$\frac{\partial H}{\partial x} = -\dot{\lambda} = 0; \qquad \text{thus } \lambda = C = \text{constant}$$

$$\frac{\partial H}{\partial u} = u + \lambda = 0; \qquad \text{thus } u = -\lambda = -C$$

$$\frac{\partial H}{\partial \lambda} = \dot{x} = u$$

with boundary conditions $\lambda(t_f) = sx(t_f)$. The Riccati equation is given by

$$\dot{p} - p^2 = 0, \qquad p(t_f) = s$$

The Ricatti equation may be written as follows:

$$\frac{dp}{p^2} = dt$$

Integrating from t to t_f we have

$$-\frac{1}{p}\bigg]_t^{t_f} = t_f - t \qquad \text{and hence} \qquad p(t) = \left[\frac{1}{s^{-1} + t_f - t}\right]x(t)$$

Consequently, the optimal $u(t)$ becomes

$$u(t) = -p(t)x(t) = \left[\frac{-1}{s^{-1} + t_f - t}\right]x(t)$$

Example 11.3.3

Consider the scalar system

$$\dot{x}(t) = -ax(t) + bu(t), \qquad x(t_0) = x_0$$

with cost function

$$J = \frac{1}{2}sx^2(t_f) + \frac{1}{2}\int_{t_0}^{t_f} u^2(t)\,dt$$

Find the optimal $u(t)$ as a function of $x(t)$.

Solution

The Hamiltonian is given by

$$H = \tfrac{1}{2}u^2 + \lambda(-ax + bu)$$

Hence, the canonical equations are

$$\frac{\partial H}{\partial x} = -\dot{\lambda} = -a\lambda$$

$$\frac{\partial H}{\partial u} = u + b\lambda$$

$$\frac{\partial H}{\partial \lambda} = \dot{x} = -ax + bu$$

with boundary conditions $\lambda(t_f) = sx(t_f)$. The Riccati equation is

$$\dot{p} - 2ap - b^2p^2 = 0, \qquad p(t_f) = s$$

and may be written as

$$\dot{p} = 2ap + b^2p^2 = 2a\left(p + \frac{b^2}{2a}p^2\right)$$

or

$$\frac{dp}{p} - \frac{\dfrac{b^2}{2a}\,dp}{\dfrac{b^2}{2a}p + 1} = 2a\,dt$$

Integrating from t to t_f, we have

$$\ln s - \ln p(t) - \ln\left(s\frac{b^2}{2a} + 1\right) + \ln\left[p(t)\frac{b^2}{2a} + 1\right] = 2a(t_f - t)$$

or

$$\ln\left[p(t)\frac{b^2}{2a} + 1\right] - \ln p(t) = \ln\left[\frac{s\dfrac{b^2}{2a} + 1}{s}\right] + 2a(t_f - t)$$

or

$$\frac{p(t)\dfrac{b^2}{2a} + 1}{p(t)} = \left[\frac{s\dfrac{b^2}{2a} + 1}{s}\right]\exp[2a(t_f - t)]$$

Solving for $p(t)$, we obtain

$$p(t) = \frac{\exp[-2a(t_f - t)]}{\dfrac{1}{s} + \dfrac{b^2}{2a}[1 - \exp[-2a(t_f - t)]]}$$

Consequently, the optimal $u(t)$ becomes

$$u(t) = -bp(t)x(t) = -b\left[\frac{\exp[-2a(t_f - t)]}{\dfrac{1}{s} + \dfrac{b^2}{2a}[1 - \exp[-2a(t_f - t)]]}\right]x(t)$$

11.4 OPTIMAL LINEAR SERVOMECHANISM OR TRACKING PROBLEM

Consider the linear time-varying system

$$\dot{\mathbf{x}}(t) = \mathbf{A}(t)\mathbf{x}(t) + \mathbf{B}(t)\mathbf{u}(t) \tag{11.4-1a}$$

$$\mathbf{y}(t) = \mathbf{C}(t)\mathbf{x}(t) \tag{11.4-1b}$$

Let $\xi(t)$ be the desired closed-loop system output vector. Then the optimal linear servomechansim or tracking problem may be stated as follows: determine a control vector $\mathbf{u}(t)$ such that the cost function

$$J = \frac{1}{2}\mathbf{e}^{\mathrm{T}}(t_f)\mathbf{S}\mathbf{e}(t_f) + \frac{1}{2}\int_{t_0}^{t_f} \left[\mathbf{e}^{\mathrm{T}}(t)\mathbf{Q}(t)\mathbf{e}(t) + \mathbf{u}^{\mathrm{T}}(t)\mathbf{R}(t)\mathbf{u}(t)\right]dt \tag{11.4-2}$$

is minimized, where $\mathbf{e}(t) = \xi(t) - \mathbf{y}(t) = \xi(t) - \mathbf{C}\mathbf{x}(t)$.

Clearly, the optimal linear servomechanism problem is a generalization of the optimal linear regulator problem. The solution of the optimal servomechanism problem will be determined using the same technique which was applied for the solution of the optimal regulator problem. Thus, we start with the Hamiltonian, defined as follows:

$$H(\mathbf{x}, \mathbf{u}, \lambda, t) = \tfrac{1}{2}\mathbf{e}^{\mathrm{T}}(t)\mathbf{Q}(t)\mathbf{e}(t) + \tfrac{1}{2}\mathbf{u}^{\mathrm{T}}(t)\mathbf{R}(t)\mathbf{u}(t) + \lambda^{\mathrm{T}}(t)[\mathbf{A}(t)\mathbf{x}(t) + \mathbf{B}(t)\mathbf{u}(t)]$$

$$\tag{11.4-3}$$

The canonical equations are

$$\frac{\partial H}{\partial \mathbf{x}} = -\dot{\lambda}(t) = \mathbf{C}^{\mathrm{T}}(t)\mathbf{Q}(t)[\mathbf{C}(t)\mathbf{x}(t) - \xi(t)] + \mathbf{A}^{\mathrm{T}}(t)\lambda(t) \tag{11.4-4a}$$

$$\frac{\partial H}{\partial \mathbf{u}} = 0 = \mathbf{R}(t)\mathbf{u}(t) + \mathbf{B}^{\mathrm{T}}(t)\lambda(t) \tag{11.4-4b}$$

$$\frac{\partial H}{\partial \lambda} = \mathbf{A}(t)\mathbf{x}(t) + \mathbf{B}(t)\mathbf{u}(t) \tag{11.4-4c}$$

and the final condition

$$\lambda(t_f) = \frac{\partial}{\partial \mathbf{x}}\left[\frac{1}{2}\mathbf{e}^{\mathrm{T}}(t_f)\mathbf{S}\mathbf{e}(t_f)\right] = \mathbf{C}^{\mathrm{T}}(t_f)\mathbf{S}[\mathbf{C}(t_f)\mathbf{x}(t_f) - \xi(t_f)] \tag{11.4-4d}$$

where use was made of the following vector and matrix properties:

$$\frac{\partial}{\partial \mathbf{x}}\left[\mathbf{q}^{\mathrm{T}}(t)\mathbf{x}(t)\right] = \mathbf{q}(t) \quad \text{and} \quad \frac{1}{2}\frac{\partial}{\partial \mathbf{x}}\left[\mathbf{z}^{\mathrm{T}}(t)\mathbf{E}(t)\mathbf{z}(t)\right] = \left[\frac{\partial \mathbf{z}^{\mathrm{T}}(t)}{\partial \mathbf{x}}\right]\mathbf{E}(t)\mathbf{z}(t)$$

where $\mathbf{E}(t)$ is a symmetric matrix. From relation (11.4-4b) and provided that $|\mathbf{R}(t)| \neq 0$ for $t \in [t_0, t_{\mathrm{f}}]$, we have

$$\mathbf{u}(t) = -\mathbf{R}^{-1}(t)\mathbf{B}^{\mathrm{T}}(t)\boldsymbol{\lambda}(t) \tag{11.4-5}$$

Assume that $\boldsymbol{\lambda}(t)$ has the form

$$\boldsymbol{\lambda}(t) = \mathbf{P}(t)\mathbf{x}(t) - \boldsymbol{\mu}(t) \tag{11.4-6}$$

Then, the control vector (11.4-5) becomes

$$\mathbf{u}(t) = -\mathbf{R}^{-1}(t)\mathbf{B}(t)[\mathbf{P}(t)\mathbf{x}(t) - \boldsymbol{\mu}(t)] \tag{11.4-7}$$

Finally

$$\mathbf{u}(t) = \mathbf{K}(t)\mathbf{x}(t) + \boldsymbol{\rho}(t)$$

where

$$\mathbf{K}(t) = -\mathbf{R}^{-1}(t)\mathbf{B}^{\mathrm{T}}(t)\mathbf{P}(t) \quad \text{and} \quad \boldsymbol{\rho}(t) = \mathbf{R}^{-1}(t)\mathbf{B}^{\mathrm{T}}(t)\boldsymbol{\mu}(t)$$

If we substitute Eq. (11.4-5) in Eq. (11.4-1), we obtain

$$\dot{\mathbf{x}}(t) = \mathbf{A}(t)\mathbf{x}(t) - \mathbf{B}(t)\mathbf{R}^{-1}(t)\mathbf{B}^{\mathrm{T}}(t)\mathbf{P}(t)\mathbf{x}(t) + \mathbf{B}(t)\mathbf{R}^{-1}(t)\mathbf{B}^{\mathrm{T}}(t)\boldsymbol{\mu}(t) \tag{11.4-8a}$$

$$\mathbf{y}(t) = \mathbf{C}(t)\mathbf{x}(t) \tag{11.4-8b}$$

Differentiate Eq. (11.4-6) to yield

$$\begin{aligned}
\dot{\boldsymbol{\lambda}}(t) &= \dot{\mathbf{P}}(t)\mathbf{x}(t) + \mathbf{P}(t)\dot{\mathbf{x}}(t) - \dot{\boldsymbol{\mu}}(t) \\
&= -\mathbf{C}^{\mathrm{T}}(t)\mathbf{Q}(t)[\mathbf{C}(t)\mathbf{x}(t) - \boldsymbol{\xi}(t)] - \mathbf{A}^{\mathrm{T}}(t)\boldsymbol{\lambda}(t) \\
&= -\left[\mathbf{C}^{\mathrm{T}}(t)\mathbf{Q}(t)\mathbf{C}(t) + \mathbf{A}^{\mathrm{T}}(t)\mathbf{P}(t)\right]\mathbf{x}(t) + \mathbf{C}^{\mathrm{T}}(t)\mathbf{Q}(t)\boldsymbol{\xi}(t) + \mathbf{A}^{\mathrm{T}}(t)\boldsymbol{\mu}(t) \quad (11.4\text{-}9)
\end{aligned}$$

where use was made of relations (11.4-4a) and (11.4-6). Substituting Eq. (11.4-8a) in Eq. (11.4-9) gives

$$\left[\dot{\mathbf{P}}(t) + \mathbf{P}(t)\mathbf{A}(t) - \mathbf{P}(t)\mathbf{B}(t)\mathbf{R}^{-1}(t)\mathbf{B}^{\mathrm{T}}(t)\mathbf{P}(t) + \mathbf{C}^{\mathrm{T}}(t)\mathbf{Q}(t)\mathbf{C}(t) + \mathbf{A}^{\mathrm{T}}(t)\mathbf{P}(t)\right]\mathbf{x}(t)$$

$$+ \left[-\dot{\boldsymbol{\mu}}(t) - \left[\mathbf{A}(t) - \mathbf{B}(t)\mathbf{R}^{-1}(t)\mathbf{B}^{\mathrm{T}}(t)\mathbf{P}(t)\right]^{\mathrm{T}}\boldsymbol{\mu}(t) - \mathbf{C}^{\mathrm{T}}(t)\mathbf{Q}(t)\boldsymbol{\xi}(t)\right] = 0$$

$$\tag{11.4-10}$$

For Eq. (11.4-10) to be valid, the coefficient of $\mathbf{x}(t)$ and the second term in Eq. (11.4-10) must simultaneously be equal to zero. This reduces Eq. (11.4-10) to the following two differential equations, together with their corresponding final conditions:

$$\dot{\mathbf{P}}(t) + \mathbf{P}(t)\mathbf{A}(t) + \mathbf{A}^{\mathrm{T}}(t)\mathbf{P}(t) - \mathbf{P}(t)\mathbf{B}(t)\mathbf{R}^{-1}(t)\mathbf{B}^{\mathrm{T}}(t)\mathbf{P}(t) = -\mathbf{C}^{\mathrm{T}}(t)\mathbf{Q}(t)\mathbf{C}(t)$$

$$\tag{11.4-11a}$$

with final condition

$$\mathbf{P}(t_{\mathrm{f}}) = \mathbf{C}^{\mathrm{T}}(t_{\mathrm{f}})\mathbf{S}\mathbf{C}(t_{\mathrm{f}}) \tag{11.4-11b}$$

and

$$\dot{\boldsymbol{\mu}}(t) + \left[\mathbf{A}(t) - \mathbf{B}(t)\mathbf{R}^{-1}(t)\mathbf{B}^{\mathrm{T}}(t)\mathbf{P}(t)\right]^{\mathrm{T}}\boldsymbol{\mu}(t) = -\mathbf{C}^{\mathrm{T}}(t)\mathbf{Q}(t)\boldsymbol{\xi}(t) \tag{11.4-12a}$$

with final condition

$$\mu(t_f) = \mathbf{C}^T(t_f)\mathbf{S}\xi(t_f) \tag{11.4-12b}$$

Consequently, we note that for the determination of the optimal control law (11.4-7) for the linear servomechanism problem it is required to solve two matrix differential equations: the Riccati equation (11.4-11), which is in fact essentially the same as the Riccati equation (11.3-11) of the linear regulator problem; and Eq. (11.4-12), which provides the part of the control vector $\mathbf{u}(t)$ which depends on the desired output $\xi(t)$ of the closed-loop system. If $\xi(t) = \mathbf{0}$ and $\mathbf{C}(t) = \mathbf{I}$, then Eq. (11.4-12) yields $\mu(t) = \mathbf{0}$ and Eq. (11.4-11) becomes identical to Eq. (11.3-11). This means that in this case the linear servomechanism problem reduces to the linear regulator problem.

Remark 11.4.1

Equation (11.4-8a), together with the canonical equation (11.4-4a), may be written as follows:

$$\begin{bmatrix} \dot{\mathbf{x}}(t) \\ \dot{\lambda}(t) \end{bmatrix} = \begin{bmatrix} \mathbf{A}(t) & -\mathbf{B}(t)\mathbf{R}^{-1}(t)\mathbf{B}^T(t) \\ -\mathbf{C}^T(t)\mathbf{Q}(t)\mathbf{C}(t) & -\mathbf{A}^T(t) \end{bmatrix} \begin{bmatrix} \mathbf{x}(t) \\ \lambda(t) \end{bmatrix} + \begin{bmatrix} \mathbf{0} \\ \mathbf{C}^T(t)\mathbf{Q}(t)\xi(t) \end{bmatrix} \tag{11.4-13}$$

If we set $\mathbf{C}(t) = \mathbf{I}$ and $\xi(t) = \mathbf{0}$ in Eq. (11.4-13), we obtain Eq. (11.4-17). The solution of Eq. (11.4-13) is given by

$$\begin{bmatrix} \mathbf{x}(t_f) \\ \lambda(t_f) \end{bmatrix} = \phi(t_f, t)\begin{bmatrix} \mathbf{x}(t) \\ \lambda(t) \end{bmatrix} + \int_t^{t_f} \phi(t_f, \sigma)\begin{bmatrix} \mathbf{0} \\ \mathbf{C}^T(\sigma)\mathbf{Q}(\sigma)\xi(\sigma) \end{bmatrix} d\sigma \tag{11.4-14}$$

Partition the matrix $\phi(t_f, t)$, as in the case of $\phi(t_f, t)$ of the linear regulator problem in relation (11.3-19), and define

$$\int_t^{t_f} \phi(t_f, \sigma)\begin{bmatrix} \mathbf{0} \\ \mathbf{C}^T(\sigma)\mathbf{Q}(\sigma)\xi(\sigma) \end{bmatrix} d\sigma = \begin{bmatrix} \mathbf{f}_1(t) \\ \mathbf{f}_2(t) \end{bmatrix}$$

Then Eq. (11.4-14) may be rewritten as follows:

$$\mathbf{x}(t_f) = \phi_{11}(t_f, t)\mathbf{x}(t) + \phi_{12}(t_f, t)\lambda(t) + \mathbf{f}_1(t) \tag{11.4-15a}$$

$$\lambda(t_f) = \phi_{21}(t_f, t)\mathbf{x}(t) + \phi_{22}(t_f, t)\lambda(t) + \mathbf{f}_2(t) \tag{11.4-15b}$$

with boundary condition

$$\lambda(t_f) = \mathbf{C}^T(t_f)\mathbf{S}[\mathbf{C}(t_f)\mathbf{x}(t_f) - \xi(t_f)] \tag{11.4-16}$$

Substituting Eq. (11.4-16) in Eq. (11.4-15b) and $\mathbf{x}(t_f)$ from Eq. (11.4-15a) in Eq. (11.4-15b) and finally solving for $\lambda(t)$, we obtain

$$\lambda(t) = \mathbf{P}(t)\mathbf{x}(t) - \mu(t) \tag{11.4-17}$$

where

$$\mathbf{P}(t) = \left[\phi_{22}(t_f, t) - \mathbf{C}^T(t_f)\mathbf{S}\mathbf{C}(t_f)\phi_{12}(t_f, t)\right]^{-1}\left[\mathbf{C}^T(t_f)\mathbf{S}\mathbf{C}(t_f)\phi_{11}(t_f, t) - \phi_{21}(t_f, t)\right] \tag{11.4-18}$$

and

$$\boldsymbol{\mu}(t) = -\left[\boldsymbol{\phi}_{22}(t_f, t) - \mathbf{C}^T(t_f)\mathbf{S}\boldsymbol{\phi}_{12}(t_f, t)\right]^{-1}\left[\mathbf{C}^T(t_f)\mathbf{S}\mathbf{C}(t_f)\mathbf{f}_1(t)\right.$$
$$\left. - \mathbf{C}^T(t_f)\mathbf{S}\boldsymbol{\xi}(t_f) - \mathbf{f}_2(t)\right] \tag{11.4-19}$$

Finally

$$\mathbf{u}(t) = \mathbf{K}(t)\mathbf{x}(t) + \boldsymbol{\rho}(t)$$

where

$$\mathbf{K}(t) = -\mathbf{R}^{-1}(t)\mathbf{B}^T(t)\mathbf{P}(t) \qquad \text{and} \qquad \boldsymbol{\rho}(t) = \mathbf{R}^{-1}(t)\mathbf{B}^T(t)\boldsymbol{\mu}(t)$$

Figures 11.5 and 11.6 give an overview of the optimal servomechanism problem. Comparing these figures with Figures 11.3 and 11.4 of the optimal regulator problem we note that here an additional input vector $\boldsymbol{\rho}(t)$ is present. The vector $\boldsymbol{\rho}(t)$ is due to $\boldsymbol{\xi}(t)$, which can be viewed as a reference vector.

Remark 11.4.2

Consider the special case where the system under control is time invariant, the weighting matrix $\mathbf{S} = \mathbf{0}$, and $t_f \rightarrow +\infty$. Then, similar results to those of Subsec. 11.3.3 can be obtained for the optimal linear servomechanism [2].

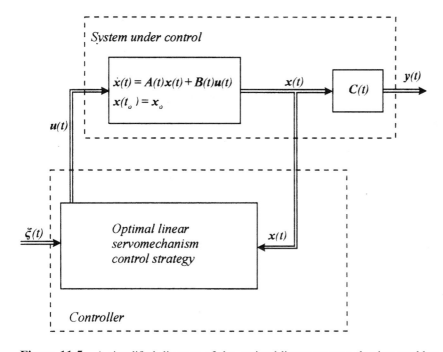

Figure 11.5 A simplified diagram of the optimal serovmechanism problem.

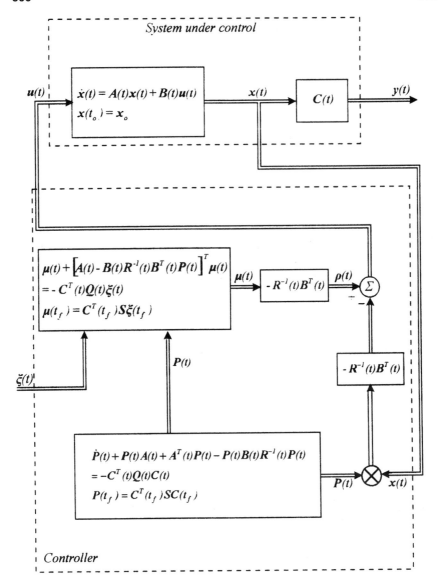

Figure 11.6 Block diagram of the optimal linear servomechanism problem.

Example 11.4.1

Consider the system

$$\begin{bmatrix} \dot{x}_1(t) \\ \dot{x}_2(t) \end{bmatrix} = \begin{bmatrix} 0 & 1 \\ 0 & 0 \end{bmatrix} \begin{bmatrix} x_1(t) \\ x_2(t) \end{bmatrix} + \begin{bmatrix} 0 \\ 1 \end{bmatrix} u(t)$$

$$\begin{bmatrix} y_1(t) \\ y_2(t) \end{bmatrix} = \begin{bmatrix} 1 & 0 \\ 0 & 1 \end{bmatrix} \begin{bmatrix} x_1(t) \\ x_2(t) \end{bmatrix}$$

with cost function

$$J = \frac{1}{2} \int_0^{t_f} \left[[\mathbf{x}(t) - \boldsymbol{\xi}(t)]^T \mathbf{Q}[\mathbf{x}(t) - \boldsymbol{\xi}(t)] + u(t)Ru(t) \right] dt$$

$$= \frac{1}{2} \int_0^{t_f} \left[[x_1(t) - \xi_1(t)]^2 + u^2(t) \right] dt$$

where

$$\mathbf{Q} = \begin{bmatrix} 1 & 0 \\ 0 & 0 \end{bmatrix} \quad \text{and} \quad R = 1$$

Find the optimal $u(t)$ as a function of $\mathbf{x}(t)$.

Solution

The Riccati equation is given by

$$\dot{\mathbf{P}} + \mathbf{P}\mathbf{A} + \mathbf{A}^T\mathbf{P} - \mathbf{P}\mathbf{B}\mathbf{R}^{-1}\mathbf{B}^T\mathbf{P} = -\mathbf{C}^T\mathbf{Q}\mathbf{C}$$

or

$$\begin{bmatrix} \dot{p}_{11} & \dot{p}_{12} \\ \dot{p}_{21} & \dot{p}_{22} \end{bmatrix} + \begin{bmatrix} 0 & p_{11} \\ 0 & p_{21} \end{bmatrix} + \begin{bmatrix} 0 & 0 \\ p_{11} & p_{12} \end{bmatrix} - \begin{bmatrix} p_{12}p_{21} & p_{12}p_{22} \\ p_{21}p_{22} & p_{22}^2 \end{bmatrix} = - \begin{bmatrix} 1 & 0 \\ 0 & 0 \end{bmatrix}$$

Since matrix $\mathbf{P}$ is symmetric, namely $p_{12} = p_{21}$, the above equation reduces to the following three nonlinear differential equations:

$$\dot{p}_{11} - p_{12}^2 = -1$$

$$\dot{p}_{12} + p_{11} - p_{12}p_{22} = 0$$

$$\dot{p}_{22} + 2p_{12} - p_{22}^2 = 0$$

with boundary conditions

$$\mathbf{P}(t_f) = \mathbf{C}^T\mathbf{S}\mathbf{C} = \mathbf{0}, \quad \text{since } \mathbf{S} = \mathbf{0}$$

To facilitate the solution of the Riccati equation, we assume that $t_f \to +\infty$. This results in the following algebraic system of nonlinear equations:

$$p_{12}^2 = 1, \quad p_{11} - p_{12}p_{22} = 0, \quad 2p_{12} - p_{22}^2 = 0$$

from which we obtain that $p_{12} = 1$, $p_{22} = \sqrt{2}$, and $p_{11} = \sqrt{2}$. Consequently,

$$\mathbf{P} = \begin{bmatrix} \sqrt{2} & 1 \\ 1 & \sqrt{2} \end{bmatrix}$$

In the sequel, we determine the vector $\boldsymbol{\mu}(t)$. We have

$$\mathbf{A} - \mathbf{B}\mathbf{R}^{-1}\mathbf{B}^T\mathbf{P} = \begin{bmatrix} 0 & 1 \\ -1 & -\sqrt{2} \end{bmatrix}$$

Hence, Eq. (11.4-12a) becomes

$$\begin{bmatrix} \dot{\mu}_1(t) \\ \dot{\mu}_2(t) \end{bmatrix} + \begin{bmatrix} 0 & -1 \\ 1 & -\sqrt{2} \end{bmatrix} \begin{bmatrix} \mu_1(t) \\ \mu_2(t) \end{bmatrix} = - \begin{bmatrix} \xi_1(t) \\ 0 \end{bmatrix}$$

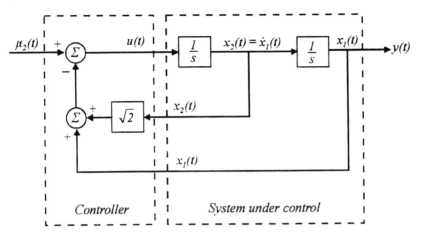

Figure 11.7 Block diagram of the closed-loop system.

When $\xi_1(t) = \gamma = $ constant, then for $t_f \to +\infty$ we further assume that $\dot{\mu}_1(t) = \dot{\mu}_2(t) = 0$, and thus we have $\mu_2 = 0.707\mu_1 = \gamma$. This solution is finally substituted in Eq. (11.4-7) to obtain the optimal control signal

$$u = -x_1 - \sqrt{2}x_2 + \mu_2$$

Figure 11.7 shows the block diagram of the closed-loop system.

11.5 PROBLEMS

1. Find the optimal vector $x(t)$ which minimizes the cost function

$$J = \int_{t_0}^{t_f} \varphi(\mathbf{x}, \dot{\mathbf{x}}, t)\, dt = \int_0^{\pi/4} \left(x_1^2 + 4x_2^2 + \dot{x}_1 \dot{x}_2 \right) dt$$

where $\mathbf{x}^T = [x_1 \quad x_2]$, and where

$$\mathbf{x}(0) = \begin{bmatrix} x_1(0) \\ x_2(0) \end{bmatrix} = \begin{bmatrix} -1 \\ 1 \end{bmatrix} \quad \text{and} \quad \mathbf{x}(\pi/4) = \begin{bmatrix} x_1(\pi/4) \\ x_2(\pi/4) \end{bmatrix} = \begin{bmatrix} 1 \\ 0 \end{bmatrix}$$

2. Find the optimal vector $x(t)$ which minimizes the cost function

$$J = \int_{t_0}^{t_f} \varphi(\mathbf{x}, \dot{\mathbf{x}}, t)\, dt = \int_0^{\pi/4} \left(x_1^2 + \dot{x}_1 \dot{x}_2 + \dot{x}_2^2 \right) dt$$

where $\mathbf{x}^T = [x_1 \quad x_2]$, and where

$$\mathbf{x}(0) = \begin{bmatrix} x_1(0) \\ x_2(0) \end{bmatrix} = \begin{bmatrix} 1 \\ 3/2 \end{bmatrix} \quad \text{and}$$

$$\mathbf{x}(\pi/4) = \begin{bmatrix} x_1(\pi/4) \\ x_2(\pi/4) \end{bmatrix} = \begin{bmatrix} 2 \\ \text{unspecified} \end{bmatrix}$$

3. Consider the system

$$\dot{\mathbf{x}}(t) = \mathbf{A}\mathbf{x}(t) + \mathbf{B}\mathbf{u}(t), \qquad \mathbf{x}(0) = \mathbf{x}_0$$

$$y(t) = Cx(t) + Du(t)$$

where

$$A = \begin{bmatrix} 0 & 1 \\ -1 & 0 \end{bmatrix}, \qquad B = \begin{bmatrix} 1 & 1 & 1 \\ 1 & 1 & -1 \end{bmatrix},$$

$$C = \begin{bmatrix} 1/2 & 1/2 \\ 1/4 & -1/4 \end{bmatrix}, \qquad D = \begin{bmatrix} 0 & 0 & 0 \\ 0 & 0 & 1 \end{bmatrix}$$

Find the optimal input vector $u(t)$ so that the output vector $y(t)$ belongs to the ellipse

$$\tfrac{1}{4} y_1^2(t) + y_2^2(t) = 1$$

or equivalently

$$y_1(t) = 2 \cos t \qquad \text{and} \qquad y_2(t) = \sin t$$

while the following cost function is minimized:

$$J = \int_0^{\pi/2} \left[\dot{u}^T(t)\dot{u}(t) + u^T(t)u(t) \right] dt$$

with

$$u(0) = \begin{bmatrix} 1 \\ -1 \\ 0 \end{bmatrix} \qquad \text{and} \qquad u(\pi/2) = \begin{bmatrix} 1 \\ -1 \\ 0 \end{bmatrix}$$

4. Find the optimal input $u(t)$ which minimizes the cost function

$$J = \frac{1}{2} \int_{t_0}^{t_f} \left\{ [x(t) - \xi(t)]^T Q(t)[x(t) - \xi(t)] + u^T(t)R(t)u(t) \right\} dt$$

for the linear time-varying system

$$\dot{x}(t) = A(t)x(t) + B(t)u(t), \qquad x(0) = x_0$$

where $\xi(t)$ is a predetermined desired state trajectory.

5. Consider the network of Figure 11.8. The capacitor's initial voltage is x_0. At $t = 0$, the switches S_1 and S_2 are closed. Find the optimal input $u(t)$ which minimizes the following cost function:

$$J = \int_0^1 \left[x^2(t) + \frac{1}{5} u^2(t) \right] dt$$

Let $RC = 1$ and consider both cases: the open- and the closed-loop system.

6. Consider the system

$$\dot{x}(t) = x(t) + u(t), \qquad x(0) = x_0$$

Find the optimal input $u(t)$ which minimizes the cost function

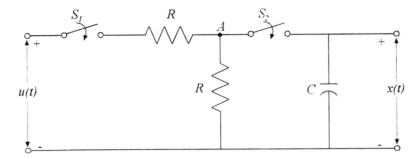

Figure 11.8

$$J = x^2(1) + \int_0^1 \left[x^2(t) + \frac{1}{8} u^2(t) \right] dt$$

for both the open- as we well as for the closed-loop system case.

7. Consider the system

$$\dot{x}(t) = ax(t) + u(t), \qquad x(t_0) = x_0$$

and the cost function

$$J = \frac{1}{2} s x^2(t_f) + \frac{1}{2} \int_{t_0}^{t_f} \left[q x^2(t) + r u^2(t) \right] dt$$

Find the optimal $u(t)$ as a function of $x(t)$.

8. Consider the system

$$\begin{bmatrix} \dot{x}_1(t) \\ \dot{x}_2(t) \end{bmatrix} = \begin{bmatrix} 0 & 1 \\ -\omega^2 & 0 \end{bmatrix} \begin{bmatrix} x_1(t) \\ x_2(t) \end{bmatrix} + \begin{bmatrix} 0 \\ 1 \end{bmatrix} u(t), \qquad \mathbf{x}(t_0) = \begin{bmatrix} x_2(t_0) \\ x_2(t_0) \end{bmatrix}$$

and the cost function

$$J = \frac{1}{2} s x_1^2(t_f) + \frac{1}{2} \int_{t_0}^{t_f} u^2(t) \, dt$$

Find the optimal $u(t)$ as a function of $x(t)$.

9. Consider the controllable and observable system

$$\dot{\mathbf{x}}(t) = \mathbf{A}(t)\mathbf{x}(t) + \mathbf{B}(t)\mathbf{u}(t)$$

$$\mathbf{y}(t) = \mathbf{C}(t)\mathbf{x}(t)$$

The response of the closed-loop system is desired to follow the response of an ideal model described by the vector differential equation

$$\dot{\mathbf{w}}(t) = \mathbf{L}(t)\mathbf{w}(t)$$

Find the optimal control vector $\mathbf{u}(t)$, as a function of $x(t)$, which minimizes the cost function

$$J = \frac{1}{2} \int_{t_0}^{t_f} \left\{ [\dot{\mathbf{y}}(t) - \mathbf{L}(t)\mathbf{y}(t)]^T \mathbf{Q}(t)[\dot{\mathbf{y}}(t) - \mathbf{L}(t)\mathbf{y}(t)] + \mathbf{u}^T(t)\mathbf{R}(t)\mathbf{u}(t) \right\} dt$$

where $\mathbf{Q}(t)$ is symmetrical, positive semidefinite and $\mathbf{R}(t)$ is symmetrical positive definite.

10. The yaw motion of a tanker is described in state space by the equations [21]

$$\dot{\mathbf{x}}(t) = \mathbf{A}\mathbf{x}(t) + \mathbf{B}\mathbf{u}(t)$$

$$\mathbf{y}(t) = \mathbf{C}\mathbf{x}(t)$$

where

$$\mathbf{x}^T(t) = [v(t) \quad r(t) \quad \psi(t)], \qquad u(t) = \delta(t), \quad \mathbf{y}^T(t) = [\psi(t) \quad v(t)],$$

$$\mathbf{A} = \begin{bmatrix} -0.44 & -0.28 & 0 \\ -2.67 & -2.04 & 0 \\ 0 & 1 & 0 \end{bmatrix}, \qquad \mathbf{B} = \begin{bmatrix} 0.07 \\ -0.53 \\ 0 \end{bmatrix}, \qquad \mathbf{C} = \begin{bmatrix} 0 & 0 & 1 \\ 1 & 0 & 0 \end{bmatrix}$$

In particular (see Figure 11.9),

$v(t) =$ the y-component of the tanker velocity
$r(t) =$ the tanker velocity
$\psi(t) =$ the axial inclination of the tanker relative to the given frame of reference
$\delta(t) =$ the rudder orientation with respect to the axial direction.

Find a state feedback control law of the form

$$\mathbf{u}(t) = \mathbf{k}^T \mathbf{x}(t)$$

so that the following cost function is minimized:

$$J = \int_0^\infty \left[\mathbf{x}^T(t)\mathbf{Q}\mathbf{x}(t) + ru^2(t) \right] dt, \qquad \text{where} \qquad \mathbf{Q} = \begin{bmatrix} 1 & 0 & 0 \\ 0 & 2 & 1 \\ 0 & 1 & 2 \end{bmatrix}$$

and $r = 3$.

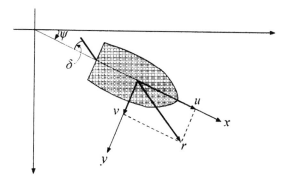

Figure 11.9

11. Consider a linear time-invariant system described by the differential equation

$$y^{(2)}(t) + 4y^{(1)}(t) + 3y(t) = u(t)$$

with zero initial conditions.

(a) Find the optimal solution $u(t) = f[y(t), \dot{y}(t), t]$ of the linear serovmechanism problem which minimizes the cost function

$$J = \int_{t_0}^{t_f} \left\{ [\dot{y}(t) - 2]^2 + [y(t) - 1]^2 + 3u^2(t) \right\} dt$$

with t_f given. Do not solve the differential equations derived in the solution procedure but put them in their simplest form and find the necessary boundary conditions for their solution.

(b) Consider the same cost function as in (a), and determine the analytical expression of the optimal input $u(t)$, for $t_f \to \infty$.

12. The simplified block diagram of a position control servomechanism is shown in Figure 11.10. Let $\beta = K/A$ and $\gamma = B/A$. Then $G(s) = \beta/s(s + \gamma)$ and the differential equation of the open system is the following: $\theta_y^{(2)}(t) + \gamma\theta_y^{(1)}(t) = \beta\theta_r(t)$. Assume that $\theta_r(t) = 1$, and that the desired output of the closed-loop system is $\xi(t) = 1$. Choose the following cost function:

$$J = \frac{1}{2} \int_0^{t_f} \left\{ q[\theta_y(t) - \xi(t)]^2 + r\theta_r^2(t) \right\} dt, \qquad \text{with} \qquad \xi(t) = 1$$

(a) Find the optimal $\theta_r(t)$ which minimizes the cost function. Do not solve the differential equations derived in the solution procedure but put them in their simplest form and find the necessary boundary conditions for their solution.

(b) Find the optimal $\theta_r(t)$, for $t_f \to \infty$. (Let $\gamma = 1$, $\beta = 2$, $q = 1$, and $r = 1$).

13. For the yaw motion control system of the tanker of Problem 10, find the optimal control input $u(t)$, which minimizes the cost function

$$J = \frac{1}{2} \int_0^\infty \left[e_1^2(t) + 0.2e_2^2(t) + u^2(t) \right] dt$$

where

$$e_1(t) = \xi_1(t) - y_1(t) = \xi_1(t) - x_3(t) = \xi_1(t) - \psi(t)$$

$$e_2(t) = \xi_2(t) - y_2(t) = \xi_2(t) - x_1(t) = \xi_2(t) = v(t)$$

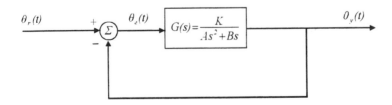

Figure 11.10

and the output of the system to follow the desired course $\xi(t)$, where

$$\xi(t) = \begin{bmatrix} \xi_1(t) \\ \xi_2(t) \end{bmatrix} = \begin{bmatrix} -2\,\text{deg} \\ 7\,\text{knots} \end{bmatrix}$$

BIBLIOGRAPHY

Books

1. BDO Anderson, JB Moore. Linear Optimal Control. Englewood Cliffs, New Jersey: Prentice Hall, 1971.
2. BDO Anderson, JB Moore. Optimal Control Quadratic Methods. Englewood Cliffs, New Jersey: Prentice Hall, 1990.
3. KJ Astrom. Introduction to Stochastic Control Theory. New York: Academic Press, 1970.
4. M Athans, PL Falb. Optimal Control. New York: McGraw-Hill, 1966.
5. AE Bryson, YC Ho. Applied Optimal Control. New York: Holsted Press, 1968.
6. AE Bryson Jr, YC Ho. Applied Optimal Control. New York: John Wiley, 1975.
7. AA Feldbaum. Optimal Control Systems. New York: Academic Press, 1965.
8. RE Kalman. The theory of optimal control and the calculus of variations. In: R Bellman, ed. Mathematical Optimization Techniques. Berkeley: University of California Press, 1963.
9. DE Kirk. Optimal Control Theory, An Introduction. Englewood Cliffs, New Jersey: Prentice Hall, 1970.
10. H Kwakernaak, R Sivan. Linear Optimal Control Systems. New York: Wiley-Interscience, 1972.
11. G Leitman. An Introduction to Optimal Control. New York: McGraw-Hill, 1966.
12. RCK Lee. Optimal Estimation, Identification and Control. Cambridge, Massachusetts: Technology Press, 1964.
13. EB Lee, L Markus. Foundations of Optimal Control Theory. New York: John Wiley, 1967.
14. FL Lewis. Optimal Control. New York: Wiley, 1986.
15. DG Luenberger. Optimization by Vector Space Methods. New York: John Wiley, 1969.
16. CW Merriam. Optimization Theory and the Design of Feedback Control Systems. New York: McGraw-Hill, 1964.
17. LS Pontryagin et al. The Mathematical Theory of Optimal Processes. New York: Interscience Publishers, 1962.
18. HA Prime. Modern Concepts in Control Theory. New York: McGraw-Hill, 1969.
19. AD Sage. Optimum Systems Control. Englewood Cliffs, New Jersey: Prentice Hall, 1968.
20. AD Sage, CC White. Optimum Systems Control. 2nd ed. Englewood Cliffs, New Jersey, 1977.

Articles

21. KJ Astrom, CG Kallstrom. Identification of ship steering dynamics. Automatica 12:9–22, 1976.

12

Digital Control

12.1 INTRODUCTION

The aim of this chapter is to introduce the reader to the modern and very promising approach of controlling systems using a computer. Our goal is to extend, as much as possible, the material covered thus far in this book for continuous-time systems to discrete-time systems. The material of this chapter is a condensed version of most of the material presented by the author in the first five chapters of his book *Digital Control Systems* [13].

12.1.1 The Basic Structure of Digital Control Systems

The basic structure of a typical digital control system or computer-controlled system or discrete-time system is shown in Figure 12.1. The system (plant or process) under control is a continuous-time system (e.g., a motor, electrical power plant, robot, etc.). The "heart" of the controller is the digital computer. The A/D converter converts a continuous-time signal into a discrete-time signal at times specified by a clock. The D/A converter, in contrast, converts the discrete-time signal output of the computer to a continuous-time signal to be fed to the plant. The D/A converter normally includes a hold circuit (for more on A/D and D/A converters see Subsecs 12.3.1 and 12.3.2). The quantizer Q converts a discrete-time signal to binary digits.

The controller may be designed to satisfy simple, as well as complex, specifications. For this reason, it may operate as a simple logic device as in programmable logic controllers (PLCs), or make dynamic and complicated processing operations on the error $e(kT)$ to produce a suitable input $u(t)$ to control the plant. This control input $u(t)$ to the plant must be such that the behavior of the closed-loop system (i.e., the output $y(t)$) satisfies desired specifications.

The problem of realizing a digital controller is mainly one of developing a computer program. Digital controllers present significant advantages over classical analog controllers. Some of these advantages are as follows:

1. Digital controllers have greater flexibility in modifying the controller's features. Indeed, the controller's features may be readily programmed.

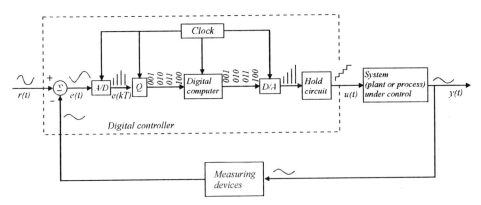

Figure 12.1 Simplified block diagram of a typical closed-loop digital control or computer-controlled system.

For classical analog controllers, any change in the characteristics of the controller is usually laborious and expensive, since it reqires changes in the structure and/or the elements of the controller.

2. Processing of data is simple. Complex computations may be performed in a fast and convenient way. Analog controllers do not have this characteristic.

3. Digital controllers are superior over analog controllers with regard to the following characteristics:
    a. Sensitivity
    b. Drift effects
    c. Internal noise
    d. Reliability

4. Digital controllers are cheaper than analog controllers.

5. Digital controllers are considerably smaller in size than analog controllers.

Nevertheless, digital controllers have certain disadvantages compared with analog controllers. The most significant disadvantage is due to the error introduced during sampling of the analog signals, as well as during the quantization of the discrete-time signals.

12.1.2 Mathematical Background

The mathematical background necessary for the study of digital control systems is the Z-transform, presented in Appendix B. The Z-transform facilitates the study and design of discrete-time control systems in an analogous way to that Laplace transform does for the continuous-time control systems. For this reason, we strongly advise that the reader becomes familiar with the material presented in Appendix B.

12.2 DESCRIPTION AND ANALYSIS OF DISCRETE-TIME SYSTEMS

The term *discrete-time systems* covers systems which operate directly with discrete-time signals. In this case, the input, as well as the output, of the system is obviously a discrete-time signal (Figure 12.2). A well-known discrete-time system is the digital

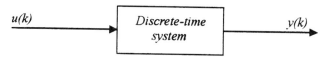

Figure 12.2   Block diagram of a discrete-time system.

computer. In this case the signals $u(k)$ and $y(k)$ are number sequences (usually 0 and 1). These types of systems, as we shall see, are described by difference equations.

The term *sampled-data systems* [1, 8] covers the usual analog (continuous-time) systems, having the following distinct characteristics: the input $u(t)$ and the output $y(t)$ are piecewise constant signals, i.e., they are constant over each interval between two consecutive sampling points (Figure 12.3). The piecewise constant signal $u(t)$ is derived from the discrete-time signal $v(kT)$ using a hold circuit (see Subsec. 12.3.2). The output $s(t)$ of the system is a continuous-time function. Let $y(t)$ be the output of the system having a piecewise constant form with $y(t) = s(t)$ at the sampling points. Then, the system having $u(t)$ as input and $y(t)$ as output, where both signals are piecewise constant in each interval, is a sampled-date system and may be described by difference equations, as shown in Subsecs 12.3.3 and 12.3.4. This means that sampled-data systems may be described and subsequently studied similarly to discrete-time systems. This fact is of particular importance since it *unifies* the study of hybrid systems, which consist of continuous-time and discrete-time subsystems (the computer-controlled system shown in Figure 12.1 is a hybrid system) using a common mathematical tool, namely the difference equations. For this reason, and for reasons of simplicity, sampled-data systems are usually addressed in the literature (and in this book) as *discrete-time systems*. It is noted that *sampled-data systems* are also called *discretized systems*.

12.2.1   Properties of Discrete-Time Systems

From a mathematical point of view, discrete-time system description implies the determination of a law which assigns an output sequence $y(k)$ to a given input sequence $u(k)$ (Figure 12.4). The specific law connecting the input and output sequences $u(k)$ and $y(k)$ constitutes the mathematical model of the discrete-time system. Symbolically, this relation can be written as follows:

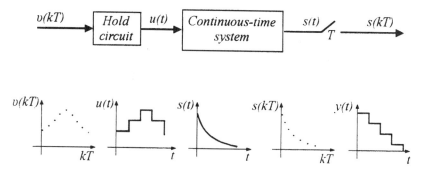

Figure 12.3   Block diagram of a sampled-data system.

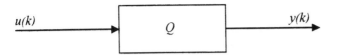

Figure 12.4 Block diagram of a discrete-time system.

$$y(k) = Q[u(k)]$$

where Q is a discrete operator.

Discrete-time systems have a number of properties, some of which are of special interest and are presented below.

1 Linearity

A discrete-time system is linear if the following relation holds true:

$$Q[c_1 u_1(k) + c_2 u_2(k)] = c_1 Q[u_1(k)] + c_2 Q[u_2(k)] = c_1 y_1(k) + c_2 y_2(k) \qquad (12.2\text{-}1)$$

for every c_1, c_2, $u_1(k)$, and $u_2(k)$, where c_1, c_2 are constants and $y_1(k) = Q[u_1(k)]$ is the output of the system with input $u_1(k)$ and $y_2(k) = Q[u_2(k)]$ is the output of the system with input $u_2(k)$.

2 Time-Invariant System

A discrete-time system is time-invariant if the following holds true:

$$Qu(k - k_0)] = y(k - k_0) \qquad (12.2\text{-}2)$$

for every k_0. Equation (12.2-2) shows that when the input to the system is shifted by k_0 units, the output of the system is also shifted by k_0 units.

3 Causality

A discrete-time system is called causal if the output $y(k) = 0$ for $k < k_0$, when the input $u(k) = 0$ for $k < k_0$. A discrete-time signal is called causal if it is zero for $k < k_0$. Hence, a system is causal if every causal excitation produces a causal response.

12.2.2 Description of Linear Time-Invariant Discrete-Time Systems

A linear time-invariant causal discrete-time system involves the following elements: summation units, amplification units, and delay units. The block diagram of all three elements is shown in Figure 12.5. The delay unit is designated as z^{-1}, meaning that the output is identical to the input delayed by a time unit.

When these three elements are suitably interconnected, then one has a discrete-time system, as, for example, the first-order discrete-time system shown in Figure 12.6. Adding the three signals at the summation point Σ, we arrive at the equation

$$y(k) + a_1 y(k - 1) = b_0 u(k) + b_1 u(k - 1) \qquad (12.1\text{-}3a)$$

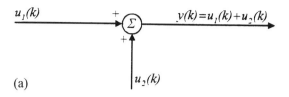

(a)

(b)

(c)

Figure 12.5   (a) Summation, (b) amplification, and (c) delay units.

Similarly, for the second-order discrete-time system shown in Figure 12.7, we obtain the equation

$$y(k) + a_1 y(k-1) + a_2 y(k-2) = b_0 u(k) + b_1 u(k-1) + b_2 u(k-2) \qquad (12.2\text{-}3b)$$

Obviously, Eqs (12.2-3a and b) are mathematical models describing the discrete-time systems shown in Figures 12.6 and 12.7, respectively. Equations (12.2-3a and b) are examples of difference equations.

There are many ways to describe discrete-time systems, as is also the case for continuous-time systems. The most popular ones are the following: the difference equations, as in Eqs (12.2-3a and b); the transfer function; the impulse response or weight function; and the state-space equations.

In presenting these four methods, certain similarities and dissimilarities between continuous-time and discrete-time systems will be revealed. There are

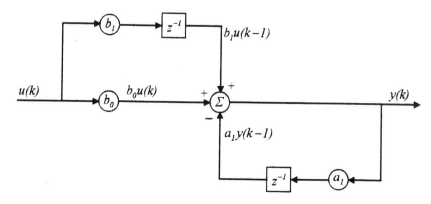

Figure 12.6   Block diagram of a first-order discrete-time system.

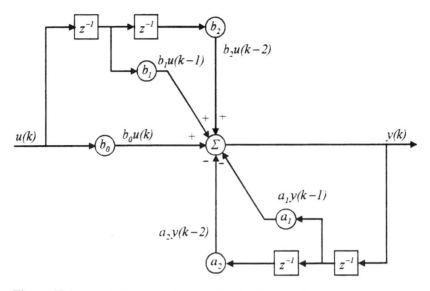

Figure 12.7   Block diagram of a second-order discrete-time system.

three basic differences, going from continuous-time to discrete-time systems: differential equations are now difference equations; the Laplace transform gives way to the Z-transform (see Appendix B); and the integration procedure is replaced by summation.

1 Difference Equations

The general form of a difference equation is as follows:

$$y(k) + a_1 y(k-1) + \cdots + a_n y(k-n) = b_0 u(k) + b_1 u(k-1) + \cdots + b_m u(k-m)$$

$$(12.2\text{-}4)$$

with initial conditions $y(-1), y(-2), \ldots, y(-n)$. The solution of Eq. (12.2-4) may be found either in the time domain (using methods similar to those for solving a differential equation in the time domain) or in the complex frequency or z-domain using the Z-transform.

2 Transfer Function

The transfer function of a discrete-time system is denoted by $H(z)$ and is defined as the ratio of the Z-transform of the output $y(k)$ divided by the Z-transform of the input $u(k)$, under the condition that $u(k) = y(k) = 0$, for all negative values of k. That is

$$H(z) = \frac{Z[y(k)]}{Z[u(k)]} = \frac{Y(z)}{U(z)}, \qquad \text{where} \qquad u(k) = y(k) = 0 \qquad \text{for} \qquad k < 0$$

$$(12.2\text{-}5)$$

The transfer function of a system described by the difference equation (12.2-4), with $y(k) = u(k) = 0$, for $k < 0$, is determined as follows: multiply both sides of Eq. (12.2-4) by the term z^{-k} and add for $k = 0, 1, 2, \ldots, \infty$, to yield:

$$\sum_{k=0}^{\infty} y(k)z^{-k} + a_1 \sum_{k=0}^{\infty} y(k-1)z^{-k} + a_2 \sum_{k=0}^{\infty} y(k-2)z^{-k} + \cdots$$

$$+ a_n \sum_{k=0}^{\infty} y(k-n)z^{-k} = b_0 \sum_{k=0}^{\infty} u(k)z^{-k} + b_1 \sum_{k=0}^{\infty} u(k-1)z^{-k} + \cdots$$

$$+ b_m \sum_{k=0}^{\infty} u(k-m)z^{-k}$$

Using the Z-transform time-shifting property given by Eq. (B.3-12a) in Appendix B and the assumption that $u(k)$ and $y(k)$ are zero for negative values of k, the above equation can be simplified as follows:

$$Y(z) + a_1 z^{-1} Y(z) + \cdots + a_n z^{-n} Y(z) = b_0 u(z) + b_1 z^{-1} U(z) + \cdots + b_m z^{-m} U(z)$$

where use was made of the definition of the Z-transform (Eq. (B.3-1)) in Appendix B. Hence, using definition (12.2-5), we arrive at the following rational polynomial form for $H(z)$:

$$H(z) = \frac{Y(z)}{U(z)} = \frac{b_0 + b_1 z^{-1} + \cdots + b_m z^{-m}}{1 + a_1 z^{-1} + \cdots + a_n z^{-n}} \qquad (12.2\text{-}6)$$

3 Impulse Response or Weight Function

The impulse response (or weight function) of a system is denoted by $h(k)$ and is defined as the output of a system when its input is the unit impulse sequence $\delta(k)$ (see Figure B.1 in Appendix B), under the constraint that the initial conditions $y(-1), y(-2), \ldots, y(-n)$ of the system are zero. The block diagram definition of the impulse response is shown in Figure 12.8. The transfer function $H(z)$ and the weight function $h(k)$ are related by the following equation:

$$H(z) = Z[h(k)] \qquad (12.2\text{-}7)$$

where $Z[f(k)]$ indicates the Z-transform of $f(k)$ defined by Eq. (B.3-1) in Appendix B.

4 State-Space Equations

State-space equations or simply state equations is a set of first-order difference equations describing high-order systems and have the form

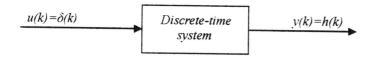

Figure 12.8 Block diagram definition of the impulse response.

$$\mathbf{x}(k+1) = \mathbf{Ax}(k) + \mathbf{Bu}(k) \tag{12.2-8a}$$
$$\mathbf{y}(k) = \mathbf{Cx}(k) + \mathbf{Du}(k) \tag{12.2-8b}$$

where $\mathbf{u}(k) \in R^m$, $\mathbf{x}(k) \in R^n$, and $\mathbf{y}(k) \in R^p$, are the input, state, and output vectors, respectively, and $\mathbf{A}$, $\mathbf{B}$, $\mathbf{C}$, and $\mathbf{D}$ are constant matrices of appropriate dimensions.

Let $\mathbf{Y}(z) = Z[\mathbf{y}(k)]$ and $\mathbf{U}(z) = Z[\mathbf{u}(k)]$. Then, the transfer function matrix $\mathbf{H}(z)$ of Eqs (12.2-8) is given by

$$\mathbf{H}(z) = \mathbf{C}(z\mathbf{I}_n - \mathbf{A})^{-1}\mathbf{B} + \mathbf{D} \tag{12.2-9}$$

The impulse response matrix $\mathbf{H}(k)$ of Eqs (12.2-8) is given by

$$\mathbf{H}(k) = Z^{-1}[\mathbf{H}(z)] = \begin{cases} \mathbf{D}, & \text{for} & k = 0 \\ \mathbf{CA}^{k-1}\mathbf{B}, & \text{for} & k > 0 \end{cases} \tag{12.2-10}$$

12.2.3 Analysis of Linear Time-Invariant Discrete-Time Systems

The problem of the analysis of linear time-invariant discrete-time systems will be treated using four different methods, where each method corresponds to one of the four description models presented in Subsec. 12.2.2.

1 Analysis Based on the Difference Equation

We present the following introductory example.

Example 12.2.1

A discrete-time system is described by the difference equation

$$y(k) = u(k) + ay(k-1)$$

with the initial condition $y(-1)$. Solve the difference equation, i.e., determine $y(k)$.

Solution

The difference equation may be solved to determine $y(k)$, using the Z-transform, as follows. Take the Z-transform of both sides of the equation to yield

$$Z[y(k)] = Z[u(k)] + aZ[y(k-1)]$$

or

$$Y(z) = U(z) + a[z^{-1}Y(z) + y(-1)]$$

and thus

$$Y(z) = \frac{z[U(z) + ay(-1)]}{z - a} = \frac{ay(-1)z}{z - a} + \frac{U(z)z}{z - a}$$

Suppose that the excitation $u(k)$ is the unit step sequence $\beta(k)$ (see Figure B.2 in Appendix B). In this case

$$U(z) = Z[\beta(k)] = \frac{z}{z - 1}$$

Then, the output $Y(z)$ becomes

$$Y(z) = \frac{ay(-1)z}{z - a} + \frac{z^2}{(z - a)(z - 1)} = \frac{ay(-1)z}{z - a} + \left[\frac{1}{1 - a}\right]\left[\frac{z}{z - 1}\right] - \left[\frac{a}{1 - a}\right]\left[\frac{z}{z - a}\right]$$

Take the inverse Z-transform (see Appendix C) to yield

$$y(k) = a^{k+1}y(-1) + \frac{1}{1-a} - \frac{1}{1-a}a^{k+1}$$

The expression for the output $y(k)$ clearly converges for $|a| < 1$. The initial condition $y(-1)$ contributes only during the transient period. The output $y(k)$, in the steady state, takes on the form

$$y_{ss}(k) = \frac{1}{1-a}$$

where $y_{ss}(k)$ denotes the steady-state value of $y(k)$. Figure 12.9 shows $y(k)$ when the initial condition $y(-1) = 0$, the input $u(k) = \beta(k)$, and $|a| < 1$.

2 Analysis Based on the Transfer Function

The input $U(z)$, the output $Y(z)$, and the transfer function $H(z)$ are related by the equation

$$Y(z) = H(z)U(z)$$

Hence

$$y(k) = Z^{-1}[Y(z)] = Z^{-1}[H(z)U(z)]$$

3 Analysis Based on the Impulse Response

The input $u(k)$, the output $y(k)$, and the impulse response $h(k)$ are related via the following convolution relation:

$$y(k) = u(k) * h(k) = \sum_{i=-\infty}^{\infty} u(i)h(k-i) \tag{12.2-11}$$

If the system is causal, i.e., if $h(k) = 0$ for $k < 0$, then relation (12.2-11) becomes

$$y(k) = \sum_{i=0}^{\infty} u(i)h(k-i) = \sum_{i=0}^{\infty} u(k-i)h(i) \tag{12.2-12}$$

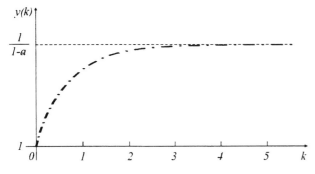

Figure 12.9 Response of system of Example 12.2.1 when $y(-1) = 0$, $u(k) = \beta(k)$, and $|a| < 1$.

If both the system and the input signal are causal, i.e., if $h(k) = 0$ and $u(k) = 0$ for $k < 0$, then Eq. (12.2-12) becomes

$$y(k) = \sum_{i=0}^{k} h(i)u(k-i) = \sum_{i=0}^{k} h(k-i)u(i) \tag{12.2-13}$$

The values $y(o), y(1), \ldots$ of the output $y(k)$ can be calculated from Eq. (12.2-13) as follows:

$$y(0) = h(0)u(0)$$
$$y(1) = h(0)u(1) + h(1)u(0)$$
$$y(2) = h(0)u(2) + h(1)u(1) + h(2)u(0)$$
$$\vdots$$
$$y(k) = h(0)u(k) + h(1)u(k-1) + \cdots + h(k)u(0)$$

or more compactly as

$$\mathbf{y} = \mathbf{Hu} = \mathbf{Uh} \tag{12.2-14}$$

where

$$\mathbf{y} = \begin{bmatrix} y(0) \\ y(1) \\ \vdots \\ y(k) \end{bmatrix}, \qquad \mathbf{u} = \begin{bmatrix} u(0) \\ u(1) \\ \vdots \\ u(k) \end{bmatrix}, \qquad \text{and} \qquad \mathbf{h} = \begin{bmatrix} h(0) \\ h(1) \\ \vdots \\ h(k) \end{bmatrix}$$

$$\mathbf{H} = \begin{bmatrix} h(0) & 0 & \cdots & 0 \\ h(1) & h(0) & \cdots & 0 \\ \vdots & \vdots & & \vdots \\ h(k) & h(k-1) & \cdots & h(0) \end{bmatrix}, \qquad \text{and}$$

$$\mathbf{U} = \begin{bmatrix} u(0) & 0 & \cdots & 0 \\ u(1) & u(0) & \cdots & 0 \\ \vdots & \vdots & & \vdots \\ u(k) & u(k-1) & \cdots & u(0) \end{bmatrix}$$

Remark 12.2.1

Equation (12.2-14) can be used for the determination of the impulse response $h(k)$ based on the input $u(k)$ and the output $y(k)$. Indeed, from Eq. (12.2-14), we have that

$$\mathbf{h} = \mathbf{U}^{-1}\mathbf{y}, \qquad \text{if} \qquad u(0) \neq 0 \tag{12.2-15}$$

The above procedure is called *deconvolution* (since it is the reverse of convolution) and constitutes a simple identification method for a discrete-time system. For more on the issue of identification see Chap. 13.

4 Analysis Based on the State Equations

Consider the state equations (12.2-8). From Eq. (12.2-8a),

$$\mathbf{x}(k+1) = \mathbf{A}\mathbf{x}(k) + \mathbf{B}\mathbf{u}(k)$$

we have that

$$\text{For } k = 0 : \mathbf{x}(1) = \mathbf{A}\mathbf{x}(0) + \mathbf{B}\mathbf{u}(0)$$

$$\text{For } k = 1 : \mathbf{x}(2) = \mathbf{A}\mathbf{x}(1) + \mathbf{B}\mathbf{u}(1) = \mathbf{A}[\mathbf{A}\mathbf{x}(0) + \mathbf{B}\mathbf{u}(0)] + \mathbf{B}\mathbf{u}(1)$$

$$= \mathbf{A}^2\mathbf{x}(0) + \mathbf{A}\mathbf{B}\mathbf{u}(0) + \mathbf{B}\mathbf{u}(1)$$

$$\text{For } k = 2 : \mathbf{x}(3) = \mathbf{A}\mathbf{x}(2) + \mathbf{B}\mathbf{u}(2) = \mathbf{A}[\mathbf{A}^2\mathbf{x}(0) + \mathbf{A}\mathbf{B}\mathbf{u}(0) + \mathbf{B}\mathbf{u}(1)] + \mathbf{B}\mathbf{u}(2)$$

$$= \mathbf{A}^3\mathbf{x}(0) + \mathbf{A}^2\mathbf{B}\mathbf{u}(0) + \mathbf{A}\mathbf{B}\mathbf{u}(1) + \mathbf{B}\mathbf{u}(2)$$

If we continue this procedure for $k = 3, 4, 5, \ldots$ we arrive at the following general expression for $\mathbf{x}(k)$:

$$\mathbf{x}(k) = \mathbf{A}^k\mathbf{x}(0) + \mathbf{A}^{k-1}\mathbf{B}\mathbf{u}(0) + \mathbf{A}^{k-2}\mathbf{B}\mathbf{u}(1) + \cdots + \mathbf{A}\mathbf{B}\mathbf{u}(k-2) + \mathbf{B}\mathbf{u}(k-1)$$

or more compactly

$$\mathbf{x}(k) = \mathbf{A}^k\mathbf{x}(0) + \sum_{i=0}^{k-1} \mathbf{A}^{k-i-1}\mathbf{B}\mathbf{u}(i) \tag{12.2-16}$$

According to Eq. (12.2-8b), the output vector $\mathbf{y}(k)$ is

$$\mathbf{y}(k) = \mathbf{C}\mathbf{x}(k) + \mathbf{D}\mathbf{u}(k)$$

or

$$\mathbf{y}(k) = \mathbf{C}\mathbf{A}^k\mathbf{x}(0) + \mathbf{C}\sum_{i=0}^{k-1} \mathbf{A}^{k-i-1}\mathbf{B}\mathbf{u}(i) + \mathbf{D}\mathbf{u}(k) \tag{12.2-17}$$

where use was made of Eq. (12.2-16)

The matrix $\mathbf{A}^k$ is called the *fundamental* or *transition* matrix of system (12.2-8) and is denoted as follows:

$$\boldsymbol{\phi}(k) = \mathbf{A}^k \tag{12.2-18}$$

The matrix $\boldsymbol{\phi}(k)$ is analogous to the matrix $\boldsymbol{\phi}(t)$ of the continuous-time systems (see Table 12.1).

Table 12.1 Comparison of the Description Methods Between Continuous-Time and Discrete-Time Systems

Description method	Continuous-time system	Discrete-time system
State-space equations	$\dot{\mathbf{x}}(t) = \mathbf{F}\mathbf{x}(t) + \mathbf{G}\mathbf{u}(t)$ $\mathbf{y}(t) = \mathbf{C}\mathbf{x}(t) + \mathbf{D}\mathbf{u}(t)$	$\mathbf{x}(k+1) = \mathbf{F}\mathbf{x}(k) + \mathbf{G}\mathbf{u}(k)$ $\mathbf{y}(k) = \mathbf{C}\mathbf{x}(k) + \mathbf{D}\mathbf{u}(k)$
Transition matrix	$\boldsymbol{\phi}(t) = e^{\mathbf{F}t}$	$\boldsymbol{\phi}(k) = \mathbf{A}^k$
L/Z-transform of transition matrix	$\boldsymbol{\phi}(s) = (s\mathbf{I} - \mathbf{F})^{-1}$	$\boldsymbol{\phi}(z) = z(z\mathbf{I} - \mathbf{A})^{-1}$
Transfer function matrix	$\mathbf{H}(s) = \mathbf{C}\boldsymbol{\phi}(s)\mathbf{G} + \mathbf{D}$	$\mathbf{H}(z) = z^{-1}\mathbf{C}\boldsymbol{\phi}(z)\mathbf{B} + \mathbf{D}$
Impulse response matrix	$\mathbf{H}(t) = \mathbf{C}\boldsymbol{\phi}(t)\mathbf{G} + \mathbf{D}\delta(t)$	$\mathbf{H}(k) = \mathbf{C}\boldsymbol{\phi}(k-1)\mathbf{G} + \mathbf{D}$ for $k > 0$ $\mathbf{H}(k) = \mathbf{D}$ for $k = 0$

The state vector $\mathbf{x}(k)$ may also be calculated from Eq. (12.2-8a) using the Z-transform as follows. Take the Z-transform of both sides of the equation to yield

$$z\mathbf{X}(z) - z\mathbf{x}(0) = \mathbf{A}\mathbf{X}(z) + \mathbf{B}\mathbf{U}(z)$$

or

$$\mathbf{X}(z) = z[z\mathbf{I} - \mathbf{A}]^{-1}\mathbf{x}(0) + [z\mathbf{I} - \mathbf{A}]^{-1}\mathbf{B}\mathbf{U}(z)$$

Taking the inverse Z-transform, we have

$$\mathbf{x}(k) = \mathbf{A}^k\mathbf{x}(0) + \mathbf{A}^{k-1} * \mathbf{B}\mathbf{u}(k) = \mathbf{A}^k\mathbf{x}(0) + \sum_{i=0}^{k-1} \mathbf{A}^{k-i-1}\mathbf{B}\mathbf{u}(i) \qquad (12.2\text{-}19)$$

Equation (12.2-19) is in agreement with Eq. (12.2-16), as expected. It is evident that the state transition matrix can also be expressed as

$$\boldsymbol{\phi}(k) = \mathbf{A}^k = Z^{-1}[z(z\mathbf{I} - \mathbf{A})^{-1}] \qquad (12.2\text{-}20)$$

A comparison between the description methods used for continuous-time and discrete-time systems is shown in Table 12.1.

Example 12.2.2

Find the transition matrix, the state and the output vectors of a discrete-time system with zero initial condition, with $u(k) = \beta(k)$ and

$$\mathbf{A} = \begin{bmatrix} 0 & 1 \\ -2 & 3 \end{bmatrix}, \qquad \mathbf{b} = \begin{bmatrix} 0 \\ 1 \end{bmatrix}, \qquad \text{and} \qquad \mathbf{c} = \begin{bmatrix} 1 \\ 1 \end{bmatrix}$$

Solution

We have

$$\boldsymbol{\phi}(z) = Z(\boldsymbol{\phi}(k)) = z(z\mathbf{I} - \mathbf{A})^{-1} = \frac{1}{z^2 - 3z + 2}\begin{bmatrix} z(z-3) & z \\ -2z & z^2 \end{bmatrix}$$

Hence

$$\boldsymbol{\phi}(k) = \begin{bmatrix} Z^{-1}\left[\dfrac{z(z-3)}{z^2-3z+2}\right] & Z^{-1}\left[\dfrac{z}{z^2-3z+2}\right] \\ Z^{-1}\left[\dfrac{-2z}{z^2-3z+2}\right] & Z^{-1}\left[\dfrac{z^2}{z^2-3z+2}\right] \end{bmatrix}$$

Using the Z-transform pairs given in Appendix C, one obtains

$$\boldsymbol{\phi}(k) = \begin{bmatrix} 2 - 2^k & 2^k - 1 \\ 2(1 - 2^k) & 2^{k+1} - 1 \end{bmatrix}$$

These results may be checked as follows. Since $\boldsymbol{\phi}(k) = \mathbf{A}^k$, it follows that $\boldsymbol{\phi}(0) = \mathbf{I}$ and $\boldsymbol{\phi}(1) = \mathbf{A}$. Moreover, from the initial value theorem it follows that $\lim_{z \to \infty} \boldsymbol{\phi}(z) = \boldsymbol{\phi}(0)$. Indeed

$$\boldsymbol{\phi}(0) = \lim_{z \to \infty} \boldsymbol{\phi}(z) = \begin{bmatrix} 1 & 0 \\ 0 & 1 \end{bmatrix} = 1$$

Since the initial conditions are zero, the state vector may be calculated as follows:

$$X(z) = (z\mathbf{I} - \mathbf{A})^{-1}\mathbf{B}u(z) = \frac{\Phi(z)}{z}\mathbf{B}u(z) = \frac{1}{(z-1)^2(z-2)}\begin{bmatrix} z \\ z^2 \end{bmatrix}$$

Using the Z-transform pairs given in the Appendix C, one obtains

$$\mathbf{x}(k) = Z^{-1}[X(z)] = \begin{bmatrix} 2^k - k - 1 \\ 2(2^k - 1) - k \end{bmatrix}$$

From this expression, it follows that $\mathbf{x}(0) = \mathbf{0}$. Finally, the output of the system is given by

$$y(k) = \mathbf{c}^T\mathbf{x}(k) = 2^k - k - 1 + (2)(2^k) - 2 - k = (3)(2^k) - 2k - 3$$

Remark 12.2.2

When the initial conditions hold for $k = k_0$, the above results take on the following general forms:

$$\Phi(k, k_0) = A^{k-k_0} \qquad\qquad (12.2\text{-}21a)$$

$$\mathbf{x}(k) = \Phi(k - k_0)\mathbf{x}(k_0) + \sum_{i=k_0}^{k-1} \Phi(k - i - 1)\mathbf{B}u(i) \qquad\qquad (12.2\text{-}21b)$$

$$y(k) = \mathbf{C}\Phi(k - k_0)\mathbf{x}(k_0) + \sum_{i=k_0}^{k-1} \mathbf{C}\Phi(k - i - 1)\mathbf{B}u(i) + \mathbf{D}u(k) \qquad\qquad (12.2\text{-}21c)$$

12.3   DESCRIPTION AND ANALYSIS OF SAMPLED-DATA SYSTEMS

12.3.1   Introduction to D/A and A/D Converters

As we have already noted in Subsec. 12.1.1, in modern control systems a continuous-time system is usually controlled using a digital computer (Figure 12.1). As a result, the closed-loop system involves continuous-time, as well as discrete-time, subsystems. To have a common base for the study of the closed-loop system, it is logical to use the same mathematical model for both continuous-time and discrete-time systems. The mathematical model uses difference equations. This approach *unifies* the study of closed-loop systems. Furthermore, it *facilitates* the study of closed-loop systems, since well-known methods and results, such as transfer function, stability criteria, controller design techniques, etc., may be extended to cover the case of discrete-time systems.

A practical problem which we come across in such systems is that the output of a discrete-time system, which is a discrete-time signal, may be the input to a continuous-time system (in which case, of course, the input ought to be a continuous-time signal). And vice versa, the output of a continuous-time system, which is a continuous-time signal, could be the input to a discrete-time system (which, of course, ought to be a discrete-time signal). This problem is dealt with using special devices called converters. There are two types of converters: D/A converters, which convert the discrete-time signals to analog or continuous-time signals (Figure

12.10a), and A/D converters, which convert analog or continuous-time signals to discrete-time signals (Figure 12.10b). The constant T is the sampling time period. It is noted that before the discrete-time signal $y(kT)$ of the A/D converter is fed into a digital computer, it is first converted to a digital signal with the help of a device called a quantizer. The digital signal is a sequence of 0 and 1 digits (see also Figure 12.1).

12.3.2 Hold Circuits

A D/A converter is actually a hold circuit whose output $y(t)$ is a piecewise constant function. Specifically, the operation of the hold circuit (Figure 12.10a) is described by the following equations:

$$y(t) = u(kT), \qquad \text{for} \qquad kT \le t < (k+1)T \qquad (12.3\text{-}1a)$$

or

$$y(kT + \xi) = u(kT), \qquad \text{for} \qquad 0 \le \xi < T \qquad (12.3\text{-}1b)$$

We will show that the operation of a hold circuit, described by Eqs (12.3-1a and b), may be equivalently described (from a mathematical point of view) by the idealized

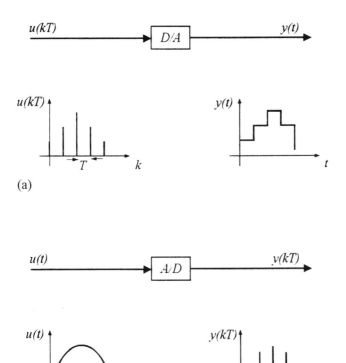

(a)

(b)

Figure 12.10   The operation of (a) D/A and (b) A/D converters.

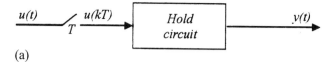

(a)

(b)

Figure 12.11 Hold circuit block diagrams: (a) hold circuit; (b) idealized hold circuit.

hold circuit shown in Figure 12.11b. Here it is assumed that the signal $u(t)$ goes through an ideal sampler δ_T, whose output $u^*(t)$ is given by the relation

$$u^*(t) = \sum_{k=0}^{\infty} u(kT)\delta(t - kT) \tag{12.-2}$$

Clearly, $u^*(t)$ is a sequence of impulse functions. Applying the Laplace transform on $u^*(t)$, we obtain

$$U^*(s) = \sum_{k=0}^{\infty} u(kT)e^{-skT} \tag{12.3-3}$$

Therefore, the output $Y(s)$ is

$$Y(s) = \left[\frac{1 - e^{-sT}}{s}\right] \sum_{k=0}^{\infty} u(kT)e^{-skT} \tag{12.3-4}$$

We would have arrived at the same result as in Eq. (12.3-4) if we had calculated $Y(s)$ of the hold circuit shown in Figure 12.11a. Indeed, since the impulse response $h(t)$ of the hold circuit is a gate function (see Figure B.3 in Appendix B), i.e., $h(t) = \beta(t) - \beta(t - T)$, the output $y(t)$ should be the convolution of $u(kT)$ and $h(t)$, i.e.,

$$y(t) = u(kT) * h(t) = \sum_{k=0}^{\infty} u(kT)[\beta(t - kT) - \beta(t - kT - T)] \tag{12.3-5}$$

Taking the Laplace transform of Eq. (12.3-5), we have

$$Y(s) = \sum_{k=0}^{\infty} u(kT)\left[\frac{e^{-skT} - e^{-s(k+1)T}}{s}\right] = \left[\frac{1 - e^{-sT}}{s}\right] \sum_{k=0}^{\infty} u(kT)^{-skT} \tag{12.3-6}$$

Equations (12.3-6) and (12.3-4) are identical. Therefore, both configurations in Figures 12.11a and b are equivalent with regard to the output $y(t)$. It is clear that the device shown in Figure 12.11b cannot be realized in practice and it is used only because it facilitates the mathematical description of the hold circuit. Figure 12.12 shows a typical output of the present hold circuit (known as a zero-order hold circuit).

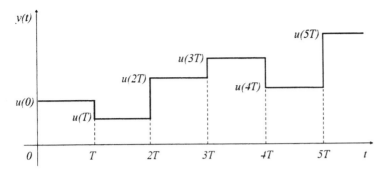

Figure 12.12   Typical output of a zero-order hold circuit.

12.3.3 Conversion of $G(s)$ to $G(z)$

To convert a continuous-time transfer function $G(s)$ to the transfer function $G(z)$ of the respective discretized system, various techniques have been proposed. In what follows, we briefly present some of the most popular techniques.

1 The Backward Difference Method

For simplicity, consider the case of a first-order system described by the transfer function

$$G(s) = \frac{Y(s)}{U(s)} = \frac{a}{s+a} \tag{12.3-7}$$

The system's differential equation is

$$y^{(1)} = -ay + au \tag{12.3-8}$$

Integrating both sides of the differential equation from 0 to t, one obtains

$$\int_0^t \frac{dy}{dt}\,dt = -a \int_0^t y\,dt + a \int_0^t u\,dt$$

Suppose that we want to determine the values of the output $y(t)$ at the sampling instants, i.e., at the points where $t = kT$. Then, the integral equation above becomes

$$\int_0^{kT} \frac{dy}{dt}\,dt = -a \int_0^{kT} y\,dt + a \int_0^{kT} u\,dt$$

Hence, we have

$$y(kT) - y(0) = -a \int_0^{kT} y\,dt + a \int_0^{kT} u\,dt \tag{12.3-9}$$

Substituting kT by $(k-1)T$ in this equation, one obtains

$$y[(k-1)T] - y(0) = -a \int_0^{(k-1)T} y\,dt + a \int_0^{(k-1)T} u\,dt \tag{12.3-10}$$

Substracting Eq. (12.3-10) from Eq. (12.3-9), we further obtain

$$y(kT) - y[(k-1)T] = -a \int_{(k-1)T}^{kT} y\,dt + a \int_{(k-1)T}^{kT} u\,dt \qquad (12.3\text{-}11)$$

Both terms on the right-hand side of Eq. (12.3-11) may be calculated approximately in various ways. If the approximation is done as shown in Figure 12.13a, i.e., by applying the backward difference method, then Eq. (12.3-11) takes the form

$$y(kT) = y[(k-1)T] - aT[y(kT) - u(kT)] \qquad (12.3\text{-}12)$$

Obviously, Eq. (12.3-12) is the equivalent difference equation of the differential equation (12.3-8). To find $G(z)$, we only need to take the Z-transform of Eq. (12.3-12) to yield

$$G(z) = \frac{Y(z)}{U(z)} = \frac{a}{\left[\dfrac{1 - z^{-1}}{T}\right] + a} \qquad (12.3\text{-}13)$$

Extending the results of the above example to the general case, we arrive at the following procedure for discretizing $G(s)$ using the backward difference method:

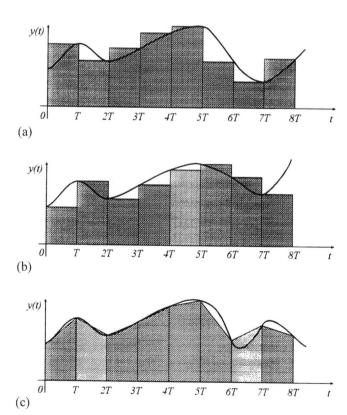

Figure 12.13 Area approximation using (a) the backward difference method, (b) the forward difference method, and (c) the Tustin or the trapezoidal method.

$$G(z) = G(s)\Big|_{s=(1-z^{-1})/T} \tag{12.3-14}$$

2 The Forward Difference Method

In this case, the approximation of the two terms of the right-hand side of Eq. (12.3-11) is done, as shown in Figure 12.13b. Working in the same way as in the previous case, we arrive at the following result for discretizing $G(s)$ using the forward difference method:

$$G(z) = G(s)\Big|_{s=\left(1-z^{-1}\right)/Tz^{-1}} \tag{12.3-15}$$

3 The Bilinear Transformation Method or Trapezoidal Method or Tustin Transformation Method

The Tustin transformation is based on the approximation of the two terms on the right-hand side of Eq. (12.3-11) using the trapezoidal rule, as shown in Figure 12.13a. This leads to the following result for discretizing $G(s)$:

$$G(z) = G(s)\Big|_{s=2/T\left[(1-z^{-1})/(1+z^{-1})\right]} \tag{12.3-16}$$

4 The Invariant Impulse Response Method

In this case, both $G(s)$ and $G(z)$ present a common characteristic in that their respective impulse functions $g(t)$ and $g(kT)$ are equal for $t = kT$. This is achieved when

$$G(z) = Z[g(kT)], \quad \text{where} \quad g(kT) = [L^{-1}G(s)]_{t=kT} \tag{12.3-17}$$

5 The Invariant Step Response Method

In this case, both $G(s)$ and $G(z)$ present the common characteristic that their step responses, i.e., the response $y(t)$ produced by the excitation $u(t) = 1$ and the response $y(kT)$ produced by the excitation $u(kT) = 1$, are equal for $t = kT$. This is achieved when

$$Z^{-1}\left[G(z)\frac{1}{1-z^{-1}}\right] = \left[L^{-1}\left[G(s)\frac{1}{s}\right]\right]_{t=kT} \tag{12.3-18}$$

where, obviously, the left-hand side of Eq. (12.3-18) is equal to $y(kT)$, whereas the right-hand side is equal to $y(t)$ at $t = kT$. Applying the Z-transform on Eq. (12.3-18), we obtain

$$G(z) = \left(1 - z^{-1}\right)Z\left[\frac{G(s)}{s}\right] = Z\left[\frac{1 - e^{-Ts}}{s}G(s)\right] = Z[G_h(s)G(s)] \tag{12.3-19}$$

where $G_h(s)$ is the transfer function of the zero-order hold circuit presented in Sec. 12.3.2.

6 Pole-Zero Matching Method

Consider the transfer function

$$G(s) = K_s \frac{(s + \mu_1)(s + \mu_2)\cdots(s + \mu_m)}{(s + \pi_1)(s + \pi_2)\cdots(s + \pi_n)}, \qquad m \leq n \tag{12.3-20}$$

Then, the pole-zero matching method assumes that $G(z)$ has the general form

$$G(z) = K_z \frac{(z + 1)^{n-m}(z + z_1)(z + z_2)\cdots(z + z_m)}{(z + p_1)(z + p_2)\cdots(z + p_n)} \tag{12.3-21}$$

where the z_i's and p_i's are "matched" to the respective μ_i's and π_i's according to the following relations

$$z_i = -e^{\mu_i T} \qquad \text{and} \qquad p_i = -e^{-\pi_i T} \tag{12.3-22}$$

The $n - m$ multiple zeros $(z + 1)^{n-m}$ which appear in $G(z)$ represent the order difference between the numerator's polynomial and the denominator's polynomial in Eq. (12.3-20). The constant K_z is calculated so as to satisfy particular requirements. For example, when we are interested in the behavior of a system at low frequencies (and this is the usual case in control systems), K_z is chosen such that $G(s)$ and $G(z)$ are equal for $s = 0$ and $z = 1$, respectively, i.e., the following relation holds:

$$G(z)\Big|_{z=1} = K_z 2^{n-m} \frac{(1 + z_1)(1 + z_2)\cdots(1 + z_m)}{(1 + p_1)(1 + p_2)\cdots(1 + p_n)} = G(s)\Big|_{s=0} = K_s \frac{\mu_1 \mu_2 \cdots \mu_m}{\pi_1 \pi_2 \cdots \pi_n} \tag{12.3-23}$$

Using Eq. (12.3-23), one can easily determine K_z.

Note that when a second-order term appears in $G(s)$, then the pole-zero matching method yields the following "matching":

$$(s + a)^2 + b^2 \Rightarrow z^2 - 2(e^{-aT} \cos bT)z + e^{-2aT}$$

Example 12.3.1

For the system with transfer function $G(s) = a/(s + a)$, find all above six equivalent descriptions of $G(z)$.

Solution

After several simple algebraic calculations, all six descriptions of $G(z)$ are found and summarized in Table 12.2.

Example 12.3.2

Consider a second-order continuous-time system having the following transfer function:

$$H(s) = \frac{b}{s(s + a)}$$

Find $H(z)$ using the invariant impulse response method.

Solution

We have

$$h(t) = L^{-1} H(s) = \frac{b}{a} L^{-1} \left[\frac{1}{s} - \frac{1}{(s + a)} \right] = \frac{b}{a} [1 - e^{-at}]$$

Table 12.2 Equivalent Discrete-Time Transfer Functions $G(z)$ of the Continuous-Time Transfer Function $G(s) = a/(s + a)$.

Method	Conversion of $G(s)$ to $G(z)$	Equivalent discrete-time transfer function
Backward difference method	$s = \dfrac{1 - z^{-1}}{T}$	$G(z) = \dfrac{a}{\left[\dfrac{1 - z^{-1}}{T}\right] + a}$
Forward difference method	$s = \dfrac{1 - z^{-1}}{Tz^{-1}}$	$G(z) = \dfrac{a}{\left[\dfrac{1 - z^{-1}}{Tz^{-1}}\right] + a}$
Tustin transformation	$s = \dfrac{2}{T}\left[\dfrac{1 - z^{-1}}{1 + z^{-1}}\right]$	$G(z) = \dfrac{a}{\left[\dfrac{2}{T}\left(\dfrac{1 - z^{-1}}{1 + z^{1}}\right)\right] + a}$
Invariant impulse response method	$G(z) = Z[g(t)]$ where $g(t) = Z^{-1}G(s)$	$G(z) = \dfrac{a}{1 - e^{-At}z^{-1}}$
Invariant step response method	$G(z) = Z\left[\dfrac{1 - e^{Ts}}{s}G(s)\right]$	$G(z) = \dfrac{(1 - e^{-aT})z^{-1}}{1 - e^{-aT}z^{-1}}$
Pole-zero matching method	$G(z) = K_z\left[\dfrac{z + 1}{z + p_1}\right]$	$G(z) = \left[\dfrac{1 - e^{-aT}}{2}\right]\left[\dfrac{1 + z^{-1}}{1 - e^{-aT}z^{-1}}\right]$

Hence

$$h(kT) = h(t)\bigg|_{t=kT} = \frac{b}{a}\left[1 - e^{-akT}\right]$$

The Z-transform of $h(kT)$ is

$$H(z) = Z[h(kT)] = \frac{b}{a}\sum_{k=0}^{\infty}\left(1 - e^{-akT}\right)z^{-k} = \frac{\dfrac{b}{a}}{1 - z^{-1}} - \frac{\dfrac{b}{a}}{1 - e^{-aT}z^{-1}}$$

or

$$H(z) = \frac{bz^{-1}\left(1 - e^{-aT}\right)}{a\left(1 - z^{-1}\right)\left(1 - e^{-aT}z^{-1}\right)}$$

Example 12.3.3

Consider a second-order continuous-time system having the following transfer function:

$$H(s) = \frac{2}{(s + 1)(s + 2)}$$

Find $H(z)$ using the invariant step response method.

Solution

We have

$$H(z) = Z[G_h(s)H(s)] = Z\left[\left[\frac{1-e^{-Ts}}{s}\right]H(s)\right] = (1-z^{-1})Z\left[\frac{H(s)}{s}\right]$$

Using the relation

$$\frac{H(s)}{s} = \frac{2}{s(s+1)(s+2)} = \frac{1}{s} - \frac{2}{s+1} + \frac{1}{s+2}$$

and the Z-transform tables of Appendix C, we obtain

$$Z\left[\frac{H(s)}{s}\right] = \frac{1}{1-z^{-1}} - \frac{2}{1-e^{-T}z^{-1}} + \frac{1}{1-e^{-2T}z^{-1}}$$

Hence

$$H(z) = \left(1-z^{-1}\right)Z\left[\frac{H(s)}{s}\right] = \left(1-z^{-1}\right)\left[\frac{1}{1-z^{-1}} - \frac{2}{1-e^{-T}z^{-1}} + \frac{1}{1-e^{-2T}z^{-1}}\right]$$

12.3.4 Conversion of Differential State Equations to Difference State Equations

Consider the continuous-time multi-input–multi-output (MIMO) open-loop system shown in Figure 12.14. Let this system be described in state space by the equations

$$\dot{\mathbf{x}}(t) = \mathbf{F}\mathbf{x}(t) + \mathbf{G}\mathbf{m}(t) \tag{12.3-24a}$$

$$\mathbf{y}(t) = \mathbf{C}\mathbf{x}(t) + \mathbf{D}\mathbf{m}(t) \tag{12.3-24b}$$

We will show that Eqs (12.3-24a and b) may be approximately written in the form of difference equations. To this end, consider the piecewise constant excitation vector $\mathbf{m}(t)$ described by

$$\mathbf{m}(t) = \mathbf{u}(kT), \qquad \text{for} \quad kT \le t < (k+1)T \tag{12.3-25}$$

Then, solving Eq. (12.3-24a) for $\mathbf{x}(t)$, we have

$$\mathbf{x}(t) = e^{\mathbf{F}t}\mathbf{x}(0) + \int_0^t e^{\mathbf{F}(t-\xi)}\mathbf{G}\mathbf{m}(\xi)d\xi \tag{12.3-26}$$

According to Eq. (12.-3-25), $\mathbf{m}(0) = \mathbf{u}(0)$ for $0 \le t < T$ and hence Eq. (12.3-26) becomes

$$\mathbf{x}(t) = e^{\mathbf{F}t}\mathbf{x}(0) + \int_0^t e^{\mathbf{F}(t-\xi)}\mathbf{G}\mathbf{u}(0)d\xi, \qquad 0 \le t < T \tag{12.3-27}$$

The state vector $\mathbf{x}(t)$, for $t = T$, will be

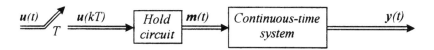

Figure 12.14 Open-loop system with a sampler and a hold circuit.

$$\mathbf{x}(T) = e^{\mathbf{F}T}\mathbf{x}(0) + \int_0^T e^{\mathbf{F}(T-\xi)}\mathbf{G}\mathbf{u}(0)d\xi \tag{12.3-28}$$

Define

$$\mathbf{A}(T) = e^{\mathbf{F}T} \tag{12.3-29a}$$

$$\mathbf{B}(T) = \int_0^T e^{\mathbf{F}(T-\xi)}\mathbf{G}d\xi = \int_0^T e^{\mathbf{F}\lambda}\mathbf{G}d\lambda, \qquad \lambda = T - \xi \tag{12.3-29b}$$

Then, for $t = T$, Eq. (12.3-27) can be simplified as follows:

$$\mathbf{x}(T) = \mathbf{A}(T)\mathbf{x}(0) + \mathbf{B}(T)\mathbf{u}(0) \tag{12.3-30}$$

Repeating the above procedure for $T \le t < 2T, 2T \le t < 3T$, etc., we arrive at the following general formula:

$$\mathbf{x}[(k + 1)T] = \mathbf{A}(T)\mathbf{x}(kT) + \mathbf{B}(T)\mathbf{u}(kT)$$

The output equation (12.3-24b) may therefore be written as

$$\mathbf{y}(kT) = \mathbf{C}\mathbf{x}(kT) + \mathbf{D}\mathbf{u}(kT)$$

Hence, the state differential equations (12.3-24) can be written as a system of difference equations, as follows:

$$\mathbf{x}[(k + 1)T] = \mathbf{A}(T)\mathbf{x}(kT) + \mathbf{B}(T)\mathbf{u}(kT) \tag{12.3-31a}$$

$$\mathbf{y}(kT) = \mathbf{C}\mathbf{x}(kT) + \mathbf{D}\mathbf{u}(kT) \tag{12.3-31b}$$

The state equations (12.3-24) and (12.3-31) are equivalent only for the time instants $t = kT$ under the constraint that the input vectors $\mathbf{m}(t)$ and $\mathbf{u}(t)$ satisfy the condition (12.3-25).

Remark 12.3.1

The transfer function matrix of the continuous-time system (12.3-24) is $\mathbf{H}(s) = \mathbf{C}(s\mathbf{I} - \mathbf{F})^{-1}\mathbf{G} + \mathbf{D}$ and the transfer function matrix of the equivalent discrete-time system (12.3-31) is $\mathbf{G}(z) = \mathbf{C}[z\mathbf{I} - \mathbf{A}(T)]^{-1}\mathbf{B}(T) + \mathbf{D}$. These two matrices are related as follows:

$$\mathbf{G}(z) = Z[G_h(s)\mathbf{H}(s)] = Z\left[\left[\frac{1 - e^{-sT}}{s}\right]\mathbf{H}(s)\right] \tag{12.3-32}$$

This means that the matrix $\mathbf{G}(z)$ is equivalent to the matrix $\mathbf{H}(s)$ in the sense of the invariant step response (see Eq. (12.3-19)).

Remark 12.3.2

Going from the continuous-time state-space description (12.3-24) to the discrete-time (sampled-data) state-space description (12.3-31), we need to determine the matrices $\mathbf{A}(T)$ and $\mathbf{B}(T)$ using the definition (12.3-29). For the determination of $\mathbf{A}(T)$ we note that it can be easily carried out using the following relation:

$$\mathbf{A}(T) = e^{\mathbf{F}t}\Big|_{t=T} = \left[L^{-1}[s\mathbf{I} - \mathbf{F}]^{-1}\right]_{t=T} \tag{12.3-33a}$$

Equation (12.3-33a) facilitates the determination of $\mathbf{B}(T)$, since, according to the definition (12.3-29b) of $B(T)$, we have that

$$\mathbf{B}(T) = \int_0^T \left[L^{-1}[s\mathbf{I} - \mathbf{F}]^{-1} \right]_{t=\lambda} \mathbf{G}d\lambda = \int_0^T \mathbf{A}(\lambda)\mathbf{G}d\lambda \qquad (12.3\text{-}33b)$$

Example 12.3.4

Consider the continuous-time system

$$\dot{\mathbf{x}}(t) = \mathbf{F}\mathbf{x}(t) + \mathbf{g}m(t)$$

$$y(t) = \mathbf{c}^T\mathbf{x}(t)$$

where

$$\mathbf{F} = \begin{bmatrix} -1 & 0 \\ 1 & 0 \end{bmatrix}, \qquad \mathbf{g} = \begin{bmatrix} 2 \\ 1 \end{bmatrix}, \qquad \mathbf{c} = \begin{bmatrix} 0 \\ 1 \end{bmatrix}$$

Find the equivalent discrete-time (sampled-data) system, i.e., find the matrix $\mathbf{A}(T)$ and the vector $\mathbf{b}(T)$.

Solution

We have

$$s\mathbf{I} - \mathbf{F} = \begin{bmatrix} s+1 & 0 \\ -1 & s \end{bmatrix}, \qquad (s\mathbf{I} - \mathbf{F})^{-1} = \begin{bmatrix} \dfrac{1}{s+1} & 0 \\ \dfrac{1}{s(s+1)} & \dfrac{1}{s} \end{bmatrix}, \qquad \text{and}$$

$$L^{-1}[s\mathbf{I} - \mathbf{F}]^{-1} = \begin{bmatrix} e^{-t} & 0 \\ 1 - e^{-t} & 1 \end{bmatrix}$$

Therefore, from Eq. (12.3-33a), we obtain

$$\mathbf{A}(T) = \left[L^{-1}[s\mathbf{I} - \mathbf{F}]^{-1} \right]_{t=T} = e^{\mathbf{F}T} = \begin{bmatrix} e^{-T} & 0 \\ 1 - e^{-T} & 1 \end{bmatrix}$$

Moreover, from Eq. (12.3-33b), we obtain

$$\mathbf{b}(T) = \int_0^T e^{\mathbf{F}\lambda}\mathbf{g}d\lambda = \int_0^T \begin{bmatrix} e^{-\lambda} & 0 \\ 1 - e^{-\lambda} & 1 \end{bmatrix} \begin{bmatrix} 2 \\ 1 \end{bmatrix} d\lambda = \int_0^T \begin{bmatrix} 2e^{-\lambda} \\ 3 - 2e^{-\lambda} \end{bmatrix} d\lambda$$

$$= \begin{bmatrix} 2(1 - e^{-T}) \\ 3T - 2(1 - e^{-T}) \end{bmatrix}$$

Example 12.3.5

Consider a harmonic oscillator with transfer function

$$G(s) = \frac{Y(s)}{U(s)} = \frac{\omega^2}{s^2 + \omega^2}$$

A state-space description of the oscillator is of the form $\dot{\mathbf{x}} = \mathbf{F}\mathbf{x} + \mathbf{g}u$, $y = \mathbf{c}^T\mathbf{x}$, where

$$\mathbf{x} = \begin{bmatrix} x_1 \\ x_2 \end{bmatrix} = \begin{bmatrix} y \\ \omega^{-1}y^{(1)} \end{bmatrix}, \qquad \mathbf{F} = \begin{bmatrix} 0 & \omega \\ -\omega & 0 \end{bmatrix}, \qquad \mathbf{g} = \begin{bmatrix} 0 \\ \omega \end{bmatrix}, \qquad \mathbf{c} = \begin{bmatrix} 1 \\ 0 \end{bmatrix}$$

Find the equivalent discrete-time (sampled-data) system of the form (12.3-31), i.e., find the matrix $\mathbf{A}(T)$ and the vector $\mathbf{b}(T)$.

Solution

We have

$$(s\mathbf{I} - \mathbf{F})^{-1} = \begin{bmatrix} \dfrac{s}{s^2 + \omega^2} & \dfrac{\omega}{s^2 + \omega^2} \\ \dfrac{-\omega}{s^2 + \omega^2} & \dfrac{s}{s^2 + \omega^2} \end{bmatrix} \quad \text{and} \quad e^{\mathbf{F}t} = \begin{bmatrix} \cos\omega t & \sin\omega t \\ -\sin\omega t & \cos\omega t \end{bmatrix}$$

Therefore

$$\mathbf{A}(T) = \left[L^{-1}[s\mathbf{I} - \mathbf{F}]^{-1} \right]_{t=T} = \begin{bmatrix} \cos\omega T & \sin\omega T \\ -\sin\omega T & \cos\omega T \end{bmatrix}$$

$$\mathbf{b}(T) = \left[\int_0^T e^{\mathbf{F}\lambda}\mathbf{g}\,d\lambda \right] = \begin{bmatrix} 1 - \cos\omega T \\ \sin\omega T \end{bmatrix}$$

12.3.5   Analysis of Sampled-Data Systems

1 Analysis Based on the State Equations

To solve Eqs (12.3-31), we take advantage of the results of Subsec. 12.2.3, since they differ only by the constant T in Eqs (12.3-31). We therefore have that the general solution of Eq. (12.3-31a) is given by

$$\mathbf{x}(kT) = \boldsymbol{\phi}[(k - k_0)T]\mathbf{x}(k_0 T) + \sum_{i=k_0}^{k-1} \boldsymbol{\phi}[(k - i - 1)T]\mathbf{B}(T)\mathbf{u}(iT) \qquad (12.3\text{-}34a)$$

and the general solution of Eq. (12.3-31b) by

$$\mathbf{y}(kT) = \mathbf{C}\boldsymbol{\phi}[(k - k_0)T]\mathbf{x}(k_0 T) + \mathbf{C}\sum_{i=k_0}^{k-1} \boldsymbol{\phi}[(k - i - 1)T]\mathbf{B}(T)\mathbf{u}(iT) + \mathbf{D}\mathbf{u}(kT)$$

$$(12.3\text{-}34b)$$

where $\boldsymbol{\phi}[(k - k_0)T]$ is the transition matrix, given by the relation

$$\boldsymbol{\phi}[(k - k_0)T] = [\mathbf{A}(T)]^{k-k_0}$$

Clearly, if we set $T = 1$ in Eqs (12.3-34a and b), then we obtain the formulas (12.2-16) and (12.2-17), respectively.

2 Analysis Based on $\mathbf{H}(kT)$

Consider the continuous-time system shown in Figure 12.15, where the two samplers are synchronized. The output vector $\mathbf{y}(kT)$, i.e., the vector $\mathbf{y}(t)$ at the sampling points $t = kT$, is

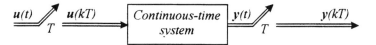

Figure 12.15   A continuous-time system with input and output samplers.

$$\mathbf{y}(kT) = \sum_{i=0}^{\infty} \mathbf{H}(kT - iT)\mathbf{u}(iT) \tag{12.3-35}$$

Equation (12.3-35) represents, as is already known, a convolution. If we set $T = 1$, then Eq. (12.3-35) is the vector form of the scalar convolution (12.2-12).

3 Analysis Based on $\mathbf{H}(z)$

If we take the Z-transform of Eq. (12.3-35), we obtain the following expression for the output vector:

$$\mathbf{Y}(z) = \mathbf{H}(z)\mathbf{U}(z) \tag{12.3-36}$$

where

$$\mathbf{Y}(z) = Z[\mathbf{y}(kT)], \qquad \mathbf{U}(z) = Z[\mathbf{u}(kT)], \qquad \text{and} \qquad \mathbf{H}(z) = Z[\mathbf{H}(kT)] \tag{12.3-37}$$

Example 12.3.6

For the system of Example 12.3.5, for $T = 1$ and $\omega = \pi/2$, find:

(a)   The transition matrix $\boldsymbol{\phi}(kT)$
(b)   The state-space vector $\mathbf{x}(kT)$
(c)   The output vector $\mathbf{y}(kT)$.

Solution

For the given values of ω and T, $\mathbf{A}$, $\mathbf{B}$, and $\mathbf{c}$ become

$$\mathbf{A} = \begin{bmatrix} 0 & 1 \\ -1 & 0 \end{bmatrix}, \qquad \mathbf{b} = \begin{bmatrix} 1 \\ 1 \end{bmatrix}, \qquad \mathbf{c} = \begin{bmatrix} 1 \\ 0 \end{bmatrix}$$

(a) $\boldsymbol{\phi}(z) = Z[\boldsymbol{\phi}(kT)] = z(z\mathbf{I} - \mathbf{A})^{-1} = \dfrac{1}{z^2 + 1}\begin{bmatrix} z^2 & z \\ -z & z^2 \end{bmatrix}$

Therefore

$$\boldsymbol{\phi}(kT) = \begin{bmatrix} Z^{-1}\left[\dfrac{z^2}{z^2 + 1}\right] & Z^{-1}\left[\dfrac{z}{z^2 + 1}\right] \\[3mm] Z^{-1}\dfrac{-z}{z^2 + 1} & Z^{-1}\left[\dfrac{z^2}{z^2 + 1}\right] \end{bmatrix}$$

Using the tables of Z-transform pairs given in Appendix C, we have

$$\phi(kT) = \begin{bmatrix} \cos\dfrac{k\pi T}{2} & \sin\dfrac{k\pi T}{2} \\ -\sin\dfrac{k\pi T}{2} & \cos\dfrac{k\pi T}{2} \end{bmatrix}$$

(b) $$\mathbf{X}(z) = (z\mathbf{I} - \mathbf{A})^{-1}\mathbf{B}U(z) = \frac{1}{z^2 + 1}\begin{bmatrix} z & 1 \\ -z & z \end{bmatrix}\begin{bmatrix} 1 \\ 1 \end{bmatrix}\begin{bmatrix} z \\ z-1 \end{bmatrix}$$

$$= \frac{1}{(z^2 + 1)(z - 1)}\begin{bmatrix} z(z+1) \\ z(z-1) \end{bmatrix}$$

Hence

$$\mathbf{x}(kT) = Z^{-1}[\mathbf{X}(z)] = \begin{bmatrix} 1 - \cos\dfrac{k\pi T}{2} \\ \sin\dfrac{k\pi T}{2} \end{bmatrix}$$

(c)   Finally, the output of the system is given by

$$y(kT) = \mathbf{c}^T\mathbf{x}(kT) = [1 \quad 0]\mathbf{x}(kT) = 1 - \cos\frac{k\pi T}{2}$$

12.4 STABILITY

12.4.1 Definitions and Basic Theorems of Stability

1 Introduction

Consider the nonlinear discrete-time system

$$\mathbf{x}(kT + T) = \mathbf{f}[\mathbf{x}(kT), kT, \mathbf{u}(kT)], \qquad \mathbf{x}(k_0 T) = \mathbf{x}_0 \qquad (12.4\text{-}1)$$

Let $\mathbf{u}(kT) = \mathbf{0}$, for $k \geq k_0$. Moreover, let $\mathbf{x}(kT)$ and $\tilde{\mathbf{x}}(kT)$ be the solutions of Eq. (12.4-1) when the initial conditions are $\mathbf{x}(k_0 T)$ and $\tilde{\mathbf{x}}(k_0 T)$, respectively. Also let the symbol $\| \cdot \|$ represent the Euclidean norm

$$\|\mathbf{x}\| = \left[x_1^2 + x_2^2 + \cdots + x_n^2\right]^{1/2}$$

We give the following definitions of stability.

Definition 12.4.1: *Stability*

The solution $\mathbf{x}(kT)$ of Eq. (12.4-1) is stable if for a given $\varepsilon > 0$ there exists a $\delta(\varepsilon, k_0) > 0$ such that all solutions satisfying $\|\mathbf{x}(k_0 T) - \tilde{\mathbf{x}}(k_0 T)\| < \varepsilon$ imply that $\|\mathbf{x}(kT) - \tilde{\mathbf{x}}(kT)\| < \delta$ for all $k \geq k_0$.

Definition 12.4.2: *Asymptotic Stability*

The solution $\mathbf{x}(kT)$ of Eq. (12.4-1) is asymptotically stable if it is stable and if $\|\mathbf{x}(kT) - \tilde{\mathbf{x}}(kT)\| \to 0$ as $k \to +\infty$, under the constraint that $\|\mathbf{x}(k_0 T) - \tilde{\mathbf{x}}(k_0 T)\|$ is sufficiently small.

In the case where the system (12.4-1) is stable in accordance with Definition 12.4.1, the point $\mathbf{x}(k_0 T)$ is called the *equilibrium point*. In the case where the system (12.4-1) is asymptotically stable, the equilibrium point is the origin $\mathbf{0}$.

2 Stability of Linear Time-Invariant Discrete-Time Systems

Consider the linear time-invariant discrete-time system

$$
\left.
\begin{aligned}
\mathbf{x}(kT + T) &= \mathbf{Ax}(kT) + \mathbf{Bu}(kT), \qquad \mathbf{x}(k_0 T) = \mathbf{x}_0 \\
\mathbf{y}(kT) &= \mathbf{Cx}(kT) + \mathbf{Du}(kT)
\end{aligned}
\right\}
\tag{12.4-2}
$$

Applying Definition 12.4.1 to the system (12.4-2), we have the following theorem.

Theorem 12.4.1

System (12.4-2) is stable according to Definition 12.4.1 if, and only if, the eigenvalues λ_i of the matrix $\mathbf{A}$, i.e., the roots of the characteristic equation $|\lambda \mathbf{I} - \mathbf{A}| = 0$, lie inside the unit circle (i.e., $|\lambda_i| < 1$), or the matrix $\mathbf{A}$ has eigenvalues on the unit circle (i.e., $|\lambda_i| = 1$) of multiplicity one.

Applying Definition 12.4.2 to the system (12.4-2), we have the following theorem.

Theorem 12.4.2

System (12.4-2) is asymptotically stable if, and only if, $\lim_{k \to \infty} \mathbf{x}(kT) = \mathbf{0}$, for every $\mathbf{x}(k_0 T)$, when $\mathbf{u}(kT) = \mathbf{0}$ $(k \geq k_0)$.

On the basis of Theorem 12.4.2, we prove the following theorem.

Theorem 12.4.3

System (12.4-2) is asymptotically stable if, and only if, the eigenvalues λ_i of $\mathbf{A}$ are inside the unit circle.

Proof

The theorem will be proved for the special case where the matrix $\mathbf{A}$ has distinct eigenvalues. The proof of the case where the eigenvalues are repeated is left as an exercise. When $\mathbf{u}(kT) = \mathbf{0}$ $(k \geq k_0)$, the state vector $\mathbf{x}(kT)$ is given by

$$
x(kT) = \mathbf{A}^{k-k_0} \mathbf{x}(k_0 T)
\tag{12.4-3}
$$

Let the eigenvalues $\lambda_1, \lambda_2, \ldots, \lambda_n$ of the matrix $\mathbf{A}$ be distinct. Then, according to the Sylvester theorem (see Sec. 2.12), the matrix $\mathbf{A}^k$ can be written as

$$
\mathbf{A}^k = \sum_{i=1}^{n} \mathbf{A}_i \lambda_i^k
\tag{12.4-4}
$$

where $\mathbf{A}_i$ are special matrices which depend only on $\mathbf{A}$. Substituting Eq. (12.4-4) in Eq. (12.4-3), we obtain

$$
\mathbf{x}(kT) = \sum_{i=1}^{n} \mathbf{A}_i \lambda_i^{k-k_0} \mathbf{x}(k_0 T)
\tag{12.4-5}
$$

Hence

$$\lim_{k \to \infty} \mathbf{x}(kT) = \left[\sum_{i=1}^{n} \mathbf{A}_i \mathbf{x}(k_0 T) \left[\lim_{k \to \infty} \lambda_i^{k-k_0} \right] \right] \tag{12.4-6}$$

From Eq. (12.4-6) it is obvious that $\lim_{k \to \infty} \mathbf{x}(kT) = \mathbf{0}$, $\forall \mathbf{x}(k_0 T)$ if, and only if, $\lim_{k \to \infty} \lambda_i^{k-k_0} = 0, \forall i = 1, 2, \ldots, n$, which is true if and only if $|\lambda_i| < 1, \forall i = 1, 2, \ldots, n$, where $|\cdot|$ stands for the magnitude of a complex number.

A comparison between linear time-invariant continuous-time systems and linear time-invariant discrete-time systems with respect to the asymptotic stability is shown in Figure 12.16.

3 Bounded-Input Bounded-Output Stability

Definition 12.4.3

A linear time-invariant system is bounded-input–bounded-output (BIBO) stable if a bounded input produces a bounded output for every initial condition.

Applying Definition 12.4.3 to system (12.4-2), we have the following theorem.

Theorem 12.4.4

The linear time-invariant system (12.4-2) is BIBO stable if, and only if, the poles of the transfer function $\mathbf{H}(z) = \mathbf{C}(z\mathbf{I} - \mathbf{A})^{-1}\mathbf{B} + \mathbf{D}$, before any pole-zero cancellation, are inside the unit circle.

	Continuous–time systems	Discrete–time systems
State–space equations	$\dot{x}(t) = Fx(t) + Gu(t)$ $y(t) = Cx(t) + Du(t)$	$x[(k+1)T] = Ax(kT) + Bu(kT)$ $y(kT) = Cx(kT) + Du(kT)$
Characteristic equation	$\|\lambda I - F\| = 0$	$\|\lambda I - A\| = 0$
Stability definition	$Re\{\lambda_i\} < 0$	$\|\lambda_i\| < 1$
Stability description	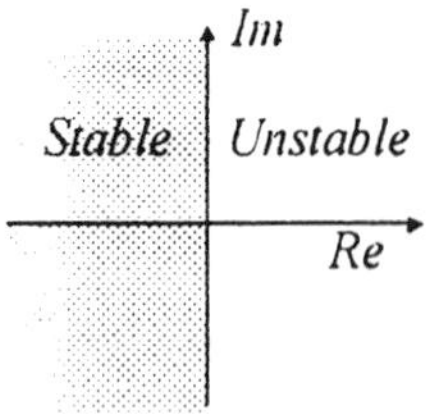	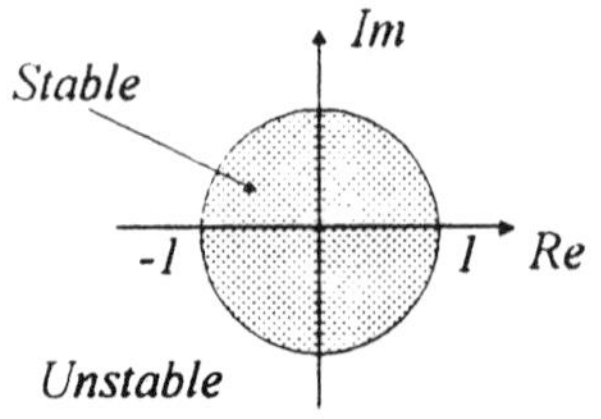

Figure 12.16 A comparison of asymptotic stability between continuous-time and discrete-time systems.

From Definition 12.4.3 we may conclude that asymptotic stability is the strongest, since it implies both stability and BIBO stability. It is easy to give examples showing that stability does not imply BIBO stability and vice versa.

12.4.2　Stability Criteria

The concept of stability has been presented in some depth in the preceding subsection. For testing stability, various techniques have been proposed. The most popular techniques for determining the stability of a discrete-time system are the following:

1.　The Routh criterion, using the Möbius transformation
2.　The Jury criterion
3.　The Lyapunov method
4.　The root locus method
5.　The Bode and Nyquist criteria

Here, we briefly present criteria 1 and 2. For the rest of the stability criteria see [3, 11, 13].

1 The Routh Criterion Using the Möbius Transformation

Consider the polynomial

$$a(z) = a_0 z^n + a_1 z^{n-1} + \cdots + a_n \tag{12.4-7}$$

The roots of the polynomial are the roots of the equation

$$a(z) = a_0 z^n + a_1 z^{n-1} + \cdots + a_n = 0 \tag{12.4-8}$$

As stated in Theorem 12.4.3, asymptotic stability is secured if all the roots of the characteristic polynomial lie inside the unit circle. The well-known Routh criterion for continuous-time systems is a simple method for determining if all the roots of an arbitrary polynomial lie in the left complex plane without requiring the determination of the values of the roots. The Möbius bilinear transformation

$$w = \frac{z+1}{z-1} \quad \text{or} \quad z = \frac{w+1}{w-1} \tag{12.4-9}$$

maps the unit circle of the z-plane into the left w-plane. Consequently, if the Möbius transformation is applied to Eq. (12.4-8), then the Routh criterion may be applied, as in the case of continuous-time systems.

Example 12.4.1

The characteristic polynomial $a(z)$ of a system is given by $a(z) = z^2 + 0.7z + 0.1$. Investigate the stability of the system.

Solution

Applying the transformation (12.4-9) to $a(z)$ yields

$$a(w) = \left[\frac{w+1}{w-1}\right]^2 + 0.7\left[\frac{w+1}{w-1}\right] + 0.1 = \frac{(w+1)^2 + 0.7(w^2-1) + 0.1(w-1)^2}{(w-1)^2}$$

$$= \frac{1.8w^2 + 1.8w + 0.4}{(w-1)^2}$$

The numerator of $a(w)$ is called the "auxiliary" characteristic polynomial to which the well-known Routh criterion will be applied. For the present example, the Routh array is

w^2	1.8	0.4
w^1	1.8	0
w^0	0.4	0

The coefficients of the first column have the same sign and, according to Routh's criterion, the system is stable. We can reach the same result if we factorize $a(z)$ into a product of terms, in which case $a(z) = (z+0.5)(z+0.2)$. The two roots of $a(z)$ are -0.5 and -0.2, which are both inside the unit circle and hence the system is stable.

2 The Jury Criterion

It is useful to establish criteria which can directly show whether a polynomial $a(z)$ has all its roots inside the unit circle instead of determining its eigenvalues. Such a criterion, which is equivalent to the Routh criterion for continuous-time systems, has been developed by Schur, Cohn, and Jury. This criterion is usually called the Jury criterion and is described in detail below.

First, the Jury table is formed for the polynomial $a(z)$, given by Eq. (12.4-7), as shown in Table 12.3. The first two rows of the table are the coefficients of the polynomial $a(z)$ presented in the forward and reverse order, respectively. The third row is formed by multiplying the second row by $\beta_n = a_n/a_0$ and subtracting the result from the first row. Note that the last element of the third row becomes zero. The fourth row is identical to the third row, but in reverse order. The above procedure is repeated until the $2n + 1$ row is reached. The last row consists of only a single element. The following theorem holds.

Theorem 12.4.5: *The Jury Stability Criterion*

If $a_0 > 0$, then the polynomial $a(z)$ has all its roots inside the unit circle if, and only if, all a_0^k, $k = 0, 1, \ldots, n-1$, are positive. If all coefficients a_0^k differ from zero, then

Table 12.3   The Jury Table

a_0	a_1	$\cdots$	a_{n-1}	a_n	
a_n	a_{n-1}	$\cdots$	a_1	a_0	
					$\beta_n = a_n/a_0$
a_0^{n-1}	a_1^{n-1}	$\cdots$	a_{n-1}^{n-1}		
a_{n-1}^{n-1}	a_{n-2}^{n-1}	$\cdots$	a_0^{n-1}		
					$\beta_{n-1} = a_{n-1}^{n-1}/a_0^{n-1}$
$\vdots$					
a_0^0			where $a_i^{n-1} = a_i^k - \beta_k a_{k-1}^k$ and $\beta_k = a_k^k/a_0^k$		

the number of negative coefficients a_0^k is equal to the number of roots which lie outside the unit circle.

Remark 12.4.1

If all coefficients a_0^k, $k = 1, 2, \ldots, n$, are positive, then it can be shown that the condition $a_0^0 > 0$ is equivalent to the following two conditions:

$$a(1) > 0 \tag{12.4-10a}$$

$$(-1)^n a(-1) > 0 \tag{12.4-10b}$$

Relations (12.4-10a and b) present necessary conditions for stability and may therefore be used to check for stability, prior to construction of the Jury table.

Example 12.4.2

Consider the characteristic polynomial $a(z) = z^3 - 1.3z^2 - 0.8z + 1$. Investigate the stability of the system.

Solution

Condition (12.4-10a) is first examined. Here $a(1) = 1 - 1.3 - 0.8 + 1 = -0.1$ and, since $a(1) < 0$, the necessary condition (12.4-10a) is not satisfied. Therefore, one or more roots of the characteristic polynomial lie outside the unit circle. It is immediately concluded that the system must be unstable.

Example 12.4.3

Consider the second-order characteristic polynomial $a(z) = z^2 + a_1 z + a_2$. Investigate the stability of the system.

Solution

The Jury table is formed as shown in Table 12.4. All the roots of the characteristics polynomial are inside the unit circle if

$$1 - a_2^2 > 0 \quad \text{and} \quad \left[\frac{1 - a_2}{1 + a_2}\right][(1 + a_2)^2 - a_1^2] > 0$$

which lead to the conditions $-1 < a_2 < 1$, $a_2 > -1 + a_1$, and $a_2 > -1 - a_1$. The stability region of the second-order characteristic polynomial is shown in Figure 12.17.

Table 12.4   The Jury Table for Example 12.4.3

1	a_1	a_2	
a_2	a_1	1	
			$\beta_2 = a_2$
$1 - a_2^2$	$a_1(1 - a_2)$		
$a_1(1 - a_2)$	$1 - a_2^2$		
			$\beta_1 = \dfrac{a_1}{1 + a_2}$
$1 - a_2^2 - \dfrac{a_1^2(1 - a_2)}{1 + a_2}$			

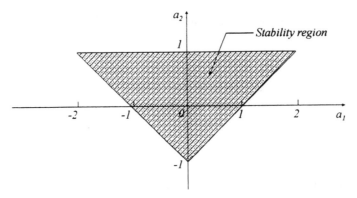

Figure 12.17 The stability region for Example 12.4.3.

Example 12.4.4

Consider the characteristic polynomial $a(z) = z^3 + Kz^2 + 0.5z + 2$. Find the values of the constant K for which all the roots of the polynomial $a(z)$ lie inside the unit circle.

Solution

The Jury table is formed as shown in Table 12.5. According to Theorem 12.4.5, all a_0^k elements are positive, except for the last element which may become positive for certain values or for a range of values of K. This last element in the Jury table may be written as

$$\tfrac{1}{3}(1-K)^2 - \tfrac{1}{3}(2K+1)^2(1-K) = \tfrac{1}{3}(1-K)\big[1-K-(2K+1)^2\big]$$
$$= \tfrac{1}{3}(1-K)(-4K^2-5K)$$
$$= \tfrac{1}{3}(1-K)(-K)(4K+5)$$

The last product of terms becomes positive if the following inequalities are true: $K > -\tfrac{5}{4}$ and $K < 0$. Hence, when $-\tfrac{5}{4} < K < 0$, the characteristic polynomial has all its roots inside the unit circle and the system under consideration is stable.

12.5 CONTROLLABILITY AND OBSERVABILITY

12.5.1 Controllability

Simply speaking, controllability is a property of a system which is strongly related to the ability of the system to go from a given initial state to a desired final state within a finite time (see Sec. 5.6). Consider the system

$$\mathbf{x}(k+1) = \mathbf{A}\mathbf{x}(k) + \mathbf{B}\mathbf{u}(k), \qquad \mathbf{x}(0) = \mathbf{x}_0 \tag{12.5-1a}$$
$$\mathbf{y}(k) = \mathbf{C}\mathbf{x}(k) \tag{12.5-1b}$$

The state $\mathbf{x}(k)$ at time k is given by Eq. (12.2-16), which can be rewritten as follows:

Table 12.5 The Jury Table for Example 12.4.4

1	K	0.5	0.5	
1	2	K	1	
$1 - 0.25$	$K - 1$	$0.5 - 0.5K$		$\beta_3 = 0.5$
$0.5 - 0.5K$	$K - 1$	0.75		
$\frac{1}{3}(1 - K)^2$	$\frac{2}{3}(1 - K)(K - 1) - (K - 1)$			$\beta_2 = \frac{2}{3}(1 - K)$
$\frac{2}{3}(1 - K)(K - 1) - (K - 1)$	$\frac{1}{3}(1 - K)^2$	0		
$\frac{1}{3}(1 - K)^2 - \frac{1}{3}(2K + 1)^2(1 - K)$				$\beta_1 = 2K + 1$

$$\mathbf{x}(k) = \mathbf{A}^k \mathbf{x}(0) + [\mathbf{B} \vdots \mathbf{AB} \vdots \cdots \vdots A^{k-1}\mathbf{B}] \begin{bmatrix} \mathbf{u}(k-1) \\ \mathbf{u}(k-2) \\ \vdots \\ \mathbf{u}(0) \end{bmatrix} \tag{12.5-2}$$

Definition 12.5.1

Assume that $|\mathbf{A}| \neq 0$. Then system (12.5-1a) is controllable if it is possible to find a control sequence $\{\mathbf{u}(0), \mathbf{u}(1), \ldots, \mathbf{u}(q-1)\}$ which allows the system to reach an arbitrary final state $\mathbf{x}(q) = \xi \in \mathbf{R}^n$, within a finite time, say q, from any initial state $\mathbf{x}(0)$.

According to Definition 12.5.1, Eq. (12.5-2) takes on the form

$$\xi - \mathbf{A}^q \mathbf{x}(0) = [\mathbf{B} \vdots \mathbf{AB} \vdots \cdots \vdots \mathbf{A}^{q-1}\mathbf{B}] \begin{bmatrix} \mathbf{u}(q-1) \\ \mathbf{u}(q-2) \\ \vdots \\ \mathbf{u}(0) \end{bmatrix} \tag{12.5-3}$$

This relation is an inhomogenous algebraic system of equations, having the control sequence $\{\mathbf{u}(0), \mathbf{u}(1), \ldots, \mathbf{u}(q-1)\}$ and the parameter q as unknowns. From linear algebra, it is known that this equation has a solution if and only if

$$\text{rank}[\mathbf{B} \vdots \cdots \vdots \mathbf{A}^{q-1}\mathbf{B} \mid \xi - \mathbf{A}^q\mathbf{x}(0)] = \text{rank}[\mathbf{B} \vdots \cdots \vdots \mathbf{A}^{q-1}\mathbf{B}]$$

This condition, for every arbitrary final state $\mathbf{x}(q) = \xi$, holds true if and only if

$$\text{rank}[\mathbf{B} \vdots \mathbf{AB} \vdots \cdots \vdots \mathbf{A}^{q-1}\mathbf{B}] = n, \qquad q \in \mathbf{N} \tag{12.5-4}$$

Clearly, an increase in time q improves the possibility of satisfying condition (12.5-4). However, the Cayley–Hamilton theorem (Sec. 2.11) states that the terms $\mathbf{A}^j\mathbf{B}$, for $j \geq n$, are linearly dependent on the first n terms (i.e., on the terms $\mathbf{B}, \mathbf{AB}, \ldots, \mathbf{A}^{n-1}\mathbf{B}$). Thus, condition (12.5-4) holds true if and only if $q = n$, i.e., if

$$\text{rank}[\mathbf{B} \vdots \mathbf{AB} \vdots \cdots \vdots \mathbf{A}^{n-1}\mathbf{B}] = n \tag{12.5-5}$$

Therefore, the following theorem has been proved.

Theorem 12.5.1

System (12.5-1a) is controllable if and only if

$$\text{rank } \mathbf{S} = n, \qquad \text{where} \qquad \mathbf{S} = [\mathbf{B} \quad \mathbf{AB} \quad \cdots \quad \mathbf{A}^{n-1}\mathbf{B}] \tag{12.5-6}$$

Here the $n \times nm$ matrix $\mathbf{S}$ is called the *controllability matrix* (see also Subsec. 5.6.1).

Remark 12.5.1

For a controllable system of order n, n time units are sufficient for the system to reach any final state $\xi = \mathbf{x}(n)$.

Example 12.5.1

Consider the system (12.5-1a), where

$$A = \begin{bmatrix} 1 & 1 \\ -0.25 & 0 \end{bmatrix}, \quad b = \begin{bmatrix} 1 \\ -0.5 \end{bmatrix}, \quad \text{and} \quad x(0) = \begin{bmatrix} 2 \\ 2 \end{bmatrix}$$

Find a control sequence, if it exists, such that $x^T(2) = [-0.5 \quad 1]$ and $x^T(2) = [0.5 \quad 1]$.

Solution

From Eq. (12.5-2), for $k = 2$, we have

$$x(2) = A^2 x(0) + Abu(0) + bu(1)$$

For the first case, where $x^T(2) = [-0.5 \quad 1]$, the above equation yields

$$\begin{bmatrix} -0.5 \\ 1 \end{bmatrix} = \begin{bmatrix} 3.5 \\ -1 \end{bmatrix} + \begin{bmatrix} 1 \\ -0.5 \end{bmatrix} [0.5u(0) + u(1)]$$

This equation leads to the scalar equation $0.5u(0) + u(1) = -4$. A possible control sequence would be $u(0) = -2$ and $u(1) = -3$.

For the second case, where $x^T(2) = [0.5 \quad 1]$, we have

$$\begin{bmatrix} 0.5 \\ 1 \end{bmatrix} = \begin{bmatrix} 3.5 \\ -1 \end{bmatrix} + \begin{bmatrix} 1 \\ -0.5 \end{bmatrix} [0.5u(0) + u(1)]$$

This equation does not possess a solution. Of course, this occurs because the system is uncontrollable, since rank $S = 1$, where

$$S = [b \; \vdots \; Ab] = \begin{bmatrix} 1 & 0.5 \\ -0.5 & -0.25 \end{bmatrix}$$

Therefore, when the system is uncontrollable, it is not possible for the state vector to reach any preassigned value.

Example 12.5.2

Consider the system (12.5-1a), where

$$A = \begin{bmatrix} 1 & 1 \\ 0 & 1 \end{bmatrix}, \quad b = \begin{bmatrix} 0 \\ 1 \end{bmatrix}, \quad \text{and} \quad x(0) = \begin{bmatrix} 0 \\ 0 \end{bmatrix}$$

Find a control sequence, if it exists, that can drive the system to the desired final state $\xi = [1 \quad 1.2]^T$.

Solution

The controllability matrix of the system is

$$S = [b \; \vdots \; Ab] = \begin{bmatrix} 0 & 1 \\ 1 & 1 \end{bmatrix}$$

Here, $|S| \neq 0$. Hence, the system is controllable and therefore there exists a control sequence that can drive the system to the desired final state $\xi = [1 \quad 1.2]^T$. The response of the system at time $k = 2$ is

$$x(2) = bu(1) + Abu(0) = \begin{bmatrix} 0 \\ 1 \end{bmatrix} u(1) + \begin{bmatrix} 1 \\ 1 \end{bmatrix} u(0) = \begin{bmatrix} u(0) \\ u(1) + u(0) \end{bmatrix}$$

For $\xi = x(2)$, we obtain $u(0) = 1$ and $u(1) = 0.2$. Thus, the desired control sequence is $\{u(0), u(1)\} = \{1, 0.2\}$.

12.5.2 Observability

Definition 12.5.2

System (12.5-1) is observable if there exists a finite time q such that, on the basis of the input sequence $\{u(0), u(1), \ldots, u(q-1)\}$ and the output sequence $\{y(0), y(1), \ldots, y(q-1)\}$, the initial state $x(0)$ of the system may be uniquely determined.

Consider the system (12.5-1). The influence of the input signal $u(k)$ on the behavior of the system can always be determined. Therefore, without loss of generality, we can assume that $u(k) = 0$. We also assume that the output sequence $\{y(0), y(1), \ldots, y(q-1)\}$ is known (for a certain q). This leads to the following system of equations:

$$\begin{bmatrix} C \\ CA \\ \vdots \\ CA^{q-1} \end{bmatrix} x(0) = \begin{bmatrix} y(0) \\ y(1) \\ \vdots \\ y(q-1) \end{bmatrix} \tag{12.5-7}$$

where use was made of Eq. (12.2-17) with $u(k) = 0$. Equation (12.5-7) is an inhomogenous linear algebraic system of equations with $x(0)$ unknown. Equation (12.5-7) has a unique solution for $x(0)$ (as is required from Definition 12.5.2) if, and only if, there exists a finite q such that

$$\mathrm{rank} \begin{bmatrix} C \\ CA \\ \vdots \\ CA^{q-1} \end{bmatrix} = n \tag{12.5-8}$$

Clearly, an increase in time q improves the possibility of satisfying condition (12.5-8). However, the Cayley–Hamilton theorem (Sec. 2.11) states that the terms CA^j, for $j \geq n$, are linearly dependent on the first n terms (i.e., on the terms $C, CA, \ldots, CA^{n-1}$). Thus, condition (12.5-8) holds true if, and only if, $q = n$, i.e., if

$$\mathrm{rank} \begin{bmatrix} C \\ CA \\ \vdots \\ CA^{n-1} \end{bmatrix} = n \tag{12.5-9}$$

Therefore, the following theorem has been proved.

Theorem 12.5.2

System (12.5-1) is observable if, and only if,

$$\mathrm{rank}\, \mathbf{R} = n, \qquad \text{where} \qquad \mathbf{R} = \begin{bmatrix} C \\ CA \\ \vdots \\ CA^{n-1} \end{bmatrix}$$

Here the $np \times n$ matrix $\mathbf{R}$ is called the *observability matrix* (see also Subsec. 5.6.3).

Remark 12.5.3

In an observable system of order n, the knowledge of the first n output values $\{\mathbf{y}(0),$ $\mathbf{y}(1), \ldots, \mathbf{y}(n-1)\}$ is sufficient to determine the initial condition $\mathbf{x}(0)$ of the system uniquely.

Example 12.5.3

Consider the system (12.5-1), where $\mathbf{u}(kT) = \mathbf{0}, \forall k$, and

$$\mathbf{A} = \begin{bmatrix} 1 & 0 \\ 1 & 1 \end{bmatrix}, \qquad \mathbf{c} = \begin{bmatrix} 0 \\ 1 \end{bmatrix}$$

The output sequence of the system is $\{y(0), y(1)\} = \{1, 1.2\}$. Find the initial state $\mathbf{x}(0)$ of the system.

Solution

The observability matrix $\mathbf{R}$ of the system is

$$\mathbf{R} = \begin{bmatrix} \mathbf{c}^T \\ \mathbf{c}^T\mathbf{A} \end{bmatrix} = \begin{bmatrix} 0 & 1 \\ 1 & 1 \end{bmatrix}$$

which has a nonzero determinant. Hence, the system is observable and the initial conditions may be determined from Eq. (12.5-7) which, for the present example, is

$$\begin{bmatrix} 0 & 1 \\ 1 & 1 \end{bmatrix}\begin{bmatrix} x_1(0) \\ x_2(0) \end{bmatrix} = \begin{bmatrix} 1 \\ 1.2 \end{bmatrix}$$

From this equation, we obtain $x_2(0) = 1$ and $x_1(0) = 0.2$.

12.5.3 Loss of Controllability and Observability Due to Sampling

As we already know from Sec. 12.3, when sampling a continuous-time system, the resulting discrete-time system matrices depend on the sampling period T. How does this sampling period T affect the controllability and the observability of the discretized system?

For a discretized system to be controllable, it is necessary that the initial continuous-time system be controllable. This is because the control signals of the sampled-data system are only a subset of the control signals of the continuous-time system. However, the controllability may be lost for certain values of the sampling period. Hence, the initial continuous-time system may be controllable, but the equivalent discrete-time system may not. Similar problems occur for the observability of the system.

Example 12.5.4

The state equations of the harmonic oscillator with $H(s) = \omega^2/(s^2 + \omega^2)$ are

$$\dot{\mathbf{x}}(t) = \begin{bmatrix} 0 & \omega \\ -\omega & 0 \end{bmatrix}\mathbf{x}(t) + \begin{bmatrix} 0 \\ \omega \end{bmatrix}u(t)$$

$$y(t) = [1 \quad 0]\mathbf{x}(t)$$

Investigate the controllability and the observability of the sampled-data (discrete-time) system whose states are sampled with a sampling period T.

Solution

The discrete-time model of the harmonic oscillator is (see Example 12.3.5)

$$\mathbf{x}(kT + T) = \begin{bmatrix} \cos \omega T & \sin \omega T \\ -\sin \omega T & \cos \omega T \end{bmatrix} \mathbf{x}(kT) + \begin{bmatrix} 1 - \cos \omega T \\ \sin \omega T \end{bmatrix} u(kT)$$

$$y(kT) = [1 \quad 0]\mathbf{x}(kT)$$

One can easily calculate the determinants of the controllability and observability matrices to yield $|\mathbf{S}| = -\sin \omega T(1 - \cos \omega T)$ and $|\mathbf{R}| = \sin \omega T$, respectively. We observe that the controllability and observability of the discrete-time system is lost when $\omega T = q\pi$, where q is an integer, although the respective continuous-time system is both controllable and observable.

12.6   CLASSICAL AND DISCRETE-TIME CONTROLLER DESIGN

The classical discrete-time controller design methods are categorized as indirect and direct techniques.

1 Indirect Techniques

Using these techniques, a discrete-time controller $G_c(z)$ is determined indirectly as follows. Initially, the continuous-time controller $G_c(s)$ is designed in the s-domain, using well-known classical techniques (e.g., root locus, Bode, Nyquist, etc.). Then, based on the continuous-time controller $G_c(s)$, the discrete-time controller $G_c(z)$ may be calculated using one of the discretization techniques presented in Subsec. 12.3.3. The indirect techniques are presented in Sec. 12.7 that follows.

2 Direct Techniques

These techniques start by deriving a discrete-time mathematical model of the continuous-time system under control. Subsequently, the design is carried out in the z-domain, wherein the discrete-time controller $G_c(z)$ is directly determined. The design in the z-domain may be done either using the root locus (Sec. 12.8) or the Bode and Nyquist diagrams (Sec. 12.9).

Special attention is given to the PID discrete-time controller design (Sec. 12.10). In Sec. 12.11, a brief description of the steady errors appearing in discrete-time systems is presented.

12.7   DISCRETE-TIME CONTROLLERS DERIVED FROM CONTINUOUS-TIME CONTROLLERS

12.7.1   Discrete-Time Controller Design Using Indirect Techniques

The practicing control engineer has often greater knowledge and experience in designing continuous-time rather than discrete-time controllers. Moreover, many practical systems already incorporate a continuous-time controller which we desire to replace by a discrete-time controller.

The remarks above are the basic motives for the implementation of indirect design techniques for discrete-time controllers mentioned in Sec. 12.6. Indirect techniques take advantage of the knowledge and the experience one has for continuous-time systems. Furthermore, in cases where a continuous-time controller is already incorporated in the system under control, it facilitates the design of a discrete-time controller.

Consider the continuous-time closed-loop control system shown in Figure 12.18 and the discrete-time closed-loop control system shown in Figure 12.19. The indirect design technique for the design of a discrete-time controller may be stated as follows. Let the specifications of the closed-loop systems shown in Figures 12.18 and 12.19 be the same. Assume that a continuous-time controller $G_c(s)$, satisfying the specifications of the closed-loop system shown in Figure 12.18, has already been determined. Then, the discrete-time controller $G_c(z)$ shown in Figure 12.19 may be calculated from the continuous-time controller $G_c(s)$ of Figure 12.18, using the discretization techniques presented in Subsec. 12.3.3.

12.7.2 Specifications of the Time Response of Continuous-Time Systems

In this section, a brief review of the specifications of the time response of the continuous-time systems is given. These specifications are useful for the material that follows and, as it is usually done, refer to the step response of a second-order system (see also Sec. 4.3).

1 Overshoot

One of the basic characteristics of the transient response of a system is the overshoot v, which depends mainly on the damping factor ζ. In the case of a second-order system, without zeros, i.e., for a system with a transfer function of the form

$$H(s) = \frac{\omega_n}{s^2 + 2\zeta\omega_n s + \omega_n^2} \qquad (12.7\text{-}1)$$

it is approximately true [see also Eq. (4.3-12)] that

$$\text{Overshoot percentage} = \%v = 100 \exp\left[\frac{-\zeta\pi}{\sqrt{1-\zeta^2}}\right] \cong 100\left[1 - \frac{\zeta}{0.6}\right] \qquad (12.7\text{-}2)$$

where ω_n is the natural frequency of the system. Therefore, for a desired overshoot percentage, the damping ratio would be

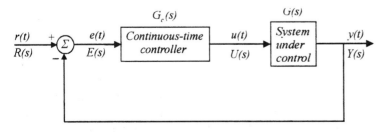

Figure 12.18 Continuous-time closed-loop control system.

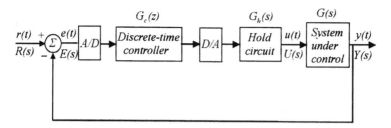

Figure 12.19   Discrete-time closed-loop control system.

$$\zeta \geq (0.6)\left[1 - \frac{\%v}{100}\right]$$ (12.7-3)

2 Rise Time

Another property which is of interest is the rise time T_r, which is defined as the time required for the response of the system to rise from 0.1 to 0.9 of its final value. For all values of ζ around 0.5, the rise time is approximately given by

$$T_r \cong 1.8/\omega_n$$ (12.7-4)

Hence, satisfying the above relation for the rise time, the natural frequency ω_n should satisfy the condition

$$\omega_n \geq 1.8/T_r$$ (12.7-5)

3 Settling Time

Finally, another significant characteristic of the response in the time domain is the settling time T_s, which is defined as the time required for the response to remain close (i.e., within a small error) to the final value. The settling time T_s is given by the relation

$$T_s = \beta/\zeta\omega_n$$ (12.7-6)

where β is a constant. It is mentioned that in the case of an error tolerance of about 1%, the constant β takes on the value 4.6, whereas in the case of an error tolerance of about 2%, the constant β takes on the value 4. Hence, if we desire that the settling time be smaller than a specified value and for an error tolerance of about 1%, then

$$\zeta\omega_n \geq 4.6/T_s$$ (12.7-7)

Remark 12.7.1

Theorem B.3.1 of Appendix B requires that the sampling frequency f be at least twice the highest frequency of the frequency spectrum of the continuous-time input signal. In practice, for a wide class of systems, the selection of the sampling period $T = 1/f$ is made using the following approximate method: let q be the smallest time constant of the system; then, T is chosen such that $T \in [0.1q, 0.5q]$.

Example 12.7.1

Consider the position control servomechanism described in Subsec. 3.13.2. For simplicity, let $L_a \simeq 0$, $K_p = 1$, and $K = A = 1$ and $B = 2$. Then, the transfer function of the motor–gear–load system becomes $G_p(s) = 1/s(s+2)$. To this servomechanism, a continuous-time controller $G_c(s)$ is introduced, as shown in Figure 12.20, which satisfies certain design requirements. Of course, if a discrete-time controller $G_c(z)$ is introduced instead of the continuous-time controller $G_c(s)$, then the closed-loop system would be as shown in Figure 12.21.

Let the continuous-time controller $G_c(s)$ satisfying the design requirements have the following form:

$$G_c(s) = K_s\left[\frac{s+a}{s+b}\right] = 101\left[\frac{s+2}{s+6.7}\right] \tag{12.7-8}$$

Then the problem at hand is to determine $G_c(z)$ of the closed-loop system shown in Figure 12.21 satisfying the same design requirements, where the sampling time $T = 0.2$ sec.

Solution

The transfer function $H(s)$ of the closed-loop system of Figure 12.20 is

$$H(s) = \frac{\Theta_y(s)}{\Theta_r(s)} = \frac{G_c(s)G_p(s)}{1 + G_c(s)G_p(s)} = \frac{101(s+2)}{s(s+2)(s+6.7) + 101(s+2)}$$

$$= \frac{101s + 202}{s^3 + 8.7s^2 + 114.4s + 202} \tag{12.7-9}$$

To find $G_c(z)$ of the closed-loop system of Figure 12.21, it is sufficient to discretize $G_c(s)$ given in Eq. (12.7-8). To this end, we will use the method of pole-zero matching presented in Subsec. 12.3.3 [relations (12.3-20)–(12.3-23)]. According to this method, $G_c(z)$ has the form

$$G_c(z) = K_z\left[\frac{z+z_1}{z+p_1}\right] \tag{12.7-10}$$

where the pole $s = -6.7$ of $G_c(s)$ is mapped into the pole $z = -p_1$ of $G_c(z)$ and the zero $s = -2$ of $G_c(s)$ is mapped into the zero $z = -z_1$ of $G_c(z)$. That is, we have that

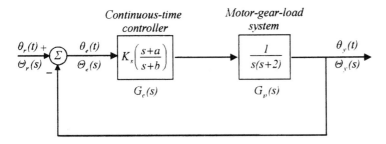

Figure 12.20   Continuous-time closed-loop control system of the position control servomechanism.

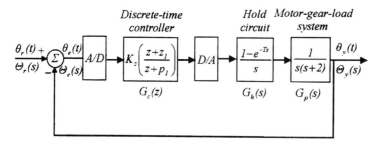

Figure 12.21 Discrete-time closed-loop control system of the position control servo-mechanism.

$$p_1 = -e^{-bT} = -e^{-6.7(0.2)} = -e^{-1.34} = -0.264$$
$$z_1 = -e^{-aT} = -e^{-2(0.2)} = -e^{-0.4} = -0.67$$

The constant K_z of Eq. (12.7-10) is calculated so that the zero frequency amplification constants of $G_c(z)$ and $G_c(s)G_h(s)$ are the same, i.e., so that the following holds (see relation (12.3-23)):

$$G_c(z = 1) = K_z \left[\frac{1 - 0.67}{1 - 0.264} \right] = G_c(s = 0)G_h(s = 0) = 101 \left[\frac{0+2}{0+6.7} \right] \left[\frac{2}{0+10} \right]$$

Thus $K_z = 13.6$. It is noted that in the relation above the value of $G_h(s)$ for $s = 0$ was taken into consideration to obtain the total zero frequency amplification for the continuous-time controller. Hence

$$G_c(z) = 13.6 \left[\frac{z - 0.67}{z - 0.264} \right] \tag{12.7-11}$$

In what follows, the responses of the closed-loop systems of Figures 12.20 and 12.21 are compared. To this end, the discrete-time transfer function of $\hat{G}(s) = G_h(s) G_p(s)$ is determined for $T = 0.2$. We have

$$\hat{G}(z) = Z\{G_h(s)G_p(s)\} = Z\left\{ \left[\frac{1 - e^{-0.2s}}{s} \right] \left[\frac{1}{s(s+2)} \right] \right\}$$

$$= (1 - z^{-1})Z\left\{ \frac{1}{s^2(s+2)} \right\} = (1 - z^{-1})Z\left\{ \frac{0.5}{s^2} - \frac{0.25}{s} + \frac{0.25}{s+2} \right\}$$

$$= \frac{z-1}{z} \left[\frac{0.1}{(z-1)^2} - \frac{0.25z}{z-1} + \frac{0.25z}{z - e^{-0.4}} \right]$$

$$= \frac{0.0176(z + 0.876)}{(z-1)(z-0.67)} \tag{12.7-12}$$

The transfer function of the closed-loop system of Figure 12.21 would be

$$H(z) = \frac{G_c(z)\hat{G}(z)}{1 + G_c(z)\hat{G}(z)} = \frac{0.239z^{-1}(1 + 0.876z^{-1})}{(1 - 0.264z^{-1})(1 - z^{-1}) + 0.239z^{-1}(1 + 0.876z^{-1})}$$

$$= \frac{0.239z^{-1} + 0.209z^{-2}}{1 - 1.025z^{-1} + 0.473z^{-2}} \tag{12.7-13}$$

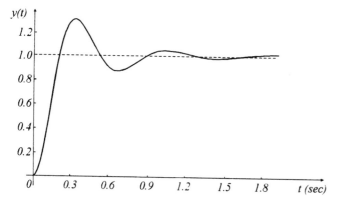

Figure 12.22   Step response of the continuous-time closed-loop system of Figure 12.20.

Figures 12.22 and 12.23 show the response of the closed-loop systems of Figures 12.20 and 12.21, respectively, where it can be seen that the two step responses are almost the same.

12.8   CONTROLLER DESIGN VIA THE ROOT LOCUS METHOD

The root locus method is a direct method for determining $G_c(z)$ and is applied as follows. Consider the closed-loop system shown in Figure 12.24. The transfer function $H(z)$ of the closed-loop system is

$$H(z) = \frac{G(z)}{1 + G(z)F(z)} \tag{12.8-1}$$

The characteristic equation of the closed-loop system is

$$1 + G(z)F(z) = 0 \tag{12.8-2}$$

For linear time-invariant systems, the open-loop transfer function $G(z)F(z)$ has the form

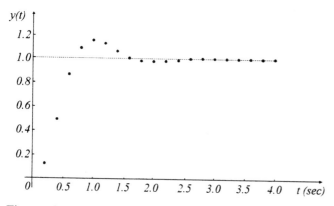

Figure 12.23   Step response of the discrete-time closed-loop system of Figure 12.21.

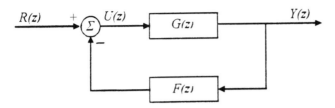

Figure 12.24   Discrete-time closed-loop system.

$$G(z)F(z) = K \frac{\prod\limits_{i=1}^{m}(z+z_i)}{\prod\limits_{i=1}^{n}(z+p_i)}$$

(12.8-3)

Substituting Eq. (12.8-3) in Eq. (12.8-2) yields the algebraic equation

$$\prod_{i=1}^{n}(z+p_i) + K \prod_{i=1}^{m}(z+z_i) = 0$$

(12.8-4)

Definition 12.8.1

The root locus of the closed-loop system of Figure 12.24 are the loci of (12.8-4) in the z-domain as the parameter K varies from $-\infty$ to $+\infty$. Since the poles $-p_i$ and the zeros $-z_i$ are, in general, functions of the sampling time T, it follows that for each T there corresponds a root locus of Eq. (12.8-4), thus yielding a family of root loci for various values of T.

The construction of the root locus of Eq. (12.8-4) is carried out using the material of Chap. 7.

The following example illustrates the construction of the root locus as the parameter K varies from 0 to $+\infty$. It also illustrates the influence on the root locus of the parameter T. Clearly, the influence of the sampling time T on the root locus is a feature which appears in the case of discrete-time systems, but not in the continuous-time systems.

Example 12.8.1

Consider the closed-loop system shown in Figure 12.25. Construct the root locus of the system for $K > 0$ and for several values of the sampling time T. Note that for $a = 2$, the system under control is the position control system presented in Example 12.7.1.

Solution

Let $\hat{G}(s) = G_h(s)G_p(s)$. Then

$$\hat{G}(s) = \left[\frac{1 - e^{-sT}}{s}\right]\left[\frac{1}{s(s+a)}\right] = \frac{1}{s^2(s+a)}\left[1 - e^{-sT}\right]$$

$$= \frac{1}{a^2}\left[\frac{a}{s^2} - \frac{1}{s} + \frac{1}{s+a}\right](1 - e^{-sT})$$

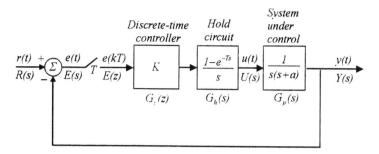

Figure 12.25 Discrete-time closed-loop control system of Example 12.8.1.

Hence

$$\hat{g}(t) = L^{-1}\{\hat{G}(s)\} = p(t) - p(t - T)\beta(t - T)$$

where

$$p(t) = L^{-1}\left\{\frac{1}{a^2}\left[\frac{a}{s^2} - \frac{1}{s} - \frac{1}{s+a}\right]\right\} = \frac{1}{a^2}\left[at - 1 + e^{-at}\right]$$

Therefore

$$\hat{g}(kT) = p(kT) - p(kT - T)\beta(kT - T)$$

Applying relation (B.3-12b) of Appendix B, we have

$$\hat{G}(z) = Z\{\hat{g}(kT)\} = P(z) - z^{-1}P(z) = \left(1 - z^{-1}\right)P(z)$$

where

$$P(z) = Z\{p(kT)\} = \frac{1}{a^2}Z\{akT - \beta(kT) + e^{-akT}\}$$

$$= \frac{1}{a^2}\left[\frac{aTz}{(z-1)^2} - \frac{z}{z-1} + \frac{z}{z-e^{-aT}}\right]$$

Therefore

$$\hat{G}(z) = Z\{\hat{G}(s)\} = Z\{G_h(s)G_p(s)\} = Z\{\hat{g}(kT)\} = \left(1 - z^{-1}\right)P(z)$$

$$= \frac{1}{a^2}\left(1 - z^{-1}\right)\left[\frac{aTz}{(z-1)^2} - \frac{z}{z-1} + \frac{z}{z-e^{-aT}}\right]$$

$$= \frac{\left(aT + e^{-aT} - 1\right)}{a^2}\left[\frac{z + \dfrac{\left[1 - aTe^{-aT} - e^{-aT}\right]}{\left[-1 + aT + e^{-aT}\right]}}{(z-1)(z-e^{-aT})}\right]$$

or

$$\hat{G}(z) = K_0\left[\frac{z + z_1)}{(z-1)(z-p_1)}\right]$$

where

$$K_0 = \frac{1}{a^2}(aT + e^{-aT} - 1), \qquad p_1 = e^{-aT}, \qquad \text{and} \qquad z_1 = \frac{1 - Tae^{-aT} - e^{-aT}}{-1 + aT + e^{-aT}}$$

Hence, the open-loop transfer function $G(z)F(z)$ for the present example is

$$G(z)F(z) = K\hat{G}(z) = K\left[\frac{K_0(z + z_1)}{(z - 1)(z - p_1)}\right]$$

Clearly, the constant K_0, the pole p_1, and the root $-z_1$ are changing with the sampling period T. For this reason, as T changes, so will the root locus of $G(z)F(z)$. Figure 12.26 presents the root loci for three different values T_1, T_2, and T_3 of T, where $T_1 < T_2 < T_3$.

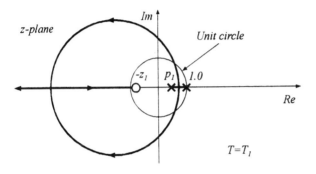

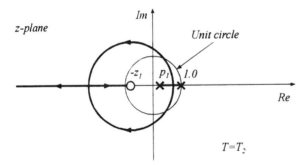

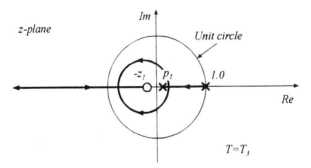

Figure 12.26 Root loci diagrams of the closed-loop system of Figure 12.25 where $T_1 < T_2 < T_3$.

Next, the special case where $a = 1$ and $T = 1$, 2, and 4 sec will be studied. For $a = 1$ and $T = 1$, the open-loop transfer function $G(z)F)z)$ becomes

$$G(z)F(z) = K\left[\frac{0.368(z + 0.718)}{(z - 1)(z - 0.368)}\right]$$

For $a = 1$ and $T + 2$, the open-loop transfer function $G(z)F(z)$ becomes

$$G(z)F(z) = K\left[\frac{1.135(z + 0.523)}{(z - 1)(z - 0.135)}\right]$$

Finally, for $a = 1$ and $T = 4$, the open-loop transfer function $G(z)F(z)$ becomes

$$G(z)F(z) = K\left[\frac{3.018(z + 0.3)}{(z - 1)(z - 0.018)}\right]$$

Figure 12.27 presents the root loci for $a = 1$ and for the three cases of $T = 1$, 2, and 4 sec. The influence of the parameter T can be observed here in greater detail than in Figure 12.26. Figure 12.27 shows that, for a fixed value of K, an increase in the sampling time T would result in a less stable closed-loop system. On the contrary, a decrease in T results in a more stable system. As a matter of fact, the more the sampling period T goes to zero ($T \rightarrow 0$), the more the behavior of the closed-loop system approaches that of the continuous-time system (here, the continuous-time closed-loop system is stable for all positive values of K). It is also noted that as the value of T increases, the critical value of K_c decreases and vice versa, where by critical value of K we mean that particular value of K where the system becomes unstable.

12.9 CONTROLLER DESIGN BASED ON THE FREQUENCY RESPONSE

12.9.1 Introduction

The well-established frequency domain design controller techniques for continuous-time systems (see Chap. 9), can be extended to cover the case of the discrete-time systems. At first, one might think of carrying out this extension by using the relation $z = e^{sT}$. Making use of this relation, the simple and easy-to-use logarithmic curves of the Bode diagrams for the continuous-time case cease to hold for the discrete-time systems (that is why the extension via the relation $z = e^{sT}$ is not recommended). To maintain the simplicity of the logarithmic curves for the discrete-time systems, we make use of the following bilinear transformation:

$$z = \frac{1 + Tw/2}{1 - Tw/2} \quad \text{or} \quad w = \frac{2}{T}\left[\frac{z - 1}{z + 1}\right] \qquad (12.9\text{-}1)$$

The transformation of a function of s to a function of z based on the relation $z = e^{sT}$ and, subsequently, the transformation of the resulting function of z to a function of w based on the relation (12.9-1), are presented in Figure 12.28. The figure shows that the transformation of the left-half complex plane on the s-plane transforms into the unit circle in the z-plane via the relation $z = e^{sT}$, whereas the unit circle on the z-plane transforms into the left-half complex plane in the w-plane, via the bilinear transformation (12.9-1).

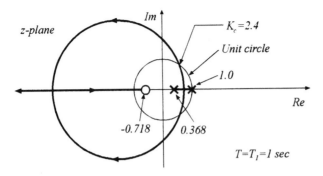

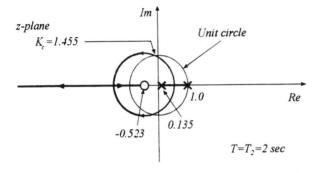

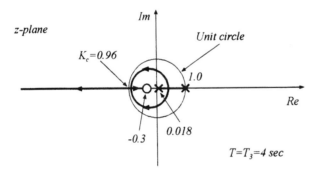

Figure 12.27 Root loci diagrams for the closed-loop system 12.25 for $a = 1$ and $T = 1, 2,$ and 4 sec.

At first sight, it seems that the frequency responses would be the same in both the s- and the w-domain. This is actually true, with the only difference that the scales of the frequencies w and v are distorted, where v is the (hypothetical or abstract) frequency in the w-domain. This frequency "distortion" may be observed if in Eq. (12.9-1) we set $w = jv$ and $z = e^{j\omega T}$, yielding

$$w\Big|_{w=jv} = jv = \frac{2}{T}\left[\frac{z-1}{z+1}\right]\Big|_{z=e^{j\omega T}} = \frac{2}{T}\left[\frac{e^{j\omega t}-1}{e^{j\omega T}+1}\right] = j\frac{2}{T}\tan\left(\frac{\omega T}{2}\right)$$

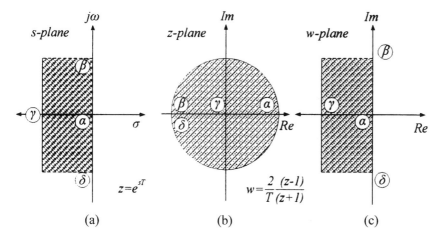

Figure 12.28 Mappings from the s-plane to the z-plane and from the z-plane to the w-plane.

Therefore

$$v = \frac{2}{T} \tan\left(\frac{\omega T}{2}\right) \tag{12.9-2}$$

Since

$$\tan\left(\frac{\omega T}{2}\right) = \frac{\omega T}{2} - \frac{(\omega T)^3}{8} + \cdots \tag{12.9-3}$$

it follows that for small values of ωT we have that $\tan(\omega T/2) \simeq \omega T/2$. Substituting this result in Eq. (12.9-2), we have

$$v \simeq \omega, \qquad \text{for small } \omega T \tag{12.9-4}$$

Therefore, the frequencies ω and v are linearly related if the product ωT is small. For greater ωT, Eq. (12.9-4) does not hold true. Figure 12.29 shows the graphical representation of Eq. (12.9-2). It is noted that the frequency range $-\omega_s/2 \le \omega \le \omega_s/2$ in the s-domain corresponds to the frequency range $-\infty \le v \le \infty$ in the w-domain, where ω_s is defined by the relation $(\omega_s/2)(T/2) = \pi/2$.

12.9.2 Bode Diagrams

Using the above results, one may readily design discrete-time controllers using Bode diagrams. To this end, consider the closed-loop system shown in Figure 12.30. Then, the five basic steps for the design of $G_c(z)$ are the following:

1. Determine $\hat{G}(z)$ from the relation $\hat{G}(z) = Z\{\hat{G}(s)\} = Z\{G_h(s)G_p(s)\}$
2. Determine $\hat{G}(w)$ using the bilinear transformation (12.9-1), yielding

$$\hat{G}(w) = \hat{G}(z)|_{z=(1+Tw/2)/(1-Tw/2)} \tag{12.9-5}$$

3. Set $w = jv$ in $\hat{G}(w)$ and draw the Bode diagrams of $\hat{G}(jv)$
4. Determine the controller $G_c(w)$ using similar techniques to those applied for continuous-time systems (see Chap. 9)

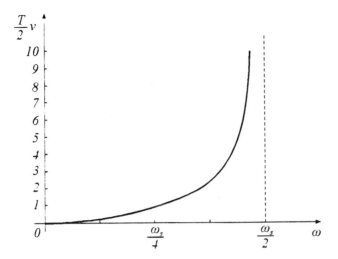

Figure 12.29 Graphical representation of relation (12.9-2).

5. Determine $G_c(z)$ from $G_c(w)$ using the bilinear transformation (12.9-1), yielding

$$G_c(z) = G_c(w)|_{w=(2/T)[(z-1)/(z+1)]} \qquad (12.9\text{-}6)$$

Note that the specifications for the bandwidth are transformed from the s-domain to the w-domain using relation (12.9-2). Thus, if for example ω_b is the desired frequency bandwidth, then the design in the w-domain must be carried out for a frequency bandwidth v_b, where

$$v_b = \frac{2}{T} \tan\left(\frac{\omega_b T}{2}\right) \qquad (12.9\text{-}7)$$

Example 12.9.1

Consider the position servomechanism shown in Figure 12.21 of Example 12.7.1. Find a controller $G_c(z)$ such that the closed-loop system satisfies the following

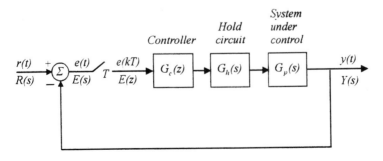

Figure 12.30 Discrete-time closed-loop system.

specifications: gain margin $K_g \geq 25$ dB, phase margin $\varphi_p \geq 70°$, and velocity error constant $K_v = 1 \, \text{sec}^{-1}$. The sampling period T is chosen to be 0.1 sec.

Solution

Let $\hat{G}(s) = G_h(s)G_p(s)$. Then

$$\hat{G}(z) = Z\{\hat{G}(s)\} = Z\left\{\frac{1 - e^{-Ts}}{s} \frac{1}{s(s+2)}\right\} = (1 - z^{-1})Z\left\{\frac{1}{s^2(s+2)}\right\}$$

$$= 0.0047z^{-1}\left[\frac{1 + 0.935z^{-1}}{(1 - z^{-1})(1 - 0.819z^{-1})}\right] = (0.0047)\frac{z + 0.935}{(z-1)(z-0.819)}$$

For $T = 0.1$ sec, the bilinear transformation (12.9-1) becomes

$$z = \frac{1 + (Tw/2)}{1 - (Tw/2)} = \frac{1 + 0.05w}{1 - 0.05w}$$

Substituting the above transformation in $\hat{G}(z)$ we have

$$\hat{G}(w) = \frac{0.0047\left(\dfrac{1 + 0.05w}{1 - 0.05w} + 0.935\right)}{\left(\dfrac{1 + 0.05w}{1 - 0.05w} - 1\right)\left(\dfrac{1 + 0.05w}{1 - 0.05w} - 0.8187\right)}$$

$$= \frac{0.5(1 + 0.00167w)(1 - 0.05w)}{w(1 + 0.5w)}$$

The gain and phase Bode diagrams of $\hat{G}(jv) = \hat{G}(w = jv)$ are given in Figure 12.31.

We choose the following form for the controller $G_c(w)$:

$$G_c(w) = K\left[\frac{1 + aw}{1 + bw}\right]$$

where a and b are constants. The open-loop transfer function is

$$G_c(w)\hat{G}(w) = K\left[\frac{1 + aw}{1 + bw}\right]\left[\frac{0.5(1 + 0.00167w)(1 - 0.05w)}{w(1 + 0.5w)}\right]$$

From the definition of the velocity error constant K_v, we have

$$K_v = \lim_{w \to 0}\left[wG_c(w)\hat{G}(w)\right] = 0.5K = 1$$

and therefore $K = 2$. The parameters a and b can be determined by applying the respective techniques of continuous-time systems (Chap. 9), which yield $a = 0.8$ and $b = 0.5$. Hence

$$G_c(w) = 2\left(\frac{1 + 0.8w}{1 + 0.5w}\right)$$

The open-loop transfer function is

$$G_c(w)\hat{G}(w) = 2\left[\frac{1 + 0.8w}{1 + 0.5w}\right]\left[\frac{0.5(1 + 0.00167w)(1 - 0.05w)}{w(1 + 0.5w)}\right]$$

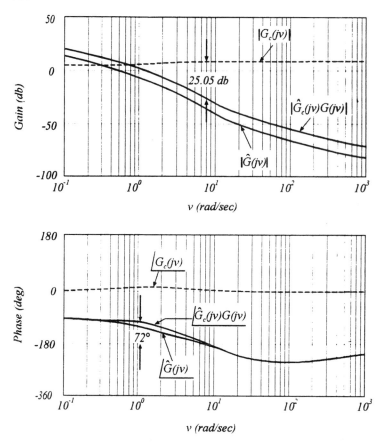

Figure 12.31 The gain and phase Bode diagrams of $\hat{G}(jv)$ of Example 12.9.1.

Checking the above results, we find that $K_g \simeq 25.05\,\text{dB}$, $\varphi_p \simeq 72°$, and $K_v = 1\,\text{sec}^{-1}$. Therefore, the closed-loop design requirements are satisfied. It remains to determine $G_c(z)$ from $G_c(w)$. To this end, we use the bilinear transformation Eq. (12.9-1) which, for $T = 0.1\,\text{sec}$, becomes

$$w = \frac{2}{T}\left[\frac{z-1}{z+1}\right] = \frac{2}{0.1}\left[\frac{z-1}{z+1}\right] = 20\left[\frac{z-1}{z+1}\right]$$

Thus, $G_c(z)$ has the form

$$G_c(z) = 2\left[\frac{1 + (0.8)(20)\left[\frac{z-1}{z+1}\right]}{1 + (0.5)(20)\left[\frac{z-1}{z+1}\right]}\right] = 3.09\left(\frac{z - 0.882}{z - 0.818}\right) = 3.09\left(\frac{1 - 0.882z^{-1}}{1 - 0.818z^{-1}}\right)$$

12.9.3 Nyquist Diagrams

Consider a closed-loop system with an open-loop transfer function $\tilde{G}(z) = Z\{G(s)F(s)\}$. Since the z- and s-domains are related via the relation $z = e^{sT}$, it follows that the Nyquist diagram of $\tilde{G}(z)$ would be the diagram of $\tilde{G}(e^{sT})$, as s traces the Nyquist path. In the z-domain, the Nyquist path is given by the relation

$$z = e^{sT}|_{s=j\omega} = e^{j\omega T} \tag{12.9-8}$$

and, therefore, the Nyquist path in the z-domain is the unit circle. Hence, to apply the Nyquist stability criterion for discrete-time systems, we draw the diagram of $\tilde{G}(e^{j\omega T})$ having the cyclic frequency ω as a parameter.

The following theorem holds.

Theorem 12.9.1

Assume that the transfer function $Z\{G(s)F(s)\}$ does not have any poles outside the unit circle. Then, the closed-loop system is stable if the Nyquist diagram of $Z\{G(s)F(s)\}$, for $z = e^{j\omega T}$, does not encircle the critical point $(-1, j0)$.

Theorem 12.9.1 is the respective (or equivalent) of the Nyquist theorem for continuous-time systems (Subsec. 8.4.3).

Clearly, the study of the stability of discrete-time closed-loop systems, as well as the design of discrete-time controllers, can be accomplished on the basis of the Nyquist diagrams, by extending the known techniques of continuous-time systems as was done in the case of the Bode diagrams in the previous subsection. This extension is straightforward and is not presented here (see for example [6]).

12.10   THE PID CONTROLLER

In discrete-time systems, as in the case of continuous-time systems (see Sec. 9.6), the PID controller is widely used in practice. This section is devoted to the study of discrete-time PID controllers. We will first study separately the proportional (P), the integral (I), and the derivative (D) controller and, subsequently, the composite PID controller.

12.10.1   The Proportional Controller

For the continuous-time systems, the proportional controller is described by the relation

$$u(t) = K_p e(t) \tag{12.10-1a}$$

and therefore

$$G_c(s) = K_p \tag{12.10-1b}$$

For the discrete-time systems, the proportional controller is described by the relation

$$u(k) = K_p e(k) \tag{12.10-2a}$$

and therefore

$$G_c(z) = K_p \tag{12.10-2b}$$

12.10.2   The Integral Controller

For the continuous-time systems, the integral controller is described by the integral equation

$$u(t) = \frac{K_{\mathrm{p}}}{T_{\mathrm{i}}} \int_{t_0}^{t} e(t)\,\mathrm{d}t \qquad\qquad (12.10\text{-}3a)$$

and therefore

$$G_{\mathrm{c}}(s) = \frac{K_{\mathrm{p}}}{T_{\mathrm{i}}s} \qquad\qquad (12.10\text{-}3b)$$

where the constant T_{i} is called the integration time constant or reset. In the case of discrete-time systems, the integral equation (12.10-3a) is approximated by the difference equation

$$\frac{u(k) - u(k-1)}{T} = \frac{K_{\mathrm{p}}}{T_{\mathrm{i}}} e(k)$$

or

$$u(k) = u(k-1) + \frac{K_{\mathrm{p}}T}{T_{\mathrm{i}}} e(k) \qquad\qquad (12.10\text{-}4a)$$

and therefore

$$G_{\mathrm{c}}(z) = \frac{K_{\mathrm{p}}T}{T_{\mathrm{i}}(1 - z)^{-1}} = \frac{K_{\mathrm{p}}Tz}{T_{\mathrm{i}}(z - 1)} \qquad\qquad (12.10\text{-}4b)$$

12.10.3 The Derivative Controller

For the continuous-time systems, the derivative controller is described by the differential equation

$$u(t) = K_{\mathrm{p}}T_{\mathrm{d}}\dot{e}(t) \qquad\qquad (12.10\text{-}5a)$$

and therefore

$$G_{\mathrm{c}}(z) = K_{\mathrm{p}}T_{\mathrm{d}}s \qquad\qquad (12.10\text{-}5b)$$

where the constant T_{d} is called the derivative or rate time constant. In the case of discrete-time systems, the differential equation (12.10-5a) is approximated by the difference equation

$$u(k) = K_{\mathrm{p}}T_{\mathrm{d}}\left[\frac{e(k) - e(k-1)}{T}\right] \qquad\qquad (12.10\text{-}6a)$$

and therefore

$$G_{\mathrm{c}}(z) = K_{\mathrm{p}}T_{\mathrm{d}}\left[\frac{1 - z^{-1}}{T}\right] = \frac{K_{\mathrm{p}}T_{\mathrm{d}}}{T}\left[\frac{z - 1}{z}\right] \qquad\qquad (12.10\text{-}6b)$$

12.10.4 The Three-Term PID Controller

Combining all the above, we have that the PID controller, for continuous-time systems, is described by the integrodifferential equation

$$u(t) = K_{\mathrm{p}}\left[e(t) + \frac{1}{T_{\mathrm{i}}}\int_{t_0}^{t} e(t)\,\mathrm{d}t + T_{\mathrm{d}}\dot{e}(t)\right] \qquad\qquad (12.10\text{-}7a)$$

and therefore

$$G_c(s) = K_p \left[1 + \frac{1}{T_i s} + T_d s \right] \tag{12.10-7b}$$

Figure 9.21 presents the block diagram of the $G_c(s)$ for the continuous-time PID controller.

In the case of discrete-time systems, the PID controller is described by the difference equation

$$u(k) = K_p \left[e(k) + \frac{T}{T_i} \sum_{i=0}^{k-1} e(i) + \frac{T_d}{T} [e(k) - e(k-1)] \right] \tag{12.10-8a}$$

where the middle term in Eq. (12.10-8a) is the solution of Eq. (12.10-4a). Hence

$$G_c(z) = K_p \left[1 + \frac{T}{T_i} \left[\frac{z}{z-1} \right] + \frac{T_d}{T} \left[\frac{z-1}{z} \right] \right] \tag{12.10-8b}$$

After some algebraic manipulations, $G_c(z)$ may be written as

$$G_c(z) = K \left[\frac{z^2 - az + b}{z(z-1)} \right] \tag{12.10-9}$$

where

$$K = K_p \left[\frac{TT_i + T_d T_i + T^2}{T_i T} \right]$$

$$a = \frac{T_i T - T_d T_i}{TT_i + T_d T_i + T^2}$$

$$b = \frac{T_d T_i}{TT_i + T_d T_i + T^2}$$

Figure 12.32 presents the block diagram of the discrete-time PID controller $G_c(z)$.

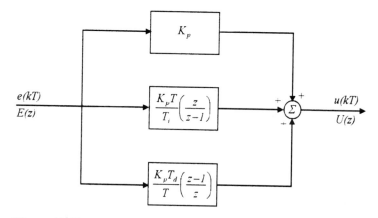

Figure 12.32 The block diagram of the discrete-time PID controller.

12.10.5 Design of PID Controllers Using the Ziegler–Nichols Methods

The Ziegler–Nichols methods for continuous-time systems have been presented in Subsec. 9.6.5. These methods can be extended directly to the case of discrete-time systems, provided that the sampling is sufficiently fast, i.e., 20 times the highest bandwidth frequency, as is normally the case in practice. If the sampling is not that fast, then the discrete-time PID controller may not produce satisfactory accurate results.

12.11 STEADY-STATE ERRORS

The subject of steady-state errors for the case of continuous-time systems was presented in Sec. 4.7. These results can readily be extended to cover the discrete-time systems case. In the sequel, we briefly cover the subject.

Consider the unity feedback discrete-time closed-loop system shown in Figure 12.33. Assume that the system under control is stable (a fact which will allow us to apply the final value theorem given by Eq. (B.3-20) of Appendix B). Define

$$\hat{G}(z) = Z\{G_h(s)G(s)\} = Z\left\{\left[\frac{1 - e^{-Ts}}{s}\right]G(s)\right\} = (1 - z^{-1})Z\left\{\frac{G(s)}{s}\right\}$$

Then, the closed-loop transfer function $H(z)$ will be

$$H(z) = \frac{Y(z)}{R(z)} = \frac{\hat{G}(z)}{1 + \hat{G}(z)} \qquad (12.11\text{-}1)$$

The error $E(z)$ is given by

$$E(z) = R(z) - B(z) = R(z) - \hat{G}(z)E(z)$$

and hence

$$E(z) = \left[\frac{1}{1 + \hat{G}(z)}\right]R(z) \qquad (12.11\text{-}2)$$

The steady-state error of $e(kT)$, denoted as e_{ss}, is defined as

$$e_{ss} = \lim_{k \to \infty} e(kT) = \lim_{z \to 1}(1 - z^{-1})E(z) \qquad (12.11\text{-}3)$$

Relation (12.11-3) is known as the final value theorem, which is defined by the relation (B.3-20) of Appendix B. If Eq. (12.11-2) is substituted in Eq. (12.11-3) then

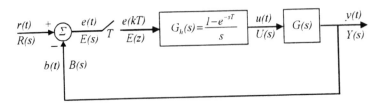

Figure 12.33   Unity feedback discrete-time closed-loop system.

$$e_{ss} = \lim_{z \to 1} \left[(1 - z^{-1}) \left[\frac{1}{1 + \hat{G}(z)} \right] R(z) \right] \qquad (12.11\text{-}4)$$

Next, we will consider three particular excitations $r(t)$: namely, the step function, the ramp function, and the acceleration function.

1 Step Function

In this case, $r(t) = 1$ or $r(kT) = 1$ and

$$R(z) = Z\{r(kT)\} = Z\{1\} = \frac{1}{1 - z^{-1}}$$

Substituting the above value of $R(z)$ in Eq. (12.11-4) yields

$$e_{ss} = \lim_{z \to 1} \left[(1 - z^{-1}) \left[\frac{1}{1 + \hat{G}(z)} \right] \left[\frac{1}{1 - z^{-1}} \right] \right] = \lim_{z \to 1} \left[\frac{1}{1 + \hat{G}(z)} \right] = \frac{1}{1 + K_p},$$

$$K_p = \lim_{z \to 1} \hat{G}(z) \qquad (12.11\text{-}5)$$

where K_p is called the *position error constant*.

2 Ramp Function

In this case $r(t) = t$ or $r(kT) = kT$, which is defined by Eq. (B.2-4) of Appendix B. Hence

$$R(z) = Z\{r(kT)\} = Z\{kT\} = \frac{Tz^{-1}}{\left(1 - z^{-1}\right)^2}$$

Substituting the value of $R(z)$ in Eq. (12.11-4) yields

$$e_{ss} = \lim_{z \to 1} \left[(1 - z^{-1}) \left[\frac{1}{1 + \hat{G}(z)} \right] \left[\frac{Tz^{-1}}{\left(1 - z^{-1}\right)^2} \right] \right] = \lim_{z \to 1} \left[\frac{T}{\left(1 - z^{-1}\right)\hat{G}(z)} \right] = \frac{1}{K_v},$$

$$K_v = \lim_{z \to 1} \left[\frac{\left(1 - z^{-1}\right)\hat{G}(z)}{T} \right] \qquad (12.11\text{-}6)$$

where K_v is called the *velocity error constant*.

3 Acceleration Function

In this case $r(t) = \frac{1}{2}t^2$ or $r(kT) = \frac{1}{2}(kT)^2$ and

$$R(z) = Z\{r(kT)\} = Z\left\{ \frac{1}{2}(kT)^2 \right\} = \frac{T^2\left(1 + z^{-1}\right)z^{-1}}{2\left(1 - z^{-1}\right)^3}$$

Substituting the value of $R(z)$ in Eq. (12.11-4) yields

$$e_{\text{ss}} = \lim_{z \to 1} \left[\left(1 - z^{-1}\right) \left[\frac{1}{1 + \hat{G}(z)} \right] \left[\frac{T^2\left(1 + z^{-1}\right)z^{-1}}{2\left(1 - z^{-1}\right)^3} \right] \right]$$

$$= \lim_{z \to 1} \left[\frac{T^2}{\left(1 - z^{-1}\right)^2 \hat{G}(z)} \right]$$

$$= \frac{1}{K_{\text{a}}}, \qquad K_{\text{a}} = \lim_{z \to 1} \left[\frac{\left(1 - z^{-1}\right)^2 \hat{G}(z)}{T^2} \right] \qquad (12.11\text{-}7)$$

where K_{a} is called the *acceleration error constant*.

It is remarked that, as in the case of continuous-time systems (see Sec. 4.7), a discrete-time system is called a *type j system* when its open-loop transfer function $\hat{G}(z)$ has the form

$$\hat{G}(z) = \frac{1}{(z - 1)^j} \left[\frac{a(z)}{b(z)} \right] \qquad (12.11\text{-}8)$$

where $a(z)$ and $b(z)$ are polynomials in z which do not involve the term $(z - 1)$.

12.12   STATE-SPACE DESIGN METHODS

The results of Chap. 10 on state-space design methods for continuous-time systems may readily be extended to the case of discrete-time systems. Indeed, the problems of pole assignment, input–output decoupling, exact model matching, and state observers may be solved in a similar way for discrete-time systems (see for example [13]).

12.13   OPTIMAL CONTROL

The results of Chap. 11 on optimal control for continuous-time systems may be extended to cover the case of discrete-time systems. However, this extension is not as easy as for the case of the results of Chap. 10. Here, one first has to present the appropriate mathematical background for the study of optimal control problems of discrete-time systems and, subsequently, use this mathematical background to solve the linear regulator and servomechanism problems.

Going from continuous-time systems, which are described by differential equations, to discrete-time systems, which are described by difference equations, a basic difference appears in the form of the cost function: the cost function for continuous-time systems is an *integral* expression, whereas for discrete-time systems it is a *summation* expression. Applying the principles of calculus of variations, one arrives at the *discrete-time Euler–Lagrange* equation. With regard to the maximum principle, an analogous treatment leads to the discrete-time Hamiltonian function, and from there to the discrete-time *canonical Hamiltonian equations*. Using this discrete-time mathematical background, one may then solve the discrete-time regulator and servomechanism problems along the same lines as those used for the case of continuous-time systems. For more details see [13].

12.14 PROBLEMS

1. A discrete-time system is described by the state-space equations (12.2-8), where

$$\mathbf{A} = \begin{bmatrix} 0 & 1 \\ -6 & -5 \end{bmatrix}, \qquad \mathbf{b} = \begin{bmatrix} 0 \\ 1 \end{bmatrix}, \qquad \mathbf{c} = \begin{bmatrix} 1 \\ 1 \end{bmatrix}, \qquad D = 0$$

   Find the transition matrix, the transfer function, and the response of the system when the input is the unit step sequence.

2. Find the transfer functions of the systems described by the difference equations

   (a) $y(k + 2) + 5y(k + 1) + 6y(k) = u(k + 1)$
   (b) $y(k + 2) - y(k) = u(k)$

3. Find the values of the responses $y(0)$, $y(1)$, $y(2)$, and $y(3)$ using the convolution method when

   (a) $h(k) = 1 - e^{-k}$ and $u(k) = \beta(k)$
   (b) $h(k) = 1 - e^{-k}$ and $u(k) = k\beta(k)$
   (c) $h(k) = \sin k$ and $u(k) = \beta(k)$
   (d) $h(k) = e^{-k}$ and $u(k) = e^{-2k}$

   where $\beta(k)$ is the unit step sequence (see Eq. (B.2-2) of Appendix B).

4. Consider the continuous-time system (12.3-24), where

$$\mathbf{F} = \begin{bmatrix} -1 & 0 \\ 0 & -2 \end{bmatrix}, \qquad \mathbf{g} = \begin{bmatrix} 1 \\ -1 \end{bmatrix}, \qquad \mathbf{c}^T = \begin{bmatrix} 0 \\ 1 \end{bmatrix}, \qquad D = 0$$

   Discretize the system, i.e., find the difference equations of the system in state space, when the sampling frequency $T = 0.1$ sec.

5. Find the equivalent discrete-time transfer function $G(z)$ of the continuous-time transfer function $G(s) = 1/[s(s + 1)]$ preceded by a zero-order hold described by $G_h(s) = (1 - e^{-Ts})/s$. Use two approaches: one making use of the Z-transform tables and the other using a time-domain analysis.

6. The block diagram of a digital space vehicle control system is shown in Figure 12.34, where G and F are constants and J is the vehicle moment of inertia (all in appropriate units). Find the discrete-time transfer function of the closed-loop system.

7. A continuous-time process described by the transfer function K/s is controlled by a digital computer, as shown in Figure 12.35.

   (a) Find the closed-loop transfer function $Y(z)/R(z)$, the disturbance-to-output transfer function $Y(z)/D(z)$, and the open-loop transfer function $Y(z)/E(z)$.
   (b) Obtain the steady-state characteristics of the system using the final value theorem.
   (c) Find the unit step response of the system for $K = 20$, $F = 5$, $K_c = 1.25$, and $T = 0.005$ sec.

8. A simplified state-space model for the altitude control (roll control) of a spacecraft is

$$\begin{bmatrix} \dot{x}_1(t) \\ \dot{x}_2(t) \end{bmatrix} = \begin{bmatrix} 0 & 1 \\ 0 & 0 \end{bmatrix} \begin{bmatrix} x_1(t) \\ x_2(t) \end{bmatrix} + \begin{bmatrix} 0 \\ 1/J \end{bmatrix} u(t) = \mathbf{F}\mathbf{x}(t) + \mathbf{g}u(t)$$

$$y(t) = \begin{bmatrix} 1 & 0 \end{bmatrix} \begin{bmatrix} x_1(t) \\ x_2(t) \end{bmatrix} = \mathbf{c}^T \mathbf{x}(t)$$

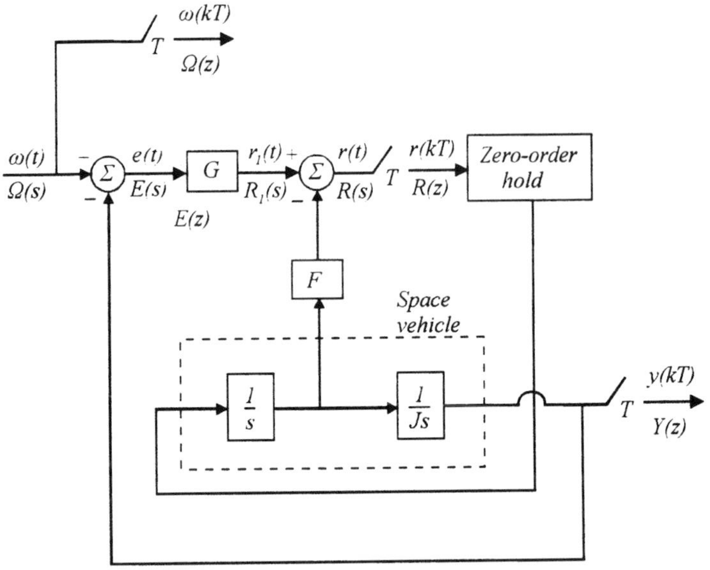

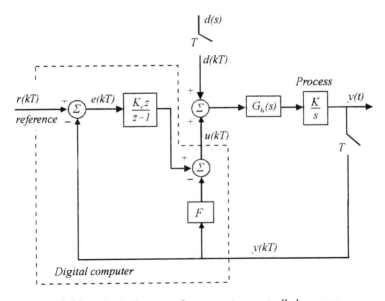

Figure 12.34 A digital space vehicle control system.

Figure 12.35 Block diagram of a computer-controlled process.

where

$$x_1 = \text{the roll of the spacecraft in rad}$$
$$x_2 = \text{the roll rate in rad/sec}$$
$$u = \text{the control torque about the roll axis produced by the thrusters in Nm}$$
$$J = \text{the moment of inertia of the vehicle about the roll axis at the vehicle}$$
$$\text{center of mass in kg m}^2$$

The transfer function relating the roll of the spacecraft to the torque input is

$$G(s) = \frac{Y(s)}{U(s)} = \frac{1}{Js^2}$$

Find the equivalent discrete-time description of the system with sampling period T.

9. Find all values of K for which the roots of the following characteristic polynomials lie inside the unit circle:

(a) $z^2 + 0.2z + K$ (b) $z^2 + Kz + 0.4$
(c) $z^2 + (K + 0.4)z + 1$ (d) $z^3 + Kz^2 + 2z + 2$
(e) $z^3 - 0.5z^2 - 0.2z + K$ (f) $z^3 + (K + 1)z^2 - 0.5z + 1$

10. A magnetic disk drive requires a motor to position a read/write head over tracks of data on a spinning disk. The motor and the head may be approximated by the transfer function

$$G(s) = \frac{1}{s(T_1 s + 1)}$$

where $T_1 > 0$. The controller takes the difference of the actual and desired positions and generates an error. This error is discretized with sampling period T, multiplied by a gain K, and applied to the motor with the use of a zero-order hold of period T (see Figure 12.36).

Determine the range of values of the gain K, so that the closed-loop discrete-time system is stable. Apply the invariant impulse response method and the Routh criterion.

11. Consider the system given in Figure 12.37. Apply the Jury criterion to determine the range of values of the gain K for which the system is stable. Assume sampling period $T = 0.1$ sec, 0.2 sec, and 1 sec.

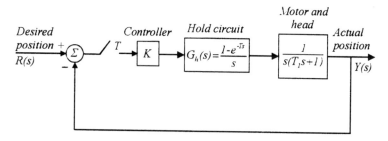

Figure 12.36 Block diagram of disk drive control system of Problem 10.

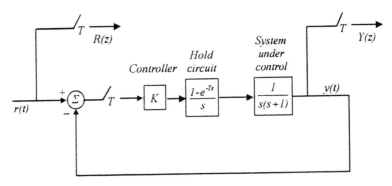

Figure 12.37 Block diagram of system of Problem 11.

12. Check the stability of the system described by

$$x_1(k+1) = x_2(k)$$

$$x_2(k+1) = 2.5x_1(k) + x_2(k) + u(k)$$

$$y(k) = x_1(k)$$

If the system is unstable, use the output feedback law $u(k) = -gy(k)$ to stabilize it. Determine the range of values of a suitable g.

13. Consider a system described by the state-space equations (12.2-8), where

(a) $\quad \mathbf{A} = \begin{bmatrix} 0 & 1 & 0 & 0 \\ 0 & 0 & 1 & 0 \\ 0 & 0 & 0 & 1 \\ -4 & -2 & 1 & 0.4 \end{bmatrix}, \quad \mathbf{B} = \begin{bmatrix} 0 \\ 0 \\ 0 \\ 1 \end{bmatrix}, \quad$ and

$\quad \mathbf{C} = \begin{bmatrix} 1 & -2 & 1 & 4 \end{bmatrix}$

(b) $\quad \mathbf{A} = \begin{bmatrix} 0 & 0 & 1 \\ 1 & 1 & 0 \\ -1 & 0 & 0.5 \end{bmatrix}, \quad \mathbf{B} = \begin{bmatrix} 1 & 2 \\ 1 & 1 \\ -1 & -0.5 \end{bmatrix}, \quad$ and

$\quad \mathbf{C} = \begin{bmatrix} 0 & 1 & 1 \\ 1 & -1 & 1 \end{bmatrix}$

(c) $\quad \mathbf{A} = \begin{bmatrix} 1 & 0 & 1 \\ 1 & 2 & 1 \\ -T & 0 & 0 \end{bmatrix}, \quad \mathbf{B} = \begin{bmatrix} 1 & 1 \\ 0 & 1 \\ 1 & 0 \end{bmatrix}, \quad$ and $\quad \mathbf{C} = \begin{bmatrix} 0 & 0 & 1 \\ 1 & 0 & 0 \end{bmatrix}$

Investigate the controllability and observability of these systems. For case (c), find the values of the sampling time T, appearing in matrix $\mathbf{A}$, which make the system controllable and/or observable.

14. Consider the continuous-time system of a rotating body described by the dynamical equation

$$\dot\omega = \frac{d^2\theta}{dt^2} = \frac{L}{J}$$

where θ is the position (angle of rotation), ω is the rate of the angle of rotation, L is the externally applied torque, and J is the moment of inertia. If $x_1 = \theta$ and $x_2 = \dot\theta = \omega$, then the state-space description is

$$\begin{bmatrix} \dot x_1 \\ \dot x_2 \end{bmatrix} = \begin{bmatrix} 0 & 1 \\ 0 & 0 \end{bmatrix}\begin{bmatrix} x_1 \\ x_2 \end{bmatrix} + \begin{bmatrix} 0 \\ 1 \end{bmatrix}\frac{L}{J} = \mathbf{F}\mathbf{x} + \mathbf{g}u$$

Obtain the discrete-time description using a zero-order hold and a sampling period T. If θ is taken to be the output, determine if this description is observable. What happens if the angular velocity ω is measured instead? Discuss the results in both cases.

15. A system is described by the state equations

$$\mathbf{x}(k+1) = \mathbf{A}\mathbf{x}(k) + \mathbf{b}u(k)$$

$$y(k) = \mathbf{c}^T\mathbf{x}(k)$$

where

$$\mathbf{A} = \begin{bmatrix} 0 & 1 \\ -2 & -3 \end{bmatrix}, \quad \mathbf{b} = \begin{bmatrix} 1 \\ 1 \end{bmatrix}, \quad \text{and} \quad \mathbf{c}^T = [1 \quad 2]$$

Determine the controllability and observability of both the open-loop and the closed-loop systems when

$$u(k) = r(k) - \mathbf{f}^T\mathbf{x}(k)$$

where $r(k)$ is some reference input and $\mathbf{f} = [f_1 \quad f_2]^T$.

16. Solve Example 12.7.1 with the following specifications:

   (a)   maximum overshoot $\leq 10\%$ and natural frequency $\omega_n = 2\,\text{rad/sec}$
   (b)   maximum overshoot $\leq 20\%$ and natural frequency $\omega_n = 6\,\text{rad/sec}$
   (c)   maximum overshoot $\leq 10\%$ and natural frequency $\omega_n = 6\,\text{rad/sec}$.

Note that here one first has to determine $G_c(s)$ satisfying these specifications and, subsequently, determine $G_c(s)$.

17. Consider the ball and beam system depicted in Figure 12.38a. The beam is free to rotate in the plane of the page about an axis perpendicular to the page, while the ball rolls in a groove along the rod. The control problem is that of maintaining the ball at a desired position by applying an input torque to the beam.

   (a)   A linear model for the system is $G(s) = 1/s^2$, as shown with a PD controller in Figure 12.38b. Obtain an equivalent discrete-time system. The sampling period is $T = 0.1$ sec.
   (b)   Design a discrete-time controller using the pole-zero matching method. Draw the unit step responses for the continuous- and the discrete-time systems.

18. Plastic extrusion is an industrial process. The extruders consist of a large barrel divided into several temperature zones with a hopper at one end and a die at the other. The polymer is fed into the barrel from the hopper and is pushed forward

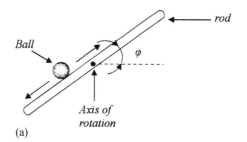

(a)

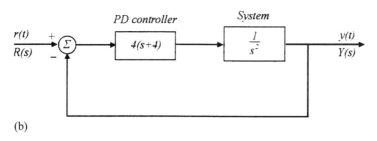

(b)

Figure 12.38   Ball and beam closed-loop control system: (a) ball and beam system; (b) block diagram of closed-loop control system.

by a powerful screw. Simultaneously, it is heated while passing through the various temperature zones set to gradually increasing temperatures. The heat produced by the heaters in the barrel, together with the heat released from the friction between the raw polymer and the surfaces of the barrel and the screw, eventually causes the polymer to melt. The polymer is then pushed out from the die. The discrete-time system for the temperature control is shown in Figure 12.39.

The transfer function relating the angular velocity of the screw and the output temperature is $G(s) = e^{-2s}/(s + 1)$, i.e., the system is of the first order, incorporating a delay of 2 sec. The sampling period $T = 1$ sec. Design a PI controller so that the dominant closed-loop poles have a damping ratio $\zeta = 0.5$ and the number of the output samples in a full cycle of the damped

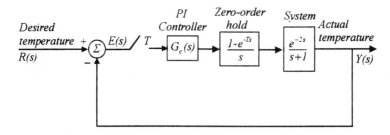

Figure 12.39   Temperature control system for plastic extrusion.

sinusoidal response is 10. Find the unit step response of the discrete-time system. Determine the velocity error coefficient K_v and the steady-state error of the output due to a ramp input.

19. A photovoltaic system is mounted on a space station in order to develop the required power. To maximize the energy production, the photovoltaic panels should follow the sun as accurately as possible. The system uses a dc motor and the transfer function of the panel mount and the motor is

$$G(s) = \frac{1}{s(s+1)}$$

An optical sensor accurately tracks the sun's position and forms a unity feedback (see Figure 12.40). The sampling period $T = 0.2\,\text{sec}$. Find a discrete-time controller such that the dominant closed-loop poles have damping ratio $\zeta = 0.5$ and there are eight output samples in a complete cycle of the damped sinusoidal response. Use the root locus method in the z-plane to determine the transfer function of the required controller. Find the unit step response of the system and the velocity error coefficient K_v.

20. In this problem, the automatic control of a wheelchair will be studied. The automatic wheelchair is specially designed for handicapped people with a disability from the neck down. It consists of a control system which the handicapped person may operate by using his or her head, thus determining the direction as well as the speed of the chair. The direction is determined by a sensor, situated on the head of the handicapped person at intervals of $90°$, so that the person may choose one of the four possible directions (motions): forward, backward, left, and right. The speed is regulated by another sensor, whose output is proportional to the movement of the head. Clearly, in the present example, the person is part of the overall controller.

For simplicity, we assume that the wheelchair, as well as the sensory device on the head, are described by first-order transfer functions, as shown in Figure 12.41a. We also assume that the time delay, which is anticipated to appear in the visual feedback path, is negligible. More specifically, we assume that $K_1 = 1$, $K_2 = 10$, $a = 1$, $b = 2$, and $F(s) = 1$. Suppose that we want to introduce a discrete-time controller to the system. Then, the closed-loop system would be as in Figure 12.41b. Find $G_c(z)$ in order for the closed-loop system to have a gain margin $K_g \geq 12\,\text{dB}$, a phase margin $\varphi_p \geq 50°$, and an error constant $K_v = 4\,\text{sec}^{-1}$. The sampling period is chosen to be 0.1 sec.

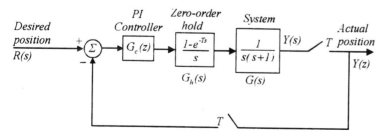

Figure 12.40   Discrete-time system for the positioning of the photovoltaic panels.

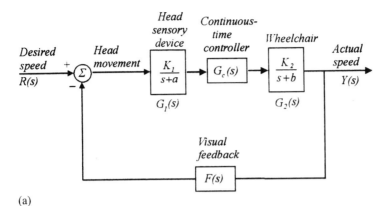

(a)

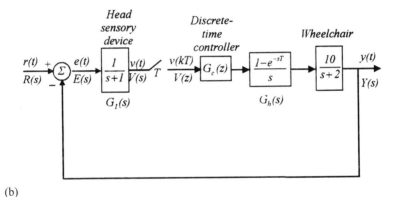

(b)

Figure 12.41 Wheelchair automatic control system: (a) wheelchair closed-loop system using a continuous time controller; (b) wheelchair closed-loop system using a discrete-time controller.

21. Construct the root locus of Example 12.8.1 for the following values of the parameters a and T: $a = 4$ and $T = 0.1, 1, 5,$ and 10 sec.

22. Consider a continuous-time open-loop system with transfer function $G(s) = a/s(s + a)$. Close the loop with unity feedback and find the position and the velocity error constants. Use a zero-order hold and unity feedback to obtain a discrete-time equivalent system. Determine the new position and velocity error constants and compare with the continuous-time case.

23. Consider the system of Figure 12.42 with unity feedback, where $G(s) = \dfrac{K}{s(s + a)}$.

   (a) Determine the transfer function $Y(z)/R(z)$ in terms of K, a, and T (sampling period).

   (b) Determine the root locus and the maximum value of K for a stable response with $T = 0.1$ sec, 0.5 sec, and 1 sec and $a = 2$.

   (c) Find the steady-state error characteristics for a unit step sequence and a ramp sequence for those error values of K and T that yield a stable system response for $a = 1$ and $a = 2$.

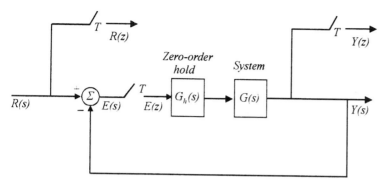

Figure 12.42 Discretized control system.

BIBLIOGRAPHY

Books

1. J Ackermann. Sampled Data Control Systems. New York: Springer Verlag, 1985.
2. KJ Astrom, T Hagglund. PID Controllers: Theory, Design and Tuning. North Carolina: Instruments Society of America, 1995.
3. KJ Astrom, B Wittenmark. Computer Controller Systems: Theory and Design. Englewood Cliffs, New Jersey: Prentice Hall, 1997.
4. JA Cadzow. Discrete-Time Systems, an Introduction with Interdisciplinary Applications. Englewood Cliffs, New Jersey: Prentice Hall, 1973.
5. JA Cadzow, HR Martens. Discrete-Time Systems and Computer Control Systems. Englewood Cliffs, New Jersey: Prentice Hall, 1970.
6. GF Franklin, JD Powell, ML Workman. Digital Control of Dynamic Systems. 2nd ed. London: Addison-Wesley, 1990.
7. CH Houpis, GB Lamont. Digital Control Systems. New York: McGraw-Hill, 1985.
8. EI Jury. Sampled Data Control Systems. New York: John Wiley, 1958; 2nd ed. Huntington, New York: Robert E Krieger, 1973.
9. T Kailath. Linear Systems. Englewood Cliffs, New Jersey: Prentice Hall, 1980.
10. P Katz. Digital Control Using Microprocessors. London: Prentice Hall, 1981.
11. BC Kuo. Digital Control Systems. Orlando, Florida: Saunders College Publishing, 1992.
12. K Ogata. Discrete-Time Control Systems. 2nd ed. Englewood Cliffs, New Jersey: Prentice Hall, 1995.
13. PN Paraskevopoulos. Digital Control Systems. London: Prentice Hall, 1996.
14. GH Perdikaris. Computer Controlled Systems, Theory and Applications. London: Kluwer Academic Publishers, 1991.
15. CL Phillips, HT Nagle Jr. Digital Control System Analysis and Design. Englewood Cliffs, New Jersey: Prentice Hall, 1984.
16. JR Ragazzini, GF Franklin. Sampled Data Control Systems. New York: McGraw-Hill, 1980.
17. M Santina, A Stubberud, G Hostetter. Digital Control System Design. Orlando, Florida: Saunders College Publishing, 1994.

Articles

18. JG Ziegler, NB Nichols. Optimum settings for automatic controllers. Trans ASME 64:759–768, 1942.

13

System Identification

13.1 INTRODUCTION

A fundamental concept in science and technology is that of mathematical modeling (see Chap. 3, particularly Secs 3.1–3.7). A mathematical model is a very useful, as well as very compact way, of describing the knowledge we have about a process or system. The determination of a mathematical model of a process or system is known as *system identification*. In control systems, a mathematical model of a process or system is in most cases necessary for the design of the controller. This becomes apparent if we recall that most of the controller design methods covered in previous chapters are based on the model of the system under control, which was assumed known. The model is also necessary for the design of adaptive and robust control systems presented in Chaps 14 and 15.

A process or a system may be described by several models, ranging from necessarily very detailed and complex microscopic models to simplistic macroscopic models which facilitate the understanding of the gross characteristics of a system's performance. Complex microscopic models require a long time to determine and they are mostly used for the detailed control of a system's performance. Between the two extremes, there exist several different types of models. Clearly, one must be able to choose the suitable type of model for each specific application.

There are basically two ways of determining a mathematical model of a system: by implementing known laws of nature or through experimentation on the process. A popular approach to obtaining a model is to combine both ways.

Mathematical models may be distinguished as parametric and nonparametric models. Parametric models obviously involve parameters: for example, the coefficients of differential or difference equations, of state equations, and of transfer functions. Nonparametric models do not involve parameters and are usually graphical representations, such as the Nyquist or Bode diagrams of a transfer function or impulse response function. This chapter refers to parametric models. For nonparametric models see for example [6]. Overall, the parametric identification problem reduces to the development of methods which give a good estimate of the parameters of the system model.

In particular, in this chapter we deal with the problem of determining mathematical models of linear, time-invariant, single-input–single-output (siso) discrete-time systems, described by difference equations. The proposed method for the identification (estimation) of the coefficients (parameters) of a difference equation is experimental and may be briefly described as follows. First, a set of N linear algebraic equations is formulated, where N is the number of measurements. From these equations, one may easily derive the *canonical equation* whose solution yields the parameter estimate $\theta(N)$, where θ is the vector parameter under identification. If an estimate of the initial conditions of the dynamic equation is also required, then $N + n$ measurements are taken and hence $N + n$ equations are produced, where n is the order of the difference equation.

An interesting feature of this chapter, is the determination of a recursive algorithm, which allows the estimation of the vector parameter θ for $N + 1$ measurements, based on the following formula:

$$\theta(N + 1) = \theta(N) + \Delta\theta = \theta(N) + \gamma(N)[y_{N+1} - \varphi^T(N + 1)\theta(N)] \qquad (13.1\text{-}1)$$

where $\gamma(N)$ and $\varphi(N + 1)$ are known vector quantities and y_{N+1} is the $N + 1$ measurement of the output y of the system. This formula shows that for the determination of $\theta(N + 1)$ one can use the previous estimate $\theta(N)$ plus a corrective term $\Delta\theta$, which is due to the new $N + 1$ measurement, instead of starting the estimation procedure right from the beginning. This algorithm facilitates the numerical part of the problem and constitutes the cornerstone notion and tool for the solution of ON-LINE identification, i.e., the identification which takes place in real time while the system is operating under normal conditions (Sec. 13.3). In contrast, when the identification procedure is desired to take place involving only the first N measurements, it is carried out off-line only once and for this reason it is called OFF-LINE identification or parameter estimation.

For relevant references to system identification see [1–23]. Most of the material of this chapter is a condensed version of the identification material reported in [15].

13.2 OFF-LINE PARAMETER ESTIMATION

13.2.1 First-Order Systems

The simple case of a first-order discrete-time system is studied first. Assume that the system under consideration is described by the differential equation

$$y(k) + a_1 y(k - 1) = b_1 u(k - 1) \qquad (13.2\text{-}1)$$

with initial condition $y(-1)$. Assume that the system (13.2-1) is excited with an input sequence $u(-1), u(0), u(1), \ldots$. As a result, the output of the system has a sequence $y(0), y(1), \ldots$. The identification problem may now be defined as follows. Given the known input sequence $u(-1), u(0), u(1), \ldots$ as well as the measured output sequence $y(0), y(1), \ldots$, find an estimate of the system's parameters a_1 and b_1 and of the initial condition $y(-1)$.

To solve the problem, we begin by writing down Eq. (13.2-1) for $N + 1$ measurements, i.e., for $k = 0, 1, 2, \ldots, N$. Consequently, we arrive at the following set of linear algebraic equations:

$$y(0) + a_1 y(-1) = b_1 u(-1) \tag{13.2-2a}$$

$$y(1) + a_1 y(0) = b_1 u(0)$$
$$y(2) + a_1 y(1) = b_1 u(1) \tag{13.2-2b}$$

$$\vdots$$

$$y(N) + a_1 y(N - 1) = b_1 u(N - 1)$$

The last N equations, i.e., Eqs (13.2-2b), are used for the estimation of the parameters a_1 and b_1. Having estimated the parameters a_1 and b_1, Eq. (13.2-2a) can be used for the estimation of the initial condition $y(-1)$. To this end, define

$$\boldsymbol{\theta} = \begin{bmatrix} a_1 \\ b_1 \end{bmatrix}, \qquad \mathbf{y} = \begin{bmatrix} y(1) \\ y(2) \\ \vdots \\ y(N) \end{bmatrix}, \qquad \text{and}$$

$$\boldsymbol{\phi} = \begin{bmatrix} \boldsymbol{\varphi}^{\mathrm{T}}(0) \\ \cdots\cdots \\ \boldsymbol{\varphi}^{\mathrm{T}}(1) \\ \cdots\cdots \\ \vdots \\ \cdots\cdots \\ \boldsymbol{\varphi}^{\mathrm{T}}(N-1) \end{bmatrix} = \begin{bmatrix} -y(0) & \vdots & u(0) \\ \cdots\cdots & \vdots & \cdots\cdots \\ -y(1) & \vdots & u(1) \\ \cdots\cdots & \vdots & \cdots\cdots \\ \vdots & \vdots & \vdots \\ \cdots\cdots & \vdots & \cdots\cdots \\ -y(N-1) & \vdots & u(N-1) \end{bmatrix} \tag{13.2-3}$$

Using these definitions, Eqs (13.2-2b) can be written compactly as

$$\mathbf{y} = \boldsymbol{\phi}\boldsymbol{\theta} \tag{13.2-4}$$

Equation (13.2-4) is an algebraic system of N equations with two unknowns. It is clear that if the known input and output sequences involve errors due to measurement or noise, then, for every input–output pair $\{u(k), y(k)\}$, there exists an error $e(k)$; thus, Eqs (13.2-2) will take on the form

$$y(k) + a_1 y(k-1) = b_1 u(k-1) + e(k), \qquad k = 0, 1, 2, \ldots, N \tag{13.2-5}$$

Consequently Eq. (13.2-4) becomes

$$\mathbf{y} = \boldsymbol{\phi}\boldsymbol{\theta} + \mathbf{e} \tag{13.2-6}$$

where $\mathbf{e}$ is the N-dimensional error vector $\mathbf{e}^{\mathrm{T}} = [e(1) \quad e(2) \quad \cdots \quad e(N)]$. For the minimization of the error vector $\mathbf{e}$, the least-squares methods can be aplied. To this end, define the following cost function

$$J = \mathbf{e}^{\mathrm{T}}\mathbf{e} = \sum_{k=1}^{N} e^2(k) \tag{13.2-7}$$

If Eq. (13.2-6) is substituted in Eq. (13.2-7), we obtain

$$J = (\mathbf{y} - \boldsymbol{\phi}\boldsymbol{\theta})^{\mathrm{T}}(\mathbf{y} - \boldsymbol{\phi}\boldsymbol{\theta})$$

Hence

$$\frac{\partial J}{\partial \theta} = -2\boldsymbol{\phi}^T(\mathbf{y} - \boldsymbol{\phi}\boldsymbol{\theta})$$

where the following formula was used (see Subsec. 2.6.2):

$$\frac{\partial}{\partial \theta}[\mathbf{A}\boldsymbol{\theta}] = \frac{\partial}{\partial \theta}[\boldsymbol{\theta}^{\mathrm{T}}\mathbf{A}^{\mathrm{T}}] = \mathbf{A}^{\mathrm{T}}$$

If we set $\partial J/\partial\theta$ equal to zero, we obtain

$$\boldsymbol{\phi}^{\mathrm{T}}\boldsymbol{\phi}\boldsymbol{\theta} = \boldsymbol{\phi}^{\mathrm{T}}\mathbf{y} \qquad\qquad (13.2\text{-}8)$$

Relation (13.2-8) is known as the *canonical equation* and has a solution when the matrix $\boldsymbol{\phi}^{\mathrm{T}}\boldsymbol{\phi}$ is invertible, in which case we have

$$\boldsymbol{\theta} = (\boldsymbol{\phi}^{\mathrm{T}}\boldsymbol{\phi})^{-1}\boldsymbol{\phi}^{\mathrm{T}}\mathbf{y} = \boldsymbol{\phi}^{\#}\mathbf{y}, \qquad \boldsymbol{\phi}^{\#} = (\boldsymbol{\phi}^{\mathrm{T}}\boldsymbol{\phi})^{-1}\boldsymbol{\phi}^{\mathrm{T}} \qquad (13.2\text{-}9)$$

where $\boldsymbol{\phi}^{\#}$ is the pseudoinverse of $\boldsymbol{\phi}$. The solution (13.2-9) minimizes the cost function (13.2-7).

The matrix $\boldsymbol{\phi}^{\mathrm{T}}\boldsymbol{\phi}$ is symmetrical and has the following form:

$$\boldsymbol{\phi}^{\mathrm{T}}\boldsymbol{\phi} = \begin{bmatrix} \displaystyle\sum_{k=0}^{N-1} y^2(k) & \displaystyle -\sum_{k=0}^{n-1} y(k)u(k) \\ \displaystyle -\sum_{k=0}^{N-1} y(k)u(k) & \displaystyle \sum_{k-0}^{N-1} u^2(k) \end{bmatrix} \qquad (13.2\text{-}10a)$$

Moreover, the vector $\boldsymbol{\phi}^{\mathrm{T}}\mathbf{y}$ has the form

$$\boldsymbol{\phi}^{\mathrm{T}}\mathbf{y} = \begin{bmatrix} \displaystyle -\sum_{k=1}^{N} y(k)y(k-1) \\ \displaystyle \sum_{k=1}^{N} y(k)u(k-1) \end{bmatrix} \qquad (13.2\text{-}10b)$$

The estimate of the parameters a_1 and a_2 is based on Eq. (13.2-9). The initial condition $y(-1)$ is estimated on the basis of Eq. (13.2-2a), which gives

$$y(-1) = \frac{1}{a_1}[b_1 u(-1) - y(0)]$$

where it is assumed that $a_1 \neq 0$ and $u(-1)$ is known.

Example 13.2.1

A discrete-time system is described by the first-order difference equation

$$y(k) + a_1 y(k-1) = b_1(k-1)$$

The input and output sequences $u(k)$ and $y(k)$, for $N = 6$, are given in Table 13.1. Estimate the parameters a_1 and b_1, as well as the initial condition $y(-1)$.

Table 13.1 Input–Output Measurements for Example 13.2.1

k	−1	0	1	2	3	4	5	6
$u(k)$	1	1	1	1	1	1	1	1
$y(k)$		1	1/2	3/4	5/8	11/16	21/32	43/64

Solution

From Eq. (13.2-10) and for $N = 6$, we have that

$$\phi^T\phi = \begin{bmatrix} \displaystyle\sum_{k=0}^{5} y^2(k) & -\displaystyle\sum_{k=0}^{5} y(k)u(k) \\ -\displaystyle\sum_{k=0}^{5} y(k)u(k) & \displaystyle\sum_{k=0}^{5} u^2(k) \end{bmatrix} = \begin{bmatrix} 3.106445 & -4.21875 \\ -4.21875 & 6 \end{bmatrix}$$

$$\phi^T y = \begin{bmatrix} -\displaystyle\sum_{k=1}^{6} y(k)y(k-1) \\ \displaystyle\sum_{k=1}^{6} y(k)u(k-1) \end{bmatrix} = \begin{bmatrix} -2.665527 \\ 3.890625 \end{bmatrix}$$

Hence

$$\theta = \begin{bmatrix} a_1 \\ b_1 \end{bmatrix} = (\phi^T\phi)^{-1}\phi^T y = \frac{1}{0.84082}\begin{bmatrix} 6 & 4.21875 \\ 4.21875 & 3.106445 \end{bmatrix}\begin{bmatrix} -2.665527 \\ 3.890625 \end{bmatrix}$$
$$= \begin{bmatrix} 0.5 \\ 1 \end{bmatrix}$$

Finally, the estimate of $y(-1)$, derived using Eq. (13.2-2a), is

$$y(-1) = \frac{1}{a_1}[b_1 u(-1) - y(0)] = 2[1 \quad -1] = 0$$

13.2.2 Higher-Order Systems

Here, the results of Subsec. 13.2.1 will be extended to cover the general case where the difference equation is of order n and has the form

$$y(k) + a_1 y(k-1) + \cdots + a_n y(k-n) = b_1 u(k-1) + \cdots + b_n u(k-n) \quad (13.2\text{-}11)$$

with initial conditions $y(-1), y(-2), \ldots, y(-n)$. In this case, the unknowns are the parameters $a_1, a_2, \ldots, a_n$, $b_1, b_2, \ldots, b_n$ and the initial conditions $y(-1)$, $y(-2), \ldots, y(-n)$. As before take $N + n$ measurements. For $k = 0$, $1, \ldots, N + n - 1$, the difference equation (13.2-11) yields the following equations:

$$y(0) + a_1 y(-1) + \cdots + a_n(y(-n) = b_1 u(-1) + \cdots + b_n(u(-n)$$
$$y(1) + a_1 y(0) + \cdots + a_n y(-n+1) = b_1 u(0) + \cdots + b_n u * -n + 1)$$
$$\vdots \qquad\qquad\qquad (13.2\text{-}12a)$$
$$y(n-1) + a_1 y(n-2) + \cdots + a_n y(-1) = b_1 u(n-2) + \cdots + b_n u(-1)$$

$$y(n) + a_1 y(n-1) + \cdots + a_n y(0) = b_1 u(n-1) + \cdots + b_n u(0)$$

$$y(n+1) + a_1 y(n) + \cdots + a_n y(1) = b_1 u(n) + \cdots + b_n u(1)$$

$$\vdots$$

$$\text{(13.2-12b)}$$

$$y(n+N-1) + a_1 y(n+N-2) + \cdots + a_n y(N-1)$$
$$= b_1 u(n+N-2) + \cdots \quad + b_n u(N-1)$$

Relation (13.2-12) has a total of $N + n$ algebraic equations: the first n equations are in Eqs (13.2-12a), whereas the remaining N equations are in Eqs (13.2-12b). Define:

$$\boldsymbol{\theta}^{\mathrm{T}} = [\, a_1 \quad a_2 \quad \cdots \quad a_n \quad b_1 \quad b_2 \quad \cdots \quad b_n \,],$$
$$\mathbf{y}^{\mathrm{T}} = [\, y(n) \quad y(n+1) \quad \cdots \quad y(n+N-1) \,]$$

$$\text{(13.2-13a)}$$

$$\boldsymbol{\phi} = \begin{bmatrix} \boldsymbol{\varphi}^{\mathrm{T}}(0) \\ \cdots\cdots\cdots \\ \boldsymbol{\varphi}^{\mathrm{T}}(1) \\ \cdots\cdots\cdots \\ \vdots \\ \cdots\cdots\cdots \\ \boldsymbol{\varphi}^{\mathrm{T}}(N-1) \end{bmatrix}$$

$$= \begin{bmatrix} -y(n-1) & \cdots & -y(0) & \vdots & u(n-1) & \cdots & u(0) \\ -y(n) & \cdots & -y(1) & \vdots & u(n) & \cdots & u(1) \\ & \vdots & & \vdots & & \vdots & \\ -y(n+N-2) & \cdots & -y(N-1) & \vdots & u(n+N-2) & \cdots & u(N-1) \end{bmatrix}$$

$$\text{(13.2-13b)}$$

Using the foregoing definitions, Eqs (13.2-12b) can be written compactly as follows:

$$\mathbf{y} = \boldsymbol{\phi}\boldsymbol{\theta} \qquad \text{(13.2-14)}$$

Based on the relation (13.2-14), the results derived for the first-order systems of Subsec. 13.2.1 can easily be extended to the higher-order case. Hence, the canonical equation for Eq. (13.2-14) takes the form

$$\boldsymbol{\phi}^{\mathrm{T}}\boldsymbol{\phi}\boldsymbol{\theta} = \boldsymbol{\phi}^{\mathrm{T}}\mathbf{y} \qquad \text{(13.2-15)}$$

and therefore

$$\boldsymbol{\theta} = \left(\boldsymbol{\phi}^{\mathrm{T}}\boldsymbol{\phi}\right)^{-1}\boldsymbol{\phi}^{\mathrm{T}}\mathbf{y} = \boldsymbol{\phi}^{\#}\mathbf{y} \qquad \text{(13.2-16)}$$

under the assumption that the matrix $\boldsymbol{\phi}^{\mathrm{T}}\boldsymbol{\phi}$ is invertible.

Example 13.2.2

Consider a discrete-time system described by the following second-order difference equation:

$$y(k+2) + \omega^2 y(k) = bu(k)$$

The input $u(k)$ and the output $y(k)$, for $N = 5$, are presented in Table 13.2. Estimate the parameters ω and b.

Solution

For $k = 0, 1, 2, 3$, we have

$$y(2) + \omega^2 y(0) = bu(0) + e(0)$$
$$y(3) + \omega^2 y(1) = bu(1) + e(1)$$
$$y(4) + \omega^2 y(2) = bu(2) + e(2)$$
$$y(5) + \omega^2 y(3) = bu(3) + e(3)$$

where $e(k)$ is the measurement error at time k. The above equations can be grouped as follows:

$$\mathbf{y} + \boldsymbol{\phi}\boldsymbol{\theta} + \mathbf{e}$$

where

$$\mathbf{y}^T = [y(2) \quad y(3) \quad y(4) \quad y(5)] = [1/2 \quad 1/3 \quad 1/4 \quad 1/5]$$

$$\boldsymbol{\phi} = \begin{bmatrix} -y(0) & 1 \\ -y(1) & 1 \\ -y(2) & 1 \\ -y(3) & 1 \end{bmatrix} = \begin{bmatrix} 0 & 1 \\ 0 & 1 \\ -1/2 & 1 \\ -1/3 & 1 \end{bmatrix}, \quad \boldsymbol{\theta} = \begin{bmatrix} \omega^2 \\ b \end{bmatrix}, \quad \mathbf{e} = \begin{bmatrix} e(2) \\ e(3) \\ e(4) \\ e(5) \end{bmatrix}$$

Using the above, we have

$$\boldsymbol{\phi}^T\boldsymbol{\phi} = \begin{bmatrix} 13/36 & -5/6 \\ -5/6 & 4 \end{bmatrix}, \quad \boldsymbol{\phi}^T\mathbf{y} = \begin{bmatrix} -23/120 \\ 77/60 \end{bmatrix},$$

$$(\boldsymbol{\phi}^T\boldsymbol{\phi})^{-1} = \frac{36}{27}\begin{bmatrix} 4 & 5/6 \\ 5/6 & 13.36 \end{bmatrix}$$

The optimum estimates of ω^2 and b are obtained from

$$\boldsymbol{\theta} = \begin{bmatrix} \omega^2 \\ b \end{bmatrix} = (\boldsymbol{\phi}^T\boldsymbol{\phi})^{-1}\boldsymbol{\phi}^T\mathbf{y} = \begin{bmatrix} 0.404 \\ 0.405 \end{bmatrix}$$

whereupon $\omega = (0.404)^{1/2} = 0.635$ and $b = 0.405$.

Table 13.2 Input–Output Data Sequence for Example 13.2.2

k	0	1	2	3	4	5
$u(k)$	1	1	1	1	1	1
$y(k)$	0	0	1/2	1/3	1/4	1/5

13.3 ON-LINE PARAMETER ESTIMATION

In many practical cases, it is necessary that parameter estimation takes place concurrently with the system's operation. This parameter estimation problem is called *on-line identification* and its methodology usually leads to a recursive procedure for every new measurement (or data entry). For this reason, it is also called *recursive identification*. In simple words, on-line identification is based on the following idea. Assume that we have available an estimate of the parameter vector θ based on N pairs of input–output data entries. Let this estimate be denoted by $\theta(N)$. Assume that $\theta(N)$ is not accurate enough and we wish to improve the accuracy using the new (the next) $N+1$ data entry. Clearly, using $N+1$ data entries, we will obtain a new estimate for θ, denoted as $\theta(N+1)$, which is expected to be an improved estimate compared with the previous estimate $\theta(N)$.

Now, it is natural to ask the following question. For the calculation of $\theta(N+1)$, do we have to estimate $\theta(N+1)$ right from the beginning, based on Eq. (13.2-16), or is there an easier way by taking advantage of the already known parameter vector $\theta(N)$? The answer to this question is that the estimate $\theta(N+1)$ may indeed be determined in terms $\theta(N)$, in accordance with the following general expression:

$$\theta(N+1) = \theta(N) + \Delta\theta \tag{13.3-1}$$

where $\Delta\theta$ is the change in $\theta(N)$ because of the new $N+1$ measurement. Expression (13.3-1) is computationally attractive, since for each new measurement we do not have to compute $\theta(N+1)$ from the beginning, a fact which requires a great deal of computation, but determine only the correction term $\Delta\theta$, which requires much less computation. Even though the calculation of the correction term $\Delta\theta$ is not always simple, for the case of linear time-invariant systems, a computationally simple expression for $\Delta\theta$ may be found, as shown below. To this end, we return to the results of Subsec. 13.2.2. Since the initial conditions $y(-1), y(-2), \ldots, y(-n)$ are not of interest in on-line identification, they are dropped from the identification procedure. We are therefore left with the canonical equation (13.2-14) which, for simplicity, will be stated in the rest of the chapter as follows:

$$\phi(N)\theta(N) = \mathbf{y}(N) \tag{13.3-2}$$

Working as usual, we obtain the following estimate for $\theta(N)$:

$$\theta(N) = \left[\phi^T(N)\phi(N)\right]^{-1}\phi^T(N)\mathbf{y}(N) \tag{13.3-3}$$

We may partition $\phi(N+1)$ and $\mathbf{y}(N+1)$ as follows:

$$\phi(N+1) = \left[\frac{\phi(N)}{\varphi^T(N+1)}\right] \quad \text{and} \quad \mathbf{y}(N+1) = \left[\frac{\mathbf{y}(N)}{y(N+1)}\right] = \left[\frac{\mathbf{y}(N)}{y_{N+1}}\right]$$

$$\tag{13.3-4}$$

where y_{N+1} indicates the last measurement $y(N+1)$ in order to avoid any confusion between the vector $\mathbf{y}(N+1)$ and the data entry y_{N+1}. Then, Eq. (13.3-2) for the $N+1$ measurements takes on the form

$$\phi(N+1)\theta(N+1) = \mathbf{y}(N+1) \tag{13.3-5}$$

Hence

$$\theta(N+1) = \left[\boldsymbol{\phi}^{\mathrm{T}}(N+1)\boldsymbol{\phi}(N+1)\right]^{-1}\boldsymbol{\phi}^{\mathrm{T}}(N+1)\mathbf{y}(N+1)$$

$$= \left[\boldsymbol{\phi}^{\mathrm{T}}(N)\boldsymbol{\phi}(N) + \varphi(N+1)\varphi^{\mathrm{T}}(N+1)\right]^{-1}\left[\boldsymbol{\phi}^{\mathrm{T}}(N)\mathbf{y}(N) + \varphi(N+1)y_{N+1}\right]$$

$$(13.3\text{-}6)$$

where use was made of Eq. (13.3-4). Equation (13.3-6) may take the form of Eq. (13.3-1) by using the following formula (known as the *matrix inversion lemma*):

$$[\mathbf{A} + \mathbf{BCD}]^{-1} = \mathbf{A}^{-1} - \mathbf{A}^{-1}\mathbf{B}\left[\mathbf{C}^{-1} + \mathbf{DA}^{-1}\mathbf{B}\right]^{-1}\mathbf{DA}^{-1} \qquad (13.3\text{-}7)$$

The foregoing equation can easily be verified. To this end, the matrix $[\mathbf{A} + \mathbf{BCD}]$ is multiplied from the left to the right-hand side of Eq. (13.3-7), to yield

$$[\mathbf{A} + \mathbf{BCD}]\left[\mathbf{A}^{-1} - \mathbf{A}^{-1}\mathbf{B}\left[\mathbf{C}^{-1} + \mathbf{DA}^{-1}\mathbf{B}\right]^{-1}\mathbf{DA}^{-1}\right]$$

$$= \mathbf{I} - \mathbf{B}\left[\mathbf{C}^{-1} + \mathbf{DA}^{-1}\mathbf{B}\right]^{-1}\mathbf{DA}^{-1} + \mathbf{BCDA}^{-1}$$

$$\quad - \mathbf{BCDA}^{-1}\mathbf{B}\left[\mathbf{C}^{-1} + \mathbf{DA}^{-1}\mathbf{B}\right]^{-1}\mathbf{DA}^{-1}$$

$$= \mathbf{I} + \mathbf{BCDA}^{-1} - \mathbf{BC}\left[\mathbf{C}^{-1} + \mathbf{DA}^{-1}\mathbf{B}\right]\left[\mathbf{C}^{-1} + \mathbf{DA}^{-1}\mathbf{B}\right]^{-1}\mathbf{DA}^{-1}$$

$$= \mathbf{I} + \mathbf{BCDA}^{-1} - \mathbf{BCDA}^{-1} = \mathbf{I}$$

Hence, the proof is completed. Now, using Eq. (13.3-7), we obtain

$$\left[\boldsymbol{\phi}^{\mathrm{T}}(N)\boldsymbol{\phi}(N) + \varphi(N+1)\varphi^{\mathrm{T}}(N+1)\right]^{-1} = \left[\boldsymbol{\phi}^{\mathrm{T}}(N)\boldsymbol{\phi}(N)\right]^{-1} = \left[\boldsymbol{\phi}^{\mathrm{T}}(N)\boldsymbol{\phi}(N)\right]^{-1}$$
$$\boldsymbol{\phi}(N+1)\left[1 + \varphi^{\mathrm{T}}(N+1)\left[\boldsymbol{\phi}^{\mathrm{T}}(N)\boldsymbol{\phi}(N)\right]^{-1}\varphi(N+1)\right]^{-1}\varphi^{\mathrm{T}}(N+1)$$
$$\left[\boldsymbol{\phi}^{\mathrm{T}}(N)\boldsymbol{\phi}(N)\right]^{-1} \qquad (13.3\text{-}8)$$

Substituting Eq. (13.3-8) in Eq. (13.3-6), we obtain

$$\theta(N+1) = \left[\boldsymbol{\phi}^{\mathrm{T}}(N)\boldsymbol{\phi}(N)\right]^{-1}\boldsymbol{\phi}^{\mathrm{T}}(N)\mathbf{y}(N) + \left[\left[\boldsymbol{\phi}^{\mathrm{T}}(N)\boldsymbol{\phi}(N) + \varphi(N+1)\varphi^{\mathrm{T}}(N+1)\right]\right.$$
$$\left. - \boldsymbol{\phi}^{\mathrm{T}}(N)\boldsymbol{\phi}(N)\right]^{-1}\Big]^{-1}\boldsymbol{\phi}^{\mathrm{T}}(N)\mathbf{y}(N) + \left[\boldsymbol{\phi}^{\mathrm{T}}(N)\boldsymbol{\phi}(N) + \varphi(N+1)\varphi(N+1)\right]^{-1}$$
$$\varphi(N+1)y_{N+1} \qquad (13.3\text{-}9)$$

The following holds true:

$$\left[\left[\boldsymbol{\phi}^{\mathrm{T}}(N)\boldsymbol{\phi}(N) + \varphi(N+1)\varphi^{\mathrm{T}}(N+1)\right]^{-1} - \left[\boldsymbol{\phi}^{\mathrm{T}}(N)\boldsymbol{\phi}(N)\right]^{-1}\right]\boldsymbol{\phi}^{\mathrm{T}}(N)\mathbf{y}(N)$$

$$= \left[\boldsymbol{\phi}^{\mathrm{T}}(N)\boldsymbol{\phi}(N) + \varphi(N+1)\varphi^{\mathrm{T}}(N+1)\right]^{-1}\left[\left[\boldsymbol{\phi}^{\mathrm{T}}(N)\boldsymbol{\phi}(N)\right] - \left[\boldsymbol{\phi}^{\mathrm{T}}(N)\boldsymbol{\phi}(N)\right]\right.$$

$$\left. + \varphi(N+1)\varphi^{\mathrm{T}}(N+1)\right]\left[\boldsymbol{\phi}^{\mathrm{T}}(N)\boldsymbol{\phi}(N)\right]^{-1}\boldsymbol{\phi}^{\mathrm{T}}(N)\mathbf{y}(N) = -\left[\boldsymbol{\phi}^{\mathrm{T}}(N)\boldsymbol{\phi}(N)\right.$$

$$\left. + \varphi(N+1)\varphi^{\mathrm{T}}(N+1)\right]^{-1}\varphi(N+1)\varphi^{\mathrm{T}}(N+1)\left[\boldsymbol{\phi}^{\mathrm{T}}(N)\boldsymbol{\phi}(N)\right]^{-1}\boldsymbol{\phi}^{\mathrm{T}}(N)\mathbf{y}(N)$$

$$= -\left[\boldsymbol{\phi}^{\mathrm{T}}(N)\boldsymbol{\phi}(N) + \varphi(N+1)\varphi^{\mathrm{T}}(N+1)\right]^{-1}\varphi(N+1)\varphi^{\mathrm{T}}(N+1)\theta(N)$$

where, in deriving the final step in the foregoing equation, use was made of Eq. (13.3-3). Using this result, Eq. (13.3-9) can be written as

$$\theta(N+1) = \theta(N) - \left[\boldsymbol{\phi}^{\mathrm{T}}(N)\boldsymbol{\phi}(N) + \boldsymbol{\varphi}(N+1)\boldsymbol{\varphi}^{\mathrm{T}}(N+1)\right]^{-1}\boldsymbol{\varphi}(N+1)\boldsymbol{\varphi}^{\mathrm{T}}(N+1)$$
$$\theta(N) + \left[\boldsymbol{\phi}^{\mathrm{T}}(N)\boldsymbol{\phi}(N) + \boldsymbol{\varphi}(N+1)\boldsymbol{\varphi}^{\mathrm{T}}(N+1)\right]^{-1}\boldsymbol{\varphi}(N+1)y_{N+1}$$

$$(13.3\text{-}10)$$

where use was made of Eq. (13.3-3). Finally, defining

$$\gamma(N) = \left[\boldsymbol{\phi}^{\mathrm{T}}(N)\boldsymbol{\phi}(N) + \boldsymbol{\varphi}(N+1)\boldsymbol{\varphi}^{\mathrm{T}}(N+1)\right]^{-1}\boldsymbol{\varphi}(N+1)$$
$$= \left[\boldsymbol{\phi}^{\mathrm{T}}(N+1)\boldsymbol{\phi}(N+1)\right]^{-1}\boldsymbol{\varphi}(N+1) \qquad (13.3\text{-}11)$$

Eq. (13.3-10) is transformed as follows:

$$\theta(N+1) = \theta(N) + \gamma(N)\left[y_{N+1} - \boldsymbol{\varphi}^{\mathrm{T}}(N+1)\theta(N)\right] \qquad (13.3\text{-}12)$$

Equation (13.3-12) is of the general form (13.3-1) sought. Unfortunately, the determination of the vector $\gamma(N)$ of Eq. (13.3-11) is numerically cumbersome since it requires inversion of the matrix $\boldsymbol{\phi}^{\mathrm{T}}(N+1)\boldsymbol{\phi}(N+1)$ at every step. To overcome this difficulty, define

$$\mathbf{P}(N) = \left[\boldsymbol{\phi}^{\mathrm{T}}(N)\boldsymbol{\phi}(N)\right]^{-1} \qquad (13.3\text{-}13)$$

Hence

$$\mathbf{P}(N+1) = \left[\boldsymbol{\phi}^{\mathrm{T}}(N+1)\boldsymbol{\phi}(N+1)\right]^{-1} = \left[\boldsymbol{\phi}^{\mathrm{T}}(N)\boldsymbol{\phi}(N) + \boldsymbol{\varphi}(N+1)\boldsymbol{\varphi}^{\mathrm{T}}(N+1)\right]^{-1}$$

Using Eq. (13.3-8), the matrix $\mathbf{P}(N+1)$ can be written as

$$\mathbf{P}(N+1) = \mathbf{P}(N) - \mathbf{P}(N)\boldsymbol{\varphi}(N+1)\left[1 + \boldsymbol{\varphi}^{\mathrm{T}}(N+1)\mathbf{P}(N)\boldsymbol{\varphi}(N+1)\right]^{-1}$$
$$\boldsymbol{\varphi}^{\mathrm{T}}(N+1)\mathbf{P}(N) \qquad (13.3\text{-}14)$$

Equation (13.3-14) offers a convenient way for calculating $\mathbf{P}(N+1)$. It is noted that the term $1 + \left[\boldsymbol{\varphi}^{\mathrm{T}}(N+1)\mathbf{P}(N)\boldsymbol{\varphi}(N+1)\right]$ is scalar, whereupon calculation of its inverse is simple. The calculation of the matrix $\mathbf{P}(N+1)$, according to the recursive formula (13.3-14), requires a matrix inversion only once at the beginning of the procedure to obtain

$$\mathbf{P}(N_0) = \left[\boldsymbol{\phi}^{\mathrm{T}}(N_0)\boldsymbol{\phi}(N_0)\right]^{-1}$$

where N_0 is the starting number of data entries. Upon computing $\mathbf{P}(N+1)$, the vector $\gamma(N)$ can easily be determined from Eq. (13.3-11), i.e., from the expression

$$\gamma(N) = \mathbf{P}(N+1)\boldsymbol{\varphi}(N+1) \qquad (13.3\text{-}15)$$

In summary, the proposed recursive algorithm is given by the following theorem.

Theorem 13.3.1

Suppose that $\theta(N)$ is the estimate of the parameters of the nth order system (13.2-11) for N data entries. Then, the estimate of the parameter vector $\theta(N+1)$ for $N+1$ data entries is given by the expression

$$\theta(N+1) = \theta(N) + \gamma(N)\left[y_{N+1} - \boldsymbol{\varphi}^{\mathrm{T}}(N+1)\theta(N)\right] \qquad (13.3\text{-}16)$$

where

$$\gamma(N) = \mathbf{P}(N+1)\boldsymbol{\varphi}(N+1) \qquad (13.3\text{-}17a)$$

and the matrix $\mathbf{P}(N + 1)$ is calculated from the recursive formula

$$\mathbf{P}(N + 1) = \mathbf{P}(N) - \mathbf{P}(N)\boldsymbol{\varphi}(N + 1)\left[1 + \boldsymbol{\varphi}^T(N + 1)\mathbf{P}(N)\boldsymbol{\varphi}(N + 1)\right]^{-1}$$
$$\boldsymbol{\varphi}^T(N + 1)\mathbf{P}(N) \tag{13.3-17b}$$

with initial conditions

$$\mathbf{P}(N_0) = \left[\boldsymbol{\varphi}^T(N_0)\boldsymbol{\varphi}(N_0)\right]^{-1} \tag{13.3-17c}$$

$$\boldsymbol{\theta}(N_0) = \mathbf{P}(N_0)\boldsymbol{\varphi}^T(N_0)\mathbf{y}(N_0) \tag{13.3-17d}$$

In Figure 13.1, a block diagram of the on-line algorithm is given. It is clear that at every step the (known) inputs are the y_{N+1}, $\boldsymbol{\varphi}^T(N + 1)$, $\boldsymbol{\theta}(N)$, and $\mathbf{P}(N)$ and the algorithm produces the new estimate $\boldsymbol{\theta}(N + 1)$. The algorithm also produces the matrix $\mathbf{P}(N + 1)$, which is used in the next step. In Figure 13.2, a more detailed block diagram of the on-line algorithm is given.

Remark 13.3.1

In Eq. (13.3-16) we observe that the correction term $\Delta\boldsymbol{\theta}$, defined in Eq. (13.3-1), is proportional to the difference $y_{N+1} - \boldsymbol{\varphi}^T(N + 1)\boldsymbol{\theta}(N)$, where y_{N+1} is the new data entry and $\boldsymbol{\varphi}^T(N + 1)\boldsymbol{\theta}(N)$ is an estimate of this new entry, based on Eq. (13.3-5) and using the latest estimate of the system parameters. Had there not existed any error in the measurements, the expected value $\boldsymbol{\varphi}^T(N + 1)\boldsymbol{\theta}(N)$ of y_{N+1} would be equal to the respective measurement value y_{N+1} and the difference between them would be zero, in which case $\boldsymbol{\theta}(N + 1) = \boldsymbol{\theta}(N)$. In other words, when there is no error in the data entries, a new entry does not add any new information, so the new estimate of $\boldsymbol{\theta}$ has exactly the same value as that of the previous estimate. Finally, it is noted that the term $\gamma(N)$ may be considered as a weighting factor of the difference term $y_{N+1} - \boldsymbol{\varphi}^T(N + 1)\boldsymbol{\theta}(N)$.

Example 13.3.1

Consider the simple case of a resistive network given in Figure 13.3. Estimate the parameter a when $u(k)$ is a step sequence, i.e., when $u(k) = 1$, for $k = 1, 2, 3, \ldots$.

Solution

The difference equation for the system is $y(k) = au(k)$. Since $u(k) = 1$, for all k, it follows that

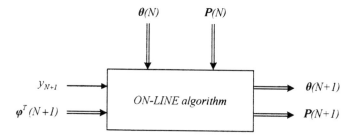

Figure 13.1 Block diagram presentation of the on-line algorithm.

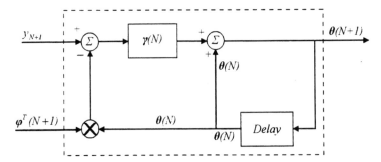

Figure 13.2 Detailed block diagram representation of the on-line algorithm.

$$y(k) = a + e(k), \qquad k = 1, 2, \ldots, N$$

To start with, we solve the problem using the following very simple technique. Define the cost function

$$J = \sum_{k=1}^{N} e^2(k) = \sum_{k=1}^{N} [y(k) - a]^2$$

Then

$$\frac{\partial J}{\partial a} = -2 \sum_{k=1}^{N} [y(k) - a] = 0$$

The foregoing equation can be written as

$$\sum_{k=1}^{N} [y(k) - a] = \left[\sum_{k=1}^{N} y(k) \right] - Na = 0$$

Hence, the estimate $a(N)$ of the parameter a is

$$a(N) = \frac{1}{N} \sum_{k=1}^{N} y(k)$$

The expression above for $a(N)$ is the mean value of the measurements $y(k)$, as was anticipated. Assume now that we have a new measurement. Then

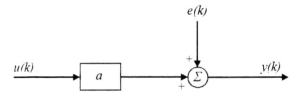

Figure 13.3 Simple case of a resistive network.

$$a(N+1) = \frac{1}{N+1}\sum_{k=1}^{N+1} y(k)$$

$$= \frac{1}{N}\sum_{k=1}^{N} y(k) - \frac{1}{N}\sum_{k=1}^{N} y(k) + \frac{1}{N+1}\sum_{k=1}^{N}[y(k) + y(N+1)]$$

$$= \frac{1}{N}\sum_{k=1}^{N} y(k) + \frac{N-(N+1)}{N(N+1)}\sum_{k=1}^{N} y(k) + \frac{1}{N+1}y(N+1)$$

$$= a(N) + \frac{1}{N+1}[y(N+1) - a(N)]$$

The foregoing equation represents the recursive algorithm (13.3-16), where $\varphi^T(N+1) = 1$, $y_{N+1} = y(N+1)$, and $\gamma(N) = 1/(N+1)$.

Now, we solve the problem using the method presented in this section. As a first step, we formulate the canonical equation

$$\boldsymbol{\phi}(N)\boldsymbol{\theta}(N) = \mathbf{y}(N)$$

where $\boldsymbol{\theta}(N) = a(N)$, $\boldsymbol{\phi}(N) = [1 \quad 1 \quad \cdots \quad 1]^T$, and $\mathbf{y}(N) = [y(1) \quad y(2) \quad \cdots \quad y(N)]^T$. The solution of the canonical equation is

$$a(N) = [\boldsymbol{\phi}^T(N)\boldsymbol{\phi}(N)]^{-1}\boldsymbol{\phi}^T(N)\mathbf{y}(N) = N^{-1}\boldsymbol{\phi}^T(N)\mathbf{y}(N) = N^{-1}\sum_{k=1}^{N} y(k)$$

We observe that $\mathbf{P}(N) = N^{-1}$. Thus Eq. (13.3-17b) yields

$$\mathbf{P}(N+1) = N^{-1} - N^{-1}[1 + N^{-1}]N^{-1} = \frac{1}{N+1}$$

whereas Eq. (13.3-17a) gives

$$\gamma(N) = \mathbf{P}(N+1)\varphi(N+1) = \frac{1}{N+1}$$

Hence, Eq. (13.3-16) becomes

$$a(N+1) = a(N) + \frac{1}{N+1}[y_{N+1} - a(N)]$$

Figure 13.4 shows the block diagram of the ON-LINE algorithm for Example 13.3.1.

Example 13.3.2

Consider the system of Example 13.3.1 with the difference that the input is not a unit step function but any other type of function. Estimate the parameter a.

Solution

The difference equation is $y(k) = au(k)$. We formulate the canonical equation

$$\boldsymbol{\phi}(N)\boldsymbol{\theta}(N) = \mathbf{y}(N)$$

where

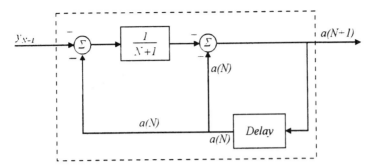

Figure 13.4 Block diagram of the on-line algorithm of Example 13.3.1.

$$\theta(N) = a(N), \qquad \phi(N) = [u(1) \quad u(2) \quad \cdots \quad u(N)]^T \qquad \text{and}$$
$$\mathbf{y}(N) = [y(1) \quad y(2) \quad \cdots \quad y(N)]^T$$

The solution of the canonical equation gives the following result:

$$a(N) = \left[\phi^T(N)\phi(N)\right]^{-1}\phi^T(N)\mathbf{y}(N) = \left[\sum_{k=1}^{N} u^2(k)\right]^{-1} \sum_{k=1}^{N} u(k)\varphi(k)$$

We observe that

$$\mathbf{P}(N) = \left[\sum_{k=1}^{N} u^2(k)\right]^{-1}$$

Hence, Eq. (13.3-17b) becomes

$$\mathbf{P}(N + 1) = \mathbf{P}(N) - \mathbf{P}(N)u(N + 1)[1 + u^2(N + 1)\mathbf{P}(N)]^{-1}u(N + 1)\mathbf{P}(N)$$
$$= \left[[1 + u^2(N + 1)\mathbf{P}(N)] - u^2(N + 1)\mathbf{P}(N)\right]\mathbf{P}(N)[1 + u^2(N + 1)\mathbf{P}(N)]^{-1}$$
$$= \mathbf{P}(N)[1 + u^2(N + 1)\mathbf{P}(N)]^{-1}$$

Therefore

$$a(N + 1) = a(N) + \gamma(N)\left[y_{N+1} - \varphi^T(N + 1)a(N)\right]$$
$$= a(N) + \mathbf{P}(N)u(N + 1)[1 + u^2(N + 1)\mathbf{P}(N)]^{-1}\left[y_{N+1} - u(N + 1)a(N)\right]$$

If $u(k) = 1$, $k = 1, 2, \ldots$, the results of Example 13.3.1 can readily be derived as a special case of the foregoing results. Indeed, since $\mathbf{P}(N) = N^{-1}$ and $u(k) = 1$, $k = 1, 2, \ldots$, the above expression for $a(N + 1)$ becomes

$$a(N + 1) = a(N) + N^{-1}\left[1 + N^{-1}\right]\left[y_{N+1} - a(N)\right]$$
$$= a(N) + \frac{1}{N + 1}\left[y_{N+1} - a(N)\right]$$

Example 13.3.3

Consider the discrete-time system described by the nonlinear difference equation

$$y(k + 1) + ay^2(k) = bu(k)$$

If the unit $u(k)$ is a unit step sequence, it results in the following output measurements:

k	0	1	2	3	4	5	6
$y(k)$	0	0.01	1.05	1.69	3.02	7.4	39.3

Determine:

(a)   An estimate of the parameters a and b.
(b)   Assume that a new output measurement $y(7) = 1082$ is available. Find the new estimates for a and b using the recursive formula.

Solution

(a) For $k = 1, 2, 3, 4, 5$ we have

$$y(2) + ay^2(1) = bu(1) + e(2)$$

$$y(3) + ay^2(2) = bu(2) + e(3)$$

$$y(4) + ay^2(3) = bu(3) + e(4)$$

$$y(5) + ay^2(4) = bu(4) + e(5)$$

$$y(6) + ay^2(5) = bu(5) + e(6)$$

where $e(k)$ is the measurement error at time k. The above equations can be written compactly as follows:

$$\mathbf{y}(N) = \boldsymbol{\phi}(N)\boldsymbol{\theta}(N) + \mathbf{e}$$

where

$$\mathbf{y}^{\mathrm{T}}(N) = [y(2) \quad y(3) \quad \cdots \quad y(6)] = [1.05 \quad 1.69 \quad 3.02 \quad 7.4 \quad 39.3]$$

$$\boldsymbol{\phi}(N) = \begin{bmatrix} -y^2(1) & u(1) \\ -y^2(2) & u(2) \\ -y^2(3) & u(3) \\ -y^2(4) & u(4) \\ -y^2(5) & u(5) \end{bmatrix} = \begin{bmatrix} -0.0001 & 1 \\ -1.1025 & 1 \\ -2.8561 & 1 \\ -9.1204 & 1 \\ -54.76 & 1 \end{bmatrix}, \qquad \boldsymbol{\theta}(N) = \begin{bmatrix} a \\ b \end{bmatrix},$$

$$\mathbf{e} = \begin{bmatrix} e(2) \\ e(3) \\ e(4) \\ e(5) \\ e(6) \end{bmatrix}$$

Using the above results, we obtain

$$\phi^T(N)\phi(N) = \begin{bmatrix} 3091.12 & -67.839 \\ -67.839 & 5 \end{bmatrix}, \qquad \phi^T(N)\mathbf{y}(N) = \begin{bmatrix} -2230.04771 \\ 52.46 \end{bmatrix}$$

$$[\phi^T(N)\phi(N)]^{-1} = \begin{bmatrix} 0.0005 & 0.0063 \\ 0.0063 & 0.2848 \end{bmatrix}$$

The optimal estimate of the parameters a and b is

$$\theta(N) = \begin{bmatrix} a \\ b \end{bmatrix} = [\phi^T(N)\phi(N)]^{-1}\phi^T(N)\mathbf{y}(N) = \begin{bmatrix} -0.7845 \\ 0.891 \end{bmatrix}$$

(b) Using the recursive equation (13.3-16), we have

$$\theta(7) = \theta(6) + \gamma(6)[y_7 - \varphi^T(7)\theta(6)]$$

where

$$\theta(6) = \begin{bmatrix} -0.7845 \\ 0.891 \end{bmatrix}, \qquad y_7 = y(7) = 1082, \qquad \varphi^T(7) = [-1544.5 \quad 1]$$

$$\gamma(6) = \mathbf{P}(7)\varphi(7) = \begin{bmatrix} -0.0006 \\ -0.008 \end{bmatrix}$$

Hence

$$\theta(7) = \begin{bmatrix} a \\ b \end{bmatrix} = \begin{bmatrix} -0.7062 \\ 1.9354 \end{bmatrix}$$

13.4   PROBLEMS

1. A system is described by the following difference equation

$$y(k+3) + a_1 y(k+2) + a_2 y(k+1) + a_3 y(k) = u(k)$$

If the input to the system is the impulse function $u(k) = \delta(k)$, it results in the following output measurements:

k	0	1	2	3	4	5	6
$y(k)$	1	0.2	-0.6	-1.2	-1.6	-1.7	-1.6

Estimate the parameters a_1, a_2, and a_3.

2. Estimate the parameters a, b, and ω of a system described by the difference equation

$$y(k) + \omega^2 y(k-2) = au(k-1) + bu(k-2)$$

given that the following measurements are available:

k	1	2	3	4	5
$u(k)$	1	1	1	1	1
$y(k)$	0	1/2	1/3	1/4	1/5

Are the estimates unique? Explain your results.

3. Estimate the parameters a, b, and c of the system

$$y(k) + ay(k-1) = bu^2(k) + cu(k-1)$$

given the information shown in Table 13.3 regarding its input and output. Make use of all the information provided in the table. The resulting identification must be unique.

4. Let a system be described by

$$y(k+2) = ay(k+1) + y(k) + bu(k)$$

(a) Estimate the parameters a and b given the following measurements:

k	0	1	2	3	4
$u(k)$	1	1	1	1	1
$y(k)$	1	0.9	0.9	0.8	0.6

(b) What are the new parameter estimates in view of the additional measurements $u(5) = 1$ and $y(5) = 0.4$?

5. The output of a given system $H(z)$ is compared with the output of a known system $H_2(z)$, as shown in Figure 13.5.
It is known that the difference equation that describes $H(z)$ is of the form

$$y(k) = au(k) + bu(k-1) + cu(k-2)$$

The equation of $H_2(z)$ is

$$y(k) = u(k) + 2u(k-1) - 3u(k-2)$$

and the feedback coefficient $f = 1$. Find the unknown parameters a, b, and c given the following measurements:

k	0	1	2	3	4	5
$r(k)$	1	1	1	1	1	1
$d(k)$	0	0	-0.5	-0.8	-1.1	-1.4

Table 13.3 Output Sequences $y(k)$ of Several Inputs for Problem 3

Input	$y(1)$	$y(2)$	$y(3)$	$y(4)$
$u_1(k) = k^2$	0	0	1	1.5
$u_2(k) = k$	0	-0.5	1	2.5
$y(k) = 1$	0	0.5	1	4

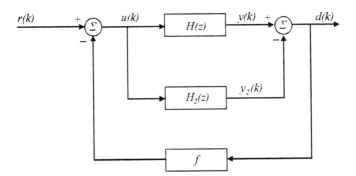

Figure 13.5   Closed-loop configuration.

REFERENCES

Books

1. KJ Astrom, B Wittenmark. Computer Controlled Systems. Englewood Cliffs, New Jersey: Prentice Hall, 1984.
2. CI Byrnes, A Lindquist (eds). Modelling, Identification and Robust Control. Amsterdam: North Holland, 1986.
3. R Calaba, K Spingarn. Control, Identification and Input Optimization. New York: Plenum Press, 1982.
4. RH Cannon Jr. Dynamics of Physical Systems. New York: McGraw-Hill, 1967.
5. P Eykhoff. System Identification. Parameter and State Estimation. New York: John Wiley & Sons, 1977.
6. P Eykhoff. Trends and Progress in System Identification. Oxford: Pergamon Press, 1981.
7. GC Goodwin, RL Payne. Dynamic System Identification: Experimental Design and Data Analysis. New York: Academic Press, 1977.
8. TC Hsia. System Identification. Boston, MA: Lexington, 1977.
9. L Ljung, T Södeström. Theory and Practice of Recursive Identification. Cambridge, MA: MIT Press, 1983.
10. L Ljung. System Identification-Theory for the User. Englewood Cliffs, New Jersey: Prentice Hall, 1987.
11. RK Mehra, DG Lainiotis (eds). System Identification. Advances and Case Studies. New York: Academic Press, 1976.
12. JM Mendel. Discrete Techniques of Parameter Estimation. New York: Marcel Dekker, 1973.
13. M Milanese, R Tempo, A Vicino (eds). Robustness in Identification and Control. New York: Plenum Press, 1989.
14. JP Norton. An Introduction to Identification. London: Academic press, 1986.
15. PN Paraskevopoulos. Digital Control Systems. London: Prentice Hall, 1996.
16. MS Rajbman, VM Chadeer. Identification of Industrial Processes. Amsterdam: North Holland, 1980.
17. AD Sage, JL Melsa. System Identification. New York: Academic Press, 1971.
18. T Söderström, P Stoica. System Identification. London: Prentice Hall, 1989.
19. H W Sorenson. Parameter Estimation. Principles and Problems. New York: Marcel Dekker, 1980.
20. CB Speedy, RF Brown, GC Goodwin. Control Theory: Identification and Optimal Control. Edinburgh: Oliver and Boyd, 1970.

21. E Walter. Identification of State Space Models. Berlin: Springer Verlag, 1982.

Articles

22. KJ Astrom, PE Eykhoff. System identification—a survey. Automatica 7:123–162, 1971.
23. Special issue on identification and system parameter estimation. Automatica 17, 1981.

14

Adaptive Control

14.1 INTRODUCTION

An *adaptive control system* is a system which adjusts automatically on-line the parameters of its controller, so as to maintain a satisfactory level of performance when the parameters of the system under control are unknown and/or time varying.

Generally speaking, the performance of a system is affected either by external perturbations or by parameter variations. Closed-loop systems involving feedback (top portion of Figure 14.1), are used to cope with external perturbations. In this case, the measured value of the output $y(kT)$ is compared with the desired value of the reference signal $r(kT)$. The difference $e(kT)$ between the two signals is applied to the controller, which in turn provides the appropriate control action $u(kT)$ to the plant or system under control. A somewhat similar approach can be used when parametric uncertainties (unknown parameters) appear in the system model of Figure 14.1. In this case the controller involves adjustable parameters. A performance index is defined, reflecting the actual performance of the system. This index is then measured and compared with a desired performance index (see Figure 14.1) and the error between the two performance indices activates the controller adaptation mechanism. This mechanism is suitably designed so as to adjust the parameters of the controller (or modify the input signals in a more general case), so that the error between the two performance indices lies within acceptable bounds.

Closer examination of Figure 14.1 reveals that two closed loops are involved: the "inner" feedback closed loop, whose controller involves adjustable parameters (upper portion of the figure); the supplementary "outer" feedback closed loop (or adaptation loop), which involves the performance indices and the adaptation mechanism (lower portion of the figure). The role of the adaptation loop is to find appropriate estimates for the adjustable controller parameters at each sampling instant.

It should be mentioned that a general definition, on the basis of which one could characterize a system as being adaptive or not, is still missing. However, it is clear that constant feedback systems are not adaptive systems. The existence of a feedback loop involving the performance index of the closed-loop system is a safe rule for characterizing a system as adaptive or not.

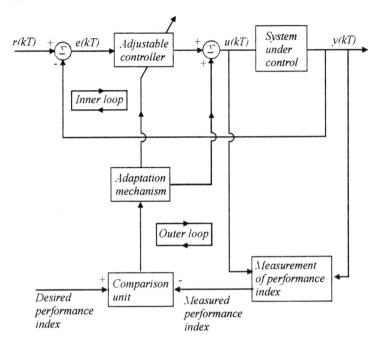

Figure 14.1 Block diagram of a general adaptive control system.

An adaptive control system is inherently nonlinear, since the controller parameters are nonlinear functions of the measured signals through the adaptation
mechanism. This is true even for the control of linear systems with unknown parameters, a fact which makes the analysis of adaptive systems very difficult. This
analysis involves the stability characteristics of the closed-loop system, the satisfaction of the performance requirements, and the convergence of the parameter estimates.

Adaptive control has been under investigation for many years. Major break-
throughs in the area have been reported in the last two decades [1–27]. Adaptive
control schemes have been applied in the paper industries, rolling mills, power
plants, motor drives, chemical reactors, cement mills, autopilots for aircrafts, missiles and ships, etc. Microprocessor advances have made it quite easy to implement
adaptive controllers and at low cost. The use of adaptive controllers may lead to
improvement of product quality, increase in production rates, fault detection, and
energy saving.

The two basic techniques to control discrete-time systems with unknown parameters are the *model reference adaptive control (MRAC)* scheme [8, 20–24] and *self-
tuning regulators (STRs)* [2, 7, 15–18, 26, 27]. These two techniques are presented in
Secs 14.3 and 14.4, respectively, and constitute a condensed version of the material
reported in Chap. 9 in [12].

In MRAC, a reference model is used explicitly in the control scheme and sets
the desired performance. Then, an appropriate on-line adaptation mechanism is
designed to adjust the controller parameters at each step, so that the output of the
system converges to the output of the reference model asymptotically, while simul-

taneously the stability of the closed-loop system is secured. In STRs, the control design and the adaptation procedure are separate. Different parameter estimators can be combined with appropriate control schemes to yield a variety of STRs. Restrictions in the structure of the models under which both methods can be applied are discussed on several appropriate occasions in the material of this chapter.

Model reference adaptive controllers can be either *direct* or *indirect*. The essential difference between them is that in direct MRAC the controller parameters are directly adjusted by the adaptation mechanism, while in indirect MRAC the adjustment of the controller parameters is made in two steps. In the first step, the control law is reparametrized so that the plant parameters appear explicitly in the control law. A relation between the controller parameters and the plant parameters is thus established. The plant parameters are adjusted by the adaptation mechanism. In the second step, the controller parameters are calculated from the estimates of the plant parameters. Direct MRAC, using the hyperstability approach for proving stability, is discussed in Sec. 14.3.

STRs can be either *explicit* or *implicit*. In explicit STRs an estimate of the explicit plant-model parameters is obtained. The explicit plant model is the actual plant model. In implicit STRs the parameters of an implicit model are estimated. The implicit model is a reparametrization of the explicit plant model. The parameters of the implicit model and those of the controller are the same; therefore, we call the plant parameters explicit or indirect and the controller parameters implicit or direct. Though of different nature and origin, a close relation between MRAC systems and STRS has been established [18, 20]. It is clear that explicit self-tuners correspond to indirect MRAC schemes, while implicit self-tuners correspond to direct MRAC schemes. Self-tuners based on pole-placement control are presented in Sec. 14.4.2.

Another approach to discrete-time MRAC is that of using Lyapunov functions to prove asymptotic stability and the satisfaction of performance requirements. An expression for the error between the output of the reference model and that of the plant is formed and then the adaptation mechanism is chosen in order to make the increments of a Lyapunov candidate function negative. This method is not developed in this chapter. A demonstration by using a simple example, can be found in [12]. The difficulty of finding an appropriate Lyapunov candidate function in the general discrete-time case restricts the use of this method. The hyperstability approach of Sec. 14.3 is preferable for discrete-time MRAC systems, while for continuous-time systems the Lyapunov design has mainly been used.

A first approach to MRAC was based on the gradient method. The parameter adaptation scheme obtained for synthesizing the adaptive loop was heuristically developed, initially for continuous-time systems and is known as the *MIT rule* [3]. A version of MRAC for discrete-time systems, based on the gradient method, is presented below.

14.2 ADAPTIVE CONTROL WITH THE GRADIENT METHOD (MIT RULE)

Consider a system with a single output $y(kT, \theta)$, where T is the sampling period and θ is the vector of unknown parameters which parametrizes the adjustable controller (hence the system's input signal is a function of θ, i.e., $u(kT, \theta)$) and the output of the system. The control objective is to follow the output $y_m(kT)$ of a reference model, in the sense that a particular performance index, involving the error $e(kT, \boldsymbol{\theta}) = y(kT, \boldsymbol{\theta}) - y_{\mathrm{m}}(kT)$, is a minimum.

Consider the quadratice performance index

$$J(kT, \boldsymbol{\theta}) = \tfrac{1}{2} e^2(kT, \boldsymbol{\theta}) \qquad (14.2\text{-}1)$$

It is obvious that to minimize the $J(kT, \boldsymbol{\theta})$, the parameter vector $\boldsymbol{\theta}$ of the adjustable controller should change in the opposite direction to that of the gradient $\partial J / \partial \boldsymbol{\theta}$. Consequently, the adaptation rule for $\boldsymbol{\theta}$, i.e., the difference equation giving the time evolution of $\boldsymbol{\theta}$ at the sampling instants, is

$$\boldsymbol{\theta}(kT + T) = \boldsymbol{\theta}(kT) - \gamma \left[\frac{\partial J(kT, \boldsymbol{\theta})}{\partial \boldsymbol{\theta}} \right] = \boldsymbol{\theta}(kT) - \gamma\, e(kT, \boldsymbol{\theta}) \left[\frac{\partial e(kT, \boldsymbol{\theta})}{\partial \boldsymbol{\theta}} \right] \qquad (14.2\text{-}2)$$

where γ is a constant positive adaptation gain. More precisely, when $\partial J(kT, \boldsymbol{\theta})/\partial \boldsymbol{\theta}$ is negative, i.e., when J decreases while $\boldsymbol{\theta}$ increases, then $\boldsymbol{\theta}$ should increase in order for J to decrease further. In the case where $\partial J(kT, \boldsymbol{\theta})/\partial \boldsymbol{\theta}$ is positive, i.e., when J and $\boldsymbol{\theta}$ increase simultaneously, then $\boldsymbol{\theta}$ should decrease in order for J to decrease further. This is achieved by the heuristic adaptation mechanism of Eq. (14.2-2). The partial derivative $\partial e(kT, \boldsymbol{\theta})/\partial \boldsymbol{\theta}$ appearing in Eq. (14.2-2) is called the system's *sensitivity derivative*. For the "MIT rule" to perform well, the adaptation gain γ should be small, since its value influences the convergence rate significantly. Moreover, it is possible for the "MIT rule" to lead to an unstable closed-loop system, since it is only a heuristic algorithm not rigidly based on stability requirements. Other performance indices are also possible.

The following example will illustrate the application of the MIT rule. This example will reveal the main problem in applying this method: namely, the necessity of using approximations to calculate the sensitivity derivatives of a certain system.

Example 14.2.1

Consider a first-order system described by the difference equation

$$y(kT + T) = -ay(kT) + bu(kT) \qquad (14.2\text{-}3)$$

where $u(kT)$ is the input and $y(kT)$ is the output. It is desired to obtain a closed-loop system of the form

$$y_{\mathrm{m}}(kT + T) = -a_{\mathrm{m}} y_{\mathrm{m}}(kT) + b_{\mathrm{m}} r(kT) \qquad (14.2\text{-}4)$$

where $r(kT)$ is a bounded reference sequence and $y_{\mathrm{m}}(kT)$ is the output of the reference model. To this end, an output feedback control law is used, having the form

$$u(kT) = -fy(kT) + gr(kT) \qquad (14.2\text{-}5)$$

Assume that the system parameters a and b are unknown. Determine the appropriate adaptation mechanism for the controller parameters f and g, using the MIT rule.

Solution

Combining Eqs (14.2-3) and (14.2-5), we obtain the closed-loop system

$$y(kT + T) = -ay(kT) + b[-fy(kT) + gr(kT)] = -(a + bf)y(kT) + bgr(kT)$$

or

$$y(kT) = -(a + bf)y(kT - T) + bgr(kT - T)$$

or

$$y(kT) = \left[\frac{bgq^{-1}}{1 + (a + bf)q^{-1}}\right] r(kT) \qquad (14.2\text{-}6)$$

where q^{-1} is the backward shift operator such that $q^{-1}y(kT) \equiv y(kT - T)$. Comparing Eqs (14.2-4) and (14.2-6), we have that in the case of known parameters a and b, the particular choice $f = (a_{\mathrm{m}} - a)/b$ and $g = b_{\mathrm{m}}/b$ leads to satisfaction of the control objective. This case is called *perfect model following*.

In the case of uncertain system parameters, we will use the "MIT rule." Here, the controller is parametrized by the adjustable parameters $f(k)$ and $g(k)$. The error $e(kT)$ between the outputs of the system and the reference model is now given by

$$e(kT) = y(kT) - y_{\mathrm{m}}(kT) = \left[\frac{bgq^{-1}}{1 + (a + bf)q^{-1}}\right] r(kT) - y_{\mathrm{m}}(kT) \qquad (14.2\text{-}7)$$

Using the foregoing expression for $e(kT)$, the system's sensitivity derivatives $\partial e/\partial g$ and $\partial e/\partial f$ can be easily determined. We have

$$\frac{\partial e(kT)}{\partial g} = \left[\frac{bq^{-1}}{1 + (a + bf)q^{-1}}\right] r(kT) \qquad (14.2\text{-}8)$$

$$\frac{\partial e(kT)}{\partial f} = -\left[\frac{b^2 g q^{-2}}{[1 + (a + bf)q^{-1}]^2}\right] r(kT) = -\left[\frac{bq^{-1}}{1 + (a + bf)q^{-1}}\right] y(kT) \qquad (14.2\text{-}9)$$

These expressions for the sensitivity derivatives cannot be used in the adaptation mechanism, since the unknown parameters a and b appear explicitly. For the present system, when perfect model following is achieved, we have that $a + bf = a_{\mathrm{m}}$. Taking advantage of this fact, the following approximate forms can be used for the sensitivity derivatives (still containing the unknown b):

$$\frac{\partial e(kT)}{\partial g} \simeq \left[\frac{bq^{-1}}{1 + a_{\mathrm{m}}q^{-1}}\right] r(kT) \qquad (14.2\text{-}10)$$

$$\frac{\partial e(kT)}{\partial f} \simeq -\left[\frac{bq^{-1}}{1 + a_{\mathrm{m}}q^{-1}}\right] y(kT) \qquad (14.2\text{-}11)$$

These sensitivity derivatives lead to the following parameter adaptation laws ("MIT rule"):

$$g(kT + T) = g(kT) - \gamma \left[\frac{q^{-1}}{1 + a_{\mathrm{m}}q^{-1}} r(kT)\right] e(kT) \qquad (14.2\text{-}12)$$

$$f(kT + T) = f(kT) + \gamma \left[\frac{q^{-1}}{1 + a_{\mathrm{m}}q^{-1}} y(kT)\right] e(kT) \qquad (14.2\text{-}13)$$

Notice here that the adaptation laws were obtained by absorbing the parameter b in the adaptation gain γ. This is done because b is unknown and should not appear in the adaptation laws; however, this requires that the sign of b is known. Then the sign of γ depends on the sign of b. The foregoing laws are initialized with arbitrary $g(0)$

and $f(0)$, which should reflect our a priori knowledge on the appropriate controller parameters which achieve model following.

Finally, the adjustable controller is given by

$$u(kT) = -f(kT)y(kT) + g(kT)r(kT) \tag{14.2-14}$$

Equations (14.2-12), (14.2-13), and (14.2-14) specify the dynamic adaptive controller being sought.

The aforementioned results can be generalized to the case of a single-input–single-output (SISO) linear system described by the difference equation

$$A(q^{-1})y(kT) = q^{-d}B(q^{-1})u(kT) \tag{14.2-15}$$

where $d \geq 1$ is the system's delay and $A(q^{-1})$ and $B(q^{-1})$ are polynomials in the backward shift operator having the form

$$A(q^{-1}) = 1 + a_1 q^{-1} + \cdots + a_{n_A} q^{-n_A} \tag{14.2-16}$$

$$B(q^{-1}) = b_0 + b_1 q^1 + \cdots + b_{n_B} q^{-n_B} \tag{14.2-17}$$

For more details, see [12], where both direct and indirect algorithms are developed.

14.3 MODEL REFERENCE ADAPTIVE CONTROL— HYPERSTABILITY DESIGN

14.3.1 Introduction

MRAC is a systematic method for controlling plants with unknown parameters. The basic scheme of an MRAC system is presented in Figure 14.2. In comparison to the general structure of an adaptive control system given in Figure 14.1, here the desired performance index is generated by means of a reference model.

The reference model is a dynamic system whose behavior is considered to be the desired (ideal) one and it is a part of the control system itself, since it appears explicitly in the control scheme. The output $y_m(kT)$ of the reference model indicates how the output $y(kT)$ of the plant should behave. Both systems are excited by the same command signal $r(kT)$. Comparing Figures 14.1 and 14.2, we observe that the desired performance index is now replaced by $y_m(kT)$ and the measured performance index by $y(kT)$.

We distinguish two control loops: the "inner" loop and the "outer" loop. The "inner" loop consists of the plant which involves unknown parameters and the adjustable controller. The "outer" loop is designed appropriately to adjust the controller's parameters so that the error $e(kT) = y(kT) - y_m(kT)$ approaches zero asymptotically, while the stability of the overall system can be proved using the so-called hyperstability approach.

Compared with techniques which involve other kinds of performance indices, the MRAC technique is characterized by high speed of adaptation. This is because a simple subtracter is needed to form the error $e(kT) = y(kT) - y_m(kT)$. This error, together with other available on-line data, is then fed to the adaptation mechanism. The parameters of the adjustable controller are modified accordingly, in order to minimize the difference between the two performance indices: namely, the desired performance index and the measured performance index.

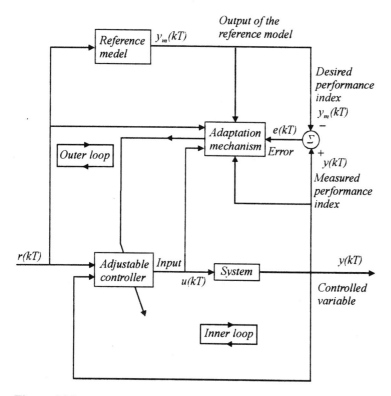

Figure 14.2 Block diagram of the model reference adaptive control (MRAC) scheme.

14.3.2 Definition of the Model Reference Control Problem

Consider a deterministic, SISO, discrete-time, linear, time-invariant systems, described by

$$A(q^{-1})y(k+d) = B(q^{-1})u(k) \quad \text{or} \quad A(q^{-1})y(k) = q^{-d}B(q^{-1})u(k) \quad (14.3\text{-}1)$$

with initial condition $y(0) \neq 0$. Here

$$A(q^{-1}) = 1 + a_1 q^{-1} + \cdots + a_{n_A} q^{-n_A} = 1 + q^{-1}A^*(q^{-1}) \qquad (14.3\text{-}2)$$

$$B(q^{-1}) = b_0 + b_1 q^{-1} + \cdots + b_{n_B} q^{-n_B} = b_0 + q^{-1}B^*(q^{-1}), \qquad b_0 \neq 0 \quad (14.3\text{-}3)$$

where q^{-1} is the backward shift operator, $d > 0$ represents the system's time delay, and $u(k)$ and $y(k)$ represent the system's input and output signals, respectively. The following three assumptions are made for the system under control:

1. The roots of $B(z^{-1})$, which are the system's zeros, are all inside the unit circle $|z| < 1$, i.e., $z_i^{n_B} B(z_i^{-1}) = 0$ with $|z_i| < 1$. Thus, the system zeros are stable and can be canceled out without leading to an unbounded control signal.
2. The system's delay d is known (this implies $b_0 \neq 0$).
3. An upper limit for the orders n_A and n_B of the polynomials $A(q^{-1})$ and $B(q^{-1})$, respectively, is given.

According to the foregoing assumptions, any change in the system's characteristics should not affect the delay d, whereas the system's zeros can move only inside the unit circle. The method is therefore valid only for minimum-phase systems.

The control objective is twofold: linear model following during tracking and elimination of any initial output disturbance during regulation. It is desirable to be able to specify the tracking and regulation objectives independently. This flexibility is crucial for certain applications. The control objectives are specified as follows.

1 Tracking

During tracking, it is desired for the plant output $y(k)$ to satisfy the equation

$$A_m(q^{-1})y(k) = q^{-d}B_m(q^{-1})r(k) \tag{14.3-4}$$

where

$$A_m(q^{-1}) = 1 + a_1^m q^{-1} + \cdots + a_{n_{A_m}}^m q^{-n_{A_m}} \tag{14.3-5}$$

$$B_m(q^{-1}) = b_0^m + b_1^m q^{-1} + \cdots + b_{n_{B_m}}^m q^{-n_{B_m}} \tag{14.3-6}$$

Here, $r(k)$ is a bounded reference sequence and the polynomial $A_m(q^{-1})$ is chosen to be asymptotically stable.

2 Regulation $(r(k) \equiv 0, y_m(k) \equiv 0)$

In regulation, the influence of any initial nonzero output $y(0) \neq 0$ (which corresponds to an impulse perturbation), should be eliminated via the dynamics defined by

$$\Gamma(q^{-1})y(k + d) = 0 \qquad \text{for} \qquad k \geq 0 \tag{14.3-7}$$

where $\Gamma(q^{-1})$ is an asymptotically stable polynomial of the designer's choice, having the form

$$\Gamma(q^{-1}) = 1 + \gamma_1 q^{-1} + \cdots + \gamma_{n_\Gamma} q^{-n_\Gamma} \tag{14.3-8}$$

Consider the following explicit reference model:

$$A_m(q^{-1})y_m(k) = q^{-d}B_m(q^{-1})r(k) \tag{14.3-9}$$

with input $r(k)$ and output $y_m(k)$. Note here that the sequence $y_m(k)$, apart from being calculated by means of the reference model (14.3-9), can also be a predefined sequence stored in memory.

It is obvious that both control objectives (i.e., tracking and regulation) can be accomplished if the control law $u(k)$ is such that

$$\bar{e}(k + d) = \Gamma(q^{-1})e(k + d) = \Gamma(q^{-1})[y(k + d) - y_m(k + d)] \equiv 0 \qquad \text{for} \qquad k \geq 0 \tag{14.3-10}$$

The error $e(k)$ is the difference between the plant and reference model outputs (*plant-model error*), i.e.,

$$e(k) = y(k) - y_m(k) \tag{14.3-11}$$

and $\bar{e}(k) = \Gamma(q^{-1})e(k)$ is the so-called *filtered error* between the plant and the reference model outputs. The error $\bar{e}(k)$ is also called the *a priori adaptation error*. The foregoing objectives will be satisfied below in the case of known or unknown parameters in the polynomials $A(q^{-1})$ and $B(q^{-1})$, in Subsecs 9.3.3 and 9.3.4, respectively, by seeking appropriate control laws.

If Eq. (14.3-10) is satisfied, then any initial plant-model error or any initial output disturbance, will converge to zero, i.e., $\lim_{k\to+\infty} e(k) = 0$. Note that if $\Gamma(q^{-1}) = 1$, one has $e(k + d) \equiv 0$, $k \geq 0$, which means that in this case the plant-model error vanishes d steps after the control input is applied. The polynomial $\Gamma(q^{-1})$ is a filtering polynomial. As is made clear in the sequel, adaptive control performance depends critically on the choice of $\Gamma(q^{-1})$.

14.3.3 Design in the Case of Known Parameters

In this subsection we assume that the plant parameters appearing in the polynomials $A(q^{-1})$ and $B(q^{-1})$ are known. Consider the general case where $d > 1$. We wish to obtain a control law $u(k)$ satisfying the control objectives and being causal, i.e., not depending on future values of the input and output. This control law we seek will therefore have the form

$$u(k) = f(y(k), y(k - 1), \ldots, u(k - 1), u(k - 2), \ldots)$$

To this end, consider the following equation, which is equivalent to Eq. (14.3-1):

$$y(k + d) = -A^*(q^{-1})y(k + d - 1) + B(q^{-1})u(k) \tag{14.3-12}$$

Next, we want to express Eq. (14.3-12) in the form

$$\Gamma(q^{-1})y(k + d) = g(y(k), y(k - 1), \ldots, u(k), u(k - 1), \ldots) \tag{14.3-13}$$

The specific form of g sought can be determined in two ways: either by repeatedly substituting $y(k + d - 1), \ldots, y(k + 1)$ in Eq. (14.3-12) generated by the same equation (Eq. (14.3-12)) delayed in time, or more easily, by directly considering the following polynomial identity (decomposition of $\Gamma(q^{-1})$):

$$\Gamma(q^{-1}) = A(q^{-1})S(q^{-1}) + q^{-d}R(q^{-1}) \tag{14.3-14}$$

with $R(q^{-1})$ and $S(q^{-1})$ appropriate polynomials. We adopt the second method for simplicity and, to this end, the results of the following remark will be useful.

Remark 14.3.1

The above identity (14.3-14) is a special case of what is referred to as the *Diophantine equation* or the *Bezout identity*. It can be proven that $\Gamma(q^{-1})$ can be uniquely factorized as in Eq. (14.3-14), where

$$S(q^{-1}) = 1 + s_1 q^{-1} + \cdots + s_{n_S} q^{-n_S} \tag{14.3-15}$$

$$R(q^{-1}) = r_0 + r_1 q^{-1} + \cdots + r_{n_R} q^{-n_R} \tag{14.3-16}$$

with $n_S = d - 1$ and $n_R = \max(n_A - 1, n_\Gamma - d)$ (see Subsec. 14.4.2 that follows for the uniqueness conditions). The coefficients of the polynomials $S(q^{-1})$ and $R(q^{-1})$ are uniquely determined by the solution of the following algebraic equation:

$$
\begin{bmatrix}
1 & & & & & & & & \\
a_1 & 1 & & & & & & & \\
a_2 & a_1 & 1 & & & & & & \\
\vdots & & & & & \mathbf{0} & & & \\
a_{d-1} & a_{d-2} & \cdots & a_1 & 1 & & & & \\
a_d & a_{d-1} & \cdots & a_2 & a_1 & 1 & & & \\
a_{d+1} & a_d & \cdots & a_3 & a_2 & 0 & 1 & & \\
a_{d+2} & a_{d+1} & \cdots & a_4 & a_3 & 0 & 0 & 1 & \\
\vdots & \vdots & & \vdots & \vdots & 0 & 0 & 0 & 1 \\
& & & & & \vdots & \vdots & \vdots & \vdots \\
& & & & & 0 & 0 & 0 & \cdots & 1 \\
& & & & & 0 & 0 & 0 & \cdots & 0 & 1
\end{bmatrix}
\begin{bmatrix}
1 \\ s_1 \\ s_2 \\ \vdots \\ s_{d-1} \\ r_0 \\ r_1 \\ r_2 \\ \vdots \\ \\ \\ r_{n_R}
\end{bmatrix}
=
\begin{bmatrix}
1 \\ \gamma_1 \\ \gamma_2 \\ \vdots \\ \gamma_{d-1} \\ \vdots \\ \\ \\ \\ \\ \\
\end{bmatrix}
$$

$$(14.3\text{-}17)$$

Returning to Eq. (14.3-13) and using Eq. (14.3-14), we express $\Gamma(q^{-1})y(k+d)$ as follows:

$$
\begin{aligned}
\Gamma(q^{-1})y(k+d) &= A(q^{-1})S(q^{-1})y(k+d) + q^{-d}R(q^{-1})y(k+d) \\
&= B(q^{-1})S(q^{-1})u(k) + R(q^{-1})y(k)
\end{aligned}
$$

$$(14.3\text{-}18)$$

Let

$$
\psi(q^{-1}) = B(q^{-1})S(q^{-1}) = b_0 + q^{-1}\psi^*(q^{-1}) = b_0 + \psi_1 q^{-1} + \cdots + \psi_{d+n_B-1}q^{-(d+n_B-1)}
$$

$$(14.3\text{-}19)$$

where

$$
\psi_1 = b_0 s_1 + b_1, \qquad \psi_2 = b_0 s_2 + b_1 s_1 + b_2, \dots, \qquad \psi_{d+n_B-1} = b_{n_B}s_{d-1}
$$

$$(14.3\text{-}20)$$

Finally, we have

$$
\Gamma(q^{-1})y(k+d) = b_0 u(k) + \psi^*(q^{-1})u(k-1) + R(q^{-1})y(k)
$$

$$(14.3\text{-}21)$$

Note that the right-hand side of Eq. (14.3-21) is the function g appearing in Eq. (14.3-13). Equation (14.3-21) can also be written as

$$
\Gamma(q^{-1})y(k+d) = \boldsymbol{\theta}^T \boldsymbol{\varphi}(k) = b_0 u(k) + \boldsymbol{\theta}_0^T \boldsymbol{\varphi}_0(k)
$$

$$(14.3\text{-}22)$$

where

$$
\boldsymbol{\theta}^T = \left[b_0 \; \vdots \; \psi_1, \dots, \psi_{d+n_B-1}, r_0, r_1, \dots, r_{n_R} \right] = \left[b_0 \; \vdots \; btheta_0^T \right]
$$

$$(14.3\text{-}23)$$

and $\boldsymbol{\varphi}(k)$ is the so-called *regression vector* having the form

$$
\begin{aligned}
\boldsymbol{\varphi}^T(k) &= \left[u(k) \; \vdots \; u(k-1), \dots, u(k-d-n_B+1), y(k), y(k-1), \dots, y(k-n_R) \right] \\
&= \left[u(k) \; \vdots \; \boldsymbol{\varphi}_0^T(k) \right]
\end{aligned}
$$

$$(14.3\text{-}24)$$

In what follows, we seek to find a control law $u(k)$, which drives the filtered plant-model error $\bar{e}(k+d) = \Gamma(q^{-1})e(k+d)$ to zero. Using Eqs (14.3-21) and (14.3-22), we have

$$\Gamma(q^{-1})e(k+d) = \Gamma(q^{-1})y(k+d) - \Gamma(q^{-1})y_m(k+d)$$
$$= b_0 u(k) + \psi^*(q^{-1})u(k-1) + R(q^{-1})y(k) - \Gamma(q^{-1})y_m(k+d)$$
$$= b_0 u(k) + \theta_0^T \varphi_0(k) - \Gamma(q^{-1})y_m(k+d) \qquad (14.3\text{-}25)$$

Solving for $u(k)$, which drives the filtered error to zero, i.e., $\bar{e}(k+d) = \Gamma(q^{-1})e(k+d) = 0$, we arrive at the desired control law:

$$u(k) = \frac{\Gamma(q^{-1})y_m(k+d) - R(q^{-1})y(k) - \psi^*(q^{-1})u(k-1)}{b_0} \qquad (14.3\text{-}26)$$

or equivalently

$$u(k) = \frac{\Gamma(q^{-1})y_m(k+d) - \theta_0^T \varphi_0(k)}{b_0} \qquad (14.3\text{-}27)$$

where use has been made of the fact that $b_0 \neq 0$. Finally, using the fact that $\psi(q^{-1}) = B(q^{-1})S(q^{-1})$, the expression for $u(k)$ becomes

$$u(k) = \frac{1}{B(q^{-1})S(q^{-1})}\left[\Gamma(q^{-1})y_m(k+d) - R(q^{-1})y(k)\right] \qquad (14.3\text{-}28)$$

From this last expression for $u(k)$, it is readily seen why the process should be minimum phase, as the system zeros appear in the denominator of the control law.

It can be seen that the control law (14.3-26), which satisfies the control objective $\Gamma(q^{-1})e(k+d) = 0$, also minimizes the quadratic performance index

$$J(k+d) = \left[\Gamma(q^{-1})[y(k+d) - y_m(k+d)]\right]^2 \qquad (14.3\text{-}29)$$

thereby assuring that $J(k+d) \equiv 0$, for $k \geq 0$.

The control scheme analyzed above for the case of known parameters is shown in Figure 14.3.

14.3.4 Hyperstability Design in the Case of Unknown Parameters

1 The Adaptation Algorithm

When the system parameters appearing in the polynomials $A(q^{-1})$ and $B(q^{-1})$ are unknown, we keep the same structure for the controller, but replace the constant b_0 and the vector θ_0 (which are now unknown) in Eq. (14.3-27), with the time-varying adjustable parameters

$$\hat{b}_0(k) \quad \text{and} \quad \hat{\theta}_0^T(k) = \left[\hat{\psi}_1(k), \ldots, \hat{\psi}_{d+n_B-1}(k), \hat{r}_0(k), \hat{r}_1(k), \ldots, \hat{r}_{n_R}(k)\right] \qquad (14.3\text{-}30)$$

This procedure is widely known in the literature as the *certainty equivalence principle*. The adjustable parameters of Eq. (14.3-30) will be appropriately updated by the adaptation mechanism. The certainty equivalence control law now becomes

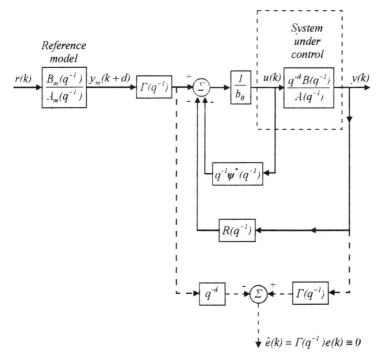

Figure 14.3 The control scheme for tracking and regulation with independent dynamics for the case of known parameters.

$$u(k) = \frac{\Gamma(q^{-1})y_m(k+d) - \hat{\theta}_0^T(k)\varphi_0(k)}{\hat{b}_0(k)} \qquad (14.3\text{-}31)$$

For convenience, we keep the same notation $u(k)$ for the certainty equivalence control law (14.3-31) as well. Expression (14.3-31) may also be written as

$$\Gamma(q^{-1})y_m(k+d) = \hat{\theta}^T(k)\varphi(k) \qquad (14.3\text{-}32)$$

where

$$\hat{\theta}^T(k) = \left[\hat{b}_0(k) \vdots \hat{\theta}_0^T(k) \right] \qquad (14.3\text{-}33)$$

In the case of unknown plant parameters, it is not possible to keep the filtered error $\bar{e}(k+d) = \Gamma(q^{-1})e(k+d)$ identically equal to zero. The design objective now changes and becomes that of finding a suitable adaptation mechanism for the adjustable parameters in Eq. (14.3-33), which will secure the asymptotic convergence of $\bar{e}(k+d)$ to zero, with bounded input and output signals. Consequently, in the case of unknown plant parameters the control objective becomes

$$\lim_{k \to +\infty} \bar{e}(k+d) = \lim_{k \to +\infty} \Gamma(q^{-1})[y(k+d) - y_m(k+d)] = 0, \qquad \forall \bar{e}(0) \neq 0,$$

$$\hat{\theta}(0) \in \mathbf{R}^{d+n_B+n_R+1}$$

$$(14.3\text{-}34)$$

with $\|\boldsymbol{\varphi}(k)\|$ bounded by all k. Then, if Eq. (14.3-34) holds, since $\Gamma(q^{-1})$ is an asymptotically stable polynomial, one could conclude that $\lim_{k \to +\infty} e(k) = 0$. That is, the plant-model error vanishes asymptotically. Using Eqs (14.3-22) and (14.3-32), the filtered plant-model error (or a priori adaptation error) $\bar{e}(k + d)$ is expressed as

$$\bar{e}(k + d) = \Gamma(q^{-1})e(k + d) = \Gamma(q^{-1})[y(k + d) - y_{\mathrm{m}}(k + d)]$$
$$= \boldsymbol{\theta}^{\mathrm{T}}\boldsymbol{\varphi}(k) - \hat{\boldsymbol{\theta}}^{\mathrm{T}}(k)\boldsymbol{\varphi}(k)$$

or

$$\bar{e}(k + d) = \left(\boldsymbol{\theta} - \hat{\boldsymbol{\theta}}(k)\right)^{\mathrm{T}}\boldsymbol{\varphi}(k) \tag{14.3-35}$$

Equivalently

$$\bar{e}(k) = \Gamma(q^{-1})e(k) = \left(\boldsymbol{\theta} - \hat{\boldsymbol{\theta}}(k - d)\right)^{T}\boldsymbol{\varphi}(k - d) \tag{14.3-36}$$

Define the *auxiliary error* $\bar{e}(k)$ as

$$\bar{\varepsilon}(k) = \left(\hat{\boldsymbol{\theta}}(k - d) - \hat{\boldsymbol{\theta}}(k)\right)^{\mathrm{T}}\boldsymbol{\varphi}(k - d) \tag{14.3-37}$$

and the *a posteriori filtered plant-model error* or *augmented error* $\varepsilon(k)$ as

$$\varepsilon(k) = \bar{e}(k) + \bar{\varepsilon}(k) = \left(\boldsymbol{\theta} - \hat{\boldsymbol{\theta}}(k)\right)^{\mathrm{T}}\boldsymbol{\varphi}(k - d) \tag{14.3-38}$$

By using the so-called hyperstability approach not presented herein, but analyzed in [12], it can be proven that the following adaptation algorithm:

$$\hat{\boldsymbol{\theta}}(k) = \hat{\boldsymbol{\theta}}(k - 1) + \mathbf{F}(k)\boldsymbol{\varphi}(k - d)\varepsilon(k) \tag{14.3-39}$$

assures that, for all $\bar{e}(0) \neq 0$ and $\hat{\boldsymbol{\theta}}(0) \in \mathbf{R}^{d + n_B + n_R + 1}$, we have

$$\lim_{k \to +\infty} \varepsilon(k) = 0 \quad \text{and} \quad \lim_{k \to +\infty} \bar{e}(k) = \lim_{k \to +\infty} e(k) = 0 \tag{14.3-40}$$

The gain matrix $\mathbf{F}(k)$ is positive definite and is generated by

$$\mathbf{F}^{-1}(k + 1) = \lambda_1(k)\mathbf{F}^{-1}(k) + \lambda_2(k)\boldsymbol{\varphi}(k - d)\boldsymbol{\varphi}^{\mathrm{T}}(k - d) \quad \text{when} \quad \mathbf{F}(0) > 0 \tag{14.3-41}$$

with

$$0 < \lambda_1(k) \leq 1 \quad \text{and} \quad 0 \leq \lambda_2(k) < 2 \quad \text{for all } k \tag{14.3-42}$$

Clearly, relation (14.3-40) states that the control objective is satisfied asymptotically.

Note that the algorithm presented above is a special case of the algorithm given by Ionescu & Monopoli in [19], who introduced the notion of the augmented error for the first time for discrete-time systems.

Remark 14.3.2

To apply the adaptation algorithm (14.3-39), an implementable form for the *a posteriori* filtered plant-model error $\varepsilon(k)$ may be derived using Eqs (14.3-32) and (14.3-39), as follows:

$$\varepsilon(k) = \bar{e}(k) + \tilde{\varepsilon}(k) = \Gamma(q^{-1})(y(k) - y_m(k)) + \left(\hat{\theta}(k - d) - \hat{\theta}(k)\right)^T \varphi(k - d)$$

$$= \Gamma(q^{-1})y(k) - \hat{\theta}^T(k - d)\varphi(k - d) + \left(\hat{\theta}(k - d) - \hat{\theta}(k)\right)^T \varphi(k - d)$$

$$= \Gamma(q^{-1})y(k) - \hat{\theta}^T(k)\varphi(k - d)$$

$$= \Gamma(q^{-1})y(k) - \hat{\theta}^T(k - 1)\varphi(k - d) - \varphi^T(k - d)\mathbf{F}(k)\varphi(k - d)\varepsilon(k) \quad (14.3\text{-}43)$$

Hence

$$\varepsilon(k) = \frac{\tilde{\varepsilon}(k)}{1 + \varphi^T(k - d)\mathbf{F}(k)\varphi(k - d)} \quad (14.3\text{-}44)$$

where

$$\tilde{\varepsilon}(k) = \Gamma(q^{-1})y(k) - \hat{\theta}^T(k - 1)\varphi(k - d) \quad (14.3\text{-}45)$$

Remark 14.3.3

During the adaptation procedure we have $\hat{b}_0(k) = 0$. To avoid division by zero in Eq. (14.3-31), if $|\hat{b}_0(k)| < \delta(\delta > 0)$ for a certain k, we repeat evaluating $\hat{\theta}(k)$ from Eq. (14.3-39), using appropriate values for $\lambda_1(k - 1)$ and $\lambda_2(k - 1)$ in Eq. (14.3-41). These values must be chosen by trial and error so that $|\hat{b}_0(k)| \geq \delta$.

The control algorithm is summarized in Table 14.1. The control scheme for tracking and regulation with independently chosen dynamics, for the case of unknown parameters, is given in Figure 14.4.

Table 14.1   The Model Reference Adaptive Control (MRAC) Algorithm

Algorithm	Equation No.
$\varphi^T(k) = [u(k) \vdots u(k - 1), \ldots, u(k - d - n_B + 1), y(k), y(k - 1), \ldots, y(k - n_R)]$	
$\qquad = [u(k) \vdots \quad \varphi_0^T(k)]$	(14.3-46)
$\hat{\theta}^T(k) = [\hat{b}_0(k) \vdots \quad \hat{\theta}_0^T(k)]$	(14.3-47)
$u(k) = \dfrac{\Gamma(q^{-1})y_m(k + d) - \hat{\theta}_0^T(k)\varphi_0(k)}{\hat{b}_0(k)}$	(14.3-48)
$\tilde{\varepsilon}(k) = \Gamma(q^{-1})y(k) - \hat{\theta}^T(k - 1)\varphi_0(k - d)$	(14.3-49)
$\varepsilon(k) = \dfrac{\tilde{\varepsilon}(k)}{1 + \varphi^T(k - d)\mathbf{F}(k)\varphi(k - d)}$	(14.3-50)
$\hat{\theta}(k) = \hat{\theta}(k - 1) + \mathbf{F}(k)\varphi(k - d)\varepsilon(k)$	(14.3-51)
$\mathbf{F}^{-1}(k + 1) = \lambda_1 \mathbf{F}^{-1}(k) + \lambda_2(k)\varphi(k - d)\varphi^T(k - d)$	(14.3-52)
$\quad$ with initial conditions $y(0)$, $\mathbf{F}(0) > 0$, $\hat{\theta}(0)$	

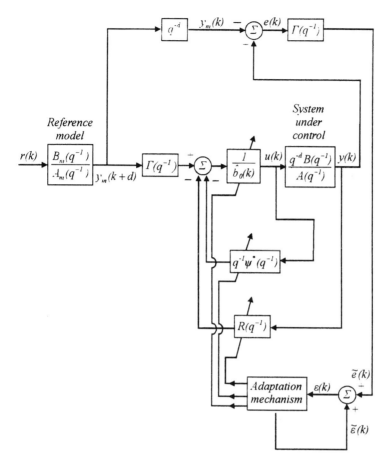

Figure 14.4 The control scheme for tracking and regulation with independent dynamics for the case of unknown parameters.

Example 14.3.1

Consider the system

$$A(q^{-1})y(k) = q^{-1}B(q^{-1})u(k), \qquad y(0) = 1$$

where $A(q^{-1}) = 1 + 2q^{-1} + q^{-2}$ and $B(q^{-1}) = 2 + q^{-1} + 0.5q^{-2}$ (asymptotically stable). It is desired to track the output of the reference model

$$y_m(k) = \left[\frac{q^{-1}B_m(q^{-1})}{A_m(q^{-1})}\right]r(k) = q^{-1}\left[\frac{1 + 0.3q^{-1}}{1 - q^{-1} + 0.25q^{-2}}\right]r(k) \qquad \text{with} \qquad y_m(0) = 2$$

The dynamics during regulation are characterized by the asymptotically stable polynomial $\Gamma(q^{-1}) = 1 + 0.5q^{-1}$.

(a) Determine a model reference control law which achieves the control objectives in the case of known parameters for $A(q^{-1})$ and $B(q^{-1})$.

(b) In the case of unknown plant parameters, determine a control law and appropriate adaptations (MRAC design) to satisfy the control objectives asymptotically.

Solution

(a) Here $d = 1$, $S(q^{-1}) = 1$, and we are looking for $R(q^{-1}) = r_0 + r_1 q^{-1}$ such that

$$\Gamma(q^{-1}) = 1 + 0.5q^{-1} = (1 + 2q^{-1} + q^{-2}) + q^{-1}(r_0 + r_1 q^{-1})$$
$$= A(q^{-1})S(q^{-1}) + q^{-d}R(q^{-1})$$

or equivalently

$$\begin{bmatrix} 1 & 0 & 0 \\ a_1 & 1 & 0 \\ a_2 & 0 & 1 \end{bmatrix} \begin{bmatrix} 1 \\ r_0 \\ r_1 \end{bmatrix} = \begin{bmatrix} 1 \\ \gamma_1 \\ 0 \end{bmatrix} \quad \text{or} \quad \begin{bmatrix} 1 & 0 & 0 \\ 2 & 1 & 0 \\ 1 & 0 & 1 \end{bmatrix} \begin{bmatrix} 1 \\ r_0 \\ r_1 \end{bmatrix} = \begin{bmatrix} 1 \\ 0.5 \\ 0 \end{bmatrix}$$

One easily obtains $r_0 = -1.5$ and $r_1 = -1$. Moreover,

$$\psi(q^{-1}) = B(q^{-1})S(q^{-1}) = 2 + q^{-1} + 0.5q^{-2} = 2 + q^{-1}(1 + 0.5q^{-1})$$
$$= b_0 + q^{-1}\psi^*(q^{-1})$$

In the case of unknown parameters, the control law is

$$u(k) = \frac{\Gamma(q^{-1})y_m(k+1) - R(q^{-1})y(k) - \psi^*(q^{-1})u(k-1)}{b_0}$$
$$= \frac{y_m(k+1) + 0.5y_m(k) + 1.5y(k) + y(k-1) - u(k-1) - 0.5u(k-2)}{2}$$

(b) In the case of unknown plant parameters, the certainty equivalence control law is

$$u(k) = \frac{\begin{array}{c} y_m(k+1) + 0.5y_m(k) - \hat{\theta}_1(k)u(k-1) - \\ \hat{\theta}_2(k)u(k-2) - \hat{\theta}_3(k)y(k) - \hat{\theta}_4(k)y(k-1) \end{array}}{\hat{\theta}_0(k)}$$

where

$$\hat{\theta}^T(k) = \left[\hat{\theta}_0(k), \hat{\theta}_1(k), \hat{\theta}_2(k), \hat{\theta}_3(k), \hat{\theta}_4(k) \right]$$

where $\hat{\theta}(k)$ is appropriately changed at each step, by using the adaptation algorithm given below. In this algorithm we let $\lambda_1(k) = 0.98$ and $\lambda_2(k) = 1$. This corresponds to a forgetting factor algorithm, as explained in the discussion of the parameter adaptation algorithm presented at the end of this section. The adaptation algorithm is

$$\mathbf{F}(0) = \frac{1}{10^{-3}} \mathbf{I}_5$$

$$\boldsymbol{\varphi}^T(k) = [u(k), u(k-1), u(k-2), y(k), y(k-1)] \quad \text{for} \quad k = 0, 1, 2, \ldots$$

$$\tilde{\varepsilon}(k) = y(k) + 0.5y(k-1) - \hat{\theta}^T(k-1)\varphi(k-1) \quad \text{for} \quad k = 1, 2, \ldots$$

$$\varepsilon(k) = \frac{\tilde{\varepsilon}(k)}{1 + \varphi^T(k-1)\mathbf{F}(k)\varphi(k-1)} \quad \text{for} \quad k = 1, 2, \ldots$$

$$\hat{\theta}(k) = \hat{\theta}(k-1) + \mathbf{F}(k)\varphi(k-1)\varepsilon(k) \qquad \text{for} \qquad k = 1, 2, \ldots$$

$$\mathbf{F}^{-1}(k+1) = 0.98\mathbf{F}^{-1}(k) + \varphi(k-1)\varphi^{\mathrm{T}}(k-1) \qquad \text{for} \qquad k = 0, 1, 2, \ldots$$

We can initialize $\hat{\theta}(k)$ with $\hat{\theta}^{\mathrm{T}}(0) = [1, 0, 0, -1, -1]$ for convenience

The particular choice made for the adaptation algorithm was guided by the objective of global asymptotic stability for the whole system (14.3-39), (14.3-41), (14.3-42), (14.3-44), i.e., asymptotic stability for any finite initial parameter error and plant-model error. Moreover, the adaptation mechanism should ensure that the error between the plant output and the ouput of the reference model tends to zero asymptotically, which is the control objective. The approach applied to satisfy the aforementioned objectives relies upon the hyperstability theory presented in [12]. Global asymptotic stability is guaranteed.

Indeed, most adaptive control schemes, after an adequate analysis, lead to an equation of the form

$$v(k) = H(q^{-1})\left(\theta - \hat{\theta}(k)\right)^{\mathrm{T}}\varphi(k-d) \tag{14.3-53}$$

where θ is an unknown parameter vector, $\hat{\theta}(k)$ is the estimate of θ resulting from an appropriate parameter adaptation algorithm, $\varphi(k-d)$ is a measurable regressor vector, $H(q^{-1})$ is a rational discrete transfer function of the form

$$H(z^{-1}) = \frac{1 + h'_1 z^{-1} + + h'_\alpha z^{-\alpha}}{1 + h_1 z^{-1} + \cdots + h_\beta z^{-\beta}} \tag{14.3-54}$$

and the measurable quantity $v(k)$ is the so-called *processed augmented error*. A particular case of Eq. (14.3-53) is given by Eq. (14.3-38), where $H(q^{-1}) = 1$ and the *a posteriori* filtered plant-model error $\varepsilon(k)$ takes the place of $v(k)$. Then, a stability theorem given in [22] provides the following appropriate adaptation mechanism sought, which makes use of $v(k)$ as the basis of the parameter update law for $\hat{\theta}(k)$:

$$\hat{\theta}(k) = \hat{\theta}(k-1) + \mathbf{F}(k)\varphi(k-d)v(k) \tag{14.3-55}$$

$$\mathbf{F}^{-1}(k+1) = \lambda_1(k)\mathbf{F}^{-1}(k) + \lambda_2(k)\varphi(k-d)\varphi^{\mathrm{T}}(k-d), \qquad \mathbf{F}(0) > 0 \tag{14.3-56}$$

with

$$0 < \lambda_1(k) \leq 1, \qquad 0 \leq \lambda_2(k) < 2, \qquad \forall k \tag{14.3-57}$$

The convergence of the plant-model error $e(k)$ to zero and the boundedness of the input and output signals can be proved.

2 Discussion of the Parameter Adaptation Algorithm

Expression (14.3-56) defines a general law for the determination of the adaptation gain matrix $\mathbf{F}(k)$ and is repeated here for convenience:

$$\mathbf{F}^{-1}(k+1) = \lambda_1(k)\mathbf{F}^{-1}(k) + \lambda_2(k)\varphi(k-d)\varphi^{\mathrm{T}}(k-d) \qquad \text{with} \qquad \mathbf{F}(0) > 0 \tag{14.3-58}$$

where $0 < \lambda_1(k) \leq 1$ and $0 \leq \lambda_2(k) < 2$. Using the matrix inversion lemma of Chapter 13 (relation 13.3-7)), the above equation may be written equivalently as follows:

$$\mathbf{F}(k+1) = \frac{1}{\lambda_1(k)} \mathbf{F}(k) - \left[\frac{\mathbf{F}(k)\boldsymbol{\varphi}(k-d)\boldsymbol{\varphi}^{\mathrm{T}}(k-d)\mathbf{F}(k)}{\lambda_1(k)\lambda_2^{-1}(k) + \boldsymbol{\varphi}^{\mathrm{T}}(k-d)\mathbf{F}(k)\boldsymbol{\varphi}(k-d)} \right] \qquad (14.3\text{-}59)$$

We note that, in general, $\lambda_1(k)$ and $\lambda_2(k)$ have opposite effects on the adaptation gain. That is, as $\lambda_1(k) \leq 1$ increases, the gain $\lambda_2(k)$ does the opposite, i.e., it decreases the gain.

Different types of adaptation algorithms are obtained by appropriate choices of $\lambda_1(k)$ and $\lambda_2(k)$, $0 < \lambda_1(k) \leq 1$, $0 \leq \lambda_2(k) < 2$. We distinguish the following choices:

1. $\lambda_1(k) \equiv 1$ and $\lambda_2(k) \equiv 0$. In this case $\mathbf{F}(k) = \mathbf{F}(0)$. This choice corresponds to an algorithm with a constant gain. It is the simplest to implement, but also the least efficient. It is convenient for the estimation of unknown constant parameters, but not for time-varying parameters.
2. $\lambda_1(k) = \lambda_2(k) \equiv 1$. This choice corresponds to a recursive least-squares algorithm with decreasing gain.
3. $\lambda_1(k) \equiv \lambda_1 < 1$ (usually $0.95 \leq \lambda_1 \leq 0.99$) and $\lambda_2(k) \equiv 1$. This choice corresponds to an algorithm with a constant forgetting factor λ_1 (it "forgets" old measurements exponentially).
4. $\lambda_1(k) < 1$ and $\lambda_2(k) \equiv 1$. This choice corresponds to a variable forgetting factor type of algorithm. Usually, $0.95 \leq \lambda_1(k) \leq 0.99$ or $\lambda_1(k+1) = \lambda_0\lambda_1(k) + 1 - \lambda_0$, with $0.95 \leq \lambda_0 \leq 0.99$ and $0.95 \leq \lambda_1(0) \leq 0.99$A. In this last case it holds true that $\lim_{k \to +\infty} \lambda_1(k) = 1$.
5. When both $\lambda_1(k)$ and $\lambda_2(k)$ are time varying, we have extra freedom in choosing the gain profiles. For example, by choosing $\lambda_1(k)/\lambda_2(k) = \alpha(k)$, we have the following expression for the trace of $\mathbf{F}(k)$:

$$\mathrm{tr}\,\mathbf{F}(k+1) = \frac{1}{\lambda_1(k)} \mathrm{tr}\left[\mathbf{F}(k) - \frac{\mathbf{F}(k)\boldsymbol{\varphi}(k-d)\boldsymbol{\varphi}^{\mathrm{T}}(k-d)\mathbf{F}(k)}{\alpha(k) + \boldsymbol{\varphi}^{\mathrm{T}}(k-d)\mathbf{F}(k)\boldsymbol{\varphi}(k-d)} \right]$$

$$(14.3\text{-}60)$$

At each step we can choose $\alpha(k) = \lambda_1(k)/\lambda_2(k)$ and then specify $\lambda_1(k)$ from Eq. (14.3-60) such that the trace of $\mathbf{F}(k)$ has a prespecified value (constant or time varying) at each step.

Remark 14.3.4

We note that when $\boldsymbol{\varphi}(k-d) = \mathbf{0}$ for a long period of time (this may happen in the steady state or in the absence of any signal in the input), using choices 3 or 4 may lead to an undesirable increase in the adaptation gain. In this case there is no change in the parameter estimates and $\mathbf{F}(k)$ will grow exponentially if $\lambda_1(k) < 1$, since in this case we have that $\mathbf{F}(k+1) = 1/\lambda_1^{-1}(k)\mathbf{F}(k)$. A new change in the set point can then lead to large changes in the parameter estimates and the plant output.

Remark 14.3.5

In practice we initialize at $\mathbf{F}(0) = (1/\delta)\mathbf{I}$, $0 < \delta \ll 1$.

14.4 SELF-TUNING REGULATORS

14.4.1 Introduction

Another important class of adaptive systems with many industrial applications is that of STRs. The block diagram of an STR is shown in Figure 14.5.

The STR is based on the idea of separating the estimation of the unknown parameters of the system under control, from the design of the controller. The control scheme consists of two loops: the "inner" loop, which involves the plant with unknown parameters and a linear feedback controller with adjustable parameters and the "outer" loop, which is used in the case of unknown plant parameters and is composed of a recursive parameter estimator and a block named "controller design."

In the case of known plant parameters, the design of the controller (i.e., the determination of its parameters as functions of the plant parameters) is carried out off-line. This controller satisfies a specific control design problem, such as minimum variance, pole placement, model following, etc. This control problem, in the context of the STRs, is called the *underlying control problem*.

When the plant parameters are uncertain, the recursive parameter estimator provides on-line estimates of the unknown plant parameters. On the basis of these estimates, the solution of the control design problem (i.e., the determination of the controller parameters as functions of the plant parameters) is achieved on-line in each step by the "controller design" block. The controller parameter estimates thus obtained, are used to recalculate the control law at each step. Apart from the fact

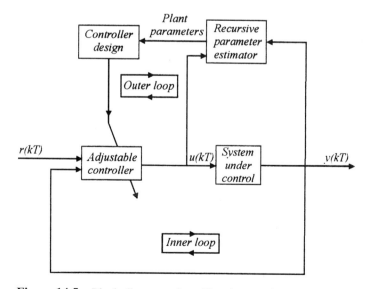

Figure 14.5 Block diagram of a self-tuning regulator (STR).

that the controller parameters are substituted by estimates obtained by the on-line solution of the control design problem, the controller structure is kept the same as in the case of known plant parameters. This is the *certainty equivalence principle*.

For the estimation of the plant parameters, various schemes can be used: least squares, recursive least squares, maximum likelihood, extended Kalman filtering, etc. Different combinations of appropriate parameter estimation methods and suitable control strategies lead to different adaptive controllers. For example, an adaptive controller based on least-squares estimation and deadbeat control was first described by Kalman in 1958, while the original STR design by Åström et al. [16] was based on least-squares estimation and the minimum-variance control problem.

The control procedure discussed above leads to an *explicit STR*, where the term explicit is used because the plant parameters are estimated explicitly. Such explicit STRs need to solve, at each step, the tedious controller design problem. It is sometimes possible, in order to eliminate the design calculations, to reparametrize the plant model, so that it can be expressed in terms of the controller parameters, which are then updated directly by the estimator. This leads to *implicit STRs*. Implicit STRs avoid controller design calculations and are based on estimates of an implicit plant model. Explicit STRs correspond to indirect adaptive control, while implicit STRs correspond to direct adaptive control.

A close relation has been established between STRs and MRAC systems, in spite of differences in their origin. Indeed, MRAC design was based on the deterministic servoproblem, while STR design was based on the stochastic regulation problem. Although the design methods of the "inner" loop and the parameter adjustments in the "outer" loop are different, direct MRAC systems are closely related to implicit STRs, while indirect MRAC systems are related to explicit STRs.

Implicit STRs are not discussed here. Explicit STRs, using the pole placement technique, are treated in Subsec. 14.4.2 that follows. Furthermore, it is explained how implicit pole-placement designs can also be derived.

14.4.2 Pole-Placement Self-Tuning Regulators

1 Pole-Placement Design with Known Parameters

The pole-placement design (chosen as the underlying control problem), can be applied for nonminimum phase systems. The procedure consists of finding a feedback law for which the closed-loop poles have desired locations. Both explicit and implicit schemes may be formulated. Explicit schemes are based on estimates of parameters in an explicit system model, while implicit schemes are based on estimates of parameters in a modified implicit system model. Similarities between MRAC and STRs will emerge.

The discussion is limited to SISO systems described by

$$A(q^{-1})y(k) = q^{-d}B(q^{-1})u(k) \tag{14.4-1}$$

where

$$A(q^{-1}) = 1 + a_1 q^{-1} + \cdots + a_{n_A} q^{-n_A} \tag{14.4-2}$$

$$B(q^{-1}) = b_0 + b_1 q^{-1} + \cdots + b_{n_B} q^{-n_B} \tag{14.4-3}$$

The polynomial $A(q^{-1})$ is thus monic, $A(q^{-1})$ and $B(q^{-1})$ are relatively prime (have no common factors), and the system's delay is $d \geq 1$. It is desired to find a controller for which the relation from the command signal $r(k)$ to the output $y(k)$ becomes

$$A_m(q^{-1})y(k) = q^{-d}B_m(q^{-1})r(k) \tag{14.4-4}$$

where $A_m(q^{-1})$ is a stable monic polynomial and $A_m(q^{-1})$ and $B_m(q^{-1})$ are relatively prime. Restrictions on $B_m(q^{-1})$ will appear in what follows.

A general structure (R–S–T canonical structure) for the controller is presented in Figure 14.6. The controller is described by

$$R(q^{-1})u(k) = T(q^{-1})r(k) - S(q^{-1})y(k) \tag{14.4-5}$$

This controller offers a negative feedback with transfer function $-S(q^{-1})/R(q^{-1})$ and feedforward with transfer function $T(q^{-1})/R(q^{-1})$. Multiplying Eq. (14.4-5) by $q^{-d}B(q^{-1})$, one obtains

$$q^{-d}B(q^{-1})R(q^{-1})u(k) = q^{-d}T(q^{-1})B(q^{-1})r(k) - q^{-d}S(q^{-1})B(q^{-1})y(k)$$

or

$$\left[A(q^{-1})R(q^{-1}) + q^{-d}B(q^{-1})S(q^{-1}) \right]y(k) = q^{-d}T(q^{-1})B(q^{-1})r(k) \tag{14.4-6}$$

Hence, the relation between $y(k)$ and $r(k)$ is given by

$$\frac{y(k)}{r(k)} = \frac{q^{-d}T(q^{-1})B(q^{-1})}{A(q^{-1})R(q^{-1}) + q^{-d}B(q^{-1})S(q^{-1})} \tag{14.4-7}$$

Relation (14.4-4), which represents the desired behavior, may be written as

$$\frac{y(k)}{r(k)} = \frac{q^{-d}B_m(q^{-1})}{A_m(q^{-1})} \tag{14.4-8}$$

Thus, the design problem is equivalent to the algebraic problem of finding $R(q^{-1})$, $S(q^{-1})$, and $T(q^{-1})$, for which the following equation holds true:

$$\frac{q^{-d}T(q^{-1})B(q^{-1})}{A(q^{-1})R(q^{-1}) + q^{-d}B(q^{-1})S(q^{-1})} = \frac{q^{-d}B_m(q^{-1})}{A_m(q^{-1})} \tag{14.4-9}$$

From the left-hand side of Eq. (14.4-9), it is evident that the system zeros ($z^{n_B}B(z^{-1}) = 0$) will also be closed-loop zeros, unless they are canceled out by corre-

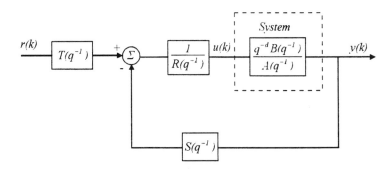

Figure 14.6 The R–S–T canonical structure used for the controller.

sponding closed-loop poles. But unstable (or poorly damped) zeros should not be canceled out by the controller, since they would lead to instability. Thus, let us factor out $B(q^{-1})$ as follows

$$B(q^{-1}) = B^+(q^{-1})B^-(q^{-1}) \qquad (14.4\text{-}10)$$

where $B^+(q^{-1})$ contains the well-damped zeros (which are canceled out) and $B^-(q^{-1})$ contains the unstable and poorly damped zeros (which are not canceled out). To obtain a unique factorization, we also require that $B^+(q^{-1})$ is a monic polynomial. From Eq. (14.4-9), it follows that the characteristic polynomial of the closed-loop system is

$$A(q^{-1})R(q^{-1}) + q^{-d}B(q^{-1})S(q^{-1}) \qquad (14.4\text{-}11)$$

The factors of this polynomial should be the desired reference model poles, i.e., the roots of $A_m(q^{-1})$, and the system zeros which can be canceled out, i.e., the roots of $B^+(q^{-1})$. Moreover, since in general, the order of the reference model (deg $A_m(q^{-1})$), is less than the order of the closed-loop system deg$(A(q^{-1})R(q^{-1}) + q^{-d}B(q^{-1})S(q^{-1}))$), there are factors in the left-hand side of Eq. (14.4-9) which cancel out. These factors correspond to a polynomial $A_0(q^{-1})$. The polynomial $A_0(q^{-1})$ is called the observer polynomial and is chosen to have well-damped roots. The appearance of this polynomial is more evident when a state-space solution to this problem is considered. In this case, the solution is a combination of state feedback and an observer. Hence, the characteristic polynomial of the closed-loop system assumes the form

$$A(q^{-1})R(q^{-1}) + q^{-d}B(q^{-1})S(q^{-1}) = B^+(q^{-1})A_m(q^{-1})A_0(q^{-1}) \qquad (14.4\text{-}12)$$

Now, since $B^+(q^{-1})$ is the divident of $B(q^{-1})$ and the polynomials $A(q^{-1})$ and $B(q^{-1})$ are relatively prime, it is clear from Eq. (14.4-12) that $B^+(q^{-1})$ should also be the divident of the polynomial $R(q^{-1})$, i.e.,

$$R(q^{-1}) = B^+(q^{-1})R_1(q^{-1}) \qquad (14.4\text{-}13)$$

Equation (14.4-12) may then be rewritten as

$$A(q^{-1})R_1(q^{-1}) + q^{-d}B^-(q^{-1})S(q^{-1}) = A_m(q^{-1})A_0(q^{-1}) \qquad (14.4\text{-}14)$$

Hence, Eq. (14.4-9) is then equivalent to the equation

$$\frac{q^{-d}B^+(q^{-1})B^-(q^{-1})T(q^{-1})}{B^+(q^{-1})A_m(q^{-1})A_0(q^{-1})} = \frac{q^{-d}B_m(q^{-1})}{A_m(q^{-1})} \qquad (14.4\text{-}15)$$

In order that the foregoing equation holds true and since $B^-(q^{-1})$ cannot be canceled out, it is clear that $B^-(q^{-1})$ must be a factor of $B_m(q^{-1})$ i.e.,

$$B_m(q^{-1}) = B^-(q^{-1})B_m^+(q^{-1}) \qquad (14.4\text{-}16)$$

and also that

$$T(q^{-1}) = A_0(q^{-1})B_m^+(q^{-1}) \qquad (14.4\text{-}17)$$

It should be evident that we are not absolutely free in the choice of $B_m(q^{-1})$, which corresponds to the specifications for the closed-loop zeros. We can choose freely the part $B_m^+(q^{-1})$ of $B_m(q^{-1})$ while Eq. (14.4-16) should be valid, otherwise there is no solution to the design problem.

It is necessary to establish conditions under which a solution for the polynomials $R_1(q^{-1})$ and $S(q^{-1})$ in Eq. (14.4-14), is guaranteed. This equation, linear in the polynomials $R_1(q^{-1})$ and $S(q^{-1})$ is a special case of the Diophantine equation (or Bezout identity), which has the general form (see also Remark 14.3.1):

$$\bar{A}(q^{-1})\bar{R}(q^{-1}) + \bar{B}(q^{-1})\bar{S}(q^{-1}) = \bar{C}(q^{-1}) \qquad (14.4\text{-}18)$$

It can be proved that the Diophantine equation (14.4-18) always has a solution for $\bar{R}(q^{-1})$ and $\bar{S}(q^{-1})$, if the greatest common factor of $\bar{A}(q^{-1})$ and $\bar{B}(q^{-1})$ is a dividend of $\bar{C}(q^{-1})$. Therefore, Eq. (14.4-14) will always have a solution for $R_1(q^{-1})$ and $S(q^{-1})$, since we have assumed that $A(q^{-1})$ and $B(q^{-1})$ are coprime and, consequently, $A(q^{-1})$ and $q^{-d}B^-(q^{-1})$ are also coprime.

Note that if a solution exists, then Eq. (14.4-18), in general, has infinitely many solutions. Indeed, if $R^0(q^{-1})$ and $S^0(q^{-1})$ are solutions of Eq. (14.4-18), then it can be easily verified that $R^0(q^{-1}) + \bar{B}(q^{-1})Q(q^{-1})$ and $S^0(q^{-1}) - \bar{A}(q^{-1})Q(q^{-1})$, with $Q(q^{-1})$ an arbitrary polynomial, are also solutions of Eq. (14.4-18). Particular solutions can be specified in several ways. Different solutions give systems with different noise rejection properties.

It can be proved that there are unique solutions to Eq. (14.4-18) if, in addition, we impose the following restriction for the solution sought:

$$\deg \bar{R}(q^{-1}) < \deg \bar{B}(q^{-1}) \qquad (14.4\text{-}19)$$

or

$$\deg \bar{S}(q^{-1}) < \deg \bar{A}(q^{-1}) \qquad (14.4\text{-}20)$$

Moreover, for the pole placement control problem, we seek particular solutions which lead to causal control laws (i.e., $\deg S(q^{-1}) \le \deg R(q^{-1})$ and $\deg T(q^{-1}) \le \deg R(q^{-1})$). Note also that it is often advantageous to keep $\deg S(q^{-1}) = \deg T(q^{-1}) = \deg R(q^{-1})$, in order to avoid an unnecessary delay in the controller.

Note that from Eq. (14.4-12) we must select either

$$\deg R(q^{-1}) = \deg A_m(q^{-1}) + \deg A_o(q^{-1}) + \deg B^+(q^{-1}) - \deg A(q^{-1}) \qquad (14.4\text{-}21)$$

or

$$\deg S(q^{-1}) = \deg A_m(q^{-1}) + \deg A_0(q^{-1}) - \deg B^-(q^{-1}) - d \qquad (14.4\text{-}22)$$

The degrees of $R(q^{-1})$ and $S(q^{-1})$ are imposed by the structure of the system and the structure of the desired closed-loop transfer function. To assure unique solutions, using Eq. (14.4-21) we must have $\deg S(q^{-1}) \le \deg A(q^{-1}) - 1$ (this results from Eq. (14.4-20)), and if we choose to satisfy Eq. (14.4-22) we must have $\deg R(q^{-1}) \le \deg B(q^{-1}) + d - 1$ (this results from Eq. (14.4-19)).

By selecting Eq. (14.4-21) or Eq. (14.4-22), possible choices for the degrees of $R_1(q^{-1})$ and $S(q^{-1})$ in Eq. (14.4-14), corresponding to unique solutions and minimum-order polynomials, are consequently given below:

$$\deg R_1(q^{-1}) = \deg A_m(q^{-1}) + \deg A_0(q^{-1}) - \deg A(q^{-1}) \qquad (14.4\text{-}23)$$

$$\deg S(q^{-1}) = \deg A(q^{-1}) - 1 \qquad (14.4\text{-}24)$$

or

$$\deg R_1(q^{-1}) = \deg B^-(q^{-1}) + d - 1 \qquad (14.4\text{-}25)$$

$$\deg S(q^{-1}) = \deg A_m(q^{-1}) + \deg A_0(q^{-1}) - \deg B^-(q^{-1}) - d \qquad (14.4\text{-}26)$$

By selecting Eqs (14.4-23) and (14.4-24), and in order to have causal control laws (that is $\deg S(q^{-1}) = \deg A(q^{-1}) - 1 \leq \deg R(q^{-1})$), relation (14.4-21) leads to

$$\deg A_0(q^{-1}) \geq 2 \deg A(q^{-1}) - \deg A_m(q^{-1}) - \deg B^+(q^{-1}) - 1 \qquad (14.4\text{-}27)$$

Relation (14.4-27) is a restriction on the degree of the observer polynomial $A_0(q^{-1})$. Moreover, requiring that $\deg T(q^{-1}) \leq \deg R(q^{-1})$ and using Eqs (14.4-17) and (14.4-21), we obtain

$$\begin{aligned}
\deg A_0(q^{-1}) + \deg B_m^+(q^{-1}) = \deg T(q^{-1}) &\leq \deg R(q^{-1}) \\
&= \deg A_m(q^{-1}) + \deg A_0(q^{-1}) + \deg B^+(q^{-1}) \\
&\quad - \deg A(q^{-1})
\end{aligned}$$

or

$$\deg A(q^{-1}) - \deg B(q^{-1}) \leq \deg A_m(q^{-1}) - \deg B_m(q^{-1}) \qquad (14.4\text{-}28)$$

The pole excess of the system should be less than the pole excess of the reference model. Condition (14.4-27) in combination with Eq. (14.4-28) guarantees that the feedback will be causal when Eqs (14.4-23) and (14.4-24) are chosen. This, in turn, implies that the transfer functions S/R and T/R will be causal.

The control algorithm, in the case of known parameters, is summarized in Table 14.2.

Table 14.2 The Pole-Placement Control Algorithm for the Case of Known Parameters*

Step 1	Factor $B(q^{-1}) = B^+(q^{-1})B^-(q^{-1})$ with $B^+(q^{-1})$ monic. Choose $A_m(q^{-1})$, $B_m(q^{-1}) = B^-(q^{-1})B_m^+(q^{-1})$ and $A_0(q^{-1})$ such that Eqs (14.4-27) and (14.4-28) are satisfied
Step 2	Select the degrees of $R_1(q^{-1})$ and $S(q^{-1})$ in order to satisfy Eqs (14.4-23) and (14.4-24) or Eqs (14.4-25) and (14.4-26). Solve $A(q^{-1})R_1(q^{-1}) + q^{-d}B^-(q^{-1})S(q^{-1}) = A_m(q^{-1})A_0(q^{-1})$ for $R_1(q^{-1})$ and $S(q^{-1})$
Step 3	Compute $R(q^{-1}) = B^+(q^{-1})R_1(q^{-1})$ and $T(q^{-1}) = A_0(q^{-1})B_m^+(q^{-1})$. The foregoing steps are executed once off-line.
Step 4	Apply the control law $$u(k) = \left[\frac{T(q^{-1})}{R(q^{-1})}\right]r(k) - \left[\frac{S(q^{-1})}{R(q^{-1})}\right]y(k) \qquad \text{at each step}$$

*Given $A(q^{-1})$ and $B(q^{-1})$ monic and $A(q^{-1})$ and $B(q^{-1})$ co prime.

2 Pole-Placement Design in the Case of Unknown Parameters

In the case of uncertain system model parameters, an STR is used on the basis of the following separation principle. Here, the unknown system parameters are estimated recursively. Based on the certainty equivalence principle, the controller is recomputed at each step using the estimated system parameters. The controller design problem (Diophantine equation) is therefore solved at each step.

The parameter estimator is based on the system model

$$A(q^{-1})y(k) = B(q^{-1})u(k - d) \tag{14.4-29}$$

or explicitly

$$y(k) + a_1 y(k-1) + \cdots + a_{n_A} y(k - n_A) = b_0 u(k - d) + b_1 u(k - d - 1) + \cdots$$
$$+ b_{n_B} u(k - d - n_B) \tag{14.4-30}$$

Introducing the parameter vector

$$\boldsymbol{\theta}^{\mathrm{T}} = [a_1, \ldots, a_{n_A}, b_0, \ldots, b_{n_B}] \tag{14.4-31}$$

and the regression vector

$$\boldsymbol{\varphi}^{\mathrm{T}}(k) = [-y(k-1), \ldots, -y(k - n_A), u(k - d), \ldots, u(k - d - n_B)] \tag{14.4-32}$$

Eq. (14.4-30) is expressed compactly as

$$y(k) = \boldsymbol{\theta}^{\mathrm{T}} \boldsymbol{\varphi}(k) \tag{14.4-33}$$

Based on the prediction model (14.4-33), the recursive least-squares estimator is described by the recursive equation

$$\hat{\boldsymbol{\theta}}(k) = \hat{\boldsymbol{\theta}}(k-1) + \mathbf{F}(k)\boldsymbol{\varphi}(k)\varepsilon(k) \tag{14.4-34}$$

with prediction error

$$\varepsilon(k) = y(k) - \boldsymbol{\varphi}^{\mathrm{T}}(k)\hat{\boldsymbol{\theta}}(k-1) \tag{14.4-35}$$

The gain matrix $\mathbf{F}(k)$ can be deduced recursively using the expression

$$\mathbf{F}(k+1) = \frac{1}{\lambda}\left[\mathbf{F}(k) - \frac{\mathbf{F}(k)\boldsymbol{\varphi}(k)\boldsymbol{\varphi}^{T}(k)\mathbf{F}(k)}{1 + \boldsymbol{\varphi}^{\mathrm{T}}(k)\mathbf{F}(k)\boldsymbol{\varphi}(k)}\right], \qquad \mathbf{F}(0) > 0 \tag{14.4-36}$$

where $0 < \lambda \leq 1$ is a *forgetting factor*. The restrictions of Remark 14.3.4 hold for Eq. (14.4-36).

In self-tuning, the convergence of the parameter estimates to the true values is of great importance. To obtain good estimates using Eq. (14.4-34), it is necessary that the process input be sufficiently rich in frequencies, or persistently exciting. The concept of *persistent excitation* was first introduced in identification problems. This states that we cannot identify all the parameters of a model unless we have enough distinct frequencies in the spectrum of the input signal. In general, when the input to a system is the result of feedback and is therefore a dependent variable within the adaptive loop, the input signal is not persistently exciting.

In the explicit STR based on the pole-placement design discussed above, the estimated parameters are the parameters of the system model. This explicit adaptive pole-placement algorithm is summarized in Table 14.3.

Table 14.3 The Explicit Adaptive Pole-Placement Algorithm for the Case of Unknown Parameters

Step 1	Estimate the model parameters in $A(q^{-1})$ and $B(q^{-1})$ using Eqs (14.4-34), (14.4-35), and (14.4-36), recursively, at each step. It is assumed that $A(q^{-1})$ and $B(q^{-1})$ have no common factors
Step 2	Factorize the polynomial $B(q^{-1})$ so that the decomposition $B^{+}(q^{-1})B^{-}(q^{-1})$ can be made ON-LINE at each step. Solve the controller design problem with the estimates obtained in step 1, i.e., solve Eq. (14.4-14) for $R_1(q^{-1})$ and $S(q^{-1})$ using $A(q^{-1})$ and $B^{-}(q^{-1})$ calculated on the basis of the estimation at step. 1. Calculate $R(q^{-1})$ and $T(q^{-1})$ from Eqs (14.4-13) and (14.4-17), respectively
Step 3	Compute the control law using Eq. (14.4-5)
Step 4	Repeat steps 1–3 at each sampling period

An implicit STR design procedure based on pole placement may also be considered. To this end, we reparametrize the system model such that the controller parameters appear. These controller parameters can then be updated directly. The proper system model structure sought is obtained by multiplying Eq. (14.4-14) by $y(k)$ to yield

$$
\begin{aligned}
A_{\mathrm{m}}(q^{-1})A_0(q^{-1})y(k) &= A(q^{-1})R_1(q^{-1})y(k) + q^{-d}B^{-}(q^{-1})S(q^{-1})y(k) \\
&= q^{-d}B(q^{-1})R_1(q^{-1})u(k) + q^{-d}B^{-}(q^{-1})S(q^{-1})y(k) \\
&= q^{-d}B^{-}(q^{-1})\big[R(q^{-1})u(k) + S(q^{-1})y(k)\big]
\end{aligned} \tag{14.4-37}
$$

The reparametrization (14.4-37), which is an implicit system model, is redundant, since it has more parameters than Eq. (14.4-29). It is also bilinear in the parameters of $B^{-}(q^{-1})$, $R(q^{-1})$, and $S(q^{-1})$. This leads to a nontrivial bilinear estimation problem. We can obtain the regulator parameters by estimating $B^{-}(q^{-1})$, $R(q^{-1})$, and $S(q^{-1})$ in Eq. (14.4-37) directly, avoiding at each step the control design problem, i.e., the solution of the Diophantine equation. This leads to a less time-consuming algorithm, in the sense that the design calculations become trivial. The implicit STR is summarized in Table 14.4.

To avoid nonlinear parametrization, Eq. (14.4-37) is rewritten equivalently as

$$
A_{\mathrm{m}}(q^{-1})A_0(q^{-1})y(k) = q^{-d}\bar{R}(q^{-1})u(k) + q^{-d}\bar{S}(q^{-1})y(k) \tag{14.4-38}
$$

where

$$
\bar{R}(q^{-1}) = B^{-}(q^{-1})R(q^{-1}) \tag{14.4-39}
$$

and

$$
\bar{S}(q^{-1}) = B^{-}(q^{-1})S(q^{-1}) \tag{14.4-40}
$$

Based on the linear model (14.4-41), it is possible to estimate the coefficients of the polynomials $\bar{R}(q^{-1})$ and $\bar{S}(q^{-1})$. However, it should be noted that, in general, this is not a minimal parametrization since the coefficients of the polynomial $B^{-}(q^{-1})$ are estimated twice. Moreover, possible common factors in $\bar{R}(q^{-1})$ and $\bar{S}(q^{-1})$ (corre-

Table 14.4 The Implicit Pole-Placement STR for the Case of Unknown Parameters

Step 1 Estimate the coefficients in $R(q^{-1})$, $B^-(q^{-1})$, and $S(q^{-1})$ recursively based on the reparametrized model (bilinear estimation problem)

$$A_m(q^{-1})A_0(q^{-1})y(k) = q^{-d}B^-(q^{-1})[R(q^{-1})u(k) + S(q^{-1})y(k)] \qquad (14.4-41)$$

Step 2 Compute the control law using the relations

$$T(q^{-1}) = A_0(q^{-1})B_m^+(q^{-1}) \qquad (14.4-42)$$

$$u(k) = \left[\frac{T(q^{-1})}{R(q^{-1})}\right]r(k) - \left[\frac{S(q^{-1})}{R(q^{-1})}\right]y(k) \qquad (14.4-43)$$

Step 3 Repeat steps 1 and 2 at each sampling period

sponding to $B^-(q^{-1})$) should be canceled out to avoid cancellation of unstable modes in the control law. The algorithm thus obtained is summarized in Table 14.5.

Example 14.4.1

Consider the system

$$A(q^{-1})y(k) = q^{-1}B(q^{-1})u(k) \qquad \text{with} \qquad y(0) = 1$$

where $A(q^{-1}) = 1 + 2q^{-1} + q^{-2}$ and $B(q^{-1}) = 2 + q^{-1} + q^{-2}$. The polynomial $B(q^{-1})$ can be factored as follows:

$$B(q^{-1}) = 2(1 + 0.5q^{-1} + 0.5q^{-2}) = B^-(q^{-1})B^+(q^{-1})$$

with $B^+(q^{-1})$ monic. The desired closed-loop behavior is given by

$$A_m(q^{-1})y(k) = q^{-1}B_m(q^{-1})r(k)$$

with

$$A_m(q^{-1}) = 1 - q^{-1} + 0.25q^{-2}$$

Table 14.5 An Alternate Implicit Pole-Placement STR for the Case of Unknown Parameters

Step 1	Using the model (14.4-41) and least-squares, estimate the coefficients of the polynomials $\bar{R}(q^{-1})$ and $\bar{S}(q^{-1})$
Step 2	Cancel out possible common factors in $\bar{R}(q^{-1})$ and $\bar{S}(q^{-1})$ in order to obtain $R(q^{-1})$ and $S(q^{-1})$
Step 3	Compute the control law using the relations $$T(q^{-1}) = A_0(q^{-1})B_m^-(q^{-1})$$ $$u(k) = \left[\frac{T(q^{-1})}{R(q^{-1})}\right]r(k) - \left[\frac{S(q^{-1})}{R(q^{-1})}\right]y(k)$$
Step 4	Repeat steps 1 and 3 at each sampling period

and

$$B_m(q^{-1}) = 1 + 0.3q^{-1} = 2(0.5 + 0.15q^{-1}) = B^-(q^{-1})B_m^+(q^{-1})$$

(a) In the case of known plant parameters, calculate the pole-placement control law
(b) In the case of unknown plant parameters, define an explicit adaptive pole-placement control scheme
(c) Repeat part (b) for an implicit adaptive pole-placement control scheme.

Solution

(a) In the case of known parameters and to satisfy Eqs (14.4-23), (14.4-24), and (14.4-27) we select deg $A_0(q^{-1}) = 0$, deg $R_1(q^{-1}) = 0$, and deg $S(q^{-1}) = 1$. Hence, $A_0(q^{-1}) = 1$, $R_1(q^{-1}) = r_0$, and $S(q^{-1}) = s_0 + s_1 q^{-1}$. We now solve the following Diophantine equation for r_0, s_0, and s_1:

$$A(q^{-1})R_1(q^{-1}) + q^{-1}B^-(q^{-1})S(q^{-1}) = A_m(q^{-1})A_0(q^{-1})$$

or

$$(1 + 2q^{-1} + q^{-2})r_0 + 2q^{-1}(s_0 + s_1 q^{-1}) = 1 - q^{-1} + 0.25q^{-2}$$

One easily obtains $r_0 = 1$, $s_0 = -1.5$, and $s_1 = -0.375$ and, consequently, $R_1(q^{-1}) = 1$ and $S(q^{-1}) = -1.5 - 0.375q^{-1}$. Note from Eq. (14.4-14) that when $A_0(q^{-1})$ is a monic polynomial, then $R_1(q^{-1})$ and $R(q^{-1})$ are restricted to being monic polynomials also. Now, $R(q^{-1}) = B^+(q^{-1})R_1(q^{-1}) = 1 + 0.5q^{-1} + 0.5q^{-2}$ and $T(q^{-1}) = B_m^+(q^{-1})A_0(q^{-1}) = 0.5 + 0.15q^{-1}$. The control law is given by

$$u(k) = \left[\frac{T(q^{-1})}{R(q^{-1})}\right]r(k) - \left[\frac{S(q^{-1})}{R(q^{-1})}\right]y(k)$$

or

$$u(k) = \left[\frac{0.5 + 0.15q^{-1}}{1 + 0.5q^{-1} + 0.5q^{-2}}\right]r(k) + \left[\frac{1.5 + 0.375q^{-1}}{1 + 0.5q^{-1} + 0.5q^{-2}}\right]y(k)$$

(b) The system model belongs to the following class of models:

$$(1 + a_1 q^{-1} + a_2 q^{-2})y(k) = (b_0 + b_1 q^{-1} + b_2 q^{-2})u(k - 1)$$

which may be rewritten as

$$y(k) = \theta^T \varphi(k)$$

where

$$\theta^T = [a_1, a_2, b_0, b_1, b_2]$$

$$\varphi^T(k) = [-y(k - 1), -y(k - 2), u(k - 1), u(k - 2), u(k - 3)]$$

The parameters a_i, b_i can be estimated on-line using the following algorithm:

$$\varphi^T(k) = [-y(k-1), -y(k-2), u(k-1), u(k-2), u(k-3)]$$

$$F(k+1) = \frac{1}{0.99}\left[F(k) - \frac{F(k)\varphi(k)\varphi^T(k)F(k)}{1 + \varphi^T(k)F(k)\varphi(k)}\right] \qquad \text{with} \qquad F(0) = \frac{1}{10^{-3}}I_5$$

$$\varepsilon(k) = y(k) - \varphi^T(k)\hat{\theta}(k-1)$$

$$\hat{\theta}(k) = \hat{\theta}(k-1) + F(k)\varphi(k)\varepsilon(k)$$

$$\hat{\theta}^T(k) = \left[\hat{a}_1(k), \hat{a}_2(k), \hat{b}_0(k), \hat{b}_1(k), \hat{b}_2(k)\right]$$

initialized for example at

$$\hat{\theta}^T(0) = [1, 0, 1, 0, 1]$$

At each step

$$\hat{B}(q^{-1}) = \hat{b}_0(k) + \hat{b}_1(k)q^{-1} + \hat{b}_2(k)q^{-2}$$

is factored as

$$\hat{B}(q^{-1}) = \hat{B}^-(q^{-1})\hat{B}^+(q^{-1})$$

where $\hat{B}^+(q^{-1})$ is chosen to be monic. Moreover, at each step, the following Diophantine equation is solved for $\hat{r}_0(k)$, $\hat{s}_0(k)$, and $\hat{s}_1(k)$:

$$\hat{A}(q^{-1})\hat{r}_0(k) + q^{-1}\hat{B}^-(q^{-1})[\hat{s}_0(k) + \hat{s}_1(k)q^{-1}] = 1 - q^{-1} + 0.25q^{-2}$$

where

$$\hat{A}(q^{-1}) = 1 + \hat{a}_1(k)q^{-1} + \hat{a}_2(k)q^{-2} \qquad \text{and} \qquad \hat{S}(q^{-1}) = \hat{s}_0(k) + \hat{s}_1(k)q^{-1}$$

Then, the following computations are made:

$$\hat{R}(q^{-1}) = \hat{R}_1(q^{-1})\hat{B}^+(q^{-1}) = r_0(k)\hat{B}^+(q^{-1})$$

$$T(q^{-1}) = B_m^+(q^{-1})$$

where

$$B_m(q^{-1}) = \hat{B}^-(q^{-1})B_m^+(q^{-1})$$

Here, $B_m^+(q^{-1})$ can be any polynomial of our choice. The control law to be applied to the system at each step is

$$u(k) = \left[\frac{T(q^{-1})}{\hat{R}(q^{-1})}\right]r(k) - \left[\frac{\hat{S}(q^{-1})}{\hat{R}(q^{-1})}\right]y(k)$$

(c) In the case of an implicit adaptive pole-placement design, the parameters of the following implicit system model are estimated using recursive least squares:

$$A_0(q^{-1})A_m(q^{-1})y(k) = q^{-1}\bar{R}(q^{-1})u(k) + q^{-1}\bar{S}(q^{-1})y(k)$$

or

$$(1 - q^{-1} + 0.25q^{-2})y(k) = [\bar{r}_0(k) + \bar{r}_1(k)q^{-1} + \bar{r}_2(k)q^{-2}]u(k-1)$$
$$+ [\bar{s}_0(k) + \bar{s}_1(k)q^{-1}]y(k-1)$$

or

$$y(k) - y(k-1) + 0.25y(k-2) = [\bar{r}_0(k), \bar{r}_1(k), \bar{r}_2(k), \bar{s}_0(k), \bar{s}_1(k)] \begin{bmatrix} u(k-1) \\ u(k-2) \\ u(k-3) \\ y(k-1) \\ y(k-2) \end{bmatrix}$$

with $\bar{R}(q^{-1}) = B^-(q^{-1})R(q^{-1})$ and $\bar{S}(q^{-1}) = B^-(q^{-1})S(q^{-1})$. Next, any common factors in $\bar{R}(q^{-1})$ and $\bar{S}(q^{-1})$ (corresponding to $B^-(q^{-1})$) are canceled out to obtain $R(q^{-1})$ and $S(q^{-1})$. One then has

$$T(q^{-1}) = B_m^+(q^{-1})$$

and the control law is given by

$$u(k) = \left[\frac{T(q^{-1})}{R(q^{-1})}\right] r(k) - \left[\frac{S(q^{-1})}{R(q^{-1})}\right] y(k)$$

The procedure described above is repeated at each sampling period.

14.5 PROBLEMS

1. Consider a system described by

$$y(k) = \theta_0 G(q^{-1})u(k)$$

   where θ_0 is an unknown parameter and $G(q^{-1})$ is a known rational function of q^{-1}. The reference model is described by

$$y_m(k) = \theta_m G(q^{-1})r(k)$$

   where θ_m is a known parameter. The controller is of the form

$$u(k) = \theta r(k)$$

   Find an adaptation mechanism for the feedforward gain θ, by using the MIT rule.

2. Consider a system described by

$$y(k) = \left[\frac{q^{-1}(0.36 + 0.28q^{-1})}{1 - 1.36q^{-1} + 0.36q^{-2}}\right] u(k)$$

   and a reference model given by

$$y_m(k) = \left[\frac{q^{-1}(0.38 + 0.24q^{-1})}{1 - 0.78q^{-1} + 0.37q^{-2}}\right] r(k)$$

   Determine an adaptive controller to achieve model following by using the MIT rule (assume that the parameters appearing in the system model are unknown).

3. Consider a system described by

$$y(k) = \left[\frac{q^{-2}\beta_0}{1 + \alpha_1 q^{-1} + \alpha_2 q^{-2}}\right] u(k)$$

where β_0, α_1, and α_2 are free parameters. The desired input–output behavior is given by

$$y_m(k) = q^{-2}\left[\frac{b_0}{1 + a_1 q^{-1} + a_2 q^{-2}}\right] r(k)$$

where $1 + a_1 q^{-1} + a_2 q^{-2}$ is an asymptotically stable polynomial.

(a) In the case where β_0, α_1, and α_2 are assumed to be known, calculate a pole-placement control law

(b) Design an implicit adaptive pole-placement algorithm in the case where β_0, α_1, and α_2 are unknown.

4. The plastic extrusion process is briefly described in Problem 18 (Sec. 12.14) of Chap. 12. The discrete-time system for the temperature control is shown in Figure 14.7. The transfer function from the screw speed (which is the main controlling variable) to the temperature of the polymer at the output is given by

$$H(s) = \frac{Ke^{-s\tau}}{\beta s + 1}$$

where K is the static gain, β is the time constant, and τ the system delay. The system delay τ is such that

$$\tau = (d-1)T + L, \qquad 0 < L < T$$

where T is the sampling period and d is a positive integer.

(a) Verify that the equivalent discrete-time transfer function, using a zero-order hold circuit, is given by

$$H(z) = \frac{z^{-d}(b_0 + b_1 z^{-1})}{1 + a_1 z^{-1}}$$

with

$$a_1 = -e^{-T/\beta}$$

$$b_0 = K\left(1 - e^{(L-T)/\beta}\right)$$

$$b_1 = Ke^{-T/\beta}\left(e^{L/\beta} - 1\right)$$

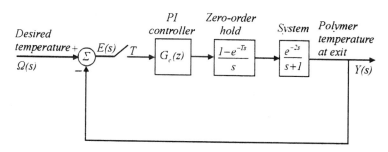

Figure 14.7   Temperature control system for plastic extrusion.

The discretized system is thus equivalently described by the difference equation

$$y(k + d) = -a_1 y(k + d - 1) + b_0 u(k) + b_1 u(k - 1)$$

It is clear that d is the discretized (sampled-date) system delay.

(b)   Choose $d = 1$ and a sampling period T such that the sampled-data system has a stable zero (i.e., choose $-b_1/b_0$ to be inside the unit circle). The minimum-phase system thus obtained is described by

$$y(k + 1) = -a_1 y(k) + b_0 u(k) + b_1 u(k - 1)$$

For this system model, determine a model reference control law which, in the case of known model parameters, satisfies the control objective

$$\Gamma(q^{-1})[y(k + 1) - y_m(k + 1)] \equiv 0 \qquad \text{for} \qquad k \geq 0$$

where $\Gamma(q^{-1}) = 1 + \gamma_1 q^{-1}$. The polynomial $\Gamma(q^{-1})$ is assumed to be asymptotically stable. The asymptotically stable reference model is given by

$$y_m(k) = \left[\frac{q^{-1}(b_0^m + b_1^m q^{-1})}{1 + a_1^m q^{-1}} \right] r(k)$$

(c)   In the case of unknown plant parameters, determine an MRAC design that satisfies the control objective of part (b) asymptotically.

5.   A system is described by the model (initially known to the designer) [22]

$$y(k) = \left[\frac{q^{-2}(1 + 0.4q^{-1})}{(1 - 0.5q^{-1})[1 - (0.8 + 0.3j)q^{-1}][1 - (0.8 - 0.3j)q^{-1}]} \right] u(k)$$

It is desired to follow the reference model

$$y_m(k) = \left[\frac{q^{-2}(0.28 + 0.22q^{-1})}{(1 - 0.5q^{-1})[1 - (0.7 + 0.2j)q^{-1}][1 - (0.7 - 0.2j)q^{-1}]} \right] r(k)$$

(a)   By choosing $\Gamma(q^{-1}) = 1$, or $\Gamma(q^{-1}) = [1 - 0.4q^{-1}]^3$, design a model reference control law, assuming the new model reference parameters known.

(b)   Suppose now that parameter changes occur in the system model. The system, after the parameter changes, is described by (the new model is assumed unknown to the designer)

$$y(k) = \left[\frac{q^{-2}(0.9 + 0.5q^{-1})}{(1 - 0.5q^{-1})[1 - (0.9 + 0.5j)q^{-1}][1 - (0.9 - 0.5j)q^1]} \right] u(k)$$

When the control objective is tracking, the changes occur at $t = 25$ sec, while when the control objective is regulation, the changes occur at $t = 0$ sec. In the case of unknown parameters determine an adaptive MRAC scheme. Simulate the behavior of the closed-loop system during tracking and during regulation, with the choices $\Gamma(q^{-1}) = 1$ or $\Gamma(q^{-1}) = [1 - 0.4q^{-1}]^3$. Use as initial parameter values for the adaptive controller those obtained from the design in the nonadaptive case.

Assume that a constant trace algorithm is used for the adaptation gain, with $\lambda_1(k)/\lambda_2(k) = 1$ and $\text{tr}\,\mathbf{F}(k) = \text{tr}\,\mathbf{F}(0)$, with $\mathbf{F}(0) = 10\mathbf{I}$.

6. For the system of Problem 4, it is desired to design an explicit STR, considering the model reference control as the "underlying control problem." Define a least-squares algorithm to estimate the plant parameters. Reparametrize the control law of Problem 4, so that the plant parameters appear explicitly. Solve the controller design problem and define the STR. Simulate the behavior of the system.

7. For the system of Problem 4, design an explicit pole-placement STR. Distinguish the cases of stable and unstable system zero.

8. Determine explicit and implicit self-tuners for the plant

$$y(k) = 0.86y(k-1) + 0.08u(k-1) + 0.06u(k-2)$$

The desired characteristic polynomial is chosen as follows:

$$A_m(q^{-1}) = 1 - 1.5q^{-1} + 0.6q^{-2}$$

Also, choose $B_m^+(q^{-1}) = 1$. Use a forgetting adaptation algorithm with $\lambda = 0.95$. Moreover, use $\mathbf{F}(0) = 100\mathbf{I}$. Simulate the behavior of the system.

REFERENCES

Books

1. BDO Anderson, RR Bitmead, CR Johnson Jr, *et al.* Stability of Adaptive Systems: Passivity and Averaging Analysis. Cambridge, Massachusetts: MIT Press, 1986.
2. KJ Åström, B Wittenmark. Computer Controlled Systems. Englewood Cliffs, New Jersey: Prentice Hall, 1984.
3. KJ Åström, B Wittenmark. Adaptive Control. New York: Addison-Wesley, 1989.
4. VV Chalam. Adaptive Control Systems: Techniques and Applications. New York: Marcel Dekker, 1987.
5. G Goodwin, K Sin. Adaptive Filtering, Prediction and Control. Englewood Cliffs, New Jersey: Prentice Hall, 1984.
6. B Egardt. Stability of Adaptive Controllers. Berling: Springer-Verlag, 1989.
7. CJ Harris, SA Billings (eds). Self Tuning and Adaptive Control: Theory and Applications. Stevenage, UK, and New York: IEE London and New York, Peter Peregrinus, 1981.
8. ID Landau. Adaptive Control: The Model Reference Approach. New York: Marcel Dekker, 1979.
9. E Mishkin, L Braun (eds). Adaptive Control Systems. New York: McGraw-Hill, 1961.
10. KS Narendra, RV Monopoli (eds). Applications of Adaptive Control. New York: Academic Press, 1980.
11. KS Narendra, A Annaswamy. Stable Adaptive Systems. Englewood Cliffs, New Jersey: Prentice Hall, 1989.
12. PN Paraskevopoulos. Digital Control Systems. London: Prentice Hall, 1996.
13. SS Sastry, M Bodson. Adaptive Control: Stability Convergence and Robustness. Englewood Cliffs, New Jersey: Prentice Hall, 1988.
14. H Ubenhauen (ed). Methods and Applications in Adaptive Control. Berlin: Springer-Verlag, 1980.

Articles

15. KJ Åström. Theory and applications of adaptive control – a survey. Automatica 19(5):471–486, 1983.
16. KJ Åström, U Gorisson, L Ljung, B Wittenmark. Theory and applications of self-tuning regulators. Automatica 13:457–476, 1977.
17. KJ Åström, B Wittenmark. On self tuning regulators. Automatica 9:185–199, 1973.
18. KJ Åström, B Wittenmark. Self tuning controllers based on pole-zero placement. IEE Proc. 127 (Part D, No. 3):120–130, May 1980.
19. T Ionescu, R Monopoli. Discrete model reference adapative control with an augmented error signal. Automatica 13:507–517, 1977.
20. ID Landau. A survey of model reference adaptive techniques – theory and applications. Automatica 10:353–379, 1974.
21. ID Landau. An extension of a stability theorem applicable to adaptive control. IEEE Trans Automatic Control AC-25(4):814–817, August 1980.
22. ID Landau, R Lozano. Unification of discrete time explicit model reference adaptive control designs. Automatica 17(4):593–611, 1981.
23. ID Landau, HM Silveiva. A stability theorem with applications to adaptive control. IEEE Trans Automatic Control AC-24(2):305–312, April 1979.
24. R Lozano, ID Landau. Redesign of explicit and implicit discrete-time model reference adaptive control schemes. Int J Control 33(2):247–268, 1981.
25. P Parks. Liapunov redesign of model reference adaptive control systems. IEEE Trans Automatic Control AC-11(3):362–367, July 1966.
26. B Wittenmark. Stochastic adaptive control methods – a survey. Int J Control 21(5):705–730, 1975.
27. B Wittenmark, KJ Åström. Practical issues in the implementation of self-tuning control. Automatica 20(5):595–605, 1984.

15

Robust Control

15.1 INTRODUCTION

Control engineers are always aware that any design of a controller based on a fixed plant model (e.g., transfer function or state space) is very often unrealistic. This is because there is always a nagging doubt about the performance specifications if the model on which the design is based deviates from the assumed value over a certain range.

Robust control refers to the control of uncertain plants with unknown disturbance signals, uncertain dynamics, and imprecisely known parameters making use of fixed controllers. That is, the problem of robust control is to design a fixed controller that guarantees acceptable performance norms in the presence of plant and input uncertainty. The performance specification may include properties such as stability, disturbance attenuation, reference tracking, control energy reduction, etc. In the case of single-input–single-output (SISO) systems this is roughly covered by concepts such as gain and phase margins (see Chap. 8). However, in the multiple-input–multiple-output (MIMO) systems case, matters become quite complicated and an easy extension of gain and phase margins is not possible. This led to new approaches and new techniques to deal with the situation. In this chapter, we report some of these new developments. We limit our presentation to linear time-invariant SISO systems for both the plant and the controller, wherein the controller configuration remains fixed.

It was pointed out in Chap. 1 that control theory is a relatively new discipline and was recognized as such only during the early 1930s. The contributions made by Nyquist [27] and Bode [2] in those days, placed this discipline on firm theoretical foundations. Furthermore, the contribution of Wiener [10] and Kolmogorov [26] on filtering and prediction of stationary processes constitutes a landmark in stochastic control theory. In all these developments the emphasis was on frequency domain techniques. In particular, Bode and Nyquist understood and appreciated the concept of robustness, which is embodied in their definitions of gain and phase margins. An important change in control theory development took place in the 1950s with the emergence of the state-space approach. The high point of this development was reached during the 1960s with the formulation and solution of what is known as the linear quadratic gaussian (LQG) problem [5]. This development was primarily inspired by the important contributions of Kalman [21–24]. The LQG approach provided a mathematically elegant method for designing feedback controllers of systems working in a noisy environment. However, it was soon realized that one of its principal drawbacks was its inability to guarantee a robust solution. Control engineers found it difficult to incorporate robustness criteria in a quadratic integral performance index used in the LQG problem. For this reason, during the early 1970s, an attempt was made to generalize some of the useful concepts such as gain and phase margins so that they are made applicable to MIMO systems [7, 8]. This brings us to the 1980s, where a variety of ideas of far-reaching significance were developed in the area of robust control. Some of these ideas are the following:

1. Use of singular values as a measure of gain in transformations [16]
2. The factorization approach in controller synthesis [3, 9]
3. Parametrization of stabilizing controllers [14, 28, 29]
4. H_∞ optimization [5, 20, 31, 32]
5. Robust stabilization and sensitivity minimization [12, 18, 25, 30]
6. Computational aspects of H_∞ optimization, such as
   (a) Interpolation methods based on Nevanlinna–Pick interpolation theory [13]
   (b) Hankel norm approach [19]
   (c) Operator–theoretic approach [17]
   (d) State space approach using separation principle [15]
7. Kharitonov theory and related approaches [1]

With all these developments, robust control gained a great momentum, and it is currently one of the most important areas of research in the field of control theory and practice. In the subsequent sections of this chapter, an introduction to the main problems and principles of robust control is presented based on certain transfer function approaches. There have been proposed many other important robust control techniques, which are not presented here due to space limitations.

15.2 MODEL UNCERTAINTY AND ITS REPRESENTATION

15.2.1 Origins of Model Uncertainty

Uncertainty in control systems may stem from different sources. Model uncertainty is one main consideration. Other considerations include sensor and actuator failures, physical constraints, changes in control purposes, loop opening and loop closure, etc. Moreover, in control design problems based on optimization, robustness issues due to mathematical objective functions not properly describing the real control problem may occur. On the other hand, numerical design algorithms may not be robust. However, when we refer to robustness in this chapter, we mean robustness with respect to model uncertainty. We also assume that a fixed linear controller is used.

Model uncertainty may have several origins. In particular, it may be caused by:

1. Parameters in a linear model, which are approximately known or are simply in error
2. Parameters, which may vary due to nonlinearities or changes in the operating conditions

3. Neglected time delays and diffusion processes
4. Imperfect measurement devices
5. Reduced (low-order) models of a plant, which are commonly used in practice, instead of very detailed models of higher order
6. Ignorance of the structure and the model order at high frequencies
7. Controller order reduction issues and implementation inaccuracies.

The above sources of model uncertainty may be grouped into three main categories.

Parametric or Structured Uncertainty

In this case the structure of the model and its order is known, but some of the parameters are uncertain and vary in a subset of the parameter space.

Neglected and Unmodeled Dynamics Uncertainty

In this case the model is in error because of missing dynamics (usually at high frequencies), most likely due to a lack of understanding of the physical process.

Lumped Uncertainty or Unstructured Uncertainty

In this case uncertainty represents several sources of parametric and/or unmodeled dynamics uncertainty combined into a single lumped perturbation of prespecified structure. Here, nothing is known about the exact nature of the uncertainties, except that they are bounded.

15.2.2 Representation of Uncertainty

Parametric uncertainty will be quantified by assuming that each uncertain parameter α is bounded within some region $[\alpha_{min}, \alpha_{max}]$. In other words, there are parameter sets of the form

$$\alpha_p = \alpha_m(1 + r_\alpha \Delta) \tag{15.2-1}$$

where α_m is the mean parameter value, $r_\alpha = (\alpha_{max} - \alpha_{min})/(\alpha_{max} + \alpha_{min})$ is the relative parametric uncertainty and Δ is any scalar satisfying $|\Delta| \leq 1$.

Neglected and unmodeled dynamics uncertainty is more difficult to quantify. The frequency domain is particularly well suited for representing this class of uncertainty, through complex perturbations, which are normalized such that $\|\Delta\|_\infty \leq 1$, where $\|\Delta\|_\infty$ is given by the following definition.

Definition 15.2.1

For a scalar complex function $\Delta(s)$, the H_∞-norm of $\Delta(s)$ is defined as

$$\|\Delta\|_\infty = \sup_\omega |\Delta(j\omega)|$$

Lumped unstructured uncertainty can easily be described in the frequency domain. In most cases, it is preferred to lump the uncertainty into a "*multiplicative uncertainty*" of the form

$$P_m : G_p(s) = G(s)[1 + w_m(s)\Delta_m(s)]; \qquad \|\Delta_m\|_\infty \leq 1 \tag{15.2-3}$$

where

P_m : set of possible linear time-invariant models (usually called *the uncertainty set*)

$G(s) \in P_m$: nominal plant model (without uncertainty)

$G_p(s) \in P_m$: perturbed plant model

$w_m(s)$: the multiplicative uncertainty weight

$\Delta_m(s)$: *any stable* transfer function which is less than or equal to 1 in magnitude at all frequencies. Some examples of allowable $\Delta_m(s)$ with $\|\Delta\|_\infty \leq 1$ are

$$\frac{1}{\beta s + 1} \quad \text{and} \quad \frac{0.1}{s^2 + 0.1s + 1}, \quad \text{where} \quad \beta > 0$$

It is worth noting at this point that the requirement for stability on $\Delta_m(s)$ may be removed if one assumes that the number of right-half plane poles in $G(s)$ and $G_p(s)$ remains unchanged.

The block diagram of Figure 15.1 represents a plant with multiplicative uncertainty.

Other less common uncertainty forms are the *additive uncertainty*, having the form

$$P_a : G_p(s) = G(s) + w_a(s)\Delta_a(s), \qquad \|\Delta_a\|_\infty \leq 1 \tag{15.2-4}$$

the *inverse multiplicative uncertainty*, having the form

$$P_{im} : G_p(s) = G(s)[1 + w_{im}(s)\Delta_{im}(s)]^{-1}; \qquad \|\Delta_{im}\|_\infty \leq 1 \tag{15.2-5}$$

and the *division uncertainty*, having the form

$$P_d : G_p(s) = G(s)[1 + w_d(s)\Delta_d(s)G(s)]^{-1}; \qquad \|\Delta_d\|_\infty \leq 1 \tag{15.2-6}$$

It is pointed out here that the additive and the multiplicative uncertainty descriptions are equivalent if at each frequency the following relation holds:

$$|w_m(j\omega)| = \frac{|w_a(j\omega)|}{|G(j\omega)|} \tag{15.2-7}$$

Next, we give some examples on representing parametric model uncertainty.

Example 15.2.1

Assume that the set of possible plants is given by

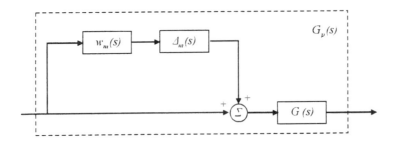

Figure 15.1   Plant with multiplicative uncertainty.

$$G_P(s) = \beta_p G_0(s), \qquad \beta_{min} \le \beta_P \le \beta_{max} \tag{15.2-8}$$

where β_p is an uncertain gain and $G_0(s)$ is a transfer function without uncertainty. Determine the multiplicative uncertainty description of Eq. (15.2-8).

Solution

Write β_p as follows

$$\beta_p = \beta_m(1 + r_\beta\Delta), \qquad |\Delta| \le 1$$

where

$$\beta_m = \frac{\beta_{min} + \beta_{max}}{2} \qquad \text{and} \qquad r_\beta = \frac{\beta_{max} - \beta_{min}}{\beta_{max} + \beta_{min}}$$

Clearly, β_m and r_β are the average gain and the relative magnitude of the gain uncertainty, respectively. Therefore, the model set (15.2-8) can be written as

$$G_P(s) = \beta_m G_0(s)[1 + r_\beta\Delta] = G(s)[1 + r_\beta\Delta], \qquad \|\Delta\| \le 1$$

The above expression for $G_p(s)$ is the multiplicative uncertainty description of the set (15.2-8).

Example 15.2.2

Assume that the set of possible plants is given by

$$G_P(s) = \frac{1}{\tau_p s + 1} G_0(s), \qquad \tau_{min} \le \tau_p \le \tau_{max} \tag{15.2-9}$$

Determine the inverse multiplicative uncertainty description of Eq. (15.2-9).

Solution

By writing

$$\tau_P = \tau_m(1 + r_\tau\Delta), \qquad |\Delta| \le 1$$

where

$$\tau_m = \frac{\tau_{min} + \tau_{max}}{2} \qquad \text{and} \qquad r_\tau = \frac{\tau_{max} - \tau_{min}}{\tau_{max} + \tau_{min}}$$

the model set (15.2-9) can be written as

$$G_p(s) = \frac{G_0(s)}{1 + \tau_m s + r_\tau\tau_m s\Delta} = \frac{G_0(s)}{1 + \tau_m s}[1 + w_{im}(s)\Delta]^{-1} = G(s)[1 + w_{im}(s)\Delta]^{-1}$$

where

$$G(s) = \frac{G_0(s)}{1 + r_m s} \qquad \text{and} \qquad w_{im}(s) = \frac{r_\tau\tau_m s}{1 + \tau_m s}$$

The above expression for $G_p(s)$ is the inverse multiplicative form of the uncertainty set (15.2-9).

Example 15.2.3

Assume that the set of possible plants is given by

$$G_p(s) = (1 + z_p s)G_0(s), \qquad z_{min} \le z_p \le z_{max} \tag{15.2-10}$$

where $G_0(s)$ is assumed to have no uncertainty. Determine the multiplicative uncertainty form of Eq. (15.2-10).

Solution

By writing

$$z_p = z_m(1 + r_z\Delta), \qquad |\Delta| \le 1$$

where

$$z_m = \frac{z_{max} + z_{min}}{2} \quad \text{and} \quad r_z = \frac{z_{max} - z_{min}}{z_{max} + z_{min}}$$

the model (15.2-10) can be written as

$$G_p(s) = (1 + z_m s + z_m r_z s\Delta)G_0(s) = (1 + z_m s)G_0(s) + z_m r_z s\Delta G_0(s)$$

$$= (1 + z_m s)G_0(s) + \frac{z_m r_z s}{1 + z_m s}(1 + z_m s)G_0(s)\Delta = G(s) + \frac{z_m r_z s}{1 + z_m s}G(s)\Delta$$

$$= G(s)[1 + w_m(s)\Delta]$$

where

$$G(s) = (1 + z_m s)G_0(s) \quad \text{and} \quad w_m(s) = \frac{z_m r_z s}{1 + z_m s}$$

The above expression for $G_p(s)$ is the multiplicative form of the uncertainty set (15.2-10).

Example 15.2.4

Consider the family of plant transfer functions

$$G_p(s) = \frac{1}{s^2 + \alpha s + 1}, \qquad 0.4 \le \alpha \le 0.8 \tag{15.2-11}$$

Determine the division uncertainty description of Eq. (15.2-11).

Solution

It is easy to see that

$$\alpha = 0.6 + 0.2\Delta, \qquad |\Delta| \le 1$$

Therefore, the model set (15.2-11) can be expressed as

$$G_p(s) = G(s)[1 + w_d(s)\Delta G(s)]^{-1}$$

where

$$G(s) = \frac{1}{s^2 + 0.6s + 1}, \qquad w_d(s) = 0.2s$$

The foregoing relation for $G_p(s)$ is the division uncertainty description of the family (15.2-11).

Although parametric uncertainty is easily represented in some simple cases, it is avoided in most cases because of the following reasons:

1. It requires large efforts to model parametric uncertainty, particularly in the case of systems with a large number of uncertain parameters.
2. In many cases, the assumptions about the model and the parameters may be inexact. However, the description of a family of systems through parametric uncertainty is very detailed and accurate.
3. In order to model uncertain systems through parametric uncertainty, the exact model structure is indispensable. Unmodeled dynamics cannot then be incorporated in this description.

We next focus our attention on the problem of describing a set of possible plants P by a single unstructured perturbation $\Delta_a(s)$ or $\Delta_m(s)$. This description can be obtained on the basis of the following steps:

1. Choose a nominal model $G(s)$. A nominal model can be selected to be either a low-order, delay-free model or a model of mean parameter values or, finally, the central plant obtained from the Nyquist plots corresponding to all of the plants of the given set P.
2. In the case of additive uncertainty, find the smallest radius $\ell_a(\omega)$, which includes all possible plants

$$\ell_a(\omega) = \max_{G_p \in P_a} |G_p(j\omega) - G(j\omega)| \qquad (15.2\text{-}12)$$

In most cases we look for a rational transfer function weight $w_a(s)$ for additive uncertainty. This weight must be chosen such that

$$|w_a(j\omega)| \geq \ell_a(\omega), \qquad \forall \omega \qquad (15.2\text{-}13)$$

and is usually selected to be of low order to simplify the design of controllers.
3. In the case of multiplicative uncertainty (which is the preferred uncertainty form), find the smallest radius $\ell_m(\omega)$, which includes all possible plants

$$\ell_m(\omega) = \max_{G_p \in P_m} \left| \frac{G_p(j\omega) - G(j\omega)}{G(j\omega)} \right| \qquad (15.2\text{-}14)$$

For a chosen rational weight $w_m(s)$, there must be

$$|w_m(j\omega)| \geq \ell_m(\omega), \qquad \forall \omega \qquad (15.2\text{-}15)$$

We next give an example of how this approach is applied in practice.

Example 15.2.5

Consider the family of plants with parametric uncertainty given by

$$P : G_p(s) = \frac{s}{s^2 + as + b}, \qquad 1 \leq a \leq 3, \qquad 2 \leq b \leq 6 \qquad (15.2\text{-}16)$$

Obtain a representation of the above set using multiplicative uncertainty with a single rational weight $w_p(s)$.

Solution

Choose a nominal model as the model of mean parameter values. That is, let

$$G(s) = \frac{s}{s^2 + 2s + 4}$$

In order to obtain $\ell_m(\omega)$, we consider three values for each of the two parameters a and b. In particular, we consider $a = 1$, 2.5, or 3 and $b = 2$, 5, or 6. With this choice, we obtain $3^2 = 9$ alternative plants. In general, this choice does not guarantee that the worst case is obtained, since the worst case may correspond to another interior point of the intervals. However, in our example, it can be shown that the worst case corresponds to the choice $a = 1$ and $b = 2$. The relative error

$$\left| \frac{G_p(j\omega) - G(j\omega)}{G(j\omega)} \right|$$

for the nine resulting $G_p(s)$, are depicted as functions of frequency in Figure 15.2. According to our analysis, the curve for $\ell_m(\omega)$ must at each frequency lie above all the curves corresponding to the nine $G_p(s)$. One can also observe that as $s \to 0$, $\ell_m(\omega) \to 1$, and as $s \to \infty$, $\ell_m(\omega) \to 0$. With these observations one must choose a simple first-order weight that approximately matches this limiting behavior, as for example

$$w_m(s) = \frac{20}{s + 20} \tag{15.2-17}$$

This weight is also depicted in Figure 15.2. It can be shown that this weight gives a good fit of $\ell_m(\omega)$, except at frequencies around $\omega = 1.5$, where $|w_m(j\omega)|$ is too small, and thus this weight does not include all possible plants. To change this such that $|w_m(j\omega)| \geq \ell_m(\omega)$, $\forall\omega$, we can work along two alternative approaches: The first approach is to augment the numerator of Eq. (15.2-17), as for example

$$w_m(s) = \frac{37}{s + 20} \tag{15.2-18}$$

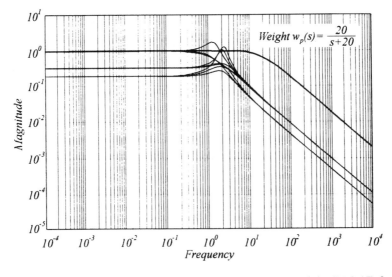

Figure 15.2 The first-order multiplicative uncertainty weight (15.2-17) for Example 15.2.5.

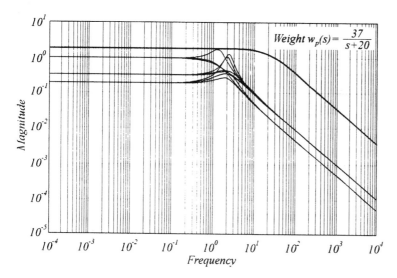

Figure 15.3 The first-order multiplicative uncertainty weight (15.2-18) for Example 15.2.5.

This multiplicative uncertainty weight is shown in Figure 15.3 and obviously includes all possible plants.

The second approach is to multiply $w_m(s)$ of the form (15.2-17) by a correction factor to lift the gain slightly at $\omega = 1.5$. Thus, for example, we obtain the following weight:

$$w_m(s) = \frac{20(2.15s + 1)}{(s + 20)(s + 1)} = \frac{43s + 20}{s^2 + 21s + 20} \tag{15.2-19}$$

This second-order multiplicative uncertainty weight is shown in Figure 15.4, and obviously includes all possible plants.

15.3 ROBUST STABILITY IN THE H_∞-CONTEXT

In the previous section we discussed how to represent model uncertainty in a mathematical context. In this section, we will derive conditions under which a system remains stable for all perturbations in an uncertainty set.

15.3.1 Robust Stability with a Multiplicative Uncertainty

In Figure 15.5, a feedback system with a plant $H(s)$, a controller $K(s)$, and a multiplicative uncertainty is presented. In what follows, our aim is to determine whether the stability of the uncertain feedback system is maintained, if there is a multiplicative uncertainty of magnitude $|w_m(j\omega)|$.

With the uncertainty present, the open-loop transfer function of the feedback system of Figure 15.5 is given by

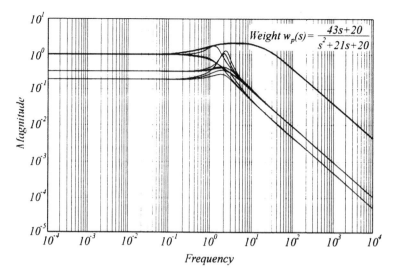

Figure 15.4   The second-order multiplicative uncertainty weight (15.2-19) for Example 15.2.5.

$$G_p(s) = H_p(s)K(s) = H(s)K(s)[1 + w_m(s)\Delta_m(s)] = G(s) + w_m(s)G(s)\Delta_m(s),$$

$$\|\Delta_m(j\omega)\|_\infty \leq 1$$

$$(15.3\text{-}1)$$

Assume that, by design, the stability of the nominal closed-loop system is guaranteed. For simplicity, we also assume tht the open-loop transfer function $G_p(s)$ is stable. To test for robust stability of the closed-loop feedback system, we use the Nyquist stability condition. Then, we obtain that robust stability, which is equivalent to the stability of the system for all $G_p(s)$, is also equivalent to the fact that $G_p(s)$ should not encircle the point $-1 + j0$, for all $G_p(s)$.

Now, consider a typical plot of $G_p(s)$ as shown in Figure 15.6. The distance from the point $1 + j0$ to the center of the disk, which represents $G_p(s)$, is $|1 + G(s)|$.

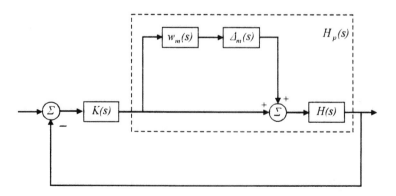

Figure 15.5   Closed-loop feedback system with multiplicative uncertainty.

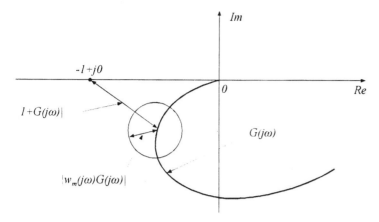

Figure 15.6   Graphical derivation of the robust stability condition through Nyquist plot.

Furthermore, the radius of this disk is $|w_m(s)G(s)|$. To avoid encirclement of $-1 + j0$, none of the disks should cover the critical point. By inspection of Figure 15.6, we conclude that the encirclement is avoided if and only if

$$|w_m(s)G(s)| < |1 + G(s)|, \forall \omega \tag{15.3-2}$$

or equivalently if and only if

$$\left| \frac{w_m(s)G(s)}{1 + G(s)} \right| < 1, \forall \omega \tag{15.3-3}$$

Definition 15.3.1

The sensitivity function designated by $S(s)$ and the complementary sensitivity function designated by $T(s)$ are defined as follows:

$$S(s) = [1 + G(s)]^{-1} = [1 + K(s)H(s)]^{-1} \quad \text{and}$$
$$T(s) = K(s)H(s)[1 + K(s)G(s)]^{-1} \tag{15.3-4}$$

The sensitivity functions $S(s)$ and $T(s)$ satisfy the relation $S(s) + T(s) = 1$.

Using Eq. (15.3-4), one can conclude that the encirclement is avoided (equivalently the robust stability condition is satisfied) if and only if

$$|w_m(s)T(s)| < 1, \forall \omega \tag{15.3-5}$$

Making use of Definition 15.2.1, we can finally conclude that robust stability under multiplicative perturbation is assumed if and only if

$$\|w_m(s)T(s)\|_\infty < 1 \tag{15.3-6}$$

It is worth noting that the robust stability condition (15.3-6) for the case of the multiplicative uncertainty gives an upper bound on the complementary sensitivity function. In other words, to guarantee robust stability in the case of multiplicative uncertainty one has to make $T(s)$ small at frequencies where the uncertainty weight exceeds 1 in magnitude.

Condition (15.3-6) is *necessary and sufficient* provided that, at each frequency, all perturbations satisfying $|\Delta_m(j\omega)| \leq 1$ are possible for the feedback system studied. If this is not the case, the condition is only *sufficient*.

An alternative, rather algebraic, way of obtaining the robust stability condition (15.3-6) is the following. Since $G_p(s)$ is assumed to be stable and the nominal closed-loop system is stable by design, then the nominal open-loop transfer function does not encircle the critical point $-1 + j0$. Conseqently, since the family of uncertain plants is norm bounded, it then follows that, if for some $G_{p1}(s)$ in the uncertain family, we have encirclement of $-1 + j0$, then there must be another $G_{p2}(s)$ in the uncertain family, which passes through $-1 + j0$ at some frequency. Therefore, to guarantee robust stability, the following condition must hold (and vice versa):

$$|1 + G_p(s)| \neq 0, \forall G_p, \forall \omega \tag{15.3-7}$$

Hence, robust stability is guaranteed if and only if

$$|1 + G(s) + w_m(s)G(s)\Delta_m(s)| > 0, \forall |\Delta_m(s)| \leq 1, \forall \omega \tag{15.3-8}$$

This last condition is most easily violated at each frequency when $\Delta_m(j\omega)$ has magnitude 1 and the phase is such that the terms $1 + G(s)$ and $w_m(s)G(s)\Delta_m(s)$ have opposite signs. Thus, robust stability is guaranteed if and only if

$$|1 + G(s)| - |w_m(s)G(s)| > 0, \forall \omega \tag{15.3-9}$$

Then, condition (15.3-6) follows easily.

We next give an example of how to check robust stability when using multiplicative perturbation.

Example 15.3.1

Consider the uncertain feedback control system of Figure 15.5. Assume that the uncertain plant transfer function is given by

$$H_p(s) = H(s)[1 + w_m(s)\Delta_m(s)]$$

where

$$H(s) = \frac{1}{s-1} \quad \text{and} \quad w_m(s) = \frac{2}{s+10}$$

while the controller $K(s)$ is a constant gain controller of the form $K(s) = 10$. Determine whether the closed-loop system is robustly stable.

Solution

For this case the complementary sensitivity function $T(s)$ is given by

$$T(s) = \frac{10}{s+9}$$

Figure 15.7 gives the magnitude of $T(s)$ as a function of the frequency, versus the magnitude of $1/w_m(s) = (s+10)/2$. From the figure, it is clear that, at each frequency, the magnitude of $1/w_m(s)$ overbounds the magnitude of $T(s)$. Hence, in our case, condition (15.3-6) is satisfied, and the closed-loop system is robustly stable.

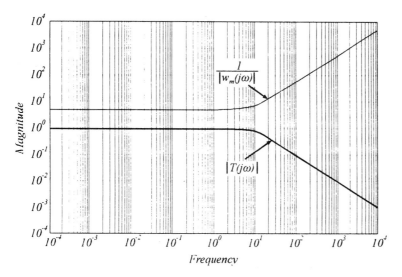

Figure 15.7 Checking robust stability with a multiplicative uncertainty, for Example 15.3.1.

15.3.2 Robust Stability with an Inverse Multiplicative Uncertainty

In this subsection a corresponding robust stability condition is derived for the case of feedback systems with inverse multiplicative uncertainty. To this end, we consider the feedback system of Figure 15.8, with a plant $H(s)$, a controller $K(s)$, and an inverse multiplicative uncertainty of magnitude $w_{im}(s)$. That is, here,

$$H_p(s) = H(s)[1 + w_{im}(s)\Delta_{im}(s)]^{-1} \tag{15.3-10}$$

Now, suppose that the open-loop transfer function $G_p(s)$ is stable and that the nominal closed-loop system is also stable. As mentioned above, robust stability is guaranteed, if encirclements of the point $-1 + j0$ are avoided, and since $G_p(s)$ belongs to a norm-bounded set, we conclude that robust staiblity is guaranteed if and only if one of the following four equivalent inequalities holds:

$$|1 + G_p(s)| > 0, \forall G_p(s), \forall \omega \tag{15.3-11a}$$

$$|1 + G(s)[1 + w_{im}(s)\Delta_{im}(s)]^{-1}| > 0, \forall |\Delta_{im}(j\omega)| \le 1, \forall \omega \tag{15.3-11b}$$

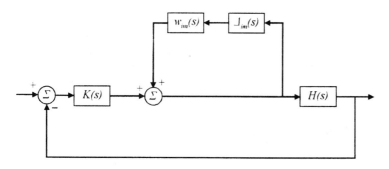

Figure 15.8 Closed-loop feedback system with inverse multiplicative uncertainty.

$$\left| \frac{1 + G(s) + w_{im}(s)\Delta_{im}(s)}{1 + w_{im}(s)\Delta_{im}(s)} \right| > 0, \forall |\Delta_{im}(j\omega)| \leq 1, \forall \omega \qquad (15.3\text{-}11c)$$

$$|1 + G(s) + w_{im}(s)\Delta_{im}(s)| > 0, \forall |\Delta_{im}(j\omega)| \leq 1, \forall \omega \qquad (15.3\text{-}11d)$$

The last condition is most easily violated at each frequency when $\Delta_{im}(j\omega)$ has magnitude 1 and the phase is such that the terms $1 + G(s)$ and $w_{im}(s)\Delta_{im}(s)$ have opposite signs. Thus, robust stability is guaranteed if and only if

$$|1 + G(s)| - |w_{im}(s)| > 0, \forall \omega \qquad (15.3\text{-}12)$$

Taking into account the definitions of the sensitivity function $S(s)$ and of the H_∞-norm, we finally obtain that robust stability with inverse multiplicative uncertainty is guaranteed if and only if

$$\|w_{im}(s)S(s)\|_\infty < 1 \qquad (15.3\text{-}13)$$

Condition (15.3-13) indicates that in order to guarantee robust stability, in the case of an inverse multiplicative perturbation, one has to make $S(s)$ small at frequencies where the uncertainty weight exceeds 1 in magnitude.

Example 15.3.2

Consider the feedback system of Figure 15.8. Assume that the uncertain plant transfer function is given by

$$H_p(s) = H(s)[1 + w_{im}(s)\Delta_{im}(s)]^{-1}$$

where

$$H(s) = \frac{1}{s - 1} \quad \text{and} \quad w_{im}(s) = \frac{s + 2.1}{3s + 0.7}$$

while the controller $K(s)$ is a PI controller of the form

$$K(s) = 1 + \frac{2}{s}$$

Determine whether the closed-loop system is robustlys table.

Solution

For this case, the sensitivity function $S(s)$ is given by

$$S(s) = \frac{s^2 - s}{s^2 + 2}$$

Figure 15.9 gives the magnitude of $S(s)$ as a function of the frequency versus the magnitude of $1/w_{im}(s) = (3s + 0.7)/(s + 2.1)$. From the figure, it is clear that, at each frequency, the magnitude of $1/w_{im}(s)$ overbounds the magnitude of $S(s)$. Hence, in our case, condition (15.3-4) is satisfied and the closed-loop system is robustly stable.

Remark 15.3.1

In the case of other well-known uncertainty descriptions, such as the additive or the division uncertainty, one can easily obtain robust stability conditions analogous to the conditions (15.3-6) and (15.3-13). Table 15.1 summarizes the robust stability tests for several commonly used uncertainty models.

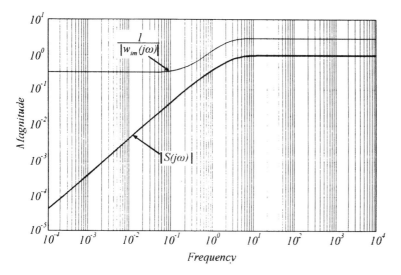

Figure 15.9 Checking robust stability with an inverse multiplicative uncertainty, for Example 15.3.2.

15.4 ROBUST PERFORMANCE IN THE H_∞-CONTEXT

In this section we study the performance of a perturbed plant. The general notion of *robust performance* is that internal stability and performance, of a specific type, should hold for all plants in a family P. Before dealing with robust performance, we study briefly the nominal performance and its relation to the sensitivity function.

15.4.1 Nominal Performance

Consider the feedback system presented in Figure 15.10. Here, $H(s)$ is the (unperturbed) plant transfer function, $K(s)$ is the controller transfer function, $r(t)$ or $R(s)$ is the reference input (command, setpoint), $d(t)$ or $D(s)$ is the disturbance (process noise), $n(t)$ or $N(s)$ is the measurement noise, $y_m(t)$ or $Y_m(s)$ is the measured output,

Table 15.1 Robust Stability Tests

Uncertainty description	Robust stability condition
Additive uncertainty $G(s) + w_a(s)\Delta_a(s)$	$\|w_a(s)K(s)S(s)\|_\infty < 1$
Multiplicative uncertainty $G(s)(1 + w_m(s)\Delta_m(s))$	$\|w_m(s)T(s)\|_\infty < 1$
Inverse multiplicative uncertainty $G(s)(1 + w_{im}(s)\Delta_{im}(s))^{-1}$	$\|w_{im}(s)S(s)\|_\infty < 1$
Division uncertainty $G(s)(1 + w_d(s)G(s)\Delta_d(s))^{-1}$	$\|w_d(s)G(s)S(s)\|_\infty < 1$

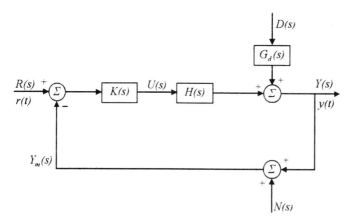

Figure 15.10   Block diagram of feedback control system with disturbance and noise.

and $u(t)$ or $U(s)$ is the control signal (actuator signal). The control error $e(t) = y(t) - r(t)$ or $E(s) = Y(s) - R(s)$ is given by

$$E(s) = [1 + K(s)H(s)]^{-1}R(s) + [1 + K(s)H(s)]^{-1}G_{\mathrm{d}}(s)D(s)$$
$$- K(s)H(s)[1 + K(s)H(s)]^{-1}N(s) \tag{15.4-1}$$

or, in terms of the sensitivity and the complementary sensitivity functions,

$$E(s) = S(s)R(s) + S(s)G_{\mathrm{d}}(s)D(s) - T(s)N(s) \tag{15.4-2}$$

For "perfect control," we want $e(t) = y(t) - r(t) = 0$. That is, we would like to have good disturbance rejection and command tracking as well as reduction of measurement noise on the plant output. This means that, for disturbance rejection and command tracking, the sensitivity function $S(s)$ must be chosen to be small in magnitude, whereas for zero noise transmission the same function must have a large magnitude, close to 1 (in this case $T(s)$ is small in magnitude). This illustrates the fundamental nature of feedback design, which always involves a trade-off among conflicting control objectives. Moreover, it illustrates that the sensitivity function $S(s)$ is a very good indicator of closed-loop performance. In particular, when considering $S(s)$ as such an indicator, our main advantage stems from the fact that it is sufficient to consider just its magnitude and not worry about its phase.

Some very common specifications in terms of $S(s)$ are listed below:

1. Maximum tracking error at prespecified frequencies
2. Minimum steady-state tracking error A
3. Maximum peak magnitude M of $S(s)$
4. Minimum bandwidth ω_{B}^{*}

Performance specifications of the above type can usually be incorporated in an upper bound, $1/|w_{\mathrm{P}}(s)|$, on the magnitude of the sensitivity function, where $w_{\mathrm{P}}(s)$ is a weight chosen by the designer. The subscript P stands for *performance*, since, as already mentioned, the sensitivity function is used as a performance indicator. Then, the performance requirement is guaranteed if and only if one of the following three equivalent inequalities holds:

$$|S(j\omega)| < 1/|w_P(j\omega)|, \forall \omega \qquad (15.4\text{-}3a)$$

$$|w_P(j\omega)S(j\omega)| < 1, \forall \omega \qquad (15.4\text{-}3b)$$

$$\|w_P(s)S(s)\|_\infty < 1 \qquad (15.4\text{-}3c)$$

A typical performance weight is the following:

$$w_P(s) = \frac{s/M + \omega_B^*}{s + \omega_B^* A} \qquad (15.5\text{-}4)$$

It can be easily seen from Eq. (15.4-4) that

1. As $s \to 0$, $S(s) \to A$.
2. As $s \to \infty$, $S(s) \to M$.
3. The asymptote of the Bode plot of the magnitude of $S(s)$ crosses $0\,\mathrm{dB}$, at the frequency ω_B^*, which is the bandwidth requirement.

Now, consider the Nyquist plot of Figure 15.11. Taking into account the definition of the sensitivity function, one can obtain from Eq. (15.4-3) that nominal performance is equivalent to

$$|w_P(j\omega)| < |1 + G(j\omega))|, \forall \omega \qquad (15.4\text{-}5)$$

At each frequency, the term $|1 + G(s)|$ is the distance of $G(s)$ from the critical point $-1 + j0$ in the Nyquist plot. Therefore, for nominal performance, $G(j\omega)$ must be at least at a distance of $|w_P(j\omega)|$ from the critical point. In other words, for nominal performance, $G(j\omega)$ must stay outside a disk of radius $|w_P(j\omega)|$, centered at $-1 + j0$. This graphical interpretation of nominal performance is depicted in Figure 15.11.

Example 15.4.1

Consider the feedback system depicted in Figure 15.10, with

$$H(s) = \frac{1}{s-1} \quad \text{and} \quad K(s) = 10$$

Let the design specifications for the closed-loop system be the following:

1. Steady-state tracking error $A = 0.2$

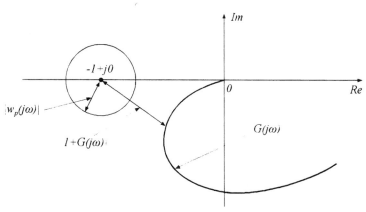

Figure 15.11 Nominal performance in the Nyquist plot.

2. Maximum peak magnitude $M = 2$ of the sensitivity function $S(s)$
3. Minimum bandwidth $\omega_B^* = 0.5\,\text{rad/sec}$

Determine whether the closed-loop system meets the nominal performance requirement.

Solution

These design specifications can be written in the form of a rational performance bound $w_P(s)$ of the form (15.4-4). In particular, for the present case, the performance bound has the form

$$w_P(s) = \frac{s+1}{2s+0.2}$$

In Figure 15.12, the magnitude of the sensitivity function $S(s)$ as a function of the frequency, versus the magnitude of $1/w_P(s) = (s+1)/(2s+0.2)$ is shown. From the figure it is apparent that, at each frequency, the magnitude of the sensitivity function $S(s)$ is bounded by the magnitude of $1/w_P(s)$. Therefore, in our case, condition (15.4-3) is satisfied, and the closed-loop system meets the nominal performance requirement.

15.4.2 Robust Performance

Clearly, for robust performance, it is sufficient to require that condition (15.4-5) is satisfied for all possible plants $G_p(s)$. In mathematical terms, robust performance is defined by one of the following two equivalent inequalities:

$$|w_P(j\omega)S_p(j\omega)| < 1, \forall S_p, \forall \omega \tag{15.4-6a}$$

$$|w_P(j\omega)| < |1 + G_p(j\omega)|, \forall G_p, \forall \omega \tag{15.4-6b}$$

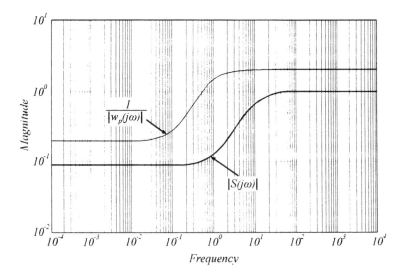

Figure 15.12   Checking nominal performance for Example 15.4.1.

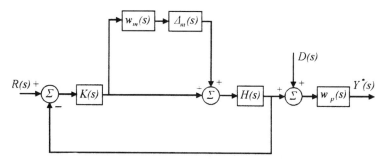

Figure 15.13 Block diagram for robust performance in the multiplicative uncertainty case.

Next, and for sake of simplicity, we focus our attention on the multiplicative uncertainty case. Figure 15.13 presents the block diagram for robust performance in the multiplicative uncertainty case. It is not difficult to see that condition (15.4-6) corresponds to the requirement $|y^*/d| < 1, \forall \Delta_m$. In this case, the set of possible open-loop transfer functions is given by

$$G_p(s) = K(s)H_p(s) = G(s)[1 + w_m(s)\Delta_m(s)] = G(s) + w_m(s)G(s)\Delta_m(s) \quad (15.4\text{-}7)$$

Now, consider the Nyquist plot of Figure 15.14. To guarantee robust performance, one must require that all possible open-loop transfer functions $G_p(j\omega)$ stay outside a disk of radius $|w_P(j\omega)|$ centered on the critical point $-1 + j0$. It is evident that $G_p(j\omega)$, at each frequency, stays within a disk of radius $w_m(j\omega)G(j\omega)$, centered on $G(j\omega)$. Therefore, from Figure 15.14, the condition for robust performance is that these two disks must not overlap. The centers of the two disks are located at a distance $|1 + G(j\omega)|$ apart. Consequently, the robust performance is guaranteed if and only if one of the following two equivalent inequalities holds:

$$|w_P(j\omega)| + |w_m(j\omega)G(j\omega)| < |1 + G(j\omega)|, \forall \omega \quad (15.4\text{-}8a)$$

$$|w_P(j\omega)[1 + G(j\omega)]^{-1}| + |w_m(j\omega)G(j\omega)[1 + G(j\omega)]^{-1}| < 1, \forall \omega \quad (15.4\text{-}8b)$$

Taking into account the definitions of the sensitivity and the complementary sensitivity functions, we may further obtain robust performance if and only if

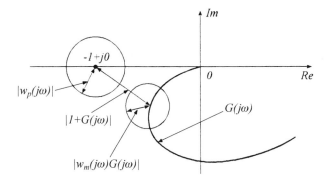

Figure 15.14 Robust performance in the Nyquist plot.

$$|w_P(j\omega)S(j\omega)| + |w_m(j\omega)T(j\omega)| < 1, \forall \omega$$

or in other words

$$\||w_P(j\omega)S(j\omega)| + |w_m(j\omega)T(j\omega)|\|_\infty < 1 \tag{15.4-9}$$

where, in deriving Eq. (15.4-9), use was made of the definition of the H_∞-norm. Relation (15.4-9) is a necessary and sufficient condition for robust performance.

An alternative (rather algebraic) way of obtaining the robust performance condition (15.4-9) is the following. According to relation (15.4-6a), robust performance is guaranteed if the maximum *weighted sensitivity* $w_P(s)S(s)$, at each frequency, is less than 1 in magnitude. This means that robust performance is assured if and only if

$$\sup_{S_p} |w_P(j\omega)S_p(j\omega)| < 1, \forall \omega \tag{15.4-10}$$

The perturbed sensitivity is

$$S_p(s) = [1 + G_p(s)]^{-1} = \frac{1}{1 + G(s) + w_m(s)G(s)\Delta(s)} \tag{15.4-11}$$

The worst-case (maximum) is obtained in the case where, at each frequency, we select $|\Delta_m(s)| = 1$, such that the signs of the terms $1 + G(s)$ and $w_m(s)G(s)\Delta_m(s)$ are opposite. In mathematical terms, we have

$$\sup_{S_p} |w_P(j\omega)S_p(j\omega)| = \frac{|w_P(j\omega)|}{|1 + G(j\omega)| - |w_m(j\omega)G(j\omega)|} = \frac{|w_P(j\omega)S(j\omega)|}{1 - |w_m(j\omega)T(j\omega)|}$$

$$\tag{15.4-12}$$

Combining relations (15.4-10) and (15.4-12) and taking into account the definition of the H_∞-norm we readily obtain the robust performance condition (15.4-9).

Condition (15.4-9) provides us with some useful bounds on the magnitude of $G(s)$. In particular, by observing that $|1 + G(j\omega)| \geq 1 - |G(j\omega)|$ and $|1 + G(j\omega)| \geq |G(j\omega)| - 1$, $\forall \omega$, we can easily see that the robust performance condition (15.4-9) is satisfied if

$$|G(j\omega)| > \frac{1 + |w_P(j\omega)|}{1 - |w_m(j\omega)|}, \forall \omega : |w_m(j\omega)| < 1 \tag{15.4-13a}$$

or if

$$|G(j\omega)| < \frac{1 - |w_P(j\omega)|}{1 + |w_m(j\omega)|}, \forall \omega : |w_P(j\omega)| < 1 \tag{15.4-13b}$$

We can also prove that the robust performance condition is satisfied if

$$|G(j\omega)| > \frac{|w_P(j\omega)| - 1}{1 - |w_m(j\omega)|}, \qquad \forall \omega : |w_m(j\omega)| < 1 \quad \text{and} \quad |w_P(j\omega)| > 1$$

$$\tag{15.4-14a}$$

or if

$$|G(j\omega)| < \frac{1 - |w_P(j\omega)|}{|w_m(j\omega)| - 1}, \qquad \forall \omega : |w_P(j\omega)| < 1, \quad \text{and} \quad |w_m(j\omega)| > 1$$

$$\tag{15.4-14b}$$

For the case of SISO systems the term $|w_P(s)S(s)| + |w_m(s)T(s)|$ is the *structured singular value* (SSV) designated by $\mu(\omega)$. With this definition, robust performance is guaranteed if and only if

$$\|\mu(\omega)\|_\infty < 1 \tag{15.4-15}$$

Example 15.4.2

Consider the feedback control system of Figure 15.13, where

$$H(s) = \frac{1}{s-1}, \quad K(s) = 10, \quad w_m(s) = \frac{2}{s+10}, \quad \text{and} \quad w_P(s) = \frac{s+1}{2s+0.2}$$

As shown in Example 15.3.1, the system is robustly stable. Moreover, in Example 15.4.1, it has been shown that the nominal plant satisfies the nominal performance specification imposed by $w_P(s)$. Determine whether the closed-loop system also satisfies the robust performance specification.

Solution

Figure 15.15 shows that the structured singular value $\mu(\omega)$ as function of the frequency. From this figure, it becomes clear that, at each frequency, the SSV is less than 1. Therefore, condition (15.4-15) (or equivalently, condition (15.4-9)) is satisfied, and robust performance of the feedback system is guaranteed.

15.4.3 Some Remarks on Nominal Performance, Robust Stability, and Robust Performance

Consider once again the block diagram of Figure 15.13 representing a feedback loop with multiplicative uncertainty and assume that the nominal closed-loop system is stable. Relations (15.4-3), (15.3-5), and (15.4-9) (or (15.4-15)) give the conditions for nominal performance, robust stability, and robust performance, respectively. From

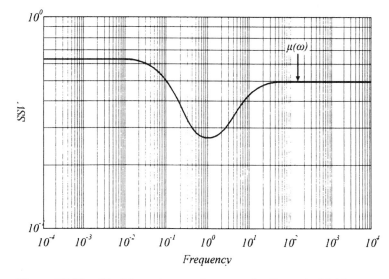

Figure 15.15   Checking robust performance for Example 15.4.2.

these conditions, it becomes clear that *the closed-loop system satisfies the robust performance specification if it satisfies the nominal performance specification and is simultaneously robustly stable*. Therefore, robust performance *is not* our primary objective for siso systems. Our *primary objectives* for siso systems are nominal performance and robust stability. However, this is not true, in general, for the MIMO system case.

From condition (15.3-5), it is clear that in order to satisfy robust stability we want, in general, to make $T(s)$ small. On the other hand, for nominal performance, we want, in general, to make $S(s)$ small. However, since $S(s) + T(s) = 1$, *we cannot make both $S(s)$ and $T(s)$ small at the same frequency*. That is, *we cannot satisfy more than 100% uncertainty and good performance at the same frequency*. This is another example of conflicting control objectives in feedback control systems.

It is worth noting at this point that robust performance can be viewed as a special case of robust stability with multiple uncertainty description. To make this clear, consider the block diagrams of Figure 15.16. The block diagram of Figure 15.16a is almost the same as that of Figure 15.13. The block diagram of Figure 15.16b represents a closed-loop feedback system with both multiplicative and inverse multiplicative uncertainties. Referring to Figure 15.16a, in order to satisfy robust performance, condition (15.4-9) must hold. We now focus our attention on Figure 15.16b. To guarantee robust stability, it is necessary and sufficient that the following relation holds

$$|1 + G_p(j\omega)| > 0, \forall G_p(j\omega), \forall \omega \tag{15.4-16}$$

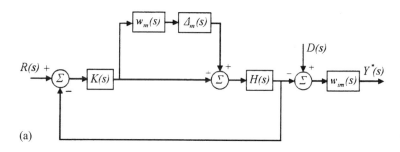

(a)

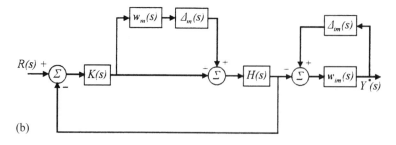

(b)

Figure 15.16 Similarity between robust stability and robust performance: (a) robust performance with multiplicative uncerainty; (b) robust performance with both multiplicative and inverse uncertainty.

Block diagram algebra shows that, from condition (15.4-16), we can readily obtain that robust stability is ensured if and only if one of the following two inequalities hold:

$$\left|1 + G(j\omega)[1 + w_{\mathrm{m}}(j\omega)\Delta_{\mathrm{m}}(j\omega)][1 - w_{\mathrm{im}}(j\omega)\Delta_{\mathrm{im}}(j\omega)]^{-1}\right| > 0, \forall \Delta_{\mathrm{m}}, \Delta_{\mathrm{im}}, \forall \omega$$

(15.4-17a)

$$\left|1 + G(j\omega) + G(j\omega)w_{\mathrm{m}}(j\omega)\Delta_{\mathrm{m}}(j\omega) - w_{\mathrm{im}}(j\omega)\Delta_{\mathrm{im}}(j\omega)\right| > 0, \forall \Delta_{\mathrm{m}}, \Delta_{\mathrm{im}}, \forall \omega$$

(15.4-17b)

The worst-case (maximum) is obtained in the case where, at each frequency, we select $|\Delta_{\mathrm{m}}(s)| = 1$, and $|\Delta_{\mathrm{im}}(s)| = 1$, such that the signs of the terms $w_{\mathrm{im}}(s)\Delta_{\mathrm{im}}(s)$ and $w_{\mathrm{m}}(s)G(s)\Delta_{\mathrm{m}}(s)$ are opposite to the sign of $1 + G(s)$. In mathematical terms, robust stability is guaranteed if and only if one of the following four equivalent inequalities holds:

$$|1 + G(j\omega)| - |G(j\omega)w_{\mathrm{m}}(j\omega)| - |w_{\mathrm{im}}(j\omega)| > 0, \forall \omega \qquad (15.4\text{-}18a)$$

$$|G(j\omega)w_{\mathrm{m}}(j\omega)| + |w_{\mathrm{im}}(j\omega)| < |1 + G(j\omega)|, \forall \omega \qquad (15.4\text{-}18b)$$

$$|G(j\omega)[1 + G(j\omega)]^{-1}w_{\mathrm{m}}(j\omega)| + |w_{\mathrm{im}}(j\omega)[1 + G(j\omega)]^{-1}| < 1, \forall \omega \qquad (15.4\text{-}18c)$$

$$|w_{\mathrm{m}}(j\omega)T(j\omega)| + |w_{\mathrm{im}}(j\omega)S(j\omega)| < 1, \forall \omega \qquad (15.4\text{-}18d)$$

Condition (15.4-18) is equivalent to condition (15.4-9), provided that $w_{\mathrm{im}}(s) \equiv w_{\mathrm{P}}(s)$.

15.5 KHARITONOV'S THEOREM AND RELATED RESULTS

This section refers mainly to the seminal theorem of Kharitonov [1], which, since its appearance in the late 1970s, has motivated a variety of powerful results (such as the sixteen-plant theorem, the edge theorem [1], etc.) for more general robustness problems.

15.5.1 Kharitonov's Theorem for Robust Stability

Kharitonov's theorem addresses robust stability of *interval polynomials with lumped uncertainty and fixed degree* of the form

$$p(s, \mathbf{a}) = \sum_{i=0}^{n} a_i s^i \qquad (15.5\text{-}1)$$

where

$$a_i \in [a_i^-, a_i^+], \qquad i = 0, 1, \ldots, n$$

where $[a_i^-, a_i^+]$ denotes the a priori known bounding interval for the ith component of uncertainty (the ith uncertain coefficient) a_i.

To describe Kharitonov's theorem for robust stability, it is first necessary to define four fixed polynomials associated with a family of interval polynomials.

Definition 15.5.1 (Khartinov Polynomials)

Associated with the interval polynomial $p(s, \mathbf{a})$ are the four fixed Kharitonov polynomials

$$p_1(s) = a_0^- + a_1^- s + a_2^+ s^2 + a_3^+ s^3 + a_4^- s^4 + a_5^- s^5 + a_6^+ s^6 + \cdots$$
$$p_2(s) = a_0^+ + a_1^+ s + a_2^- s^2 + a_3^- s^3 + a_4^+ s^4 + a_5^+ s^5 + a_6^- s^6 + \cdots$$
$$p_3(s) = a_0^+ + a_1^- s + a_2^- s^2 + a_3^+ s^3 + a_4^+ s^4 + a_5^- s^5 + a_6^- s^6 + \cdots$$
$$p_4(s) = a_0^- + a_1^+ s + a_2^+ s^2 + a_3^- s^3 + a_4^- s^4 + a_5^+ s^5 + a_6^+ s^6 + \cdots$$

The Kharitonov polynomials are easily constructed by inspection, as in the following example.

Example 15.5.1

Consider the following interval polynomial with fixed degree 6:

$$p(s, \mathbf{a}) = [2, 3]s^6 + [1, 8]s^5 + [3, 12]s^4 + [5, 6]s^3 + [4, 7]s^2 + [9, 11]s + [6, 15]$$

Derive the four Kharitonov polynomials.

Solution

For this case, the four Kharitonov polynomials are

$$p_1(s) = 6 + 9s + 7s^2 + 6s^3 + 3s^4 + s^5 + 3s^6$$
$$p_2(s) = 15 + 11s + 4s^2 + 5s^3 + 12s^4 + 8s^5 + 2s^6$$
$$p_3(s) = 15 + 9s + 4s^2 + 6s^3 + 12s^4 + s^5 + 2s^6$$
$$p_4(s) = 6 + 11s + 7s^2 + 5s^3 + 3s^4 + 8s^5 + 3s^6$$

We are now in position to present the celebrated Kharitonov's theorem.

Theorem 15.5.1

An interval polynomial $p(s, \mathbf{a})$ with invariant degree n is robustly stable if and only if its four associated Kharitonov polynomials are stable.

We next give an application example of Kharitonov's theorem.

Example 15.5.2

Consider the following interval polynomial with fixed degree 5:

$$p(s, \mathbf{a}) = [1, 3]s^5 + [3, 6]s^4 + [4, 7]s^3 + [5, 9]s^2 + [3, 4]s + [2, 5]$$

Determine if this interval polynomial is robustly stable.

Solution

In this case, the four Kharitonov polynomials are

$$p_1(s) = 2 + 3s + 9s^2 + 7s^3 + 3s^4 + s^5$$
$$p_2(s) = 5 + 4s + 5s^2 + 4s^3 + 6s^4 + 3s^5$$
$$p_3(s) = 5 + 3s + 5s^2 + 7s^3 + 6s^4 + s^5$$
$$p_4(s) = 2 + 4s + 9s^2 + 4s^3 + 3s^4 + 3s^5$$

According to Kharitonov's theorem, in order to guarantee robust stability of the given interval polynomial, it is sufficient to test the stability of the above four Kharitonov polynomials. This can be accomplished by using an algebraic stability criterion, e.g., the Routh criterion. Applying the Routh criterion in the four Kharitonov's polynomials, we obtain the following four Routh tables:

1. Polynomial $p_1(s)$:

$$
\begin{array}{c|ccc}
s^5 & 1 & 7 & 3 \\
s^4 & 3 & 9 & 2 \\
s^3 & 4 & 7/3 & 0 \\
s^2 & 29/4 & 2 & 0 \\
s^1 & 107/77 & 0 & \\
s^0 & 2 & & \\
\end{array}
$$

2. Polynomial $p_2(s)$:

$$
\begin{array}{c|ccc}
s^5 & 3 & 4 & 4 \\
s^4 & 6 & 5 & 5 \\
s^3 & 3/2 & 3/2 & 0 \\
s^2 & -1 & 15/2 & 0 \\
s^1 & 51/4 & 0 & \\
s^0 & 15/2 & & \\
\end{array}
$$

3. Polynomial $p_3(s)$:

$$
\begin{array}{c|ccc}
s^5 & 1 & 7 & 3 \\
s^4 & 6 & 5 & 5 \\
s^3 & 37/6 & 13/6 & 0 \\
s^2 & 107/37 & 5 & 0 \\
s^1 & -909/107 & 0 & \\
s^0 & 5 & & \\
\end{array}
$$

4. Polynomial $p_4(s)$:

$$
\begin{array}{c|ccc}
s^5 & 3 & 4 & 4 \\
s^4 & 3 & 9 & 2 \\
s^3 & -5 & 2 & 0 \\
s^2 & 51/5 & 2 & 0 \\
s^1 & 152/51 & 0 & \\
s^0 & 2 & & \\
\end{array}
$$

Since the polynomials $p_2(s)$, $p_3(s)$, and $p_4(s)$ are unstable, the interval polynomial $p(s, \mathbf{a})$ of the present example is not robustly stable.

Kharitonov's test for robust stability can signficantly be simplified if the interval polynomial studied is of degree 5, 4, or 3. In these cases, the Kharitonov polynomials necessary for performing the test are 3, 2, or 1 in number, respectively, as against the four polynomials of the general case. More precisely, we have the following propositions.

Proposition 15.5.1

An interval polynomial $p(s, \mathbf{a})$ with invariant degree 5 is robustly stable if and only if the Kharitonov polynomials $p_1(s)$, $p_2(s)$, and $p_3(s)$ are stable.

Proposition 15.5.2

An interval polynomial $p(s, \mathbf{a})$ with invariant degree 4 is robustly stable if and only if the Kharitonov polynomials $p_2(s)$ and $p_3(s)$ are stable.

Proposition 15.5.3

An interval polynomial $p(s, \mathbf{a})$ with invariant degree 3 is robustly stable if and only if the Kharitonov polynomial $p_3(s)$ is stable.

Although Kharitonov's theorem is a very important result in the area of robustness analysis, it has several limitations. The most important limitations are the following:

(a)   Kharitonov's theorem is applicable only in problems for which the stability region is the open left-half plane. In other words, Kharitonov's theorem cannot be applied in the case of discrete-time systems.

(b)   Kharitonov's theorem is applicable only in the case of interval polynomials whose coefficients vary independently. In the more general case of interval polynomials of the form

$$p(s, \mathbf{a}) = \sum_{i=0}^{n} f_i(a_1, a_2, \ldots, a_m)s^i \tag{15.5-2}$$

where $f_i(a_1, a_1, \ldots, a_m), i = 1, 2, \ldots, n$ are multilinear functions of the uncertain coefficients $a_1, a_2, \ldots, a_m$, Kharitonov's theorem fails to give an answer to the question of the robust stability of interval polynomials of the form (15.5-2).

Much research effort has been devoted to removing these limitations. An important such effort is the celebrated Edge theorem, which will not be presented here since it is beyond the scope of this book. The interested reader may refer to [1] for a detailed discussion of this important theorem.

15.5.2   The Sixteen-Plant Theorem

Here, we generalize the analytical results presented in the previous subsection, in order to develop a technique for the design of robustly stabilizing compensators. In particular, we focus our attention on the design of proper first-order compensators of the form

$$F(s) = \frac{K(s - z)}{s - p} \tag{15.5-3}$$

which robustly stabilize a strictly proper interval plant family of the form

$$G(s, \mathbf{a}, \mathbf{b}) = \frac{A(s, \mathbf{a})}{B(s, \mathbf{b})} \; \frac{\displaystyle\sum_{j=0}^{m} a_j s^j}{\displaystyle s^n + \sum_{i=0}^{n-1} b_i s^i}, \qquad m < n \tag{15.5-4}$$

$$a_j \in [a_j^-, a_j^+], \qquad j = 0, 1, \ldots, m, \qquad b_i \in [b_i^-, b_i^+], \qquad i = 0, 1, \ldots, n-1$$

with $[a_j^-, a_j^+]$ and $[b_i^-, b_i^+]$ denoting the a priori known bounding intervals for the jth and the ith components of uncertainties a_j and b_i, respectively.

We say that the compensator $F(s)$ of the form (15.5-3) robustly stabilizes the interval plant family (15.5-4), if for all $a_j \in [a_j^-, a_j^+]$, $j = 0, 1, \ldots, m$ and $b_i \in [b_i^-, b_i^+]$, $i = 0, 1, \ldots, n-1$, the resulting closed-loop polynomial

$$p_c(s, \mathbf{a}, \mathbf{b}) = K(s-z)A(s, \mathbf{a}) + (s-p)B(s, \mathbf{b}) \tag{15.5-5}$$

is Hurwitz (i.e., its roots lie in the open left-half plane). $F(s)$ is then called a robust stabilizer of the interval plant family (15.5-4).

The focal point of this subsection is the following question. Given the interval plant family (15.5-4), with a compensator interconnected as a Figure 15.17, under what conditions can we establish robust stability of the closed-loop system, under a "small" finite subset of systems corresponding to the extreme members of the family (15.5-4)?

We next try to answer this question. To this end, we introduce the Kharitonov polynomials for the numerator and the denominator of Eq. (15.5-4). For the numerator, let

$$A_1(s) = a_0^- + a_1^- s + a_2^+ s^2 + a_3^+ s^3 + a_4^- s^4 + a_5^- s^5 + a_6^+ s^6 + \cdots$$
$$A_2(s) = a_0^+ + a_1^+ s + a_2^- s^2 + a_3^- s^3 + a_4^+ s^4 + a_5^+ s^5 + a_6^- s^6 + \cdots$$
$$A_3(s) = a_0^+ + a_1^- s + a_2^- s^2 + a_3^+ s^3 + a_4^+ s^4 + a_5^- s^5 + a_6^- s^6 + \cdots$$
$$A_4(s) = a_0^- + a_1^+ s + a_2^+ s^2 + a_3^- s^3 + a_4^- s^4 + a_5^+ s^5 + a_6^+ s^6 + \cdots$$

and for the denominbtor let

$$B_1(s) = b_0^- + b_1^- s + b_2^+ s^2 + b_3^+ s^3 + b_4^- s^4 + b_5^- s^5 + b_6^+ s^6 + \cdots$$
$$B_2(s) = b_0^+ + b_1^+ s + b_2^- s^2 + b_3^- s^3 + b_4^+ s^4 + b_5^+ s^5 + b_6^- s^6 + \cdots$$
$$B_3(s) = b_0^+ + b_1^- s + b_2^- s^2 + b_3^+ s^3 + b_4^+ s^4 + b_5^- s^5 + b_6^- s^6 + \cdots$$
$$B_4(s) = b_0^- + b_1^+ s + b_2^+ s^2 + b_3^- s^3 + b_4^- s^4 + b_5^+ s^5 + b_6^+ s^6 + \cdots$$

By taking all combinations of the $A_i(s)$, $i = 1, 2, 3, 4$, and $B_k(s)$, $k = 1, 2, 3, 4$, we obtain the 16 Kharitonov plants

$$G_{ik}(s) = \frac{A_i(s)}{B_k(s)} \tag{15.5-6}$$

for $i, k = 1, 2, 3, 4$. For these extreme plants, when it is said that $F(s)$ stabilizes $G_{ik}(s)$, we understand that the closed-loop polynomial

$$p_{c,ik}(s) = K(s-z)A_i(s) + (s-p)B_k(s) \tag{15.5-7}$$

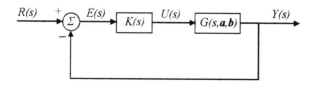

Figure 15.17 Interval plant family interconnected with a first-order compensator.

is asymptotically stable. We are now able to state the so-called *sixteen-plant theorem*.

Theorem 15.5.2

A proper first-order compensator of the form (15.5-3) robustly stabilizes the interval plant family (15.5-4), if and only if it stabilizes all of the 16 Kharitonov plants $G_{ik}(s)$, $i, k = 1, 2, 3, 4$.

We next give some examples regarding the 16 Kharitonov plants and the application of the sixteen-plant theorem as a design tool.

Example 15.5.3

Consider the interval plant family

$$G(s, \mathbf{a}, \mathbf{b}) = \frac{[2, 5]s^2 + [6, 9]s + [3, 11]}{s^3 + [4, 6]s^2 + [1, 8]s + [5, 7]}$$

Determine one proper first-order compensator that can robustly stabilize the family and one that cannot.

Solution

The 16 Kharitonov plants associated with this family are

$$G_{11}(s) = \frac{2s^2 + 6s + 11}{s^3 + 4s^2 + s + 7}, \qquad G_{12}(s) = \frac{2s^2 + 6s + 11}{s^3 + 6s^2 + 8s + 5}$$

$$G_{13}(s) = \frac{2s^2 + 6s + 11}{s^3 + 6s^2 + s + 5}, \qquad G_{14}(s) = \frac{2s^2 + 6s + 11}{s^3 + 4s^2 + 8s + 7}$$

$$G_{21}(s) = \frac{5s^2 + 9s + 3}{s^3 + 4s^2 + s + 7}, \qquad G_{22}(s) = \frac{5s^2 + 9s + 3}{s^3 + 6s^2 + 8s + 5}$$

$$G_{23}(s) = \frac{5s^2 + 9s + 3}{s^3 + 6s^2 + s + 5}, \qquad G_{24}(s) = \frac{5s^2 + 9s + 3}{s^3 + 4s^2 + 8s + 7}$$

$$G_{31}(s) = \frac{5s^2 + 6s + 3}{s^3 + 4s^2 + s + 7}, \qquad G_{32}(s) = \frac{5s^2 + 6s + 3}{s^3 + 6s^2 + 8s + 5}$$

$$G_{33}(s) = \frac{5s^2 + 6s + 3}{s^3 + 6s^2 + s + 5}, \qquad G_{34}(s) = \frac{5s^2 + 6s + 3}{s^3 + 4s^2 + 8s + 7}$$

$$G_{41}(s) = \frac{2s^2 + 9s + 11}{s^3 + 4s^2 + s + 7}, \qquad G_{42}(s) = \frac{2s^2 + 9s + 11}{s^3 + 6s^2 + 8s + 5}$$

$$G_{43}(s) = \frac{2s^2 + 9s + 11}{s^3 + 6s^2 + s + 5}, \qquad G_{44}(s) = \frac{2s^2 + 9s + 11}{s^3 + 4s^2 + 8s + 7}$$

With a particular first-order compensator, say $F(s) = (s - 1)/(s + 1)$, it can easily be verified that the closed-loop polynomial, associated with the Kharitonov plant $G_{32}(s)$, is given by

$$p_{c,32}(s) = (s - 1)(5s^2 + 6s + 3) + (s + 1)(s^3 + 6s^2 + 8s + 5)$$
$$= s^4 + 12s^3 + 15s^2 + 10s + 2$$

It is not difficult to show that the controller $f(s) = (s - 1)/(s + 1)$ cannot robustly stabilize the given interval plant family. Indeed, with this controller, the closed-loop polynomial $p_{c,11}(s)$ is given by

$$p_{c,11}(s) = (s - 1)(2s^2 + 6s + 11) + (s + 1)(s^3 + 4s^2 + s + 7)$$
$$= s^4 + 7s^3 + 9s^2 + 13s - 4$$

Since there is a sign change in the coefficients of $p_{c,11}(s)$, the polynomial is not stable. Therefore, $F(s) = (s - 1)/(s + 1)$ cannot robustly stabilize the entire interval plant family.

On the other hand, as can easily be checked by performing 16 Routh tests for closed-loop polynomials of the 16 Kharitonov plants, the PI controller of the form

$$F(s) = 1 + \frac{1}{s}$$

robustly stabilizes the given interval plant family.

Example 15.5.4

Although the sixteen-plant theorem is stated as an analysis tool, it can be quite easily used in a synthesis context as well. Explain how this is possible.

Solution

Consider the interval plant family

$$G(s, \mathbf{a}, \mathbf{b}) = \frac{[1, 1.5]s + [0.5, 1]}{s^3 + [2, 3]s^2 + [1, 2]s + [3, 4]}$$

Suppose that one wants to find a robustly stabilizing PI controller of the form

$$F(s) = K_1 + K_2/s \tag{15.5-8}$$

for the above family. Note that a PI controller of the form (15.5-8) is the special case of a first-order compensator of the form (15.5-3), for which $p = 0$, $K = K_1$, and $z = -K_2/K_1$.

To deal with this problem, we first construct the 16 Kharitonov plants associated with the given interval plant family. These are the following:

$$G_{11}(s) = \frac{s + 0.5}{s^3 + 2s^2 + s + 4}, \qquad G_{12}(s) = \frac{s + 0.5}{s^3 + 3s^2 + 2s + 3}$$

$$G_{13}(s) = \frac{s + 0.5}{s^3 + 3s^2 + s + 3}, \qquad G_{14}(s) = \frac{s + 0.5}{s^3 + 2s^2 + 2s + 4}$$

$$G_{21}(s) = \frac{1.5s + 1}{s^3 + 2s^2 + s + 4}, \qquad G_{22}(s) = \frac{1.5s + 1}{s^3 + 3s^2 + 2s + 3}$$

$$G_{23}(s) = \frac{1.5s + 1}{s^3 + 3s^2 + s + 3}, \qquad G_{24}(s) = \frac{1.5s + 1}{s^3 + 2s^2 + 2s + 4}$$

$$G_{31}(s) = \frac{1.5s + 0.5}{s^3 + 2s^2 + s + 4}, \qquad G_{32}(s) = \frac{1.5s + 0.5}{s^3 + 3s^2 + 2s + 3}$$

$$G_{33}(s) = \frac{1.5s + 0.5}{s^3 + 3s^2 + s + 3}, \qquad G_{34}(s) = \frac{1.5s + 0.5}{s^3 + 2s^2 + 2s + 4}$$

$$G_{41}(s) = \frac{s+1}{s^3 + 2s^2 + s + 4}, \qquad G_{42}(s) = \frac{s+1}{s^3 + 3s^2 + 2s + 3}$$

$$G_{43}(s) = \frac{s+1}{s^3 + 3s^2 + s + 3}, \qquad G_{44}(s) = \frac{s+1}{s^3 + 2s^2 + 2s + 4}$$

With the PI compensator of the form (15.5-8), the closed-loop polynomials associated with the 16 Kharitonov plants can be easily obtained. For example, for the Kharitonov plant $G_{42}(s)$, the associated closed-loop polynomials, is found to be

$$\begin{aligned} p_{c,42}(s) &= s(s^3 + 3s^2 + 2s + 3) + (K_1 s + K_2)(s+1) \\ &= s^4 + 3s^3 + (2+K_1)s^2 + (3 + K_1 + K_2)s + K_2 \end{aligned}$$

Now, using the polynomial $p_{c,42}(s)$, we generate the Routh table

s^4	1	$2 + K_1$	K_2
s^3	3	$3 + K_1 + K_2$	0
s^2	$a_1(K_1, K_2)$	K_2	
s^1	$a_2(K_1, K_2)$	0	
s^0	K_2		

where

$$a_1(K_1, K_2) = \frac{3 + 2K_1 - K_2}{3} \qquad \text{and} \qquad a_2(K_1, K_2) = 3 + K_1 + K_2 - \frac{3K_2}{a_1(K_1, K_2)}$$

In a similar manner one can generate Routh tables for the remaining 15 Kharitonov plants. Using the 16 Routh tables our next steps is to enforce positivity in each of the first columns. As a result, we obtain inequalities involving K_1 and K_2. For example, for the Routh table corresponding to the Kharitonov polynomial $p_{c,42}(s)$, positivity of the first column leads to the conditions

$$K_1 > \frac{K_2 - 3}{2} \qquad \text{and} \qquad 2K_1^2 - K_2^2 + K_1 K_2 + 9K_1 - 3K_2 + 9 > 0 \qquad (15.5\text{-}9)$$

It is not difficult to plot the set of the gains K_1 and K_2, satisfying condition (15.5-9). This set is shown in Figure 15.18, for the range $0 < K_1 < 70, 0 < K_2 < 70$. The set of stabilizing PI controllers for each of the remaining 15 Kharitonov plants can be obtained in a similar way. Then, the desired set of robustly stabilizing controllers for the interval plant family is obtained as the cross-section of the above 16 particular stabilizing sets. In our case, the set K of robust PI stabilizers of the form (15.5-8) is shown in Figure 15.19. This set is obviously nonempty, and the given interval plant family is robustly stabilizable. To stabilize the interval plant family, we can choose any pair (K_1, K_2) which belongs to the set K. For example, a robust PI stabilizer is given by

$$F(s) = 20 + \frac{10}{s}$$

Before closing this section, we point out that the sixteen-plant theorem can be extended to the more general class of compensators

$$F(s) = \frac{K(s-z)}{s^q(s-p)}, \qquad q > 1$$

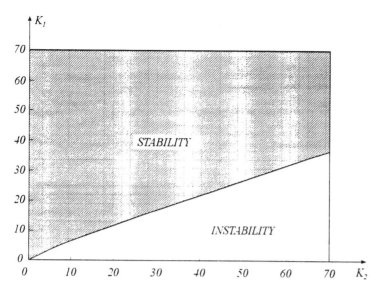

Figure 15.18   The set of stabilizing PI controllers for $G_{42}(s)$.

Unfortunately, the sixteen-plant theorem does not hold in the case of more general classes of compensators, or in the case of more general classes of interval plant families. In these cases, one may use other important results (such as the "sixty-four polynomial approach" or the "4^k polynomial approach") to characterize the stability of an interval plant family, by performing tests in a finite number of characteristic plants (see [4] for details). It should, however, be noted that in these situations, the issue of the computational effort needed to find such a compensator

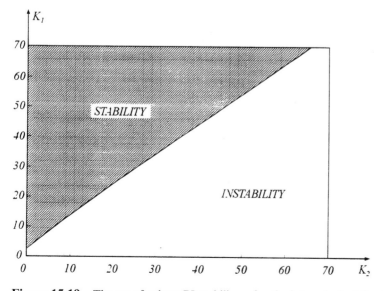

Figure 15.19   The set of robust PI stabilizers for the interval plant family $G(s, \mathbf{a}, \mathbf{b})$.

is of paramount importance, since the number of parameters entering $F(s)$ may be very large.

PROBLEMS

1. Consider the set of plants with parametric uncertainty given by relation (15.2-9). Show that the additive form of the uncertainty set (15.2-9) is given by

$$G_p(s) = G(s) + w_a(s)\Delta_a, \qquad |\Delta_a| \le 1$$

where

$$G(s) = \frac{G_0(s)}{1 + \tau_m s}, \qquad w_a(s) = -\frac{r_\tau \tau_m s}{(1 + \tau_m s)^2}$$

and where

$$\tau_m = \frac{\tau_{min} + \tau_{max}}{2}, \qquad r_\tau = \frac{\tau_{max} - \tau_{min}}{\tau_{max} + \tau_{min}}$$

2. Consider the set of plants with parameters uncertainty given by

$$P : G_p(s) = \frac{3(s+1)}{(as+1)(bs+1)}, \qquad a_{min} \le a \le a_{max}, \qquad b_{min} \le b \le b_{max}$$

Show that the above set of plants can be set in the following inverse multiplicative uncertainty form

$$P : G_p(s) = G(s)[1 + w_{im}(s)\Delta]^{-1}, \qquad |\Delta| \le 1$$

where

$$G(s) = \frac{s+1}{(a_m s + 1)(b_m s + 1) + r_a r_b a_m b_m s^2}$$

$$w_{im}(s) = \frac{[r_a a_m (b_m s + 1) + r_b b_m (a_m s + 1)]s}{(a_m s + 1)(b_m s + 1) + r_a r_b a_m b_m s^2}$$

and where

$$a_m = \frac{a_{min} + a_{max}}{2}, \qquad b_m = \frac{b_{min} + b_{max}}{2}$$

$$r_a = \frac{a_{max} - a_{min}}{a_{max} + a_{min}}, \qquad r_b = \frac{b_{max} - b_{min}}{b_{max} + b_{min}}$$

3. Consider the family of plants with parametric uncertainty given by

$$P : G_p(s) = \frac{3(as+1)}{(2s+1)(bs+1)^2}, \qquad 1 \le a \le 2, \qquad 2 \le b \le 3$$

Suppose that we want to obtain a multiplicative uncertainty description of the above family. Plot the smallest radius $\ell_m(\omega)$ and approximate it by a rational transfer function weight $w_m(s)$. Show that two good choices for the multiplicative uncertainty weight are

$$w_{\mathrm{m}}(s) = \frac{5s + 1}{2s + 3} \qquad \text{and} \qquad w_{\mathrm{m}}(s) = \frac{s^2 + 3s + 0.01}{0.7s^2 + 3s + 1}$$

Also, show that the approximation of $\ell_{\mathrm{m}}(\omega)$ by

$$w_{\mathrm{m}}(s) = \frac{5s + 1}{2s + 7} \qquad \text{or} \qquad w_{\mathrm{m}}(s) = \frac{s^2}{2s^2 + 3s + 1}$$

is not good enough to represent in the multiplicative uncertainty form the given family of plants.

4. Consider the uncertain feedback control system of Figure 15.5. Assume that the uncertain plant transfer function is given by

$$H_{\mathrm{p}}(s) = H(s)[1 + w_{\mathrm{m}}(s)\Delta_{\mathrm{m}}(s)]$$

where

$$H(s) = \frac{s + 2}{s^2 - 2s + 1} \qquad \text{and} \qquad w_{\mathrm{m}}(s) = \frac{3s + 1}{2s + 10}$$

while the controller $K(s)$ is a PI controller of the form

$$K_1(s) = 5 + \frac{10}{s}$$

Show that the closed-loop systems is robustly stable. Repeat the test with the PI controller of the form

$$K_2(s) = 5 - \frac{10}{s}$$

and verify that, in this case, the closed-loop system is not robustly stable.

5. Consider the feedback system of Figure 15.8. Assume that the uncertain plant transfer function is given by

$$H_{\mathrm{p}}(s) = H(s)[1 + w_{\mathrm{im}}(s)\Delta_{\mathrm{im}}(s)]^{-1}$$

where

$$H(s) = \frac{s + 2}{s^2 - 2s + 1} \qquad \text{and} \qquad w_{\mathrm{im}}(s) = \frac{s + 1}{2s + 5}$$

while the controller $K(s)$ is a PI controller of the form

$$K(s) = 3 + \frac{5}{s}$$

Show that the closed-loop system is robustly stable.

6. Consider the feedback system of Figure 15.10, with

$$H(s) = \frac{s + 3}{s^2 - 2s + 3} \qquad \text{and} \qquad K(s) = 2 + \frac{3}{s}$$

The design specifications for the closed-loop system are

(a) Steady-state tracking error $A = 0$
(b) Maximum peak magnitude $M = 2$ of the sensitivity function $S(s)$
(c) Minimum bandwidth $\omega_{\mathrm{B}}^* = 0.2 \, \mathrm{rad/sec}$

Show that the closed-loop system satisfies the above performance specifications.

7. Consider the feedback control system of Figure 15.13, with

$$H(s) = \frac{s+3}{s^2 - 2s + 3}, \qquad K(s) = 3 - \frac{4}{s}, \qquad w_{\mathrm{m}}(s) = \frac{3s+1}{2s+10},$$

$$w_{\mathrm{P}}(s) = \frac{s+0.4}{s}$$

Show that the closed-loop system does not satisfy the robust performance specification. Show that the same is true if the controller has the form $K(s) = 7$ (i.e., the form of a constant gain controller). Repeat the test with the controller $K(s) = 1 + 5/s$, and verify that in this case the closed-loop system meets the robust performance specification.

8. Consider the interval polynomial family with fixed degree

$$p(s, \mathbf{a}) = [3, 4.5]s^6 + [5, 8]s^5 + [6, 8]s^4 + [7, 11]s^3 + [4, 5]s^2 + [1, 5]s + [2, 13]$$

Construct the four Kharitonov polynomials related to this family. Is this polynomial family robustly stable? Repeat the same test with the interval polynomial family

$$p(s, \mathbf{a}) = [2, 7]s^5 + [8, 10]s^4 + [4, 7]s^3 + [4, 5]s^2 + [3, 5]s + [9, 11]$$

9. Consider the interval plant family

$$G(s, \mathbf{a}, \mathbf{b}) = \frac{[2, 3]s + [1, 3]}{s^2 + [2, 4]s + [1, 2.5]}$$

Show that the first-order compensator of the form

$$F(s) = \frac{s-1}{s-2}$$

cannot robustly stabilize the given interval plant family. Show that the entire family can be robustly stabilized by the use of a simple gain controller of the form

$$F(s) = K, \qquad K > 0$$

10. Consider the interval plant family

$$G(s, \mathbf{a}, \mathbf{b}) = \frac{[1, 2]s + [2, 3]}{s^2 + [2, 5]s + [2, 6]}$$

Let a first-order compensator of the form

$$F(s) = \frac{s + K_1}{s + K_2}$$

be connected to the given plant family, as suggested in Figure 15.17. Find the ranges of K_1 and K_2 for which the closed-loop system is robustly stable. Show, by using any two-variable graphic (e.g., the version provided by Matlab), that the range of K_1 and K_2 has the form shown in Figure 15.20.

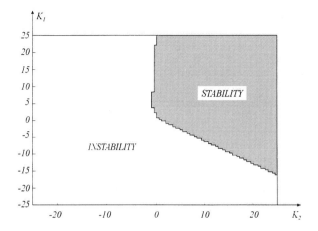

Figure 15.20   The set of robust stabilizers for Problem 10.

REFERENCES

Books

1. BR Barmish. New Tools for Robustness of Linear Systems. New York: MacMillan, 1994.
2. HW Bode. Network Analysis and Feedback Amplifier Design. Princeton, New Jersey: Van Nostrand, 1945.
3. FM Callier, CA Desoer. Multivariable Feedback Systems. New York: Springer-Verlag, 1982.
4. TE Djaferis. Robust Control Design: A Polynomial Approach. Boston, MA: Kluwer Academic Publishers, 1995.
5. BA Francis. A Course in H_∞ Control. Berlin: Springer-Verlag, 1987.
6. AGJ McFarlane (ed). Frequency Response Methods in Control Systems. New York: IEEE Press, 1979.
7. HH Rosenbrock. State Space and Multivariable Theory. London: Nelson, 1970.
8. HH Rosenbrock. Computer-Aided Control System Design. New York: Academic Press, 1974.
9. M Vidyasagar. Control System Synthesis—A Factorization Approach. Cambridge, MA: MIT Press, 1985.
10. N Wiener. Extrapolation, Interpolation and Smoothing of Stationary Time Series with Engineering Applications. New York: Wiley, 1949.

Articles

11. MA Athans. The role and the use of the stochastic linear-quadratic-gaussian problem. IEEE Trans Automatic Control AC-16:529–552, 1971.
12. BC Chang, JB Pearson. Optimal disturbance rejection in linear multivariable systems. IEEE Trans Automatic Control AC-29:880–887, 1984.
13. PH Delsarte, Y Genin, Y Kamp. The Nevanlinna–Pick problem in circuit and system theory. Int J Circuit Th Appl 9:177–187, 1981.
14. CA Desoer, RW Liu, J Murrary, R Saeks. Feedback system design: the fractional representation approach to analysis and synthesis. IEEE Trans Automatic Control AC-25:339–412, 1980.

15. JC Doyle, K Glover, PP Khargonekar, BA Francis. State-space solutions to standard H_2 and H_∞ control problems. IEEE Trans Automatic Control AC-34:831–847, 1989.

16. JC Doyle, G Stein. Multivariable feedback design: concepts for a classical modern synthesis. IEEE Trans Automatic Control AC-26:4–16, 1981.

17. BA Francis, JW Helton, G Zames. H_∞ optimal controllers for linear multivariable systems. IEEE Trans Automatic Control AC-29:888–920, 1984.

18. BA Francis, G Zames. On H_∞ optimal sensitivity theory for SISO systems. IEEE Trans Automatic Control AC-29:9–16, 1984.

19. K Glover. All optimal Hankel norm approximation of linear multivariable systems and their L_∞ error bounds. Int J Control 39:1115–1193, 1984.

20. JW Helton. Worst case analysis in the frequency domain: the H_∞ approach to control. IEEE Trans Automatic Control AC-30:1154–1170, 1985.

21. RE Kalman. Contribution to the theory of optimal control. Bol Socied Mathematica Mexicana, pp 102–119, 1960.

22. RE Kalman. Mathematical description of linear dynamical systems. SIAM J Control 1:151–192, 1963.

23. RE Kalman. When is a linear system optimal. ASME Trans Series D, J Basic Eng 86:51–60, 1964.

24. RE Kalman. Irreducible realizations and the degree of rational matrices. SIAM J Appl Math 13:520–544, 1965.

25. H Kimura. Robust stabilization of a class of transfer functions. IEEE Trans Automatic Control AC-29:778–793, 1984.

26. AN Kolmogorov. Interpolation and extrapolation of stationary random sequences. Bull Acad Sci USSR Vol 5, 1941.

27. N Nyquist. Regenerative theory. Bell Syst Tech J January issue, 1932.

28. DC Youla, JJ Bongiorno, CN Lu. Modern Wiener–Hopf design of optimal controllers. Part I: The single input case. IEEE Trans Automatic Control AC-21:3–14, 1976.

29. DC Youla, HA Jabar, JJ Bongiorno. Modern Wiener–Hopf design of optimal controllers, Part II. The multivariable case. IEEE Trans Automatic Control AC-21:319–338, 1976.

30. M Vidyasagar, H Kimura. Robust controllers for uncertain linear multivariable systems. Automatica 22:85–94, 1986.

31. G Zames. Feedback and optimal sensitivity: model reference transformations, multiplicative seminorms and approximate inverses. IEEE Trans Automatic Control AC-26:301–320, 1981.

32. G Zames, BA Francis. Feedback, minimax sensitivity and optimal robustness. IEEE Trans Automatic Control AC-28:585–601, 1983.

16

Fuzzy Control

16.1 INTRODUCTION TO INTELLIGENT CONTROL

In the last two decades, a new approach to control has gained considerable attention. This new approach is called *intelligent control* (to distinguish it from *conventional* or traditional control) [1]. The term conventional control refers to theories and methods that are employed to control dynamic systems whose behavior is primarily described by differential and difference equations. Thus, all the well-known classical and state-space techniques in this book fall into this category.

The term "intelligent control" has a more general meaning and addresses more general control problems. That is, it may refer to systems which cannot be adequately described by a differential/difference equations framework but require other mathematical models, as for example, discrete event system models. More often, it treats control problems, where a qualitative model is available and the control strategy is formulated and executed on the basis of a set of linguistic rules [2, 5, 10, 12, 14, 16, 33–36]. Overall, intelligent control techniques can be applied to ordinary systems and more important to systems whose complexity defies conventional control methods.

There are three basic approaches to intelligent control: *knowledge-based expert systems, fuzzy logic, and neural networks*. All three approaches are interesting and very promising areas of research and development. In this book, we present only the fuzzy logic approach. For the interested reader, we suggest references [6] and [9] for knowledge-based systems and neural networks.

The fuzzy control approach has been studied intensively in the last two decades and many important theoretical, as well as practical, results have been reported. The fuzzy controller is based on fuzzy logic. Fuzzy logic was first introduced by Zadeh in 1965 [33], whereas the first fuzzy logic controller was implemented by Mamdani in 1974 [17]. Today, fuzzy control applications cover a variety of practical systems, such as the control of cement kilns [3, 19], train operation [32], parking control of a car [27], heat exchanger [4], robots [23], and are in many other systems, such as home appliances, video cameras, elevators, aerospace, etc.

In this chapter, a brief introduction to fuzzy control is presented (see also Chap. 11 in [11]). This material aims to give the reader the heuristics of this approach to control, which may be quite useful in many practical control problems, but treats the theoretical aspects in an introductory manner only. Furthermore, we hope that this material will inspire further investigation, not only in the area of fuzzy control but also in the more general area of intelligent control. For further reading on the subject of fuzzy control, see the books [1–16] and articles [17–36] cited in the Bibliography.

16.2 GENERAL REMARKS ON FUZZY CONTROL

A principal characteristic of fuzzy control is that it works with linguistic rules (such as "if the temperature is high then increase cooling") rather than with mathematical models and functional relationships. With conventional control, the decisions made by a controller are a rigid "true" or "false." Fuzzy control uses fuzzy logic, which is much closer in spirit to human thinking and natural language than conventional control systems. Furthermore, fuzzy logic facilitates the computer implementation of imprecise (fuzzy) statements.

Fuzzy logic provides an effective means of capturing the approximate and inexact nature of the real world. To put it simply, the basic idea in fuzzy logic, instead of specifying a truth or falsehood, 0 or 1, etc., is to exert a gradual transition depending on the circumstances. For example, an air conditioning unit using conventional control recognizes room temperature only as warm, when the temperature is greater than 21°C, and cold, when the temperature is less than 21°C. Using fuzzy control, room temperature can be recognized as cold, cool, comfortable, warm, or hot and, furthermore, if this temperature is increasing or decreasing. On the basis of these fuzzy variables, a fuzzy controller makes its decision on how to cool the room.

In Figure 16.1a–c the fuzzy notion of cold, hot, and comfortable are presented in graphical form. The magnitude of these graphical representations lies between 0 and 1. The whole domain of fuzzy variables referring to the notion of temperature may be constructed by adding other variables such as cool, warm, etc., as shown in Figure 16.1d.

16.3 FUZZY SETS

A nonfuzzy set (or class) is any collection of items (or elements or members) which can be treated as a whole. Consider the following examples:

1. The set of all positive integers less than 11. This is a finite set of 10 members, i.e., the numbers $1, 2, 3, \ldots, 9, 10$. This set is written as $\{1, 2, \ldots, 9, 10\}$.
2. The set of all positive integers greater than 4. This set has an infinite number of members and can be written as $x > 4$.
3. The set of all humans having four eyes. This set does not have any members and is called an empty (or null) set.

In contrast to nonfuzzy (or crisp) sets, in a fuzzy set there is no precise criterion for membership. Consider for example the set *middle-aged* people. What are the members of this set? Of course, babies or 100-year-old people are not middle-aged people! One may argue that people from 40 to 60 appear to be in the set of middle-aged people! This may not, however, hold true for all people, in all places and at all

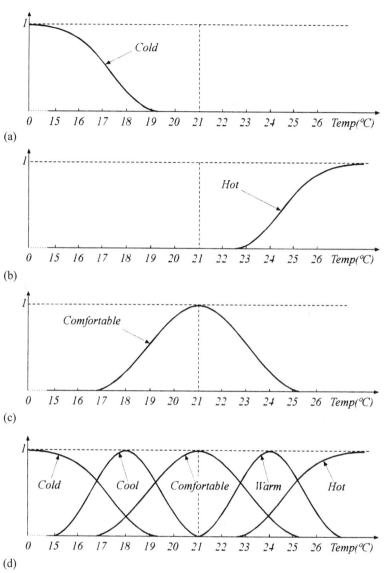

Figure 16.1 The fuzzy notion of the variable temperature.

times. For example, centuries ago, in most countries, the mean life expectancy was around 50 (this is true today for certain underdeveloped countries). We may now ask: are the ages 32, 36, 38, 55, 58, 60, and 65 members of the set of middle-aged people? The answer is that the set "middle-aged people" is a fuzzy set, where there is no precise criterion for membership and depends on time, place, circumstances, on the subjective point of view, etc. Other examples of fuzzy sets are intelligent people, tall people, strong feelings, strong winds, bad weather, feeling ill, etc.

To distinguish between members of a fuzzy set which are more probable than those which are less probable in belonging to the set, we use the *grade of membership*, denoted by μ, which lies in the range of 0 to 1, meaning that as μ gets closer to 1, the grade of membership becomes higher. If $\mu = 1$, then it is certainly a member and, of course, if $\mu = 0$, then it is certainly not a member.

The elements of a fuzzy set are taken from a *universe*. The universe contains all elements. Consider the examples:

(a) The set of numbers from 1 to 1000. The elements are taken from the universe of all numbers.

(b) The set of tall people. The elements are taken from the universe of all people.

If x is an element of a fuzzy set, then the associated grade of x with its fuzzy set is described via a *membership function*, denoted by $\mu(x)$. There are two methods for defining fuzzy sets, depending on whether the universe of discourse is discrete or continuous. In the discrete case, the grade of membership function of a fuzzy set is represented as a vector whose dimension depends on the degree of discretization. An example of a discrete membership function, referring to the fuzzy set *middle* $= (0.6/30, 0.8/40, 1/50, 0.8/60, 0.6/70)$, where the universe of discourse represents the ages $[0, 100]$. This is a bell-shaped membership function. In the continuous case, a functional definition expresses the membership function of a fuzzy set in a functional form. Some typical examples of continuous membership functions are given below:

$$\mu(x) = \exp\left[-\frac{(x - x_0)^2}{2\sigma^2}\right] \tag{16.3-1}$$

$$\mu(x) = \left[1 + \left(\frac{x - x_0}{\sigma}\right)^2\right]^{-1} \tag{16.3-2}$$

$$\mu(x) = 1 - \exp\left[-\left(\frac{\sigma}{x_0 - x}\right)^\zeta\right] \tag{16.3-3}$$

where x_0 is the point where $\mu(x)$ is maximum (i.e., $\mu(x_0) = 1$) and σ is the standard deviation. Expression (16.3-1) is the well-known standard Gaussian curve. In expression (16.3-3) the exponent ζ shapes the gradient of the sloping sides.

To facilitate our understanding further, we refer to Figure 16.2, where the very simple case of the fuzzy and nonfuzzy interpretation of an old man is given. In the nonfuzzy or crisp case, everyone older than 70 is old, whereas in the fuzzy case the transition is gradual. This graphical presentation reveals the distinct difference between fuzzy and nonfuzzy (or crisp) sets. Next, consider the membership functions given in Figure 16.3. Figure 16.3a presents the membership function $\mu(x)$, defined as follows:

$$\mu(x) = \begin{cases} 1, & 6 \leq x \leq 10 \\ 0, & \text{otherwise} \end{cases} \tag{16.3-4}$$

The nonfuzzy (or crisp) membership function is unique. The corresponding fuzzy membership function may have several forms: triangular (16.3b), bell-shaped curve (16.3c), trapezoidal or flattened bell-shaped (16.3d), etc.

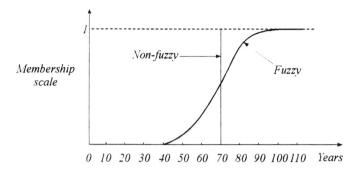

Figure 16.2 The fuzzy and nonfuzzy interpretation of an old man. In the nonfuzzy case, anyone older than 70 is old; in the fuzzy case, the transition is gradual.

In fuzzy sets, the variable x may be algebraic, as in relations (16.3-1)–(16.3-3), or it may be a *linguistic variable*. A linguistic variable takes on words or sentences as values. This type of a value is called a *term set*. For example, let the variable x be the linguistic variable "age." Then, one may construct the following term: {very young, young, middle age, old, very old}. Note that each term in the set (e.g., young) is a fuzzy variable itself. Figure 16.4 shows three sets: young (Y), middle age (M), and old (O). Figure 16.5 shows four sets: young (Y), very young (VY), old (O), and very old (VO). We say that the sets "young" and "old" are *primary sets*, whereas the sets "very young" and "very old" are derived from them.

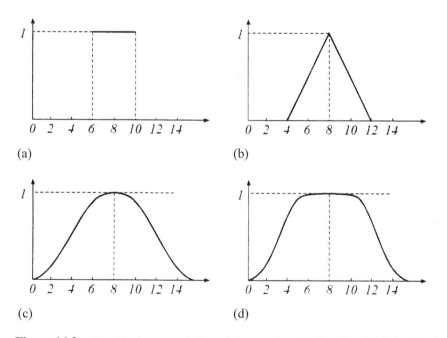

Figure 16.3 Graphical representation of the membership function (10.3-4): (a) crisp membership funciton; (b) triangular membership function; (c) bell-shaped membership function; (d) trapezoidal membership function.

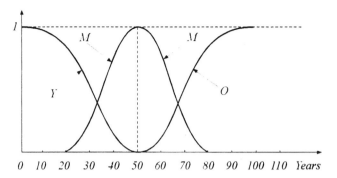

Figure 16.4 The membership curves of the three sets: young (Y), middle age (M), and old (O).

16.4 FUZZY CONTROLLERS

The nonfuzzy (crisp) PID controller has been presented in Chap. 9 (continuous-time) and in Chap. 12 (discrete-time), where it was pointed out that this type of controller has many practical merits and has become the most popular type of controller in industrial applications. The same arguments hold true for the fuzzy PID controller and, for this reason, we focus our attention on this controller. We will examine, in increasing order of complexity, the fuzzy proportional (FP), the fuzzy proportional–derivative (FPD), and the fuzzy proportional–derivative plus integral (FDP + I) controller.

1 The FP Controller

Let $e(k)$ be the input and $u(k)$ be the output of the controller, respectively. The input $e(k)$ is the error

$$e(k) = r(k) - y(k) \tag{16.4-1}$$

where $r(k)$ is the reference signal and $y(k)$ is the output of the system (see any closed-loop figure in the book, or Figure 16.9 in Sec. 16.5 that follows). Then

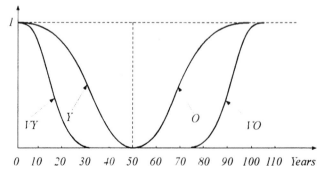

Figure 16.5 The membership curves of the four sets: very young (VY), young (Y), old (O), and very old (VO).

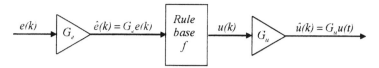

Figure 16.6   The fuzzy proportional (FP) controller.

$$u(k) = f(e(k)) \tag{16.4-2}$$

that is, the output of the controller is a nonlinear function of $e(k)$. A simplified diagram of Eq. (16.4-2) is given in Figure 16.6. In comparison with Figure 12.32 of Chap. 12, where only one tuning parameter appears (the parameter Kp), for FP controllers we have two tuning parameters, namely the parameters G_e and G_u, which are the error and controller output gains, respectively. The block designated as "rule base" is the heart of the fuzzy controller whose function is explained in Secs 16.5 and 16.6 below.

2 The FPD Controller

Let $ce(k)$ denote the change in the error (for continuous-time systems, $ce(k)$ corresponds to the derivative of the error de/dt). An approximation to $ce(t)$ is given by

$$ce(k) = \frac{e(k) - e(k-1)}{T} \tag{16.4-3}$$

where T is the sampling time. The block diagram of the FPD controller is given in Figure 16.7. Here, the output of the controller is a nonlinear function of two variables, namely the variables $e(k)$ and $ce(k)$, i.e.,

$$u(k) = f(e(k), ce(k)) \tag{16.4-4}$$

Note that here we have three tuning gains (G_e, G_{ce}, and G_u), as compared with the crisp PI controller, which has only two (see Sec. 12.10).

3 The PFD+I Controller

It has been shown that it is not straightforward to write rules regarding integral action. Furthermore, the rule base involving three control actions (proportional, derivative, and integral) simultaneously becomes very large. To circumvent these difficulties, we separate the integral action from the other two actions, resulting in an FPD + I controller, as shown in Figure 16.8. In the present case the output of the

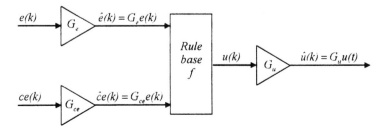

Figure 16.7   The fuzzy proportional–derivative (FPD) controller.

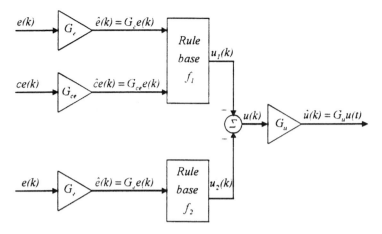

Figure 16.8 The fuzzy proportional–derivative and integral (FPD + I) controller.

controller $u(k)$ is a nonlinear function of three variables, namely the variables $e(k)$, $ce(k)$, and $ie(k)$, i.e.,

$$u(k) = f(e(k), ce(k), ie(k)) \qquad (16.4\text{-}5)$$

where $ie(k)$ denotes the integral of the error. Since the integral action has been separated from the proprotional and derivative actions, relation (16.4-5) breaks down to two terms:

$$u(k) = u_1(k) + u_2(k) = f_1(e(k), ce(k)) + f_2(ie(k)) \qquad (16.4\text{-}6)$$

Note that here we have four tuning parameters (G_e, G_{ce}, G_{ie}, and G_u) as compared with the crisp PID controller of Sec. 12.10, which has only three.

16.5 ELEMENTS OF A FUZZY CONTROLLER

A simplified block diagram of a fuzzy controller incorporated in a closed-loop system is shown in Figure 16.9. The fuzzy controller involves four basic operations, namely the *fuzzification interface*, the *rule base*, the *inference engine*, and the *defuz-*

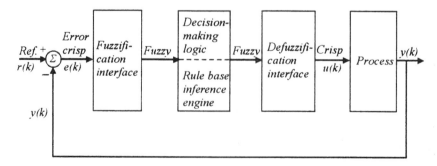

Figure 16.9 Basic configuration of a closed-loop system involving an FLC, wherein the fuzzy and crisp data flow is identified.

zification interface. A brief explanation of these four elements is given below. A more detailed explanation is given in the next four sections.

1 Fuzzification Interface

Here, the crisp error signal $e(k)$ is converted into a suitable linguistic fuzzy set.

2 Rule Base

The rule base is the heart of a fuzzy controller, since the control strategy used to control the closed-loop system is stored as a collection of *control rules*. For example, consider a controller with three inputs e_1, e_2, and e_3 and output u. Then, a typical control rule has the form

$$\text{if } e_1 \text{ is } A, e_2 \text{ is } B, \text{ and } e_3 \text{ is } C, \text{ then } u \text{ is } D \qquad (16.5\text{-}1)$$

where A, B, C, and D are linguistic terms, such as very low, very high, medium, etc. The control rule (16.5-1) is composed of two parts: the "if" part and the "then" part. The "if" part is the input to the controller and the "then" part is the output of the controller. The "if" part is called the *premise* (or *antecedent* or *conditon*) and the "then" part is called the *consequence* (or *action*).

3 Inference Engine

The basic operation of the interference engine is that it "infers," i.e., it deduces (from evidence or data) a logical conclusion. Consider the following example described by the logical rule, known as *modus ponens*:

Premise 1: If an animal is a cat, then it has four legs.
Premise 2: My pet is a cat.

Conclusion: My pet has four legs.

Here, premise 1 is the base rule, premise 2 is the fact (or the evidence or the data), and the conclusion is the consequence.

The inference engine is a program that uses the rule base and the input data of the controller to draw the conclusion, very much in the manner shown by the above *modus ponens* rule. The conclusion of the inference engine is the fuzzy output of the controller, which subsequently becomes the input to the defuzzification interface.

4 Defuzzification Interface

In this last operation, the fuzzy conclusion of the inference engine is defuzzified, i.e., it is converted into a crisp signal. This last signal is the final product of the fuzzy logic controller (FLC), which is, of course, the crisp control signal to the process.

The above four operations are explained in greater detail in the sections that follow.

16.6   FUZZIFICATION

The fuzzification procedure consists of finding appropriate membership functions to describe crisp data. For example, let speed be a linguistic variable. Then the set $T(\text{speed})$ could be

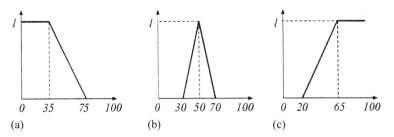

Figure 16.10   The membership function for the term set T(speed) = {slow, medium, fast}. (a) Slow speed (shouldered); (b) medium speed (triangular); (c) high speed (shouldered).

$$T(\text{speed}) = \{\text{slow, medium, fast}\} \tag{16.6-1}$$

On a scale from 0 to 100, slow speed may be up to 35, medium speed could be from 30 to 70, and high speed could be from 65 to 100. The membership functions for the three fuzzy variables may have several shapes. Figure 16.10 shows some membership functions for each of the three fuzzy variables.

Other examples of fuzzification of crisp data have already been presented in Figures 16.1, 16.2, 16.3b–d, 16.4 and 16.5.

16.7   THE RULE BASE

The most usual source for constructing linguistic control rules are human experts. We start by questioning experts or operators using a carefully prepared questionaire. Using their answers, a collection of if–then rules is established. These rules contain all the information regarding the control of the process. Note that there are other types of sources for constructing the rule base, such as control engineering knowledge, fuzzy models, etc. [20].

The linguistic control rules are usually presented to the end-user in different formats. One such format has the verbal form of Table 16.1 which refers to the two-input one-output controller of Figure 16.7 for the control of the temperature of a room. Here, the controller inputs e and ce refer to the error and change in error, respectively, whereas the variable u refers to the output of the controller. This format involves the following five fuzzy sets: zero (Z), small positive (SP), large positive (LP), small negative (SN), and large negative (LN). Clearly, the set of if–then rules presented in Table 16.1 is an example of a linguistic control strategy applied by the controller in order to maintain the room temperature close to the desired optimum value of 21°C.

In Figure 16.11 the *graphical representation* of the five fuzzy sets Z, SP, LP, SN, and LN is given. Using Figure 16.11, the *graphical forms* of the nine rules of Table 16.1 are presented in Figures 16.12 and 16.13.

16.8   THE INFERENCE ENGINE

The task of the inference engine is to deduce a logical conclusion, using the rule base. To illustrate how this is performed, we present three examples, in somewhat increasing order of complexity.

Table 16.1   Verbal Format of If–Then Rules

Rule 1	If $Z(e)$ and $Z(ce)$, then $Z(u)$
Rule 2	If $SP(e)$ and $Z(ce)$, then $SN(u)$
Rule 3	If $LP(e)$ and $Z(ce)$, then $LN(u)$
Rule 4	If $SN(e)$ and $Z(ce)$, then $SP(u)$
Rule 5	If $LN(e)$ and $Z(ce)$, then $LP(u)$
Rule 6	If $SP(e)$ and $SN(ce)$, then $Z(u)$
Rule 7	If $SN(e)$ and $SP(ce)$, then $Z(u)$
Rule 8	If $SP(e)$ and $SP(ce)$, then $LN(u)$
Rule 9	If $LP(e)$ and $LP(ce)$, then $LN(u)$

Example 16.8.1

Consider a simple one rule fuzzy controller, having the following rule:

Rule: If e_1 is slow and e_2 is fast, then u is medium

The graphical representation of the rule involving the membership functions of the three members slow, fast, and medium is given in Figure 16.14a. Determine the fuzzy control u.

Solution

To determine the fuzzy control u, we distinguish the following two steps:

Step 1

Consider the particular time instant k. For this time instant, let the fuzzy variable e_1 have the value 25 and the fuzzy variable e_2 the value 65, both on the scale 0–100. Through these points, two vertical lines are drawn, one for each column, intersecting the fuzzy sets e_1 and e_2 (Figure 16.14b). Each of these two *intersection points* (also called *triggering points*) has a particular μ, denoted as $\mu_{e_i}^k$, $i = 1, 2$. This results in the following:

First column (fuzzy variable e_1): $\mu_{e_1}^k = 0.5$
Second column (fuzzy variable e_2): $\mu_{e_2}^k = 0.2$

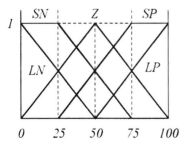

Figure 16.11   Graphical representation of the five fuzzy sets: zero (Z), small positive (SP), large positive (LP), small negative (SN), and large negative (LN).

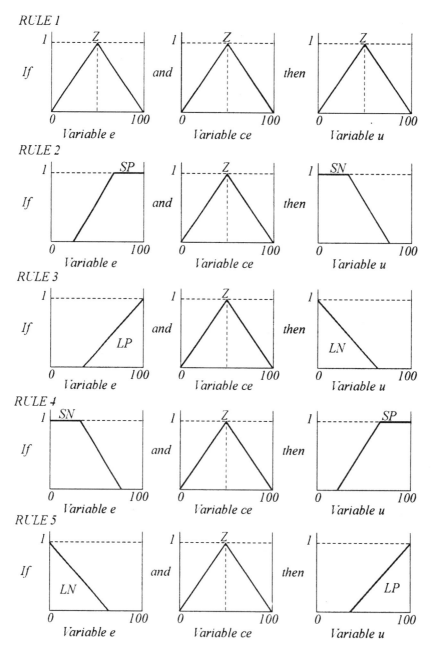

Figure 16.12 Graphical forms of rules 1–5 of Table 16.1.

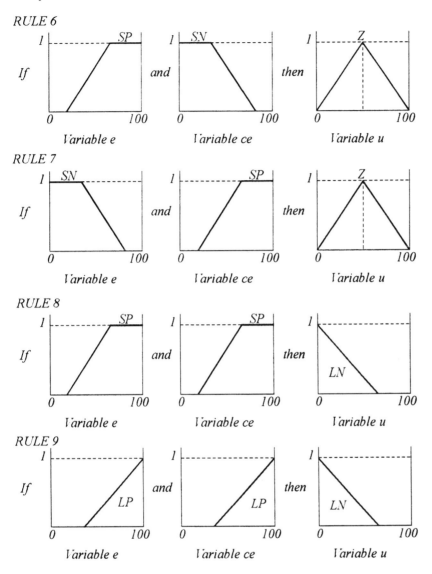

Figure 16.13 Graphical forms of the rules 6–9 of Table 16.1.

Next, determine the value of s_1^k, defined as follows:

$$s_1^k = \min\left\{\mu_{e_1}^k, \mu_{e_2}^k\right\} = \min\{0.5, 0.2\} = 0.2 \tag{16.8-1}$$

This completes the first step, i.e., the determination of the s_1^k. Note that s_1^k is related only to the "if" part of the rule. This step is depicted in Figure 16.14b. Clearly, when $e_1, e_2, \ldots, e_n$ variables are involved in the "if" part of the rule and there is a total of r rules, then

$$s_p^k = \min\left\{\mu_{e_1}^k, \mu_{e_2}^k, \ldots, \mu_{e_n}^k\right\}, \qquad p = 1, 2, \ldots, r \tag{16.8-2}$$

where p indicates the particular rule under consideration.

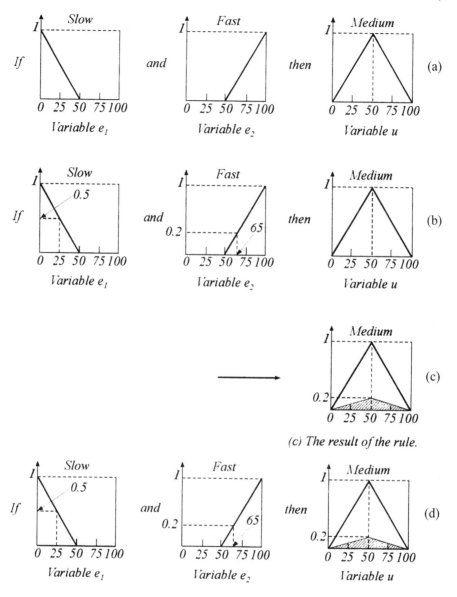

Figure 16.14 The inference procedure for Example 16.8.1. (a) Graphical representation of membership functions; (b) determination of the triggering points and of the $\mu_{\min}$; (c) the result of the rule; (d) compact representation of the three figures (a), (b), and (c).

Step 2

The second step is the most important step in the inference engine, since it deduces the *result of the rule* for the particular instant of time k. One way to deduce this result is to multiply the fuzzy variable u (third column) by s_1^k. The resulting curve is the fuzzy control sought, and constitutes the *result of the rule*. This curve is the shaded area depicted in Figure 16.14c.

To state this procedure more formally, let $\mu_1(u)$ and $\lambda_1^k(u)$ denote the membership functions of the given output fuzzy set u (first row, last column of Figure 16.14) and of the curve depicted in Figure 16.14c, respectively. Then, $\lambda_1^k(u)$ is given by the following expression:

$$\lambda_1^k(u) = \left[\min\left\{\mu_{e_1}^k, \mu_{e_2}^k\right\}\right]\left[\mu_1(u)\right] = \left[s_1^k\right]\left[\mu_1(u)\right] \tag{16.8-3}$$

For the general case, where n variables are involved in the "if" part of the rule and there is a total of r rules, we have the following expression:

$$\lambda_p^k(u) = \left[\min\left\{\mu_{e_1}^k, \ldots, \mu_{e_n}^k\right\}\right]\left[\mu_p(u)\right] = \left[s_p^k\right]\left[\mu_p(u)\right], \qquad p = 1, 2, \ldots, r \tag{16.8-4}$$

where p indicates the particular rule under consideration.

In practice, the above two steps are presented compactly, as shown in Figure 16.14d. Clearly, the two steps are repeated for all desirable instants of time k in order to construct u for the particular time interval of interest.

Example 16.8.2

Consider a fuzzy controller that is to apply a control strategy described by the following three if–then rules:

Rule 1: If e_1 is negative and e_2 is negative, then u is negative
Rule 2: if e_1 is zero and e_2 is zero, then u is zero
Rule 3: If e_1 is positive and e_2 is positive, then u is positive.

The graphical representation of the three rules involving the membership functions of the three fuzzy members negative, zero, and positive is given in Figure 16.15. (To facilitate the presentation of the method, the members positive and negative are actually crisp. A fuzzy presentation is given in Figure 16.24 of Problem 2 of Sec. 16.12.). Determine the fuzzy control u.

Solution

Making use of the results of Example 16.8.1, we carry out the first two steps, as follows:

Step 1

Consider the particular time instant k. For this time instant, let the fuzzy variable e_1 have the value 25 and the fuzzy variable e_2 also have the value 25, all in the scale 0–100. Through these points, two vertical lines are drawn, one for each column. These vertical lines intersect the fuzzy curves at different *triggering points*, having a particular μ. This results in the following:

First column: in rule 1, $\mu_{e_1}^k = 1$; in rule 2, $\mu_{e_1}^k = 0.5$; and in rule 3, $\mu_{e_1}^k = 0$
Second column: in rule 1, $\mu_{e_2}^k = 1$; in rule 2, $\mu_{e_2}^k = 0.5$; and in rule 3, $\mu_{e_2}^k = 0$

Next, determine, s_p^k, $p = 1, 2, 3$, using definition (16.8-2) to yield:

For rule 1: $s_1^k = \min\left\{\mu_{e_1}^k, \mu_{e_2}^k\right\} = \min\{1, 1\} = 1$
For rule 2: $s_2^k = \min\left\{\mu_{e_1}^k, \mu_{e_2}^k\right\} = \min\{0.5, 0.5\} = 0.5$
For rule 3: $s_3^k = \min\left\{\mu_{e_1}^k, \mu_{e_2}^k\right\} = \min\{0, 0\} = 0$

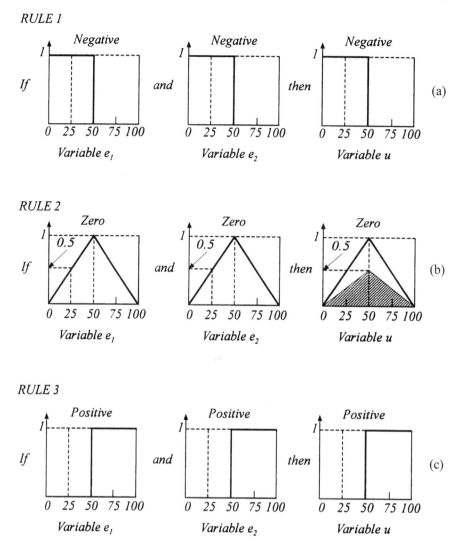

Figure 16.15   The inference procedure for Example 16.8.2. (a) Graphical representation of rule 1; (b) graphical representation of rule 2; (c) graphical representation of rule 3.

It is evident that only rules 1 and 2 have nonzero contribution on u, while rule 3 plays no part in the final value of u. Furthermore, rule 1 is seen to be dominant, while rule 2 plays a seconday role.

Step 2

Now multiply s_p^k of each rule with the corresponding curve of the third column, using definition (16.8-4). The result of this product is a curve (shaded area) for each of the three variables shown in the third column.

Since the present example involves more than one rule, as compared with Example 16.8.1, the following extra step is needed.

Step 3

The above procedure is the implementation of each one of the rules 1, 2, and 3 via their corresponding fuzzy curves of each of the elements appearing in each rule. The next step is to take the *result of each rule*, which is the shaded area in the third column, and "unite" them together, as shown in Figure 16.16a. A popular method to construct the shaded area of Figure 16.16a is as follows. We take the *union* of the three shaded areas of the third column of Figure 16.15. Then, the actual control of u is the set (envelope) of these "united" (superimposed) areas, shown in Figure 16.16a.

Strictly speaking, the term "union" refers to two or more sets and is defined as the maximum of the corresponding values of the membership functions. More specifically, if we let $\lambda^k(u)$ be the membership function of the fuzzy control u depicted in Figure 16.16, then $\lambda^k(u)$ is evaluated as follows:

$$\lambda^k(u) = \max\left\{\lambda_1^k(u), \lambda_2^k(u), \lambda_3^k(u)\right\} \tag{16.8-5}$$

where $\lambda_1^k(u)$, $\lambda_2^k(u)$, and $\lambda_3^k(u)$ are the membership functions of the *result of each rule*. For the general case of r rules, Eq. (16.8-5) becomes

$$\lambda^k(u) = \max\left\{\lambda_1^k(u), \ldots, \lambda_r^k(u)\right\} \tag{16.8-6}$$

Now consider another instant of time k: let both fuzzy variables in the first and second column of Figure 16.15 have the value 75 on a scale from 0 to 100. Then, following the same procedure, one may similarly construct the corresponding third column of Figure 16.15. The resulting u for this second case is the shaded area in Figure 16.16b.

Example 16.8.3

Consider a fuzzy controller that is to apply a control strategy described by the following two if–then rules:

Rule 1: If e_1 is positive and e_2 is zero, then u is negative.
Rule 2: If e_1 is zero and e_2 is zero, then u is zero.

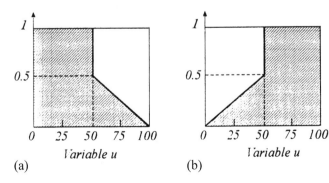

(a)

(b)

Figure 16.16 Fuzzy control signal graphical construction for Example 16.8.2. (a) Actual control signal u when triggering at 25; (b) actual control signal u when triggering at 75.

The graphical representation of the two rules involving the membership functions of the three fuzzy members positive, zero, and negative is given in Figure 16.17. Determine the fuzzy control u.

Solution

Using the results of examples 16.8.1 and 16.8.2, we have the following steps.

Step 1

Consider the particular time instant k. For this time instant, let the fuzzy variable e_1 have the value of 60 and the fuzzy variable e_2 have the value of 25. Through these points, two vertical lines are drawn, one for each column. These vertical lines intersect the fuzzy curves at different triggering points, having a particular μ. This results in the following:

First column: in rule 1, $\mu_{e_1}^k = 0.75$ and in rule 2, $\mu_{e_1}^k = 0.5$

Second column: in rule 1, $\mu_{e_2}^k = 0.4$ and in rule 2, $\mu_{e_2}^k = 0.5$

Next, determine s_p^k, $p = 1, 2$, using definition (16.8-2) to yield:

For rule 1: $s_1^k = \min\{\mu_{e_1}^k, \mu_{e_2}^k\} = \min(0.75, 0.4\} = 0.4$

For rule 2: $s_2^k = \min\{\mu_{e_1}^k, \mu_{e_2}^k\} = \min\{0.5, 0.5\} = 0.5$

Step 2

Multiply s_p^k of each rule with the corresponding curve of the third column, according to definition (16.8-4). The resulting curves are the fuzzy curves of the output u (third column, shaded areas).

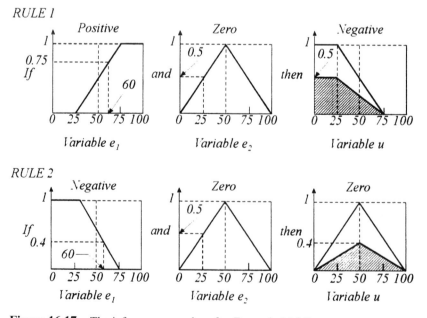

Figure 16.17 The inference procedure for Example 16.8.3.

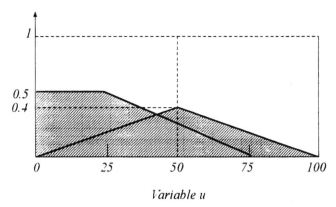

Figure 16.18 Fuzzy control signal graphical construction for Example 16.8.3.

Step 3

Using definition (16.8-6), the envelope of the actual fuzzy control *u* is constructed by superimposing the two shaded areas of the third column to yield the curve shown in Figure 16.18.

Remark 16.8.1

We may now make the following remark, regarding the overall philosophy of an FLC. To estimate the fuzzy control signal at each instant of time *k*, the FLC works as follows. Each rule contributres an "area" (i.e., the shaded areas in the last column in Figure 16.14 or 16.15 or 16.17). This area describes the output *u* of the controller as a fuzzy set. All these areas are subsequently superimposed in the manner explained above (see Figure 16.16 or 16.18), to give the fuzzy set of the ouput *u*. The envelope of this total area is the *final conclusion* of the interference engine for the instant of time *k*, deduced using the rule base. One can conclude, therefore, that the end product of the inference engine, given in Figure 16.16 or 16.18, is a rule base result, where *all* rules are *simultaneously* taken into consideration.

The very last action in an FLC is, by using Figure 16.16 or 16.18, to determine the crisp values for the control signal, which will serve as an input to the process. This is defuzzification, which is explained below.

16.9 DEFUZZIFICATION

There are several methods for defuzzification. A rather simple method is the center of area method, which is defined as follows:

$$u = \frac{\sum_i \mu(x_i)x_i}{\sum_i \mu(x_i)} \tag{16.9-1}$$

or

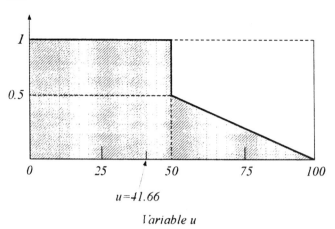

Figure 16.19   Defuzzified value of control signal.

$$u = \frac{\int \mu(x) d \, dx}{\int \mu(x) \, dx} \qquad (16.9\text{-}2)$$

where u is the crisp function sought, x_i is the member of the set, and $\mu(x_i)$ is the associated membership function. Clearly, expressions (16.9-1) and (16.9-2) correspond to the discrete- and continuous-time cases, respectively.

Example 16.9.1

Consider the shaded area in Figure 16.16a. Calculate the center of area of this shaded area using Eq. (16.9-1).

Solution

We have:

$$\sum_i \mu(x_i)x_i = \text{The shaded area in Figure 16.16a}$$

$$= (50) + (0.5)(0.5)(50) = 50 + 12.5 = 62.5$$

$$\sum_i \mu(x_i) = 1 + 0.5 = 1.5$$

Hence $u = 62.5/1.5 = 41.66$. Therefore, the defuzzification procedure yields the crisp value of u, which is depicted in Figure 16.19.

There are other types of defuzzification, such as the mean of maximum, first of maxima, last of maxima, etc. For more information on these techniques see [9, 20].

16.10   PERFORMANCE ASSESSMENT

Up to now, no systematic procedures for the design of an FLC have been proposed (such as root locus, Nyquist plots, pole placement, stability tests, etc.). The basic difficulty in developing such procedures is the fact that the rule base has no mathematical description. As a consequence, it is not obvious how the rules and gains affect the overall performance of the closed-loop system.

The problem of stability of a closed-loop system incorporating an FLC essentially remains an unsolved question, even though increasing research results have appeared recently in the literature. For linear time-variant systems with a known transfer function or state-space model, if the describing function approach is applied and together with the Nyquist plot, one may reach some safe conclusions regarding the stability margins of the systems.

To evaluate the performance of FLCs a theoretical approach has been proposed, which yields a partial evaluation performance. This approach refers to the *integrity* of the rule base and aims at securing the accuracy of the rule base. One way to investigate integrity is to plot the input and output signal of the FLC controller. Comparison of those two waveforms provides some idea of the integrity of the rule base. Clearly, the objective of this investigation is to study the behavior of the control system and, if it is not satisfactory, to suggest improvements.

16.11 APPLICATION EXAMPLE: KILN CONTROL

The kilning process in the manufacturing of cement has attracted much attention from the control viewpoint and, indeed, was one of the first applications of fuzzy control in the process industry [3, 19, 25]. The kilning process is one of the most complex industrial processes to control and has, until the advent of fuzzy control, defied automatic control; However, human operators can successfully control this process using rules, the result of years of experience. These rules now form the basis for fuzzy control, and there are many successful applications worldwide.

Briefly, in the kilning process a blend of finely ground raw materials is fed into the upper end of a long, inclining, rotating cylinder and slowly flows to the lower end, while undergoing chemical transformation due to the high temperatures produced by a flame at the lower end. The resultant product, clinker, constitutes the major component of cement. A measure of the burning zone temperature at the lower end of the rotary kiln can be obtained indirectly by measuring the torque of the motor rotating the kiln, whereas a measure of the quality of the end product is its free lime content (FCAO). These two quantities (or process output measurements) are essential in specifying the fuel feed to the kiln (i.e., the control strategy).

The block diagram of Figure 16.20 is a simplified controller for the kilning process. There are two inputs e_1 and e_2 and one output u, defined as follows [25]:

e_1 = change in kiln torque drive (DELTQUE or ΔTQUE)
e_2 = free lime content (FCAO)
u = output fuel rate (DELFUEL or ΔFUEL)

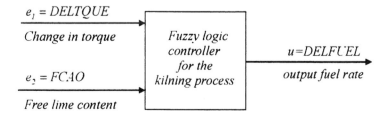

Figure 16.20 Simplified controller for the kilning process.

where Δ stands for change. The corresponding ranges of e_1, e_2, and u are $(-3, 0, 3)$, $(0.3, 0.9, 1.5)$, and $(-0.2, 0, 0.2)$, respectively, where the middle number indicates the center of the fuzzy membership function.

The rule base is composed of nine if–then rules, as shown in Table 16.2. The graphical representation of this rule base is shown in Figures 16.21 and 16.22, where each row represents a rule. The first column represents the membership function for the first input e_1 = change in kiln drive torque (DELTQUE or ΔTQUE), the second column the membership function of the second input e_2 = free lime content (FCAO), and the third column the membership function of the output u = output fuel rate (DELFUEL or ΔFUEL). Clearly, in this example we have three sets with their corresponding members as follows:

First set: ΔTQUE (ZERO (ZE), NEGATIVE (NE), POSITIVE (PO))
Second set: FCAO (LOW (LO), OK (OK), HIGH (HI))
Third set: ΔFUEL (LARGE POSITIVE (LP), MEDIUM POSITIVE (MP),
    SMALL POSITIVE (SP), NO CHANGE (NC), SMALL NEGATIVE (SN),
    MEDIUM NEGATIVE (MN), LARGE NEGATIVE (LN)).

To determine the fuzzy controller output at some particular instant k, assume that DELTQUE and FCAO are $e_1 = -1.2\%/\text{hr}$ and $e_2 = 0.54\%/\text{hr}$, respectively. Thus for the first controller input variable e_1 (corresponding to the first column) a vertical line centered at $-1.2\%/\text{hr}$ is drawn to intercept the fuzzy sets for the change in kiln drive torque for every rule. Likewise a vertical line, centered at $0.54\%/\text{hr}$, is drawn to intercept the fuzzy sets for the second input e_2, free lime, for every rule.

To obtain the final fuzzy output u of the controller, for this particular instant of time k, we follow the procedure presented in the examples of Sec. 16.8. We have:

Table 16.2   The Nine If–Then Rules for the Kiln Process FLC

Rule 1	If DELTQUE is zero and FCAO is low, then DELFUEL is medium negative
Rule 2	If DELTQUE is zero and FCAO is OK, then DELFUEL is zero
Rule 3	If DELTQUE is zero and FCAO is high, then DELFUEL is medium positive
Rule 4	If DELTQUE is negative and FCAO is low, then DELFUEL is small positive
Rule 5	If DELTQUE is negative and FCAO is OK, then DELFUEL is medium positive
Rule 6	If DELTQUE is negative and FCAO is high, then DELFUEL is large positive
Rule 7	If DELTQUE is positive and FCAO is low, then DELFUEL is large negative
Rule 8	If DELTQUE is positive and FCAO is OK, then DELFUEL is medium negative
Rule 9	If DELTQUE is positive and FCAO is high, then DELFUEL is small negative

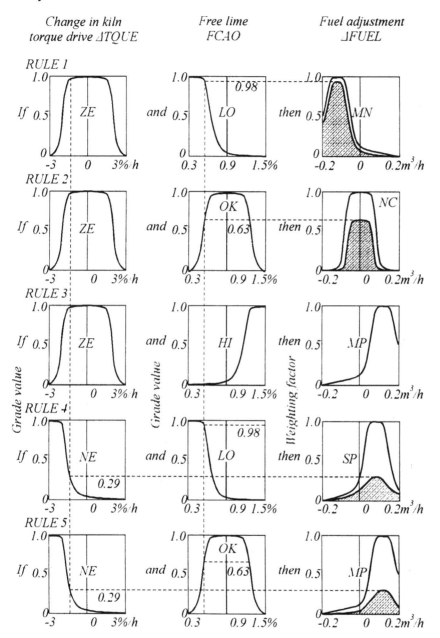

Figure 16.21 Graphical representation of the fuzzy logic interpretation of the control rules 1–5 (Table 16.2) for the kilning process.

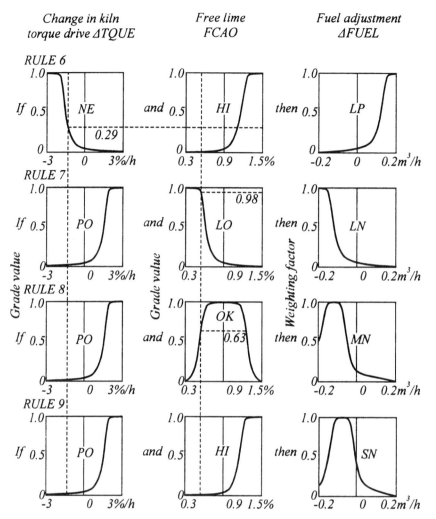

Figure 16.22 Graphical representation of the fuzzy logic interpetation of the control rules 6–9 (Table 16.2) for the kilning process. (Reproduced by the kind permission of FLS Automation, Denmark.)

Step 1

Using the relationship (16.8-2) for the case of a two-input controller with nine rules at the kth time instant, we have that for each rule

$$s_p^k = \min\left\{\mu_{e_1}^k, \mu_{e_2}^k\right\} = \min\left\{\mu_{\text{DLTQUE}}^k, \mu_{\text{FCAO}}^k\right\} \qquad (16.11\text{-}1)$$

As a result, the set of minima at this time instant, which is a measure of the strength or contribution of each rule on the final decision, is

$$\left\{s_1^k, s_2^k, s_3^k, s_4^k, s_5^k, s_6^k, s_7^k, s_8^k, s_9^k\right\} = \{0.98, 0.63, 0, 0.29, 0.29, 0, 0, 0, 0\}$$

It is evident here that only rules 1, 2, 4, and 5 have a nonzero contribution, the remainder playing no part in the final decision. Furthermore, rule 1 is seen to be dominant, while rule 2 has a significant contribution. In contrast, rules 4 and 5 have only a small contribution.

Step 2

To determine the contribution to u of each rule, i.e., to determine $\lambda_p^k(u), p = 1, 2, \ldots, 9$, we apply relation (16.8-4), i.e., the relation

$$\lambda_p^k(u) = \left[s_p^k\right]\left[\mu_p(u)\right] \tag{16.11-2}$$

As a result, the nine curves in the third column of Figures 16.21 and 16.22 are produced.

Step 3

To determine the final fuzzy control u, *simultaneously* taking into account all nine rules, we make use of Eq. (16.8-6) to yield

$$\lambda^k(u) = \max\left\{\lambda_1^k(u), \ldots, \lambda_9^k(u)\right\} \tag{16.11-3}$$

The fuzzy set $\lambda^k(u)$ is given in Figure 16.23.

Finally, we defuzzify $\lambda^k(u)$ by obtaining the centroid of this resultant output fuzzy set $\lambda^k(u)$. The final crisp output to the fuel actuator at this sampling instant is the center of the area (COA) of the envelope of the resultant ouput fuzzy set $\lambda^k(u)$, and is calculated to be (see Figure 16.23)

$$\Delta\text{FUEL}(k) = -0.048\,\text{m}^3/\text{hr}$$

It is clear that this procedure must be repeated at every sampling instant k. The sequence of these control decisions is then the desired rule-based *control strategy*. For more details see [3, 25].

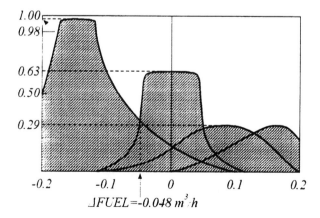

Figure 16.23 The graphical representation of the control output u.

PROBLEMS

1. In Example 16.8.2, construct the curves of the third column for the case $k = k_2$, where the fuzzy variables have the value 75.
2. Solve Example 16.8.2 when the graphical representation of the membership functions of the three fuzzy members positive, zero, and negative are as in Figure 16.24.
3. Solve Example 16.8.3 when the graphical representation of the membership functions of the three fuzzy positive, zero, negative are as in Figure 16.25.
4. A fuzzy controller is to apply a control strategy described by the following three if–then rules:

    Rule 1: If the temperature is low and the pressure is zero, then the speed is low.
    Rule 2: If the temperature is medium and the pressure is low, then the speed is medium.
    Rule 3: If the temperature is high and the pressure is high, then the speed is high.

    The ranges of the variables are: temperature of 0 to 100°C, pressure from 0 to 10 lb, and the speed from 0 to 100 m/sec.

    (a) Describe the temperature, pressure, and speed by graphical representation as fuzzy sets.
    (b) Determine the three rules using the above fuzzy sets.
    (c) For the instant of time k, the values of temperature and pressure are 30°C and 5 lb, respectively. Determine the fuzzy output (speed) set.
    (d) Determine the crisp value by defuzzifying the above fuzzy output set.

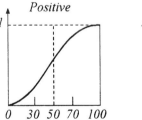
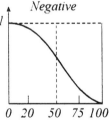
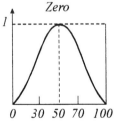

Figure 16.24   The membership functions for Problem 2.

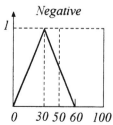
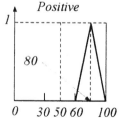
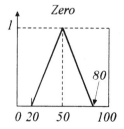

Figure 16.25   The membership functions for Problem 3.

BIBLIOGRAPHY

Books

1. PJ Antsaklis, KM Passino (eds). An Introduction to Intelligent and Autonomous Control. Boston: Kluwer, 1992.
2. D Driankov, H Hellendoorn, M Reinfrank. An Introduction to Fuzzy Control. Berlin: Springer-Verlag, 1993.
3. LP Homblad, JJ Ostergaard. Control of a cement kiln by fuzzy logic. In: MM Gupta and Sanchez, eds. Fuzzy Information and Decision Processes. Amsterdam: North-Holland, 1982.
4. JJ Ostergaard. Fuzzy logic control of a heat exchanger process. In: MM Gupta, GN Saridis, BR Gaines, eds. Fuzzy Automata and Decision Processes. Amsterdam: North-Holland, 1977.
5. M Mizumoto, S Fukami, K Tanaka. Some methods of fuzzy reasoning. In: MM Gupta, RK Ragade, RR Yager, eds. Advances in Fuzzy Set Theory Applications. New York: North Holland, 1979.
6. CJ Harris, CG Moore, M Brown. Intelligent Control: Aspects of Fuzzy Logic and Neural Nets. London: World Scientific, 1990.
7. A Kaufman. Introduction to the Theory of Fuzzy Sets. New York: Academic Press, 1975.
8. RE King. Computational Intelligence in Control Engineering. New York: Marcel Dekker, 1999.
9. B Kosko. Neural Networks and Fuzzy Systems. Englewood Cliffs, New Jersey: Prentice Hall, 1992.
10. PM Larsen. Industrial application of fuzzy logic control. In: EH Mamdani, BR Gains, eds. Fuzzy Reasoning and Its Applications. London: Academic Press, 1981.
11. PN Paraskevopoulos. Digital Control Systems. London: Prentice Hall, 1996.
12. W Pedrycz. Fuzzy Control and Fuzzy Systems. New York: John Wiley & Sons, 1993.
13. GN Saridis. Self-Organizing Control of Stochastic Systems. New York: Marcel Dekker, 1977.
14. M Sugeno (ed.). Industrial Applications of Fuzzy Control. Amsterdam: North Holland, 1985.
15. RR Yager, S Ovchinnikov, RM Tong, HT Nguyen. Fuzzy Sets & Applications: Selected Papers by LA Zadeh. New York: John Wiley & Sons, 1987.
16. HJ Zimmerman. Fuzzy Set Theory and Its Applications. Boston: Kluwer, 1993.

Articles

17. S Assilian, EH Mamdani. An experiment in linguistic synthesis with a fuzzy logic controller. Int J Man Machine Studies 7:1–13, 1974.
18. WJM Kickert, HR van Nauta Lemke. Application of a fuzzy controller in a warm water plant. Automatica 12:301–308, 1976.
19. RE King. Expert supervision and control of a large-scale plant. J Intelligent Systems and Robotics 5:167–176, 1992.
20. CC Lee. Fuzzy logic in control systems: fuzzy logic controller. IEEE Trans on Systems, Man and Cybernetics 20:404–435, 1990.
21. EH Mamdani. Application of fuzzy algorithms for the control of a simple dynamic plant. Proc IEE 121:1585–1588, 1974.
22. EH Mamdani. Application of fuzzy logic to approximate reasoning. IEEE Trans Computers 26:1182–1191, 1977.
23. NJ Mandic, EM Scharf, EH Mamdani. Practical application of a heuristic fuzzy rule-based controller to the dynamic control of a robot arm. IEE Proc D 132:190–203, 1985.

24. S Murakami, F Takemoto, H Fulimura, E Ide. Weldline tracking control of arc welding robot using fuzzy logic controller. Fuzzy Sets and Systems 32:221–237, 1989.
25. JJ Ostergaard. FUZZYII: the new generation of high level kiln control. Zement Kalk Gips (Cement-Lime-Gypsum) 43:539–541, 1990.
26. TJ Procyk, EH Mamdani. A linguistic self-organizing process controller. Automatica 15:15–30, 1979.
27. M Sugeno, T Murofushi, T Mori, T Tatematsu, J Tanaka. Fuzzy algorithmic control of a model car by oral instructions. Fuzzy Sets and Systems 32:207–219, 1989.
28. KL Tang, RJ Mulholland. Comparing fuzzy logic with classical controller design. IEEE Trans Systems Man and Cybernetics 17:1085–1087, 1987.
29. SG Tzafestas. Fuzzy systems and fuzzy expert control: an overview. The Knowledge Engineering Review 9:229–268, 1994.
30. BAM Wakileh, KF Gill. Use of fuzzy logic in robotics. Computers in Industry 10:35–46, 1988.
31. T Yamakawa, T Miki. The current mode fuzzy logic integrated circuits fabricated by the standard CMOS process. IEEE Trans Computers 35:161–167, 1986.
32. S Yasunobu, S Miyamoto, H Ihara. Fuzzy control for automatic train operation system. Proceedings IFAC/IFIP/IFORS Int Congress on Control in Transportation Systems, Baden-Baden, 1983.
33. LA Zadeh. Fuzzy sets. Information and Control 8:338–353, 1965.
34. LA Zadeh. Outline of a new approach to the analysis of complex systems and decision processes. IEEE Trans SMC 3:43–80, 1973.
35. LA Zadeh. Making computer think like people. IEEE Spectrum pp 26–32, 1984.
36. LA Zadeh. The concept of a linguistic variable and its application to approximate reasoning. Information Sci 8:43–80, 1975.

Appendix A

Laplace Transform Tables

Table A.1 Laplace Transform Properties and Theorems

Properties or theorems	$f(t)$	$F(s)$
1 Definition of the Laplace transform	$f(t)$	$\displaystyle\int_0^\infty f(t)e^{-st}\,dt$
2 Definition of the inverse Laplace transform	$\dfrac{1}{2\pi j}\displaystyle\int_{c-j\omega}^{c+j\infty} F(s)e^{st}\,ds$	$F(s)$
3 Linearity	$c_1 f_1(t) + c_2 f_2(t)$	$c_1 F_1(s) + c_2 F_2(s)$
4 First derivative	$\dfrac{df(t)}{dt}$	$sF(s) - f(0)$
5 Second derivative	$\dfrac{d^2 f(t)}{dt^2}$	$s^2 F(s) - sf(0) - f^{(1)}(0)$
6 nth derivative	$\dfrac{d^n f(t)}{dt^n}$	$s^n F(s) - s^{n-1}f(0) - \cdots - f^{(n-1)}(0)$
7 Integral	$\displaystyle\int_0^t f(t)\,dt$	$\dfrac{F(s)}{s}$
8 Integral	$\displaystyle\int_{-\infty}^t f(t)\,dt$	$\dfrac{F(s)}{s} + \dfrac{f^{(-1)}(0)}{s}$
9 Double integral	$\displaystyle\int_{-\infty}^t \int_{-\infty}^t f(t)(dt)^2$	$\dfrac{F(s)}{s^2} + \dfrac{f^{(-1)}(0)}{s^2} + \dfrac{f^{(-2)}(0)}{s}$
10 nth time integral	$\underbrace{\displaystyle\int_\infty^t \cdots \int_\infty^t}_{n\text{ times}} f(t)(dt)^n$	$\dfrac{F(s)}{s^n} + \dfrac{f^{(-1)}(0)}{s^n} + \dfrac{f^{(-2)}(0)}{s^{n-1}} + \cdots$ $+ \dfrac{f^{(-n)}(0)}{s}$
11 Time scaling	$f(at)$	$\dfrac{1}{a}F\!\left(\dfrac{s}{a}\right)$

(continued)

Table A.1 (*continued*)

Properties or theorems	$f(t)$	$F(s)$
12 Shift in the frequency domain	$e^{-at}f(t)$	$F(s+a)$
13 Shift in the time domain	$f(t-a)u(t-a)$	$e^{-at}F(s)$
14 Multiplication of a function by t	$tf(t)$	$-\dfrac{\mathrm{d}}{\mathrm{d}s}F(s)$
15 Division of a function by t	$\dfrac{f(t)}{t}$	$\displaystyle\int_s^\infty F(a)\,\mathrm{d}a$
16 Multiplication of a function by t^n	$t^n f(t)$	$(-1)^n \dfrac{\mathrm{d}^n}{\mathrm{d}s^n}F(s)$
17 Division of a function by t^n	$\dfrac{f(t)}{t^n}$	$\underbrace{\displaystyle\int_s^t \cdots \int_s^t F(s)(\mathrm{d}s)^n}_{n\,\text{times}}$
18 Convolution	$\displaystyle\int_0^t h(t-\tau)u(\tau)\mathrm{d}\tau$	$H(s)U(s)$
19 The initial value theorem	$\lim\limits_{t\to 0} f(t)$	$\lim\limits_{t\to\infty} sF(s)$
20 The final value theorem	$\lim\limits_{t\to\infty} f(t)$	$\lim\limits_{t\to 0} sF(s)$

Remark A.1.1 In the properties 8, 9, and 10, the constant $f^{(-k)}(0)$ is defined as follows:

$$f^{(-1)}(0) = \int_{-\infty}^0 f(t)\,\mathrm{d}t, \, f^{(-2)}(0) = \int_{-\infty}^0 \int_{-\infty}^0 f(t)\mathrm{d}t^2, \text{ etc.}$$

Table A.2 Laplace Transform Pairs

SN	$F(s) = L[f(t)]$	$f(t) = L^{-1}[F(s)]$	Remarks
1	1	$\delta(t)$	
2	s	$\delta^{(1)}(t)$	
3	s^n	$\delta^{(n)}(t)$	n is a positive integer
4	$\dfrac{1}{s}$	$u(t)$	
5	$\dfrac{1}{s^2}$	t	
6	$\dfrac{1}{s^n}$	$\dfrac{t^{n-1}}{(n-1)!}$	n is a positive integer
7	$\dfrac{1}{s^{1/2}}$	$\dfrac{1}{(\pi t)^{1/2}}$	
8	$\dfrac{1}{s^{n+1/2}}$	$\dfrac{2^n t^{n-1/2}}{1 \cdot 3 \cdot 5 \cdots (2n-1) \cdot \pi^{1/2}}$	n is a positive integer
9	$\dfrac{1}{s+a}$	e^{-at}	
10	$\dfrac{1}{(s+a)^2}$	te^{-at}	
11	$\dfrac{1}{(s+a)^n}$	$\dfrac{t^{n-1}e^{-at}}{(n-1)!}$	n is a positive integer
12	$\dfrac{1}{s^2+a^2}$	$\dfrac{1}{a}\sin at$	
13	$\dfrac{1}{(s^2+a^2)^2}$	$\dfrac{1}{2a^3}(\sin at - at\cos at)$	
14	$\dfrac{1}{s^2-a^2}$	$\dfrac{1}{a}\sinh at$	
15	$\dfrac{1}{(s+a)(s+b)}$	$\dfrac{e^{-at}-e^{-bt}}{b-a}$	
16	$\dfrac{1}{(s+a)(s+b)(s+c)}$	$\dfrac{-(c-b)e^{-at}-(a-c)e^{-bt}-(b-a)e^{-ct}}{(b-a)(c-b)(a-c)}$	
17	$\dfrac{(s+a)}{(s+b)(s+c)}$	$\dfrac{(a-b)e^{-bt}-(a-c)e^{-ct}}{(c-b)}$	
18	$\dfrac{s}{s^2+a^2}$	$\cos at$	

(*continued*)

Table A.2 (*continued*)

SN	$F(s) = L[f(t)]$	$f(t) = L^{-1}[F(s)]$	Remarks
19	$\dfrac{s}{s^2 - a^2}$	$\cosh at$	
20	$\dfrac{s+a}{(s+a)^2 + b^2}$	$e^{-at}\cos bt$	
21	$\dfrac{b}{(s+a)^2 + b^2}$	$e^{-at}\sin bt$	
22	$\dfrac{1}{s^2(s+a)}$	$\dfrac{1}{a^2}(e^{-at} + at - 1)$	
23	$\dfrac{1}{(s+a)^2(s+b)}$	$\dfrac{1}{(b-a)^2}\left[[(b-a)t - 1]e^{-at} + e^{-bt}\right]$	
24	$\dfrac{1}{s(s^2 + a^2)}$	$\dfrac{1}{a^2}(1 - \cos at)$	
25	$\dfrac{s}{(s+a)(s+b)}$	$\dfrac{1}{b-a}(be^{-bt} - ae^{-at})$	
26	$\dfrac{1}{s(s+a)^2}$	$\dfrac{1}{a^2}(1 - (at + 1)e^{-at})$	
27	$\dfrac{1}{s(s+a)(s+b)}$	$\dfrac{be^{-at} - ae^{-bt}}{ab(b-a)} + \dfrac{1}{ab}$	
28	$\dfrac{1}{s^2(s^2 + a^2)}$	$\dfrac{1}{a^3}(at - \sin at)$	
29	$\dfrac{1}{s^4 - a^4}$	$\dfrac{1}{2a^3}(\sinh at - \sin at)$	
30	$\dfrac{s+b}{s^2 + a^2}$	$\dfrac{\sqrt{a^2 + b^2}}{a}\sin(at + \theta)$	$\theta = \tan^{-1}\left(\dfrac{a}{b}\right)$
31	$\dfrac{s}{(s^2 + a^2)^2}$	$\dfrac{t}{2a}\sin at$	
32	$\dfrac{s^2}{(s^2 + a^2)^2}$	$\dfrac{1}{2a}(at \cos at + \sin at)$	
33	$\dfrac{s^n}{(s^2 + a^2)^{n+1}}$	$\dfrac{t^n \sin at}{n!2^n a}$	
34	$\dfrac{s+b}{s(s+a)^2}$	$\dfrac{b}{a^2} + \left[\dfrac{a-b}{a}t - \dfrac{b}{a^2}e^{-at}\right]$	
35	$\dfrac{s+c}{s(s+a)(s+b)}$	$\dfrac{c-a}{a(a-b)}e^{-at} + \dfrac{c-b}{b(b-a)}e^{-bt} + \dfrac{c}{ab}$	

Table A.2 (*continued*)

SN	$F(s) = L[f(t)]$	$f(t) = L^{-1}[F(s)]$	Remarks
36	$\dfrac{s^2 - b^2}{(s^2 + b^2)^2}$	$t \cos bt$	
37	$\dfrac{s}{(s^2 + a^2)(s^2 + b^2)}$	$\dfrac{\cos at - \cos bt}{b^2 - a^2}$	$a \neq b$
38	$\dfrac{s}{(s + a)(s + b)(s + c)}$	$-\dfrac{a\mathrm{e}^{-at}}{(b - a)(c - a)} - \dfrac{b\mathrm{e}^{-bt}}{(a - b)(c - b)}$ $-\dfrac{c\mathrm{e}^{-ct}}{(a - c)(b - c)}$	
39	$\dfrac{s^2}{(s + a)(s + b)(s + c)}$	$\dfrac{a^2\mathrm{e}^{-at}}{(b - a)(c - a)} + \dfrac{b^2\mathrm{e}^{-bt}}{(a - b)(c - b)}$ $+ \dfrac{c^2\mathrm{e}^{-ct}}{(a - c)(b - c)}$	
40	$\dfrac{s}{(s + a)(s + b)^2}$	$-\dfrac{a\mathrm{e}^{-at} + (b(a - b)t - a)\mathrm{e}^{-bt}}{(a - b)^2}$	
41	$\dfrac{s^2}{(s + a)(s + b)^2}$	$\dfrac{a^2\mathrm{e}^{-at} + (b^2(a - b)t + b^2 - 2ab)\mathrm{e}^{-bt}}{(a - b)^2}$	
42	$\dfrac{1}{(s + a)(s^2 + b)^2}$	$\dfrac{1}{a^2 + b^2}\left[\mathrm{e}^{-at} - \dfrac{1}{b}\sqrt{a^2 + b^2}\,\sin(bt + \theta)\right]$	$\theta = \tan^{-1}\left(\dfrac{a}{b}\right)$
43	$\dfrac{s}{(s + a)(s^2 + b^2)}$	$\dfrac{-1}{a^2 + b^2}\left[\mathrm{e}^{-at} - \dfrac{1}{a}\sqrt{a^2 + b^2}\,\sin(bt + \theta)\right]$	$\theta = \tan^{-1}\left(\dfrac{a}{b}\right)$
44	$\dfrac{s^2}{(s + a)(s^2 + b^2)}$	$\dfrac{a^2}{a^2 + b^2}\left[\mathrm{e}^{-at} - \dfrac{b}{a^2}\sqrt{a^2 + b^2}\,\sin(bt - \theta)\right]$	$\theta = \tan^{-1}\left(\dfrac{a}{b}\right)$
45	$\dfrac{1}{s[(s + a)^2 + b^2]}$	$\dfrac{1}{a^2 + b^2}\left[1 - \dfrac{b}{a}\sqrt{a^2 + b^2}\,\mathrm{e}^{-at}\sin(bt + \theta)\right]$	$\theta = \tan^{-1}\left(\dfrac{a}{b}\right)$
46	$\dfrac{1}{s^2[(s + a)^2 + b^2]}$	$\dfrac{1}{a^2 + b^2}\left[t - \dfrac{2a}{a^2 + b^2} + \dfrac{1}{b}\mathrm{e}^{-at}\sin(bt + \theta)\right]$	$\theta = \tan^{-1}\left(\dfrac{a}{b}\right)$
47	$\dfrac{1}{s(s^2 + a^2)^2}$	$\dfrac{1}{a^4}(1 - \cos at) - \dfrac{1}{2a^3}t\sin at$	
48	$\dfrac{1}{s^4 - a^4}$	$\dfrac{1}{2a^2}(\cosh at - \cos at)$	
49	$\dfrac{s^2}{s^4 - a^4}$	$\dfrac{1}{2a}(\sinh at + \sin at)$	

(*continued*)

Table A.2 (*continued*)

SN	$F(s) = L[f(t)]$	$f(t) = L^{-1}[F(s)]$	Remarks
50	$\dfrac{1}{\sqrt{s^2 + a^2}}$	$J_0(at)$	J_0: Bessel function
51	$\dfrac{1}{s - \ln b}$	b^t	
52	$\ln \dfrac{s+a}{s+b}$	$\dfrac{1}{t}(\mathrm{e}^{-bt} - \mathrm{e}^{-at})$	

Appendix B

The Z-Transform

B.1 INTRODUCTION

It is well known that the Laplace transform is a mathematical tool which facilitates the study and the design of linear time-invariant continuous-time control systems (see Chap. 2). The reason for this is that it transforms the *differential equation* which describes the system under control to an *algebraic equation*. The corresponding technique for the discrete-time systems is the Z-transform. Indeed, the Z-transform facilitates significantly the study and the design of linear time-invariant discrete-time systems since it transforms the *difference equation* which describes the system under control to an *algebraic equation*. Since the study of an algebraic equation is much easier than that of a difference equation, the Z-transform has been extensively used as a basic study and design tool for discrete-time systems.

This appendix is devoted to the Z-transform, covering the basic theory together with several examples. More specifically, we begin the appendix with the definitions of certain basic discrete-time signals. Subsequently, we present the definitions and some basic properties and theorems of the Z-transform. Finally, the inverse Z-transform is defined and some illustrative examples are presented.

B.2 THE BASIC DISCRETE-TIME CONTROL SIGNALS

In this section we present the definitions of the following basic discrete-time signals: the unit pulse sequence, the unit step sequence, the unit gate sequence, the ramp sequence, the exponential sequence, the alternating sequence, and the sine sequence. These signals are very important for control applications.

1 The Unit Pulse Sequence

The unit pulse sequence is designated by $\delta(k - k_0)$ and is defined as follows:

$$\delta(k - k_0) = \begin{cases} 1, & \text{for } k = k_0 \\ 0, & \text{for } k \neq k_0 \end{cases} \tag{B.2-1}$$

The graphical representation of $\delta(k - k_0)$ is given in Figure B.1.

Figure B.1   The unit pulse sequence $\delta(k - k_0)$.

2 The Unit Step Sequence

The unit step sequence is designated by $\beta(k - k_0)$ and is defined as follows:

$$\beta(k - k_0) = \begin{cases} 1, & \text{for } k \geq k_0 \\ 0, & \text{for } k < k_0 \end{cases} \tag{B.2-2}$$

The graphical representation of $\beta(k - k_0)$ is given in Figure B.2.

3 The Unit Gate Sequence

The unit gate sequence is designated by $g_\pi(k) = \beta(k - k_1) - \beta(k - k_2)$ and is defined as follows:

$$g_\pi(k) = \begin{cases} 1, & \text{for } k_1 \leq k \leq k_2 \\ 0, & \text{for } k < k_1 \text{ and for } k > k_2 \end{cases} \tag{B.2-3}$$

Figure B.3 shows the graphical representation of $g_\pi(k)$. The unit gate sequence is usually used to zero all values of another sequence outside a certain time interval. Consider, for example, the sequence $f(k)$. Then, the sequence $y(k) = f(k)g_\pi(k)$ becomes

$$y(k) = f(k)g_\pi(k) = \begin{cases} f(k), & \text{for } k_1 \leq k \leq k_2 \\ 0, & \text{for } k < k_1 \text{ and for } k > k_2 \end{cases}$$

4 The Ramp Sequence

The ramp sequence is designated by $r(k - k_0)$ and is defined as follows:

$$r(k - k_0) = \begin{cases} k - k_0, & \text{for } k \geq k_0 \\ 0, & \text{for } k < k_0 \end{cases} \tag{B.2-4}$$

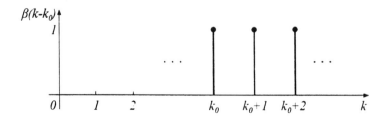

Figure B.2   The unit step sequence $\beta(k - k_0)$.

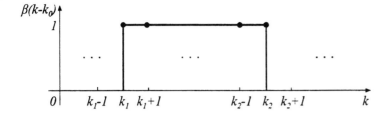

Figure B.3   The unit gate sequence $g_\pi(k) = \beta(k - k_1) - \beta(k - k_2)$.

Figure B.4 shows the graphical representation of $r(k - k_0)$.

5 The Exponential Sequence

The exponential sequence is designated by $g(k)$ and is defined as follows:

$$g(k) = \begin{cases} a^k, & \text{for } k \geq 0 \\ 0, & \text{for } k < 0 \end{cases} \tag{B.2-5}$$

Figure B.5 shows the graphical representation of $g(k) = a^k$. Clearly, when $a > 1$ the values of $g(k)$ increase as k increases, whereas for $a < 1$ the values of $g(k)$ decrease as k decreases. For $a = 1$, $g(k)$ remains constantly equal to 1. In this last case, $g(k)$ becomes the unit step sequence $\beta(k)$.

6 The Alternating Sequence

The alternating sequence is designated by $\varepsilon(k)$ and is defined as follows:

$$\varepsilon(k) = \begin{cases} (-1)^k, & \text{for } k \geq k_0 \\ 0, & \text{for } k < k_0 \end{cases} \tag{B.2-6}$$

Figure B.6 shows the graphical representation of $\varepsilon(k)$.

7 The Sine Sequence

The sine sequence is defined as follows:

$$f(k) = \begin{cases} A \sin \omega_0 k, & \text{for } k \geq 0 \\ 0, & \text{for } k < 0 \end{cases} \tag{B.2-7}$$

Figure B.7 shows the graphical representation of $f(k) = A \sin \omega_0 k$, with $\omega_0 = 2\pi/12$.

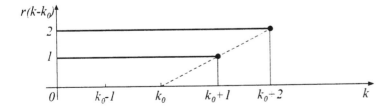

Figure B.4   The ramp sequence $r(k - k_0)$.

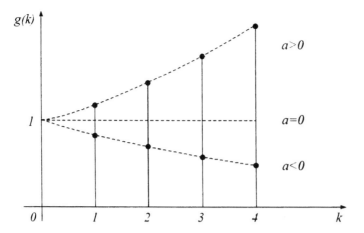

Figure B.5 The exponential sequence $g(k) = a^k$.

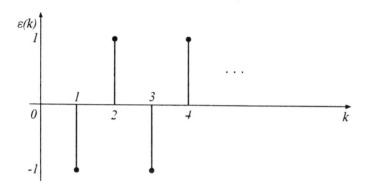

Figure B.6 The alternating sequence $\varepsilon(k)$.

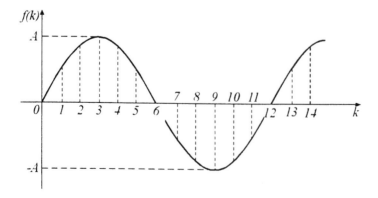

Figure B.7 The sine sequence $f(k) = A \sin \omega_0 k$.

B.3 THE *Z*-TRANSFORM

B.3.1 Introduction to the *Z*-Transform

The *Z*-transform of a discrete-time function $f(k)$ is designated by $F(z)$ and is defined as follows:

$$F(z) = Z[f(k)] = \sum_{k=-\infty}^{\infty} f(k)z^{-k} \tag{B.3-1}$$

If the discrete-time function $f(k)$ is causal, i.e., $f(k) = 0$ for $k < 0$, then the definition (B.3-1) becomes

$$F(z) = Z[f(k)] = \sum_{k=0}^{\infty} f(k)z^{-k} \tag{B.3-2}$$

In practice, the discrete-time sequence is usually produced from a continuous-time function $f(t)$. The conversion of $f(t)$ to $f(kT)$, where T represents the time distance between two points of $f(kT)$, is achieved through a sampler, as is shown in Figure B.8. The sampler is actually a switch which closes instantly and with frequency $f_s = 1/T$. The resulting output $f(kT)$ represents a discrete-time function with amplitude equal to the amplitude of $f(t)$ at the sampling instants kT, $k = 0, 1, 2, \ldots$.

The following theorem refers to the criteria for choosing the sampling frequency $f_s = 1/T$ (this issue was first investigated by Nyquist, but Shannon gave the complete proof of the theorem).

Theorem B.3.1

Let f_1 be the highest frequency in the frequency spectrum of $f(t)$. Then, for $f(t)$ to be recovered from $f(kT)$, it is necessary that $f_s \geq 2f_1$.

It is noted that the function $f(t)$ may be reproduced from $f(kT)$ using a hold circuit (see Sec. 12.3) in series with a low-frequency filter which smooths out the form of the signal.

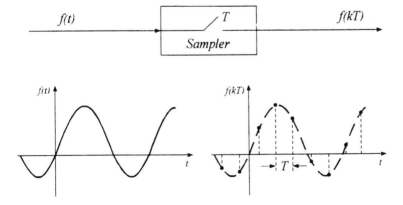

Figure B.8 The operation of a sampler.

Let $\delta_T(t)$ denote the infinite sequence of unit pulse functions (or Dirac functions) shown in Figure B.9. In addition, let $f^*(t)$ be the following function:

$$f^*(t) = f(t)\delta_T(t) = \sum_{k=-\infty}^{\infty} f(kT)\delta(t - kT) \qquad \text{where} \qquad \delta_T(t) = \sum_{k=-\infty}^{\infty} \delta(t - kT)$$

(B.3-3)

When $f(t)$ is causal (and this is usually the case), Eq. (B.3-3) becomes

$$f^*(t) = \sum_{k=0}^{\infty} f(kT)\delta(t - kT)$$

(B.3-4)

The Laplace transform of Eq. (B.3-4) is

$$F^*(s) = L[f^*(t)] = \sum_{k=0}^{\infty} f(kT) \int_0^{\infty} \delta(t - kT)e^{-st}dt = \sum_{k=0}^{\infty} f(kT)e^{-kTs}$$

(B.3-5)

The Z-transform of $f(kT)$ is

$$F(z) = Z[f(kT)] = \sum_{k=0}^{\infty} f(kT)z^{-k}$$

(B.3-6)

If we use the mapping

$$z = e^{Ts} \qquad \text{or} \qquad s = T^{-1}\ln z$$

(B.3-7)

then

$$F(z) = F^*(s)\Big|_{s=T^{-1}\ln z}$$

(B.3-8)

Equation (B.3-8) shows the relation between the sequence $f(kT)$ and the function $f^*(t)$ described in the z- and s-domain, respectively. A continuous-time system with input $f(t)$ and output $f^*(t)$ is called an ideal sampler (see, also, Sec. 12.3).

The inverse Z-transform of a function $F(z)$ is denoted as $f(kT)$ and is defined as

$$f(kT) = Z^{-1}[F(z)] = \frac{1}{2\pi j}\oint F(z)z^{k-1}dz$$

(B.3-9)

Equations (B.3-6) and (B.3-8) constitute the Z-transform pair.

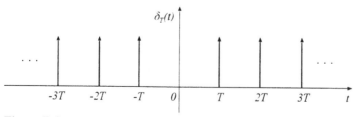

Figure B.9 The Dirac functions $\delta_T(t)$.

B.3.2 Properties and Theorems of the Z-Transform

1 Linearity

The Z-transform is linear, i.e., the following relation holds:

$$Z[c_1 f_1(kT) + c_2 f_2(kT)] = c_1 Z[f_1(kT)] + c_2 Z[f_2(kT)]$$
$$= c_1 F_1(z) + c_2 F_2(z) \tag{B.3-10}$$

where c_1 and c_2 are constants and

$$F_i(z) = Z[f_i(kT)], \qquad i = 1, 2 \tag{B.3-11}$$

Proof

Apply the Z-transform definition (B.3-5) to relation (B.3-10) to yield

$$Z[c_1 f_1(kT) + c_2 f_2(kT)] = \sum_{k=0}^{\infty} [c_1 f_1(kT)z^{-k} + c_2 f_2(kT)z^{-k}]$$

$$= c_1 \sum_{k=0}^{\infty} f_1(kT)z^{-k} + c_2 \sum_{k=0}^{\infty} f_2(kT)z^{-k}$$

$$= c_1 F_1(z) + c_2 F_2(z)$$

2 Shift in the Time Domain

The discrete-time functions $f(kT - \sigma T)$ and $f(kT + \sigma T)$ are actually the function $f(kT)$, shifted σT time units to the right and to the left, respectively. From definition (B.3-6), we have

(a) $$Z[f(kT - \sigma T)] = \sum_{k=0}^{\infty} f(kT - \sigma T)z^{-k} = \sum_{m=-\sigma}^{\infty} f(mT)z^{-m}z^{-\sigma}$$

$$= z^{-\sigma} \left[\sum_{m=0}^{\infty} f(mT)z^{-m} + \sum_{m=-\sigma}^{-1} f(mT)z^{-m} \right]$$

$$= z^{-\sigma} F(z) + \sum_{m=-\sigma}^{-1} f(mT)z^{-(\sigma+m)} \tag{B.3-12a}$$

(b) $$Z[f(kT - \sigma T)\beta(kT - \sigma T)] = \sum_{k=0}^{\infty} f[(k-\sigma)T]\beta[(k-\sigma)T]z^{-k}$$

$$= z^{-\sigma} \left[\sum_{k=0}^{\infty} f[(k-\sigma)T]\beta[(k-\sigma)T]z^{-(k-\sigma)} \right]$$

$$= z^{-\sigma} \sum_{m=-\sigma}^{\infty} f(mT)\beta(mT)z^{-m}$$

$$= z^{-\sigma} \sum_{m=0}^{\infty} f(mT)z^{-m}$$

$$= z^{-\sigma} F(z) \tag{B.3-12b}$$

(c) $Z[f(kT + \sigma T)] = \displaystyle\sum_{k=0}^{\infty} f(kT + \sigma T)z^{-k} = z^{\sigma} \sum_{k=0}^{\infty} f[(k + \sigma)T]z^{-(k+\sigma)}$

$$= z^{\sigma} \sum_{m=\sigma}^{\infty} f(mT)z^{-m}$$

$$= z^{\sigma} \left[\sum_{m=0}^{\infty} f(mT)z^{-m} - \sum_{m=0}^{\sigma-1} f(mT)z^{-m} \right]$$

$$= z^{\sigma} \left[F(z) - \sum_{m=0}^{\sigma-1} f(mT)z^{-m} \right]$$

$$= z^{\sigma} \left[F(z) - \sum_{k=0}^{\sigma-1} f(kT)z^{-k} \right] \qquad \text{(B.3-12c)}$$

where in the last step, we have set $m = k$.

From the foregoing equations, the following special cases are obtained:

$$Z[f(t + T)] = zF(z) - zf(0) \qquad \text{(B.3-13a)}$$

$$Z[f(t + 2T)] = z^2 F(z) - z^2 f(0) = zf(T) \qquad \text{(B.3-13b)}$$

$$Z[f(t - T)] = z^{-1}F(z) + f(-T) \qquad \text{(B.3-13c)}$$

$$Z[f(t - 2T)] = z^{-2}F(z) + z^{-1}f(-T) + f(-2T) \qquad \text{(B.3-13d)}$$

3 Change in the z-Scale

Consider the function $a^{\mp t} f(t)$. Then, according to definition (B.3-6), it follows that

$$Z[a^{\mp t} f(t)] = \sum_{k=0}^{\infty} a^{\mp kT} f(kT)z^{-k} = \sum_{k=0}^{\infty} f(kT)[a^{\pm T}z]^{-k} = F(a^{\pm T}z) \qquad \text{(B.3-14)}$$

4 The Z-Transform of a Sum

Consider the finite sum

$$\sum_{i=0}^{k} f(iT)$$

This sum represents the summation of the first $k + 1$ terms of the sequence $f(kT)$. Defining

$$g(kT) = \sum_{i=0}^{k} f(iT), \qquad g[(k - 1)T] = \sum_{i=0}^{k-1} f(iT), \dots$$

the discrete-time function $g(kT)$ may be described by the following difference equation:

$$g(kT + T) = g(kT) + f(kT + T)$$

Applying the Z-transform to this difference equation yields

$$z[G(z) - g(0)] = G(z) + z[F(z) - f(0)]$$

where use was made of Eq. (B.3-13a). Since $g(0) = f(0)$, this relation becomes

$$G(z) = Z\left[\sum_{i=0}^{k} f(iT)\right] = \left[\frac{z}{z-1}\right]F(z) \tag{B.3-15}$$

5 Multiplication by k

Consider the discrete-time function $f(kT)$. Then, the Z-transform of the function $kf(kT)$ is

$$Z[kf(kT)] = \sum_{k=0}^{\infty} kf(kT)z^{-k} = z\sum_{k=0}^{\infty} f(kT)[kz^{-k-1}] = -z\sum_{k=0}^{\infty} f(kT)\frac{\mathrm{d}z^{-k}}{\mathrm{d}z}$$

$$= -z\frac{\mathrm{d}}{\mathrm{d}z}\left[\sum_{k=0}^{\infty} f(kT)z^{-k}\right] = -z\frac{\mathrm{d}}{\mathrm{d}z}F(z) \tag{B.3-16}$$

6 Convolution of Two Discrete-Time Functions

Consider the causal discrete-time functions $f(kT)$ and $h(kT)$. The convolution between these two functions is designated by $y(kT) = f(kT) * h(kT)$ and is defined as follows:

$$y(kT) = f(kT) * h(kT) = \sum_{i=0}^{\infty} f(iT)h(kT - iT) = \sum_{i=0}^{\infty} h(iT)f(kT - iT)$$

The Z-transform of the function $y(kT)$ is

$$Y(z) = Z[y(kT)] = Z[f(kT) * h(kT)] = \sum_{k=0}^{\infty}\left[\sum_{i=0}^{\infty} f(iT)h(kT - iT)\right]z^{-k}$$

$$= \sum_{k=0}^{\infty}\left[\sum_{i=0}^{\infty} h(iT)f(kT - iT)\right]z^{-k}$$

Reversing the summing order, we have

$$Y(z) = \sum_{i=0}^{\infty} h(iT)\sum_{k=0}^{\infty} f(kT - iT)z^{-k} = \sum_{i=0}^{\infty} h(iT)z^{-i}\sum_{k=0}^{\infty} f(kT - iT)z^{-(k-i)}$$

$$= \left[\sum_{i=0}^{\infty} h(iT)z^{-i}\right]\left[\sum_{m=0}^{\infty} f(mT)z^{-m}\right]$$

Since $f(mT)$ is a causal function, i.e., $f(mT) = 0$ for $m < 0$, it follows that

$$Y(z) = \left[\sum_{i=0}^{\infty} h(iT)z^{-i}\right]\left[\sum_{m=0}^{\infty} f(mT)z^{-m}\right] = H(z)F(z) \tag{B.3-17}$$

7 Discrete-Time Periodic Functions

A discrete-time function $f(kT)$ is called periodic with period p if the following relation holds:

$$f(kT) = f(kT + pT) \qquad \text{for every } k = 0, 1, 2, \ldots$$

Let $F_1(z)$ be the Z-transform of the first period of $f(kT)$, i.e., let

$$F_1(z) = \sum_{k=0}^{p-1} f(kT)z^{-k}$$

Then the Z-transform of the periodic function $f(kT)$ is

$$Z[f(kT)] = F(z) = Z[f(kT + pT)] = z^p\left[F(z) - \sum_{k=0}^{p-1} f(kT)z^{-k}\right]$$

$$= z^p[F(z) - F_1(z)]$$

where relation (B.3-12c) was used. Hence

$$F(z) = \left[\frac{z^p}{z^p - 1}\right]F_1(z) \tag{B.3-18}$$

8 Initial Value Theorem

The following relation holds:

$$f(0) = \lim_{z \to \infty} F(z) \tag{B.3-19}$$

Proof

The Z-transform of $f(kT)$ may be written as

$$F(z) = \sum_{k=0}^{\infty} f(kT)z^{-k} = f(0) + f(T)z^{-1} + f(2T)z^{-2} + \cdots$$

Taking the limits of both sides of the above equation, as $z \to \infty$, we immediately arrive at the relation (B.3-19).

9 Final Value Theorem

The following relation holds:

$$\lim_{k \to \infty} f(kT) = \lim_{z \to 1}(1 - z^{-1})F(z) \tag{B.3-20}$$

under the assumption that the function $(1 - z^{-1})F(z)$ does not have any poles outside or on the unit circle.

Proof

Consider the Z-transform of $f(kT + T) - f(kT)$:

$$Z[f(kT + T) - f(kT)] = \lim_{m \to \infty} \sum_{k=0}^{m}[f(kT + T) - f(kT)]z^{-k}$$

Using Eqs (B.3-6) and (B.3-13a), we obtain

$$zF(z) - zf(0) - F(z) = \lim_{m \to \infty} \sum_{k=0}^{m} [f(kT + T) - f(kT)]z^{-k}$$

or

$$(1 - z^{-1})F(z) - f(0) = \lim_{m \to \infty} \sum_{k=0}^{m} [f(kT + T) - f(kT)]z^{-k-1}$$

Taking the limits on both sides of the above equation, as $z \to 1$, we obtain

$$\lim_{z \to 1}(1 - z^{-1})F(z) - f(0) = \lim_{m \to \infty} \sum_{k=0}^{m} [f(kT + T) - f(kT)]$$
$$= \lim_{m \to \infty} \{[F(T) - f(0)] + [f(2T) - f(T)]$$
$$+ \cdots + [f(mT + T) - f(mT)]\}$$
$$= \lim_{m \to \infty} [-f(0) + f(mT + T)] = -f(0) + f(\infty)$$

Hence

$$\lim_{k \to \infty} f(kT) = \lim_{z \to 1}(1 - z^{-1})F(z)$$

All the foregoing properties and theorems are summarized in Appendix C.

Example B.3.1

Find the Z-transform of the impulse sequence $\delta(kT - \sigma T)$.

Solution

Using definition (B.3-6), we have

$$Z[\delta(kT - \sigma T)] = \sum_{k=0}^{\infty} \delta(kT - \sigma T)z^{-k} = z^{-\sigma}$$

Example B.3.2

Find the Z-transform of the step sequence $\beta(kT - \sigma T)$.

Solution

Here

$$Z[\beta(kT - \sigma T)] = z^{-\sigma}Z[\beta(kT)] = z^{-\sigma} \sum_{k=0}^{\infty} \beta(kT)z^{-k} = z^{-\sigma}\left[\sum_{k=0}^{\infty}(z^{-1})^k\right]$$
$$= z^{-\sigma}\left[\frac{1}{1 - z^{-1}}\right] = \frac{z^{-\sigma+1}}{z - 1}$$

where use was made of property (B.3-12c) and of the relation

$$\sum_{k=0}^{\infty} z^i = \frac{1}{1 - z}, \qquad |z| < 1$$

Example B.3.3

Find the Z-transform of the exponential sequence $g(kT) = a^{kT}$.

Solution

Here

$$Z[g(kT)] = \sum_{k=0}^{\infty} a^{kT} z^{-k} = \sum_{k=0}^{\infty} (a^T z^{-1})^k = \frac{1}{1 - a^T z^{-1}} = \frac{z}{z - a^T}$$

$$\text{for} \quad |az^{-1}| < 1$$

Example B.3.4

Find the Z-transform of the ramp sequence $r(kT - \sigma T)$.

Solution

Here

$$Z[r(kT - \sigma T)] = z^{-\sigma} Z[r(kT)] = z^{-\sigma} \sum_{k=0}^{\infty} r(kT) z^{-k} = Tz^{-\sigma} \sum_{k=0}^{\infty} k z^{-k}$$

$$= Tz^{-\sigma}(-z) \frac{d}{dz} Z[\beta(kT)] = -Tz^{-\sigma+1} \frac{d}{dz} \left[\frac{z}{z-1} \right] = \frac{Tz^{-\sigma+1}}{(z-1)^2}$$

where use was made of the property (B.3-16).

Example B.3.5

Find the Z-transform of the alternating sequence $\varepsilon(kT) = (-1)^{kT}$.

Solution

Here

$$Z[\varepsilon(kT)] = \sum_{k=0}^{\infty} (-1)^{kT} z^{-k} = \sum_{k=0}^{\infty} [(-1)^T z^{-1}]^k = \frac{1}{1 - (-1)^T z^{-1}} = \frac{z}{z - (-1)^T}$$

Example B.3.6

Find the Z-transform of the function $y(kT) = e^{bkT}$.

Solution

Setting $a = e^b$ in Example B.3.3, we obtain

$$Z[e^{bkT}] = \frac{z}{z - e^{bT}}$$

Example B.3.7

Find the Z-transform of the functions $f(kT) = \sin \omega_0 kT$ and $f(kT) = \cos \omega_0 kT$.

Solution

Here

$$Z[e^{j\omega_0 kT}] = \frac{z}{z - e^{j\omega_0 T}} = \frac{z}{z - \cos\omega_0 T - j\sin\omega_0 T}$$

$$= \frac{z(z - \cos\omega_0 T + j\sin\omega_0 T)}{(z - \cos\omega_0 T)^2 + \sin^2\omega_0 T}$$

$$= \left[\frac{z(z - \cos\omega_0 T)}{z^2 - 2z\cos\omega_0 T + 1}\right] + j\left[\frac{z\sin\omega_0 T}{z^2 - 2z\cos\omega_0 T + 1}\right]$$

Since $e^{j\theta} = \cos\theta + j\sin\theta$, it follows that

$$Z[\cos\omega_0 kT] = \frac{z(z - \cos\omega_0 T)}{z^2 - 2z\cos\omega_0 T + 1}$$

$$Z[\sin\omega_0 kT] = \frac{z\sin\omega_0 T}{z^2 - 2z\cos\omega_0 T + 1}$$

Example B.3.8

Find the Z-transform of the function $f(kT) = e^{-bkT}\sin\omega kT$.

Solution

Setting $a = e^b$ in Eq. (B.3-14), we have

$$Z[e^{-bkT} f(kT)] = F(e^{bT} z)$$

Using the results of Example B.3.7, we obtain

$$Z[e^{-bkT}\sin\omega_0 kT] = \frac{ze^{bT}\sin\omega_0 T}{z^2 e^{2bT} - 2ze^{bT}\cos\omega_0 T + 1}$$

$$= \frac{ze^{-bT}\sin\omega_0 T}{z^2 - 2ze^{-bT}\cos\omega_0 T + e^{-2bT}}$$

B.4   THE INVERSE Z-TRANSFORM

The determination of the inverse Z-transform (as in the case of the inverse Laplace transform) is usually based upon the expansion of a rational function $F(z)$ into partial fraction expansion whose inverse transform can be directly found in the tables of the Z-transform pairs given in Appendix C. It is noted that in cases where the numerator of $F(z)$ involves the term z, it is more convenient to expand into partial fraction expansion the function $F(z)/z$, instead of $F(z)$ and, subsequently, determine $F(z)$ from the relation $z[F(z)/z]$.

It is also noted that there are several other techniques for the determination of the inverse Z-transform, as for example the method of the continuous fraction expansion, the direct implementation of the definition of the inverse Z-transform given by Eq. (B.3-9), and others. The method of partial fraction expansion appears to be computationally simpler over the other methods, and for this reason it is almost always used for the determination of the inverse Z-transform.

Example B.4.1

Find the inverse Z-transform of the function

$$F(z) = \frac{-3z}{(z-1)(z-4)}$$

Solution

Expanding $F(z)/z$ into partial fraction expansion, we obtain

$$\frac{F(z)}{z} = \frac{-3}{(z-1)(z-4)} = \frac{1}{(z-1)} - \frac{1}{(z-4)}$$

and hence

$$F(z) = \frac{z}{(z-1)} - \frac{z}{(z-4)}$$

From the table of the Z-transform pairs (Appendix C), we find that

$$Z^{-1}\left[\frac{z}{z-1}\right] = \beta(kT) \qquad \text{and} \qquad Z^{-1}\left[\frac{z}{z-4}\right] = 4^k$$

where $T = 1$.
   Hence

$$f(kT) = Z^{-1}[F(z)] = \beta(kT) - 4^k = 1 - 4^k$$

Example B.4.2

Find the inverse Z-transform of the function

$$F(z) = \frac{z(z-4)}{(z-2)^2(z-3)}$$

Solution

Expanding $F(z)/z$ into partial fraction expansion, we obtain

$$\frac{F(z)}{z} = \frac{1}{z-2} + \frac{2}{(z-2)^2} - \frac{1}{z-3}$$

and hence

$$F(z) = \frac{z}{z-2} + \frac{2z}{(z-2)^2} - \frac{z}{z-3}$$

Since for the case $T = 1$

$$Z^{-1}\left[\frac{z}{z-2}\right] = 2^k, \qquad Z^{-1}\left[\frac{2z}{(z-2)^2}\right] = k2^k, \qquad \text{and} \qquad Z^{-1}\left[\frac{z}{z-3}\right] = 3^k$$

it follows that

$$f(kT) = Z^{-1}[F(z)] = 2^k + k2^k + 3^k = (k+1)2^k + 3^k$$

Example B.4.3

Find the inverse Z-transform of the function

$$F(z) = \frac{2z^3 + z}{(z-2)^2(z-1)}$$

Solution

Expanding $F(z)/z$ into partial fraction expansion, we obtain

$$\frac{F(z)}{z} = \frac{9}{(z-2)^2} - \frac{1}{z-2} + \frac{3}{z-1}$$

and hence

$$F(z) = \frac{9z}{(z-2)^2} - \frac{z}{z-2} + \frac{3z}{z-1}$$

Since for $T = 1$

$$Z^{-1}\left[\frac{z}{z-2}\right] = 2^k, \quad Z^{-1}\left[\frac{z}{(z-2)^2}\right] = k2^{k-1}, \quad \text{and} \quad Z^{-1}\left[\frac{z}{z-1}\right] = 1$$

it follows that

$$f(kT) = Z^{-1}[F(z)] = 9k2^{k-1} - 2^k + 3$$

Example B.4.4

Find the inverse Z-transform of the function

$$F(z) = \frac{z^2}{z^2 - 2z + 2}$$

Solution

Examining the form of the denominator $z^2 - 2z + 2$ we observe that $F(z)$ may be the Z-transform of a function of the type $e^{-akT}(c_1 \sin \omega_0 kT + c_2 \cos \omega_0 kT)$, where c_1 and c_2 are constants. To verify this observation, we work as follows. The constant term 2 is equal to the exponential e^{-2aT}, in which case $a = -(\ln 2)/2T$. The coefficient -2 of the z term must be equal to the function $-2e^{-aT}\cos \omega_0 T$, in which case $\cos \omega_0 T = 1/\sqrt{2}$ and $\omega_0 = \pi/4T$. Consequently, the denominator of $F(z)$ can be written as follows:

$$z^2 - 2z + 2 = z^2 - 2z\,e^{-\left(\frac{-\ln 2}{2T}\right)T}\cos\left(\frac{\pi}{4T}\right)T + e^{-2\left(\frac{-\ln 2}{2T}\right)T}$$

The numerator can be written as $z^2 = (z^2 - z) + z$. Since

$$e^{-\left(\frac{-\ln 2}{2T}\right)T}\cos\left(\frac{\pi}{4T}\right)T = e^{-\left(\frac{-\ln 2}{2T}\right)T}\sin\left(\frac{\pi}{4T}\right)T = 1$$

it follows that the numerator $(z^2 - z) + z$ may be written as follows:

$$z^2 = \left[z^2 - ze^{-\left(\frac{-\ln 2}{2T}\right)T}\cos\left(\frac{\pi}{4T}\right)T\right] + ze^{-\left(\frac{-2}{2T}\right)T}\sin\left(\frac{\pi}{4T}\right)T$$

Hence, the function $F(z)$ may finally be written as

$$F(z) = \frac{z^2 - z}{z^2 - 2z + 2} + \frac{z}{z^2 - 2z + 2}$$

$$= \frac{z^2 - ze^{-\left(\frac{-\ln 2}{2T}\right)T} \cos\left(\frac{\pi}{4T}\right)T}{z^2 - 2ze^{-\left(\frac{-\ln 2}{2T}\right)T} \cos\left(\frac{\pi}{4T}\right)T + e^{-2\left(\frac{-\ln 2}{2T}\right)T}}$$

$$+ \frac{ze^{-\left(\frac{-\ln 2}{2T}\right)T} \sin\left(\frac{\pi}{4T}\right)T}{z^2 - 2ze^{-\left(\frac{-\ln 2}{2T}\right)T} \cos\left(\frac{\pi}{4T}\right)T + e^{-2\left(\frac{-\ln 2}{2T}\right)T}}$$

From the table of the Z-transform pairs (Appendix C), it follows that

$$f(kT) = Z^{-1}[F(z)] = e^{-\left(\frac{-\ln 2}{2T}\right)kT}\left[\cos\left(\frac{\pi}{4T}\right)kT + \sin\left(\frac{\pi}{4T}\right)kT\right]$$

$$= e^{\frac{-\ln 2}{2}k}\left[\cos\frac{\pi k}{4} + \sin\frac{\pi k}{4}\right] \qquad \text{for} \qquad T = 1$$

Appendix C

Z-Transform Tables

Table C.1 Properties and Theorems of the Z-Transform

Property or theorem	$f(kT)$	$F(z)$
1 Definition of Z-transform	$f(kT)$	$\displaystyle\sum_{k=0}^{\infty} f(kT)z^{-k}$
2 Definition of the inverse Z-transform	$\displaystyle\frac{1}{2\pi j}\oint F(z)z^{k-1}\,dz$	$F(z)$
3 Linearity	$c_1 f_1(kT) + c_2 f_2(kT)$	$c_1 F_1(z) + c_2 F_2(z)$
4 Shift to the left (advance)	$f(kT + \sigma T)$	$\displaystyle z^{\sigma}\left(F(z) - \sum_{k=0}^{\sigma-1} f(kT)z^{-k}\right)$
5 Shift to the right (delay)	$f(kT - \sigma T)$	$z^{-\sigma} F(z)$
6 Change in z-scale	$a^{\mp kT} f(kT)$	$F(a^{\pm T} z)$
7 Change in kT-scale	$f(mkT)$	$F(z^{-m})$
8 Multiplying by k	$k f(kT)$	$\displaystyle -z\frac{d}{dz}F(z)$
9 Summation	$\displaystyle\sum_{k=0}^{m} f(kT)$	$\displaystyle\frac{z}{z-1}F(z)$
10 Convolution	$f(kT) \times h(kT)$	$F(z)H(z)$
11 Periodic function	$f(kT) = f(kT + pT)$	$\displaystyle\frac{z^p}{z^p - 1}F_1(z)$
12 Initial value theorem	$f(0)$	$\displaystyle\lim_{z\to\infty} F(z)$
13 Final value theorem	$\displaystyle\lim_{k\to\infty} f(kT)$	$\displaystyle\lim_{z\to 1}(1 - z^{-1})F(z)$

723

Table C.2 Z-Transform Pairs

SN	$f(kT)$	$F(s) = \int_0^\infty f(t)\mathrm{e}^{-st}\mathrm{d}t$	$F(z) = \sum_{k=0}^{\infty} f(kT)z^{-k}$
1	$\delta(kT - aT)$	e^{-aTs}	z^{-a}
2	$\delta(kT)$	1	1 or z^{-0}
3	$\beta(kT - aT)$	$\dfrac{\mathrm{e}^{-aTs}}{s}$	$\dfrac{z^{-a+1}}{z-1}$
4	$\beta(kT)$	$\dfrac{1}{s}$	$\dfrac{z}{z-1}$
5	$kT - aT$	$\dfrac{\mathrm{e}^{-aTs}}{s^2}$	$\dfrac{Tz^{-a+1}}{(z-1)^2}$
6	kT	$\dfrac{1}{s^2}$	$\dfrac{Tz}{(z-1)^2}$
7	$\dfrac{1}{2!}k^2T^2$	$\dfrac{1}{s^3}$	$\dfrac{T^2z(z+1)}{2(z-1)^3}$
8	$\dfrac{1}{3!}k^2T^3$	$\dfrac{1}{s^4}$	$\dfrac{T^3z(z^2+4z+1)}{6(z-1)^4}$
9	$\dfrac{1}{m!}k^mT^m$	$\dfrac{1}{s^{m+1}}$	$\displaystyle\lim_{a\to 0}\dfrac{(-1)^m}{m!}\dfrac{\partial^m}{\partial a^m}\left[\dfrac{z}{z-\mathrm{e}^{-aT}}\right]$
10	a^{kT}	$\dfrac{1}{s - T\ln a}$	$\dfrac{z}{z - a^T}$

(continued)

11	e^{-akT}	$\dfrac{1}{s+a}$	$\dfrac{z}{z-\mathrm{e}^{-aT}}$
12	kTe^{-akT}	$\dfrac{1}{(s+a)^2}$	$\dfrac{Tz\mathrm{e}^{-aT}}{(z-\mathrm{e}^{-aT})^2}$
13	$\dfrac{k^2T^2}{2}\mathrm{e}^{-akT}$	$\dfrac{1}{(s+a)^3}$	$\dfrac{T^2\mathrm{e}^{-aT}\,z}{2(z-\mathrm{e}^{-aT})^2}+\dfrac{T^2\mathrm{e}^{-2aT}\,z}{(z-\mathrm{e}^{-aT})^3}$
14	$\dfrac{k^m T^m}{m!}\mathrm{e}^{-akT}$	$\dfrac{1}{(s+a)^{m+1}}$	$\dfrac{(-1)^m}{m!}\dfrac{\partial^m}{\partial a^m}\left[\dfrac{z}{z-\mathrm{e}^{-aT}}\right]$
15	$1-\mathrm{e}^{-akT}$	$\dfrac{a}{s(s+a)}$	$\dfrac{(1-\mathrm{e}^{-aT})z}{(z-1)(z-\mathrm{e}^{-aT})}$
16	$kT-\dfrac{1-\mathrm{e}^{-akT}}{a}$	$\dfrac{1}{s^2(s+a)}$	$\dfrac{T}{(z-1)^2}-\dfrac{1-\mathrm{e}^{-aT}}{a(z-1)(z-\mathrm{e}^{-aT})}$
17	$\sin\omega_0 kT$	$\dfrac{\omega_0}{s^2+\omega_0^2}$	$\dfrac{z\sin\omega_0 T}{z^2-2z\cos\omega_0 T+1}$
18	$\cos\omega_0 kT$	$\dfrac{s}{s^2+\omega_0^2}$	$\dfrac{z(z-\cos\omega_0 T)}{z^2-2z\cos\omega_0 T+1}$
19	$\sinh\omega_0 kT$	$\dfrac{\omega_0}{s^2-\omega_0^2}$	$\dfrac{z\sinh\omega_0 T}{z^2-2z\cosh\omega_0 T+1}$
20	$\cosh\omega_0 kT$	$\dfrac{s}{s^2-\omega_0^2}$	$\dfrac{z(z-\cosh\omega_0 T)}{z^2-2z\cosh\omega_0 T+1}$

Table C.2 *(continued)*

	$f(kT)$	$F(s)=\int_0^\infty f(t)e^{-st}\,dt$	$F(z)=\sum_{k=0}^\infty f(kT)z^{-k}$
21	$\cosh \omega_0 kT - 1$	$\dfrac{\omega_0^2}{s(s^2-\omega_0^2)}$	$\dfrac{z(z-\cosh\omega_0 T)}{z^2-2z\cosh\omega_0 T+1}-\dfrac{z}{z-1}$
22	$1-\cos\omega_0 kT$	$\dfrac{\omega_0^2}{s(s^2+\omega_0^2)}$	$\dfrac{z}{z-1}-\dfrac{z(z-\cos\omega_0 T)}{z^2-2z\cos\omega_0 T+1}$
23	$e^{-akT}-e^{-bkT}$	$\dfrac{b-a}{(s+a)(s+b)}$	$\dfrac{z}{z-e^{-aT}}-\dfrac{z}{z-e^{-bT}}$
24	$(c-a)e^{-akT}+(b-c)e^{-bkT}$	$\dfrac{(b-a)(s+c)}{(s+a)(s+b)}$	$\dfrac{(c-a)z}{z-e^{-aT}}-\dfrac{(b-c)z}{z-e^{-bT}}$
25	$1-(1+akT)e^{-akT}$	$\dfrac{a^2}{s(s+a)^2}$	$\dfrac{z}{z-1}-\dfrac{z}{z-e^{-aT}}-\dfrac{aTze^{-aT}}{(z-e^{-aT})^2}$
26	$b-be^{-akT}+a(a-b)kTe^{-akT}$	$\dfrac{a^2(s+b)}{s(s+a)^2}$	$\dfrac{bz}{z-1}-\dfrac{bz}{z-e^{-aT}}+\dfrac{a(a-b)Tze^{-aT}}{(z-e^{-aT})^2}$
27	$e^{-bkT}-e^{-akT}+(a-b)kTe^{-akT}$	$\dfrac{(a-b)^2}{(s+b)(s+a)^2}$	$\dfrac{z}{z-e^{-bT}}-\dfrac{z}{z-e^{-aT}}+\dfrac{(a-b)Tze^{-aT}}{(z-e^{-aT})^2}$
28	$e^{-akT}\sin\omega_0 kT$	$\dfrac{\omega_0}{(s+a)^2+\omega_0^2}$	$\dfrac{ze^{-aT}\sin\omega_0 T}{z^2-2ze^{-aT}\cos\omega_0 T+e^{-2aT}}$

29	$e^{-akT}\cos\omega_0 kT$	$\dfrac{s+a}{(s+a)^2+\omega_0^2}$	$\dfrac{z^2 - ze^{-aT}\cos\omega_0 T}{z^2 - 2ze^{-aT}\cos\omega_0 T + e^{-2aT}}$
30	$e^{-bkT} - e^{-akT}\sec\theta\cos(\omega_0 kT - \theta),$ where $\theta = \tan^{-1}\left[\dfrac{b-a}{\omega_0}\right]$	$\dfrac{(a-b)^2+\omega_0^2}{(s+b)[(s+a)^2+\omega_0^2]}$	$\dfrac{z}{z-e^{-bT}} - \dfrac{z^2 - ze^{-aT}\sec\theta\cos(\omega_0 T + \theta)}{z^2 - 2ze^{-aT}\cos\omega_0 T + e^{-2aT}}$
31	$1 - e^{-akT}\sec\theta\cos(\omega_0 kT + \theta),$ where $\theta = \tan^{-1}\left[-\dfrac{a}{\omega_0}\right]$	$\dfrac{a^2+\omega_0^2}{s[(s+a)^2+\omega_0^2]}$	$\dfrac{z}{z-1} - \dfrac{z^2 - ze^{-aT}\sec\theta\cos(\omega_0 T + \theta)}{z^2 - 2ze^{-aT}\cos\omega_0 T + e^{-2aT}}$
32	$b - be^{-akT}\sec\theta\cos(\omega_0 kT + \theta),$ where $\theta = \tan^{-1}\left[\dfrac{a^2+\omega_0^2 - ab}{b\omega_0}\right]$	$\dfrac{(a^2+\omega_0^2)(s+b)}{s[(s+a)^2+\omega_0^2]}$	$\dfrac{bz}{z-1} - \dfrac{b[z^2 - ze^{-aT}\sec\theta\cos(\omega_0 T + \theta)]}{z^2 - 2ze^{-aT}\cos\omega_0 T + e^{-2aT}}$

Index